6th International Conference on Software Engineering & Computer Systems

IOP Conference Series: Materials Science and Engineering
Volume 769

Kuantan, Malaysia
25 - 27 September 2019

ISBN: 978-1-7138-4707-6
ISSN: 1757-8981

Printed from e-media with permission by:

Curran Associates, Inc.
57 Morehouse Lane
Red Hook, NY 12571

Some format issues inherent in the e-media version may also appear in this print version.

Institute of Physics
Dirac House, Temple Back
Bristol BS1 6BE UK

Phone: 44 1 17 929 7481
Fax: 44 1 17 920 0979

techtracking@iop.org

Additional copies of this publication are available from:

Curran Associates, Inc.
57 Morehouse Lane
Red Hook, NY 12571 USA
Phone: 845-758-0400
Fax: 845-758-2633
Email: curran@proceedings.com
Web: www.proceedings.com

TABLE OF CONTENTS

IOP Conf. Series: Materials Science and Engineering **769** (2020) 011001 doi:10.1088/1757-899X/769/1/011001

Preface

The **6th International Conference on Software Engineering & Computer Systems (ICSECS'19)** organised by **Faculty of Computing, Universiti Malaysia Pahang**, is a biennial conference series that serves as a forum for international community of researchers, practitioners and industries in subfield of Software Engineering and Computer Systems. This year it held in Vistana Kuantan City Center, Kuantan, Pahang, Malaysia from 25-26 September 2019 with the theme of ***IR 4.0: Eyeing The Future***. We are profound to promote the conference to a wider community by contributing the matters of public concern and future applications in the fields of software engineering and computer system with IR 4.0.

Three keynote speakers were invited based on their expertise areas. The first keynote speaker was Professor Dr. Ts. Kamal Zuhairi bin Zamli from Universiti Malaysia Pahang. As a well-known professor in *software testing*, he delivers a speech on "Optimization via Meta/Hyper Heuristics: Achievements & Prospects for the Future". The second keynote speaker was Professor Dr. Yuko Hatano from University of Tsukuba, Japan. She delivered her keynote speech on "Supercomputing by the Oakforest of University of Tsukuba Regarding the Flow in Porous Media". The last speaker was Dr. Ishkandar bin Baharin, the President of Malaysia Robotics and Automation Society (MyRAS). His speech ws on "Bridging the Robotics and Automation Gap for IR 4.0 Adoption for local SMEs and the role of Local Academia".

This proceeding book is a compilation of selected papers that are submitted and presented in the conference. The content of the proceeding book is structured based on 4 conference tracks. The track has emphasized on the following topics include, but not limited to:

Track 1: Internet of Things, Big Data and Autonomous Robots

Track 2: Simulation and Virtual/Augmented Reality

Track 3: Additive Manufacturing and System Integration

Track 4: Cyber-Security and Cloud Computing

The editors would like to extend their gratitude to all reviewers and program committees for the hard works. They would also like to thank to all presenters and participants for sharing their research and innovation outcomes in this conference. In addition, their special appreciation is dedicated to the sponsors, namely Malaysian Communications and Multimedia Commission, UMP Holdings, UMP Advance Education, Purnama Sdn. Bhd., Optik Prima Sdn. Bhd., KARAR Solutions Sdn. Bhd., Brilliant Grid Resources and Kafe Bonda Habsah Universiti Malaysia Pahang. Last but not least, the editors would like to thank to all for their enormous supports in this conference.

IOP Conf. Series: Materials Science and Engineering **769** (2020) 011001 doi:10.1088/1757-899X/769/1/011001

Statement of Peer Review:

"All papers published in this volume of *IOP Conference Series: Materials Science and Engineering* have been peer reviewed through processes administered by the Editors. Reviews were conducted by expert referees to the professional and scientific standards expected of a proceedings journal published by IOP Publishing."

ICSECS'19

INTERNATIONAL CONFERENCE ON SOFTWARE ENGINEERING & COMPUTER SYSTEMS

6th INTERNATIONAL
CONFERENCE ON

Universiti Malaysia PAHANG
ICSECS'19
6th INTERNATIONAL CONFERENCE ON SOFTWARE ENGINEERING & COMPUTER SYSTEMS
25 - 26 S
Vistana
Kuantan
IR 4
Organised

IOP Conf. Series: Materials Science and Engineering **769** (2020) 011004 doi:10.1088/1757-899X/769/1/011004

Peer review statement

All papers published in this volume of *IOP Conference Series: Materials Science and Engineering* have been peer reviewed through processes administered by the proceedings Editors. Reviews were conducted by expert referees to the professional and scientific standards expected of a proceedings journal published by IOP Publishing.

Industry 4.0: Eyeing The Future via Simulation

Aisyah Ibrahim, Tuty Asmawaty Abdul Kadir, Adzhar Kamaludin

Faculty of Computing, University Malaysia Pahang
Kuantan, Pahang

Corresponding author's email : aisyah.ibrahim@gmail.com

Abstract. Industry 4.0 is a future agenda. Currently, Industry 4.0 has paved the way in many industries around the world. Many initiatives have been put forward and assessed to ensure Industry 4.0 is becoming a reality. Although widely discussed in the both academic literatures and industrial publications, Industry 4.0 is encapsulated with many terms since it's inception. Some of the terms are not self-explained and often overlapped or even contradicted with similar terms in different perspectives. To demystify the issue, this paper begins with a brief overview of Industry 4.0 background including its importance and readiness. It reviews not only the technological concept with the support of nine pillars of prominent technologies, but also including the human resources as the driver behind the Industry 4.0. Dedicated section on how simulations can contribute to Industry 4.0 in ways where simulation applications can be combined with each Industry 4.0 pillars, including human resources are also presented. This paper is written in general view of Industry 4.0 so that it is easier to relate with not only with manufacturing sectors (which where Industry 4.0 was originally intended for) but also with other sectors.

1. Introduction

Industry 4.0 seems to be at the forefront of the discussion in both industry and academia. The topic is widely discussed even though it is still a non-consensual concept [1]–[3]. Industry 4.0 brings a significant long-term strategic impact on global industrial development. Industry 4.0 does not only confined to manufacturing sector but also impacted other sectors as well such as healthcare [4], agriculture [5], food industry [6], energy management [7], education [8] and perhaps many more.

The novelty in Industry 4.0 is not just in new technology but also include a various mixed combination of available technology in a new way [9]. For this, Industry 4.0 has raised high expectations, but to date, not all have been met yet. One way of doing the planning and assessing situation is through modeling and simulation. Simulation involves a process of imitating the system behavior [10]. Depending on the motivation of usage, various methodologies and techniques can be applied in simulation for the evaluation of complex systems [10]. Afterall, simulation is considered as one of the key pillars in industry 4.0. Therefore, how simulation can support in Industry 4.0 is reviewed too in this paper. Although most literature regarding Industry 4.0 are heavily emphasising on manufacturing sectors, this paper is intended to maintain the focus on the general view of Industry 4.0 so that it is easier to relate with other sectors too.

The remainder of this paper is organized as follows. Section 2 provides an overview of Industry 4.0 background, concepts, importance and readiness. Section 3 presents on how simulations can contribute to Industry 4.0. Lastly, this paper ends with closing remarks in Section 4.

2. Industry 4.0

2.1. Background and concept

Generally, the term 'Industrie 4.0' or 'Industry 4.0' is referred to as 'The Fourth Industrial Revolution' [3], [8]. To date, Industry 4.0 is still a new concept compared to the first three industrial revolutions which have spanned more than 200 years. The first industrial revolution or 'The Age of Steam'[11], began in United Kingdom [12], [13]. At the time, water and steam were used to power mechanical engine systems in ships, trains, production plants and manufacturing factories [12]. The second industrial revolution or 'The Age of Electricity', was driven by electrical energy. The movement started in Europe and US[13]. In this era, the application of electrical devices and internal combustion engines were used to support industries for mass productions [2], [11]. The third industrial revolution was driven by the Programmable Logic Controllers (PLC) and computerised IT systems and automation for digitalisation [9], [11]. Automated machines and computer applications were more widely used across the world in Industry 3.0 compared to Industry 2.0 [12]. Thus, this era is known as 'The Information Age'. As Industry 3.0 is matured, it becomes the foundation for Industry 4.0 to move forward, which is now, it is paving it's way to reality [2], [11]. A brief overview of the four Industrial Revolutions evolution is shown in Figure 1.

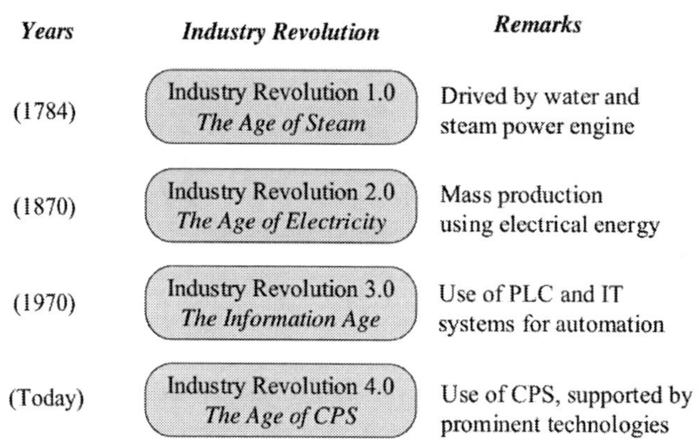

Figure 1 Overview of four Industrial Revolutions evolution, adapted from [2], [11]

Looking back at history, 'Industry 4.0' denotation was first coined by the German government as key initiatives that marked the new beginning of the next industrial revolution [9], [14]. Originally, Industry 4.0 is meant for Cyber-physical Production Systems (CPS) applications for Smart Factory [9]. The ideas are to use smart monitoring systems in production processes to support decision making, to lower down machine maintenance costs and to increase the competitiveness of German's industry. Besides technological aspects, human capital is another Industry 4.0 driver [15], [16]. Human labor ship will be assigned where human expertise is needed. Since the first announced in 2011 in Hanover Fair, Industry 4.0 has become the central focus of future industrial development [9], [11], [14].

Technological wise, there are six principles of intelligent production systems for Industry 4.0 development and deployment in the future have been recognised [16]. The six principles identified are illustrated in pictorial form in Figure 2(a).

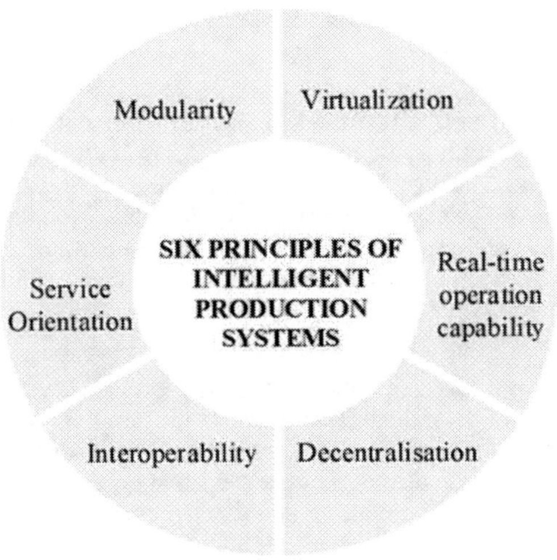

(a) Six principles of intelligent production systems for Industry 4.0. [16]

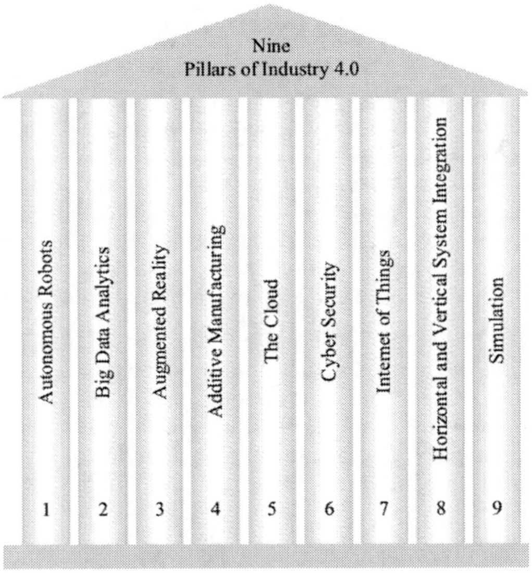

(b) The nine pillars that support Industry 4.0. [2], [17]

Figure 2 The principles of intelligence production systems and the nine pillars of Industry 4.0 in pictorial form.

Each principle is explained as follows:

- Interoperability – allows information exchange between machines and machines, between machines and processes, and between machines and humans (through interface) in both physical and virtual worlds.
- Real-time operation capability – allows factory or business operation in real-time manners with instantaneous data acquisition and processing. Thus, this can encourage real-time decision makings as well.
- Virtualisation – allows remote traceability and monitoring of all operation processes.
- Decentralisation – allows smart application modules to work in a decentralised manner to improve production processes.
- Service Orientation – deploys service-oriented software architectures coupled with the IoT concept.
- Modularity – allows flexibility to change machine tasks easily where production processes can be designed according to the demand and where coupling and decoupling of modules in production can be done quickly.

Based on the six principles in Figure 2(a), Industry 4.0 is supported by the nine pillars of prominent technologies in Figure 2(b). All nine of them constitute the critical enabler of industry 4.0 implementation. These are the core which allows connections of machines, products, systems and people. Each pillar is explained briefly as follows:

1) Autonomous Robots:
With CPS, there was speculation that factories of the future would operate autonomously where less human intervention is needed especially on the tasks that are beyond human's capabilities and the monotonous and repetitive tasks. [18]. With CPS, the tasks will be taken in charge by autonomous robots[9]. It will be no longer necessary for humans to perform tedious, repetitive activities. This will

help free up workforces and redirect them to manoeuvre the robots or do the tasks that robots cannot perform instead.

2) Big Data Analytics
Once IoT is fully implemented, a gargantuan amount of non-real-time data and real-time data will be generated. These 'data goldmines' has potentials to generate valuable insights. Many industries will need to organize, manage and figure out how to analyze in order to benefits from their data to suit their needs[1]. One way of doing this is through Big Data Analytics. Analytics for big data may include data minings, machine learnings, artificial intelligence (AI) techniques, applied statistics and perhaps many more [19], [20]. In fact, there also needs for more advanced data science and data analytics techniques to be developed and employed for better decision makings [11].

3) Augmented Reality
The augmented reality applications serve in various purposes ranging from visual arts and games to medical, military, navigation, sports and educations. For industrial purposes, the applications are narrowed down to e-commerce, administration, training and marketing. Augmented reality can be an essential element in the intelligent industry for operation executions. Together with augmented reality, real-time information is supplied to workers to improve decision making and work procedures [2].

4) Additive Manufacturing
Additive Manufacturing is the technologies that can build a new 3D object or a recreation copy of any other 3D object. This process is done by adding layer-upon-layer of material at a time. This technology plays an essential role in creating new materials. For instance, in production planning where evidently, it helps to visualise the material's creation process and advanced further by the creation of prototypes [17].

5) The Cloud
Cloud-based applications serve as a critical aspect for storage, connection and communication in digital connectivity and digital production. Three prominent Cloud applications are Software as a Service, Platform as a Service, and Infrastructure as a Service [21]. These Clouds can be either public, private or hybrid mode. Each has a variety of speed. On the application side of smart factory, "Digital production" offers a concept of having the connections of different devices to the same Cloud for information sharing. For example, administrators can store their data in the Clouds. And using the Clouds, they also can implement new systems to monitor the operations and to control processes. Meanwhile from the business perspectives, the Clouds also can offer a platform for digital connectivity through the concept of information sharing. For example, information with collaborators or customers. As results, this can lead to a transparent business ecosystem in competitive market environment.

6) Cybersecurity
Transparency can expose a company to vulnerabilities. Thus, information security and privacy protection are among top priorities in Industry 4.0. However, current existing technologies, may not be enough for future industrial applications. Cybersecurity has to comply to suit the company's safety, security regulations and standard requirements. Network security, information security and data privacy protection are critical aspects that future should take account before implementing IoT technologies and services in Industry 4.0 [22]. Access to production-related data and services has to be administrable to protect the company's intellectual property or intellectual know-how [9]. Besides that, it is also important to avoid illegal access to production systems, especially the ones that can cause environmental or economic damage and harm to humans.

7) Internet of Things (IoT)
Internet is the backbone of data communication in Industry 4.0 [21]. The introduction of information and communication systems for IoT into industrial network leads to a steep rise in the degree of automation and data sharing. When everything is connected, everything can communicate with one another. With digital product memories, a collection of data can be recorded in all phases of the product life cycle and can be saved for analysis. The information needs to be organized in all aspects according to territories, storages and aspects of usage such as Cloud, cybersecurity, simulation, augmented reality

and perhaps many more[17]. With full IoT capabilities, many can reap the benefits of smart power grids, smart healthcare, smart logistics, and perhaps many more smart paradigms[21].

8) Horizontal and Vertical System Integration
Horizontal and vertical systems integration promotes a chain of collaborative systems. The chain may involve multiple combinations of units or departments. The combinations can be between engineering, production, suppliers, marketing, and supply chain operations and perhaps many more. In order to integrate the systems, there are two aspect that must be considered. First, is the levels of automation and second, is the information flow [17]. By far, there are three categories of integration identified in the literatures [2], [17] namely:

- horizontal integration across the entire value creation network: this means the information integration of business operation occurs at the same level of the business's value chain, either in a similar industry or different ones.
- vertical integration and networked manufacturing systems: this means the information integration of business operation involve across different level of the business's value chain, either upstream manner or downstream manner or both.
- end-to-end engineering across the entire product life cycle: this means the information and data can be obtained and analysed throughout the production process. This features is beneficial for quick decision makings.

9) Simulation
Simulation helps to evaluate the effects and impacts of actions in safe mode. It enhances learning and decision makings. The changes of behaviors in the configuration of machines, process flow and plant designs can be observed in the simulation environment. The effectiveness of the changes can be tested without them being realized in real-world [2], [17]. As Industry 4.0 is a complex environment, it is not a surprise that new paradigms is highly needed to support Industry 4.0 [23]. For this, models for simulation are expected to be smart, can perform self-correction, can be a learning system that is able to adapt behavior based on changing conditions and past experiences [24]. Furthermore, future simulation can be used for semi-autonomous problem solving [24]. There are also initiatives to combine analytical with simulation tools. In doing so, decision makers can evaluate options to take the best decisions.

2.2. Why Industry 4.0 is Important?

In term of a global perspective, Industry 4.0 significantly can influence economic growth and increases nation economic prosperity. Here are several reasons why Industry 4.0 is important in terms of business operations.

- Industry 4.0 encourages industrial processes to change for the better. For examples in manufacturing systems and supply chains. It promotes a new way of doing business or perhaps a new business value, especially for traditional business model organizations [1], [13], [25].
- Industry 4.0 cultivates decentralized and digitalized production in business operation [9], [15], [26].
- Industry 4.0 implementation help to increase competitive advantage. This can be achieved through the successful digital business model implementation and technology creation [27].
- With robots and automated applications, tasks completion are better and faster by machines rather than by people [9][13].
- The use of cutting edge technologies in Industry 4.0 such as digital chains, smart systems, centralized management and the industrial Internet will speed up innovations as new business models can be implemented much faster [9], [14], [27].

- In Industry 4.0, it is believed that things will become smarter, more reliable, and more autonomous. This will increase the added-value of businesses, especially on products and services [13].

2.3. Are we ready for Industry 4.0?

Smart systems, smart operations, smart productions, humans capitals in Industry 4.0 were identified as the peak priorities. To date, Industry 4.0 is still far in the future. Many ideas are still in the realm of fiction, but some are on its way to reality. Governments, industries and universities worldwide embrace the transformation era of Industry 3.0 and welcome Industry 4.0. There has been an increasing trend and research attention on the nine pillars [9], [11], [14], [22]. Efforts have been made to reap the benefits from what the new industrial revolution wave could offer.

By doing so, Industry 4.0 exposes established companies to new competitive challenges [18]. On the operational level, there is the need to explore more the possibilities and benefits of adopting or adapting the concept. Currently, most industries are facing the lacking of applicable frameworks for the implementation of Industry 4.0 [18].

On the business level, Industry 4.0 does not only concern about the issues on the technical sides but also on the management sides [22]. Many industries are not yet ready or mature for Industry 4.0. Some are still evaluating their readiness and capabilities towards implementing Industry 4.0. Decisions made have to properly analysed and studied before putting substantial investment in upgrading the business model to fit the Industry 4.0 concept. For some industries, this is a long way to go, and for some industries, the concept is still unknown. Many are aware that adopting and adapting Industry 4.0 comes with risks and uncertainties. The process involves many aspects and types of difficulties and challenges. These may stem from scientific sides, technological sides, and economic challenges, social problems as well as political issues [1].

On the national level, besides Germany, several other countries around the world were also have made their movements towards industry 4.0 [14], [27]. The countries include US, French, United Kingdom, some European countries under the European Commission, and some Asian countries such as South Korea, China, Japan, Singapore and Malaysia too. Table 1 is showing the national initiatives of several countries adapted from [14] with an extended update on Malaysia [28].

On the human aspects, most national Industry 4.0 initiatives by the countries mentioned are taking account humans resources as the centre of production. People is regarded as another Industry 4.0 catalyst [15]. Instead of looking at technology applications alone, it is worth pay attention at the people as the future depends on how the people are moving on the journey within the next few years or decades ahead [17]. Tasks slowly will be taken over by machines and robots. Most manual labours will be replaced by trained professionals or knowledgeable workers [16]. A knowledgeable worker is portrayed by World Economic Forum as a person who can deal with complex problem, think and judge critically, able to coordinate with others, creative and master digital skills [8].

Apart from human resources, politicians are among the important actors in shaping and supporting Industry 4.0 [15]. They play an essential role in including the different players in industries. One way of doing this is by creating a shared understanding of a future situation in economic sectors. Success stories are believed to be one way of getting business-minded people and entrepreneurs on board.

Table 1. Government's National Initiatives on Industry 4.0 in several countries [14]

Year	Country	Industry 4.0 Plans	
		National Plan	**Aims**
2011	USA	Advanced Manufacturing Partnership	to ensure the US is ready to lead the next generation of manufacturing
2012	Germany	'High-Tech Strategy 2020' Action Plan	to gain competitive advantage, with billions euros worth of high-end technologies and developments for German's manufacturing
2013	French	'La Nouvelle France Industrielle'	focuses on 34 sector-based initiatives in France's industrial policy priorities
2013	UK	'Future of Manufacturing'	to provide a refocused and rebalanced policy for supporting the growth of UK manufacturing till 2050
2014	The European Commission	Pubic-Private Partnership (PPP) on 'Factories of the Future (FoF)'	providing financial support up to 80 billion euros under European Economic Recovery Plan from 2014 to 2020
2014	South Korea	'Innovation in Manufacturing 3.0'	contains strategies for a new leap of Korean's future manufacturing industries.
2015	China	'Made in China 2025'	prioritises ten fields in the manufacturing sector to accelerate the informatisation and industrialisation in China
2015	Japan	5th Science and Technology Basic Plan	for realising its world-leading 'Super Smart Society.'
2016	Singapore	RIE 2020 Plan (Research, Innovation, and Enterprise)	focuses on eight vital industries within the advanced manufacturing and engineering domain, which sets nineteen millions of Singaporean dollars.
2018	Malaysia	Industry4WRD [28]	to cultivate conducive ecosystem for Industry 4.0 technologies, and to ensure manufacturing sectors to be ready for Industry 4.0

3. How Simulation Can Help?

Computer simulation is considered as a safe platform to play around, learn, test assumptions and study and impact of particular actions chosen. The trends in simulation development indicated that simulation has moved to non-stand-alone paradigm instead using the former the traditional stand-alone paradigm [23]. To date, some of modeling and simulation approaches can be connected to various data sources, and output destinations [23], [24]. Some can be controlled and modified via user-friendly front-ends or other applications [23], [24]. For other cases, modeling and simulation is no longer relying on single approach but also taking account to integrate with other approaches as well such as optimizations, machine learnings, data minings etc.

In simulation, "system" is referring to "what, from the real world, is being simulated" [29]. A system can be referred to anything such as people, machines or resources. Real-world system is modeled as close as possible and simulated to support testing, learning and decision makings. In relation to Industry 4.0, every system is linked to one another. Smart systems of Industry 4.0 are "systems of systems" not only from the holistic views but in detail views too. The systems and the systems related need to work smoothly in a proper way in order for overall systems to be working effectively [30]. Due to these reasons, smart systems really need simulation technology to design and assess the applicability before the real implementation takes place. This is because to test in real world would be very costly. The side effects would be unbearable and irreversible. Through simulation-based decision support tools, users or decision makers will be able to seek solution development, validation and testing for systems and individual elements of systems [23]. For this, simulation can be combined with the other eight pillars, including extra section on simulation and human resources. For examples:

a) Simulation and Autonomous Robots – Simulation can help to design the kinematics and dynamics of robotic manipulators. Such systems are not easy to be developed and could not be realized in real world easily. Only simulations on system-level can gain the necessary insights [30]. Autonomous robots are complex. It is not easy to understand reality with all of its complexity. One way of doing this is by building artificial objects and dynamically act out roles with them to realise intelligent systems. There are two identified categories of simulations tools for robotic systems: (1) tools based on general simulation systems and (2) special tools for robot systems. The former comes together with specialized modules and libraries, including user interfaces. The purpose is to ease robot systems development for simulation. Whereas the latter is for advanced usage and multitasking tasks. This may involve off-line programming, designing robot work cells, kinematic and dynamic analysis, mechanical design and perhaps many more. They can be specialized for special types of robots. Among the common tools used for simulation of the autonomous robots are Discrete event simulation systems, MATLAB/Simulink, Modelica modeling language [30]. Some are provided by robot manufacturers such as KUKA.Sim, V-REP, and GAZEBO, Emulate3D and perhaps many more [30].

b) Simulation and Big Data Analytics – Under Big Data paradigm, data-model can be built upon abundant data such as from real-time devices or information systems [24]. The "data model" is then can be simulated to understand problems and solve complex problems to support decision makings. Heavy computations, high complexity, and difficulties to model feasibly are among challenges faced in this domain. But for Industry 4.0, these open opportunities for discoveries of new modelling and simulation paradigms [23], [30]. This can be new in-depth analysis, new way of processing especially on large-scale data processing. At the same time, innovation of existing modeling and simulation methodology also can be enhanced and explored further.

c) Simulation and Augmented Reality – Both can be a great tool when combined together, especially for training, upskilling, learning purposes. Augmented reality can help to increase people‘s understanding. This is evidently shown in a study by [31], where augmented reality was used as a medium for displaying the simulation results and analysis. Another purpose for combining simulation and augmented reality is for The Teaching Factory concept. This concept is close to “Education 4.0” where Teaching Factory is aimed to polish students through real industrial problems [8].

d) Simulation and Additive Manufacturing - Simulation is important in additive manufacturing. There are needs for the implementation of computationally efficient techniques to predict, to study and to assess varying characteristics of newly produced components. Among common simulation used are Monte Carlo simulation, Cellular Automata, Discrete Element Method and perhaps many more.

e) Simulation and the Cloud - Companies can retrieve the data from the Cloud and carry on doing simulation without having to worry about enterprise-level software deployments, server upgrades, hardware and software licensing or other issues. If they are able to run real-time simulations of their entire factory or company’s operation in the Cloud and they could configure their simulation environment whether to include new machines, new staffs or not, and getting the answer in real-time. Thus, faster decision can be made whether the additions will make a difference to productivity and profitability or not.

f) Simulation and Cybersecurity – With everything is interconnected to the network, the system inside a factory is vulnerable for attacks. Simulation applications for cybersecurity can help to assess the risk in cyber network and the network behaviour.

g) Simulation and IoT. One of critical application in this area is for planning IoT. This because to build effective and smart services can be quite complicated. IoT networks are complex. IoT connectivity is influenced by the location where they are going to be situated geographically, network communication capabilities, and device availability for distributions. Thus, IoT simulation is necessary for planning and risk assessment. To name a few issues: capacity planning, what-if simulation and analysis, proactive management and support for many specific security-related evaluations. Agent-Based Simulation and Discrete Event Simulation are among perfect simulation paradigms that are usually used[10]. Aside of being applied in IoT, there is also an emerging technology regarding simulation and IOT such as Internet of Simulation, which is an extension of internet technologies in simulations [24].

h) Simulations and Horizontal and Vertical System Integrations. Designing Industry 4.0 systems is tedious due to its high dimensionality and complexity. Simulation of inventory, logistics and transport, and usage history of products can help to positively influence the production processes. The complexity of the innovation systems lead to the development of modern approaches of new modeling approach. Thus simulation is treated as tool for further designing, creating, and modifying real innovative systems of different levels of organization under the conditions of Industry 4.0.

i) Simulation and human resources - Every revolution creates worries over job losses. This happens to Industry 4.0 as well. Although robots and machine automations are handy, but at the same time it creates insecurities. At the other end, in some sense, they are perceived as threat to human employability. Due to this reason, human resources in Industry 4.0 is widely discussed in the both industries and academic literatures. Many efforts and initiatives have been made to analyse the consequences of Industry 4.0. The efforts cover not only on operational, indirect management tasks and activities but also on the social impacts on national and international basis [32]. Based on literatures, evidently, simulation can help to effectively plan for human resources management. For example, a workforce plan can be developed in a way that can satisfy organization's or national's needs in meeting the objectives for recruitment, deployment and promoting right employees at the right places at the right times.

4. Concluding Remarks

Industry 4.0 promotes a new era of opportunities, challenges with uncertainties in a highly dynamic competitive environment. This paper reviews the concepts of Industry 4.0 including the nine pillars technologies that support Industry 4.0 conception to reality. Although many efforts have been laid to welcome Industry 4.0 including the governments (national level), not many industries are ready to embrace the idea of Industry 4.0 to the fullest. As aforementioned, these may come from many factors which may include scientific, technological, and economic challenges, social problems as well as political issues as well. As one of the nine pillars that support Industry 4.0, simulation plays a vital role. Simulations are proven to work well for designing overlapping system functions, for decision makings, learnings, trainings and many more. In order to achieve this, several aspects must be considered such as suitable tools selections, combination methodologies, data(s), and techniques used. With extremely complex system of Industry 4.0, simulation is very much needed to create a room for safe system execution, hence prevent unnecessary complications and losses. That is the main reason why simulation is considered one of the pillars for Industry 4.0.

Acknowledgement

This paper is financially supported by research grant (RDU140310) under Universiti Malaysia Pahang. Special thanks to Assoc. Prof Dr Hamdan Daniyal for the support given during the documentation of this paper.

References

[1] M. Mohamed, "Challenges and benefits of Industry 4.0: An overview," *Int. J. Supply Oper. Manag.*, vol. 5, no. 3, pp. 256--265, 2018.

[2] S. Vaidya, P. Ambad, and S. Bhosle, "Industry 4.0 - A glimpse," *Procedia Manuf.*, vol. 20, pp. 233–238, 2018.

[3] M. Hermann, T. Pentek, and B. Otto, "Design principles for industrie 4.0 scenarios," in *Proceedings of the Annual Hawaii International Conference on System Sciences*, 2016, vol. 2016-March, pp. 3928–3937.

[4] G. Manogaran, C. Thota, and D. Lopez, "Big data security intelligence for healthcare Industry

4.0," *Cybersecurity Ind. 4.0*, pp. 103--126, 2017.

[5] I. Zambon, M. Cecchini, G. Egidi, M. G. Saporito, and A. Colantoni, "Revolution 4.0: Industry vs. agriculture in a future development for SMEs," *Processes*, vol. 7, no. 1, p. 36, 2019.

[6] A. Luque, M. E. Peralta, A. De Las Heras, and A. Córdoba, "State of the Industry 4.0 in the Andalusian food sector," *Procedia Manuf.*, vol. 13, pp. 1199–1205, 2017.

[7] M. Faheem *et al.*, "Smart grid communication and information technologies in the perspective of Industry 4.0: Opportunities and challenges," *Comput. Sci. Rev.*, vol. 30, pp. 1–30, 2018.

[8] D. Mourtzis, A. Vasilakopoulos, E. Zervas, and N. Boli, "Manufacturing system sesign using simulation in metal industry towards Education 4.0," *Procedia Manuf.*, vol. 31, pp. 155–161, 2019.

[9] R. Drath and A. Horch, "Industrie 4.0: hit or hype?," *IEEE Ind. Electron. Mag.*, vol. 8, no. 2, pp. 56–58, 2014.

[10] G. D'Angelo, S. Ferretti, and V. Ghini, "Simulation of the Internet of Things," *2016 Int. Conf. High Perform. Comput. Simulation, HPCS 2016*, pp. 1–8, 2016.

[11] L. Da Xu, E. L. Xu, and L. Li, "Industry 4.0: State of the art and future trends," *Int. J. Prod. Res.*, vol. 56, no. 8, pp. 2941–2962, 2018.

[12] I. Garbie, *Sustainability in manufacturing enterprises: Concepts, analyses and assessments for Industry 4.0*. Springer, 2016.

[13] A. C. Pereira and F. Romero, "A review of the meanings and the implications of the Industry 4.0 concept," in *Procedia Manufacturing*, 2017, vol. 13, pp. 1206–1214.

[14] Y. Liao, F. Deschamps, E. de F. R. Loures, and L. F. P. Ramos, "Past, present and future of Industry 4.0 - A systematic literature review and research agenda proposal," *Int. J. Prod. Res.*, vol. 55, no. 12, pp. 3609–3629, 2017.

[15] M. Türkeş, I. Oncioiu, H. Aslam, A. Marin-Pantelescu, D. Topor, and S. Căpuşneanu, "Drivers and barriers in using Industry 4.0: A perspective of SMEs in Romania," *Processes*, vol. 7, no. 3, p. 153, 2019.

[16] M. Y. Santos *et al.*, "A big data system supporting Bosch Braga Industry 4.0 strategy," *Int. J. Inf. Manage.*, vol. 37, no. 6, pp. 750–760, 2017.

[17] J. A. Saucedo-Martínez, M. Pérez-Lara, J. A. Marmolejo-Saucedo, T. E. Salais-Fierro, and P. Vasant, "Industry 4.0 framework for management and operations: a review," *J. Ambient Intell. Humaniz. Comput.*, vol. 9, no. 3, pp. 789–801, 2018.

[18] E. Mueller, X. L. Chen, and R. Riedel, "Challenges and requirements for the application of Industry 4.0: a special insight with the usage of Cyber-Physical System," *Chinese J. Mech. Eng. (English Ed.*, vol. 30, no. 5, pp. 1050–1057, 2017.

[19] M. J. Liberatore and W. Luo, "The analytics movement: Implications for operations research," *Interfaces (Providence).*, vol. 40, no. 4, pp. 313–324, 2010.

[20] C. Holsapple, A. Lee-Post, and R. Pakath, "A unified foundation for business analytics," *Decis. Support Syst.*, vol. 64, pp. 130–141, 2014.

[21] L. Thames and D. Schaefer, "Industry 4.0: an overview of key benefits, technologies, and challenges," in *Cybersecurity for Industry 4.0*, Springer, 2017, pp. 1–33.

[22] A. Sanders, C. Elangeswaran, and J. Wulfsberg, "Industry 4.0 implies lean manufacturing: research activities in industry 4.0 function as enablers for lean manufacturing," *J. Ind. Eng. Manag.*, vol. 9, no. 3, pp. 811–833, 2016.

[23] B. Rodič, "Industry 4.0 and the new simulation modelling paradigm," *Organizacija*, vol. 50, no. 3, pp. 193–207, 2017.

[24] A. G. Uriarte, A. H. C. Ng, and M. U. Moris, "Supporting the lean journey with simulation and optimization in the context of Industry 4.0," in *Procedia Manufacturing*, 2018, vol. 25, pp. 586–593.

[25] J. M. Müller, D. Kiel, and K. I. Voigt, "What drives the implementation of Industry 4.0? The role of opportunities and challenges in the context of sustainability," *Sustain.*, vol. 10, no. 1, 2018.

[26] H. Lasi, P. Fettke, H.-G. Kemper, T. Feld, and M. Hoffmann, "Industry 4.0," *Bus. Inf. Syst. Eng.*, vol. 6, no. 4, pp. 239--242, 2014.
[27] E. Oztemel and S. Gursev, "Literature review of Industry 4.0 and related technologies," *J. Intell. Manuf.*, no. June, 2018.
[28] M. of I. and I. Trade, "National Policy on Industry 4.0." [Online]. Available: https://www.miti.gov.my/index.php/pages/view/industry4WRD?mid=559. [Accessed: 01-Aug-2019].
[29] J. Banks, "Introduction to simulation," in *2000 Winter Simulation Conference Proceedings*, 2000, no. 1, pp. 9–16.
[30] M. Schluse, M. Priggemeyer, L. Atorf, and J. Rossmann, "Experimentable digital twins-streamlining simulation-based systems engineering for Industry 4.0," *IEEE Trans. Ind. Informatics*, vol. 14, no. 4, pp. 1722–1731, 2018.
[31] I. Karlsson, J. Bernedixen, A. H. C. Ng, and L. Pehrsson, "Combining augmented reality and simulation-based optimization for decision support in manufacturing," in *2017 Winter Simulation Conference (WSC)*, 2017, pp. 3988–3999.
[32] L. Bonekamp and M. Sure, "Consequences of Industry 4.0 on human labour and work organisation," *J. Bus. Media Psychol.*, vol. 6, no. 1, pp. 33–40, 2015.

A Combined Weighting for the Feature-Based Method on Topological Parameters in Semantic Taxonomy Using Social Media

Ali Muttaleb Hasan, Noorhuzaimi Mohd Noor, Taha Hussein Rassem, Ahmed Muttaleb Hasan and Waleed A.Hammood
Faculty of Computing, Universiti Malaysia Pahang, 26300 Gambang, Kuantan, Pahang Darul Makmur, Malaysia

E-mail: nhuzaimi@ump.edu.my

Abstract. The textual analysis has become most important task due to the rapid increase of the number of texts that have been continuously generated in several forms such as posts and chats in social media, emails, articles, and news. The management of these texts requires efficient and effective methods, which can handle the linguistic issues that come from the complexity of natural languages. In recent years, the exploitation of semantic features from the lexical sources has been widely investigated by researchers to deal with the issues of "synonymy and ambiguity" in the tasks involved in the Social Media like document clustering. The main challenges of exploiting the lexical knowledge sources such as [1]WordNet 3.1 in these tasks are how to integrate the various types of semantic relations for capturing additional semantic evidence, and how to settle the high dimensionality of current semantic representing approaches. In this paper, the proposed weighting of features for a new semantic feature-based method as which combined four things as which is "Synonymy, Hypernym, non-taxonomy, and Glosses". Therefore, this research proposes a new knowledge-based semantic representation approach for text mining, which can handle the linguistic issues as well as the high dimensionality issue. Thus, the proposed approach consists of two main components: a feature-based method for incorporating the relations in the lexical sources, and a topic-based reduction method to overcome the high dimensionality issue. The proposed method approach will evaluated using WordNet 3.1 in the text clustering and text classification.

Keywords: component, text mining, text classification, sentiment analysis, semantic representation, weighting, topological parameter, Non-taxonomy.

1. Introduction

The proportion of the query and user-produced content in many websites, social media, and online services, such as Twitter, Facebook, YouTube and LinkedIn increases the quality of opinion and information obtainable from the internet. Therefore, the mining of data currently begun utilized to handle huge amounts of data all in name of defending the native land. Different kinds of information retrieval, including healthcare records, network traffic, and financial transactions being subject to mining with a view to creating profiles, constructing models of social network, and detecting communications by terrorists. This activity generates serious concerns on privacy and has creation of particularity certification data mining techniques. In text mining, the documents have clusters, for data mining is about looking for patterns in data. Furthermore, text mining is looking also for modality in the text, it's the processing of analyzing the text to extraction the information that is useful for nominated purposes,

[1] https://wordnet.princeton.edu

compared with the kind of data we have been taking about. In this paper, the motivation of this aim paper for trying to extract the vector of text mining [1]. In addition, the NLP is the aim of the techniques in AI is to give an income to computers the capacity to realize natural language processing NLP. Thus, the major problem in this is understanding how to represent natural language texts semantically in an efficient and effective manner [2]. Social networking that is Web-based (otherwise known as long-range informal communication) happens to be an online stage used by individuals for purpose of building interpersonal organizations or social relations with other individuals who are comparable and interesting, exercises, institute or veritable associations. There is some regular components among different web-based social networking accessible on the web. They are Web 2.0 based web applications, embolden the advancement of online informal platforms by associating the profile of a client with those of different people, and additional bunches. Therefore, to take the information extraction to the next level, the extracted information can be in a subsequent step to understand how to make rules for extracting the text information that is characteristic of the content of itself. These rules might predict the values of definite slot filters from the rest of the text. In certain tightly constrained situations, like internet job posting for computing-related jobs, or post from friend feed by politics to relate with the post, these all treasure based on the database as what applies to extraction of the information. We have proposed in this paper, the weighting of feature method to combine four things as which "[2]Synonymy, [3]Hypernym, [4]Non-taxonomy, and [5]Glosses"; this research proposed a new sematic representation approach of texting mining to reduce the high dimensionality relations in social Media. To capturing more relevance features between two words.

The remaining part of the paper arranged as follows: Section 2, semantic representation. While in section 3, Topological of parameter discussed. In section 4, Definition of terms. In section 5, structural-based methods discussed. In section 6, Semantic Representation Approach discussed. While section 7, result and dissection. In section 8, Conclusion.

2. Semantic representation

In cognitive psychology, semantic representation defined as a topic, which studies the techniques of using the mental lexicon to interpret the meaning of words in natural languages. In computational linguistics and artificial intelligence, semantic representation deals with how word-meaning can extracted from the lexical source database or corpus. Therefore, the most extensive model which implements distinct semantic representation is WordNet as a lexical source [1],[2] it uses a semantic representation to compute the similarity of two words, a network model for representing huge number with nouns, and verbs in English. In WordNet 3.1, "Both nouns and verbs have their own semantic relations as well as their own organization based on the role they play in constructing linguistic messages. These relations and organization that created depending on the relations, which relevant within a specific class of words. For nouns, relations, such as hierarchical relations, synonymy, and part-whole relations usually play the most significant roles. In a number of studies, they introduce the semantic representation for various methods [2]. For verbs, instead, dominant is the taxonomy (hierarchical relations related to specificity in manner), Synonym, is-a relation, holonym, meronym, and antonyms. Some evidence is compatible with a different role of relations such as hyponym and antonym for nouns and verbs.

2.1 Feature-based on semantic taxonomy

The 'Is-A' relations are a semantic representation formulated in every concepts/word construct on the semantic features in lexical resources like WordNet [3, 4]. To use by representing the knowledge in the lexical source as the semantic taxonomy. The semantic taxonomy is a network among concepts in the

[2]https://github.com/alimuttaleb/Ali-Muttaleb/blob/master/Synonym.txt
[3] https://github.com/alimuttaleb/Ali-Muttaleb/blob/master/Hypernym
[4] https://github.com/alimuttaleb/Ali-Muttaleb/blob/master/Non-taxonomy
[5] https://github.com/alimuttaleb/Ali-Muttaleb/blob/master/Glosses

IOP Conf. Series: Materials Science and Engineering **769** (2020) 012002 doi:10.1088/1757-899X/769/1/012002

lexicon, in which nodes of representing concepts and each edge has represented as a hypernyms and hyponyms relations. These relations are a relation that organizes nominal and verbal synsets as the inheritance semantic taxonomy. Figure 1 shows the part of feature-based of semantic taxonomy of the WordNet 3.1.

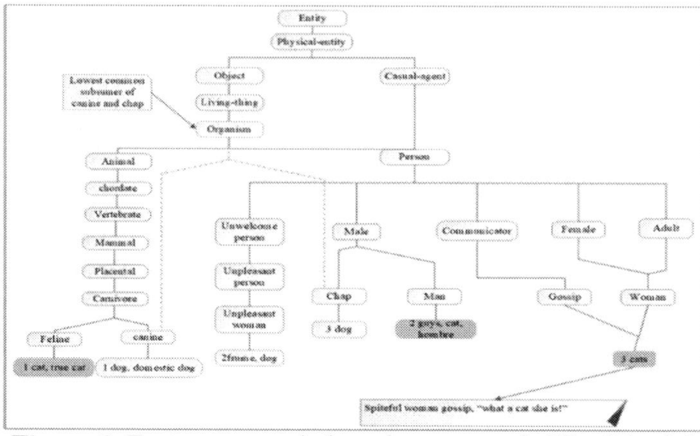

Figure 1. Feature semantic-based taxonomy in WordNet 3.1

In the semantic taxonomy, there are several terms, which will used during this research:

- The root of the taxonomy it is a node of the entity, which superordinate all nodes in the taxonomy. In figure.1, root an entity.
- Hyponymy parent concept, it is a function, which returning the direct hypernyms "*parents*" in the concept, for example, hyponymy (*Chap, Canine*) =*Organism*.
- Subsumes (C) (or ancestors (C)) that will returning the hyponyms in concept C in a recursively way, for example, subsumes (*Chap, Canine*) = (*Organism, living thing, Object, Physical Entity, entity*).
- Hyponym or child concept, as which returned the direct hyponym (*Children*) in the concepts, for example, hyponym (*Person*) = (*Unwelcome person, Male, Communicator, Female, Adult*).
- Descendant (C) returning the hyponyms of a concept c in a recursively way, for example, (*Person*) = (*Unwelcome person, Male, Communicator, Female, Adult, Unpleasant person, Chap, Man, Gossip, Woman, Unpleasant woman, 3 dog, 2 guy, cat, hombre, 3 cats, 2frume, dog*).
- Least common subsumes LCS (C1, C2): it is a function, which takes as input two concepts (C1 and C2) and returns the least common subsumes of C1, C2, for example, LCS (*Chap, Canine*) = *Organism*.
- Path (C1, C2): It is a set containing the nodes (sometimes edges) from the concept C1 to the concept C2.
- Length (C1, C2): It is the number of nodes in the shortest path from the concept C1 to the concept C2.
- Depth (C): It is the length shortest path from the root to the concept C.

2.2 Pre-processing

Text pre-processing is fundamental in many NLP applications and tasks, as which depend on the unstructured knowledge sources for extracting information or manipulating the textual parts. The main aim of this step is to prepare the unstructured knowledge sources such as a collection of documents. The document in the corpus may include some non-informative and noisy terms that lead to the wrong process in natural language processing applications. For this reason, the corpus requires the pre-processing step to handle the noisy terms [5]. Since the distributional-based semantic representation approach depends on the collection of documents to extract the semantics from the text, it is a required pre-processing step. The pre-processing step is implemented to achieve three functions: tokenization, removing stop-words and stemming [6-8]. For each document, the tokenization applied to segment the text in document text into the basic linguistic units called tokens. The second function is removing stop-words, which aimed to exclude the stop-word from the document before representing the semantics [9]. The stop-words such as the, of, and for may lead to the poor semantic representation of the words [10]. For example, the text "*a machine for performing calculations automatically*" contains two stop-words *a* and *for* which are non-descriptive terms in text and have to be excluded. The final function of the pre-processing step is the stemming, which is responsible for converting the different forms of the words to their single source or a lexeme. This function is to avoid repeating the representation of the same lexeme.

For instance, the words books and book have to represent one time only. There are several stemmers for English texts such as the Java implementation of Porter's stemmer.

1. Building Co-occurrence for association model - In distributional of Co-occurrence model methods to build semantic representation matrix of the words. Generally, the vector space model used for the semantic representation of the concepts/words using the information in the corpus. This model is to construct the word-context matrix that contains the words as the rows and the contextual relationship in which that words occurrences as the columns. Formally, for a given corpus C, let w= w_n is a group of words, which appear in whole corpus, and T= {T1, T2,..., Tm} is that set of contexts, the semantic representation of these words is represented as the co-occurrences matrix as shown in Table 1.

Table 1. Weighting for taxonomy

	T1	T2	....	....	....	Tm
W1	f(1,1)	f(1,2)	...	...	...	f(1,m)
W2	f(2,1)	f(2,2)	...	...	...	f (2,m)
:			...	...	...	...
Wn	f(n,1)	f(n,2)	...	...	...	f(n,m)

Each vector in this matrix reveals the representation of one word, in which the elements of this vector are the weights that correspond to each context. In this regard, the context can defined as the environment of words that uses to extract the semantics of words. Different distributional methods used different contexts to represent the corpus. The context of a word can be one of the following types: direct surrounding words, -nword with +nword, sentence, paragraph or document. In all types except document, the elements t1, t2,..., tm are single words or multiword expressions. For example, the word *animal* can represented in the sentence "*The dog and cat are a type of animals which can live with people*" using the context sentence by a *dog, cat, live* and *people,* in which other words are stop-words.
2. Weighting - In this section, we weighting each element in the semantic representation of a word according to relevance, in which the element with a high weighting semantically associated with the concept of the word. The f (i, j) matrix is basic weighting (frequency of words occurs the neighbour with the context) of each context j to represent the semantics of the word i. The weight can be measured by a different formula such as TF-IDF [11], and mutual information.

3. Topological of parameter

In this phase, to proposing the novel knowledge approach method for representing semantics in textual documents in the text mining tasks. It includes the main contribution of the current research which will prove that the topological parameters (such as, depth and density) are useful as an upstream dimensional reduction algorithm to the traditional representation of the textual documents in text mining tasks. This sub-phase tries to overcome the limitations in Social media, as which Facebook, LinkedIn, and tweeter this work, when people post something in politic in this case, will show many of feature in the concepts have like the same relevant ambiguity, (such as, synonymy, ambiguity, and high dimensionality) of the text mining tasks. The main aim of the present methodology is to investigate exploitation of knowledge in lexical sources for text mining tasks, the lexicon contain have a several semantic features such as: Taxonomical relation, meronym relation, synonyms, IS-A relations, textual definitions, glosses definitions, etc. The main challenge is how to select the informative features that improve semantic representation of concepts. Thus, the knowledge-based semantic representation of documents by replacing the words in the document usually leads to very huge number of features. This due to the number of concepts in the lexical sources is extremely large the number of concepts in smaller lexicon (WordNet 3.1) is around 117,659. Thus, the feature-based method is used to represent knowledge in a given lexical source as a taxonomy using "is-a" relations. Then, each concept is semantically representing by a set of feature recursively extracted from a taxonomy. In addition, the feature-based method is how to incorporate the features to capturing more semantic evidence, in the lexical source, by using topological parameter to reduce these features from the weighting of words such as depth, density, descendants and ancestors. All features of concepts have the same relevance based on feature semantic method that most promising topological parameters that could be exploiting by constant weighting assumption. Figure 2 show the processing of semantic representation based on the feature-based method.

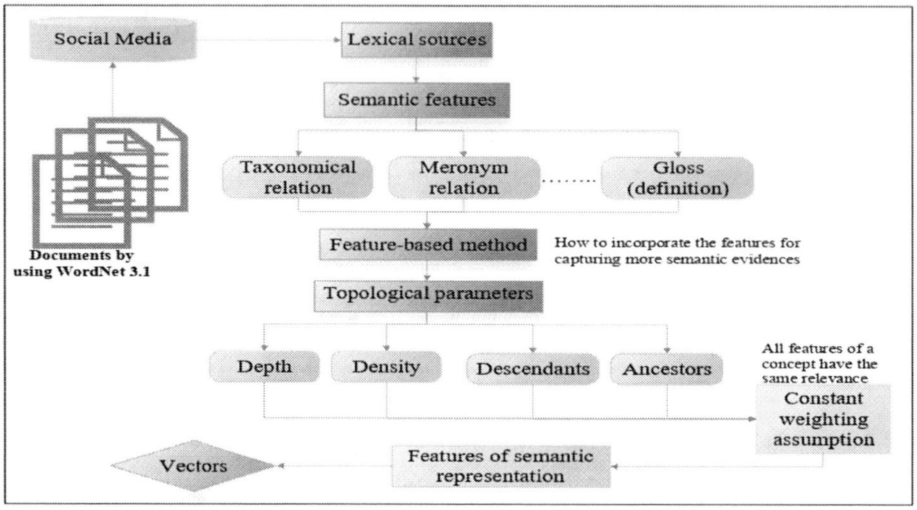

Figure 2. The methodology of illustration of the research issues

3.1 Depth of synset

The weighting of depth is a mechanism used to reduce the features from the concepts in a semantic taxonomy the weighting of the taxonomy. The depth of synset and taxonomy to the concepts indicate, by depth (synset) is commonly the numbering of nodes straight the protracted path between synset and root of the entity. The depth is considerable in taxonomy-based on semantic representations. However, the words in the synset will reduce by using the depth to find the nearest words in the concepts to give specific words/concepts. The depth of concepts in WordNet 3.1 ranged between 1 and 20. The weight concepts or synset will determine by taxonomy like how many of descendant's node and how many of ancestor's node awhile arrange the depth of all words. In Fig 3 shows how the depth accounts the words to find the relevance features by hyponym and hypernym.

To find the possible paths of a synset to its top most hypernym:

```
from nltk.corpus import wordnet
>>> from nltk.corpus import wordnet as wn
>>> max(max(len(hyp_path) for hyp_path in ss.hypernym_paths()) for ss in
wn.all_synsets())
>>> wn.synset('dog.n.1')
Synset('dog.n.01')
>>> wn.synset('dog.n.1').hypernym_paths()
[[Synset('entity.n.01'), Synset('physical_entity.n.01'), Synset('object.n.01'),
Synset('whole.n.02'), Synset('living_thing.n.01'), Synset('organism.n.01'),
Synset('animal.n.01'), Synset('chordate.n.01'), Synset('vertebrate.n.01'),
Synset('mammal.n.01'), Synset('placental.n.01'), Synset('carnivore.n.01'),
Synset('canine.n.02'), Synset('dog.n.01')], [Synset('entity.n.01'),
Synset('physical_entity.n.01'), Synset('object.n.01'), Synset('whole.n.02'),
Synset('living_thing.n.01'), Synset('organism.n.01'), Synset('animal.n.01'),
Synset('domestic_animal.n.01'), Synset('dog.n.01')]]
```

To find the maximum of one synset:

```
>>> max(len(hyp_path) for hyp_path in wn.synset('dog.n.1').hypernym_paths())
```

Figure 3. Code of features by hyponym and hypernym

3.2 Density of synset

In a semantic taxonomy, the domestic density of a synset node indicate to the numbering of connexion nodes or the number of its child nodes. Based on the assumption that two concepts are semantically close if they densely connected locally. However, density used as a topological parameter to locate the specific of a concept. In the WordNet 3.1 the description of the semantic relations between words in synset, one such as a relationship in is-a relationship, at which connect as a Hyponym and Hypernym in synset to give the more specific concept of the words.

IOP Conf. Series: Materials Science and Engineering **769** (2020) 012002 doi:10.1088/1757-899X/769/1/012002

4. Definition of terms

The terms of deflation, it is worth defining some of the terms, which related to the topic of the current research. The following sub-sections provide the definitions of the terms relating to computational semantic analysis.

4.1 Semantic Relation

Semantic relation is a link between two words/concepts reveals the relevance of their semantics. Semantic relation has been exploiting in natural language processing applications such as WSD (*Word Sense Disambiguation*) [12] Query expansion and suggestion. There are many types of semantic relations such as:

- Synonyms: it is equivalence relations. Two words are synonyms if their meaning is similar such as *car* and *automobile*.
- Is-A relation: includes hypernyms and hyponyms. It has called also a hierarchical relation. For example, *fruit* is a hypernym of *apple*; *apple* is a hyponym of *fruit*.
- Holonym: the concept C1 is a holonym of the concept C2 if C2 is a part of C1. For example, the computer is holonym of *RAM*.
- Meronym: the concept C1 is a meronym of the concept C2 if c1 is a part of C2. For example, the finger is meronym of *hand*. The *Cheese*, *beef*, *tomato*, and *bread* are meronyms of the *burger*.

To clarity, more based on our episode of our implementation, as which show the synonyms of words.

Class mapping synonyms

```
public class Synonyms {
private final FileProcess fp=new FileProcess();
private Map<Integer,String> Synsets = new HashMap< >();
private final   Map<String,Set<Integer>> WordToSynset = new HashMap<>(); public   Synonyms ()
String pth="data/wn/synonyms";
Synsets=fp.OpenMapFile(pth);
BuildWordTosynset();
public Map<Integer,String> getSynsets()
return Synsets;
public Map<String,Set<Integer>>  getWordSynsets()
return WordToSynset;
public String getSynset(int t)
return Synsets.get(t).trim();
public Set<Integer > getSynsetsForWord(String s)
return WordToSynset.get(s.trim().toLowerCase());
private void BuildWordTosynset(){
for(Map.Entry<Integer,String> e: Synsets.entrySet())
String sx1[]=e.getValue().toLowerCase().split(";");
System.out.println(sx+"\t"+sx1.length);
for(int i=0;i<sx1.length;i++)
String stemmedWord =sx1[i].trim()   ;
if (WordToSynset.containsKey(stemmedWord))
Set<Integer> sid=WordToSynset.get(stemmedWord);
sid.add( (e.getKey()));
WordToSynset.put(stemmedWord, sid);
```

4.2 Semantic "Is-a" relation

The semantic of "is-a" relation in the reviewing previous of literature studies of the current research, the recent works about using the lexical sources for representing the textual documents in text mining task focus on three topics. The first topic focus on how the semantic information can be organized to provide the efficient semantic measures by introducing new approach for representation [13, 14], proposing new measure [15, 16] and combining the semantic relations [17] to ameliorate the performance of measuring in semantic relatedness SR. The second topic focuses on using the existing approaches and measures to examine them by using the comparative evaluation among approaches or measures or between lexical sources. The third topic focuses on handling the semantic knowledge acquisition issues to enhance the quality and quantity of the lexical resources by integrating two resources. In this paper, the feature-based method based on representing the semantic of concepts as a set of mixed attributes from the semantic relations and glosses in the knowledge sources. The semantic relations include the "is-a" 'hypernym-hypernym' relations, holonym, inverse gloss, and meronym. Thus, the semantic of the *car* represented as the set of features containing the following terms: vehicle, convertible, accelerator, train, and cable *car*. The performance of the feature-based method depends on the factors, which related to the fidelity, continuity, and balance of the knowledge sources such as WordNet 3.1. Although there are a number of knowledge sources that have created to contain the semantic features of the concepts in the language, they include only the classical semantic features.

5. Structural-based methods

The structural of method refer to the "semantic similarity" SS measures, which are depend on the features of the semantic taxonomy (ST), of the knowledge sources to weighting the features. There are several features in the semantic taxonomy that can used for measuring semantic similarity between two concepts, such as, depth, ancestors and path. In the previous study, many ways have used to classify these types of semantic measures. In this work, these measurements are clustering depending using weighting features to measures the semantic similarity (SS) into two main types: direct and indirect.

5.1 Direct Measures

The direct measures refer to the semantic measures, which directly exploited the structural weighting of features to compute the similarity between two concepts. The structural features are length; depth of concepts, depth of semantic taxonomy, and least common subsumer (LCS). Table 2 shows a summary of direct semantic measures.

Table 2. The summary of direct semantic measures

Author	Semantic measures	Features
[18]	$Sim_{Rada}(c_1, c_2) = \frac{1}{\delta(c_1, c_2) + 1}$	Length
[19]	$Sim_{LCh}(c_1, c_2) = -\log\left(\frac{\delta(c_1, c_2) + 1}{2 \times D}\right)$	Length, depth of semantic taxonomy
[20]	$Sim_{WuP}(c_1, c_2) = \frac{2 \times d(LCS(c_1, c_2))}{\delta(c_1, c_2) + 2 \times d(LCS(c_1, c_2))}$	LCS, length, depth
[21]	$Sim_{Li}(c_1, c_2) = e^{-\alpha \times \delta(c_1, c_2)} \times \frac{e^{\beta h} - e^{-\beta h}}{e^{\beta h} + e^{-\beta h}};$ $h = d(LCS(c_1, c_2))$ *where*	LCS, length, depth
[22]	$Sim_{Liu}(c_1, c_2) = \frac{\beta \times d(LCS(c_1, c_2))}{\beta \times d(LCS(c_1, c_2)) + \alpha \times \delta(c_1, c_2)}$	LCS, length, depth
[23]	$Dist_{Mubaid}(c_1, c_2) = \log((\delta(c_1, c_2) - 1)^{\alpha} CSpec(c_1, c_2)^{\beta} + k)$ $CSpec(c_1, c_2) = D - depth(LCS(c_1, c_2))$D is the depth of the semantic taxonomy	LCS, length, depth, depth of semantic taxonomy
[24]	$Sim(c_1, c_2) = -\log_2 \frac{\lvert T(c_1) \cup T(c_2)\rvert - \lvert T(c_1) \cap T(c_2)\rvert}{\lvert T(c_1) \cup T(c_2)\rvert}$ $T(c_i) = \{c_j \in C: c_j \text{ is a subsumer of } c_i\} \cup \{c_i\}$	Subsumes

5.2 Indirect Measures

The indirect measures refer to the semantic measures, which exploit the structural features to quantify the information content (IC) of a given concept and then use the IC to measure the semantic similarity between two concepts. These measures have classified in the previous researches as the IC-based measures. The IC based on previous researchers shown in Table 3.

Table 3. Information content

Author	IC metrics	Features
[25]	$IC_{Seco}(c) = 1 - \frac{\log(desc(c) + 1)}{\log(Max_{nodes})}$	Descendants, between two nodes in the semantic relation
[26]	$IC_{Zhou}(c) = k\left(1 - \frac{\log(desc(c) + 1)}{\log(Max_{nodes})}\right) + (1 - k)\left(\frac{\log(depth(c))}{\log(Max_{depth})}\right)$	Descendants, number of nodes in the semantic taxonomy, depth of concept, and depth of taxonomy
[27]	$IC_{Sanchez}(c) = -\log\left(\frac{\frac{\lvert Leaves(c)\rvert}{\lvert subsumers(c)\rvert} + 1)}{Max_{Leaves} + 1}\right)$	Leaves of concept subsumes, and number of leaves in the semantic taxonomy

[28]	$IC_{SanBatet}(c) = -\log(\frac{commonness(c)}{commonness(Root)})$ $commonness(c) = \begin{cases} \frac{1}{\lvert subsumers(c)\rvert} & where\ c\ is\ a\ leaf \\ \sum_{c' \in Leaves(c)} \frac{1}{\lvert subsumers(c')\rvert} & otherwise \end{cases}$	Leaves of concept subsumes, and number of leaves in the semantic taxonomy
[29]	$IC_{Taieb}(c) = \left(\sum_{c' \in V} Score(c')\right) \times AvgDepth(c)$ $Score(c) = \left(\sum_{a \in Hyper(c)} \frac{depth(a)}{Desc(a)}\right) \times Hypo(c)$	Descendants, depth of concept, hypernyms, ancestors
[15]	$IC_{Meng}(c) = \frac{\log(depth(c))}{\log(Max_{depth})} \times \left(1 - \frac{\log\left(\sum_{c' \in Hypo(c)} \frac{1}{depth(c')} + 1\right)}{\log(Max_{nodes})}\right)$	hyponyms of the concept c, depth of concept, number of nodes, depth of semantic taxonomy
[4]	$IC_{Aouicha}(c) = \sum_{c' \in Subgraph\ (c)} Score(c')$ $Score(c) = \left(\sum_{c' \in Hyper(C)} \left(-\log\left(\frac{1}{depth(c')}\right) \times Term(c')\right)\right) \times Term(c)$ $Term(c) = 1 - \frac{\log(HypoInfo(c) + 1)}{\log(maxHypoInfo)}$ $HypoInfo(c) = \sum_{c' \in descendants(c)} \frac{1}{depth\ (c')}$	depth, ancestors, hypernyms, descendants, and hyponyms

6. Semantic representation approach

After compiling the literature survey on semantic representation approaches, semantic measures, and knowledge-based text mining tasks, to develop and improve the semantic representation approach for text mining tasks/applications. In the semantic representation approach, we proposed a weighted semantic representation approach to handle the constant weighting assumption in feature-based methods using topological parameters. The researchers showed the numbers of depending on the techniques for Words/concepts on the classification. During our research, we have tried to consider the most recently confirmed efficient approaches that we have found the result of our investigation. In this section, we have tried to presenting a study on some research papers to give a view of three mechanization to grade the feature of concepts.

6.1 Text Classification Application

In text classification, we proposed a new weighting to extract the important features from the concepts by the taxonomy of semantic representation. Instead of using texting or word relation, we proposed using weighting to extract the features from concepts in an incorporate the features for capturing more semantic evidence all features of concepts have the same relevance constant weighting assumption The concept of semantic representation classify features by the feature-based method. A system based on the PM has been implemented and tested. The experimental results show that the proposed method works as an effective text classify.

1) KNN hash mapping - In this section, the hash mapping in KNN is very efficient in many tasks to reduction high dimensionality and pressuring the database. In spite of the fact that a lot of approaches have been developed to learn the dataset depending on hash functions in recent year, the hash mapping will select the depth and destiny depends on the taxonomy of weighting the features to output a good performance with agreeable statistical to challenge problems. Based on the observation the features that extremely related to the KNN classification accuracy, this paper a novel the feature- based on KNN-based on supervised hashing method. We proposed weighting on a feature-based method to the assumption the semantic representation to method the proximity relationships inveterate in training data.

2) Naïve Bayes - In this section, there are a few approaches that based on machine learning has presented. We will consider the method to be machine learning based on features selection. The interesting in the machine learning communication on the semantic representation based on semantic relation problem. Starting from simple statistical Naive Bayes-based methods, we describe methods using semantic representation fields.

6.2 Sentiment analysis

The sentiment analysis over the Social Media offers organizations a quick and efficient way to monitor the publics' feelings towards their brand, business, directors, etc. A wide range of features and methods for training sentiment classify for Social Media datasets have researched in recent years with varying results. In this paper, we introduce a new weighting feature-based approach to sentiment the analysis words in Social Media based on the measure of semantic taxonomy to reduce the noisy and irrelevant features. For each extracted features from the concept, we add its semantic concept (e.g. "politic") as an additional feature and measure the correlation of the representative concept with negative/positive sentiment.

6.3 Clustering Unsupervised

In the clustering of unsupervised feature selection using a combined weighting of a feature-based method to measure the similarity as measures described. The structural-based methods refer to the semantic, which based on the features of semantic taxonomy of the knowledge source. Although the units in the semantic representation that taxonomically extracted, the topological metrics (e.g. depth and density) ignored when computing the semantic similarity.

1. K-means - This method of vectors, originally from features based method, which is prevalent for clustering the features after made from semantic taxonomy. K-means clustering aims to root M documents perception into k clusters. The algorithm has a loose relationship to the k-nearest neighbour classifier, a prevalent machine learning technique for classification that is used with k-means. Applying the 1-nearest neighbour classifier the cluster centres obtained by k-means classifies new data into the existing clusters.

2. Hierarchical clustering - In data mining, the statistical of hierarchical clustering features method to combine the relations between two words on concepts, also called hierarchical cluster or hierarchical clustering analysis.

7. Result & discussion

In the semantic of the relation is described the exploited the computing the relatedness degree the taxonomy of the WordNet 3.1 for the comparison, the WordNet (Christiane Fellbaum, 1998), is a lexical of the database as a development from the University of the Princeton in 1985. Is the most of resource used to computing the semantic relations of the task, it's free and the document resource that is structure the semantics of the lexical contents of the English language. The WordNet 3.1 has organized the set of the synonyms called the synset, each of synset corresponds to the concepts are noun and verbs are organized in "is-a" of the taxonomy. The version is 3.1, the WordNet (Is-A) taxonomy assigned the nouns contains are 82,192 synset. In our comparison show the result between the proposed method and the previous methods using three of method as which taxonomy relation, non- taxonomy relation and glosses. In the significance of the measuring as using the ρ of the null of the hypothesis when it became true. In the most of the empirical analysis, is a result of the yielding the r value and the ρ are considered the statistical significance, in these results the considered statistically significant when the μ is a value of these result, to consider the highly statistically significant. In measuring the semantic similarity, there are two cases of the statistical of the significance are used to compare the similarity of values. The statistical of the significance of the feature-based method is different between the proposed method and human judgment in the three of the measures are (taxonomy relations, Non-taxonomy relations and Glosses), each one has the hypothesis is that the scores of the similarity. Table IV show the statistical of the significance feature-based method based on the semntic represeantion of the proposed method in the taxonomical of the relations.

Table 4. The statistical of the significance of feature-based method different between the proposed Method and human judgment in the taxonomy of relations.

Features		**Rada**	**Wu**	**Li**	**LC**	**Sebti**	**Sánchez**	**Meng**	**Taieb**	**Aouicha**	**Our result**
Taxonomy Relations	r	0.66	0.75	0.81	0.75	0.82	0.82	0.86	0.76	0.82	**0.82**
	ρ	0.75	0.76	0.74	0.75	0.78	0.74	0.77	0.76	0.76	**0.80**
	μ	0.70	0.75	0.77	0.75	0.80	0.78	0.81	0.76	0.79	**0.81**
Non-Taxonomy Relations	**Our PM**	**0.82**									
Glosses	**Non-zero**	**96%**									

8. Conclusions

In this paper, we proposed a new weighting of features based method, in this paper has described the weighting of features selection that is followed to locate the research objective and to coincide the scope of the feature based on the proposed method. Next, the details about the research phases are given, which consist of four phases starting with compiling of the literature review and defining the research problem followed by the collection of the text mining in social media and preparing it for the next phase. Based on these findings, using WordNet 3.1 as background knowledge for text clustering: First, experimenters should consider the nature of the dataset in hand and the diversity of topics before decide to use WordNet 3.1 for measuring similarity of relations. Second, the WordNet 3.1 structure does not seem to support the application of similarity measures. However, WordNet 3.1 can be better exploited by capturing specific types of relations such as "Synonymy, Hypernym, non-taxonomy, and Glosses". The statistical of the significance of feature-based method shows the proposed method 0.82 and non-zero 0.96%, is better from the other human judgment based on our methods as which (Taxonomy relations, Non-taxonomy relations and Glosses). To concerns, this feature-based method, which aims to achieve the research objectives and finally the evaluation measurements are highlighted. Each sub-phase of the design and development phase is constructed to achieve one objective of the research, which in turn tries to overcome some challenges and problems that are observed in the related works.

Acknowledgment

This work is supported by the Universiti Malaysia Pahang (UMP) in Research Grant UMP RDU1803141, PGRS190398.

References

[1] K. M. Hermann, D. Das, J. Weston, and K. Ganchev, "Semantic frame identification with distributed word representations," in *Proceedings of the 52nd Annual Meeting of the Association for Computational Linguistics (Volume 1: Long Papers)*, 2014, vol. 1, pp. 1448-1458.

[2] A. Saif, M. J. Ab Aziz, and N. Omar, "Reducing explicit semantic representation vectors using Latent Dirichlet Allocation," *Knowledge-Based Systems,* vol. 100, pp. 145-159, 2016.

[3] M. B. Aouicha, M. A. Hadj Taieb, and A. B. Hamadou, "Taxonomy-based information content and wordnet-wiktionary-wikipedia glosses for semantic relatedness," *Applied Intelligence,* journal article pp. 1-37, 2016.

[4] M. B. Aouicha, M. A. H. Taieb, and M. Ezzeddine, "Derivation of "is a" taxonomy from Wikipedia Category Graph," *Engineering Applications of Artificial Intelligence,* vol. 50, pp. 265-286, 2016.

[5] T. H. Rassem, N. M. Makbol, A. M. Hasan, S. S. M. Zaki, and P. Girija, "Restoring the missing features of the corrupted speech using linear interpolation methods," in *AIP Conference Proceedings*, 2017, vol. 1891, no. 1, p. 020119: AIP Publishing.

[6] A. M. Hasan, T. H. Rassem, and M. Karimah, "Pattern-Matching Based for Arabic Question Answering: A Challenge Perspective," *Advanced Science Letters,* vol. 24, no. 10, pp. 7655-7661, 2018.

[7] A. M. Hasan and L. Q. Zakaria, "QUESTION CLASSIFICATION USING SUPPORT VECTOR MACHINE AND PATTERN MATCHING," *Journal of Theoretical & Applied Information Technology,* vol. 87, no. 2, 2016.

[8] Q. Al-Tashi and A. M. Hasan, "WORD SENSE DISAMBIGUATION: A REVIEW," *Southern Connecticut State University, Hilton C. Buley Library,* vol. 1,2, pp. pp. 20-458, 2019. google scholar

[9] S. Fodeh, B. Punch, and P.-N. Tan, "On ontology-driven document clustering using core semantic features," *Knowledge and information systems,* vol. 28, no. 2, pp. 395-421, 2011.

[10] A. M. Hasan, T. H. Rassem, and M. Noorhuzaimi, "Combined Support Vector Machine and Pattern Matching for Arabic Islamic Hadith Question Classification System," in *International Conference of Reliable Information and Communication Technology*, 2018, pp. 278-290: Springer.

[11] K. Chen, Z. Zhang, J. Long, and H. Zhang, "Turning from TF-IDF to TF-IGM for term weighting in text classification," *Expert Systems with Applications,* vol. 66, pp. 245-260, 2016.

[12] X. Chen, Z. Liu, and M. Sun, "A unified model for word sense representation and disambiguation," in *Proceedings of the 2014 Conference on Empirical Methods in Natural Language Processing (EMNLP)*, 2014, pp. 1025-1035.

[13] S. Hassan, *Measuring semantic relatedness using salient encyclopedic concepts*. University of North Texas, 2011.

[14] M. A. H. Taieb, M. B. Aouicha, and A. B. Hamadou, "Computing semantic relatedness using Wikipedia features," *Knowledge-Based Systems,* vol. 50, pp. 260-278, 2013.

[15] L. Meng, J. Gu, and Z. Zhou, "A new model of information content based on concept's topology for measuring semantic similarity in WordNet," *International Journal of Grid and Distributed Computing,* vol. 5, no. 3, pp. 81-94, 2012.

[16] D. Sánchez, M. Batet, and D. Isern, "Ontology-based information content computation," *Knowledge-Based Systems,* vol. 24, no. 2, pp. 297-303, 2011.

[17] G. Pirró, "A semantic similarity metric combining features and intrinsic information content," *Data & Knowledge Engineering,* vol. 68, no. 11, pp. 1289-1308, 2009.

[18] R. Rada, H. Mili, E. Bicknell, and M. Blettner, "Development and application of a metric on semantic nets," *IEEE transactions on systems, man, and cybernetics,* vol. 19, no. 1, pp. 17-30, 1989.

[19] C. Leacock and M. Chodorow, "Combining local context and WordNet similarity for word sense identification," *WordNet: An electronic lexical database,* vol. 49, no. 2, pp. 265-283, 1998.

[20] Z. Wu and M. Palmer, "Verbs semantics and lexical selection," in *Proceedings of the 32nd annual meeting on Association for Computational Linguistics*, 1994, pp. 133-138: Association for Computational Linguistics.

[21] Y. Li, Z. A. Bandar, and D. McLean, "An approach for measuring semantic similarity between words using multiple information sources," *IEEE Transactions on knowledge and data engineering,* vol. 15, no. 4, pp. 871-882, 2003.

[22] H. Liu, H. Bao, and D. Xu, "Concept vector for semantic similarity and relatedness based on wordnet structure," *Journal of Systems and software,* vol. 85, no. 2, pp. 370-381, 2012.

[23] H. Al-Mubaid and H. A. Nguyen, "A cluster-based approach for semantic similarity in the biomedical domain," in *Engineering in Medicine and Biology Society, 2006. EMBS'06. 28th Annual International Conference of the IEEE*, 2006, pp. 2713-2717: IEEE.

[24] M. Batet, S. Harispe, S. Ranwez, D. Sánchez, and V. Ranwez, "An information theoretic approach to improve semantic similarity assessments across multiple ontologies," *Information Sciences,* vol. 283, pp. 197-210, 2014.

[25] N. Seco, T. Veale, and J. Hayes, "An intrinsic information content metric for semantic similarity in WordNet," in *ECAI*, 2004, vol. 16, p. 1089.

[26] Z. Zhou, Y. Wang, and J. Gu, "A new model of information content for semantic similarity in WordNet," in *Future Generation Communication and Networking Symposia, 2008. FGCNS'08. Second International Conference on*, 2008, vol. 3, pp. 85-89: IEEE.

[27] D. SáNchez and M. Batet, "A semantic similarity method based on information content exploiting multiple ontologies," *Expert Systems with Applications,* vol. 40, no. 4, pp. 1393-1399, 2013.

[28] M. A. H. Taieb, M. B. Aouicha, and A. B. Hamadou, "Ontology-based approach for measuring semantic similarity," *Engineering Applications of Artificial Intelligence,* vol. 36, pp. 238-261, 2014.

[29] M. A. H. Taieb, M. B. Aouicha, and A. B. Hamadou, "A new semantic relatedness measurement using WordNet features," *Knowledge and information systems,* vol. 41, no. 2, pp. 467-497, 2014.

IOP Conf. Series: Materials Science and Engineering **769** (2020) 012003 doi:10.1088/1757-899X/769/1/012003

A Preliminary Component Model for IoT

Arbaayah Kamarudin[1], Ng Keng Yap[1], Mohamed Abdullahi Ali[1] and Lim Swee Phang[1]

[1]Department of Software Engineering and Information Systems, Faculty Computer Science and Information Technology, Universiti Putra Malaysia

gs49499@student.upm.edu.my, kengyap@upm.edu.my, mrmoo2030@gmail.com and www.limsp95@gmail.com

Abstract. The enthusiasm of IT entrepreneurs in producing Internet of Things (IoT) systems is undeniable as currently, the number of connected devices is enormously increasing. Many research has been done to efficiently develop IoT systems. IoT systems are usually engineered from scratch. IoT component models have been introduced but lack of generic development framework or model that supports high reusability and loose coupling in dealing with the heterogeneous devices that can hinder its development. Thus, an IoT component model is proposed. Meta-modelling has been used to define the component model where the specific interaction and composition standard in a component are abstracted. IoT component model is intended to develop a prototype for IoT development. With this IoT prototype, IoT system developers will not need to develop everything from scratch every time, as generic components can be reused even when it is applied in different domains or during system enhancement is required. Smart home IoT system has been selected as a case study to evaluate our prototype tool. In this study, we provide an alternative way to develop IoT software in component-based software engineering method. A prototype has also been developed to assist reusability and reduce coupling between modules.

1. Introduction

The exponential growth of connected objects that can participate in the Internet of Things (IoT) ecosystem is inevitable as nowadays we witness the tremendous progress of interconnection between digital and physical devices [2]. Examples of physical devices, such as sensors, lights, fans, watches, cameras, alarms, medical devices, transport devices, and any other numerous devices- are aimed to be connected over a network so that they can communicate with each other and make decisions intelligently, anytime and anywhere.

IoT components are IoT sub-systems that interact with each other in an IoT system and can be reused in a new environment [2]. It allows developers to develop the IoT systems from existing IoT components rather than from scratch. The reusability of IoT components benefits the developers and organizations in many ways such as to decrease bugs and defects in the systems, to shorten the development time and cost, and to increase productivity.

IoT component model is used to define IoT components and its standard of interaction and composition of IoT systems [3]. Similar to other software component models, the IoT component model uses port connector type architectural units as components, with method calls through port-to-port connections as the mechanism to compose components.

To develop IoT systems from scratch requires tremendous effort as developers must have an in-depth understanding of technologies and the new application domain where so many software components have to be engineered from scratch especially when the systems have to deal with a huge number of sensors and actuators [9].

IoT systems have been adopted by many companies adopting monitoring, fault detection, and diagnosis of manufacturing systems (MFDD) [7]. This hinders component reusability even though their goals are the same – to achieve monitoring, fault detection, and diagnosis goals. The similar issue has been addressed in [5] whereby no unified and interoperable system which could be reused and redeployed in future smart cities.

There is no generic development framework or model for such IoT systems, which cause the producing of application codes tangled with very specific scenarios or algorithms [5]. Lack of proper abstractions for application development is also one of the factors [5].

We found that most IoT systems apparently share 3 common component types, namely sensor, actuator, and controller. All of them are driven by software. Sensors and actuators, in particular, have their own common functions, i.e. sensors are used to read data from the environment; whilst actuators are used to actuate physical response to the environment, either mechanically or electronically. This phenomenon seems to hold for any kind of sensors and actuators. Although these devices may work in different modes, such as toggling, continuous, intermittent, we believe that they can be reused if they are made generic and highly configurable.

In this paper, we propose an alternative way to develop IoT systems that emphasize on reusability and loose coupling. This is achieved by defining an IoT component model with generic and highly configurable component types. We classify components into 3 main types in general, namely sensor, actuator, and software. The sensor component can be instantiated to drive any kind of hardware sensor, similarly, the actuator component can be instantiated to drive any actuator hardware. The software component can be instantiated to encapsulate any control logic. A set of these instances will be composed to build an IoT system.

Abstraction and generalization are keys at large in defining the IoT component model. For instance, the sensor component is the type that generalizes the data reading function for all kinds of sensors. The data reading mode and data type are the intrinsic attributes of the sensor component, in which we believe they should be configured in their respective instance to suit the underlying sensor hardware and application purpose. For example, for temperature sensor hardware, the sensor (software) component instance can be configured to read a decimal value from the sensor hardware every second in one application; or read an integer value every 10 minutes in another application. The underlying sensor hardware may also vary even for the same configuration.

This paper proposes an IoT component model that considers reusability and loose coupling. Meta-modelling was used to define the component model. The IoT component model acts as the basis to develop a modelling tool prototype for IoT system. With this tool in place, when the performing system enhancement or developing a new system in different application domains, IoT system developers can reuse the components instead of developing it from scratch.

The remainder of the paper is structured as follows: Section 2 reviews existing research and related work. Section 3 presents the proposed IoT component model. Section 4 demonstrates the development of the IoT prototype tool. Section 5 discusses a case study. Section 6 presents discussion. Finally, Section 7 concludes our research.

2. Literature

Table 1 compares literature related to IoT architecture and determines whether software component models were defined or applied in these researches. We found that most of the literature suggests IoT system development without any software component model defined, there are prone to the adoption of SOA, Cloud API, or web services. Although [12] suggests component-based development, the authors did not generalize components like what we propose herein. Therefore, the component model is not IoT specific.

Table 1. Comparison for various IoT Architecture

Papers	Type of architecture	Component model
[4]	4 stages of IoT - Stage 1, where sensors, actuators, or any devices produce, receive and process the signals collected, Stage 2 where the data collected from the sensors aggregated and converted into digital stream, Stage 3 where the digitized data is being processed and Stage 4 where the data is forwarded to physical data centres or cloud-based systems.	No component model suggested
[16]	Integrated platform for heterogeneous systems which consists of IoT middleware and information bus. Heterogeneous systems consist of sensors to collect data to be supplied to the middleware.	No component models suggested systems as the resources of the data.
[13]	Layered architecture consists of 4 major layers. Smart Objects layer contains sensors and actuators, Networking Infrastructure layer for SaaS to intermediate smart objects with Information Service layer. Application layer supports various application such as security, home automation, and energy.	No component models suggested. Smart Objects layer contains sensors and actuators but not encapsulated in components.
[12]	The sensor node software architecture for this framework is designed in a multi-layered approach, which is the Application Layer (APL) and the Operating System Abstraction Layer (OAL). Components in each layer can be combined as one unit of concrete sensor node's software architecture.	There are component models but not specifically for sensors and actuators. Sensors and actuators as nodes.
[15]	Sensors are represented as nodes which designed in a distributed architecture.	No component models suggested.
[11]	Core microservice to control and coordinate other microservices. Device microservice contains the sensors and actuators.	No component models suggested.
[8]	The framework basically consists of sensors and data collection area, cloud-based backend, domain-specific software components and applications, and services area.	No component models suggested.
[3]	Semantic engine applied to IoT and smart cities, where they designed an entire workflow/semantic engine to semantically interpret and reason over data by providing services hiding semantic web technologies to IoT developers and end-users.	No component models suggested, only the semantics which limited to smart cities.

The study in [14] identifies top 10 industries investing in sensors, which are Energy and Mining, Power and Utilities, Automotive, Industrial, Hospitality, Healthcare, Retail, Entertainment, Technology, and Financial Services. These are the most affected areas by the expeditious dispersion of IoT technology, yielding numerous sensors to be developed.

Smart city is one of the examples of IoT application that society dreams of. For instances, smart surveillance, smart alarm, automated transportation, smart energy management systems, water distribution are IoT applications for smart cities [1]. [1] proposed a unified semantic engine for IoT and smart cities that they analyzed from web technologies implemented in IoT systems. However, the components of the semantic engine are difficult to be reused in other application domains as they are very specific to smart cities.

Apart from that, a generic microservice system framework for IoT application has been proposed in [10] that designed with a core service to control and coordinate other microservices. However, there is no component model adopted in this architecture to support reusability of the microservices.

IOP Conf. Series: Materials Science and Engineering **769** (2020) 012003 doi:10.1088/1757-899X/769/1/012003

3. Proposed Component Model

A Component model for IoT system has been proposed in this paper where the key component types are sensor and actuator. The component model is defined in meta-model as in Figure 1. This IoT component model defines generic IoT components. It also defines appropriate generic elements, composition rules to build an IoT system. In IoT systems, generic component types are sensor, actuator, and software.

In the meta-model, *Component* is the main class in the proposed component model, *Sensor*, *Actuator*, and *Software* are inherited from the component. A *Component* should have required ports, provided ports and may have CU (Computation Unit). A provided port has reference to required port, controllers and parameters when the component is composed with other components. A *controller* can be of types: *branching, sequence, and iteration*. An *Iteration* has a *For* and *Break* elements. A controller has provided ports, and required ports. A *for* has a *CounterLoop*, an *IterativeLoop*, and *Condition* elements. A *branching* can be of type: *Switch* or *If*. A *Switch* has a *Break* and a *Case*. An *If* has a *Condition*. A *software* component should have provided ports and required ports. However, a *sensor* and *actuator* should have a provided port with required port absent.

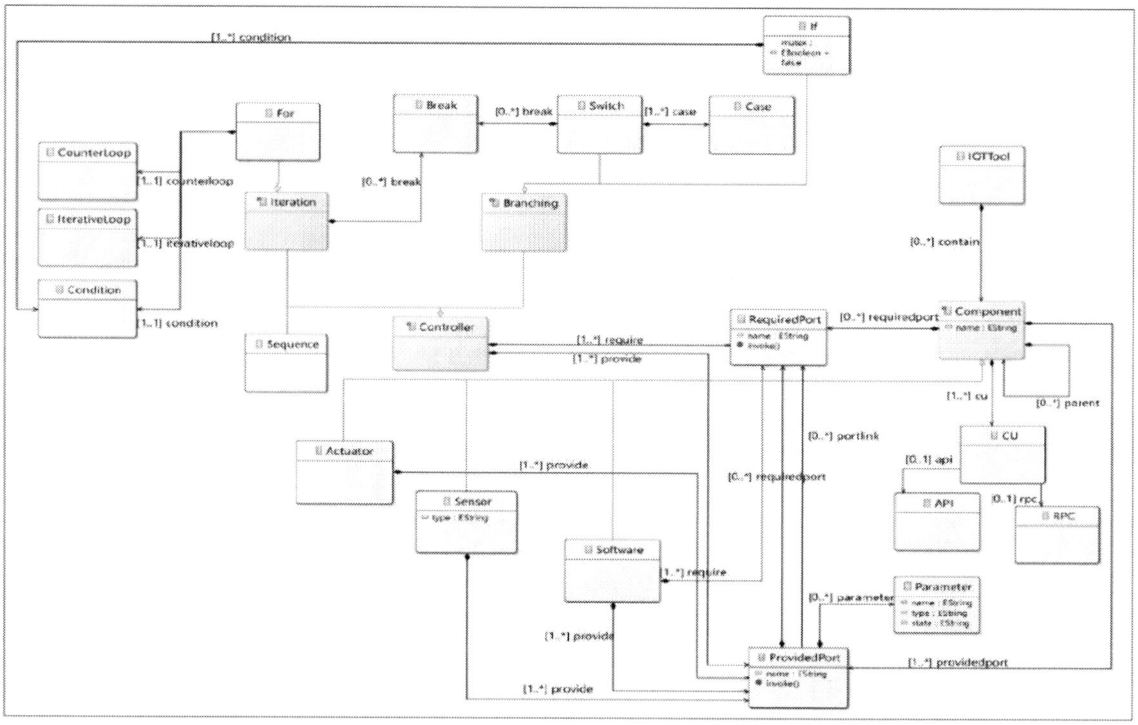

Figure 1. Meta-Model on Proposed IoT Component Model

4. Development IoT Tool

This IoT system modelling tool prototype is developed with Obeo Designer software. Obeo Designer is a graphical modelling workbench; we used it to facilitate the development of this tool.

The GUI of the development tool shown in Figure 2 provides component drag and drop to reduce the developer's effort to do coding. The palette section on the right of the screen consists of a list of components to create IoT component instances, its port, and associations. The components can be dragged from the palette and dropped into the canvas in the middle of the screen.

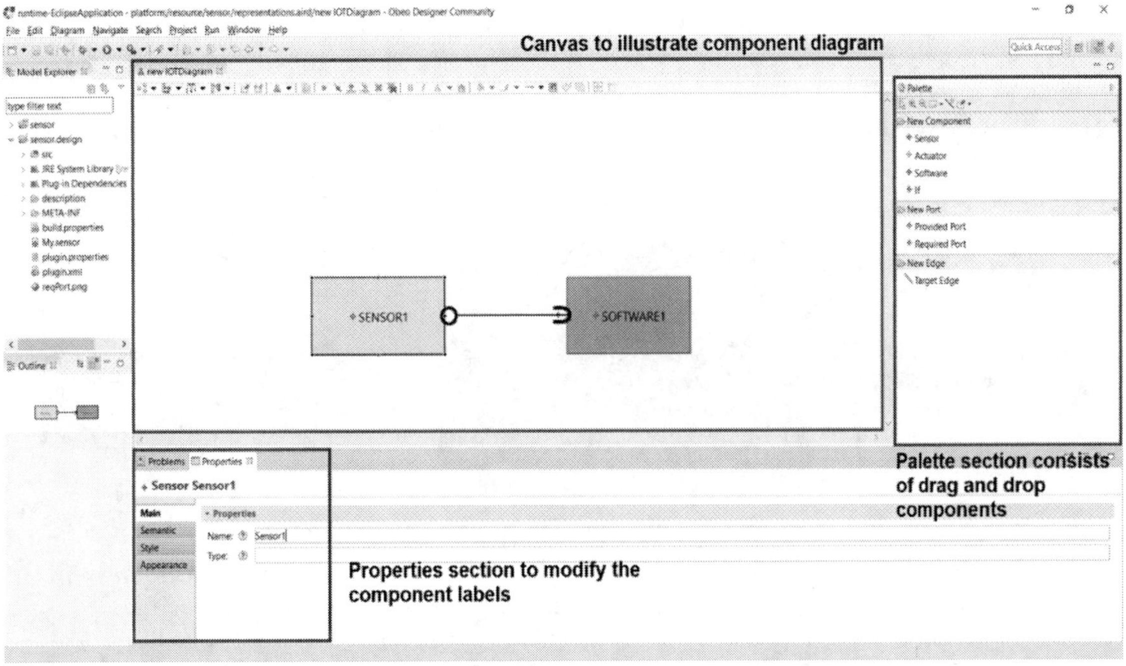

Figure 2. GUI of IoT Development Tool

5. Case Study

A smart home system [17] has been selected as a case study to evaluate the proposed IoT component model and its development tool. Two quality attributes were measured in this evaluation which namely: reusability of the components and loose coupling between components.

The selected case study allows us to describe the functions of the system and how it was implemented from scratch. The system presented is for monitoring and controlling smart home environment. As mentioned in [17], a smart home referred to a Internet-connected home or eHome as an environment for living that has highly advanced automatic systems.

The proposed smart home system mentioned in [17] consists of applications built on top of an IoT infrastructure. The main functions are:

5.1. Alert

The smart home system is able to sense the environment and sent alerts to users accordingly. The alert contains information related to the environmental data, which may include the levels of different gases in the environment, temperature, humidity, light, and intensity. An alert may be sent by email, text message, tweets or any social media on a regular basis at a predefined time.

5.2. Monitor

The smart home system is capable of monitoring its surrounding using sensors in general. It keeps track of every activity in a smart home where any further action can be taken or a decision to be made accordingly.

5.3. Control

The smart home system allows user to control different activities, which may include switching on and off the lights, air conditioners and other appliances, lock and unlock the doors, open or close the windows and doors and so on. They are actuators in general.

5.4. Intelligence

The smart home system can automatically make a decision upon the occurrence of various events. It can identify and react according to changing conditions and events. This is software-driven in general.

Figure 3 shows how Smart home system [17] was developed using FLIP. FLIP [17] is an open-source IoT platform, which has been used to develop the smart home system in the case study, which requires a lot of coding knowledge, in this case, C and Python.

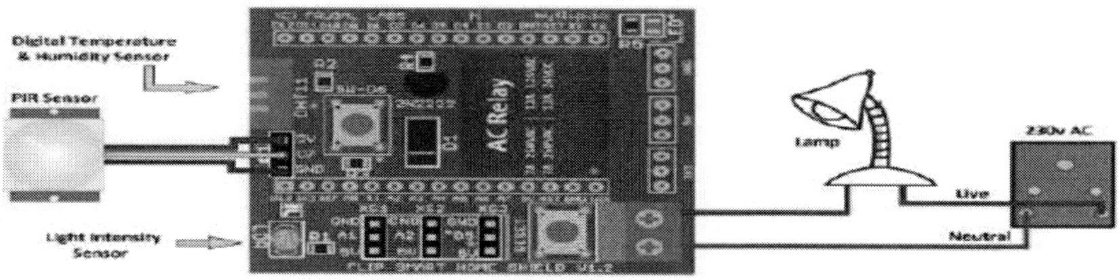

Figure 3. Architecture for Smart Home System [17]

Developers need to learn the programming skill to write codes. If the codes are written by experienced developers, it may be difficult to reuse as the codes are very system-specific.

This paper shows an alternative way to develop the smart home system based on the proposed IoT prototype tool as in Figure 4. The smart home system gathers sensor data from the environment and responds back to the environment through actuator according to the rules that have been set.

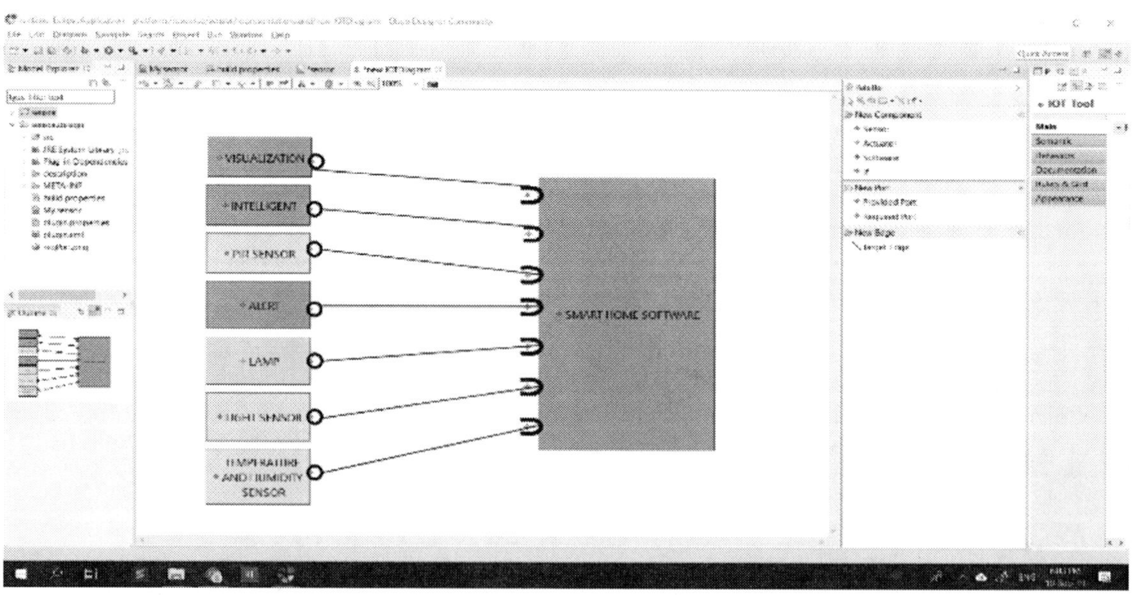

Figure 4. Smart Home System using Proposed IoT Prototype Tool

In Figure 4, boxes in blue represent sensor component, red represent software components and green represent actuator. The smart home software component is the main software component with three sensor types namely: light sensor, PIR sensor, temperature, and humidity sensor, which will gather sensor data from the environment.

The actuator called lamp will perform on/off tasks. The intelligent component is a software component that contains logics, which will automatically make decisions upon the occurrence of various events. The alert component is another software component, which is able to send alerts to the user

IOP Conf. Series: Materials Science and Engineering **769** (2020) 012003 doi:10.1088/1757-899X/769/1/012003

accordingly. As the components are independent of each other, the coupling between components is reduced. This promotes the reusability of the components.

Generic sensor and actuator components can be configured, and incrementally composed to the system. As the system evolves, components can also be replaced to update or upgrade the system. The generic sensor and actuator components are highly reusable.

6. Discussion

We do not claim that our approach is a silver bullet to IoT system development. Nevertheless, it does provide an alternative practical way for building IoT systems. We advocate that common component types should be generalized for reuse. This idea has been duly implemented as the IoT component model and subsequently the modelling tool prototype. We also demonstrated how these generic components can be reused and configured according to the IoT system requirements using a smart home case study.

Our modelling tool prototype provides an intuitive visual modelling view (refer Figure 4) which will inherently reduce the learning curve for developers. However, the modelling tool prototype is still under development. The component model is also in its preliminary stage and requires further refinement. Upon completion, the code generator is expected to generate code to compose components according to the system design model. This will significantly reduce IoT system development cost and time, and subsequently increase productivity.

Currently, most of the IoT systems are developed by coding and calling the web services as programming API. Web services provide essential distributed computing power on the cloud to complement the computing power on edge devices [8]. The typical example of IoT systems that require intensive central data processing from distributed edge devices is weather forecast. In our component model, web services can be implemented and encapsulated as the Computation Unit (CU) in our generic component.

7. Conclusion

In this paper, we have proposed an IoT component model to generalize common component types across different IoT systems, namely sensor, actuator, and software. The component model was defined using the meta-modelling technique. Then, we implemented a component-based software modelling tool according to the component model to evaluate how practical the component model is.

Although the component model is in its preliminary stage, whereby the prototype is only capable of modelling systems with a subset of controllers and the code generator is still incomplete, we could boldly conclude that the prototype has improved the productivity of IoT system development, at least in system design.

System design can be done using the drag-and-drop feature provided by the modelling tool prototype. This has also been demonstrated using a smart home system. The use of generic reusable and configurable components for sensors and actuators renders the IoT systems built using this model loosely coupled.

We have demonstrated that system design can be achieved by the drag-and-drop feature provided by the modelling tool prototype with a smart home system case study. The use of generic reusable and configurable components for sensors and actuators has also rendered the smart home system loosely coupled.

8. Acknowledgement

The authors of this work gratefully acknowledge Ministry of Education (MOE), Malaysia for their financial support under Fundamental Research Grant Scheme (FRGS) (Ref:FRGS/1/2015/ICT01/UPM/02/6).

References

[1] A. Gyrard and M. Serrano 2015 A Unified Semantic Engine for Internet of Things and Smart Cities: From Sensor Data to End-Users Applications *IEEE International Conference on Data Science and Data Intensive Systems* Sydney, NSW pp. 718-725.
[2] Ayala, I., Amor, M., Fuentes, L., & Troya, J.M 2015 A Software Product Line Process to Develop Agents for the IOT *Sensors* Pg 15640-15660.
[3] B. Councill, G. T. Heineman Definition of a Software Component and Its Elements Available from: http://heim.ifi.uio.no/~frank/inf5040/CBSE/Component-Based_Software_Engineering_-_ch1.pdf.
[4] Bharani 2019 What is IOT (Internet of Things)? IOT Architecture Explained *Available from: https://www.edureka.co/blog/what-is-iot/* Acesssed 9-10-2019.
[5] C.M. de Farias, I.C. Brito, L. Pirmez, F.C. Delicato, P.F. Pires, T.C. Rodrigues, I.L. dos Santos, L.F.R.C. Carmo, T. Batista 2017 COMFIT: A development environment for the internet of things *Future Generation Computer Systems*. 75 128–144.
[6] Flip Smart Home Kit n.d. Retrieved from http://myflip.io/Pages/SmartHome.html.
[7] I. Yen, S. Zhang, F. Bastani and Y. Zhang 2017 A Framework for IOT-Based Monitoring and Diagnosis of Manufacturing Systems *2017 IEEE Symposium on Service-Oriented System Engineering (SOSE)* San Francisco CA pp. 1-8.
[8] Im, J., Kim, S., & Kim, D 2013 IoT mashup as a service: cloud-based mashup service for the Internet of things. In 2013 IEEE International Conference on Services Computing pp. 462-469.
[9] Kamienski, C., Jentsch, M., Eisenhauer, M., Kiljander, J., Ferrera, E., Rosengren, P., Thestrup, J., Souto, E., Andrade, W.S., & Sadok, D.F 2017 Application development for the Internet of Things: A context-aware mixed criticality systems development platform *Computer Communications 104* 1-16.
[10] L. Sun, Y. Li and R. A. Memon 2017 An open IOT framework based on microservices architecture *China Communications* vol. 14, no. 2, pp. 154-162.
[11] Lengyel, L., Ekler, P., Ujj, T., Balogh, T., & Charaf, H 2015 SensorHUB: an IOT Driver Framework for Supporting Sensor Networks and Data Analysis *International Journal of Distributed Sensor Networks*, 109.
[12] Nguyen, X. T., Tran, H. T., Baraki, H., & Geihs, K 2015 FRASAD: A Framework for Model-driven IOT Application Development *IEEE 2nd World Forum on Internet of Things (WF-IOT)* 387 – 392.
[13] R. G. Anvekar and R. M. Banakar 2017 IOT application development: Home security system *2017 IEEE Technological Innovations in ICT for Agriculture and Rural Development* Chennai pp. 68-72.
[14] S. Bauer, F. Burkitt, C. Curran, L. Gioia 2014 Sensing the Future of The Internet of Things,*PwC 6th Annual Digital IQ*.
[15] S. S. Siddula, P. Babu and P. C. Jain 2018 Water Level Monitoring and Management of Dams using IOT *2018 3rd International Conference On Internet of Things: Smart Innovation and Usages* Bhimtal pp. 1-5.
[16] T. Joseph, R. Jenu, A. K. Assis, V. A. S. Kumar, P. M. Sasi and G. Alexander 2017 IOT middleware for smart city: An integrated and centrally managed IOT middleware for smart city *2017 IEEE Region 10 Symposium* Cochin pp. 1-5.
[17] T. Malche, P. Maheshwary 2017 Internet of Things (IoT) For Building Smart Home System International conference on I-SMAC (IoT in Social, Mobile, Analytics and Cloud) 65-70.

Preserving the Topology of Self-Organizing Maps for Data Analysis: A Review

B Yusob[1], Z Mustaffa[1] and S M Shamsuddin[2]
[1]Faculty of Computing, College of Computing and Applied Sciences, Universiti Malaysia Pahang, 26300 Kuantan, Pahang, Malaysia
bariahyusob@ump.edu.my
[2]School of Computing, Faculty of Engineering, Universiti Teknologi Malaysia, 81310 Johor Bahru, Johor, Malaysia
mariyam@utm.my

Abstract. In Kohonen's Self-Organizing Maps (SOM) algorithm, preserving the map structure to represent the real input patterns appears to be a significant process. Misinterpretation of the training samples can lead to failure in identifying the important features that may affect the outcomes generated by the SOM model. This paper presents detail explanation on SOM learning algorithm and its applications. Some issues related to SOM's architecture are also discussed, namely the formulation of training data from input samples, and the Best Matching Unit (BMU) identification for better visualization of large datasets, and improvement made to the SOM algorithm.
***Keywords**: Self-organizing maps, data analysis, computational intelligence*

1. Introduction

The self-organizing map (SOM) refers to a data visualization technique initiated by Professor Teuvo Kohonen [1]. This technique decreases the dimensions of data using self learning method. SOM has often been applied in the early phase of data analysis, particularly in the input data preparation, in addition to applications in numerous fields.

The conventional SOM network is composed of two phases; training and testing. SOM learning occurs during the training phase, while the process of identifying the output occurs during the testing phase. In the training phase, a sample of training data are produced by converting the input data into normalized values. Next, the Best Matching Unit (BMU) is recognized as the unit that holds the nearest distance value to the normalized input features. All neurons captured within the neighborhood of the BMU are re-calculated so that their values are reflected in the input features, to preserve the mapping topology by training the weights to simulate the actual features of the datasets.

Several major issues in the SOM learning algorithm may eventually lead to the failure of identifying of BMU [2]. The first issue refers to the pre-processing of input data for training. Missing important knowledge during data pre-processing can affect the performance of the SOM model. Next, the second issue is misrepresenting the datasets that could unable the identification of potential BMU, and therefore, produce poor mapping topology. Since SOM incorporates an iterative process, the learning process is lengthened and may generate irrelevant outputs [2]. Thus, in order to ensure the correct interpretation of data, quality measurement is required for the map topology. According to Hamel [3], quality measurement should cover two aspects: 1) mapping in the input data, and 2) the topological quality of the map. So far, most studies have only covered one aspect; either the input data or the topological structure of the map. This seems to reduce the accuracy of the SOM output.

In order to address these issues, enhancement to the SOM learning algorithm has been proposed in many prior studies. Many researchers have improved the SOM model by modifying the SOM map structures, such as lattice and map dimensions, to preserve the input data in the map topology [4][5]. Apart from that, improvements were made in SOM learning parameters itself, including initialization of weight values, learning rate, and neighborhood function [6][7]. Some studies have integrated other techniques, such as statistical and intelligent approaches as solutions to deal with issues linked with SOM, for example, Particle Swarm Optimization [8], neural gas [9], and Support Vector Machine [10].
In this paper, we study the SOM algorithm and to identify issues related to its architecture. We focus on the it's task as data analysis tools in various domains. In Section 2 of this paper, brief description on the standard SOM algorithm is presented. Various past works on SOM research in data analysis are mentioned in Section 3. Followed by issues identified in Section 4. This paper ends with a conclusion and justification for the need of future improvement to the SOM algorithm.

2. The Standard Kohonen's Self-Organizing Maps

Over the years, the SOM algorithms have emerged to be an efficient tool for data exploration due to its simplicity, practical approach, and its ability to visualize input data from higher to lower dimensions [11]. The standard Kohonen's SOM is comprised of single-layer neurons located on a regular low-dimensional lattice, usually one-dimension (1D) or two-dimension (2D). Each neuron k is represented by an n-dimensional weight vector $w_k = \{w_{k1}, \cdots, w_{kn}\}$, where n refers to the dimension of input data.

The standard Kohonen's SOM learning algorithm is given as follows:

i) Initialization of weights, radius, learning rate, and number of iteration.
ii) Input unit is randomly chosen from training samples.
iii) BMU is identified based on the closest unit with the selected input unit using distance calculation. A common formula is the Euclidean distance, as follows:

$$Euclid = \sqrt{\sum_{i=0}^{i=n}(V_i - W_i)^2} \tag{1}$$

where V denotes input unit, while W represents weight unit.

iv) Identification of neighborhood based on radius using Gaussian function. During training, radius is exponentially decreased via exponential decay:

$$\sigma(t) = \sigma_0 \exp\left(-\frac{t}{\lambda}\right), t = 1,2,3,\ldots \tag{2}$$

where σ_0 = radius at t_0, λ = time constant, and t = number of iteration.

v) Weight values for BMU and all units within the neighborhood are updated as follows:

$$W(t+1) = W(t) + L(t)(V(t) - W(t))$$
$$L(t) = L_0 \exp\left(-\frac{t}{\lambda}\right), t = 1,2,3,\ldots \tag{3}$$

where, $W(t+1)$: weight for $t = 1$, $L(t)$: learning rate at time t, and V : input unit.

When updating weights, the exponential decay function is used to decrease the learning rate value. This learning rate is decreased exponentially with time.

The closer a unit moves to the BMU, the bigger the weight values for update. In order to achieve the above function, the following equation is used:

$$\Theta(t) = \exp\left(-\frac{dist^2}{2\sigma^2(t)}\right), t = 1,2,3,\ldots \tag{4}$$

where, $dist$: distance between BMU and the closest unit, and σ :radius.

The weights are updated as follows:

$$W(t+1) = W(t) + \Theta(t)L(t)\big(V(t) - W(t)\big) \tag{5}$$

vi) Step (ii) is repeated until the maximum number of iteration is attained.

3. SOM for Data Exploration and Analysis

SOM has been widely used in data mining for many types of real domains. Most of the domains have huge volumes of complex and multi-dimensional data [12]. These domains include classification and clustering, web mining, business and finance, industrial process, biological and environmental data, image segmentation, image processing, and surface reconstruction, pattern recognition, and for general data visualization involving universal datasets [12].

One of the most important features of SOM is that it neither requires prior knowledge of the datasets nor correlation among data dimensions when using SOM for data analysis and visualization. SOM automatically learns the given datasets by generating a map topology that resembles the input data, and throughout the learning process, the map topology is preserved. The map consists of *n x m* weight units that are randomly initialized at the beginning of the network development. The preservation process occurs through BMU identification, which is the weight unit that best reflects the input data. In order to identify the BMU, SOM would calculate the distance between weight and input units using distance measures. For better map topology quality, all weight units within the neighborhood of BMU are updated to preserve the input features.

4. Architecture Aspects of SOM

The important feature of SOM that makes it a powerful visualization tool lies in its characteristic that preserves its topology. Several issues in Kohonen's SOM demand consideration to ascertain its functions, namely its network structure, identification of BMU, initialization of neighborhood radius, pre-processing of input data to train data formation, learning rate, and number of iterations during training [5]. In Yusob [5], the structure of Kohonen's SOM was built for a classification task, wherein enhancement was made on its topological structure. A hexagonal lattice with improved neighborhood area calculation was proposed and exceptional outcomes were recorded.

However, initialization of training data sample for SOM learning in the pre-processing phase of the SOM network, and BMU identification seem to remain as issues that are yet to be addressed. Thus, the following sections depict the related information regarding improving the pre-processing phase in SOM network, as well as identification of BMUs in Kohonen's SOM for better and faster visualization of multidimensional and multivariate large datasets.

4.1. Pre-Processing of Input Data for Training Sample

In SOM, only training data are needed due to its capability in processing and analyzing input data to determine its features and characteristics. Pre-processing input data refers to transformation of datasets into a standard form so that any analysis model, particularly neural networks, can process the datasets in a more accurate and efficient manner. Some methods have been applied in past studies to transform input data, which is also called normalization, including statistical or z-score, min-max, sigmoidal, energy, and principle component normalization [13] [14].

Haimoudi [2] highlighted the importance of training datasets in SOM learning. Missing of important knowledge during data pre-processing is a critical issue that can affect the success of the proposed SOM model. Linear pre-processing method lengthens the learning process and yields irrelevant outcomes.

4.2. Identification of Best Matching Unit (BMU)

Upon determining the unit closest to the input, distance between all weights and input unit is calculated. The weight that produces minimum distance is declared as the "winner" or the BMU due to the minimum dissimilarity between the weight and the input unit, which shows that the weight has become more similar to the input unit. If more units hold similar value of distance, BMU is chosen randomly among these units. Some techniques that calculate distances are Euclidean, Manhattan, and Chebyshev [5].

Euclidean distance, which is 2-norm, is generalization derived from Pythagoras theorem for area with more than two coordinates. Equation 6 presents the formula for Euclidean distance.

$$\text{Euclidean distance, } d_e = \sqrt{\sum_{i=1}^{p}(x_i - y_i)^2} \tag{6}$$

Manhattan distance, which is also known as city block distance, is a 1-norm distance. It is based on a car travelling in a city block. The journey forms a network where distance is calculated from the starting point, through the stopping points, and until the final destination (final point). The formula for Manhattan distance is as follows:

$$\text{Manhattan distance, } d_m = \sqrt{\sum_{i=1}^{p}|x_i - y_i|} \tag{7}$$

Chebyshev distance is the maximum value distance that computes definite distance between the coordinates of a pair of objects. Such distance is used for ordinal and quantitative variables, which is suitable for crucial processing time.

$$\text{Chebyshev distance, } d_{cij} = \max_k |x_{ij} - x_{jk}| \tag{8}$$

4.3. Quality Measurement of Map Preservation

SOM is a powerful technique used for visualization and to analyses high dimensional and complex datasets. Kiviluoto [15] reported a few standards to assess the quality of a map. The first measure shows how constant the structure is; inferring that nearby input features should be similar to the weights as well as the targeted output features. The second standard includes the complexity of the mapping. Maps with high complexity in terms of its dimensions have more values to be processed. Lastly, the third standard is the correctness of the mapping that reflects the likelihood of the input set. Other than that, other methods that define the quality of topology preservation [16] are topographic error, distortion, classification accuracy, mean square quantization error, and goodness of map.

4.4. Growing Grid in SOM Topology

Based on the conventional SOM learning algorithm, learning parameters are initialized to prepare for the training process. The number of neurons, map size, dimensions, and other parameters are fixed, and usually do not change regardless of the nature or behavior of the studied datasets. This causes difficulty in visualizing the data, provided that prior data are absent while dealing with large datasets. It is also difficult to obtain optimal network topology for each training iteration (or epoch) when the SOM network architecture is fixed and pre-defined before initiating SOM network training.

In order to address this issue, some researchers have proposed the growing map, for example, a 2x2 map that grows dynamically by adding a whole row or column of nodes at the boundary nodes [8].

5. Conclusion

The main function of SOM is map preservation. SOM preserves the characteristics of input data by ensuring that the trained weight units are computed to achieve the same value as the input data as much as possible. This process covers from BMU identification to updating the units within the BMU neighborhood. BMU refers to the individual weight unit that resembles the input data the most. In SOM, before learning the input data, the input data have to be transformed into uniformed units that can be processed by SOM. This is where the pre-processing phase becomes crucial. Conventional SOM usually uses the normalization method to standardize the datasets into a range value between 0 and 1. The normalization method that can generate good training samples is essential in SOM training, since SOM relies solely on good training samples to yield better results for data clustering. It is indeed an integral initiative to generate high quality training samples so that SOM can preserve its map topology for better data visualization.

This paper has presented reviews on issues related to SOM particularly in its architecture. From our findings, most of the recent researches in SOM have paid attention towards improving BMU identification through the generation of good training samples. However, most of the present real problem scenarios involve large and complex data, while some provide insufficient data for processing. Even with improvised SOM models, longer time is required to process, especially for BMU identification. Therefore, a conclusion from the related studies is that there is a need to enhance the existing SOM technique for data analysis problem, particularly within the classification task. The new improved SOM model should not only be able to identify BMU correctly, but also must be able to learn faster. This can be done by enhancing the SOM learning algorithm focusing on the structure of the map such as lattice to reduce the map's quantization error during training.

Acknowledgments
The work is funded by the Ministry of Higher Education FRGS under Project ID FRGS/1/2018/ICT02/UMP/03/2: RDU190169. The authors would like to thank all reviewers for their helpful comments, and Universiti Malaysia Pahang for the support.

References

[1] Kohonen, T. (1995). *Self-organizing maps*. Berlin, Heidelberg. Springer. ISBN: 3-540-58600-8.

[2] Haimoudi, E.K., Cherrat, L. Abdoun, O. and Ezziyyani, M. (2016). Practical Application of the Data Preprocessing Method for Kohonen Neural Networks in Pattern Recognition Tasks. In: The Sixth International Conference on Advances in Information Mining and Management (IMMM2016). Pp. 38-44.

[3] Hamel, L. (2016). SOM Quality Measures: An Efficient Statistical Approach. In: Merenyi, E., Mendenhall, M.J. and O'driscoll, P. (eds). *Advances in Self-Organizing Maps and Learning Vector Quantization*, Proceedings of the 11th International Workshop WSOM 2016, Houston, Texas, USA, January 6–8, 2016. Pp. 49-59.

[4] Ahmad, R. and Kim, D.H. (2015). An Extended Self-Organizing Map based on 2-opt algorithm for solving symmetrical Traveling Salesperson Problem. In: *Neural Computing and Applications*: 26:987–994. DOI 10.1007/s00521-014-1773-z.

[5] Yusob, B. (2007). *Pembaikan Struktur Kekisi Heksagon Dalam Pembelajaran Rangkaian Kohonen*. Masters Thesis. Universiti Teknologi Malaysia.

[6] Natita, W., Wiboonsak, W. and Dusadee, S. (2016). Appropriate Learning Rate and Neighborhood Function of Self-organizing Map (SOM) for Specific Humidity Pattern Classification over Southern Thailand. In: *International Journal of Modeling and Optimization*, Vol. 6, No. 1. Pp. 61-65.

[7] Fidae, H., Zakariae, E.N., Abdelatif, E.S. and Mohamed, E. (2015). Preprocessing phase for initializing the PRSOM architecture. In: *2015 Intelligent Systems and Computer Vision (ISCV)*. IEEE. Pp. 1-4. DOI: 10.1109/ISACV.2015.7106190.

[8] Forkan, F. (2009). *Algoritma Rangkaian Kohonen-Swarm Bagi Penjanaan Permukaan*. Master's -Thesis. Universiti Teknologi Malaysia.

[9] Vergara, J.R., Estévez, P.A., and Serrano, A. (2016). Segment Growing Neural Gas for Nonlinear Time Series Analysis. In: Merenyi, E., Mendenhall, M.J. and O'driscoll, P. (eds). *Advances in Self-Organizing Maps and Learning Vector Quantization*, Proceedings of the 11th International Workshop WSOM 2016, Houston, Texas, USA, January 6–8, 2016. Pp. 107-117.

[10] Mudali, D., Biehl, M., Leenders, K.L., and Roerdink, J.B.T.M. LVQ and SVM Classification of FDG-PET Brain Data. In: Merenyi, E., Mendenhall, M.J. and O'driscoll, P. (eds). *Advances in Self-Organizing Maps and Learning Vector Quantization*, Proceedings of the 11th International Workshop WSOM 2016, Houston, Texas, USA, January 6–8, 2016. Pp. 205-215.

[11] Cottrell M., Olteanu M., Rossi F., Villa-Vialaneix N. (2016) Theoretical and Applied Aspects of the Self-Organizing Maps. In: Merényi E., Mendenhall M., O'Driscoll P. (eds) Advances in Self-Organizing Maps and Learning Vector Quantization. *Advances in Intelligent Systems and Computing*, vol 428. Springer, Cham.

[12] Xu, Y., Xu, L., and Chow, T.W.S. (2011). PPoSOM: A new variant of PolSOM by using probabilistic assignment for multidimensional data visualization. In: *Neurocomputing*. 74. Pp. 2018-2027.

[13] Awang, S., Sulaiman, J., Mohd, N.N.L, Bayuaji, L. (2017). Comparison of accuracy performance based on normalization techniques for the features fusion of face and online signature. In: *Advanced Science Letters*, 2017. Volume 23, Issue 11. Pp. 11233-11236

[14] Yusob, B., Shamsuddin, H.S.M, and Hamed, H.N.A. (2013). Spiking Self-organizing Maps for Classification Problem. In: Procedia Technology. Volume 11, 2013. pp. 57-64. ISSN: 2212-0173. https://doi.org/10.1016/j.protcy.2013.12.162.

[15] Kiviluoto, K. (1996). Topology preservation in self-organizing maps. In: *IEEE Neural Networks Council (Eds.), Proceedings of International Conference on Neural Networks (ICNN'96)*. Vol. 1. New Jersey, USA. Pp. 294–299.

[16] Mariette, J. and Villa-Vialaneix, N. (2016). Aggregating Self-Organizing Maps with Topology Preservation. In: Merényi E., Mendenhall M., O'Driscoll P. (eds) Advances in Self-Organizing Maps and Learning Vector Quantization. *Advances in Intelligent Systems and Computing*, vol 428. Springer, Cham.

IOP Conf. Series: Materials Science and Engineering **769** (2020) 012005 doi:10.1088/1757-899X/769/1/012005

Implementation of the Greedy Algorithm to determine the nearest route Search in distributing food production

Hasanul Fahmi[1], Muhammad Zarlis[2], Erna B Nababan[3], Poltak Sihombing[4]
Program Studi Doktor Ilmu Komputer, Fakultas Ilmu Komputer dan Teknologi Informasi, Universitas Sumatera Utara, Jl. Universitas No.9A, Medan-Indonesia

***Abstract*.** Distribution is the activity of allocating goods from producers to customers within allocated time. Managing product distribution is very important for the survival of a company, yet improper planning and management will cause the production is stacked in the warehouse thus this will cause unhealthy financial cycles and turnover to the company itself. Nowadays many companies produce foodstuffs but one of the major problems faced by the companies are the delays in distributing food production. It is found that the common difficulty faced by the food production companies in distributing the product due to long travelling distance, traffic congestion, natural disasters, ambiguous routes as well as unstructured routes. To overcome the problems, greedy algorithm is implemented. The algorithm able to identify the optimal pathway and time consumption by finding the closest distance, followed by the utmost distance. Therefore this will reduce the cost of distribution and other company resources. The results of this study show that the algorithm identifies Route 5 as the best route. This is because it will go through 5 cities, namely A-B-C-F-G-H with distance of 648KM. Although it is considered as long travelling distance, on the other hand the algorithm suggests that 648KM is the optimum distance to travelling to city H.

1. Introduction

In order to achieve the objectives and objectives of the company in the field of marketing, each company conducts distribution activities of raw goods and also finished goods. Channelling is an activity to manufacture goods production to the hands of consumers at the right time. Some understanding of distribution channels relating to distribution channels by experts in their field as follows: Early research suggests that "distribution channels are a series of organizational participants that perform all the functions needed to deliver products/services from sellers to end buyers". Thus, the conclusion can be taken distribution is the flow of goods from the manufacturer to the consumer (Sorenson, Hinds, & Kiesler, 2006).

Distribution is a support for every company because when the distribution of Lancer then production can continue to run and no build-up of production goods in the warehouse that cause harm to the company(Lakhani & Panetta, 2007), Nowadays, there are very many companies engaged in production of food as well as beverages.

The results of observations that have been done in one of the companies apparently there is a route that is unclear, causing a delay in delivery and stacking of production goods are used so that it will cause a very significant loss to the company. Nowadays almost in every company both engaged in the production of food and beverages in distributing products with less clear and irregular routes (Touretzky & Hinton, 1988). So this is what triggers interest for researchers to apply the algorithm greedy where this algorithm can help the problems that occur in the company so that it can minimize the distribution costs incurred by looking for Closest distance first then the farthest distance.

In previous research conducted by Sulistiorini and Mahmudy about "application of Evolution Strategies algorithm for the optimization of two stage goods distribution". The genetic algorithm is applied in the case of two-stage goods distribution to optimize the distribution route of goods, to perform the technique of optimizing route distribution used genetic algorithms by using crossover techniques with one cut point crossover, Mutations with Exchange mutation and selection using elitism selection by assuming 2 factories with 5 distributors and 10 agents represented using a permutation representation. The final result is the optimum distribution path with the right capacity. However, it is not able to provide a solution to determine the fastest route with minimum cost(Sulistiorini & Mahmudy, 2015).

2. Methodology

2.1 Greedy Algorithm

In the previous research conducted by greedy algorithm is a type of algorithm that uses a problem solving approach by finding the maximum value while at each step. The maximum value of temporary is known as the local maximum. In most cases, the greedy algorithm will not produce the optimal solution, and the greedy algorithm usually provides a solution that approaches the optimum value in a fast enough time (Soltys, 2012).

The principle of the Greedy algorithm is to take every opportunity that exists at the moment, without regard to the future consequences. The Greedy algorithm forms a solution from step by step, and at every step should be made the best decision in determining the choice (Caruana & Freitag, 2014). At each step of the algorithm Greedy take the best option that can be gained at that moment without regard to future consequences (Bird, 1992). Figure 1 describes the concept of A directional graph from point A to B

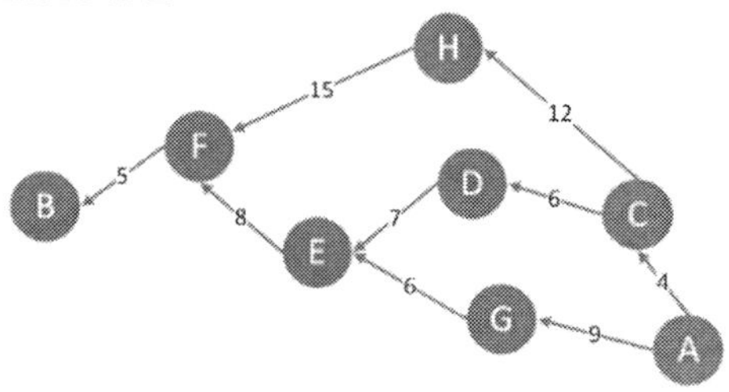

Figure 1 Directional Graph from point A to B

In previous research conducted by J. Webster & M. Moshkov to find the shortest distance from A to B, a greedy algorithm will execute steps like the following(Webster & Moshkov, 2015):

1. Visit one point on the graph, and grab the whole point that can be visited from the present point.
2. Find the local maximum to the next point.
3. Mark the graph now as the graph you have visited, and move to a predefined local maximum.
4. Go back to step 1 until the destination point is obtained.

If you have to communicate the above steps in graph A to B earlier then we will get a move as follows:

1. Start from the starting point (A). Take the whole point that can be visited.

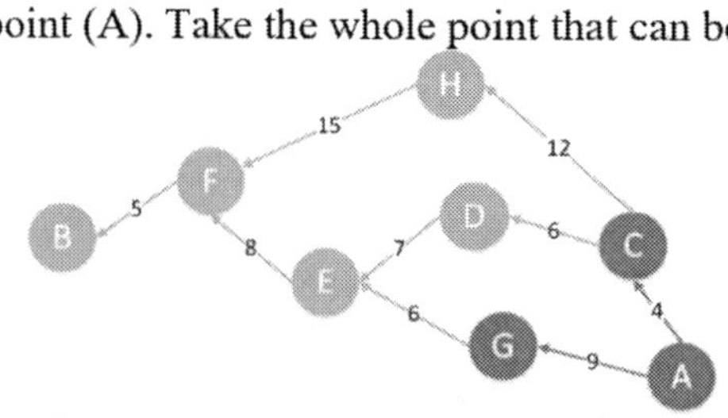

Figure 2 Greedy Steps

2. Local maximum is to C, because the distance to C is the closest.
3. Mark A as the point that has been visited, and move to C.
4. Take the whole point that can be visited from C.

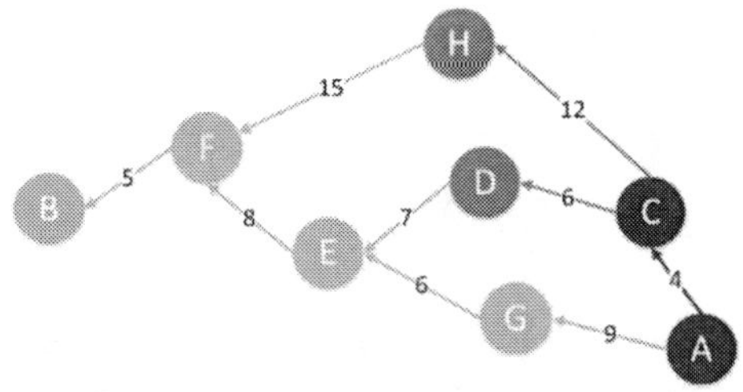

Figure 3 The Second Step Of Greedy

5. Local maximum is to D, with a distance of 6.
6. Mark C as the point that has been visited, and move to D.

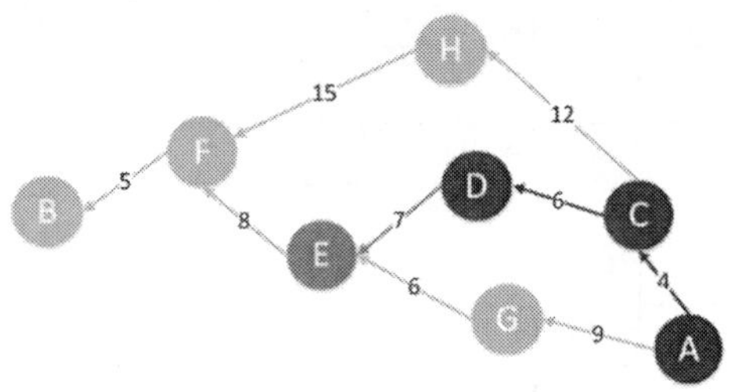

Figure 4 The Third Step Greedy

7. (Next step is handed over to the reader as exercise)

In this research will be conducted analysis on the search of product distribution route to 5 distributors and 10 agents with attention to distance and duration of inter-city travel so that distribution time can be utilized efficiently.

3. Result and Discussion

Researchers have represented distributing their products to many cities. The city that is closest to the position of the home industry we illustrate is the town of Lubuk Pakam. Then the city that often ask for our products are the high cliff, Pematangsiantar, Medan, Binjai, Kabanjahe, Sidikalang and ACEH. The distribution path is depicted with the following nodes,

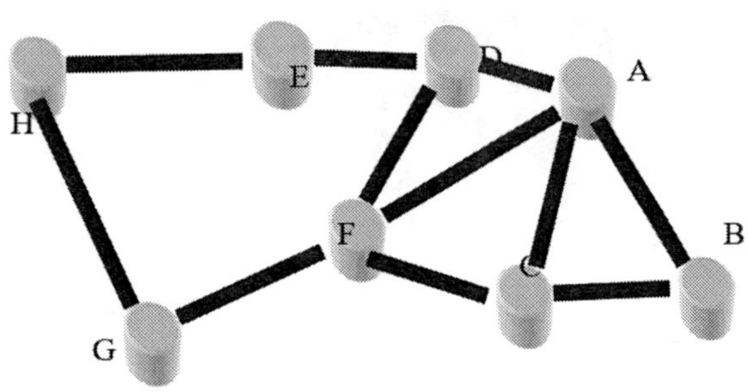

Figure 5 Distribution Path Illustration

Table 1 City name and code

Num	City	Code
1	Lubuk Pakam	A
2	Tebing Tinggi	B
3	Pematang Siantar	C
4	Medan	D

5	Binjai	E
6	Kabanjahe	F
7	Sidikalang	G
8	NAD	H

Table 2 Distance between cities

Num	City code	Distance
1	A - B	40 km
2	A – C	69 km
3	A – D	23 km
4	A – F	65 km
5	B – C	46 km
6	C – F	100 km
7	F – G	76 km
8	G – H	386 km
9	D – E	42 km
10	D – F	75 km
11	E – H	399 km

Based on the Greedy algorithm, the route starts from the first node to the node that is expected to be optimal, thus there are several routes found. These routes are described with the following nodes:

A. Route 1

The first route algorithm starts from City A to H City with distribution line that passes 2 cities

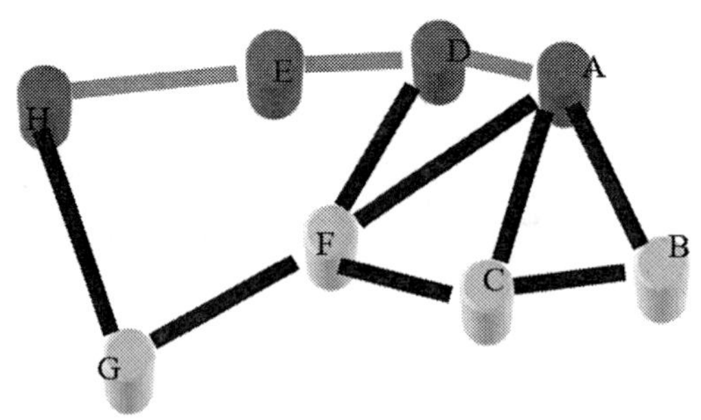

The Total distance from city A to City H is:
Rute1 = 23 km + 42 km + 399 km
= 464 km
The city passed is A-D-E-H.

B. Route 2

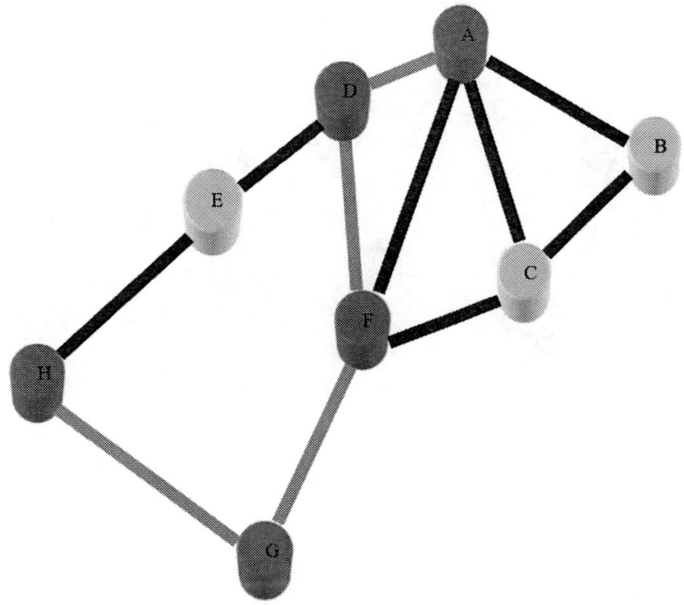

The Total distance from city A to City H is:
Rute2 = 23 km + 75 km + 76 km + 386 km
= 560 km
The skipped city was A-D-F-G-H.

C. Route 3

The third route begins from City A to the city of H with a distribution line that passes 2 cities.

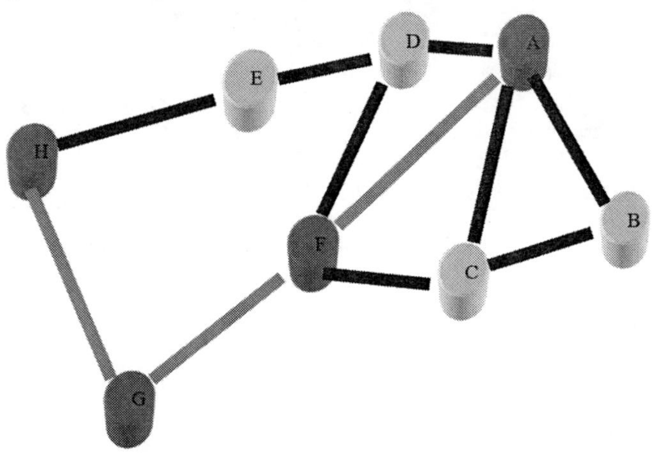

The Total distance from city A to City H is:
Rute3 = 65 km + 76 km + 386 km
= 527 km
The city being skipped was A-F-G-H

D. Route 4

The fourth route algorithm starts from City A to H City with distribution line that passes through 3 cities.

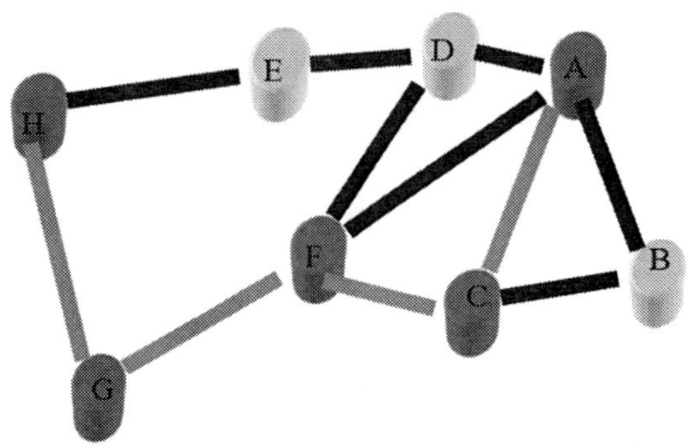

The Total distance from city A to City H is:
Rute4 = 69 km + 100 km + 76 km + 386 km
= 631 km
The skipped city was A-C-F-G-H.

E. Route 5

The fifth route algorithm starts from City A to H City with distribution line that passes 4 cities

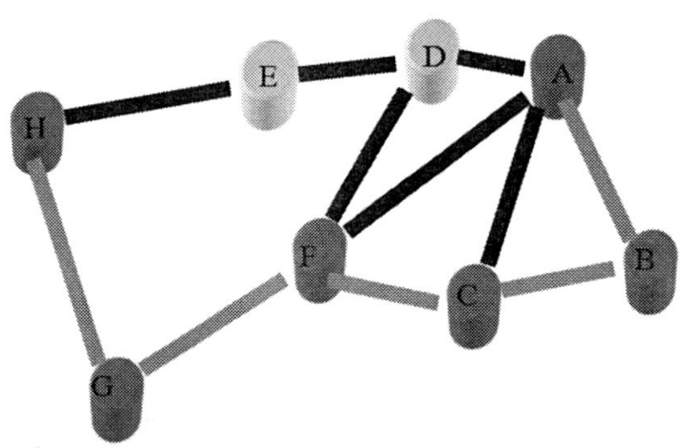

The Total distance from city A to City H is:
Rute5 = 40 km + 46 km + 100 km + 76 km + 386 km
= 648 km
The skipped city was A-B-C-F-G-H.
From the search results of the above route, obtained 5 distribution route data such as the table below:

Table 3 Route search results

Num	Route	Pathway	Distance
1	Route 1	A-D-E-H	464 km
2	Route 2	A-D-F-G-H	560 km
3	Route 3	A-F-G-H	527 km
4	Route 4	A-C-F-G-H	631 km
5	Route 5	A-B-C-F-G-H	648 km

Based on the concept of the Greedy algorithm, the most popular pathway is Route 5 which passes through the city of A-B-C-F-G-H which is considered as the best track. With a distance of 648 km.

IOP Conf. Series: Materials Science and Engineering **769** (2020) 012005 doi:10.1088/1757-899X/769/1/012005

4. Conclusion

The use of the GREEDY algorithm in the process of distributing food production obtained very good results, where the distance to distribute foods from the city of A-H is only 648KM and passing through five cities only. There are still 2 cities that are passed by, yet greedy algorithm able to decide that the 2 cities are not worth passing.

References

[1]Bird, R. S. (1992). Two greedy algorithms. *Journal of Functional Programming*. https://doi.org/10.1017/S0956796800000368

[2]Caruana, R., & Freitag, D. (2014). Greedy Attribute Selection. In *Machine Learning Proceedings 1994*. https://doi.org/10.1016/b978-1-55860-335-6.50012-x

[3]Lakhani, K. R., & Panetta, J. A. (2007). The Principles of Distributed Innovation. *Innovations: Technology, Governance, Globalization*. https://doi.org/10.1162/itgg.2007.2.3.97

[4]Soltys, M. (2012). Greedy Algorithms. In *An Introduction to the Analysis of Algorithms*. https://doi.org/10.1142/9789814271424_0002

[5]Sorenson, O., Hinds, P. J., & Kiesler, S. (2006). Distributed Work. *Administrative Science Quarterly*. https://doi.org/10.2307/3094928

[6]Sulistiorini, R., & Mahmudy, W. F. (2015). Penerapan algoritma evolution strategies untuk optimasi distribusi barang dua tahap. *Jurnal Mahasiswa*.

[7]Touretzky, D. S., & Hinton, G. E. (1988). A distributed connectionist production system. *Cognitive Science*. https://doi.org/10.1016/0364-0213(88)90029-8

[8]Webster, J. G., & Moshkov, M. (2015). Greedy Algorithms. In *Wiley Encyclopedia of Electrical and Electronics Engineering*. https://doi.org/10.1002/047134608x.w8262

IOP Conf. Series: Materials Science and Engineering **769** (2020) 012006 doi:10.1088/1757-899X/769/1/012006

A Model Of Factors Influencing Users' Adoption Of Internet Of Things Services: A Case Study Of Iraqi Educational Institutions

Hayder Salah Hashim[1], Zainab Amin Al-Sulami[1]

[1]Basrha University, Basrha-Iraq

Hayder_alasadi@uobasrah.edu.iq, zainab.abduljabba@uobasrah.edu.iq

Abstract. In the revolution of information technology, the appearance of Internet of things (IoT) is considered a significant evolution that can enhance the efficiency and prosperity of daily activities. In this context, a few people are aware of the important benefits that can be gained from using IoT services. Generally, previous researchers have studied the technical aspects of IoT, such as architectural dimensions, models, wireless sensor networks and attribute-based signature. The aim of this study is to identify the factors that influence users' adoption of IoT services at an educational institution in Iraq. The stratified sampling techniques were used to collect data from 221 users (students and lecturers) at university of Basra. Meanwhile, results corroborate that social influence is the most important factor in relation to the behavioral intention (BI) to use IoT services follow by effort expectancy, security and, performance expectancy. Facilitating conditions and BI have an effective influence on the use behavior of IoT services. Ultimately, contributions and recommendations were demonstrated.

Keywords: Internet of things, Higher Education Institutions, UTAUT, Security, TAM, DOI

1. Introduction

The appearance of (IoT) was introduced by Ashton in 1999 [1].It is a networking of electronic connection that enables the gathering and transferring of information amongst physical objects [2].Moreover, it refers to the outcomes of development, such as new IP protocol versions, broadband and combination of nanotechnology with many electronic devices (i.e. mobile phones), in the telecommunication field. Consequently, IoT is created on the basis of the idea of simultaneously integrating electronic devices that are connected within a network with other users [3]. Ultimately, IoT services are growing intensively when the devices of users, enterprises and stakeholders are connected to the Internet network [4].

Despite the increasing number of IoT services and advanced features that IoT offers, such as the maximum connection efficiency of users worldwide, a few people are aware about the important benefits of using these services. In addition, people have a slight perception regarding their motivation to continue using IoT services in daily life. In fact, many people are apprehensive about these services due to serious concerns, such as security breach [4].

The majority of the literature has covered the technical aspects of IoT usage. Moreover, previous studies have focused on architectural dimensions[5], wireless sensor network and attribute-based signature[6].Meanwhile, other relevant research has discussed other issues, such as descriptions, concepts, business models, opportunities and challenges [5],[7],[8],[9].Finally, previous studies have neglected the aspect of IoT users and their perception about the technology.

This research aims to discover the factors that influence the adoption of IoT services at higher institutions in Iraq. The technology acceptance model (TAM) was applied in order to evaluate the perception of users regarding the adoption of IoT[1]. Conversely, the acceptance unified theory and use of technology (UTAUT) was confirmed to be influential and capable in interpreting the difference in the acceptance of technology more than TAM and previous models[10],[11].They also corroborated that UTAUT model is unlike the TAM model was originally proposed to investigate the user adoption of technology [12],[13],[10]. However, despite the high explanatory power, UTAUT conceptualization was argued due to the absence of security factors. Current researches have confirmed the importance of security factors in the decision of adoption technology [14], [15].

Literary, developed countries captured the majority of the studies on IoT services .This study investigates the adoption of IoT services by users at educational institutions in Iraq, which is the perfect place for IoT. In addition, extensive international and domestic investment and economic growth coupled with limited legacy infrastructure set the stage for a digital revolution in the Iraq. The application of the technology in the country could help in minimising waste resources and increase the efficiency of educational institutions. Thus, this study will be conducted in educational institutions in Iraq to examine the ability of these organisations to apply the technology and utilise its benefits.

The present study conducted to discover the factors that influence the adoption of IoT services by users at higher educational institutions in Iraq. Meanwhile, this research is structured as follows. The neglected areas, discussions that demonstrated in the literature review, as well as the proposed aims that presented and highlighted in the Introduction section. And the third section explains the research models and the development of hypothesis.

Furthermore, the fourth part discusses the research methodology, and the fifth part demonstrates and discusses the results of this study. Ultimately, the last section derives the conclusion, limitations and, directions for advanced research.

2. Review of literature

2.1 Iot

Formulation of IoT depends on the concept of the integration of devices with embedded sensors, which permit them to connect simultaneously via an Internet interface. Although traditional Internet services connect people through communication and information sharing, IoT connection requires a vast network to enable communication amongst objects, machines and internal sensors. Conceptually, IoT can be defined as the process of aggregating network capacities, intelligence and smart objects that interact autonomously with users. Technically, it is composed of three components, namely, hardware, infrastructure and application and services [3],[16].Hence, IoT consists of a huge web that is created from physical objects, such as sensors and machines, that establishes standard communication between them and within remote databases [9].Furthermore, IoT provides users with the ability to manage and control their electrical machines, such as heating and lighting, security and lighting systems from anywhere through their computers and mobile devices [5].Ultimately, IoT must be distinguished from traditional Internet services. That is, IoT has the ability to remotely interact with machines and objects instead of merely providing online services.

2.2 Iot In educational institutions

The interface between IoT and education can be traced back to 1999 when the Auto-ID Centre of the Massachusetts Institute of Technology first proposed the IoT concept. IoT has brought a huge transformation to the educational community. In the era of rapid technology development, top management and lecturers in educational institutions are challenged to change their thought processes on teaching and educating the new generations of students. Moreover, pressure is increased on the education institutions to upgrade their educational methods and respond to the technological changes in the marketplace. In this essence and through IoT, educational institutions should redirect its strategy to enhance teaching, learning and researching activities. In addition, IoT can motivate students to be responsive and interactive with lecturers in such a manner that enables lecturers to obtain the best understanding and communicating space with students [17].For instance, IoT can facilitate an interactive English teaching course that teaches students based on individual differences and enhance their capacities to be creative

2.3 UTAUT

The UTAUT model was developed to accumulate previous interrelated models, such as the motivational model (MM),TAM, social cognitive theory and , diffusion of innovation theory (DOI) [10].The UTAUT is composed of six main factors, namely, performance expectancy (PE), effort expectancy (EE), social influence (SI), facilitating conditions (FCs), behavioural intention (BI) and use behaviour. In addition, The UTAUT contains four main moderators which are age, experience, voluntariness and gender. Table 1 provides the main factors of the UTAUT and previous models from which these factors are derived.

Table 1. Main factors of the UTAUT

Factor	Previous Models
Performance Expectancy	Perceived usefulness (TAM/TAM2/ C-TAM-TPV) extrinsic motivation (MM); job-fit (MPCU); Relative Advantage (DOI).
Effort expectancy	Perceived easy to use (TAM/TAM2); Complexity (MPCU/DOI).
Social influence	Subjective norm (TRA, TAM2, TPD/DTPB, C-TAM-TPD); Social factors (MPCU); Image (DOI).
Facilitating Conditions	Perceived behavioral control (TPB/DTPE, C-TAM-TPB); Facilitating conditions (MPCU); Compatibility (DOI).

Previous studies have described the UTAUT as a new theoretical model, which strengthens the inadequacies of previous models[12],[14].The exploratory power of the UTAUT model, which can reach up to 70%, is higher than those of previous models [12],[18],[19].

The majority of previous studies have used the TAM or DOI to discover factors that influence the adoption of IoT[1],[9].The literature focuses less on the use of the UTAUT model to determine the factors that influence the adoption of IoT. Moreover, the UTAUT was developed to assess individual acceptance of a new technology. It was applied in many areas, such as social networking sites[20] near field communication for mobile devices and services [21]and cloud computing by universities[14]. However, the use of the model in IoT studies remains limited. Ultimately, this research used the UTAUT model interrelated with security factor.

3. Conceptual model And research hypotheses

3.1 Conceptual model

The UTAUT model deploys in this research as conceptual model due to its main role to investigate user adoption of technology. This research also incorporates the factor security (SEC). Issues in SEC are crucial for ensuring the successful adoption of an appropriate IoT. In addition, researchers criticised the UTAUT and TAM because they excluded SEC from their models [22],[23]. Figure 1 presents the conceptual model of this study.

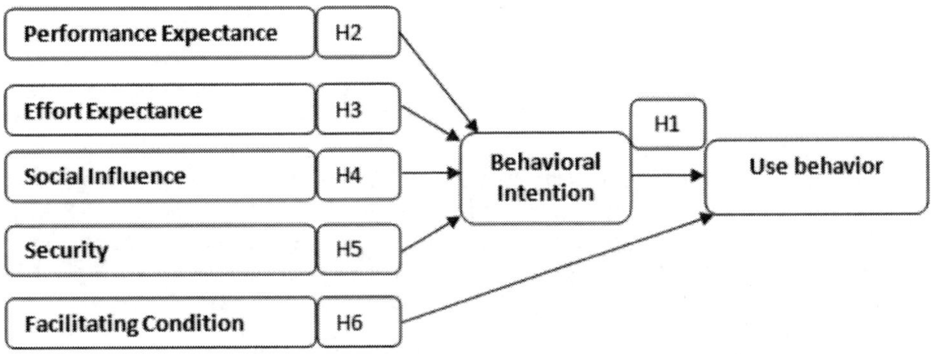

Figure 1. Conceptual model for user adoption

3.2 Research hypotheses

BI and use behavior are the dependent factors of this research. The UTAUT model indicated a direct connection between BI and use behaviour [10].Meanwhile, [9] corroborated that BI strongly influences the real usage of IoT services. Regarding this research, BI is predicted to influence the use behaviour of IoT. Thus, the following can be hypothesised:

H1: BI exerts appositive influence on the usage behaviour of IoT services.

The first factor of the UTAUT is PE, which is also a critical factor for user adoption. In this study, the PE factor has important role in DOI as relative advantage and has the same important in TAM as perceived usefulness [10],[24],[25]. In addition, it represents the users' thought process about how a specific technology improves its performance upon use. For instance, IoT technologies can save time and effort and speed up the processes in retail stores. Users positively perceive these improvements in services. Apparently, IoT technologies enhance adoption rates by enhancing services in users' ordinary life. Empirically, studies in the IoT field have proved that PE influences users' BI to adopt IoT technology [1],[26]. In this study, users' intention to adopt IoT services is expected to increase if they realised the advantages of using this technology. Hence, we propose the following hypothesis.

H2: PE has a positive influence of users' BI to use IoT services.

The EE factor has important role in DOI as complexity and has the same important role in TAM as perceived ease of use [10],[25]. It represents users' perceived efforts when they apply IoT technologies. To encourage users to adopt IoT services and technologies, users must perceive that IoT services are easy to apply without any bold obstacles and efforts. Further, studies have proved that the perceived ease of use in the TAM and UTAUT positively influences the BI of users to adopt the IT technology [14],[24],[27]. Hence, we propose the following hypothesis:

H3: EE has a positive influence on users' BI to use IoT services.

The social aspect plays an important role in how decisions are formulated and executed. The acceptance of any innovative technology is an essential factor to assess [9]. Generally, when a new technology is initially used, most users lack trusted information about this technology. Therefore, users rely on the social network context, such as opinions of friends, family and classmates[24].Therefore, users' intention to adopt any technology is influenced by social impact, judgement and evaluations from other people. For instance, SI is argued to exert an important impact on the privacy information of users[9]. In terms of IoT technology, many researchers have argued that SI has a significant impact on users' acceptance of IoT technology[1],[24].In this study, the user intention to adopt the IoT services is predicted to be influenced by users who have previously used this technology. Hence, we propose the following hypothesis:

H4: SI has a positive impact on BI of user to use IoT services.

The UTAUT was criticised for not contain factors like SEC in the conceptualization model [14],[19]. Recent researches argued that security factor have strong impact on users decisions to adopt IoT services [7],[8].Security is the degree of data confidentiality, authentication and access control within an IoT network[8]. Generally , security factor play an important role in information systems field .Particularly, security also play the same important role in the IoT services domain.[28].Consequently, several researchers have conclude same results and affirmed the importance of security for the adoption of IoT [7],[8],[9]. Therefore, the present study assumes that the SEC factor exerts a positive influence on the BI to adopt IoT services. Thus, we propose the following hypothesis:

H5: High SEC has a positive influence on users' BI to use IoT services.

The FC factor proved to have a direct influence on user behaviour[10]. FCs is described as certain conditions for users to adapt technology [12]. Thus, FCs represents the required resources, capability and relevant information that users must acquire and perform successful behaviour. Users require having essential skills to the IoT services and technology. In addition, FCs has the same role as FCs in (MPCU)[10], and compatibility in DOI. Therefore, on the basis of the above discussion, the present study assumes that the FC factor exerts a positive influence on user behaviour to adopt IoT services. Thus, we propose the following hypothesis:

H6: FCs has a positive influence on the use behaviour of IoT services.

The above-mentioned hypotheses were proved empirically to discover factors that influence users' acceptance of IoT services. Thus, the current research deploys specific questionnaire items to confirm the proposed model hypotheses and factors that could affect the acceptance of user to adapt IoT services. Table 2 illustrates the selected questionnaire items that could be used to prove the proposed hypotheses (H1– H6).

Table 2. Questionnaire items based on model factors

Factors\Authors	Items	Questions
Performance Expectancy [9] ,[29]	5	Using IoT services improves my work/life performance. Using IoT services helps me obtain information that is useful for my work/life. Using IoT services enhances my work/life efficiency. Using IoT services enables me to accomplish my work/life more quickly. *Overall, using IoT services creates significant learning spaces.
EE [14]	5	Using IoT services is easy. Learning to use IoT services is easy. Using IoT services is not a burden during transition. Using IoT services does not require considerable mental efforts. Overall, I find doing my day-to-day work/study easy with IoT services.
SI [14]	6	I use IoT services because my friends use IoT services. I use IoT services because my family members use IoT services. My professors have been supportive of using IoT services. I use IoT services because the mass media

		encourages the use of IoT services. Word-of-mouth stimulates me to use IoT services. The university has supported the use of IoT services.
FCs [14]	5	I have the hardware and software for using IoT services. I have the skills and knowledge for using IoT services. When I have problems using IoT services, someone can help me solve them. I always have electricity supply to use IoT services. I have Internet connection to utilize IoT services.
SEC [9],[14]	5	Using IoT services is associated with extreme uncertainty. Using IoT services service would lead to a loss of SEC. Service stability is a determinant of using IoT services. My decision to use IoT services exposes me to security concerns. Overall, considerable security is involved in using IoT services.
BI [9],[14]	4	I hope to use IoT services. I intend to switch from the traditional use of service to IoT services. I intend to increase my use of IoT services in the near future. I intend to recommend my friends/family members to use IoT services in the future.
Use Behaviour [14]	5	Managing my devices using IoT services is easy. *I am regularly using IoT services to control devices. *I am regularly using IoT services to manage devices. I rely on IoT services to control my devices. I have used IoT services in my university.

4. Research methodology

In this study, population refers to all users of IoT services in Basra University, which is composed of33,567 students and 2,848 academic staff according to the university website[30]. Students pertain to individuals who are pursuing diplomas and Bachelor, Masters and PhD degrees, and academic staff refers to the lecturers.

A sample refers to a group that is selected from the population for representation [31]. Proportional stratified sampling was used to obtain the number of samples from each group of users. Meanwhile, the sample must be representative of the entire population. However, the current population is composed of two groups with different characteristics. Thus, we select the stratum as groups of the population. In addition, a randomly selected sampling technique is used for each group. Finally, the sample size is 379, of which 349 were students and 30 were academic staff [32].

4.1 Research instrument

The data were collected by questionnaire survey. The questionnaire consists of two parts. Part A consists of demographics, such as age, gender, occupation, qualification and general information regarding the Internet, computers and IoT. Meanwhile, part B consists of questions regarding the proposed factors of this study. The scale of five point likert was used with the range (1 for strongly disagree ,2 for disagree, 3 for neutral, 4 for agree , and 5 for strongly agree).

4.2 Questionnaire validity and reliability

Questionnaire validity was confirmed by consulting four experts from Basra University, Mustansiriya University and Tenaga National University. The experts evaluated and adjusted the cohesion and interconnections of the questions. In addition, reliability was examined by conducting a pilot test on 38 students and academic staff. Cronbach's alphas for the reliability of the pilot test and final collected data were within acceptable thresholds. Table 3 provides the reliability results.

Table 3. Questionnaire reliability

Factor	Items	Reliability values (221)
PE	5	**0.78**
EE	5	**0.79**
SI	6	**0.73**
FCs	5	**0.72**
SEC	5	**0.74**
BI	5	**0.71**
Use behaviour	5	**0.72**

5. Data collection

Student and academic staff received the questionnaires via e-mail. Further, the respondents were given approximately five weeks to fill in the questionnaires. Finally, questionnaire data were collected, and 221 usable sets were entered into SPSS V 22.0.

5.1 Demographics of respondents

Demographic data were obtained from users in Basra University (students and lecturers). In total, 221 respondents were involved to answer the questionnaire items. Most of the respondents were aged 21 to 40 years (126 or 57%).In addition, the respondents were composed of 170 students (76.9%), and of which121 have bachelor degrees (54.8%). The level of computer and Internet knowledge of the respondents was generally positive.

5.2 Hypothesis testing

5.2.1 Correlation analysis. This research used correlation analysis, and correlations between proposed factors should be tested for regression analysis. The purpose for this notion is correlation can pinpoint whether multicollinearity exists or not[33]. Table 4 shows the Pearson correlation coefficient. Between factors, coefficients that are less than 0.90 indicate that multicollinearity is a problem [31].

Table 4. Correlation values

N=221	PE	EE	SI	FCs	SE	C BI	Use behaviour
PE	1	0.50	0.40	0.21	0.30	0.24	**0.25**
EE	0.50	1	0.64	0.48	0.51	0.57	**0.69**
SI	0.40	0.64	1	0.60	0.49	0.65	**0.67**
FCs	0.21	0.48	0.60	1	0.42	0.53	**0.59**
SEC	0.30	0.51	0.49	0.42	1	0.26	**0.52**
BI	0.24	0.57	0.65	0.53	0.26	1	**0.70**
Use behaviour	0.25	0.69	0.67	0.59	0.52	0.70	**1**

5.2.2 Regression analysis. The research hypotheses were tested using regression analysis. Table 5 provides the results, which indicate that the highest predictors of BI are SI followed by EE, SEC and PE. The table also reveals that use behaviour is influenced by FCs and BI. Moreover, the regression coefficient indicates the size effect between independent and dependent factors, which is denoted as (B) or beta. Sig refers to the P-value, which must be less than 0.05 to consider the hypothesis significant. The T statistic is the result of dividing B by the Std. error.

Table 5. Coefficient of regression analysis

M		**Unstandardised coefficients**		**T**	**Sig.**
		B	*Std. error*		
BI	(Constant)	0.55	0.35	1.57	0.16
	PE	0.16	0.07	2.28	0.02
	EE	0.18	0.06	3.03	0.02
	SI	0.21	0.09	2.33	0.00
	SEC	0.17	0.04	4.25	0.00
Use behaviour	(Constant)	−0.25	0.24	−1.05	0.46
	FCs	0.68	0.43	1.73	0.01
	BI	0.35	0.06	5.83	0.02

The results claimed that the influence of BI on use behaviour is significant (B=0.35, P-value<0.05).Therefore, H1 is supported. The effect of PE on BI was affirmed to be significant (B=0.16, P-value <0.05).Therefore, H2 is supported. The UTAUT final factors, which are EE (B=0.18, P-value<0.5), SI (B=0.21, P-value<0.05) and high SEC (B= 0.17, P-value <0.05) were confirmed to be significant. Therefore, H3, H4 and H5 are supported.

In addition, the results of regression analysis illustrates that FCs have a significant influence on the use behaviour of IoT services. The results also contended that FCs significantly affected the use behaviour of IoT services. Thus, H6 is supported.

Table 6 summarises the regression analysis of the proposed model. The table illustrates the value of R-squared in the model, which clarifies the predictive power of the model. It shows that the four factors (PE, EE, SI and SEC) that are linked to BI explain 49.6% of variance in BI towards the use of IoT services, while BI and FCs explain 55.1% of variance in use behaviour.

Table 6. Regression values of the proposed model

Model	*R*	*R-square*	*Adjusted R-square*	*Std. error of estimate*
BI (D.V.)	0.70	0.496	0.487	**1.593**
Use behaviour (D.	0.74	0.551	0.547	**2.232**

Table 7 illustrates the hypothesis testing outline. The beta value pinpoints the size of influence beside the P-value of the relationship. In addition, the research hypotheses were confirmed to be acceptable (P-value less than 0.05).

Table 7. Hypothesis testing outline

Hypotheses	**Beta**	**P-value**	**Remarks**
H1: **BI** positively influence the use behaviour of IoTservices.	0.35	0.02	**Supported**
H2: PE positively influence users' **BI** to use IoT services.	0.16	0.02	**Supported**
H3: EE positively influences users' **BI** to use IoT services.	0.18	0.02	**Supported**
H4: SI has a positive influence on users' **BI** touse IoT services.	0.21	0.00	**Supported**
H5: SEC has a positive influence on **BI** to adopt IoT services.	0.17	0.00	**Supported**
H6: FCs positively influence the use behaviour of IoT services.	0.68	0.01	**Supported**

Figure 2 presents the model of adoption IoT services for higher educational institutions in Iraq. The figure depicts that the four factors of the UTAUT, along SEC, are significant factors for the adoption of IoT services. FCs is the most powerful factor in the model, with a coefficient of 0.68. This result could be recognized due to the lack of internet availability as well as the traditional IT infrastructure in Iraq.

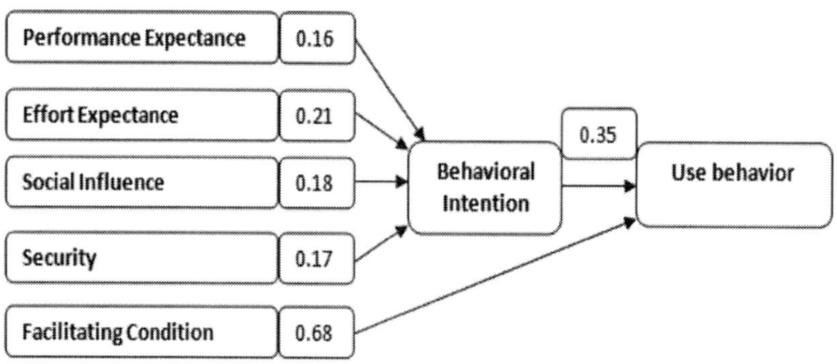

Figure 2. IoT adoption model for higher education institutions in iraq

6. Contribution

This study provided conceptual and realistic contributions. From a conceptual perspective, the model is developed and tested for IoT services in developing countries. In addition, the researches in IoT are in the early stages, and the majority of previous researches are technical. However, this study investigated the adoption of IoT from the behavioural and attitudinal perspectives. In addition, this study added value to the UTAUT model by testing the validity of this model in terms of IoT. Further, the incorporation of SEC was according to the criticism of the UTAUT model. In this regard, this research filled the gap.

In practice, top management and stakeholders can employ the results to improve the use and acceptance of IoT. FCs is the most powerful factor that directly influence use behaviour. The decision makers should establish the proper environment to aid in the evolvement of IoT in educational institutions in Iraq. In addition, awareness about the SI of IoT services, especially for university lecturers and students, should also be increased. Users must be provided with practical courses on the mechanisms of IoT services, and universities must support IoT services and facilitate the accessibility of internet services for students and lecturers. Ultimately, the Internet should be supplied in universities for all the users in as a consistent technology.

7. Recommendation for future work

The previous researches in IoT services showed that UTAUT model is merely used. In addition, the literature suggested that the concept of IoT services is relatively new. Thus, numerous academic works are required. On the basis of the results, the following recommendations are made.

IoT services are still a new and growing concept. Thus, further studies are required in this field. The literature review validated that the majority of the studies used the TAM as theoretical adoption theory, whereas studies on the UTAUT are few. Therefore, we recommend that future studies deploy the UTAUT because it has the most exploratory power comparing with other models. Researchers can mix between models to derive other pertinent factors that affect the adoption of IoT.

The respondents of the present study were students and academic staff. Future studies should focus on each group of target respondents to delineate issues in relation to the adoption of IoT services. In addition, future studies can incorporate non-academic staff.

The study was conducted on all faculty staff of Basra University. Future work could focus on one or a group of faculties that share certain characteristic. The derived results are linked to individual users, which mean that the organizational aspects need to examine in the future researches. In addition, this research conducted for the higher educational institutions. Therefore, business organisations need to test the proposed model in the future works. In addition, the university could utilise a hybrid cloud to maximize the security level and enhance the adoption of IoT services.

Lastly, on the basis of the R-squared of the BI, the factors that affect BI could explain 49.6% of the variation of BI. We recommend that future studies increase the percentage of variance by incorporating new factors that are related to the country, such as trust, culture and tradition, knowledge of users or university incentive to use IoT services.

8. Conclusion

This research was accomplished in Basra University. The purpose was to discover the factors that influence the adoption of IoT services at higher institutions in Iraq. Moreover, the study reviewed the literature and developed the proposed model. Meanwhile, the UTAUT was employed as a theoretical adoption model. With SEC as an independent factor, UTAUT factors, such as PE, EE, FC, and SI, were deemed to influence use behavior and BI of IoT services.

The respondents were composed of students and lecturers. Data were analyzed by SPSS V 22.0. The results verified that SI has the highest level of influence on the BI to adopt IoT services. This factor was followed by EE, SEC and PE. The FC was more effective than BI as a driver of use behaviour. This result was mainly due to technological infrastructure, which was more important than intention. The pre-requisites for using IoT services are Internet connection, networks and electric supply. Although these pre-requisites are available in Iraq, they require further enhancements.

Authorities and decision makers in higher educational institutions are advised to use IoT services and technology and increase awareness regarding benefits and usability.

References

[1] Gao, L. and Bai, X., 2014. A unified perspective on the factors influencing consumer acceptance of internet of things technology. *Asia Pacific Journal of Marketing and Logistics*, *26*(2), pp.211-231.

[2] Bauer, H., Patel, M. and Veira, J., 2014. The Internet of Things: Sizing up the opportunity.

Retrieved from: McKinsey at http://www. mckinsey.com/insights/high_tech_telecoms_internet/the_internet_of_things_sizing_up_the_opportunity.

[3] Gómez, J., Huete, J.F., Hoyos, O., Perez, L. and Grigori, D., 2013. Interaction system based on internet of things as support for education. Procedia Computer Science, 21, pp.132-139.

[4] Greenough, J., 2015. The'Internet of Things' will be the world's most massive device market and save companies billions of dollars. Business Insider, New York, accessed June, 2, p.2017.

[5] Gubbi, J., Buyya, R., Marusic, S. and Palaniswami, M., 2013. Internet of Things (IoT): A vision, architectural elements, and future directions. Future generation computer systems, 29(7), pp.1645-1660.

[6] Su, J., Cao, D., Zhao, B., Wang, X. and You, I., 2014. ePASS: An expressive attribute-based signature scheme with privacy and an unforgeability guarantee for the Internet of Things. Future Generation Computer Systems, 33, pp.11-18.

[7] Whitmore, A., Agarwal, A., & Da Xu, L. (2015). The Internet of Things—A survey of topics and trends. *Information Systems Frontiers*, *17*(2), 261-274.

[8] Sicari, S., Rizzardi, A., Grieco, L. A., & Coen-Porisini, A. (2015). Security, privacy and trust in Internet of Things: The road ahead. *Computer Networks*, *76*, 146-164.

[9] Hsu, C. L., & Lin, J. C. C. (2016). An empirical examination of consumer adoption of Internet of Things services: Network externalities and concern for information privacy perspectives. *Computers in Human Behavior*, *62*, 516-527.

[10] Venkatesh, V., M.G. Morris, G.B. Davis, F.D. Davis. (2003). User acceptance of information technology: Toward a unified view. *MIS Quarterly*, 27(3): 425–478.

[11] Davis, F.D., Bagozzi, R.P. and Warshaw, P.R. (1989). "User acceptance of computer technology: a comparison of two theoretical models", *Management Science*, Vol. 35 No. 8, pp. 982-1003.

[12] Mathur, S.K., T.V. Dhulla. (2014). Factors Influencing Professionals' Decision for Cloud Computing Adoption.

[13] Gangwar, H., H. Date, A.D. Raoot. (2014). Review on IT adoption: insights from recent technologies. *Journal of Enterprise Information Management*, *27*(4): 488-502.

[14] Hashim, H. S., & Hassan, Z. B. (2015). Factors That Influence The Users' Adoption Of Cloud Computing Services At Iraqi Universities: An Empirical Study. *Australian Journal of Basic and Applied Sciences*, 9(27), 379-390.

[15] Khalilzadeh, J., Ozturk, A. B., & Bilgihan, A. (2017). Security-related factors in extended UTAUT model for NFC based mobile payment in the restaurant industry. *Computers in Human Behavior*, *70*, 460-474.

[16] Gu, Y., & Liu, Q. (2013). Research on the application of the internet of things in reverse logistics information management. *Journal of Industrial Engineering and Manag*

[17] Jorge, Juan, Oscar, Luis, & Daniela. (2013). Interaction System Based on Internet of Things as Support for Education. *Procedia Computer Science 21*, (2013) 132 – 139

[18] Cao, Y., X. Bi, L. Wang. (2013). A Study on User Adoption of Cloud Storage Service in China: A Revised Unified theory of Acceptance and Use of Technology Model. *In Information Science and Cloud '*

[19] Alharbi, S.T. (2014). Trust and Acceptance of Cloud Computing: A Revised UTAUT Model. *In Computational Science and Computational Intelligence (CSCI), 2014 International Conference on* (2: 131-134). IEEE.

[20] Kaba, B., & Touré, B. (2014). Understanding information and communication technology behavioral intention to use: Applying the UTAUT model to social networking site adoption by young people in a least developed country. Journal of the Association for Information Science and Technology, 65(8), 1662-1674.

[21] Chen, K. Y., & Chang, M. L. (2013). User acceptance of 'near field communication'mobile phone service: an investigation based on the 'unified theory of acceptance and use of technology'model. The Service Industries Journal, 33(6), 609-623.

[22] Alotaibi, M.B. (2014). Exploring Users' attitudes And Intentions Toward The Adoption Of Cloud Computing In Saudi Arabia: An Empirical Investigation. Journal of Computer Science, 10(11): 2315- 2329.

[23] Singh, S., D. Chand. (2014). Trust evaluation in cloud based on friends and third party's recommendations. In Engineering and Computational Sciences (RAECS), 2014 Recent Advances in (pp): 1-6. IEEE.

[24] Venkatesh, V., Thong, J. and Xu, X. (2012). "Consumer acceptance and use of information technology: extending the unified theory of acceptance and use of technology", MIS Quarterly, Vol. 36 No. 1, pp. 157-178.

[25] Junadi, S. (2015). A model of factors influencing consumer's intention to use e-payment system in Indonesia.

[26] Mital, M., Chang, V., Choudhary, P., Papa, A., & Pani, A. K. (2017). Adoption of Internet of Things in India: A test of competing models using a structured equation modeling approach. Technological Forecasting and Social Change

[27] Lian, J.W. (2015). Critical factors for cloud based e-invoice service adoption in Taiwan: An empirical study. International Journal of Information Management, 35(1): 98-109.

[28] Phaphoom, N., X. Wang, S. Samuel, S. Helmer, P. Abrahamsson. (2015). A survey study on major technical barriers affecting the decision to adopt cloud services. Journal of Systems and Software, 103: 167-181.

[29] Hassan, H., Nasir, M. H. M., Khairudin, N., & Adon, I. (2017). Factors Influencing Cloud Computing Adoption In Small And Medium Enterprises. Journal Of Ict, 16(1), 21-41.

[30] http://www.uobasrah.edu.iq/

[31] Ary, D., Jacobs, L. C., Irvine, C. K. S., & Walker, D. (2013). Introduction to research in education. Cengage Learning.

[32] Krejcie, R. V., & Morgan, D. W. (1970). Determining sample size for research activities. Educ Psychol Meas.

[33] Pallant, J. (2010). SPSS Survival Manual.

IOP Conf. Series: Materials Science and Engineering **769** (2020) 012007 doi:10.1088/1757-899X/769/1/012007

Big data: definition, characteristics, life cycle, applications, and challenges

Hiba Basim Alwan[1] and Ku Ruhana Ku-Mahamud[2]

[1] Department of Computer Engineering, Al-Mansour University College, 10068 Al-Andalus Sq., Baghdad, Iraq
[2] School of Computing, Universiti Utara Malaysia, 06010 Sintok, Kedah, Malaysia

hiba.basim@muc.edu.iq

Abstract. Any data set contains large volumes of information and complex data is called Big Data (BD). BD is unlike other traditional data sets, so it requires special processing to manage it. BD faces many challenges starting from data capture through to the final results. BD exists in many subject areas such as business, governments, sciences, healthcare and transport. Thus it touches peoples' lives in many aspects. BD is considered as the most important topic and requires good understanding in order to be fully utilized. This paper presents the basic information of BD which includes its properties and applications. Descriptions and examples of BD and its categories are elaborated upon. The BD architectural establishment is presented followed by the conclusion of the importance of BD.

1. Introduction

One of the greatest trendy ideas these days is Big Data (BD). Everyone speaks about BD as can be seen in the media. Governments and businesses attempt to use and implement BD to their benefits [1]. The term BD was not known until in the middle of 2011. Like cloud computing the term has been implemented from product sellers to huge scale outsourcing and cloud service suppliers to powerfully encourage their offerings [2]. But what actually is BD?

Lisa [3] defines BD as a group of data from conventional and digital bases inside/outside the enterprise that characterizes a basis for continuing detection and analysis. Another definition of BD is found in [4] which defined BD as a quantity of data that is extremely relative and cannot be managed through the usual methods while [5] defined BD as not only one technology, but a group of old and modern technologies that assist businesses to obtain actionable perceptions. BD has large quantities of dissimilar data which allows processing in real time analysis and response. BD can be defined as a giant size of data [6]. Within BD, analyzing, visualizing or any other processing can be done. From the above definitions, it can be concluded that if data cannot be stored or processed by a common system's capabilities or exceed a common system's capabilities then these data are considered BD. The powerful services in this modern domain are continuously growing volumes of data and the advances in technologies are always able to mine the data for commercial purposes [4].

The unexpected increase of BD like a modern resource of knowledge has encouraged business decision-makers to generate decisions more quickly and to proactively locate environmental alterations [7].

BD requires the study and thinking about both technical and business needs. There are people who need to investigate technological specifics, whereas others need to know the cost-effective of BD equipment usage. Applying a BD setting will need an architectural and business method and lots of planning [5, 8].

To manage BD, data scientists are needed because there is an immense amount of data available where, in the past, there were no algorithms able to manage BD. In the past, large amount of data could not be stored. Now,

Exabyte storage and the tools needed to manipulate BD are available and not expensive. Data virtualization and efficient preservation of BD are now using cost efficient cloud storage [5].

BD technology is an essential progress track in the area of Internet science and technology. It has been broadly evaluated and progressed entirely around the globe and has been used in various areas of manufacturing as well as life [9].

There are a lot of advantages in utilizing BD technologies. BD could produce benefits like maximizing organizational productivity, informing strategic positioning, improved client services, recognizing and evolving modern products and services if utilized in an efficient way [10, 11]. Other advantages of utilizing BD is to improve marketing, automated decision making, descriptions of client behaviors, better profit on investments, quantification of dangers and market trending, understanding of commercial changing, planning and predicting, recognition of client activities beginning with click streams as well as manufacture income expanding [12, 13].

According to the knowledge of the authors, there is no single paper that gathers the basic concepts of BD which include BD definition, types of data, technologies to deal with BD, characteristics of BD, life cycle of BD, architecture of BD, applications of BD, its platforms, challenges of BD, limitations to implement BD, as well as motivations to doing BD research. Thus, this paper tries to fill in the gaps with the aim to provide the understanding of BD in a simple and easy way.

In addition to the introductory section, Section 2 presents the types and properties of BD followed by applications of BD in Section 3. Challenges in performing BD research are presented in Section 4. Finally, Section 5 presents the conclusion.

2. Types and proprieties of big data

There are three main categories of data within BD. The categories are structured data, semi structured data and unstructured data [1, 14].

Structured Data usually denotes data that have a described length and format. Such data includes string and dates. The majority of experts conclude that this category of data occupies about one quarter of the available data. It is frequently kept in a database and are created by machine or human. Structured data created by human include input data, like human's names and human's age, click-stream data game's moves data, while structured data created by machine include sensor data (log in data to website, data selling port, and commercial data) [5].

Unstructured Data are data that are unable to keep in a particular formula. Usually, unstructured data comprises three quarter of any company's data. Unstructured data may be located everywhere and are created by machine or human. Those created by machine include satellite images, scientific data, pictures as well as video and sensor data. Unstructured data created by humans include texts in the social media, web site and cellular phone [5, 15].

Semi Structured Data are the data that are not categorized as structured and unstructured data. This type of data do not essentially adapt as stable representation but may contain uncomplicated values [5, 15].

The 5Vs is used to describe BD [9, 16] which is also known as the characteristics of BD that relates to volume, velocity, variety, veracity and value [17, 18, 19].

Volume which is the quantity of data. This characteristic is foremost in human minds when handling BD. Many businesses have huge volumes of data that are archived in the form of logs, but do not have the ability to manage them. The benefit obtained from the ability to handle huge quantities of data to produce information is the most important desirability of BD analytics [17]. The huge volume of data can be helpful for businesses, but it affects the retrieval and analytic procedures as it is time consuming because of the calculation processes [8].

Velocity is the speed of the data that during handling and also the speed at which the data are generated. Thus, this implies the quickness that has to be considered in managing, storing and analyzing the data. Each second of each day, hundreds of hours of video are uploaded on YouTube and more than 200 million emails were sent via Gmail.

Variety refers to the range of data types and sources. The group to which BD is assigned is also an important feature that requires to be recognized by the data analysts, because BD is not usually structured data and it is not usually easy to maintain in a relational database. The difficulty of keeping and analyzing BD will maximize the difficulty of handling structured and unstructured data as 90% of created data are unstructured data.

Veracity is related to the truth of data which is important for precision in analysis. It is impossible to ensure that all data are 100% accurate when managing huge volume of data with great velocity as well as variety. The quality of the data will vary and precision of analysis depends on the data veracity of the data resource.

Value is the importance of the data importance and this is a very significant feature in BD. The possible value of BD is substantial. However, without suitable access its value cannot be exploited.

Table 1 has been provided by [20] to show the summary of 4 characteristics of BD.

Table 1. Summary of the 4V's.

	Volume	Velocity	Variety	Veracity
Description	The amount of data created is huge compared with standard data resources	Data are being created very speedily, at a progression that never ends and at which data is converted into perception	Data capture from various resources like machines, people, operations both from outside and inside the company	Quality and origin of data
Attribute	Exabyte, Zettabyte, Yottabyte, etc.	Batch, Near/Real time streams	Grade of structures difficulties	Reliability, totality, integrity, uncertainty
Driver	Maximized data resource, higher solution, sensors, scalable infrastructure	Enhanced linking competitive benefit pre-calculated information	Mobile, social media, video, genomics, Internet of Things	Fee, requires traceability and explanation

The life cycle for BD consists of five phases. These phases are data collection, date cleaning, data classification, data modelling and data delivery. Data collection phase includes the gathering and storing of data from different resources. In the data cleaning phase, management of the confirmation whether there are some unwanted items within the data or missing values are performed. Data classification will classify the data according to their types either structured, semi-structured or unstructured. In the data modeling phase, analysis of data is performed and the result is the clustered data for objectives. Finally, data delivery phase involves the creation of reports based the results of the modelling phase. These phases are depicted in figure 1 [5, 21].

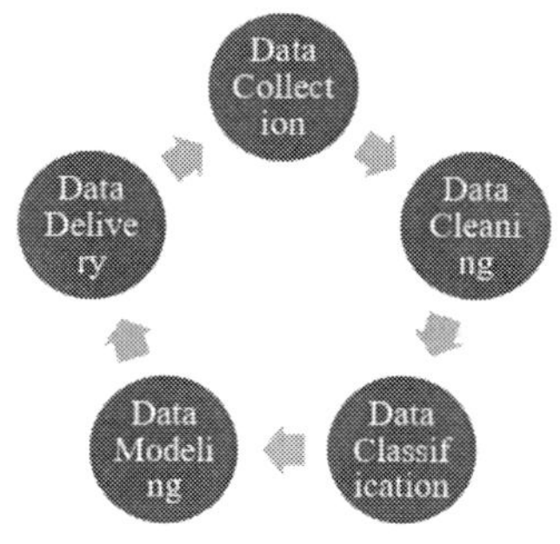

Figure 1. Life cycle of BD.

Several questions need to be considered in establishing any BD architecture. These questions relate to how much data will a company require to handle BD now and in the future and how will a company usually require handling data in actual time or near to actual time? Other technical issues that need to be resolved are the speed and accuracy of the data. The elements required in any BD architectural establishment is shown in figure 2 [5].

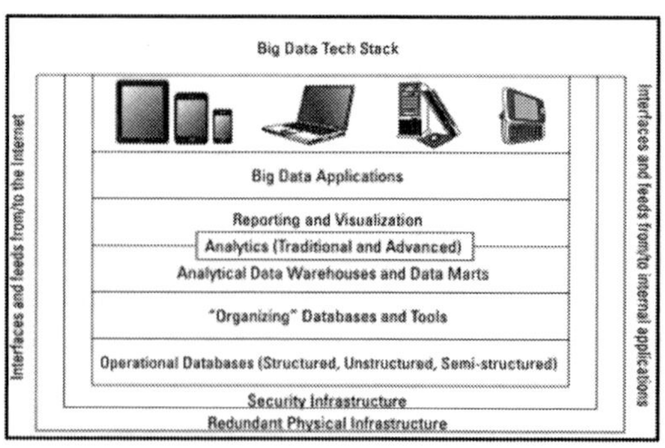

Figure 2. Big Data Architecture.

IOP Conf. Series: Materials Science and Engineering **769** (2020) 012007 doi:10.1088/1757-899X/769/1/012007

3. Applications of big data

Organizations are uncertain as to how data can be used and when to utilize the data. Applications are invented particularly to gain benefit of the exclusive features of BD especially in the fields of health care, manufacturing and city planning [5]. Figure 3 compares the different applications based on the 3V's (variety, velocity, and volume) [22].

Application	Variety	Velocity	Volume
Banking and securities	Low	Very high	Very high
Media and entertainment	Very high	High	Very high
Healthcare providers	High	Very high	Moderate
Education	Low	Very low	Very low
Manufacturing and natural resources	High	Very high	Very high
Government	Very high	High	Very high
Insurance	Moderate	Moderate	Moderate
Retail and wholesale trade	Moderate	High	High
Transportation	Moderate	High	High
Energy and utilities	Moderate	High	High

Figure 3. Comparison of different data attributes in BD applications.

Other applications of BD as reported in [15] are: smart grid, e-health, Internet of Thing (IoT), public services, transportation and logistics, and political services and government monitoring.

Smart Grid Case: Smart grids involve the forever progressively huge size of data [23]. This is a critical area that requires real-time monitoring for almost all of its operations. This is achieved through connected devices that form the whole of the infrastructure. BD analytic helps to provide an insight like identifying deteriorating critical equipment on national grid that exhibit abnormal behavior like faulty transformers. In such cases, proactive measure and the best line of preventive and maintenance actions can be deployed thereby saving cost and optimizing operation.

E-health: Health related policies are among various fields that adapt and benefited from BD. Patients monitoring sensors, laboratory data, medical history of patients with different ailments are among few various sources of useful data that if utilized properly can aid into personalized medication, help policy makers to provide adaptive health care policies and it can also be utilized to reduce general hospital operational running cost and enhance service delivery.

Internet of Things: This is another area that benefited much from DB is IoT due to variety of interconnected objects that consume, generate and share different types of data. Objects in IoT can be anything that is capable of being connected to and be accessed online.

Public Services: Public services like public water systems are putting in place sensors to monitor consumption, illegal connection and leakage on pipelines in order to benefit from real time monitoring of infrastructures. This helps to reduce manpower needed for monitor facilities which resulted into timely intervention when needed and helps in rendering efficient service to public.

Transportation and Logistics: Transportation sector is among the most prominent areas where application of BD cannot be over emphasized. Near field communication devices like radio frequency identification enable transporters to have fleet equipped with capable sensors attached to commuting vehicles. This enable administrators to efficiently plan and manage delivery routes, access to up-to-date track record of employees, have the ability to monitor fleet in a real-time manner, and access to up-to-date pattern of passengers commuting behaviour which enables optimised planning and effective management.

Political Facilities and Government Observing: Several governments are mining social data to observe political trends as well as analyze community opinions. Governments may utilize BD systems to enhance the usage of scarce resources and services. For example, data from sensors placed on various public infrastructure like water pipes can be used to determine consumption rate of different regions and several zones of a city which can result into provision of required quantity of the needed resource.

Big organizations (like Yahoo, Google, and Facebook) required to develop modern tools that could permit them for storing, accessing, and analyzing massive amount of data close to real time. Platforms like MapReduce,

Hadoop, YARN, Oozie, Flume, Hive, HBase, Apache Pig, Apache Spark, Sqoop, Zoo keeper, and Big Table are modern creation of data management tools that can be used to analyze huge amount of data effectively and timely [5].

MapReduce: Is one of the data processing options that can be executed on Hadoop [15, 24]. MapReduce has been design to handle and schedule data processing job and cluster assignment effectively. The main advantage of MapReduce is its simplification of processing huge volume of data which is achieved through effective computing resource sharing mechanism via parallel processing procedure. The effectiveness of MapReduce is achieved through its capability for distributing task through many available mapping nodes. The quantity of maps is often concluded through the entire size of the inputs which is the entire number of blocks of the input files. Once the computation task on various distributions is accomplished, extra task named as "reduce" then gathers all the processed results to produce a complete processed solution. This approach facilitates better load balancing, maximizes the quantity of reduces, maximizes load equalization and minimizes the number of breakdowns [25]. Figure 4 shows data flow within MapReduce [5].

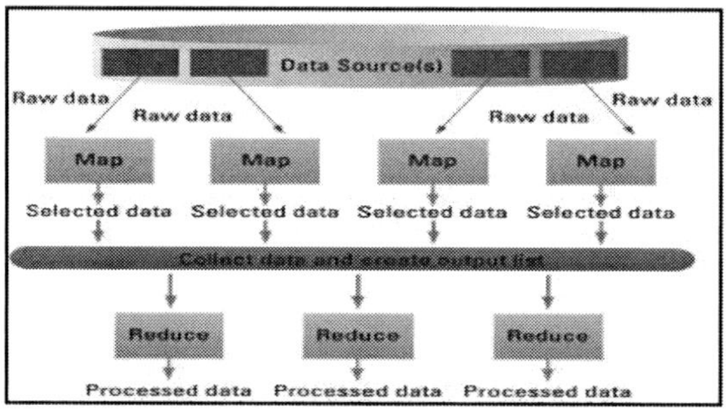

Figure 4. Data flow in MapReduce.

Big Table: This was created by Google to act as a distributed saving system that aimed to process extremely huge amount of structured data using secure servers. Data is organized as a table that has tuples and attributes. Big table differs from classical relational database in various ways. It is a spare, distributed, and permanently multi-dimensional saving map [5].

Hadoop: There are different stories about the name of Hadoop. [26] stated that Hadoop is an acronym name stand for **H**ighly **A**rchived **D**istributed **O**bject **O**riented **P**rogramming. [27] stated that Hadoop is an elephant in the room. It is not an acronym but a yellow elephant toy for the creator's son of it which is the Google's engineer Doug Cutting. However, Apache Hadoop is a hug scalable storage platform created to manage huge amount of data sets through hundreds and thousands of calculating nodes that work concurrently [2, 17, 28]. Hadoop is also considered as an Apache handled software structure resulting from Big table and MapReduce. Hadoop permits functions to be established on MapReduce to execute big clusters of productized hardware. Hadoop is the basis of the calculating architecture providing Yahoo!'s company. Hadoop is intended to simultaneous process data through computation nodes that speed up the calculation and minimize latency. Two main elements in Hadoop are (1) huge scalable distributed file system in providing petabytes of data and (2) huge mountable MapReduce that calculates outcomes in batch. Figure 5 illustrates how to map Hadoop clusters into hardware [5].

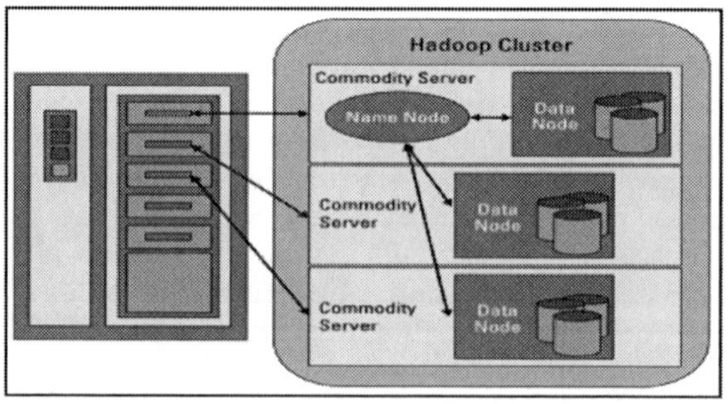

Figure 5. Mapping Hadoop clusters into hardware.

4. Challenges in big data research

Big Data Mining: BD mining presents a lot of desirable chances but with immense difficulties. The complexities depend on various stages involving data taken, storage, seeking, sharing, analysis, administration and visualization [15].

Big Data Management: BD management aims to provide a dependable clean data through various means of data gathering from huge volumes of different types of data sources like company, government as well as private/public sectors. This is achieved by various processing tasks that include preprocessing, processing and other related activities such as encrypting the data for security, confidentiality and dependability [26]. Certainly, suitable data management is the basis for BD analytics [15].

Big Data Recovery and Storage: Storage in BD is accomplished via virtualization where it processes huge sets of data from sensors, media, videos, transaction data from e-businesses, mobile signal coordinates. Many corporations manage data in huge volume through utilizing instruments such as NoSQL, Apache Drill, Horton Works, SAMOA, IKANOW, Hadoop, MapReduce, and Grid Gain [15]. Large volume storage facilities and faster I/O speeds enables improvement in working with BD. Thus, access to data should be quick and simple for on-time analysis. Previously, continuing data were replaced through utilizing Hard Disk Drive (HDD). The well-known major drawback of HDD is having slower input/output performance. Improvements in storage devices like Solid State Drive (SSD) can minimize the problems but they are not been completely utilized. HDDs are gradually being replaced by SSDs, and other improvements like pulse-code modulation which are also on the increase [22].

Big Data Processing: BD processing analyses the huge volume of BD in petabyte, exabyte, and zettabyte depending on whichever batch management or batch management is best [26].

Data Visualization: The major aim of data visualization is to present the data efficiently and sufficiently through utilizing several charts. Data visualization poses a challenge for BD applications because the huge volume and dimension of the data. Thus, there will be a need to re-test the method under which the BD is pictured. Structure and usefulness of presented data are of paramount importance to permit demonstration of knowledge which is unseen in non-simple large-scale data sets. The structured data organized in tables and its associated characteristics are necessary for informative analysis [22].

Data Transmission: Once the communication foundation is extremely big, the system's data transfer capability is bound and blocked in a cloud circulated framework. Here cloud improvement is substituted via a cloud data feed as its improved form [22].

Big Data Security: It is a challenge to ensure guaranteed safety and security of big data. This is attributed to many factors like incompetent instruments, public and private databases. In distributed programming structures, the safety challenge begins when huge amounts of personal information are kept in a database which is not encoded in standard form. Saving the data in the hand of disgruntled and unreliable persons add to the extra complexity of data security. The challenge of data security also surfaced when migrating or updating from similar and/or different data-specific instruments. Occasionally, data thieves and system thieves gather a publicly accessible BD collection, copy it and keep it in a device like a USB drive, hard disk or laptop. Therefore, when keeping of data is maximized from one level to a multi-storage levels, the safety level should also be maximized [26].

Data Curation: This area specifically contains several sub-areas, like validation, documentation, supervision, security, recovery and demonstration. The existing database management tools cannot manage BD. Data warehouse and data marts have been used to manage big data sets in a suitably structured approach. These methods follow data frameworks which are built on structured query language. These days NoSQL is utilized in BD due to the four Vs of BD [22].

Big Data Cleaning: This challenge involves five phases (cleaning, aggregation, encoding, storage and access) which are not modern and are used in conventional information handling. The challenge in BD is how to handle the difficulties of BD's nature (velocity, volume, variety, veracity and value) and operate it in a distributed situation with a combination of functions. Nevertheless, information resources may include noises, errors or incomplete data. The issue is how to clean large data and how to resolve which data is countable, and which data is beneficial [15].

Big Data Aggregation: This issue is related to synchronizing external data resources and distributed BD platforms (involving applications, repositories, sensors, networks) with the internal infrastructure of a company. Furthermore, it is not enough to analyze the data created inside companies. In mining important insight and information, it is necessary to move towards an advance of not collecting only internally generated data, but required tools should be put in place to collect both internal data and external data resources. External data could contain third-party resources, information about market fluctuations, weather predicting and traffic conditions, data from social networks, customer comments and citizen feedback [15].

Big Data Imbalance: The issue in classifying an imbalanced data set has received a lot of attention. Real world applications with diverse distributions can be categorized into two major groups. The first is the under-presented group which is characterized with insignificant number of data-points (this type is also known as minority or positive group). The second group has significant number of data-points (also known as common or negative group). The identification of positive group has significant importance in many areas like medical diagnosis, software defects detection, finances, drug discovery or bioinformatics. The traditional Machine Learning (ML) techniques cannot be implemented on imbalanced data sets. This is because the prototype building is founded on global privilege measures which by default favours the majority instances group thereby disregarding the significance of the minority group [15].

Big Data Analytical: BD brings a lot of challenges on how to extract meaning out of this large, ever-increasing voluminous data. For example, data analysis allows a company to obtain important vision and observe the patterns that may positively or negatively influence their businesses. Other data-driven applications require additional real-time analysis, like social networks, biomedicine, astronomy, and intelligent transport systems. Therefore, advanced algorithms and efficient approaches of data mining are required to obtain correct outcomes, to control the changes in different areas and be able to have future predictions with real-time responsiveness. One of the challenges is how to guarantee the timeliness of responses when the data volume is huge. Examples of the complexities observed when performing existing analytical solutions are ML, deep learning, incremental approaches, as well as granular computing [15]. The main issue of data analytic found with BD is related with the volume of data. Timeliness is the highest priority for some BD applications. The main test for BD applications is to guarantee timeliness of responses when the data being processed is huge [22].

Big Data Machine Learning: The primary aim of ML is to discover knowledge from either organized or unorganized data. ML is presently serving as the backbone of many applications that relies and produce part of big data composition, ranging from search engines, recognition systems, aeronautics, and military to mention a few [15].

BD is an innovation that will fundamentally change the method that information is grouping, keeping, monitoring, and spent by users which in turn will change the method of doing work. Several of the motivations in doing BD research are [22]:

Changing from Classical Relational Database Management System (RDBMS): This system is utilized by several enterprise information technology corporations and is still used now by many information technology enterprises. Today, data are unstructured and non-clustered and NoSQL keeps all the data with no clustering and describes them into the framework opposite to RDBMS which keeps data in suitable structures or tables.

Managing Unstructured Data: BD has the capability to manage structured as well as unstructured data. Along with data variety characteristics, BD is about text or numbers (alphanumeric fields), and unstructured data. Therefore, by utilizing NoSQL, BD can manage unstructured data.

Real Time Data Processing: In future, information systems will need an ability to manage increasingly huge volumes of data, where the velocity of BD created is presented. An expression "near real time", is often utilized with existing time of information systems, but it is not suitable. Real time data management involves the facility of managing online data or sensor information as they are created.

Most Data are either User or Machine Created: Earlier, most data were created internally in the firewall of an enterprise. However, present data are created either through end users or machine created, which are external to the bounds of the firewall of the enterprise.

There are many limitations in BD and some of these limitations can be summarized as follows [29]: firstly, the needed data are not always available because of: (i) data are simply not available; (ii) there is trouble with the holding phases; and (iii) various data platforms are shown to be interoperable. Secondly, the main core of BD is pattern recognition. Result from pattern analyzing is significant because it will demonstrate the problem of security, risk, and types of crime. Present approaches of data mining will not be able to handle BD. Lastly, BD can be used to predicate new data because it is established on old data which consist of previous patterns.

5. Conclusion

It can be concluded that BD means any quantity of data even if the data is structured, unstructured, or semi-structured that cannot fit into a processing system. Thus, BD will need special tools and technologies to handle it and can be characterized by the term 5Vs. If a company needs to model its BD and gain benefits from it, the company needs to design the architecture for its BD which will require answers to questions related to the nature of the company. BD has many applications fields in life and has many platforms to handle it. In reality, BD has many challenges and limitations to implement it, as well as there are several motivations in performing BD research.

IOP Conf. Series: Materials Science and Engineering **769** (2020) 012007 doi:10.1088/1757-899X/769/1/012007

References

[1] José T and Juan R 2018 Data learning from big data *Statistics and Probability Letters* **136** 15-19
[2] Mitchell I, Locke M, Wilson M and Fuller A 2012 *The white book of big data* (UK: Fujitsu Services Ltd.)
[3] Lisa A 2013 *Big data marketing* (New Jersey: John Wiley & Sons, Inc.)
[4] Bernard M 2016 *Big data in practice: How 45 successful companies used big data analytics to deliver extraordinary results* (New Jersey: John Wiley & Sons, Inc.)
[5] Judith H, Alan N, Fern H and Marcia K 2013 *Big data for dummies* (New Jersey: John Wiley & Sons, Inc.)
[6] Maria B, Liyana S and Elaheh Y 2019 Big data adoption: state of the art and research challenges *Information Processing and Management* **56**
[7] Alessandro M, Sally D, Maureen M, Lee Q, David W, Lyndon S and Ana C 2018 Big data, big decisions: the impact of big data on board level decision making *Journal of Business Research* **93** 67–78
[8] Michele I, Elio M, Giuseppe M, Mario M and Carlo Z 2020 Fast and effective big data exploration by clustering *Future Generation Computer Systems* **102** 84-94
[9] Yinghao Y, Meilin W, Shuhong Y, Jarvis J and Qing L 2019 Big data processing framework for manufacturing *Procedia CIRP* **83** 661-64
[10] Jaime C, Pankaj S, Unai G, Erkki J and David B 2017 A big data analytical architecture for the Asset Management *Procedia CIRP* **64** 369-74
[11] Manyika J, Chui M, Brown B, Bughin J, Dobbs R, Roxburgh C and Hung B 2012 A big data: the next frontier for innovation, competition, and Productivity (McKinsey Global Institute)
[12] Seref S and Duygu S 2013 Big data: a review *Proc. Int. Conf. on Collaboration Technoloies and Systems (CTS)* (San Diego: CA/ USA IEEE) p 42
[13] Philip R 2011 Big Data Analytics *TDWI Best Practices Report*
[14] Karim M 2019 State of the art in big data applications in microgrid: a review *Advanced Engineering Informatics* **42**
[15] Ahmed O, Fatim-Zahra B, Ayoub A and Samir B 2018 Big data technologies: a survey *Journal of King Saud University-Computer and Information Sciences* **30** 431-48
[16] Ada B, Devis B, Valeria A, Massimiliano G and Alessandro M 2019 A relevance-based approach for big data exploration *Future Generation Computer Systems* **101**, 51–69
[17] Ishwarappa K and Anuradha J 2015 A brief introduction on big data 5 vs characteristics and hadoop technology *Procedia Computer Science* **48** 319-24
[18] Jean-Louis M and Soraya S 2016 *Big data, open data and data development* (ISTE Ltd and John Wiley & Sons, Inc.)
[19] Abdulkhaliq A, Vlad K and Michael B 2017 Addressing barriers to big data *Business Horizons* **60** 285-92
[20] https://courses.cognitiveclass.ai/courses/coursev1:BigDataUniversity+BD0101EN+2016_T2/courseware/407a9f86565c44189740699636b4fb85/12eab34ec218468995e4d06566ef4a32
[21] Archenaa J and Mary E 2015 A survey of big data analytics in healthcare and government *Procedia Computer Science* **50** 408-13
[22] Kirtida N and Abhijit J 2017 Role of big data in various sectors *Proc. Int. Conf. on* IoT in Social, Mobile, Analytics and Cloud (I-SMAC) (Palladam/India IEEE) p 117
[23] Tom W, Nanlin J, Peter F and Joshua T 2019 A big data platform for smart meter data analytics *Computers in Industry* **105** 250–59
[24] https://www.ibm.com/analytics/us/en/technology/hadoop/mapreduce/#what-is-mapreduce.
[25] https://hadoop.apache.org/docs/r1.2.1/mapred_tutorial.html.
[26] Saraladevi B, Pazhanirajam N, Victer P, Saleem M S and Dhavachelvan P 2015 Big data and hadoop-a study in security perspective *Procedia Computer Science* **50** 596-601
[27] https://www.sas.com/en_us/insights/big-data/hadoop.html.
[28] https://www.ibm.com/analytics/us/en/technology/hadoop/.
[29] Dennis B, Erik S and Bart S 2017 Big data and security policies: towards a framework for regulating the phases of analytics and use of big data *Computer Law & Security Review* **33** 309-23

IOP Conf. Series: Materials Science and Engineering **769** (2020) 012008 doi:10.1088/1757-899X/769/1/012008

A Geofencing-based Recent Trends Identification from Twitter Data

M. Saef Ullah Miah[1], M. Sadid Tahsin[2], Saiful Azad[1,3,*], Gollam Rabby[1], M. Sirajul Islam[2], Shihab Uddin[2], M. Masuduzzaman[2]

[1]Faculty of Computing, University Malaysia Pahang, 26300 Gambang, Kuantan, Pahang, Malaysia

[2]American International University-Bangladesh, Ka-66/1, Kuratoli Road, Kuril, Khilkhet, Dhaka-1229, Bangladesh

[3]IBM CoE, UMP, 26300 Gambang, Kuantan, Pahang. Malaysia

E-mail: saifulazad@ump.edu.my (*corresponding author)

August 2019

Abstract. For facilitating users from information overloading by finding recent trends in twitter, several techniques are proposed. However, most of these techniques need to process extensive data. Therefore, in this paper, a geofencing-based recent trends identification technique is proposed, which acquires data based on a geofence. Afterwards, they are cleaned and the weight of these tweet data is calculated. For that, the frequency of tweet texts and hashtags are taken into account along with a boosting factor. Thereafter, they are ranked to recommend recent trends to the user. This proposed technique is applied in developing a system using Java and python. It is compared with other relevant systems, where it demonstrates that the performance of the proposed system is comparable. Over and above, since the proposed system integrates geofencing feature, it is more preferable over other systems.

1. Introduction

Since microblogging social media platforms including Twitter, FriendFeed, Dailybooth, and Tumblr have become a popular communication and information search tool, the social content recommendation has risen to a new dimension [1]. Among these platforms, twitter data are preferred by most of the researchers or industries due to its large volume of users — around 321 million claimed monthly users [2] — which produce a large volume of data. Although, these data provide us with a "gold-mine" of real-world information; they lead us to an enormous problem, called information overload [3]. Due to this, the most important, interesting, and relevant tweets remain unnoticed by the users. For more details, if an active user follows 80 users on an average, s/he may receive around 1000 tweets on an average [4]. Among these, many are personal or irrelevant tweets that are unworthy of reading.

To tackle this issue, twitter has introduced the hashtag — a word or a phrase without spaces prefixed with the hash symbol; and hence, the name. It facilitates to spreads trendy topics quickly among millions of users. Again, there remain several other important trends unidentified due to improper or no tagging. To overcome this issue, several researchers have proposed various solutions, e.g., URL recommendations for twitters, organizing trending topics in the user's timeline, recommending followers and tweets, personalized recommendation of twitter tweets [5, 6, 7, 8, 9]. However, only a few techniques are proposed that identifies trends based on geofencing — a virtual perimeter for a real-world geographic area — which are taken into account during experimental evaluations in Section 2. Again, most of these techniques do not take both hashtag and twitter text into consideration for discovering recent trends.

Therefore, in this paper, a geofencing-based recent trends identification technique is proposed utilizing twitter data. In the proposed technique, both hashtags and twitter texts are examined to identify the recent trends. The objective of this work is to assist people by offering them the most relevant and interesting trends based on geofencing; and thus, save valuable searching time for those from the ocean of data. In addition, the proposed technique also ranks the identified trends to facilitate the selection.

The rest of the subsequent sections of this paper are organized as follows. The most relevant techniques to the proposed technique are discussed in Section 2 with their known limitations and/or features to establish the necessity of proposing a new such technique. The details of the proposed technique are mentioned in Section 3. Afterwards, the performance of the proposed technique is evaluated with the relevant experiments in Section 4. This paper ends with concluding remarks in Section 5.

2. Related Works

Emotion identification employing hashtag remains an important research area to investigate. In [10], the authors endeavor to identify the emotions of different human being during crisis period. They collected tweets of various subjects after the hurricane sandy, which is considered as the deadliest, the most destructive, and the strongest hurricane of the 2012 Atlantic hurricane season. This research focuses only on analyzing the tweets of the victims to identify their emotions over the crisis. Again, in [13], another research is conducted to identify students' emotions employing twitter data. Generally, it has been observed that students cannot express their emotions to their teacher. Therefore, they prefer social media as their platform to express their emotions. In this paper, the students' tweets are acquired and data mining techniques are applied for identifying their emotions. However, the proposed technique would analyze the tweets to find the contemporary trends of a certain geographical location.

For identifying the emotions and moral relationship between them, twitter data are analyzed in [11]. Here, the authors gathered a vast amount of data from the social media. Later, they are transformed into a meaningful data. Afterwards, they are analyzed for better understanding of the audiences' emotions even before an event occurs. However, in this paper, a vast amount of data is gathered to identify the most relevant current trends which is happening in a virtual geographical location or geofence.

Twitter data are also used to identify human behavior over the social network. In [12], a such study is performed where data from Cyworld are acquired. Afterwards, they are analyzed to identify until what extent people nowadays are involved in social media and how they behave over the social network. However, the proposed technique crawls over twitter tweets to analyze trend behavior over the social network.

In [5], a technique is proposed that can identify and recommend personalized trends to the users to aid them from searching a huge number of tweets. This proposed technique utilized term frequency and inverse document frequency (tf-idf) to identify the most important term along with collaborative filtering. However, although our proposed technique utilizes tf-idf to identify the important terms, hashtag also plays an important role in identifying the recent trends. Again, the geofencing-based identification, which is necessary for the reducing number of data as well as the processing time and focus on user interest reign, make the proposed technique distinguish over other existing technique.

3. Proposed Technique

The proposed technique is divided into three phases, namely *i*)data acquisition, *ii*) recent trend identification, and *iii*) recommending recent trends and topic-based searching. All these phases are explained in details below.

3.1. Data Acquisition

Since the proposed technique identifies the recent trends based on geofencing, data must be acquired from the Twitter accordingly. Generally, tweet data are global; therefore, it is necessary to acquire data in such a way that it enables geofencing. In this proposed technique, tweet data are acquired using tweepy library, which is a python library that can acquire real time data. Afterwards, the desired location is set before acquiring the data. All the data are acquired in json format in a . json file. Then, they are converted to .CSV (Comma Separated Value) file and stored in a MySql database.

3.2. Recent Trends Identification

The process of recent trends identification starts with fetching tweets from the database, where tweets in English is only taken into account. Afterwards, stopwords, emojis, and URLs are removed from the tweet data. Let us called this as "tweet text" to distinguish it from the tweet data.

These tweet texts are further examined to identify the hashtags since they are incorporated to facilitate users in finding messages with a specific theme or content. It is possible to identify many recent trends analyzing these hashtags as demonstrated in [1]. In the proposed technique, they also play an important role in identifying the

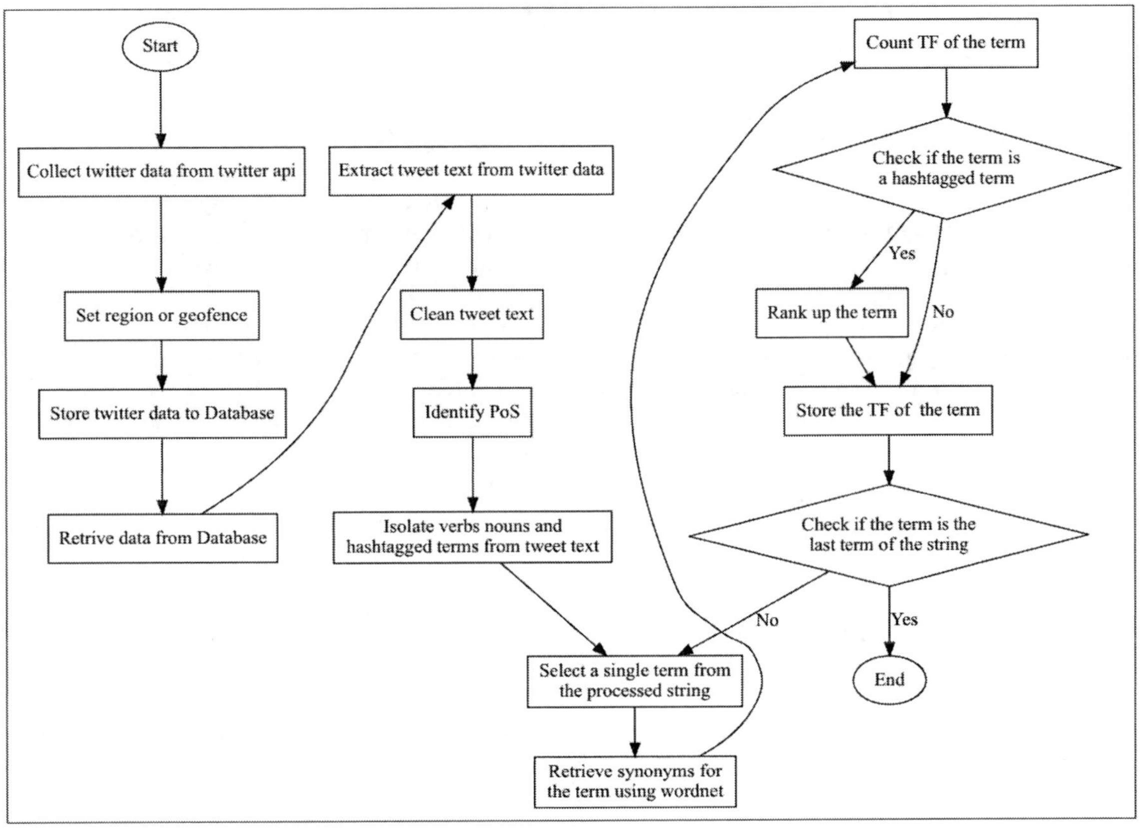

Figure 1: Steps of the proposed technique

recent trends along with other factors, which are mentioned below. Hashtags are easy to identify in a tweet text since all of them are structured and start with the "#" sign. After discovering all the hashtags from a tweet text, they are removed and saved in a list, λ_h.

Afterwards, all the nouns and verbs are identified and extracted from a tweet text; and for that, the wordnet library [17] is utilized. Later, only these nouns and verbs are considered as the representative of the entire tweet text and saved in a list, λ_s. Afterwards, the frequency of each tweet text is calculated to contribute in finding the recent trends. In the proposed technique, it is performed in the following way. Each extracted tweet text is passed to a function where a key is generated and associated with it and checked the existence of that key in a list, λ_k, which saved two-tuple data — the key and its associated count. If it is a unique key, it is appended to the list with initializing the count to 1; otherwise, the count is increased by 1. The same procedure is also followed for the hashtags in λ_h.

Then, to identify the recent trends, a matching between the hashtags and the tweeter texts are performed. Since the structure of a tweet text is different (i.e., a sentence) from that of hashtags; therefore, a substring-based matching is conducted. If

Geofencing-based Recent Trends Identification

Algorithm 1 Algorithm for finding the trends from twitter data

```
String tweets = "", hash-tags = "", all_tweets
for (i ← 0 to last row of the tweet table)
str[i] ← str[i].removeAll(comma,dots , colon etc)
str[i] ← str[i].removeAll(extra white space to single space)
new_str[array] ← str[i].split('single space')
for (j ← 0 to last index of new_str)
if new_str[j].startsWith("@http") or new_str[j].startsWith("http") then
    new_str[j] ← ""
else
    tweets+ = new_str[j]
end if
if new_str[j].startsWith("#") then
    λ_h+ = new_str[j]
else
    tweets+ = new_str[j]
end if
end for
nt[array] ← tweets.split(singlespace)
for (m ← 0 to last index of new_tweets)
if nt[m] is 'Noun' then
    nt[m] ← synonym[0].(nt[m])
    λ_s+ = nt[m]
else
    λ_s+ = ""
end if
if nt[m] is 'Verb' then
    nt[m] ← synonym[0].(nt[m])
    λ_s+ = nt[m] + "."
else
    λ_s+ = ""
end if
end for
sentence[array] ← λ_s.split(fullstop)
for (k ← 0 to last index of sentence)
generate unique Key for similar sentence
if same unique Key then
    count+ = 1
    λ_s(key, count)
else
    count ← 1
    λ_s(key, count)
end if
end for
for (k ← 0 to last index of λ_h)
generate unique Key for similar hashtags
if same unique Key then
    count+ = 1
    λ_h(sentence, count)
else
    count ← 1
    λ_h(hashtag, count)
end if
end for
for (p ← 0 to last λ_s)
tw[array] ← λ_s[p].sentence.split(singlespace)
for (q ← 0 to last λ_h)
if HTL.hashtag.matches(λ_s[p].sentence) then
    λ_s[p].count *= α
end if
end for
end for
end for
```

a matched found, the count of that tweet text is recalculated as follows:

$$\kappa_t' = \kappa_t \times (\alpha \times \kappa_h) \tag{1}$$

where, κ_t is the previous frequency count, κ_t' is the new frequency count, κ_h is the count of the hashtag, and α is a constant, which is utilized as the boosting factor. The value of α must be more than 1. As could be observed by Equation 1 is that when a tweet text or a substring of a tweet text is matched with a hashtag, its frequency is increased at least α times of the existing count. The justification of this increment is that it has been observed that hashtags are the most influential terms in a tweet data that carry recent trend related information as demonstrated in [1]. All these steps of the proposed technique are illustrated in Algorithm 1.

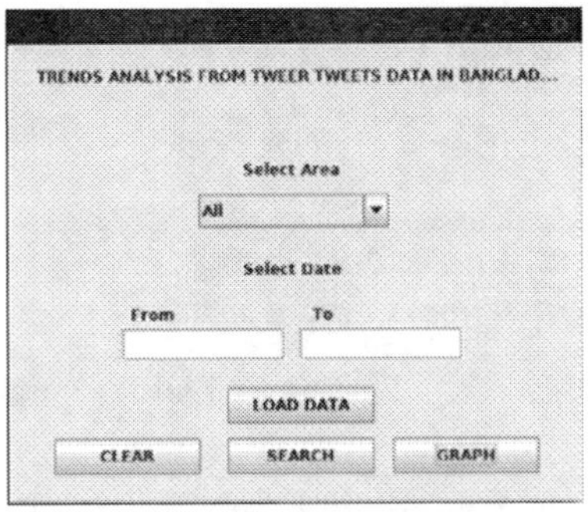

(a) Java based GUI for the trends analysis system

(b) GUI for providing search terms

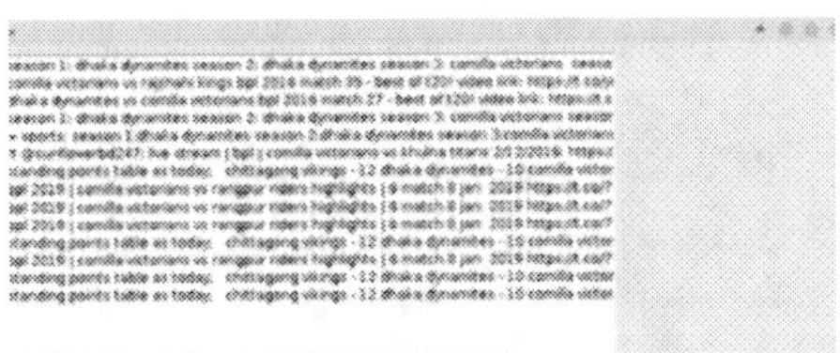

(c) Data found from database related to the search term

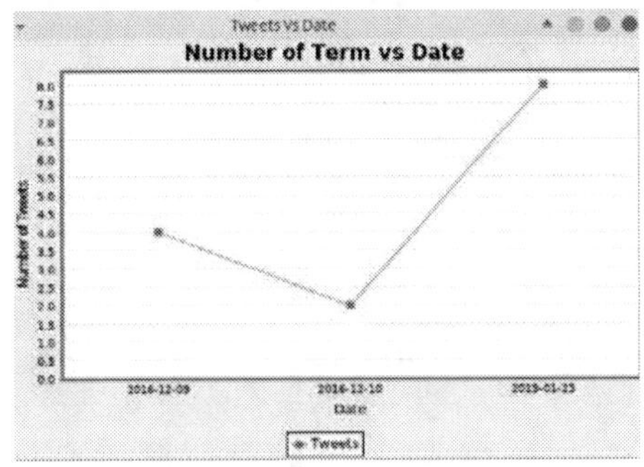

(d) Graphical representation of search term's occurrences

Figure 2: Four print-screens of the implemented system that is based on the geofencing-based proposed technique.

3.3. Recommending Recent Trends and Topic-based Searching

After calculating frequencies of all the tweet texts, they are ranked based on their frequency counts. Among them, only top-N trends are recommended to the users. For that, any advanced sorting technique would be adequate. However, in our case, a quick sort is utilized since its time complexity is $O(nlog(n))$.

Again, in the developed system employing the proposed technique, an option of searching any specific topic is also integrated. For that, a topic needs to be provided in the system. Afterwards, a search is initiated to match the topic with the tweets. Then it provides date wise count of that specific word.

4. Experimental Evaluation

A system has been implemented following Algorithm 1. Using the system, a user can search a topic or all the recent trends between a date or any term related within an epoch can be visualized as demonstrated in Figure 2. The system was developed in Java and python for the cross platform advantage; thus, this system can be employed in any platform. Once the data is loaded into the system, it takes a minimal time to

Table 1: Comparison between Three popular system and our system

	keyhole	**tweetreach**	**tweetbinder**	**proposed system**
Search Type	hash tag based	hash tag and keyword based	hash tag and keyword based	hash tag, keywords and geo location based
Licensing	paid and free with minimal feature	paid and free with minimal feature	paid and free with minimal feature	will be open-sourced after publication of the paper
Accuracy	90% on free version and stored data	95% on free version and stored data	95% on free version and stored data	95% on stored data with geo location, hashtags and keywords

identify and visualize the data.

To find out the effectiveness of the developed system, it is compared with other three most relevant existing systems, namely keyhole [14], tweetreach [15], and tweetbinder [16]. Here, it is noteworthy to mention that all these systems are freemium systems, i.e., some features of these systems are free and some of them are premium.

For our experimental evaluation, a location has been chosen for acquiring data for a period of time where some important events have taken place. In this paper, Bangladesh is chosen as the geofencing location. Again, we have tested different keywords in the free versions of the aforementioned. The results are noted in Table 1. As it could be observed from the table is that tweetreach, tweetbinder, and the proposed system demonstrate comparable performance, i.e., 95% accuracy, which is well over keyhole system with 90% accuracy. However, since the proposed system integrates geofencing for finding the recent trends, it requires only limited data to store and process; and thus, preferable over other techniques. Full system implementation can be found at the following repository hosted on github.com. Repository link: https://github.com/ping543f/A-Geofencing-based-Recent-Trends-Identification-from-Twitter-Data.git

5. Conclusions

This paper presents a geofencing-based recent trends identification technique using twitter data. In the proposed technique, data are acquired based on the geofence. Afterwards, they are cleaned and the weight of each tweet text is calculated based on the frequency of itself and substring matching of hashtags. Then, they are ranked

Geofencing-based Recent Trends Identification

to recommend to the user. The proposed technique is applied to developed a system in Java and python and is compared with other relevant systems. The experimental results demonstrate that the performance of the proposed system is comparable with other compared systems. However, due to its geofencing feature, it is more preferable over other systems.

Acknowledgement

This work has been partially supported by the RDU grant, RDU180359, of University Malaysia Pahang, Malaysia.

Referencing

[1] Eriko Otsuka, Scott A. Wallace, and David Chiu. *A hashtag recommendation system for twitter data streams.* Computational Social Networks, 3(3), 2016.

[2] Hamza Shaban. *Twitter reveals its daily active user numbers for the first time* The washington Post, 2019.
`https://www.washingtonpost.com/technology/2019/02/07/twitter-reveals-its-daily-active-user-numbers-first-time/?noredirect=on&utm_term=.778eba392257`

[3] C. Borgs, J. Chayes, B. Karrer, B. Meeder, R. Ravi, R. Regans, and A. Sayedi. *Game-theoretic models of information overload in social networks.* Algorithms and models for the Web-Graph, Springer, 146-161, 2010.

[4] Z. Qu and Y. Liu. *Interactive group suggesting for twitter.* In Proceedings of the 49th Annual Meeting of the Association for Computational Linguistics: Human Language Technologies: short papers, 2:519523, 2011.

[5] Vineeth Rakesh, Dilpreet Singh, Bhanukiran Vinzamuri, Chandan K. Reddy. *Personalized Recommendation of Twitter lists Using Content and Network Information* In Proceedings of the Eight International AAAI Conference on Weblogs and Social Media, 2014.

[6] F. Abel, Q. Gao, G.-J. Houben, and K. Tao. *Analyzing temporal dynamics in twitter profiles for personalized recommendations in the social web.* In Proceedings of the 3rd International Web Science Conference, ACM, 2011.

[7] M. G. Armentano, D. Godoy, and A. Amandi. *Topology-based recommendation of users in microblogging communities.* Journal of Computer Science and Technology, 27(3):624634, 2012.

[8] M. S. Bernstein, B. Suh, L. Hong, J. Chen, S. Kairam, and E. H. Chi. , E. H. 2010. *Eddi: interactive topic-based browsing of social status streams.* In Proceedings of the 23rd annual ACM symposium on User interface software and technology, ACM, 303312, 2010.

[9] J. Hannon, M. Bennett, and B. Smyth. *Recommending twitter users to follow using content and collaborative filtering approaches.* In Proceedings of the fourth ACM conference on Recommender systems, ACM, 199206, 2010.

[10] J. Brynielsson, F. Johansson, C. Jonsson, and A. Westling. *Emotion classification of social media posts for estimating peoples reactions to communicated alert messages during crises.* Security Informatics, 2014.

[11] D. M. Haughton, J. J. Xu, D. J. Yates, and X. Yan. *Introduction to Data Analytics and Data Mining for Social Media Minitrack.* Hawaii International Conference on System Sciences, 2016.

[12] Y.-Y. Ahn, S. Han, H. Kwak, S. Moon, and H. Jeong. *Analysis of topological characteristics of huge online social networking services.* In Proceedings of the 16th international conference on World Wide Web. ACM, 2007.

[13] X. Chen, M. Vorvoreanu, and K. Madhavan. *Mining social media data for understand-ing students learning experiences.* IEEE Transactions on Learning Technologies, 7(3): 246-259, 2014.

Geofencing-based Recent Trends Identification

[14] *Whatever happens in the world shows up on Twitter*, 2019.
texttthttps://keyhole.co/twitter-analytics/

[15] *Free Twitter Analytics Report from Union Metrics*, 2019.
`https://tweetreach.com/`

[16] *Analyze, classify, and display twitter and Instagram content*, 2019.
`https://www.tweetbinder.com/`

[17] George A. Miller. *WordNet: A Lexical Database for English.* Communications of the ACM, 38(11):39-41, 1995.

[18] *Hale, S. A. (2014) Global Connectivity and Multilinguals in the Twitter Network. In Proceedings of the 2014 ACM Annual Conference on Human Factors in Computing Systems, ACM (Montreal, Canada).*

IOP Conf. Series: Materials Science and Engineering **769** (2020) 012009 doi:10.1088/1757-899X/769/1/012009

An Improved Mapping Pattern for Digital Watermarking using Hilbert-Peano Pattern

Mohamad Nazmi Nasir, Syifak Izhar Hisham, Mohd Faizal Ab Razak

Faculty of Computing, Universiti Malaysia Pahang, 26300 Gambang, Pahang, Malaysia

syifakizhar@ump.edu.my

Abstract. In the present, watermark plays an important part in all copyrighted content whether it is physical or digital. The importance of watermark in digital data increases as the usage of internet in transferring data increases day by day as the technologies improves. The main objective of using digital watermark is to apply the highest security as possible to the digital image and in the same time to reduce as most disturbance as possible that can be visible to the human eye. This research proposed a hybrid mapping pattern which is Hilbert-Peano Curve to try to achieve the highest security as possible from alteration. This proposed digital watermark uses the Least Significant Bit (LSB) to store the watermark data, which consist of parity bits as the authentication data and recovery data. This is to reduce the disturbance to the original image as much as possible as it might alter the colour of the original image. The results are compared with the existing watermark, Hilbert-LSB to analyse which pattern is suitable to be implemented with the watermarking scheme. The results of this research shows Hilbert-Peano pattern can contribute better performance in terms of watermark data security, time taken to embed watermark data, and the imperceptibility level if compared to the original Hilbert mapping pattern.

1. Introduction

In the process of sharing information on the internet, security play a big role in ensuring the data transferred from the sender arrives to the receiver without any disturbance from an unknown entity that might be stealing any information from the transferred data. All of the data on the internet should be kept secret from the others except the sender and servers since that the data may be security-sensitive [1]. The data also requires authentication to provide confirmation on the authenticity of the data. The data also must be confirmed that it does not been tampered by any outside entity.

The most known data security techniques can be classified into two categories: Digital watermarking and steganography. Digital watermarking is a capability for inserting information known as watermark, into an image which later can be extracted or observed for various purposes such as identification and authentication [2-7]. Besides providing security, this technique can also be utilized to gain information about the source, owner, distributor or creator of an image or a document. The second category is steganography which focuses more on bandwidth of the hidden messages while hiding a message, file or an image within another message, file or image but in the aspect of watermarking, the watermark robustness is the key performance parameter [8-9].

Two fragile watermarking schemes SPIRAL-LSB and HILBERT-LSB for authentication have been developed with the special feature of embedding [10-13]. They applied unique pattern of numbering to decide the watermark bit locations. These schemes have extended the current technology of fragile watermarking. The watermarked images age proven theoretically and experimentally as good and qualified to use in clinical diagnosis. The accepted noise for human sight is 32dB, and they managed to get below this value. However, the current watermarking pattern took some time to embed the watermark into the images. Thus, the research is to test whether the proposed watermarking pattern is faster than the current watermarking pattern available.

2. Hilbert-Peano Authentication Scheme

2.1. *Generating the Hilbert-Peano Pattern*

2.1.1. Divide the Hilbert-Peano Pattern. Based on the Hilbert-Peano pattern, the pattern is generated by using two basic elements, corresponding to the connected curve. Figure 1 shows the two basic elements of the pattern.

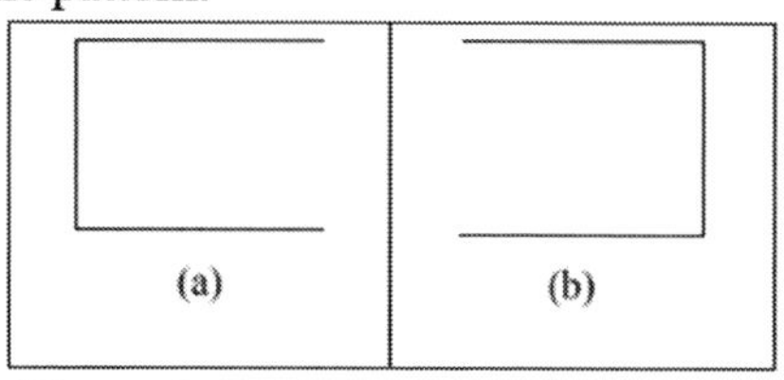

Figure 1. Basic generated elements

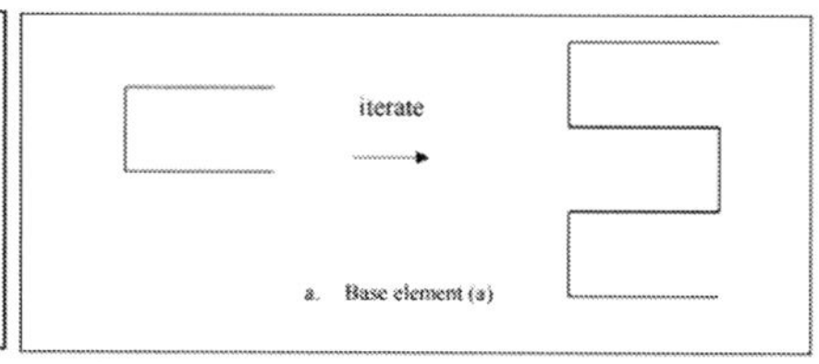

Figure 2. Generation of the pattern

2.1.2. Iterative Matrix. The Hilbert-Peano pattern generating element is simple but according to its characteristics, as the number of iterations increases, the number of the iteration of the infinity curve will traverse the area within the scope of all of the points. Based on the curve of each iteration characteristics, the amplification of iterative matrix is applied to describe the curve generation process, as demonstrated in Figure 2. The generation of the curve for the first iteration will be as in Figure 2.

2.1.3. The Connection Between the Generated Curve. According to the characteristics of the Hilbert-Peano pattern, the curve has one entrance and one exit meaning that the repeated traversal area must demand the curve to be continuous at all squares. By the n number of iterations of the Hilbert-Peano pattern after it has been translated to a $6n \times 6n$ iterative matrix according to the rules.

2.2. *Numbering and Embedding the Watermark Bits*

The detailed embedding algorithm of this authentication watermarking are described in [10-13]. Before the embedding process, all the blocks in the image are numbered and mapped with the mapping pattern to determine the location and mark the blocks as the watermarked data to avoid redundancy in the embedding process. By using a specific style of numbering and mapping of each block, it could assure the top performance of the authentication system by spreading the numbered data as far as possible from the original location. The numbering of the block is determined on the user whether they want to divide the block into 6 by 6 smaller blocks or 8x8 even smaller blocks.

The blocks of the original images are labelled as C and the blocks of watermarked image as H. The proposed mapping pattern starts with numbering C using the Hilbert-Peano method and map it using the pseudorandom way. The numbering then will start at a random pixel and then followed by moving to the next block. All of the block then will be embedded and numbered.

In the process of generating the Hilbert-Peano pattern, it follows a recursive algorithm as follows:
Step 1: Set the block size as C,
Step 2: Set k as the key number,
Step 3: Calculate the number of blocks to get the total number of columns and rows,
Step 4: Choose the starting point then start the block numbering,

Step 5: After the block numbering is completed, each block then will be mapped using the mapping pattern presented in [10-13].

3. Experimental Results and Discussion

3.1. Imperceptibility by Visual

The embedded watermark in medical images are not clearly visible to the naked eye at it is embedded in the Least Significant Bit (LSB). From Table 1, it is obvious to say there is no difference between the original and watermarked images. Therefore, to measure the imperceptibility of the watermarked images accurately, Peak Signal to Noise Ratio (PSNR) and Mean Square Error (MSE) is preferred for measurement.

Table 1: Comparison between the Original and Watermarked Images

Original Image	Watermarked Image	Original Image	Watermarked Image

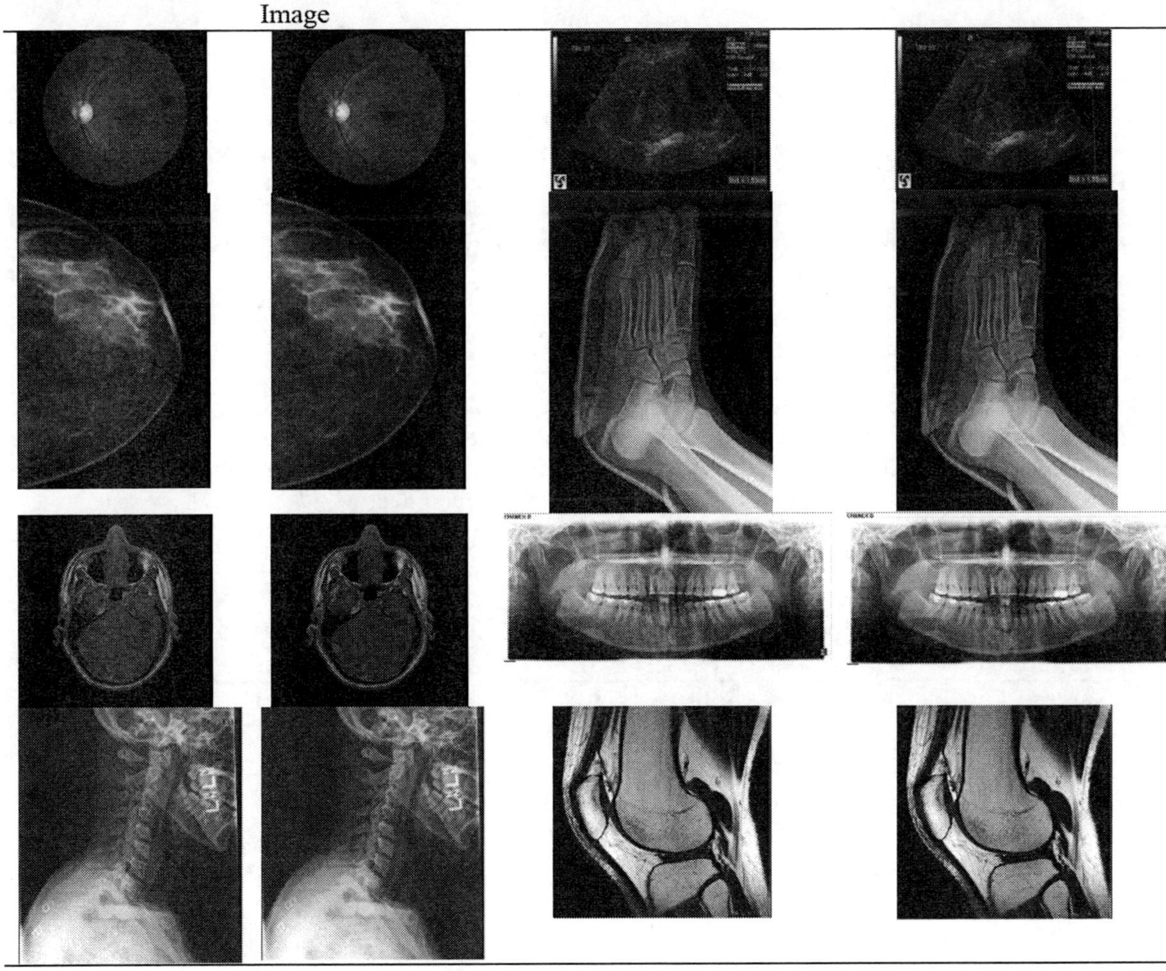

3.2. Peak Signal to Noise Ratio (PSNR)

PSNR is the ratio between a power of the signal noise and the signal's maximum power.
It is used to measure the quality of the images that has been embedded watermark into it. If the PSNR value is high, it means that the image is almost similar with the original image. Hence, each image will be embedded into two watermarking pattern and the value of the PSNR will be compared.

Table 2 shows the PSNR values obtained from the two watermarking pattern that is used in this research, Hilbert-LSB (11, 14) and Hilbert-Peano.

Table 2: PSNR Values between Hilbert-LSB and Hilbert-Peano watermarking pattern

Figure	PSNR of Hilbert-LSB	PSNR of Hilbert-Peano
Sample 1	58.9348	58.9406
Sample 2	58.8268	58.8912
Sample 3	53.9764	53.6794
Sample 4	58.3968	58.3872
Sample 5	55.4978	55.5138
Sample 6	58.4014	58.4297
Sample 7	58.6310	58.6431
Sample 8	58.6851	58.6851
Average Value	**57.6463**	**57.6687**

Table 2 shows the result of PSNR values gained from both the image watermarking schemes on the medical images that is tested on 8 different samples. The amount of PSNR value is calculated based on the ration between the power of distorting noise that affects the quality of the watermarked image and the maximum possible value of a signal.

Based on the average value of PSNR, the Hilbert watermarking scheme is lower by 2 decimal places if compared to the proposed scheme. As we know, the lower the PSNR value, the better the security of the image but based on these results, there was not much difference in PSNR value in comparison between both of the values.

3.3. Mean Square Error (MSE)

Mean Square Error (MSE) defines as the cumulative squared error between the original image and the embedded image. The lower the MSE value, the lesser the error found in the watermarked image meaning that the alteration on the embedded image is less significant to the naked eye. The same sample images will be tested on the MSE and the outcome is recorded.

Table 3 shows the MSE values obtained from the two watermarking pattern that is used in this research.

Table 3: MSE Values between Hilbert-LSB and Hilbert-Peano watermarking pattern

Figure	MSE Value of Hilbert-LSB	MSE Value of Hilbert-Peano Scheme
Sample 1	0.0830	0.0831
Sample 2	0.0839	0.0852
Sample 3	0.2787	0.2787
Sample 4	0.0943	0.0941
Sample 5	0.1827	0.1834
Sample 6	0.0933	0.0940
Sample 7	0.0889	0.0891
Sample 8	0.0880	0.0880
Average Value	**0.1241**	**0.1244**

3.4. Embedding Processing Time

The embedding process is the process of inserting the watermark into the original image. Hilbert Pattern is a single type pattern and Hilbert-Peano is a hybrid pattern as it combines two patterns into one.

Table 4 shows the time taken to embed the watermark into the images. The images are varying in sizes as they are different medical images. The size of the images also affected the time taken for the watermark to embed into the images.

Table 4: Time taken comparison between Hilbert-LSB and Hilbert-Peano watermarking pattern.

Image Sample	Image Size	Hilbert-LSB	Hilbert-Peano
		Time Taken	Time Taken
Sample 1	537 KB	45.9219	37.7813
Sample 2	645 KB	55.8750	44.8906
Sample 3	65.0 KB	5.6563	4.4844
Sample 4	1.43 MB	510.7969	408.6719
Sample 5	301 MB	351.9219	338.8594
Sample 6	101 MB	303.0781	275.1406
Sample 7	960 KB	375.8125	336.0625
Sample 8	5.39 KB	13.8750	13.0313
Average Value		**207.8672**	**182.36525**

In the table, the time taken differs and the bigger the size of the image, the longer time is needed to embed the watermark. From the table, it is clear that the Hilbert pattern is taking a bit longer time to embed as compared to the Hilbert-Peano pattern. This is because of the embedding process of the Hilbert watermarking scheme requires a 2-bit information authentication data size and 7 bits for the recovery information for each pixel as compared to the Hilbert-Peano watermarking scheme which is less.

4. Conclusion

The proposed watermarking scheme has proved better in terms of the time processing of embedding the watermark into the medical images. The quality of the PSNR and the MSE value was not significantly different meaning that the quality of the image is not too much disturbed by the watermark. This proves that even though the proposed watermark is a hybrid, it does not affect much on the watermarked image. The proposed watermarking can be used if it involves a large amount of data in order to reduce the processing time for image watermarking.

Acknowledgments:
This research work is supported by RDU1803163 supported by Universiti Malaysia Pahang. The data in this paper can be accessed in the website of FKom, UMP at [http://fskkp.ump.edu.my].

References

[1] Singh, G. (2018). A review of secure medical image watermarking. IEEE International Conference on Power, Control, Signals and Instrumentation Engineering, ICPCSI 2017, 3105–3109. https://doi.org/10.1109/ICPCSI.2017.8392297
[2] Araghi, T. K., Manaf, A. B. A., Zamani, M., & Araghi, S. K. (2016). A Survey on Digital Image Watermarking Techniques in Spatial and Transform Domains. International Journal of Advances in Image Processing Techniques– IJIPT, 3(1), 6–10.
[3] Mantoro, T., Ayu, M. A. & Priyatna, M. A. (2018). Security Scheme based Dual-Layer Fragile Watermarking for Medical Image. 2018 Third International Conference on Informatics and Computing (ICIC), Palembang, Indonesia, pp. 1-5.
[4] Rajput, V. & Ansari, I. A. (2019). Image tamper detection and self-recovery using multiple median watermarking. Multimedia Tools and Applications, pp 1–17.
[5] Ma, B., Li, B., Wang, X., Wang, C., Li, J., & Shi, Y. (2019). A code division multiplexing and block classification-based real-time reversible data-hiding algorithm for medical images.

Journal of Real-Time Image Processing, August 2019, Volume 16, Issue 4, pp 857–869.

[6] Liu, X., Lou, J., Fang, H., Chen, Y., Ouyang, P., Wang, Y., Zou, B., & Wang, L. (2019). A Novel Robust Reversible Watermarking Scheme for Protecting Authenticity and Integrity of Medical Images," in IEEE Access, vol. 7, pp. 76580-76598.

[7] Rakhmawati, L., Wirawan, W. & Suwadi, S. (2019). A recent survey of self-embedding fragile watermarking scheme for image authentication with recovery capability. EURASIP Journal on Image and Video Processing, December 2019, 2019:61

[8] Alias, N. & Ernawan, F. (2019). Multiple watermarking technique based on rdwt-svd and human visual characteristics. Journal of Theoretical and Applied Information Technology, Volume 97, Issue 14, Pages 3980-3939.

[9] Sharma, N. & Batra, U. (2017). A Review on Spatial Domain Technique Based on Image Steganography. 2017 International Conference on Computing and Communication Technologies for Smart Nation (IC3TSN), Gurgaon, 24-27. 10.1109/IC3TSN.2017.8284444

[10] Ernawan, F. (2019). Tchebichef image watermarking along the edge using YCoCg-R color space for copyright protection. International Journal of Electrical & Computer Engineering (2088-8708) 9 (3).

[11] Hisham, S. I., Zain, J. M., Arshad, N. W., & Liew, S. C. (2015). HILBERT-LSB-C as authentication system for color medical images. 2015 4th International Conference on Software Engineering and Computer Systems, ICSECS 2015: Virtuous Software Solutions for Big Data, Langkawi, 15–20. https://doi.org/10.1109/ICSECS.2015.7333114

[12] Hisham, S. I., Muhammad, A. N., Zain, M. J. & Badshah, G. (2013). Localization watermarking for authentication of text images in Quran with spiral manner numbering. 2013 Taibah University International Conference on Advances in Information Technology for the Holy Quran and Its Sciences, Madinah, 24-29.

[13] Hisham, S. I., Muhammad, A. N., Zain, M. J., Badshah, G. & Arshad, N. W. (2013). Digital watermarking for recovering attack areas of medical images using spiral numbering. 2013 International Conference on Electronics, Computer and Computation (ICECCO), Antalya. 285-288.

[14] Hisham, S. I., Zain, M. J., Arshad, N. W., Liew, S. C., Johari, N. H. & Badshah, G. (2014). Authentication system for medical images using Hilbert numbering. 2014 International Conference on Computer, Communications, and Control Technology (I4CT), Melaka. 198-202.

Improvement of the POLISAS e-Request Transcript System using Query Optimization

Mohd Hariz bin Ibrahim[1], Suryanti Awang[2], Kohbalan Moorthy[2]
[1]Kolej Komuniti Paya Besar, Kementerian Pendidikan Malaysia, BT 19, Jalan Maran, 26300 Gambang, Kuantan, Pahang, Malaysia.

[2]Soft Computing & Intelligent Systems Research Group (SPINT), Faculty of Computing, Universiti Malaysia Pahang, 26300, Kuantan, Pahang, Malaysia

hariz@kkpbe.edu.my

Abstract. This paper is about the improvement of the E-Request Transcript system used by POLISAS to be a systematic and effective system. The improvement is implemented to solve the current problem, whereby, students need to access the E-Request Transcript system according to the graduation session in individual file. This situation makes it difficult for the administrators to access more than one E-request transcript system if the application is from different sessions. It is because of the data-access issues to generate reports. To achieve the system improvement, an integrated database using the query optimization technique is implemented. Next, the query statement is analysed to produce an optimum performance. Research results show that the technique is able to improve the query time processing, thus resolving the issues discussed. A user acceptance test for this system consists of a black box test and a beta test has been deployed. A black box test was conducted to test the main menu functions in the E-Request transcript and is found to be able to succeed. Furthermore, the beta test is carried out using a survey form distributed to all students who have graduated. Survey questionnaire showed that 57% of respondents were satisfied using this system and 14% were very satisfied with this system. All respondents have provided a good scaling rate for all items assessed. For overall system assessment, the mean score obtained is 3.78 which is interpreted as a good rate based on the scale of interpretation used. In conclusion, this system has met the needs of consumers as well as simplifying their business based on surveys conducted.

Keywords - Integrated; Transcript; Optimization Technique; Query Optimization

1. Introduction

Information systems (IS) involved the process of collection, creation and distribution of useful data in organizational settings that are built based on the integration hardware, software and telecommunications networks. The interrelated components of IS works together to obtain and distribute information. It also processes, stores and distributes the information to help an organization to coordinate, analyze, control and visualize processes of an organization. This also contributes to effective decision making of an organization. There is a clear difference between IS and business processes, which is, IS a system that assists in controlling business processes.

Integrated Information Systems (IIS) consists of several aspects of Management Information Systems that can be integrated. These aspects can be integrated separately or in a combination. Some choices need be made by the organization between developing a new IS or improving an existing one. Thus, the IIS implementation can be a tedious task.

Various researchers have conducted IS assessment in the past. Assessed IS Integration management business Continuity [1]. [2] Assessed the modelling in IS integration, and [3] assessed the elements and functions integrable in integrated management systems. These efforts have contributed to better understanding of assessment issues for IIS, but fall short of looking at the IIS assessment in a broad perspective. Thus far, the assessment is focused more on the technical aspect or on the product itself, but has not covered all domains in IS integration [4].

E-request transcript (ERT) is an application of IS that is currently being used by graduates of Politeknik Sultan Haji Ahmad Shah (POLISAS) and staff at the Examination Unit in POLISAS [5]. The application involves the process of requesting forms and verification letters related to graduation. The graduating student is considered as a client who uses this system and the administrator is required to handle the process of managing the system. An innovation project called E-Request Transcripts (ERT) was produced to replace the manual application form. By using ERT, students are able to initiate online application and student information data can be updated quickly [5]. With ERT, managing production academic transcript and graduate verification letter becomes faster and systematic.

Effective ERT implementation is dependent on an understanding of the goals and benefits of system integration required. The integration is regarded as a progressive effort towards a better handling of information. Other efforts that lead to a better integration are the linking data between different session, integration of relevant data in a central database or platform, the integration of information on graduate and the opportunity to integrate data in personal level. People do not always benefit from the abundance of natural resources, if the man does not know to manage it [6]. From the database resources, ERT needs to serve as the best integration technique for solved the management issue.

2. Literature Review

2.1. Case study (E-Request Transcript System (ERT))

In total, the POLISAS Examination Unit is responsible for the management of data transcripts application at POLISAS. Among the key functions of the organization are [5]:

i. Collect and process academic transcripts application data from students before preparing the academic transcripts to students.
ii. Ensure data transcripts of academic transcripts and data records can be tracked quickly.
iii. Providing systematic services online.

Saved transcript data includes student-ending data, data application, pending application data, rejection data and processed data. To ensure that the above functions are well implemented, the data needs to be stored and maintained by a database system. Therefore, a database system called the E-Request Transcript System (ERT) has been developed for the management and maintenance of such data.

2.2 E-Request Transcript System (ERT)

An innovative project named E-Request Transcript was previously created to replace the use of the manual application forms. Students only need to apply online and student information data can be updated quickly. With E-Request Transcript, the issuance of Completed Letter of Accreditation and Academic Transcripts becomes faster and systematic. The E-Request Transcript system has been adopted by the examination unit since 2013 has greatly assisted the management of the examination unit, POLISAS [5].

2.3 Query processing

Structured Query Language or SQL is an interactive programming language used to retrieve information from databases and update data in databases. It provides a natural language syntax for constructing query statements, defining and manipulating data in databases. Questions are in the form of instruction languages that provide functions such as selection, entry, updating and data access and are more commonly known as SQL query statements [14] [15] [16].

The ultimate goal of SQL statement processing is to transform queries written in high-level languages to more accurate and efficient implementation strategies that are expressed in lower-level languages and implement those strategies in re-attaining the required data [17]. An important aspect of query processing is query optimization. There are various equivalent conversion strategies for every high-level query. Therefore, the purpose of query optimization is to choose one of the strategies that can minimize the use of resources [18]. Typically query optimization is implemented to reduce the overall execution time of the query, ie the total execution time for all separate operations that perform the query [18]. However, resource utilization can also be viewed as the response time of the query. In a research [18], they focus on maximizing the number of parallel operations in reducing the reaction time of query execution.

Query processing can be divided into the following four phases: decomposition (comprising decomposition and validation), optimization, code generation and implementation [19]. The next section briefly explains the first phase of decomposition, before focusing more on the second phase of query optimization.

Query optimization is a function of many relational database management systems in which multiple query plans for satisfying a query are examined. The query at each level of data integration and mapping components to the existing or a new scheme could influence the existing, and new attributes [20]. Here, a good query plan is identified. This may or may not be the absolute best strategy because there are many ways of doing plans. There is a trade-off between the amount of time spent figuring out the best plan and the amount of time used in running the plan. Different qualities of database management systems have different ways of balancing between these two [21]. Query optimization requires a lengthy query execution time.

Therefore, it is necessary to find new SQL technologies used to reduce a query execution time during the optimization, according to the basic operations of query such as scanning, aggregation, join algorithms [22]. This challenge can be addressed by reducing the number of queries by using strings, joins, aggregation and grouping all the relational [20].

2.4 Integration operation

Basically, the Integration operation is closely related to the product of the carteisan. The carteisan product of the two relations was encapsulate each tuple in the first relationship with each tuple in the second relationship and the combined result was stored in a new relationship [11]. For example, the results for the cartel product operations on the relations of X and Y, with x and y the number of tuples, are the relations (x X y) of the tuple. Consider the following example relations. The carteisan product of X relation (Table 2.2) and Y (Table 2.3) is the relation of Z [11]. Relation X (Table 2.2) and Y (Table 2.3) is the Relation of Z (Table 2.4)

Table 2.2: Relation X

StudID	ProgCode
7891	2
6033	3
1332	2

Table 2.3 : Relation Y

StudID	ProgCode
1	1000
2	5000
3	10000

Table 2.4 : Relation Z = X x Y

StudID	ProgCode	ProgCode	SumProg
7891	2	1	1000
7891	2	2	5000
7891	2	3	10000

3011	3	1	1000
3011	3	3	10000
1224	2	1	1000
1224	2	2	5000
1224	2	3	10000
7055	3	1	1000
7055	3	2	5000
7055	3	3	10000
3011	3	2	5000

Based on the concept of relationship, X and Y relationships are written as

$$X X_{X(\alpha)\theta y(\beta)} Y \tag{1}$$

where $x(\alpha)\theta y(\beta)$ defines the combined conditions. The operator, ş, represents the operator for the combined condition where the value of the operator must always be true between the attributes $x(\alpha)$ and $y(\beta)$ of the X and Y relationships. The join conditions usually consist of simple conditions and are connected to a logical operator [7] [8].

In general, it is arguably the combination is equivalent to the product of the cartels, where the combined result of the two relationships is a subset of cartels productions of two relationships [10].

$$X \bigcap_{X(\alpha)\theta y(\beta)} Y \equiv X_{x(\alpha)} X Y_{y(\beta)} \tag{2}$$

The combined result of X and Y relationships with attribute x and y, is the relation Z with (x + y) attributes. The Z relationship has one tuple for each tuple pair of X and Y that meet the combined requirements. The result of Z relationship can be defined as follows:

$$Z = \{z | z = xy \wedge x \in X \wedge y \in Y \wedge x(\alpha) y(\beta)\} \tag{3}$$

This relationship is a subset of the product X and Y.

2.4.1 Equijoin, In line with the optimization objective of minimizing the timing of data response and query execution, the implementation of this combined method can be implemented to reduce access to the relationship and ultimately achieve the ultimate goal of optimization techniques [9].

The combined commonality between R_1 and R_2 can be defined as follows. For each couple tuple t_{r1} and t_{r2}, where $t_{r1} \in R_1, t_{r2} \in R_2$ if $t_{r1}[\alpha] = t_{r2}[\beta]$, then tuple t_{r1} and t_{r2} was merged into one tuple. The overlap between the combined tuples was deleted. The equation for the combination of equality can be written as follows:

$$R_1 \bigcap_{r1(\alpha)=r2(\beta)} R_2 \tag{4}$$

$$R' = \{t | t = r1r2 \wedge r1 \in R1 \wedge r2 \in R_2 \wedge t_{r1}(\alpha) = t_{r2}(\beta)\} \tag{5}$$

$$Z = X \bigcap_{x(\mathrm{ProgCode}) = y(\mathrm{ProgCode})} Y \tag{6}$$

Table 2.5 : Equijoin result

No	ProgCode	ProgCode
7891	2	5000
3011	3	10000
1224	2	5000
7055	3	10000

where R' is the new relationship for R1 and R2 combined. In this combination the operators used are equality operators (=). Consider Table 2.2 and Table 2.3, the combined result of X and Y where $x(\mathrm{ProgCode}) = y(\mathrm{ProgCode})$ is the relationship Z as follows on Table 2.5 [11]:

3. Methodology

Although many methods have been proposed by integrated relationships within the database, they still have weaknesses in handling query statements that involve data access in large numbers of records [12] [13]. Therefore, this research produces an optimization technique to correct the problems discussed. The technique is known as the Logical Scheme Optimization of the Relationship Database. It is based on the equijoin method introduced by Priti and Margaret [10].

For the implementation of the technique, only a few entities in the ERT database are involved, such as *tblpelajar@transkripdis2014*, *tblpelajar@transkripjun2015*, *tblpelajar@transkripdis2015*, *tblpelajar@transkripjun2016*, *tblpelajar@transkripdis2016*. Entities that stores information related data POLISAS students. The following sections describe the attributes and data stored in each entity.

3.1 Entities tblpelajar

Basically, the primary key for the entity *tblpelajar @ transkripdis2015* is the same as the primary key for the entity *tblpelajar @ transkripdis2014, tblpelajar @ transkripjun2015 tbltranskripjun2016* and *tbltranskripdis2016*. Because of these factors, these entities joined and create a new entity. The purpose of the merger is to minimize data overlap and reduce the number of entities, thereby reducing the reaction time of a given query. The joint of these entities described in the following sections.

Figure 3.1 describes the implementation of the Integration Table Optimization techniques on the E-Request Transcript System database. As mentioned earlier, only a few entities are involved in the implementation of *tblpelajar@transkripdis2014*, *tblpelajar@ transkripjun2015*, *tblpelajar@transkripdis2015*, *tblpelajar@transkripjun2016* and *tblpelajar @transkripdis2016* represented tblpelajar_A and tblpelajar_B. With the implementation of the equiljoin Optimization, all entities are combined and the attributes used for comparison are Nopend, Nric, Name, prog. Tuple values for learners are combined across sessions with tuple learners who are different sessions if the values for the attributes are equivalent. Next, new entity (tblpelajar_C) was classified into multiple entities based on SESI attributes.

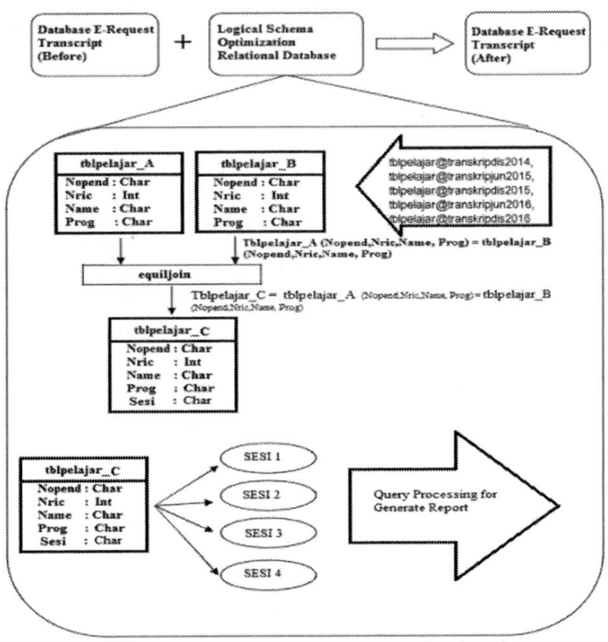

Fig 3.1: Implementation Optimization Technique of Logical Schema Optimization Relational Database to ERT.

In this implementation, the merged entity is tblpelajar for each session represented tblpelajar_A entities and entity is tblpelajar_B, while the main key for both attributes are used as terms of a combination of Nopend, Nric, Name and Prog. These attributes are combined and stored in a new entity namely tblpelajar_C (Figure 3.1) and created a new attribute of SESI. Whereas, the value of each attribute is checked, if the value is equal, then, it combined and stored in the new entity.

As explained, the goal of the research is to improve the time of data access in the database. With the implementation of this process, the number of tuples in the referenced entity can be reduced. Additionally, it can also minimize the number of referred entities. With this, the goal of research can be achieved and the problem of investigation can be improved.

4. Result and Discussion

Table 4.1 shows the number of tuples for each entity before and after the optimization technique is implemented. If observed in the table, the number of tuples before the implementation of optimization techniques is greater than after the optimization technique is implemented. The results of the combined results can be explained based on the following formula :

$$R \bullet \min(R_1, R_2) \tag{7}$$

where the number of tuples in the joined result entity (R3) must be greater or equal to the minimum tuple number between R1 or R2. R1 and R2 are the entities to be joined. Data loss and integrity issues occur if the joined results are not eligible for the formula.

Table 4.1 : Number of tuples per entity

Entity (R_j)	^{ERT}before (Without Optimization)	^{ERT}after (With Optimization)
	Number of Tuple (n)	Number of Tuple (n)
Entity R_1 (tblpelajar_A)	3500	-
Entity R_2 (tblpelajar_B)	36900	-
Entity R_3 (tblpelajar_C)	-	3650

4.1 Query Optimization Comparison Testing

This implementation only uses ORACLE query optimizer and does not change any of the basic rules in selecting the most optimal way of implementation. For each query statement, there are several possible executions that can be produced. The execution time for each possible execution are used as a criterion for determining and selecting the optimum implementation possibilities. In this paper, ORACLE SQL Analyze is used to test the execution time for each possible execution. Based on the information obtained, the most optimal implementation possibilities can be determined and selected for use as a result of obtaining the desired results.

Output for P1, P2, P3 and P4 statements is the number of requests required. To get the result, SUM aggregate function is used in each query statement. Each tuple on the affected entity such as tblmohondetail was be reviewed. If the session attribute value meets the specified predicate condition, then the value of the application was summed. The predicate conditions for each statement P1, P2, P3 and P4 vary, depending on the time interval of the desired output. For P1 statements, the desired output is the number of transcripts for a year, while P2, and P4 are the total application counts for 2 years, 3 years and 4 years respectively. For the GROUP BY statement, the compilation process is performed to get the result of the SUM function. The next section discussed comparative results in selecting and determining the likelihood of execution that results in the most optimal time.

4.1.1 Query P1 (Number of transcripts application per year)

Table 4.2 and 4.3 describes the average execution time for each plan ie plan 1, plan 2 and plan 3. As explained, the most optimal execution possibility is based on the average time taken for each execution.

IOP Conf. Series: Materials Science and Engineering **769** (2020) 012010 doi:10.1088/1757-899X/769/1/012010

Based on the comparison graph in Figure 4.6, plan 3 produces the optimal time for P1 expression. Therefore, there is a possibility of this execution to be selected for execution in producing the required reports ie a transcript application report for a year.

Table 4.2 Query Time for P1 1s – 6s

	1(s)	2(s)	3(s)	4(s)	5(s)	6(s)
Plan 1	0.12	0.12	0.07	0.09	0.10	0.11
Plan 2	0.07	0.09	0.06	0.06	0.06	0.08
Plan 3	0.04	0.04	0.06	0.06	0.06	0.06
Plan 4	0.04	0.04	0.06	0.05	0.06	0.06

Table 4.3 Query Time for P1 7s-10s

	7(s)	8(s)	9(s)	10(s)	Average (s)
Plan 1	0.09	0.06	0.07	0.08	0.091
Plan 2	0.04	0.06	0.06	0.04	0.062
Plan 3	0.06	0.04	0.06	0.04	0.052
Plan 4	0.06	0.06	0.07	0.06	0.056

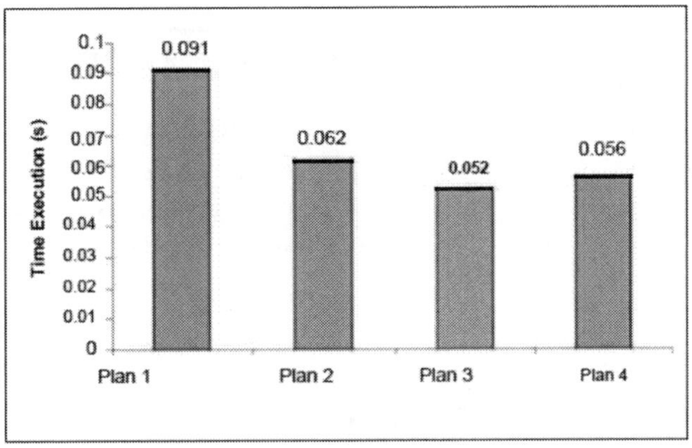

Figure 4.1 Comparison Graph for Query P1

4.1.2 Query P2 (Number of transcripts requested for 2 years)

This statement is for the re-access of transcript application data for an interval of 2 years. The average execution time for each possible execution is as in Table 4.4. Based on the average time obtained, plan 4 produces the fastest reaction time compared to plans 1, 2, 3 and 4. Therefore, the probability of execution was selected in the implementation of data retrieval. Figure 4.2 illustrates the comparison of the execution time of each plan.

Table 4.4 Query Time for P2 1s – 6s

	1(s)	2(s)	3(s)	4(s)	5(s)	6(s)
Plan 1	0.17	0.15	0.14	0.12	0.15	0.15
Plan 2	0.15	0.14	0.14	0.14	0.15	0.16
Plan 3	0.14	0.14	0.12	0.14	0.15	0.14
Plan 4	0.14	0.12	0.12	0.14	0.10	0.12

Table 4.5 Query Time for P2 7s-10s

	7(s)	8(s)	9(s)	10(s)	Average (s)
Plan 1	0.15	0.12	0.12	0.15	0.142
Plan 2	0.14	0.14	0.14	0.10	0.140
Plan 3	0.12	0.14	0.12	0.12	0.133
Plan 4	0.12	0.12	0.10	0.09	0.117

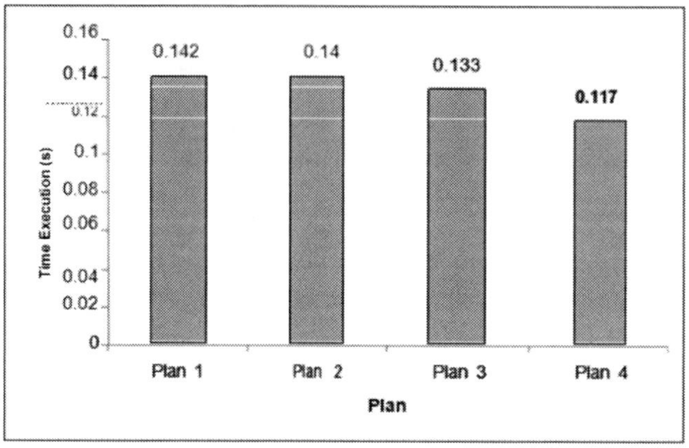

Figure 4.2 Comparison Graph for Query P2

4.1.3 Query P3 (Number of transcripts requested for 3 years)

Table 4.6 and 4.7 illustrates the average execution time for each possible execution of P3 query, while Figure 4.3 is a comparison graph of the execution time of each plan. Based on the graph, plan 3 produces the optimum execution time. Therefore, the plan is selected for the data recovery process for the P3 query statement.

Table 4.6 Query Time for P2 1s – 6s

	1(s)	2(s)	3(s)	4(s)	5(s)	6(s)
Plan 1	0.25	0.20	0.23	0.22	0.22	0.22
Plan 2	0.21	0.17	0.18	0.18	0.20	0.21
Plan 3	0.21	0.15	0.15	0.15	0.17	0.12
Plan 4	0.17	0.19	0.17	0.18	0.17	0.17

Table 4.7 Query Time for P2 7s-10s

	7(s)	8(s)	9(s)	10(s)	Average (s)
Plan 1	0.18	0.20	0.18	0.18	0.208
Plan 2	0.17	0.21	0.18	0.18	0.189
Plan 3	0.15	0.14	0.17	0.15	0.156
Plan 4	0.17	0.18	0.19	0.15	0.174

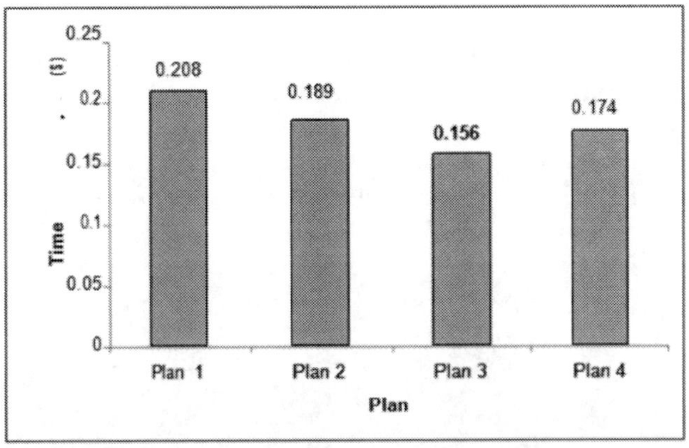

4.1.3 Query P4 (Number of transcripts requested for 4 years)

The average execution time for each possible execution of P4 query is described in Table 4.6. The results show that plan 3 produces the fastest execution time and is illustrated in the comparison graph of the implementation time in Figure 4.9 Based on the decision 3 plan selected to implement this question.

Table 4.8 Query Time for P2 1s – 6s

	1(s)	2(s)	3(s)	4(s)	5(s)	6(s)
Plan 1	0.33	0.25	0.29	0.28	0.26	0.33
Plan 2	0.26	0.23	0.24	0.23	0.26	0.26
Plan 3	0.22	0.20	0.26	0.23	0.23	0.22
Plan 4	0.21	0.21	0.25	0.23	0.26	0.21

Table 4.9 Query Time for P2 7s-10s

	7(s)	8(s)	9(s)	10(s)	Average (s)
Plan 1	0.24	0.20	0.25	0.21	0.21
Plan 2	0.25	0.23	0.25	0.26	0.28
Plan 3	0.23	0.23	0.23	0.23	0.21
Plan 4	0.23	0.21	0.21	0.23	0.23

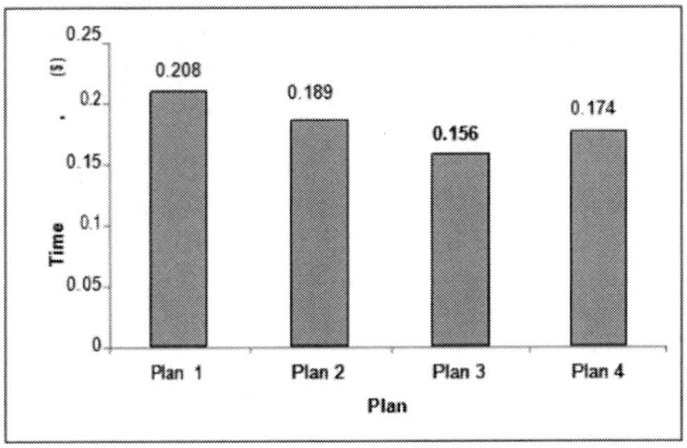

Figure 4.9 Comparison Graph for Query for P4

4.1.4 Comparison Execution Time of Plan

Figure 4.10 shows the execution time of the plan for each query statement for the number of transcript applications. Question P1 represents the number of transcripts for a year, P2 inquiry represents the number of transcripts for 2 years, the P3 question represents the total number of applications for 3 years, while the P4 query represents the total number of applications for 4 years. Plan 3 results shows the most optimal execution time of P1, P3 and P4 queries. While Plans 4 produced the optimum implementation time for P2 inquiries.

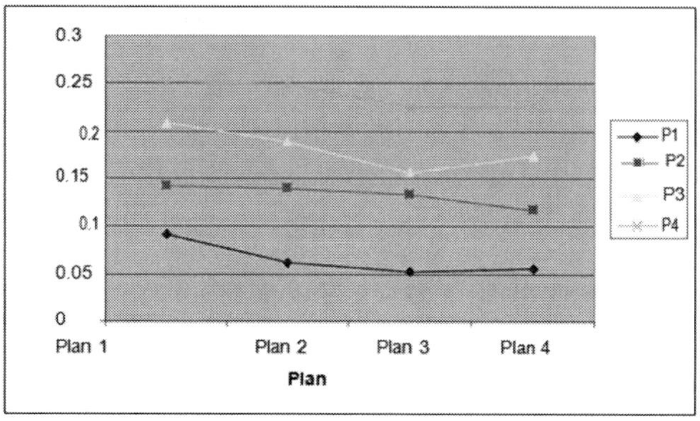

Figure 4.10 Comparative Graph Execution Time For Statement InquiriesThe Number Of Applications A Transcript

The following table 4.7 describes the comparison of execution times before and after optimization. The execution time before optimization is obtained based on the average time in the table above. While the execution time after optimization is obtained based on the probable execution testing discussed in the previous section. Each of the implementation possibilities selected for P1, P2, P3 and P4 statements was compared to the execution time before optimization.

Table 4.7 Comparison Time of Query Execution P1, P2, P3 and P4

	P1	P2	P3	P4
Before Optimization (s)	0.782	0.955	1.415	1.768
After Optimization (s)	0.052	0.117	0.156	0.225
Percent Enhancement (%)	93.4	87.7	89.0	87.3

For P1 statements, the execution time after optimization is faster than before that of 0.052 seconds. In addition, implementation time after optimization of P2, P3 and P4 statements is also faster before optimization (Refer Table 4.8). Comparative results obtained are illustrated using the graph in the following figure 4.11. Based on the graph shows that there is a percentage increase of about 89%. From the results it shows that the technique produced is more effective to fix the problems discussed.

Based on the results obtained, there is a percentage increase in response time for each query statement. Some timetable execution time shows that the response time after optimization is better than before optimization. From the results it shows that the optimization technique of the resulting table can speed up the next reaction time to improve the problem discussed.

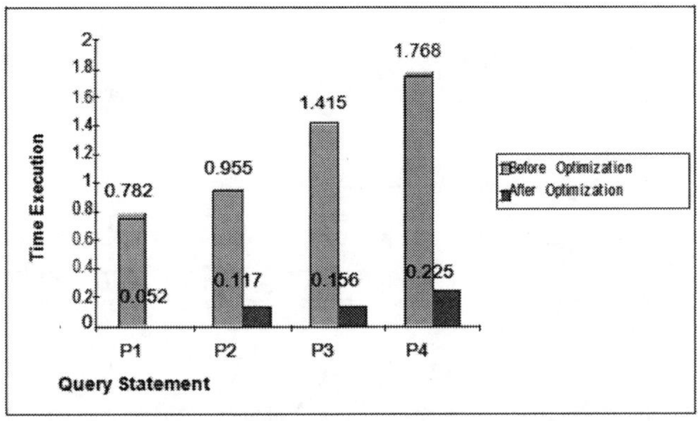

Figure 4.11 Comparison Graph P1, P2, P3 and P4

Black Box Testing is used to test the specific functions of the designed software. In order to test the e-request transcript system, black box testing is used to test the system function and beta test is used to find out the users' response to the e-request transcript system. This is conducted through data obtained by using questionnaire as a research instrument.

Beta testing is an objective test that is tested directly to the field, using a questionnaire in the effort to gather the user's response to the system that is built. The questionnaires were distributed using the Simple Random Sampling technique in which sample members of the population were randomly assigned without regard to the strata present in the population. The data was collected from 20 system users consisting of POLISAS students and 5 management users consisting of staff of the examination unit. They are given a brief explanation of this test method. First of all, they need to give their first opinion of the site. Then they were given ample time to crawl the site to get an overview of how the site operated. They were also given about 20 minutes to perform an evaluation process. Upon completion of the completed site testing, they were asked to fill out a questionnaire to answer some questions based on their experience on the site. Based on table 4.12, the questionnaire consisted of 6 main aspect questions with scoring system using Likert scale measurement technique. Listed below are the indicators and description of each indicator included in the Likert Scale.

Table 4.12 Likert score for user acceptance test

EXPLANATION	LIKERT SCORE
Very Dissatisfied	1
Dissatisfied	2
Neutral	3
Satisfied	4
Very Satisfied	5

As a result of the testing, the average user satisfaction on the ERT System is within the range of satisfactory level. Table 4.13 shows the user rating on the ERT System.

Table 4.13 User rating on the ERT System

Bil.	User Interface	Usability	System Usefulness	System Response	Overall Satisfaction	Security
1.	5	4.25	3.8	4	4.3	3
2.	4	4	4.6	5	5	4
3.	3.5	4	4	3	4	3
4.	3	3.75	4	4	4.3	4
5.	4	4.25	4.4	3.5	4	4
6.	4.5	4.25	4.2	5	4	5
7.	5	3.75	4.2	4	4.6	4
8.	4	3	3.6	4	4	3
9.	4	3.25	3.4	3	3	2
10.	3.5	2.25	2.6	3	3	3
11.	4	4	3.6	5	4	4
12.	3	3.25	3.4	4	3.6	4
13.	4	4	4.2	4.5	3.6	4
14.	4	3.5	4.2	5	4.3	5
15.	3.5	3.25	2.8	3.5	2.6	3
16.	4	4	4	4	4	4
17.	3	2.25	2.6	3.5	3	3
18.	4	4	3	3.5	3.6	4
19.	5	3.25	4	3.5	4	4
20.	3	3.5	3.2	4	2.6	3
Average	3.9	3.59	3.69	3.95	3.78	3.65

5. Conclusion

The objectives of the study outlined in this paper are as follows:

i. To study and analyse the weakness on databases of the existing system.
ii. To design and develop new databases and function on an E-request transcript application.
iii. To validate and verify the proposed solution based on query optimization technique.

The extent to which the above objectives are achieved are assessed based on the results that are obtained throughout the development of the project. It follows current database analysis in case studies of ERT system discussed to meet the first objective of the study. The E-Request Transcript application was developed by upgrading the database, using the concept of logical schema optimization relation database for equijoin method to meet the second objective of the study. From the results of the execution query time, the optimization technique was appropriate with the study frame to fulfil the third objective of the study.

In general, the resulting optimization techniques involve the restructuring of relationships and tables in the database. Therefore, this technique is suitable for use by any relational database with less normal and irregular schedule structure. Key points for each involved table need to be determined to avoid data loss after optimization is implemented.

In determining the effectiveness of the enhanced system, black box testing is used to test the system function and beta test is used to find out the user's response to the e-request transcript system, by doing the questionnaire. The test involved integration testing, system testing and user acceptance testing that have been enhanced successfully. The result of the functional testing of the menu that was set up in blackbox testing found that the menus functioned. Result of user acceptance test shows that overall user satisfaction in using the E-Request Transcript system and security aspects of the system are satisfying. A user acceptance test for this system consists of a black box test and a beta test has been deployed. A black box test was conducted to test the main menu functions in the E-Request transcript and is found to be able to succeed. Furthermore, the beta test is carried out using a survey form distributed to all students who have graduated. Survey questionnaire showed that 57% of respondents were satisfied using this system and 14% were very satisfied with this system. All respondents have provided a good scaling rate for all items assessed. For overall system assessment, the mean score obtained is 3.78 which is interpreted as a good rate based on the scale of interpretation used. In conclusion, this system has met the needs of consumers as well as simplifying their business based on surveys conducted. In conclusion, this system has met the needs of consumers as well as simplifying their business based on surveys conducted.

6. Acknowledgments

We would like to thank Universiti Malaysia Pahang for supporting this work under the RDU190315.

References

[1] An, X., Lemen, C., & Liuqi, Y. (2010). An Integration Management System for Managing Business Continuity, Records and Knowledge. Information Technology and Applications (IFITA), 2010 International Forum on, 2(April), 4–5. https://doi.org/10.1109/IFITA.2010.

[2] Igamberdiev, M., Grossmann, G., Selway, M., & Stumptner, M. (2016). An integrated multi-level modeling approach for industrial-scale data interoperability. Software and Systems Modeling, 1–26. https://doi.org/10.1007/s10270-016-0520-6

[3] Nunhes, T. V., Motta Barbosa, L. C. F., & de Oliveira, O. J. (2016). Identification and analysis of the elements and functions integrable in integrated management systems. Journal of Cleaner Production, 142. https://doi.org/10.1016/j.jclepro.2016.10.147.

[4] Ruzaini, A. A. (2013). Assessment Instrument For Integrated Information System Success In Institution Of Higher Learning. University Teknology Malaysia..

[5] Mohd Hariz, I. (2014). E-MOHON TRANSKRIP. In Icompex14 National Innovation and Invention Competition trough Exihibition 2014 (p. 2). Politeknik Sultan Abdul Halim Mu'adzam Shah.

[6] Solihadin, O., & Bandung, Y. (2016). A Case Study on Integration of Conventional Learning System and ICT-based Learning for Job Training System. 2016 International Conference on Information Technology Systems and Innovation (ICITSI) Bandung – Bali, October 24 – 27, 2016, (1).

[7] Alexandrov, A., Bergmann, R., Ewen, S. et al (2014). The Stratosphere platform for big data analytics. The VLDB Journal (2014) 23: 939. https://doi.org/10.1007/s00778-014-0357-y.

[8] Khayyat,Z., Ilyas,1.F., Jindal, A., (2015) BigDansing: A System for Big Data Cleansing. Proceeding SIGMOD '15 Proceedings of the 2015 ACM SIGMOD International Conference on Management of Data Pages 1215-1230.

[9] Breß S., Heimel M., Siegmund N., Bellatreche L., Saake G. (2014) GPU-Accelerated Database Systems: Survey and Open Challenges. In: Hameurlain A. et al. (eds) Transactions on Large-Scale Data- and Knowledge-Centered Systems XV. Lecture Notes in Computer Science, vol 8920. Springer, Berlin, Heidelberg.

[10] Priti M and Margaret H.E (1992). Join Processing in Relational Databases. ACM Computing Surveys, 24(1):63-113.

[11] Norazrin K (2006). Optimization Of Logical Scope Of Coordination Data: Hydrological Information System Malaysia. Universiti Teknologi Malaysia : Master Thesis.

[12] C, Coronel., S, Morris (2016). DATABASE SYSTEMS, Design, Implementation and Management (12nd ed.) [Online]. Available: https://www.academia.edu/.

[13] A, Dennis., B.H. Wixom., D. Tegerdan (2015, Feb). System Analysis & Design, An Object-Oriented Approach with UML (5nd ed.) [Online]. Available: https://www.academia.edu/.

[14] Campagna, G., Ramesh, R., Xu, S., Fischer, M., & Lam, M. S. (2017). Almond: The Architecture of an Open, Crowdsourced, Privacy-Preserving, Programmable Virtual Assistant. Www, 341–350. https://doi.org/10.1145/3038912.3052562

[15] Abiteboul, S., Bourhis, P., & Vianu, V. (2015). Highly Expressive Query Languages for Unordered Data Trees. Theory of Computing Systems, 57(4), 927–966. https://doi.org/10.1007/s00224-015-9617-5

[16] Torlak, E., & Bodik, R. (2013). A lightweight symbolic virtual machine for solver-aided host languages. Proceedings of the 35th ACM SIGPLAN Conference on Programming Language Design and Implementation - PLDI '14, 530–541. https://doi.org/10.1145/2594291.2594340

[17] Chilana, P. K., Singh, R., & Guo, P. J. (2016). Understanding Conversational Programmers: A Perspective from the Software Industry. Proceedings of the 2016 CHI Conference on Human Factors in Computing Systems - CHI '16, 1462–1472. https://doi.org/10.1145/2858036.2858323

[18] Nguyen, T. H., Yadav, A., An, B., Tambe, M., & Boutilier, C. (2015). Regret-based Optimization and Preference Elicitation for Stackelberg Security Games with Uncertainty. Proceedings of the 14th International Conference on Autonomous Agents and Multiagent Systems (AAMAS 2015), 756–762.

[19] Steuwer, M., Fensch, C., Lindley, S., & Dubach, C.(2015). Generating Performance Portable Code using Rewrite Rules: From High-Level Functional Expressions to High-Performance OpenCL Code. ICFP: ACM SIGPLAN International Conference on Functional Programming, 205–217. https://doi.org/10.1145/2784731.2784754

[20] Kadadi, A., Agrawal, R., Nyamful, C., & Atiq, R. (2015). Challenges of data integration and interoperability in big data. Proceedings - 2014 IEEE International Conference on Big Data, IEEE Big Data 2014, 38–40. https://doi.org/10.1109/BigData.2014.7004486

[21] Wafaa, A. S., Tiun, S., Ahmed, A. S., Awang, S., & Al-Khaleefa, A. S. (2018). Word sense disambiguation using hybrid swarm intelligence approach. PloS one, 13(12), e0208695.

[22] Al-Saffar, A., Awang, S., Tao, H., Omar, N., Al-Saiagh, W., & Al-bared, M. (2018). Malay sentiment analysis based on combined classification approaches and Senti-lexicon algorithm. PloS one, 13(4), e0194852.

IOP Conf. Series: Materials Science and Engineering **769** (2020) 012011 doi:10.1088/1757-899X/769/1/012011

Indoor occupancy estimation using carbon dioxide concentration and neural network with random weights

[1]Muhammad Faris Ramli, [1]Kishendran Muniandy, [1]Asrul Adam, [2]Ahmad Fakhri Ab. Nasir, [3]Mohd Ibrahim Shapiai

[1]College of Engineering, Universiti Malaysia Pahang, Gambang, Pahang, Malaysia

[2]Faculty of Manufacturing and Mechatronics Engineering Technology, Universiti Malaysia Pahang, Gambang, Pahang, Malaysia

[3]Malaysia-Japan International Institute of Technology, University Technology Malaysia, Kuala Lumpur, Malaysia

m.farisbinramli@gmail.com, kishenmuniandy@gmail.com,asrul@ump.edu.my, afakhri@ump.edu.my, md_ibrahim83@utm.my

Abstract. This study presents the indoor occupancy estimation using carbon dioxide concentration and neural network with random weights (NNRW). The utilization of carbon dioxide concentration is as an alternative to overcome the limitation of existing techniques, such as dependency to favourable lighting condition and camera position. Whereas, NNRW provides a generalized and fast learning speed classification. In this study, MH-Z19 sensor is used to acquire carbon dioxide concentration and the NNRW is a multiclass estimation method. The numbers of the occupants are divided into three different classes, which are 15 occupants, 30 occupant and 50 occupant classes. Result indicates that the NNRW classifier has obtained training and testing accuracy, about 100 percent and 52 percent, respectively.

1. Introduction

Humans have been known to reside inside their shelter since the dawn ages for the reason of safety and comfort. Human comfort can be influenced by four factors, which are wind, temperature, radiation and humidity. In that term, obtaining an optimum human comfort inside a building is a paramount important for most aspect of buildings development. A mechanical ventilation system has been introduced to do so. Heating, ventilation and air conditioning (HVAC) system is a system that can provide three out of those four factors. HVAC is a system that ventilates the heating and air conditioning in a closed area. The system is important as them responsible for providing acceptable indoor air quality and thermal comfort to the occupants. Overall, it provides a condition that is comfortable for the humans inside a room. However, current system unable to operate optimally as centralized air conditioning has led to energy wasting due to its inability to operate based on the presence of people time and zone [1]. An optimum HVAC system can save energy up to 15% to 25% if it is set to respond based on indoor occupancy data [1].

Input data on predicting the indoor occupancy can be obtained by two main methods, in which 1) multi camera pattern techniques (using camera) and terminal based techniques and 2) terminal based

IOP Conf. Series: Materials Science and Engineering **769** (2020) 012011 doi:10.1088/1757-899X/769/1/012011

techniques. Terminal based techniques rely on the human movements to the appliances such as keyboard and mouse. However, both methods are flawed in nature as the previous depended on lighting conditions and the camera location [2], while the second techniques cannot detect occupants that are not using the selected devices. Others sensors have also been used to estimate the number of occupancy including pyroelectric infrared (PIR) and ultrasound sensors. But these sensors can only detect presence of humans, but not estimate the number of humans inside the room [3]. These limitations have led to the implementation of environmental parameters such as humidity, temperature and carbon dioxide concentration. Since humans naturally exhaled carbon dioxide, the occupancy in a closed area can be obtained through carbon dioxide (CO_2) concentration [4]. It is also non-intrusive and non-terminal based, which is alternatively solved the problem encountered by existing techniques [3].

CO_2 concentration for indoor is usually between 350 to 2500 ppm [1]. In terms of CO_2 sensor, there are two types of sensor. First is chemical CO_2 sensor. The second one is Non-Dispersive Infrared (NDIR) CO_2 sensor. Chemical CO_2 sensor is noted to have a short life span and unsuitable in practical world [5]. NDIR sensor is noted to have higher accuracy and durability. The sensor in concern of the method is NDIR based. MH-Z14A PWM NDR Infrared carbon dioxide sensor [6], MQ135 sensor [7], B-530 sensor, H-550 NDIR sensor [5], K30 sensor [1, 8] and MH-Z16 NDIR sensor [9] have all been tested by recent research and have the ability to correctly monitor CO_2 concentration in indoor setting. Several occupancy prediction methods were used to predict the number of occupants. Extreme Learning Machine [3], Artificial Neuron Network [10], and Decision Tree [11] have all been tested by recent researchers.

In this study, carbon dioxide sensor, MH-Z19, is used to measure the concentration of carbon dioxide in a closed area. Next, NNRW classifier is implemented to estimate the number of occupants in a closed area in between three groups (15,30, and 50 occupants). This information, in future, can be used in controlling regulating HVAC system in a closed area. This will lead to a more efficient energy usage.

2. Methodology

2.1. Electrical schematic

The MH-Z19 carbon dioxide sensor will be connected to the Arduino Nano. The Arduino Nano will then be powered by USB connection with laptop. The connection of the sensor with the Arduino is as in Table 1 and Figure 1.

Table 1. MH-Z19 carbon dioxide sensor connection with Arduino Nano.

MH-Z19	**Arduino Nano**
PWM	*D05*
GND	*GND*
Vin	*5V*

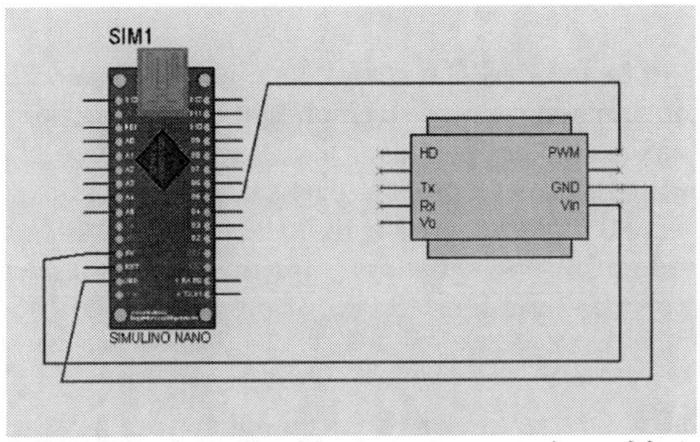

Figure 1.MH-Z19 carbon dioxide sensor connection with Arduino Nano.

2.2. Data collection

The data collection was done inside a classroom sized 6.75 meter times 9.00 meters. The data collection will be done during the class is occupied and when it is vacant. The time for data collection is shown in Table 2. Student count is done through observing. There will be three different numbers of students, which are 15 students, 30 students and 50 students. The sensor will be put at the back of the room to minimise interruption for the occupants learning process. The windows of the class are locked down to minimise air flow going in and out of the classroom. The classroom door was also closed. The air conditioner and air ventilating were ensuring to be in functioning during the data collecting. The data collection will occur for every second for at least 30 minutes.At least three set of data were obtained for each classes to ensure consistent reading. However, the effect of students going in and out of the classroom and the best sensor position for occupancy prediction is out of scope for this experiment. In theory, human occupancy should affect the reading of the sensor.

Table 2. Data collection schedule.

Time	Day	Number of occupants
8.00 AM – 10.00 AM	Monday	15
8.00 AM – 10.00 AM	Tuesday	15
2.00 PM – 4.00 PM	Tuesday	30
8.00 AM – 10.00 AM	Thursday	30
2.00 AM – 4.00 PM	Thursday	50

Figure 2. Classroom where the study conducted

2.3. Sorting of the data

The raw data is obtained from the Arduino Serial Monitor. The data is copied manually and then sorted in Microsoft Excel in ascending order. A graph will be plotted from the data with PPM reading as y-axis and seconds as x-axis. Another Microsoft Excel file is created in order to sort the data again. This time divided into three part, number of data, data sample and class. There will be 300 rows with 50 data per row, representing 50 input. The row will be labelled into three different classes (class 0 for 15 occupants, class 1 for 30 occupants, and class 2 for more than 30 occupants/50 occupants). Each class will have 100 rows of data, totalling 15,000 of datasets. The file will then be analysed and saved.

Two m.file will be created. One m.file will be detailing and setup the NNRW. Another m.file will execute the training and predicting the result. The Excel file will then be link with MATLAB for training purposes. The hidden neuron will be set to 800 while the activation function is set as manipulated variable. 225 datasets will be used for training purposes while another 75 will be used for predicting purposes. The NNRW will predict whether the dataset is in class 0, class 1 or class 2. This is illustrated in Figure 3. The result will be displayed after running ended on the Command Window. The step was repeated with changing the number of hidden nodes (200, 400, 600, 800) for each activation function (Sigmoid, Sine, Hardlim, Triangular basis and Radial).

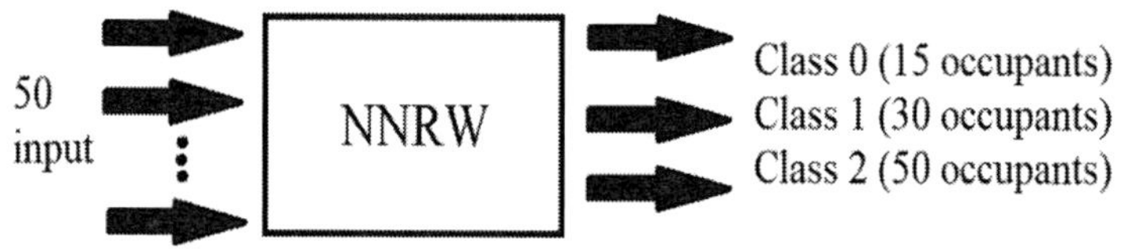

Figure 3. Sorting of the data.

2.4. Classification of the data

Neural Network with Random Weight (NNRW) is a non-iterative algorithm that have high convergence rate, fast training speed and produce certain model, difference than the traditional ANN, such as BP-based method. It was first proposed in 1992. One of the biggest differences between NNRW and BP-based method is the fact that its output weight is obtained analytically while the hidden weights are randomly selected. [12] have proposed a generalized version of NNRW, called ELM. ELM is proven to have universal approximation capability and have the ability to approximate any continuous target function with probability one under certain condition. It also has allowed nonlinear piecewise continues random hidden nodes to be used. This includes but is not limited to Sigmoid nodes, Wavelet and etc.

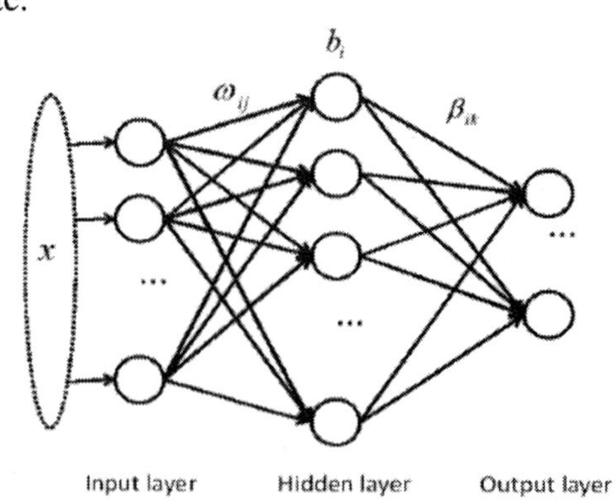

Figure 4. NNRW structure. Source: [13]

As an overview, NNRW, such as in Figure 4, consists of three layer, input layer, hidden layer and output layer. ωmean the weight between the input layer and hidden layer. $b$mean the threshold of hidden nodes, while β represent the output weights. There is an activation function in the hidden layer which performs nonlinear transformation. That activation includes but is not limited to Sigmoid, Sine, Hardlim, Triangular basis and Radial basis function. It is basically Equation 1;

$$H\beta = T \quad (1)$$

Which means;
H= Hidden layer output matrix of the neural network
B = L*m matrix of the output weights
T= N*m matrix of target
m = The number of output neurons

NNRW classifier of a given unknown sample, x will produce output function which is:

$$fc(x)=hc(x)\beta \quad (2)$$

3. Result and discussion

3.1. Carbon dioxide reading inside an occupant area

Figure 5 shows that there is a stark difference in term of reading for indoor area that is packed with 15, 30 and 50 occupants. On average, one person will increase the reading by 26 PPM. On average, 15 occupants will have 1000 PPM reading, 30 occupants will have 1400 PPM reading and 50 occupants having 1500 PPM reading.

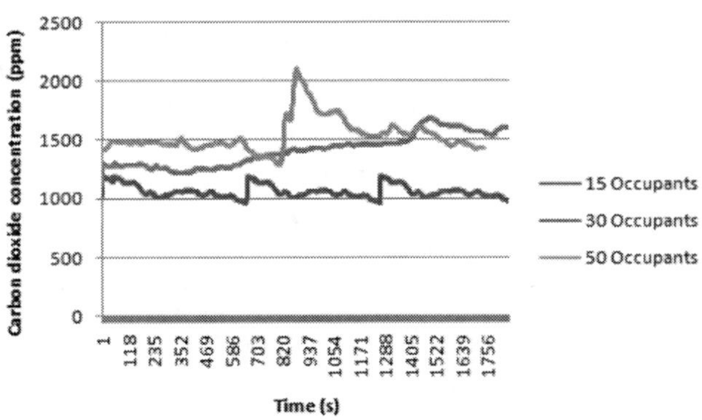

Figure 5. MH-Z19 reading comparison in indoor area.

3.2. Pattern of the data

The data collected were divided into three different classes, class 1 for 15 occupants, class 2 for 30 occupants, and class 3 for 50 occupants. Pattern of average, mode, median, standard deviation, minimum value (Min) and maximum value (Max) were tabulated on Table 3 and Figure 6. Based on the table, the minimum value is 567 PPM and the highest reading is 1706 PPM. The mode for class 0 is 698 PPM, 1251 PPM for class 1, and 1482 for class 2. On average, class 0 will have reading at 699.8706 PPM, class 1 at 1199.6910 PPM while class 2 have average reading at 1452.1320 PPM. Mode for class 0 is 698.0000 PPM, class 1 is 1251.0000 PPM while class 2 at 1482.0000 PPM. Median for class 0 is 698.0000 PPM, class 1 is 1219.0000 PPM and class 2 is 1462.0000 PPM. The standard deviation for class 0 is 34.8844, class 1 is 117.9111 while class 2 is 81.1304.

Table 3. Pattern of the data according to the classes

Classes	0	1	2
Average	699.8706	1199.6910	1452.1320
Mode	698.0000	1251.0000	1482.0000
Median	698.0000	1219.0000	1462.0000
Standard Deviation	34.8844	117.9111	81.1304
Min	567.0000	969.0000	1115.0000
Max	937.0000	1452.0000	1706.0000

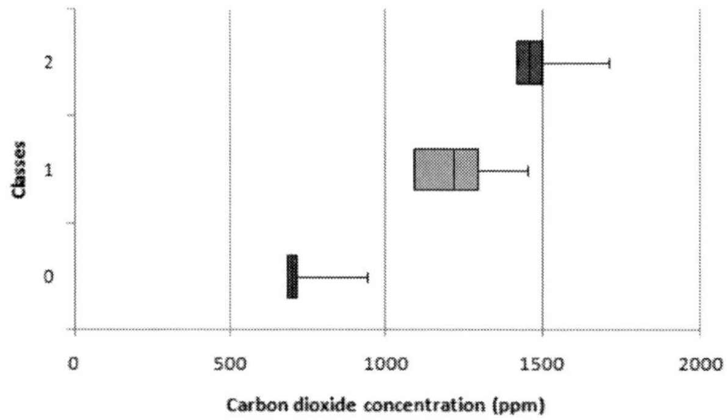

Figure 6. Pattern of the data according to the classes.

3.3. Predicting occupancy accuracy (comparison between activation function)
As illustrated on Table 4, sin activation function produces the highest accuracy, in term of training and testing. The training is higher 45.66833% compare to the second highest, Sig activation function. The testing accuracy is higher 10.50467% compare to the second highest, also Sig activation function. Tribas, on the other hand, produce the worst result in both area (36.80392% for training and 33.88235% for testing).

Table 4. Predicting accuracy (comparison between activation function)

Activation Function	Training	Testing
Sin	100.0000	52.67137
Sig	54.33167	42.16667
Hardlim	53.27538	41.52744
Tribas	36.80392	33.88235
Radbas	41.12289	34.98423

3.4. Predicting occupancy accuracy (comparison between number of hidden nodes)
Changing the number of hidden nodes gives affect the result differently for each activation function. It increases the training and testing accuracy for sig, hardlim and radbas activation function. It causes the training accuracy to fluctuate for tribas activation function. It did not give any significant differences when tested with sin activation function. This is illustrated in Table 5 till Table 9.

IOP Conf. Series: Materials Science and Engineering **769** (2020) 012011 doi:10.1088/1757-899X/769/1/012011

Table 5. Predicting accuracy for Sin activation function (comparison between number of hidden nodes)

Number of hidden nodes	Training	Testing
200	100.00000	52.42167
400	100.00000	52.57210
600	100.00000	51.70002
800	100.00000	52.67137

Table 6. Predicting accuracy for Sig activation function (comparison between number of hidden nodes)

Number of hidden nodes	Training	Testing
200	42.193460	37.416025
400	47.1565800	39.639646
600	51.4934700	40.960785
800	54.3316675	42.166666

Table 7. Predicting accuracy for Hardlim activation function (comparison between number of hidden nodes)

Number of hidden nodes	Training	Testing
200	41.96111	37.03541
400	46.49768	39.28624
600	49.99418	40.60465
800	53.27538	41.52744

Table 8. Predicting accuracy for Radbas activation function (comparison between number of hidden nodes)

Number of hidden nodes	Training	Testing
200	31.78055	30.80000
400	35.40015	32.26577
600	38.84685	34.94595
800	41.12289	34.98423

Table 9. Predicting accuracy for Tribas activation function (comparison between number of hidden nodes)

Number of hidden nodes	Training	Testing
200	34.13399	33.21421
400	36.70173	33.46135
600	35.74194	33.41577
800	36.80385	33.88235

4. Conclusion

Overall, the objective of the study managed to be achieved. The study managed to estimate the number of people in a closed area using carbon dioxide sensor by measuring the carbon dioxide concentration in a closed area. It also managed to find the best result on two parameters, activation functions and number of hidden nodes using NNRW. It was tested with sin, sig, hardlim, radbas and tribas activation functions. It was later on checked with 200, 400, 600 and 800 hidden nodes. The best result obtained when it were tested with sin activation function and 800 hidden nodes, where it achieved 100% training accuracy and 52% testing accuracy respectively. The result cannot be higher due to the fact that the testing and prediction is done using raw data.

Acknowledgement

This research is funded by University Technology Malaysia (UTM) Collaborative Research Grant (CRG) (RDU182401). Also thanks to Faculty of Manufacturing and Mechatronics Engineering Technology and College of Engineering, UMP for supporting this research work through the utilization of faculty's equipments.

References

[1] Subramany, R. (2013). *Wireless Sensor Network for HVAC Control.* (Master of Science), University of Florida.

[2] Zou, H., Zhou, Y., Yang, J., & Spanos, C. J. (2018). Device-free occupancy detection and crowd counting in smart buildings with WiFi-enabled IoT. *Energy and Buildings, 174*, 309-322. doi:10.1016/j.enbuild.2018.06.040

[3] Jiang, C., Masood, M.K, Soh, Y.C, Li, H. (2016). Indoor occupancy estimation from carbon dioxide.

[4] Jin. M, B.-L. N., Weekly K, Spanos C & Bayen A. (2015). Sensing by Proxy: Occupancy Detection Based on Indoor CO2 Concentration.

[5] Kwon, J., Ahn, G, Kim, G, Kim, J. C, Kim, H. (2009). A Study on NDIR-based CO2 Sensor to apply Remote Air Quality Monitoring System. *ICROS-SICE.*

[6] Tu, Z.-X., Hong, Cheng-chen & Feng Hao (2017). EMACS Design and Implementation of Indoor Environment Monitoring and Control System. *IEEE Computer Society.*

[7] Varzaru, G., Zarnescu, A, Ungurelu, R, Iordache, A,G. (2017). A cost-effective solution for remote Carbon Dioxide indoor monitoring. *ECAI 2017.*

[8] Yasuda, T., Yonemura, S., & Tani, A. (2012). Comparison of the characteristics of small commercial NDIR CO2 sensor models and development of a portable CO2 measurement device. *Sensors (Basel), 12*(3), 3641-3655. doi:10.3390/s120303641

[9] He, Z., Hao, G, Chen, Y. (2018). Remote Monitoring System for Plant Growth Environment. doi:10.13031/aim.201800750

[10] Khazaei, B., Shiehbeigi A. & Kani A.R.H.M.A. (2018). Modeling indoor air carbon dioxide concentration using artifcial neural network. *International Journal of Environmental Science and Technology.* doi:10.1007/s13762-018-1642-x

[11] Soki, S., Hiroshi, M., & Rajith, A. (2018). Real-time Optimized HVAC Control System on top of an IoT Framework. *International Conference on Fog and Mobile Edge Computing.*

[12] Huang, G.-B., Zhu, Q.-Y., & Siew, C.-K. (2006). Extreme learning machine: Theory and applications. *Neurocomputing, 70*(1-3), 489-501. doi:10.1016/j.neucom.2005.12.126

[13] Cao, W., Gao, J, Ming, Z, & Cai, S. (2017). Some Tricks in Parameter Selection for Extreme. *IOP Conference Series: Material Science and Engineering.* doi:10.1088/1757-899X/261/1/012002

Dengue Innovation: A Sustainability Approach for Preventing and Controlling of Dengue Diseases Outbreaks via IoT Technology

Mustafa Man[1], Wan Aezwani Wan Abu Bakar[2], Lim Chee Hwa[3], Wan Nural Jawahir Wan Mohd Yusoff[1], Mustafa Afenddi Mat Nor[1] and Mohd. Irfan Hakim Mohd. Noor[4]

[1, 4] Faculty of Ocean Engineering Technology and Informatics, Universiti Malaysia Terengganu, 21030 Kuala Terengganu, Terengganu, Malaysia.
[2] Faculty of Informatics & Computing, Universiti Sultan Zainal Abidin, Besut Campus, 22200 Besut, Terengganu, Malaysia.
[3] One Team Network, 1-1B, Block 6, Jalan Pahat 15/H, Section 15, 40200 Shah Alam, Selangor.

E-mail: wanaezwani@unisza.edu.my and mustafaman@umt.edu.my

Abstract. Dengue is a mosquito-borne viral disease that is surging and becomes worldwide serious issues. Currently, due to lack of specific treatment or vaccine against dengue, many forms of vector control remains a key strategy for dengue fever prevention. In response to the issue, this paper introduces an AedesTech Mosquitoes Home System (AMHS) that is equipped with a special trade secret formulation or also called as Insecticide Growth Regulator (IGR) and combines with Internet of Things (IoT) technology for vector control. An IGR is a pheromone-like liquid formulation that is a chemical that serves to stimulate and have sexually attract the male and female that will attract and lure adult female mosquitoes to lay eggs in them, and the adult mosquitoes will soon die after lying eggs. Those eggs that already laid with the chemical will not hatch or die at an expected of 99% rate, or go beyond the pulp level. Our sustainable approach offers for two (2) novel features. First, to monitor the collection of Aedes eggs in our MHS, we develop a mobile apps called AedesTech Apps (ATA) for Aedes eggs auto-counting and data collection are stored in cloud based server. Second, to monitor the level of liquid formulation (IGR) in AMHS, we utilize an Arduino water based sensor to real time track the IGR level status and we called as Intelligent Mosquito Home System (iMHS). Our iMHS will automatically alert the user when refill is needed. Both technologies are related to IR 4.0 that could be a new approach for preventing and monitoring of dengue outbreaks. This sustainability approach are environmental friendly and cost saving that provide the solution to the world's global issue of dengue. The results by our ATA tool indicate a satisfactory figure with only 0.01 to 0.04 percent of error rate compared to existing ICount and EggCounter tools. Meanwhile in our iMHS with water based sensor, testing results reveal that the battery lifetime solely depends upon the setting of interval time of sending data to cloud server. Results show for a half an hour interval, battery could last for a half of a month whereas for an hour interval time set, it could last for a month.

Keywords. AedesTech Mosquitoes Home System (AMHS), IoT, AedesTech Apps, Insecticide Growth Regulator (IGR)

1. Introduction

Aedes aegypti, the main vector borne diseases such as dengue and other epidemiologically significant viruses such as Chikugunya, Yellow fever, and Zika [1]. It is a domestic endophilic mosquito that has expanded its habitable range in recent decades and will likely continue to spread [2]. The Aedes mosquitoes have 4 life stages: egg, larva, pupa and adult. Mosquitoes can live and reproduce inside and outside the home. The entire life cycle, from an egg to an adult, takes approximately 8-10 days. Pupae develop into adult flying mosquitoes in 2-3 days. Dengue hemorrhagic fever (DHF) or dengue fever (DF) cases are on the increase. According to [3], in year 2018, 80615 DHF/DF cases were reported with 147 mortalities in Malaysia. Aedes aegypti and Aedes albopictus are two (2) major vector mosquitoes of DHF and DF. Vector control is one of the most effective approaches in combating DHF or DF. As DHF and DF can be transmitted transovarially [4-5] vector control of Aedes species in the immature stages is very important. Although chemical control of adult mosquitoes provides quick knockdown (mortality), resistance of mosquitoes to insecticides has been widely reported. Resistance of Aedes aegypti to organophosphate (temephos, malathion, fenthion and propoxur) [6-7], permethrin [8], DDT and dieldrin [9] have been well documented.

A *pyriproxyfen*, an insect growth regulator (IGR) also has been reported to be effective against vector mosquitoes such as aegypti [10], Culex [11-12] and Anopheles [13]. The pyriproxyfen mimics a natural hormone in insects and disrupts their growth. It is a type of insect growth regulator that affects mostly young insects and eggs. The pyriproxyfen affects many types of insects, including fleas, cockroaches, ticks, ants, carpet beetles, and mosquitoes. It has been registered for use in pesticides by the Environmental Protection Agency (EPA) since 1995 [14]. The pyriproxyfen prevent the immature mosquito to be emerged to adult. However, pyriproxyfen products such as Sumilarv 0.5 (pyriproxyfen 0.5% w/w) does not kill larvae but has extended residual activity that inhibits adult emergence in vector mosquitoes.

This paper presents two (2) novel approaches in preventing and controlling dengue outbreaks. First, our AedesTech mobile Apps (ATA) that able to scan and capture images of Aedes eggs and transfer for an automated image processing system to first segregate between Aedes eggs with non Aedes eggs and then, count its total for a further data analytic process. Second, our intelligent AedesTech Mosquito Home System (iMHS) with the embedded water based sensor in our AedesTech Mosquito Home System (AMHS) to track for the liquid formulation, Insecticide Growth Regulator (IGR) status, equipped in our AMHS. Our approaches offer for an efficient automated counting of Aedes eggs while reducing manpower cost to check and change for IGR refill in each particular AMHS.

2. Methodology of Dengue Control Using AedesTech Mosquito Home System (AMHS)

AedesTech Mosquito Home System (AMHS) as shown in Fig. 1, which consist of devices designed to attract Aedes mosquitoes and target container-breeding mosquitoes by providing an attractive water source for gravid females seeking oviposition sites. Females attracted to the device are killed and reproduction is eliminated. AMHS is an auto-dissemination device, consisting of a regular ovitrap containing pyriproxyfen, to contaminate gravid female mosquitoes during oviposition, allowing them to disseminate the chemical to other Aedes oviposition sites. Auto-dissemination is a ‘pull’ and ‘push’ technology which allows mosquito control professionals to treat larval habitats in a timely and economical fashion [15]. The objective of the AMHS is by using the “Lure & Kill Technology” to control & eliminate mosquitoes during all development stages, be it as eggs, larvae, pupae or adults, and thus to reduce the overall mosquito population. A total of 400 units of AMHS were set up surrounding the trial site, outside each of the residential house, at the fence, gate or at the front door. The AMHS’s refill was refilled at monthly basis. The AMHS installation began in September 2018, for a period of one year until August 2019 at X sites in Kedah.

Figure 1. AedesTech Mosquito Home System (AMHS) and Water Based Attractant Refill

2.1 Design of AMHS

AMHS is designed to suit the green technology environment and sleek enough to be installed in the outdoor or inner sites or location. The device measured 19.7cm (height) x 14.6cm (width) and consists 500ml of 40 ppm PPF treated water. Egg collection tissue will be placed in inner wall of AMHS. The interior skeleton of AMHS is depicted in Fig. 2. The idea comes prior to examining the details of mature adult mosquitoes, which will lay eggs in 5 or more breeding sites. Adult mosquitoes will ensure the survival of their eggs to the larvae level and eventually become adult mosquitoes. As a result of this research, the Mosquito Home System has created a formulation (IGR) that contains 500ml of 40 ppm PPF treated water. It pollutes the new reproductive site, when the adult mosquitoes lay eggs in the Mosquito Home System, they will remove the formulas and kill the entire larvae in the breeding ground. The IGR is placed upside down in the AMHS and the "lure and kill" solution is allowed to fill the AMHS slowly, which can last for a minimum of one month.

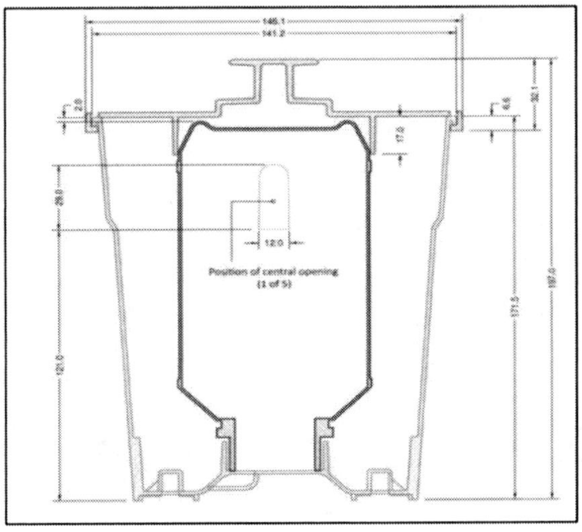

Figure 2. Interior skeleton of AMHS

3. Development of Aedestech Apps (ATA) for Mosquitoes Eggs Counting

Through this project, the mobile AedesTech Apps (ATA) is developed to collect data related to Aedes mosquito eggs directly from AMHS and stored in a cloud database. For this purpose, a rectangular piece of tissue is clipped to similar size of transparency slides and both materials are circularly attached around the inner part of AMHS. The reason is to allow for a slowly absorption of the IGR into the tissue that allow for breeding process of female Aedes mosquitoes. The tissue is then taken out to be scanned by ATA and transfer the digitized images to the Excel file for an auto counting for Aedes eggs via MATLAB AutoCountMe [16] system. Through this system, the exact images of Aedes eggs are identified from non Aedes eggs through digital image processing technology to be developed. A new algorithm will be proposed to expedite the process of counting the mosquito eggs directly using only mobile devices or mobile phones either offline or online. Refer to Fig. 3 for the overall project framework. The large data from various installations of the whole AMHS in geographically apart locations are then collected and compiled into the central cloud database for further numerous data mining and forecasting analysis. Example of the analysis would be to projection of the mosquito population rate for a region or location and it can extend to the parties for follow-up actions and comparative results that can be utilized for the manufacture decision [17].

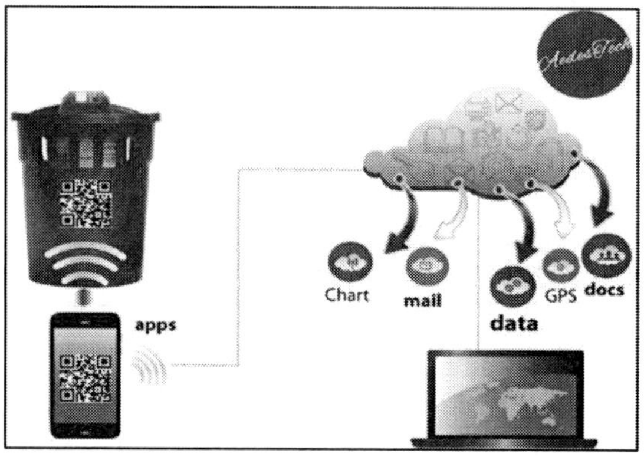

Figure 3. ATA Development Framework

The unique feature that ATA offers is for an in-situ data collection. All digitized images of an Aedes eggs are collected and auto-counting using this apps. Refer to Fig. 4 for an ATA user interface. The ATA Standard Operating Procedures are as follows:

1. *Login and Password (By Location of the user access)*
2. *Setup Pot or MHS trap and labeling using QR or Bar Coded as identification for each trap*
3. *Setup and configure the location of the trap using Global Positioning System (GPS)*
4. *Click the camera inside the apps and scan the QR-Coded then select the image and click the scan and count the Aedes eggs.*
5. *Next save in cloud or in the mobile if not internet access is available*
6. *Statistical and data visualization by each AMHS trap will visualize automatically.*
7. *By using this apps will also can retrieve and done several data mining for the related data based on Aedes eggs image captured.*

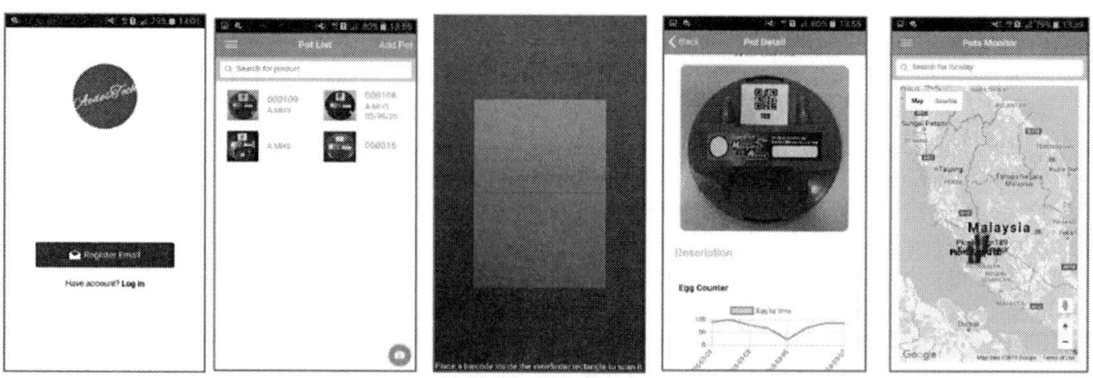

Figure 4. User Interfaces of AedesTech Apps

4. Development of iMHS: Intelligent Mosquitoes Home System Using IoT Sensor

The iMHS design framework relies upon 3-tier architecture comprises of Client Application (Presentation tier), Business layer (Application tier) and Data Layer (Data tier). The Presentation tier includes User Interface part that interfacing with client. The Application tier is the middle part between client and data layers where logic operations such as validation, calculation, and data related operations take place. The role is to ease of communication between client and data layers. Then, the Data tier consists of actual database that includes methods such as insert, delete, update and get data from/to database.

4.1 Hardware Component

The three (3) basic components of iMHS devices include the microcontroller, a water level sensor to perform the intended task and the source of power, an AA battery. The first component, an IoT based microcontroller called as ESP8266 [18-19]. It is a Wi-Fi embedded microcontroller that utilizes standard everyday Wi-Fi band frequency at 2.4 GHz. The second component of iMHS, an Arduino water level sensor [20]. This type of sensor is an easy to use water level sensor that is effective for water level drop detection. It offers for an easy completion of the water conversion to analog signals, and the simulated value of the output can be read directly by the Arduino development board, to achieve the water level alarm. Lastly, the third component refers to an AA battery [20] or also called a double A or Mignon (French for "cute" or "adorable") battery, a standard size single cell cylindrical dry battery. The AA batteries are common in portable electronic devices that compose of a single electrochemical cell which may be either a primary battery (disposable) or a rechargeable battery. Detailed hardware components are tabulated in Table 1 along with the connection of the circuit is illustrated in Fig. 5. The internal architecture and the actual implementation of our Arduino water level sensor in AMHS is depicted in Fig. 6. We use MySQL Database Management System (DBMS) and PHP 7 for page scripting.

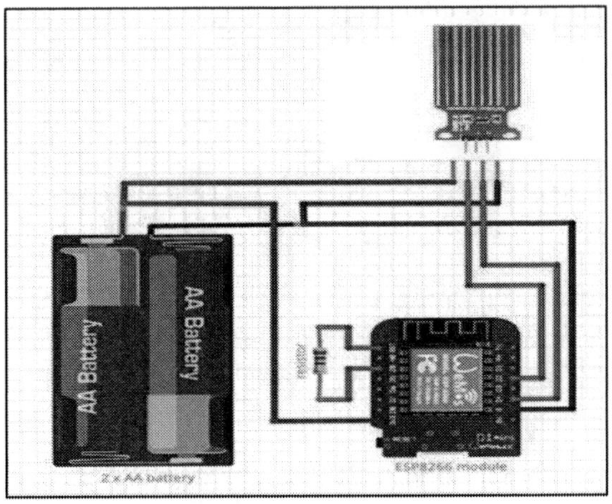

Figure 5. iMHS Circuit Board Design

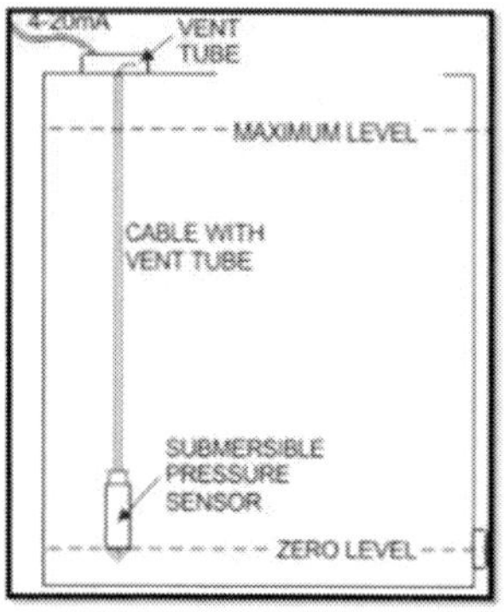

Figure 6. Installation of water level sensor in AMHS

5. Result and Discussion

5.1 Result for ATA

For the evaluation of ATA, we used a dataset provided by [21] together with their ICount tool. There are five images taken from a camera called `macro images' and five microscopic images called `macro image'. Fig. 7 shows samples of `macro image' and `micro image', respectively. We compare our proposed method with two tools that available for free download which is ICount [21] and EggCounter [22]. As a gold standard, ImageJ was used to manually count the eggs to confirm the accuracy of each tool. Table 1 shows the number of eggs obtained by our proposed tool, ICount and EggCounter using `macro image'. From the table, it is found that EggCounter unable to detect even a single egg.

(a) Macro Image

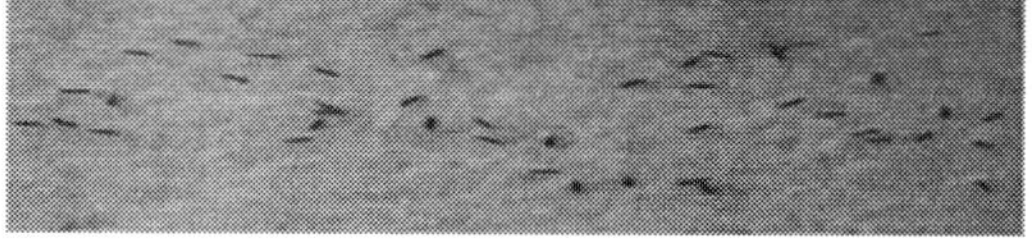

(b) Micro Image

Figure 7. Macro and Micro Image of Aedes Eggs

The results obtained by ICount are comparable to our tool, but it needs user intervention. Different parameters setting will produce different results which eventually can lead to biased or unreliable estimation. The proposed tool contrarily to ICount where a user loads the image and then the eggs are automatically counted. The results presented by the proposed tool are satisfactory with only 0.01 to 0.04 percent of error.

Table 1: Comparison between ImageJ, the ATA method, ICount and EggCounter using Macro Images

Dataset	ImageJ	ATA Method	% Error	ICount	% Error	Egg Counter	% Error
IMG_225bis	153	155	0.01	146	0.05	0	100
IMG_2253bis	479	479	0.04	482	0.01	0	100
IMG_2256bis	348	352	0.01	344	0.01	0	100

IMG_2257bis	221	213	0.04	315	0.03	0	100
IMG_2267bis	358	363	0.01	334	0.07	0	100

5.2 Result of iMHS

In iMHS, the issue arise regarding the battery power source lifetime. We run several tests to Arduino water based sensor with the AA battery in two different interval of data sent to cloud database. The results are as depicted in Table 2. Implementation results reveal that the battery lifetime solely depends upon the setting of interval time of sending data to cloud server. Changing in interval time proportionately resulting in changing the lifetime of battery power energy. The iMHS overall results show for a half an hour interval, battery could last for a half of a month whereas for an hour interval time set, it could last for a month.

Table 2: Results of battery lifetime in different interval

Interval time (in second) of Data Send	Battery lifetime (in month) (AA Battery)
1800	0.5
3600	1

The Arduino water level sensor in iMHS project is set to read in three (3) different level status i.e. low, medium and high. We have set different distances in centimeter (cm) for each status. The main interface of iMHS is illustrated in Fig. 8. Each location of iMHS devices is viewed by its longitude and latitude, while the detailed of each longitude and latitude location can be browsed through Google map.

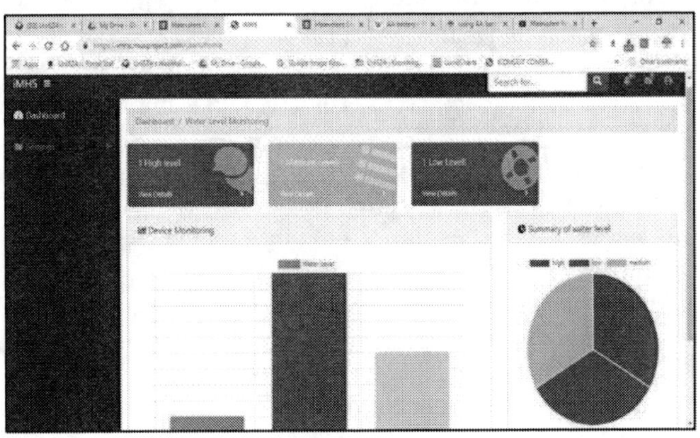

Figure 8. User interface of iMHS

6. Conclusion and Future Works

The development of AMHS, Aedestech Apps (ATA) and iMHS is considered to successfully embed the Internet of Things (IOT) technology for our dengue innovation project related to Industrial Revolution (IR) 4.0. This research collaboration has transferred the knowledge and technology between universities as well as its industry partner. As a recapitulation, AMHS offers for effective alternative in controlling dengue. The ATA provides effective way in getting the input images of Aedes egg to further process of auto-counting of Aedes eggs send to cloud storage for extensive analysis. And lastly with our iMHS, the cost for labor to manually check and track for the emptied IGR is significantly reduced and cut-off through the real-time IGR status visualization. The main objective of this project for reducing manpower usage in controlling the dengue disease at the national as well as international level. The concept of auto-detection device can be applied for other industry domain that might as well create new business opportunities in the near future.

Acknowledgment

We thanks all our research institutions-industry collaborators i.e. UniSZA, UMT, UniKL and One Team Network (OTN) Sdn. Bhd. under matching grant, UniSZA/2018/PKP/01 with project code R0047-R001. A sincere gratitude goes to the MHS co-founder, One Team Network Sdn. Bhd. As well as for supporting our work in reviewing for spelling errors and synchronization consistencies and also for the constructive comments and suggestions.

References

[1] Hemingway, J., Beaty, B. J., Rowland, M., Scott, T. W., & Sharp, B. L. "The Innovative Vector Control Consortium: improved control of mosquito-borne diseases." *Trends in parasitology*, *22*(7), 2006. 308-312.

[2] Buchman, A. Stephanie G., Ming L., Igor A., Shin-Hang L., Shin-Wei W., Chun-Hong C. "Broad Dengue Neutralization in Mosquitoes Expressing an Engineered Antibody." *CELL-D-19-01515* (2019).

[3] Seng C. M., and Jute N. "Breeding of Aedes aegypti (L.) and Aedes albopictus (Skuse) in urban housing of Sibu town, Sarawak". Southeast Asian Journal of Tropical Medicine and Public Health. 1994, Sep;25:543.

[4] Han Lim L., Mustafakamal I., and Rohani A. "Does transovarial transmission of dengue virus occur in Malaysian Aedes aegypti and Aedes albopictus?." Southeast Asian Journal of Tropical Medicine & Public Health 28.1 (1997): 230-232.

[5] Lee, H. L. "Environmental friendly approaches to mosquito control." Mosquitoes and Mosquito-borne Diseases. Ng, FSP and Yong, HS (eds.). Kuala Lumpur: Academy of Science Malaysia (2000): 223-233.

[6] World Health Organization. Dengue haemorrhagic fever: diagnosis, treatment, prevention and control. World Health Organization, 1997. George G. P., and Taylor C. E. "Genetic and biological influences in the evolution of insecticide resistance." Journal of economic entomology 70.3 (1977): 319-323.

[7] Malcolm C. A., and Wood R. J. "Location of a gene conferring resistance to knockdown by permethrin and bioresmethrin in adults of the BKPM3 strain of Aedes aegypti." Genetica 59.3 (1982): 233-237.

[8] World Health Organization. Test procedures for insecticide resistance monitoring in malaria vectors, bio-efficacy and persistence of insecticides on treated surfaces: report of the WHO informal consultation, Geneva, 28-30 September 1998. No. WHO/CDS/CPC/MAL/98.12. Geneva: World Health Organization, 1998.

[9] Loh P. Y., and Yap H. H. Laboratory Studies on the Efficacy and Sublethal Effects of an Insect Growth Regulator, Pyriproxyfen (S-31183) Against Aedesaegypti (Linnaeus). 1989.

[10] Kamimura K., and Arakawa R. "Field evaluation of an insect growth regulator, pyriproxyfen, against Culex pipiens pallens and Culex tritaeniorhynchus." Medical Entomology and Zoology 42.3 (1991): 249-254.

[11] Schaefer C. H., and Mulligan F. S. "Potential for resistance to pyriproxyfen: a promising new mosquito larvicide." Journal of the American Mosquito Control Association 7.3 (1991): 409-411.

[12] Okazawa T. "Research & Reports." Journal of the American Mosquito Control Association 7.4 (1991): 604-607.

IOP Conf. Series: Materials Science and Engineering **769** (2020) 012012 doi:10.1088/1757-899X/769/1/012012

[13] Pyriproxyfen, General Fact Sheets. Available: http://npic.orst.edu/factsheets/pyriprogen.html

[14] Unlu I, Suman D. S., Wang Y, Klingler K., Faraji A., and Gaugler R. "Effectiveness of autodissemination stations containing pyriproxyfen in reducing immature Aedes albopictus populations." Parasites & vectors 10, no. 1 (2017): 139.

[15] Wan Yussof, Wan Nural Jawahir Hj, Mustafa Man, Muhammad Suzuri Hitam, Abdul Aziz K. Abdul Hamid, Ezmamarul Afreen Awalludin, and Wan Aezwani Wan Abu Bakar. "Wavelet-based Auto-Counting Tool of Aedes Eggs." In *Proceedings of the 2018 International Conference on Sensors, Signal and Image Processing*, pp. 56-59. ACM, 2018.

[16] Man, Mustafa, Wan Aezwani Wan Abu Bakar, Wan Nural Jawahir Wan Mohd Yusoff, and Mustafa Afenddi Mat Nor. "AedesApps: Image Processing Algorithm on Aedes Eggs Auto-Counting Mobile Apps."

[17] NodeCMU Documentation. Available: https://nodemcu.readthedocs.io/en/master/

[18] Patnaik Patnaikuni, D. R. "A Comparative Study of Arduino, Raspberry Pi and ESP8266 as IoT Development Board." International Journal of Advanced Research in Computer Science 8, no. 5 (2017).

[19] AA Battery. Available: https://en.wikipedia.org/wiki/AA_battery.

[20] G. P. Nayaka, KV. Pai, G. Santhosh, J. Manjanna, 'Dissolution of cathode active material of spent Li-ion batteries using tartaric acid and ascorbic acid mixture to recover Co. Hydrometallurgy", 2016 May 1;161:54-7.

[21] Gaburro J., Duchemin J.B., Paradkar P.N., Nahavandi S., and Bhatti A., "Assessment of ICount software, a precise and fast egg counting tool for the mosquito vector Aedes aegypti". Parasites & vectors, 2016. 9(1), p.590.

[22] Mollahosseini, A., Rossignol, M., Pennetier, C., Cohuet, A., dos Anjos, A., Chandre, F. and Shahbazkia, H.R., 2012. A user-friendly software to easily count Anopheles egg batches. Parasites & vectors, 5(1), p.122.

Reducing the Throughput Time for Patient Flow in Emergency Department: Simulation and Modelling Overview

Nidal Hamza[1], Mazlina Abdul Majid[1], Hael Al-Bashiri[1]

[1]Faculty of Computing (FKom) College of Computing and Applied Sciences, Universiti Malaysia Pahang Lebuhraya Tun Razak 26300 Gambang, Kuantan, Pahang Darul Makmur.

Corresponding author e-mail: nidalhamza@hotmail.com.

Abstract-Satisfaction of patient considered as a main issue of quality of service in the healthcare sector. Typically, this satisfaction depends on the services quality provided by hospitals. Emergency Department (ED), as a critical department in the hospital, has a complicated registration system that may lead to increase the patient throughput time. Thus, to minimize this growing in the throughput time, numerous simulation and modelling, in the literature, have been developed and introduced. However, the throughput time in ED still represent in issue need for improvement to increase the ED performance. Therefore, in this paper, the main objective is providing an overview related to the characteristics and significance of current simulation and model techniques. As a result, in the ED realistically, integrating Agent-Based Simulation (ABS), Desecrate Event Simulation (DES), and System Dynamic (SD) techniques has been preferred as the solution to modelling the patient flow in ED and in turn may lead to decrease the throughput time. The proactive and independent characteristics of aforementioned techniques can contribute to the good representation the patients flow and their throughput time in ED.

Keywords: *Department Emergency, Simulation, Patient Flow, Throughput Time.*

1. Introduction

Simulation, as a high significant technique, has been developed to simulate the stochastic behaviour of sophisticated systems[1, 2]. The simulation models are overwhelmingly constructed to analyse adjustments on a system[3]. For example, the health care field is one of the most important service sectors, the service level is an key issue that has often been analysed depends on studies of simulation because the popularity of that system is too complicated [4]. Furthermore, many optimization methods correlated to health care field cost have a vast chances to be studied[5-8]. Currently, decision-making processes strategic and operative can depend on further accurate estimating [9, 10] , which can be achieved by using simulation modelling. The decision-maker grows a pearl of wisdom into the behaviour of the system, towards dissimilar possible options before committing efforts and resources, based on the result of that simulation studies [11]. Moreover, simulation and modelling are good ways to emulate the actual systems. They are applied in several sectors, such as simulation of technology for optimization the performance, safety, education, engineering, testing, video games and healthcare sector. Due to data analysis alone cannot give vision into healthcare systems that are quickly evolving into complex and dynamic systems, the simulation and modelling in healthcare sector is thriving[12]. Thus, the concept of enhancing the healthcare quality in

recently years of service has been frequently argued in patient satisfactions. Providers of healthcare attempt to introduce better treatment to patients. The patients come back to the hospital depend on the fulfilled with their experience of services [13]. Due to throughput time as a major element of patient satisfaction/dissatisfaction various researches have regarded patient in this term [14, 15]. Additionally, in [8], the relevant significance of numerous patient satisfaction factors are listed as well as locating the service throughput time to be one of the key factors. The significant healthcare indicator outcomes is the patient satisfaction and it plays a key role in enhancing the quality of healthcare for patients[11]. Consequently, growth the quality level and the efficiency of healthcare units without producing side effects to the health of staffs is the main goal of a healthcare system design via further pragmatic approach[16].
In many existing review studies, the effectiveness of modelling and simulation in healthcare departments are discussed, such as; ED, operating suits, nursing units, ambulatory patient care and auxiliary services[17]. In spite of this, the existing review studies fall short of sufficiently covering the simulation for the patient's throughput time in the ED with respect to its concept and the existing techniques. Therefore, this paper aims to give an overview on simulating the patient's throughput time in the ED.
The paper is organized as follows. Section II provides simulation of emergency department. Section III described the simulation technique for ED to improvement the performance by reduce the patient's throughput time. the conclusions and future sets are covered in the last section.

2. Simulation of patient flow in Emergency Department

Simulation is an extremely sturdy method that has been usage to learning the random probability behaviour of service systems and complicated manufacturing [1]. The ED is a medical treatment capability, specializing in critical status of patients who come without any previous appointment. The main anxiety according to the systems of healthcare is the issue of department of emergency overcrowding that has been extensively studied especially in the USA [18, 19]. The overcrowding in the ED is a thoughtful problematic, which has a serious effect on the safety of patient and reduce the survival opportunity of severe patients in some instances because of long throughput time. Subsequently, the whole national healthcare system and its competence to offer the minimal service in any country is endangered [19, 20].

Furthermore, the simulation broadly used to model the composite systems such as ED. Simulation is approved to emulate the present case of any procedure. Hence, the simulation allows decision-makers to emulate the actions of the system and offers a test bed to measure the operations and managerial policies changes and asses several options[21]. Meanwhile, the simulation technique used in ED, Agent-Based Simulation (ABS), Discrete-Event Simulation(DES), and System Dynamic(SD) are the furthermost employed tools for modelling and analysing systems regarding to the interests of users and the specify task addressed. DES is a tool for measuring the delivery structures efficiency, estimating changes in patient flow and testing the efficiency of resource in staffing. SD concern of construction on behaviour. SD is commonly used for higher level problems instead of addressing individual transactions like strategic managing controls, policy changes, or decision-making. ABS is relying on a "bottom-up" structure for the delivery of emergent phenomena based on individual interactions of resource units. Fig.1.1 illustrate the basic process of patient flow in ED[1]. The patient arrival to ED by walk-in or ambulance. Then, walk-in, the patient goes to registration or waiting room. Next, the triage nurse give assist for patient body condition and go to treatment, examination, or waiting room. Then, the examination takes the sample for patient and send the result for treatment. Finally, the treatment gives the patient report to end of diagnosis discharge, transfer to another hospital, or ward. In case of ambulance arrival, the patient goes directly to treatments part[1, 8, 22, 23].

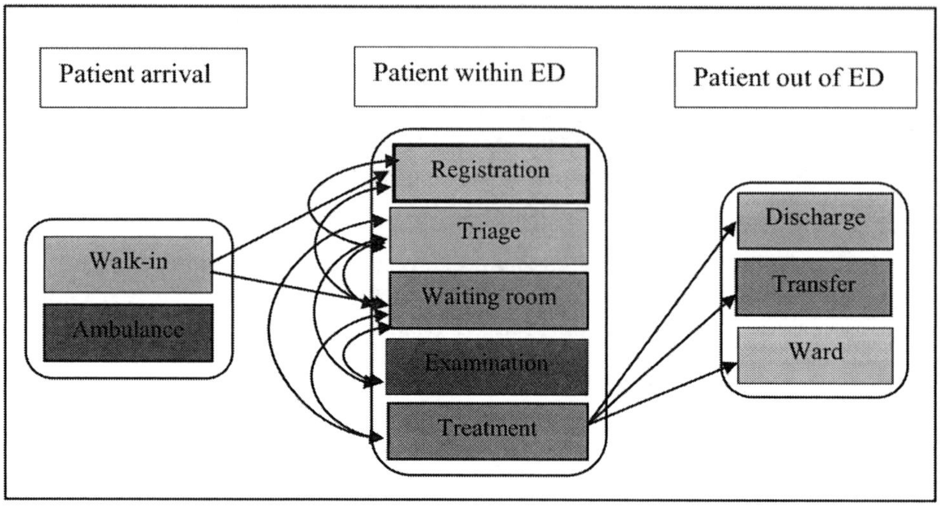

Figure 1. The Process of Patient Flow in Emergency Department

3. Simulation Technique for ED

Many simulation techniques for ED have been developed in the previous works. Thus, in this section, the author/s successful to discuss the common and powerful works introduced. In the work [24], the authors presented lengthy of ED length of Stay (LOS) related with patient experience and poorer medical outcomes. At the hospital community, the trauma patients were experiencing expanded LOS in commensurate with their medical diagnosis. The goal was to find if the modifications of functioning in the patient flow would decrease the LOS for trauma patients. Another work [25], introduced an agent-based simulation modelling in the ED. In a conventional approach, manager assigns the resources, like nurses, receptionist, doctors, and so on, to various units rely on the personal experience or by using decision-making tools. In that work, each agent of staff, based on their observation in their respective sections, took a part in the process of assigning resources, which gave the system the benefit of utilizing all the obtainable human resources through the work day by being assigned to a various unit. The yields of each simulation were number of patients who leave the emergency department without being attended, number of deaths, waiting time, length of stay, and total number of discharged patients from the ED.

An explored effect on throughput times of patient and resource of nursing request from the fast track additional unit in the ED using a queue-based Monte Carlo simulation in MATLAB [26] . in this model, the principles of queueing theory and expanded the discrete event simulation to account for time-based arrival rates are integrated. Furthermore, the nursing resource demand and ED occupancy are analysed and modelled using the Emergency Severity Index (ESI) patients' levels, rather than the number of beds in the department. Simulation results specified that the extension of a separate fast track with an additional nurse reduced overall median wait times.

In another research [18], the authors reported that in order to improve the processes of the ED, there is need to consider the various factors and targets in dynamic and unpredictable environment. This could therefore make the whole decision-making process more challenging[9, 27, 28]. As such, a possible way through it is to utilize different tools and techniques to facilitate the decision-making process. A novel approach in healthcare which combined DES, Simulation Based Multi-Objective Optimization, and Data Mining techniques were introduced. This technique was applied to ease the analysis of the system in a Swedish ED. Overall, the result of the system helped to proffer a suitable way to decision-makers in designing rules which could significantly reduce the length of stay and waiting times for patients in the ED.

Throughout the literature by [16], the authors derive a DES model to examine the patient walk-in clinics where patients are assisted without schedules. A case study is argued that a considers a walk-in clinic located in central Texas. The computational study provides useful insights that are appropriate to any walk-in health care services. In this study, presenting the application of Business Process Model (BPM) in healthcare part using the Business Process Model and Notation (BPMN), together with a multidimensional ABS of resources multidimensional organizational network and geographical. In the work [29], a technique for improving the performances at an ED in a large hospital has been developed, where the long throughput times and unstable utilization produce issues for patients and staff of ED. This technique, simulates patient flow in the ED firstly and then, finds bottlenecks that give rise to uselessness in the performance of ED. In the simulation model, patient arrival is assumed to be non-homogenous and the operation of medical tests such as MRI, CT scan, pathology testing, laboratory testing, ultrasonography, and radiology are detailed, and virtual queues of patients' specimens are considered separately from patient queues.

The research work by [25] , proposed approach for integration of DES and ABS to decrease the throughput time of patient and is the first try to apply this way to resolve the issue throughput time on orthopaedic departments. Based on the gathered data, patient behaviours are modelled and assimilated into a massive ABS. The introduced approach is an assistance analysing and modifying orthopaedic department procedures, allows us to provides more reliable results and consider far more details.

Other than that, [23] are extending application domain of meta-models into decision-making in the EDs by developing a DES model combined with suitable meta-models. This is used as a novel decision support system to improve the patients flow and relieve congestion by changing the number of ED resources (i.e., the number of receptionists, nurses, residents, and beds). This new tool could be used for decision-making in operational, tactical, and strategic levels. Experimental results with the current ED budget verify that after using the resource allocation obtained from the proposed model, the total throughput time of patients is reduced. However, according. Cassettari et al in [30] used the SD simulation and modelling in a variety of ways related to emergency medicine. SD has been used for crowding problem which is a common target to uncover essential flaws in an ED teamwork scheme. SD approaches have extensive efficacy on decision-making, which it seems highly applicable to emergency medicine.

Zhao et al.[31] , introduce a method for detecting the bottleneck in the ED, and to improve them through the use of benchmarking and design of experiments (DOE) in simulation model. Notably, overcrowding in the ED is well known to be an increasingly critical challenge to public healthcare. The bottlenecks that influence overcrowding in the ED have been identified in order to improve the flow of patients. Originality/value of integration of benchmarking and DOE in simulation modelling proposed in this research presents a potential time-saving process for identifying bottlenecks in the operations of the ED. This approach introduced by The work by [32] showed that about 50% or more patients may be attended to in a "fast track" process in comparison with the conventional procedures in many some ED. However, most of the researches on ED fast track processes were based on evidence without actually making use of a competent decision tool depict the applicability of the obtained results. On the contrary, an agent-based simulation tool was proposed by [33] to assess fast track treatment (FTT) in an ED. The simulation results provided detailed information for the FTT implementation procedures at the ED in order to reduce patient waiting time.

4. Conclusion and Future Sets

The findings of this research paper discussed the techniques with throughput time in the ED used to model patient emergency. Several techniques have been used individual or integrated to model the patient flow to save throughput time in the ED. DES is the greatest common and most regularly applied in many studies to model patient emergency throughput time. Whereas, SD effected on strategic decision-making, controls, management, or policy changes. Otherwise, the combination of ABS and DES is the newest and that is becoming more useful method. Nevertheless, there is a few works have been found so far, that modelling patient's throughput time in the hospital by using the techniques of decision-making. As a future work, the authors will be going to use the simulation technique to standardised more EDs and validate the result adaptability. The optimization for the ED performance, simulation model may be integrated with optimization methods. Additionally, several healthcare sub systems, such as EDs, out patient service units, and in patient units have to be integrated and simulated.

Acknowledgements

This research is financially supported by Universiti Malaysia Pahang (UMP) under the research grant Nos. RDU190371. Appreciation to the reviewers and editors for considering this paper.

Reference

[1] N. Hamza, M. A. Majid, K. Adam, and N. A. A. Bakar, 2019, "A Review on Simulation and Modelling for Patient Flow in Emergency Department," in *IOP Conference Series: Materials Science and Engineering*, p. 012037.

[2] M. A. Majid, K. Z. Zamli, and M. A. I. Fakhreldin, *2015,*"Modelling a Complex Human Centre Queuing System for Enhancing the Capability of Agent Based Simulation," in *Proceedings of the International Conference on Data Engineering (DaEng-2015)*, 2019, pp. 383-391.

[3] J. A. Paul and L. Lin, 2012, "Models for improving patient throughput and waiting at hospital emergency departments," *The Journal of emergency medicine,* vol. 43, pp. 1119-1126.

[4] D. Daldoul, I. Nouaouri, H. Bouchriha, and H. Allaoui, 2018 ,"A stochastic model to minimize patient waiting time in an emergency department," *Operations Research for Health Care,* vol. 18, pp. 16-25.

[5] F. Hujainah, R. B. A. Bakar, M. A. Abdulgabber, and K. Z. Zamli, 2018,"Software Requirements Prioritisation: A Systematic Literature Review on Significance, Stakeholders, Techniques and Challenges," *IEEE Access,* vol. 6, pp. 71497-71523.

[6] Z. Liu, D. Rexachs, F. Epelde, and E. Luque, 2017,"A simulation and optimization based method for calibrating agent-based emergency department models under data scarcity," *Computers & Industrial Engineering,* vol. 103, pp. 300-309.

[7] K. Wylie, J. Crilly, G. Toloo, G. FitzGerald, J. Burke, G. Williams, *et al.*, 2015,"Emergency department models of care in the context of care quality and cost: A systematic review," *Emergency Medicine Australasia,* vol. 27, pp. 95-101.

[8] E. Alhroob, M. F. Mohammed, C. P. Lim, and H. Tao, 2019, "A Critical Review on Selected Fuzzy Min-Max Neural Networks and Their Significance and Challenges in Pattern Classification," *IEEE Access,* vol. 7, pp. 56129-56146.

[9] F. Hujainah, R. B. A. Bakar, and M. A. Abdulgabber , 2019,"StakeQP: A semi-automated stakeholder quantification and prioritisation technique for requirement selection in software system projects," *Decision Support Systems,* vol. 121, pp. 94-108.

[10] W. Abo-Hamad and A. Arisha, 2013,"Simulation-based framework to improve patient experience in an emergency department," *European Journal of Operational Research,* vol. 224, pp. 154-166, .

[11] A. B. Tehrani, S. R. Feldman, F. T. Camacho, and R. Balkrishnan,2011, "Patient satisfaction with outpatient medical care in the United States," *Health Outcomes Research in Medicine,* vol. 2, pp. e197-e202.

[12] K. Freeman and S. A. Denham, 2008,"Improving patient satisfaction by addressing same day surgery wait times," *Journal of PeriAnesthesia Nursing,* vol. 23, pp. 387-393.

[13] G. M. Eilers, "Improving Patient Satisfaction With Waiting Time,2004," *Journal of American College Health,* vol. 53, pp. 41-48, 2004/07/01.

[14] T. R. Rohleder, P. Lewkonia, D. P. Bischak, P. Duffy, and R. Hendijani, 2011,"Using simulation modeling to improve patient flow at an outpatient orthopedic clinic," *Health care management science,* vol. 14, pp. 135-145.

[15] M. H. Bernhart, I. Wiadnyana, H. Wihardjo, and I. Pohan, 1999,"Patient satisfaction in developing countries," *Social science & medicine,* vol. 48, pp. 989-996.

[16] M. Sbayou, Y. Bouanan, G. Zacharewicz, J. Ribault, and J. François,2017, "DEVS modelling and simulation for healthcare process application for hospital emergency department," in *Proceedings of the 50th Annual Simulation Symposium*, p. 4.

[17] L. D. Richardson and U. Hwang, 2001,"America's Health Care Safety Net Intact or Unraveling?," *Academic Emergency Medicine,* vol. 8, pp. 1056-1063.

[18] A. G. Uriarte, E. R. Zúñiga, M. U. Moris, and A. H. Ng, 2017,"How can decision makers be supported in the improvement of an emergency department? A simulation, optimization and data mining approach," *Operations Research for Health Care,* vol. 15, pp. 102-122.

[19] S. Trzeciak and E. P. Rivers,2013, "Emergency department overcrowding in the United States: an emerging threat to patient safety and public health," *Emergency medicine journal,* vol. 20, pp. 402-405.

[20] J. Jun, S. H. Jacobson, and J. R. Swisher,1999, "Application of discrete-event simulation in health care clinics: A survey," *Journal of the operational research society,* vol. 50, pp. 109-123.

[21] N. Hamza, M. Abdul Majid, and M. Abdullateef ,2018,"IoT Cloud Simulation Managing Resources: An Overview," in *National Conference for Postgraduate Research 2018 (NCON-PGR 2018),* UMP Gambang, p. 7.

[22] F. Hujainah, R. B. A. Bakar, B. Al-haimi, and M. A. Abdulgabber, 2018,"Stakeholder quantification and prioritisation research: A systematic literature review," *Information and Software Technology,* vol. 102, pp. 85-99.

[23] H. Hajjarsaraei, B. Shirazi, and J. Rezaeian, 2018,"Scenario-based analysis of fast track strategy optimization on emergency department using integrated safety simulation," *Safety science,* vol. 107, pp. 9-21.

[24] M. Yousefi, M. Yousefi, F. Fogliatto, R. Ferreira, and J. Kim, 2018,"Simulating the behavior of patients who leave a public hospital emergency department without being seen by a physician: a cellular automaton and agent-based framework," *Brazilian Journal of Medical and Biological Research,* vol. 51.

[25] C. Kittipittayakorn and K.-C. Ying, 2016,"Using the integration of discrete event and agent-based simulation to enhance outpatient service quality in an orthopedic department," *Journal of Healthcare Engineering,* vol. 2016.

[26] K. Fitzgerald, L. Pelletier, and M. A. Reznek, 2017,"A Queue-Based Monte Carlo Analysis to Support Decision Making for Implementation of an Emergency Department Fast Track," *Journal of healthcare engineering,* vol. 2017.

[27] M. H. Yarmohammadian, F. Rezaei, A. Haghshenas, and N. Tavakoli, 2017,"Overcrowding in emergency departments: a review of strategies to decrease future challenges," *Journal of research in medical sciences: the official journal of Isfahan University of Medical Sciences,* vol. 22.

[28] A. Nahhas, A. Awaldi, and T. Reggelin, 2017,"Simulation and the Emergency Department Overcrowding Problem," *Procedia Engineering,* vol. 178, pp. 368-376.

[29] A. Gharahighehi, A. S. Kheirkhah, A. Bagheri, and E. Rashidi,2016, "Improving performances of the emergency department using discrete event simulation, DEA and the MADM methods," *Digital Health,* vol. 2, p. 2055207616664619.

[30] L. Cassettari, R. Mosca, A. Orfeo, R. Revetria, F. Rolando, and J. B. Morrison, 2013,"A System Dynamics Study of an Emergency Department Impact on the Management of Hospital's Surgery Activities," in *SIMULTECH*, pp. 597-604.

[31] Y. Zhao, Q. Peng, T. Strome, E. Weldon, M. Zhang, and A. Chochinov, 2015,"Bottleneck detection for improvement of Emergency Department efficiency," *Business Process Management Journal,* vol. 21, pp. 564-585.

[32] A. Kaushal, Y. Zhao, Q. Peng, T. Strome, E. Weldon, M. Zhang*, et al.*, 2015,"Evaluation of fast track strategies using agent-based simulation modeling to reduce waiting time in a hospital emergency department," *Socio-Economic Planning Sciences,* vol. 50, pp. 18-31.

[33] F. Zeinali, M. Mahootchi, and M. M. Sepehri, 2015,"Resource planning in the emergency departments: A simulation-based metamodeling approach," *Simulation Modelling Practice and Theory,* vol. 53, pp. 123-138.

IoT-Based Smart Environment Monitoring System for Air Pollutant Detection in Kuantan, Pahang, Malaysia

Nor Saradatul Akmar Zulkifli[1], Mohammad Ridwan Satrial[2], Mohd Zamri Osman[3], Nor Syahidatul Nadiah Ismail[4], Muhammad Rusydi Muhammad Razif[5]

[1,2,3]Soft Computing & Intelligent System (SPINT), Faculty of Computing, Universiti Malaysia Pahang, 26300 Gambang, Kuantan, Pahang, Malaysia

[4]Systems Network & Security (SYSnet), Faculty of Computing, Universiti Malaysia Pahang, 26300 Gambang, Kuantan, Pahang, Malaysia

[5]Cybernatics Research Group, Faculty of Engineering Technology, Universiti Tun Hussein Onn Malaysia, Hab Pendidikan Tinggi Pagoh, KM 1, Jalan Panchor, 84600 Panchor, Johor, Malaysia

saradatulakmar@ump.edu.my

Abstract. Air pollution is one of the pollutions that release harmful chemicals and if not controlled, it will affect the human respiratory system. Currently, there are two Air Pollution Monitoring system located in Kuantan district to record the pollution's reading once it detects the air pollutant however, the origin of the pollution cannot be detected by the system. As a solution, a Smart Environment Monitoring System for Air Pollution Using Internet of Things (IoTs) is developed to overcome the weaknesses. This project involved both hardware and software developments, in which hardware development includes the Arduino Uno, Carbon Monoxide (MQ 7), smoke/gas (MQ 135) sensors, a buzzer and LCD display. This system is design so that the current reading can be viewed by admin in a real-time value. Besides, they can also view the data for every 8 hours. These data can be printed in PDF format. In addition, when the reading reached unhealthy value, it will warn the community via its alarm system embedded in this system. However, this system is subject to limitation especially on data delays from sensors to database. As only one existing monitoring system that located at Beserah, thus this new feature will add a sub-sensor that located at the designated area which potentially expose to the air pollution.

Keywords—Internet of Things, Air Pollution Monitoring, Carbon Monoxide, Sensors, GSM Module, Arduino

1. Introduction

The key measures to be closely observed in real time for today's urban environment is the quality of air. This is because, it has a predominant impact on human health, safety and comfort. Air pollution can cause preterm births in newborns and Alzheimer's in the elderly. Air pollution is the release of harmful chemicals into the atmosphere causes discomfort to humans as well as cause damage to other living organisms such as food crops or the natural environment [1-2]. One of the main causes for air pollution in Malaysia is the emission of the combustion of fossil fuels vehicles. However, bauxite mining has been growing concern in Malaysia nowadays especially in the parts of state of Pahang in which, it covered with layers of red dust. Additionally, this bauxite sediment also pollutes the sea along the coast of Kuantan, which is one of the tourist attraction beaches near Teluk Cempedak. Practically, the existing air pollutant detection system in Malaysia are not capable to detect the source of the pollutants. It can only detect the minimum and maximum value of the air quality within its zone only. Therefore, to overcome this problem, the existing systems need to be upgrade and it should be re-locating at the potential areas to enhance its functionality.

There are some issues from the existing system that currently uses by the Department of Environment (DoE). First, the current systems use too many sensors which is not significantly used in the current environment. Second, there are only two air pollution monitoring stations in Kuantan which is at Beserah and Indera Mahkota. This is due to the high cost needed to build and develop the device. Third, the location of the current system is not too specific. The two main stations that locate in SMK Indera Mahkota and SMK Pelabuhan is the insignificant place for activities that give an impact to air pollution. There are few existing systems that had been developed previously which is Air Pollution Index (API) Real Time Monitoring System [3], Arduino-Based Air Quality Monitoring [4] and IoT Based Air Pollution Monitoring System Using Arduino [5]. Norhafizah et.al [3] developed the API Real-Time Monitoring System in purposed to detect and monitor the dust in the industrial area at Parit Raja, Johor. The authors had integrated the dust sensor and LCD shield with Arduino Uno. This system only allowed data transmitted into the laptop via USB cable and data collected were viewed using Microsoft Excel.

On the other hand, Haro Prasaath [4] had developed an Arduino-based Air Quality Monitoring by using MQ 135 sensor which is sensitive to the natural gas and Liquified Petroleum Gas (LPG). This device embeds the MQ 135 sensors, LCD Screen and Wi-Fi Modules with an Arduino Uno Microprocessor. All data is stored at http://thingspeak.com in which, ThingSpeak is a tool that let the user to store all sensor data at cloud. Moreover, this device only used one (1) gas sensor which is the MQ 135 sensor. This system integrates Wi-Fi Modules as their network connector [9-12]. However, these devices need to be installed near the Wi-Fi hotspot. In addition, this system only displays all the data in the ThingSpeak website.

IoT-Based Air Pollution Monitoring System using Arduino developed by Darshan Mirajkar [5] used only smoke sensor and specially design to detect and alerted any air pollution causing from open burning. Authors used Wi-Fi Module, Smoke Sensor, LCD Display and buzzer which integrated into Arduino Uno as their microprocessor. This system allowed the data transmitted from sensor to the local database by Wi-Fi Module and the data will be shows in the monitoring page. For outdoor air pollution monitoring, instead being the low-cost network module, Wi-Fi Module is not suitable to be used. Besides, local database like XAMPP is not suitable for this device since local database can only connect with Arduino for Local Area Network (LAN). This device development did not include any function of log in page for admin and the system can't generate any report from the data.

The main objective of this study is to develop a low-cost smart environment monitoring system for air pollution using Internet of Things (IoT). This report is organized into four (4) parts. Section II briefly describe the methodology used in this work, meanwhile Section III presents the detailed explanation of its results and discussion. At the end of this report, the development of this system is being concluded.

2. Methodology

This proposed system was developed based on Rapid Application Development (RAD) model. RAD is a combination model of different organized methods with prototyping and joint application development procedures to speed advancement of the system [6]. It aims to build application faster using an iterative development process as shown in Figure 1. Proponents of RAD claim that it increases productivity, reduces delivery time and gains high usage because of the extent of user involvement in the development [7].

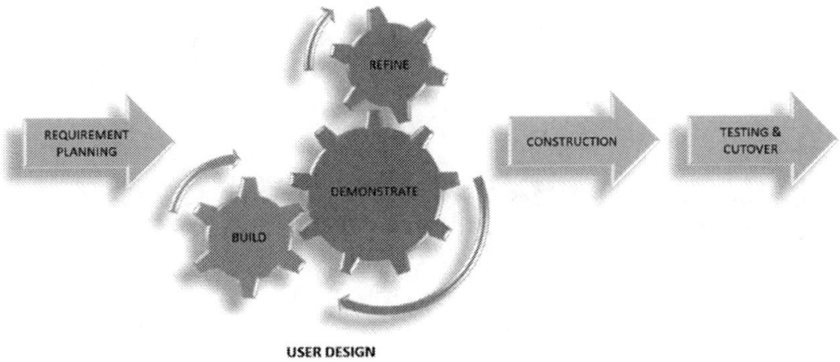

Figure 1. RAD Methodology Phase

2.1. Requirement Planning

It is important to determine the needs of the study before doing the system. These needs include the establishment of the problem statement, objectives, data and functional requirements and identification of hardware and software used in the implementation. Figure 2 shows the general architecture of this proposed system in which two nodes will be implemented in two different location given as Node A and Node B.

2.1.1. Data Requirement. There are two sensors attached to the Arduino Uno R3, which are the air quality sensor MQ-135 and carbon monoxide (CO) sensor MQ-7. These sensors provide the input for the proposed system. The MQ-135 sensor is used widely as a gas sensor for air quality. This sensor is used to detect or measure the air quality which consist of chemical substances such as Ammonia (NH3), Nitrogen Oxide (NOx), Alcohol, Benzene, Smoke and Carbon Dioxide (CO2). Whereas, the MQ-7 sensor module allows for the sensing of CO concentration in the air. It can detect the CO gas concentration from anywhere between 20 and 2000ppm. The output of the system can be seen through android mobile phone, online data display and storage as well as buzzer for red alert notification.

2.1.2. Software Requirement. C Programming language is used as the main programming language for the code development of the system. It is used to program the Arduino Uno by using Arduino IDE as its framework. Besides, several modules were also integrated with the system to realize the system's functionalities. On the other side, sublime is using to code the website development with Hypertext Preprocessor (PHP), Hypertext Markup Language (HTML) and CSS JavaScript as its main language. This website development is incorporate with MySQL database which allow the connectivity between GSM Module and the database to ensure all data from the sensor can be transmitted and stored to the database.

2.1.3. Hardware Requirement. The core development of this IoT device consist of an Arduino Uno R3, which is a microcontroller board with 20 digital input/output pins that work as a processing device for this system. Due to its flexibility and ability to respond to sensors and inputs, this Arduino can interact with a large array of outputs. Figure 3 shows the circuit diagram used for each node in the proposed system. For the prototype of this proposed system, 4 AAA battery were used as the power supply.

IOP Conf. Series: Materials Science and Engineering **769** (2020) 012014 doi:10.1088/1757-899X/769/1/012014

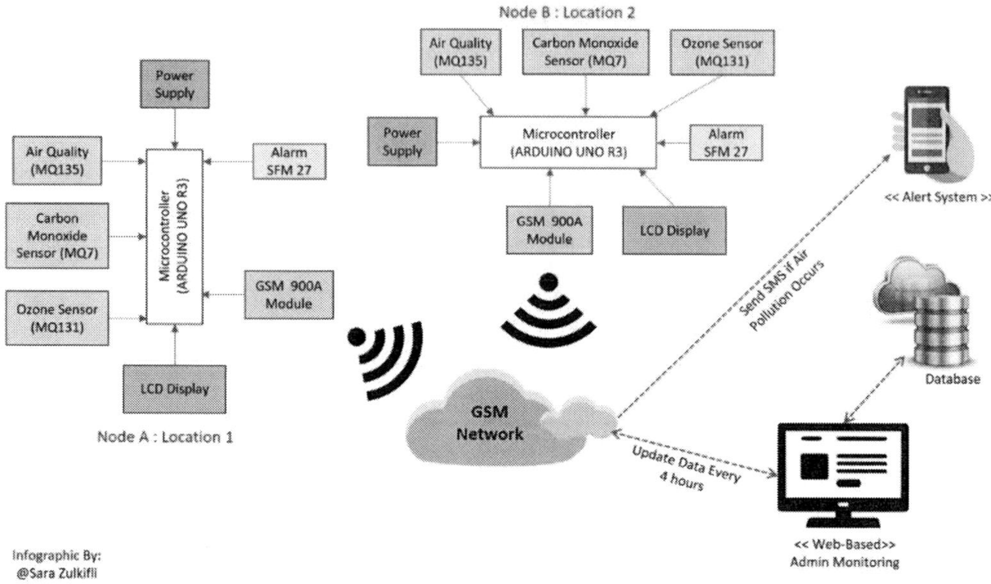

Figure 2. General Architecture of Proposed System

The important key of this IoT device is the sensor. A sensor is a device that detects or measures a physical environment and respond to it. The output of these sensors is sent directly to online MySQL database through GSM SIM 900A. GSM SIM 900A is a GSM Module that work to establish communication between a computer and a GSM-GPRS system which work together with a buzzer for indicator as an additional hardware to this IoT device [14].

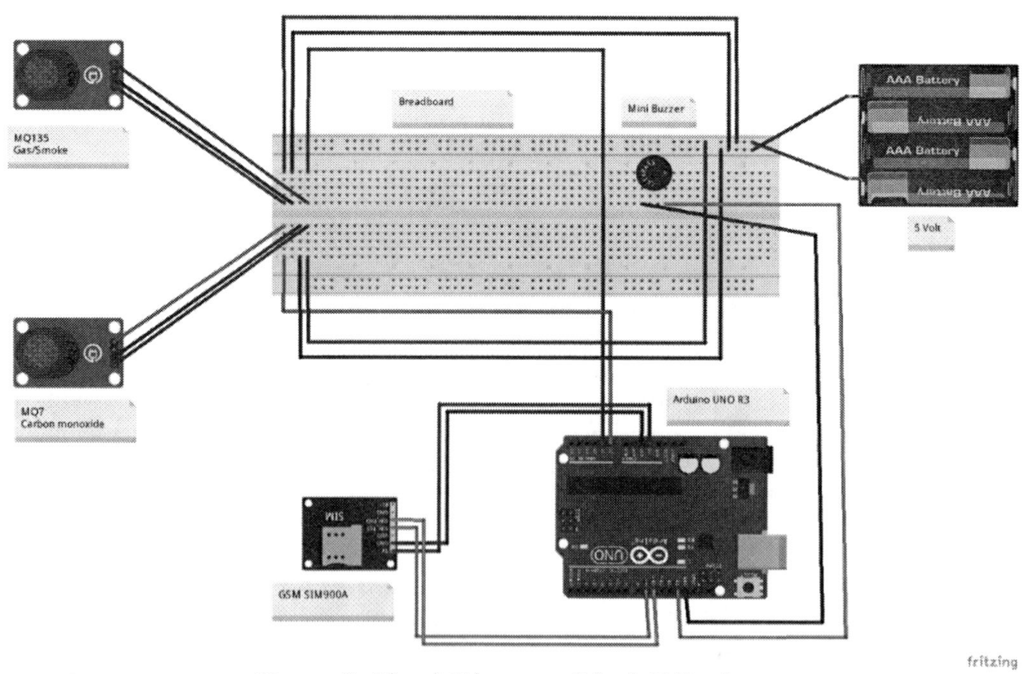

Figure 3. Circuit Diagram of the IoT Devices

2.2. User Design. User design is created based on requirements collected in the early phase. In this phase, model and prototype that represent all system processes including inputs and outputs are developed. A detailed user design includes how the system's flow, from beginning until the end, how many users include and how the systems architecture looks like. This phase enables users to see to general flow of the system, understand the process flow and modify the working model of the system that meets their needs. Figure 2 demonstrate the general architecture of the proposed system which integrates the sensors, mini buzzer, GSM Module and cloud storage with the Arduino Uno R3. The outcomes from this phase will deliver to the next phase which is the construction phase.

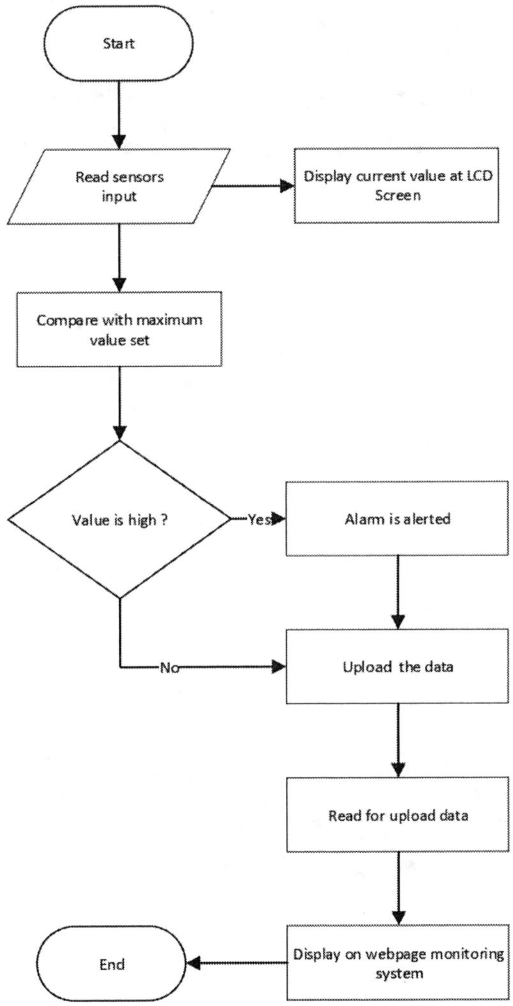

Figure 4. Flow Chart of the Proposed System

2.3 Construction. In construction phase, the detailed gritty diagram of the system models is completed with the system realization, framework and imperative documentation designed work for the proposed application. During the period time of construction, three (3) phases should be done which is the establishment of hardware and software. It is important to configure all hardware integrated in the system and program the Arduino Uno board based on the proposed design to ensure the system works properly. In addition, the GPRS GSM Module connection should be configured properly in order to enable the storage of data acquisition into the cloud and being display in the website and android applications. The last part in construction phase was an android notification to the community to alert them about the pollution. The process cycle between the user design phase and construction until final prototype is created. Figure 4 shows the flow of the proposed system.

2.4 Testing and Cutover. To acquire a complete system and test the system functionalities, both software and hardware were combined in this system. In this phase, the hardware prototype and the web-site design in its own server as well as the android applications were tested together to ensure all sensors, alert notification and buzzer work properly based on the setup function. In this proposed system, the level of air pollution was set in the construction step accordingly as indicated in Table 1. In this system, when the sensor detected the level of hazardous pollution, mini buzzer and notification will alert from the android application known as CleanKuantan2.

Table 1. Pin to Pin Details from Arduino to Hardware

Arduino	Sensors/Hardware	
A0	MQ 7	A0
A1	MQ 135	A0
9	GSM	5Tx
10		5Rx
13	Buzzer	Positive

3. Result and Discussion

In this study, three (3) experiments were conducted which is the connectivity test of sensors with Arduino Uno R3, network connection of GPRS GSM SIM900A with cloud storage and web-site design, generation of PDF report from web site and integration test of cloud storage between web sites and android application.

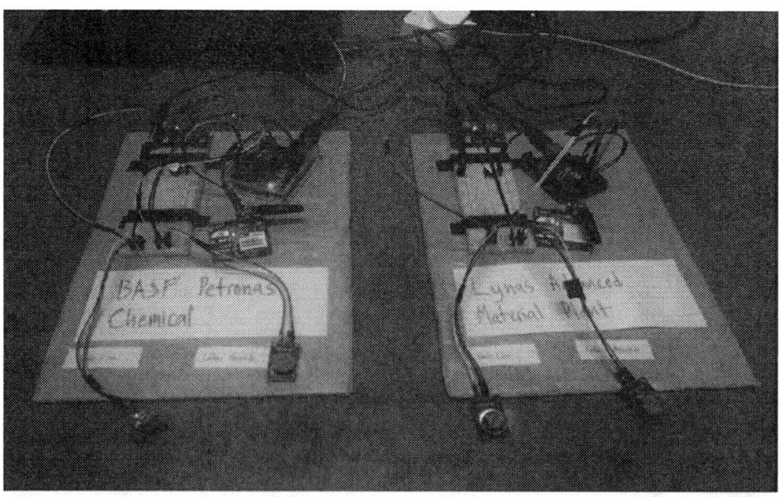

Figure 5. Connectivity of Sensors and the Arduino Uno R3 and GSM Module

In the first experiment, the connectivity of MQ-135 and MQ-7 sensors with the Arduino Uno R3 and GSM Module are shown in Figure 5 were tested. Two nodes for two different station were developed to test the concurrent data transmitted in the same time into the cloud storage. The connections of the sensors to the Arduino Uno R3 board are shows in Table 1, where MQ 135 is connected to pin A1 in Arduino and MQ 7 to pin A0. Both Transmitted, Tx and Receiver, Rx pins for GSM SIM900A were

IOP Conf. Series: Materials Science and Engineering **769** (2020) 012014 doi:10.1088/1757-899X/769/1/012014

connected to digital pin 9 and 10, respectively. Initially, Arduino IDE was used to setup the connection of all sensors and buzzer to the microprocessor Arduino Uno R3 as shows in Figure 6. Then, the Arduino Uno R3 was connected to the port to compile the programming into the Arduino. This is shows in Figure 7. It is important to check either the port connected is right or not. To ensure the right programming being compile into the right microcontroller. Once the compiler successfully done, system will show a notification as depicted in Figure 8.

Figure 9 display the code used to establish the connection in GSM SIM900A to the cloud network. This establishment allowed data to be transmitted to the cloud database. And from here, data will be retrieved and display in web-based monitoring system and android application. Figure 10 shows the serial monitor interface which is used to display the connection between Arduino Uno and GSM SIM900A Module.

Figure 12 shows the PDF report which can be downloaded by the DOE as reference report for further analysis especially when the device detected a pollution at the designated area. On the other hand, community which equipped with an android application known as CleanKuantan2 also get an alert notification regarding the pollution area through the apps. This alert notification helps the community to take early precautious.

Table 2. Air Quality Index (API) Level

API Reading	Status	Color
0 - 50	Good	Blue
51 - 100	Moderate	Green
101 - 200	Unhealthy	Yellow
201 - 300	Very Unhealthy	Orange
> 300	Dangerous	Red

```
 // Carbon Monoxide
 lcd.begin(16, 2);
 lcd.print("Carbon Monoxide");
 finalVal=(float)val/1024*5.0;
 lcd.setCursor(0, 2);
 lcd.print(finalVal, 1);//Print the value to serial port
 lcd.println(" PPM");
 delay(3000);

 //Smoke
 lcd.begin(16, 2);
 lcd.print("Smoke");
 finalSV=(float)sensorValue/1024;
 lcd.setCursor(0, 1);
 lcd.print(finalSV, 2);
 lcd.println(" PPM");
 delay(3000);

if (finalVal > 100 || finalSV > 100)
 {
   alertBuzzer();
   }
   else
   {

     }
}
```

Figure 6. Code Programming for sensors, buzzer and LCD screen

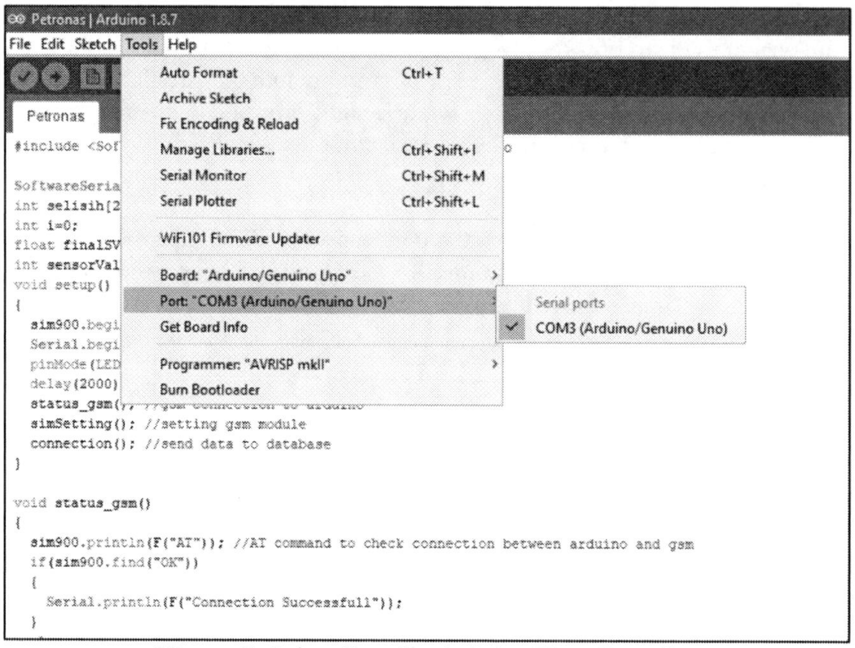

Figure 7. Select Port for Arduino Uno Compiler

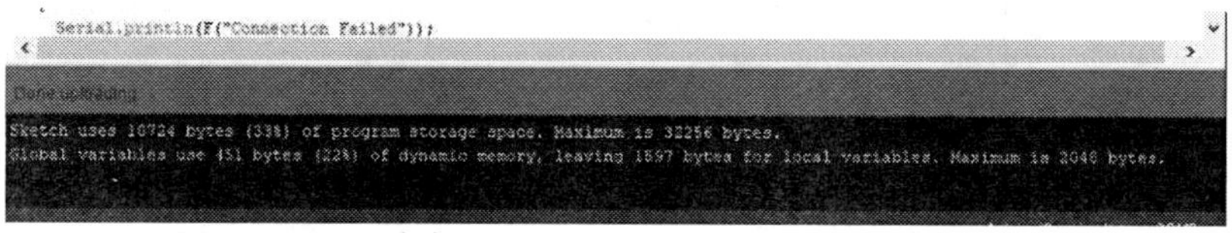

Figure 8. Arduino IDE notify the compiling process

```
void koneksi()
{
  sim900.println(F("AT+CIPSTART=\"TCP\",\"lienawan.xyz""\",80")); //memulai koneksi dengan server
  delay(1000);
  Serial.println(sim900.readString());
  if(sim900.find("OK"))
  {
    Serial.println(F("proses data"));
  }
  String link = "GET /input.php?co="+String(finalVal)+"&smoke="+String(finalSV)+" HTTP/1.1\r\nHost: lienawan.xyz\r\n\r\n";

  sim900.print(F("AT+CIPSEND=")); //mengirim request data
  sim900.println(link.length()); //mengirim panjang tautan request data
  delay(500);
  if(sim900.find(">"))
  {
    sim900.print(link); //mengirim tautan ke server
    if(sim900.find("SEND OK"))
    {
      delay(100);
      while (sim900.available())
      {
        Serial.println(sim900.readString()); //feedback data (sesuai dengan keluaran di browser)
      }
      sim900.println(F("AT+CIPCLOSE")); //stop koneksi
    }
    else
    {
```

Figure 9. Code for Network Connection in GSM SIM900A

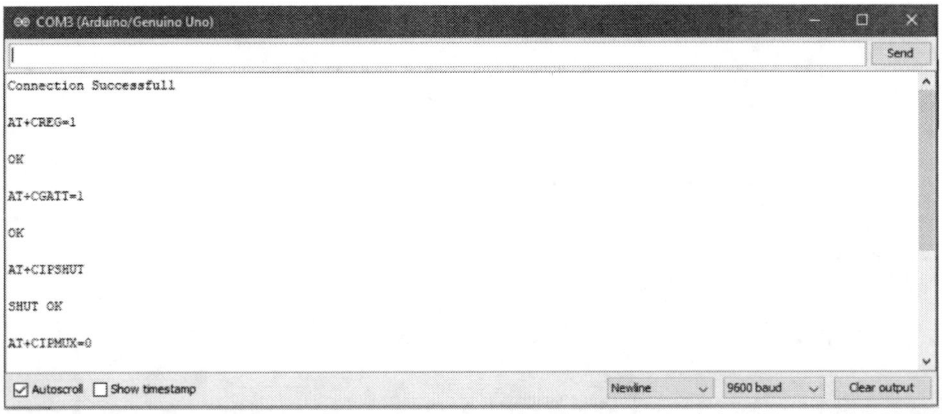

Figure 10. Serial Monitor Interface shows the connection between Arduino Uno and GSM SIM900A

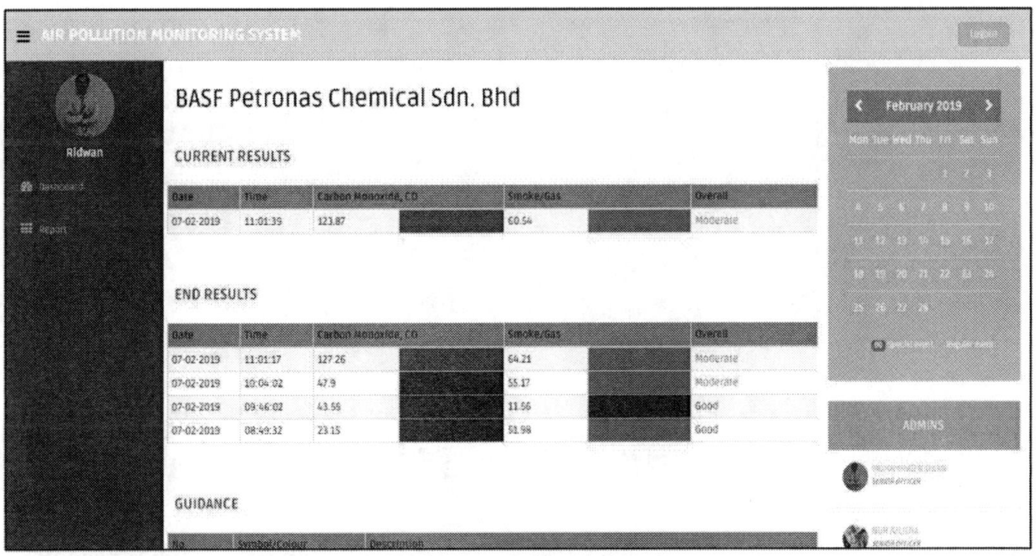

Figure 11. Web-Based Cloud Storage

IOP Conf. Series: Materials Science and Engineering **769** (2020) 012014 doi:10.1088/1757-899X/769/1/012014

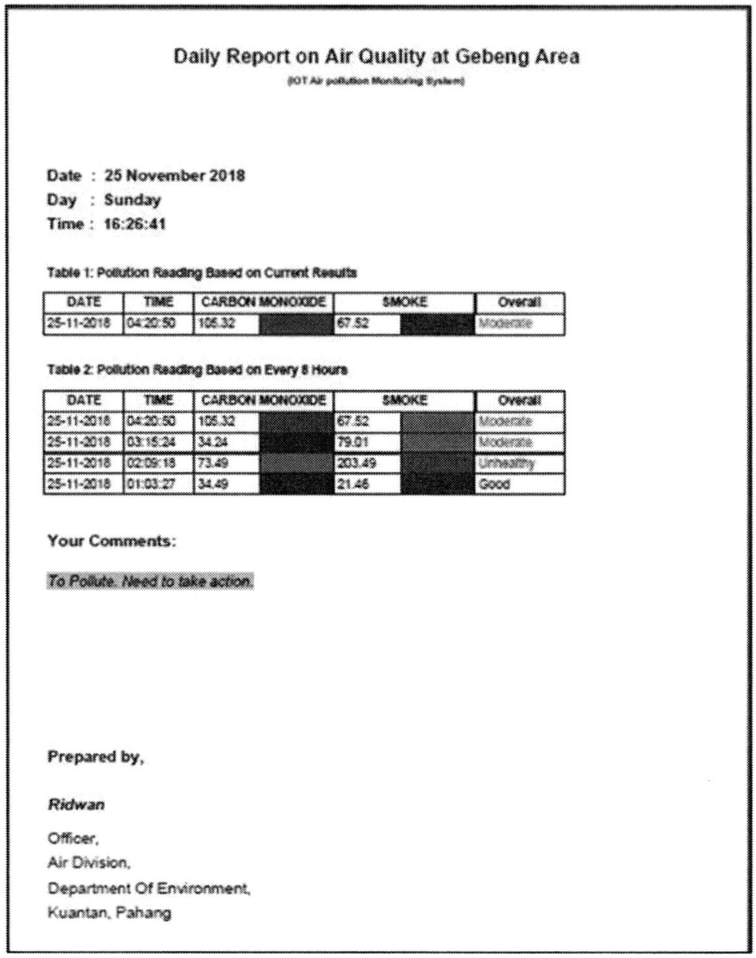

Daily Report on Air Quality at Gebeng Area

(IOT Air pollution Monitoring System)

Date : 25 November 2018
Day : Sunday
Time : 16:26:41

Table 1: Pollution Reading Based on Current Results

DATE	TIME	CARBON MONOXIDE		SMOKE		Overall
25-11-2018	04:20:50	105.32		67.52		Moderate

Table 2: Pollution Reading Based on Every 8 Hours

DATE	TIME	CARBON MONOXIDE		SMOKE		Overall
25-11-2018	04:20:50	105.32		67.52		Moderate
25-11-2018	03:15:24	34.24		79.01		Moderate
25-11-2018	02:09:18	73.49		203.49		Unhealthy
25-11-2018	01:03:27	34.49		21.46		Good

Your Comments:

To Pollute. Need to take action.

Prepared by,

Ridwan

Officer,
Air Division,
Department Of Environment,
Kuantan, Pahang

Figure 12. PDF Report generated by DOE when the device detected pollution at the designated area

4. Conclusion

In this paper, it has shown the development of the proposed solution for air pollution monitoring system which integrates GSM Modules, air quality sensor and Arduino Uno to form the Internet of Things (IoT)-based system. This system is expected to improve the current systems that had been used by the Department of Environment (DoE), Kuantan, Pahang. This monitoring system exhibit several strengths in which it can help DoE to monitor the air quality from area that exposed to get the pollution especially in the exact industrial area.

This monitoring system also providing a real-time web-based cloud application that allow the DoE to monitor, update and display the air quality data on each industrial site. If unhealthy air quality is detected in certain industrial area, it will alert both industrial and DoE through alert notification indicated using buzzer. Compared to existing system developed by N.A Zakaria [8], this device transmit data to the cloud storage using GSM Sim900. Instead of using thinkspeak.com as the web-based cloud storage, we develop our own website which can be used by DOE to monitor all stations. This smart monitoring system is easy to install and maintain as it requires very low technical skills and knowledge for device handling. However, this monitoring device still have small problem in providing a real-time to the web-based and Arduino applications [13-14]. Thus, further research in this domain should be explored to enhance the quality of data transmitted and to ensure that data display on the web-based and android application is in a real time.

Acknowledgement

A high appreciation to Faculty of Computing, Universiti Malaysia Pahang, under grant RDU 1703287.

References

[1] A. Hasfazilah, Y. Ahmad Shukri and R. Nor Azam, "PM_{10} Analysis for Three Industrialized Areas using Extreme Value," Sains Malaysiana vol. 44 (2), 2015. pp. 175 -185.

[2] Mateen, F.J., and Brook R.D, "Air Pollution as an emerging global risk factor for Stroke," JAMA, 305 (12), 2011. pp.1240-1241

[3] Ruslan,N et al., "Air Pollution Index (API) Real Time Monitoring System. University Tun Hussein Onn. Thesis. 2015.

[4] Prasaath, H., "Arduino Based Air Quality Monitoring IOT Project," Engineer Garage. 2017

[5] Mirajkar, D. "IoT based Air Pollution Monitoring System Using Arduino". 2015.

[6] Idesis,S. "What is Rapid Application Development?". 2017

[7] Coleman,G and Verbruggen,R. "A quality software process for rapid application development,". 1998

[8] N.A Zakaria, Z.Z.Abidin, N.Harun et. al, "Wireless Internet of Things-Based Air Quality Device for Smart Pollution Monitoring", International Journal of Advanced Computer Science and Applications, Vol. 9, No.11, 2018, pp.65-69.

[9] AA Shahidan, N Fisal, Nor Syahidatul N Ismail, Farizah Yunus, Sharifah HS Ariffin., "Data recovery in wireless sensor networks using network coding," Jurnal Teknologi, vol.73 (3), 2015.

[10] Farizah Yunus, Nor Syahidatul N Ismail, Sharifah HS Ariffin, SK Syed Yusof, "A rate Control Model of MPEG-4 Encoder for Video Transmission over Wireless Sensor Network," International Journal of Communication Networks and Information Security, vol 11(1), 2019. Pp.42-51.

[11] NSA Zulkifli, FK Che Harun, NS Azahar, "Centralized heart rate monitoring telemetry system using ZigBee Wireless Sensor Network," Proceeding of 2012 IEEE-EMBS International Conference on Biomedical and Health Informatics, 2012, pp. 265-268.

[12] NSA Zulkifli, FK Che Harun, NS Azahar,"XBee Wireless Sensor Networks for Heart Rate Monitoring in Sport Training," 2012 International Conference on Biomedical Engineering (ICoBE), IEEE, 2012, pp.441-444.

[13] D. Swati, B.M. Rajasekhara, H.G Amir et. al., "Internet of Things Mobile-Air Pollution Monitoring System (IoT-Mobair)," IEEE Internet of Things Journal, vol 6 (3), June 2019, pp.5577-5584.

[14] M.M. Shahkhir, A.B. Rohani, W.D. Wan Isni, "Indoor Localization Estimation Techniques in Wireless Sensor Network," 2018 IEEE International Conference on Automatic Control and Intelligent Systems, 2018, pp. 148-154.

IOP Conf. Series: Materials Science and Engineering **769** (2020) 012015 doi:10.1088/1757-899X/769/1/012015

Unstructured Malay Text Analytics Model in Crime

Rosmayati Mohemad[1], Nazratul Naziah Mohd Muhait[1], Noor Maizura Mohamad Noor[1] & Zulaiha Ali Othman[2]

[1]Faculty of Ocean Engineering Technology & Informatics, Universiti Malaysia Terengganu, 21030 Kuala Nerus, Terengganu, MALAYSIA.
[2]Faculty of Information Science & Technology, Universiti Kebangsaan Malaysia, 43600 UKM Bangi, Selangor, MALAYSIA.

rosmayati@umt.edu.my

Abstract. The concern in crime prevention to realize a safe environment for public's welfare is one of the central agenda to government authorities. The increasing volume of crime reports with unstructured data that are heavily loaded with text presents the biggest challenge to human experts in analysing the data efficiently as it contains lots of hidden and valuable information such as modus operandi and common crime patterns. The capacity limitations in human information processing demand an effective data analytics model to uncover these hidden and potentially meaningful information from large-scale of data especially in Malay language for improving analysis and supporting decision-making in crime investigation. The objective of this paper is to propose Unstructured Malay Text Analytics Model to support serial crime investigation based on modus operandi patterns. The significance output of this paper is to assist experts especially Royal Malaysia Police in crime investigation based on modus operandi patterns analysis as one of the efforts in reducing rate of crime index.

1. Introduction

The concern in crime prevention to realize a safe environment for public's welfare is one of the central agenda to government authorities in many nations. The government of Malaysia, under a dynamic plan known as Government Transformation Program (GTP) has lined a target to reduce the rate of crime index or reported crime as one of the national key areas [1]. According to Royal Malaysia Police as reported in [2], despite of the decrement in total crime index in Malaysia from year 2009 to 2012, the property offences crimes collectively contribute to approximately 80% to 82% of all reported index crime. The highest rate of crime against properties has generated considerable fear within community, making public perceptions towards safety as a challenge issue that demanding critical attention at both local and national level [2][3][4].

Crime reports are documents detailing a critical information on a crime or a problematic incident to investigators to start an investigation and lead to a number of criminals arrested and cases cleared [5]. The increasing volume of crime reports with unstructured data that are heavily loaded with natural language text presents the biggest challenge to human experts in analyzing the data efficiently as it contains lots of hidden and valuable information such as modus operandi patterns. Modus operandi is the set of habits and characteristics of criminals, as well as patterns of crime operation which become stereotyped [6][7]. Unlike the structured data stored in a database, unstructured data written in free text is difficult to understand by mature database techniques and such rigid manner is insufficient to reveal a great deal of potentially useful information [8][9]. Relying on manual analysis to discover implicit information buried in the vast amount of such crime reports is a tedious, complex, time and labor intensive [5][9][10]. Thus, the capacity limitations in human information processing and database demand an effective data analytic model to uncover these hidden and potentially meaningful information from large-scale and high-dimensional of data especially in Malay language for improving analysis and supporting decision-making in crime investigation.

Several data analytics model when dealing with unstructured text have shown significant potential to support intelligence analysis and decision-making in many domains such as healthcare crime investigation [11], education [12], manufacturing [13], banking [14] and others by integrating proper techniques in natural language processing, text mining and visualization. Even though studies relating to text-based processing for English language are extensive, insufficient studies have been conducted to work on Malay language due to its morphological variants. The Malay language is rich in colloquial, idiomatic expressions and literary allusions. Although the Malay language is widely used within the South East Asia region by more than 200 million people, it has become one of the less resourced languages in the world. Due to this, a limited number of computational linguistic researches were found related to Malay language. In crime investigation itself, there is no analytics tool available for processing Malay text. Thus, unstructured Malay text analytics model is necessary to develop an automatic approach to accurate analyze the inherent structural properties of modus operandi and crime patterns from Malay text crime reports which are essential for both crime investigation and the development of operational strategies to prevent crime.

The remaining of this paper is organized as follows. Section two discusses about the state-of-the-art of crime analysis pattern and the arising issues in the domain. The subsequent section outlines the research methodology framework. Meanwhile, the proposed unstructured Malay text analytics model is presented in the next section. Finally, the last section concludes with a summary and future research directions.

2. Related work

Crime analysis is implemented within law enforcement organizations that referring to a set of systematic and analytical processes for identifying and analyzing patterns and trends in crime data that valuable to police agencies and communities [15][16]. It can be achieved through identifying the relevant information such as modus operandi patterns, hence useful to support criminal investigation and crime prevention. Modus operandi is a unique way a criminal acts during a crime that can be classified into set of habits or behavior of working, spatial and temporal behavior and method of operating or functioning to successfully accomplish a crime [17]. Analyzing and mapping modus operandi patterns from unstructured data such as crime reports and records are important factors in crime investigation when the investigators attempt to find correlation between cases. Series of studies have been done to link spatial and temporal behavior similarity [18][19][20], criminal behavior similarity [21] and crime process [7]. However, such manual approach that requires the involvement of crime experts' domain knowledge is a challenging and time consuming with the increasing volume of unstructured text. In such a way, effective solutions using information technologies and computational approaches are needed to automate the processes.

In the past decade, the exponential increase in the creation of text analytics tools for supporting crime analysis has shown concerted efforts in handling of public-safety systems. With text analytics tools, the massive volume of unstructured crime data can be extracted, visualized and analyzed for further decision-making processes. Text analytics is the process of applying algorithms in order to analyze sets of data and extract useful and unknown patterns, relationships and information [22]. Furthermore, text analytics are used to extract previously unknown useful, valid and hidden patterns and information from large data sets as well as to detect important relationships among the stored variables. Therefore, text analytics have a significant impact on current research development especially in the crime analysis [23]

[8]. However, these text analytics tools were designed for processing English-like data. There are limited number of researches for processing Malay language.

Malay is not only a native language for Malaysia but also one of the languages used in Indonesia, Brunei, Singapore, and southern Thailand. The Malay language possesses its own unique structure and grammar [24]. A limited number of computational linguistic researches that was found on Malay text analytics. There are several studies have been conducted in text summarization [24][25], sentiment analysis [26][27][28], text normalization [29] and information retrieval [30]. Most of these studies suffer from a public Malay corpus that serves as the source of data, which currently distributed and unintegrated.

3. Research methodology framework

The overall research methodology framework is depicted in Figure 1. It consists of several phases, including theoretical study, data collection, identification of modus operandi indicator, extraction of modus operandi and classification and association. These phases contain methodological details that enables the formulation of Malay text analytics model.

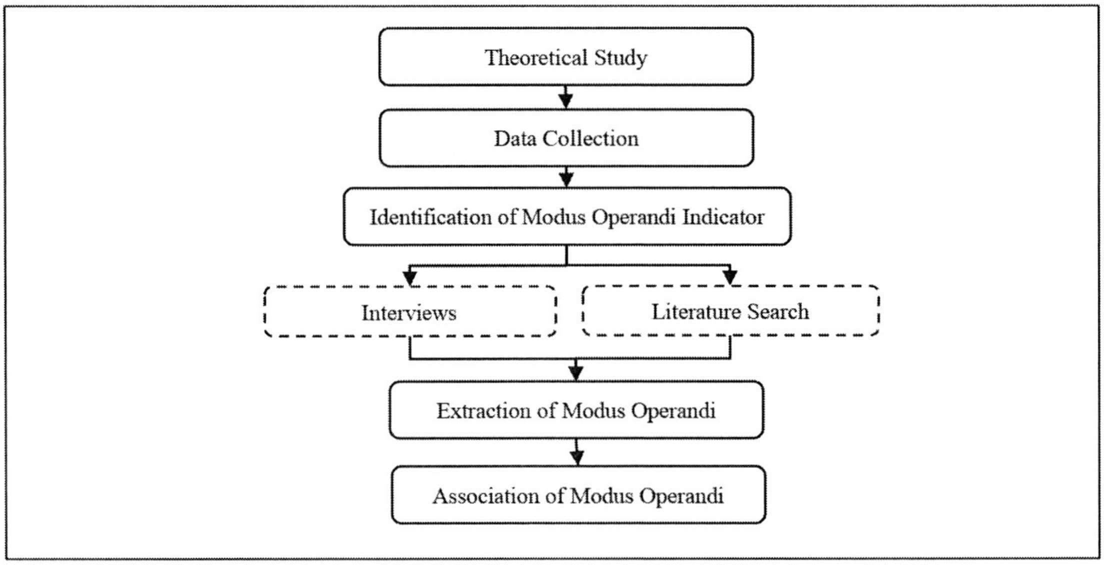

Figure 1. Research Methodology Framework

3.1. Theoretical study

Above of all, literature study is performed as the theoretical study. In this phase, the past and current literatures related to text analytics are collected, comprehended and analysed for identifying the area of concern and problems. Comprehensive literature study on existing text analytics model is conducted to find out about the purposes, gaps and concrete justification of designing unstructured Malay text analytics model.

3.2. Data collection

The data used in this study is collected from real-world house-breaking crime reports of Malay Royal Police Department in between year 2010 to 2013. The corpus contains 100,383 crime reports in Malay language. The details distribution of data is summarized in Table 1. The format of data is originally in excel documents.

Table 1. Summary of house-breaking crime reports dataset

Year	Number of reports
2010	26,777
2011	27,650
2012	24,122
2013	21,834

3.3. Identification of modus operandi indicator
The list of modus operandi indicator is crucial to indicate the relevant information that potentially need to be extracted. Thus, relevant modus operandi indicator need to be identified and constructed as a corpus. There are two methods that have been applied to recognize the potential modus operandi patterns, which are in-formal interview and literature search.

3.3.1. Interview. The main purpose of conducting an extensive interview is to understand the basic modus operandi indicators that are frequently used in crime investigation, which leads to the criminal detection. Therefore, in-depth information is gathered by consulting domain experts of Malaysia Royal Police Department. Potential related key phrases are identified and linked based on the input given by the experts.

3.3.2. Literature search. This method is carried out to identify potential modus operandi indicators in previous researches. Related journals and articles in crime domain are reviewed, focusing on house-breaking crime investigation.

3.4. Extraction of modus operandi
The process of extracting information that potentially relate to modus operandi is done automatically based on the identified modus operandi patterns. Therefore, analysis on relevant information extraction algorithms for Malay named entity recognition, criminal behavior recognition and crime process recognition are done. The analysis is divided into three steps; analysis of data, analysis of processing rules and analysis of required output.

3.4.1. Analysis of data. The collected data is observed in order to interpret its types, format and morphology. In this step, the data is also prepared for pre-processing. Several pre-processing techniques for Malay text such as tokenization, stop word and special characters removal and stemming are studied and tested with the collected data.

3.4.2. Analysis of processing rules. This step involves the analysis on information extraction algorithms for Malay text using rule-based, statistical approach and machine learning. Latent semantic analysis is also studied. All these algorithms are compared, developed and tested in order to identify its strength and weakness. Based on these existing algorithms, an improved algorithm is designed for processing Malay text.

3.4.3. Analysis of required output. The possible output presentation is observed and determined. In this step, the evaluation methods are also identified to measure the accuracy and efficiency.

3.5. Association of modus operandi.
Potential algorithms to associate and link the extracted modus operandi patterns and for the classification of similar modus operandi patterns are studied and analysed. Among potential algorithms are apriori,

logistic regression, support vector machine, k-nearest neighbor, neural network and random forest. Rapid Miner tool will be using to process the extracted data.

4. The proposed unstructured Malay text analytics model

In this section, the proposed unstructured Malay text analytics model is presented. The model as shown in Figure 2 contains four components; pre-processing, modus operandi feature extraction, modus operandi association and classification, and visualization dashboard.

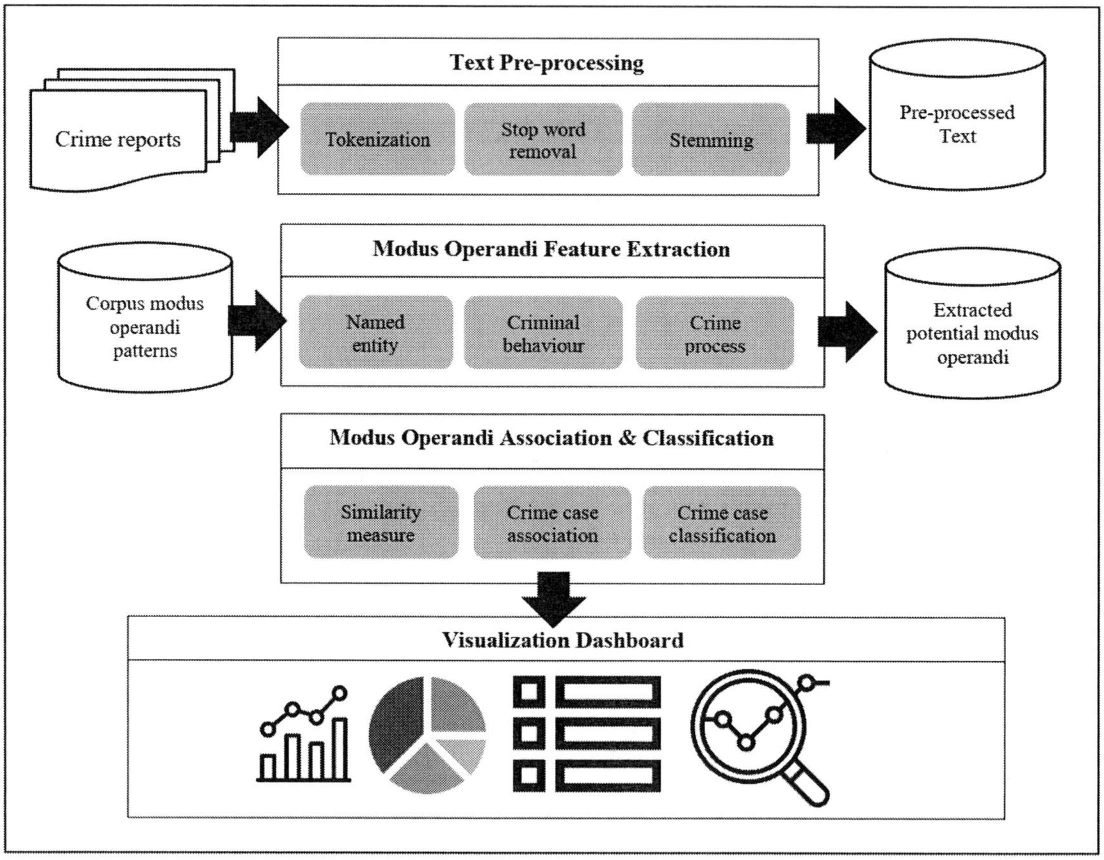

Figure 2. The model of unstructured Malay text analytics in crime.

4.1. Text pre-processing

Text pre-processing is one of the crucial component to clean and transform the text data for quantitative analysis. Among of the multi-step processes involved are tokenization, stop word removal and stemming.

4.2. Modus Operandi feature extraction

List of potential modus operandi is extracted based on the identified modus operandi indicators. Modus operandi features are categorized into named entity, crime behavior and crime process. Named entity aims to recognize nouns that related to names of person, number of criminals, tools used by the criminal, missing properties, spatial and temporal behavior. Meanwhile, criminal behavior denotes on how the criminal acts during the crime such as criminal disguise, the way criminal breaks through the premise, the way criminal acts towards victim and the way criminal rob the property. Crime process is represented by the series of actions performed and the object characteristics involved during crime.

4.3. Modus Operandi Association & Classification
This component aims to extract plausible associations between crimes based on modus operandi patterns, grouping similar crimes based on similar criminal behavior and crime process, eliciting co-criminal chain and potential list of suspects based on spatial-temporal behavioral similarity. Here, machine learning algorithms that are extensively employed for association and classification and fit to the problems are applied.

4.4. Visualization Dashboard
Visualization dashboard is used to display various types of visualization for further analysis in order to assist investigators for better decision-making.

5. Conclusion
Malaysia Ministry of Home Affairs Strategic Plan 2015-2020 and Government Transformation Program (GTP) Roadmap under the National Key Result Area (NKRA) have initiated a plan to reduce the rate of crime index at 5% annually by 2020. Therefore the proposed unstructured Malay text model is expected to provide actionable information for the police management to make a better decision in crime investigation and assist domain experts (Royal Police Department) to develop and implement more effective law enforcement policies and crime prevention programs. Next, further analysis on algorithms for Malay information extraction and algorithms on how to associate the extracted modus operandi to assist decision-making is expected to be done in order for designing an improved algorithm. In addition, prototype of the proposed model is expected to be developed.

Acknowledgement
This research is supported by Ministry of Education Malaysia, under Fundamental Research Grant Scheme (FRGS) with vote number 59541. The authors would also like to acknowledge Royal Police Department Malaysia for their full support to this research.

References
[1] Prime Minister Department, *Government Transformation Programme - The Roadmap - Executive Summary*. Putrajaya, Malaysia: Performance Management and Delivery Unit, Prime Minister Department, 2010.
[2] Y. K. Teh, "The Royal Malaysia Police has got its strategy wrong: Laws alone do not bring down crime rates," *Int. J. Police Sci. Manag.*, vol. 17, no. 1, pp. 9–16, 2015.
[3] M. A. Mohit and H. M. H. Elsawahli, "Crime and Housing in Kuala Lumpur: Taman Melati Terrace Housing," *Asian J. Environ. Stud.*, vol. 2, no. 2, pp. 53–63, 2017.
[4] S. Ishak, "Perceptions of People on Police Efficiency and Crime Prevention in Urban Areas in Malaysia," *Economics*, vol. 4, no. 5, pp. 243–248, 2016.
[5] C.-H. Ku and G. Leroy, "A decision support system: Automated crime report analysis and classification for e-government," *Gov. Inf. Q.*, vol. 31, no. 4, pp. 534–544, 2014.
[6] C. Bannel and D. V Canter, "Linking commercial burglaries by modus operandi: tests using regression and ROC analysis," in *Applications of Geographical Offender Profiling*, D. Canter and D. Youngs, Eds. 2017.
[7] Y.-S. Li and M.-L. Qi, "An approach for understanding offender modus operandi to detect serial robbery crimes," *J. Comput. Sci.*, vol. 36, p. 101024, 2019.
[8] R. N. Zaeem, M. Manoharan, Y. Yang, and K. S. Barber, "Modeling and analysis of identity threat behaviors through text mining of identity theft stories," *Comput. Secur.*, vol. 65, pp. 50–63, 2017.
[9] Y.-H. Tseng, Z.-P. Ho, K.-S. Yang, and C.-C. Chen, "Mining term networks from text collections for crime investigation," *Expert Syst. Appl.*, vol. 39, no. 11, pp. 10082–10090, 2012.
[10] W. Jentner, D. Sacha, F. Stoffel, G. Ellis, L. Zhang, and D. A. Keim, "Making Machine Intelligence Less Scary for Criminal Analysts: Reflections on Designing a Visual Comparative Case Analysis Tool," *Vis. Comput.*, vol., no., pp. 1–17, 2018.

[11] T. S. Yange, H. A. Soriyan, and O. O, "A Schematic View of the Application of Big Data Analytics in Healthcare Crime Investigation," *J. Heal. Informatics Africa*, vol. 4, no. 1, pp. 32–41, 2017.
[12] A. R. Cano, B. Fernández-Manjón, and Á. J. García-Tejedor, "GLAID: Designing a Game Learning Analytics Model to Analyze the Learning Process in Users with Intellectual Disabilities," in *Serious Games, Interaction and Simulation*, C. V. de Carvalho, P. Escudeiro, and A. Coelho, Eds. Springer, Cham, 2017, pp. 45–52.
[13] S.-J. Shin, J. Woo, and S. Rachuri, "Predictive Analytics Model for Power Consumption in Manufacturing," *Procedia CIRP*, vol. 15, pp. 153–158, 2014.
[14] N. Sun, J. G. Morris, J. Xu, X. Zhu, and M. Xie, "iCARE: A framework for Big Data-based Banking Customer Analytics," *IBM J. Res. Dev.*, vol. 58, no. 5/6, pp. 4:1-4:9, 2014.
[15] D. V. Rohini and P. Isakki, "Crime Analysis and Mapping through Online Newspapers: A survey," in *2016 International Conference on Computing Technologies and Intelligent Data Engineering (ICCTIDE'16)*, 2016, pp. 1–4.
[16] J. B. LeBlanc, J. Elder, C. W. Bruce, and R. B. Santos, "Definition and Types of Crime Analysis," 2014.
[17] J. E. Douglas, A. W. Burgess, A. G. Burgess, and R. K. Ressler, *Crime Classification Manual*, Second Edi. Jossey-Bass A Wiley Imprint, 2006.
[18] D. Johnson, "The space/time behaviour of dwelling burglars: Finding near repeat patterns in serial offender data," *Appl. Geogr.*, vol. 41, pp. 139–146, 2013.
[19] L. Markson, J. Woodhams, and J. W. Bond, "Linking Serial Residential Burglary: Comparing the Utility of Modus Operandi Behaviours, Geographical Proximity, and Temporal Proximity," *J. Investig. Psychol. Offender Profiling*, vol. 7, no. 2, pp. 91–107, 2010.
[20] K. J. Bowers and S. D. Johnson, "Who Commits Near Repeats? A Test of the Boost Explanation," *West. Crim. Rev.*, vol. 5, no. 3, pp. 12–24, 2004.
[21] H. Chi, Z. Lin, H. Jin, B. Xu, and M. Qi, "A decision support system for detecting serial crimes," *Knowledge-Based Syst.*, vol. 123, pp. 88–101, 2017.
[22] A. Ittoo, L. M. Nguyen, and A. van den Bosch, "Text analytics in industry: Challenges, desiderata and trends," *Comput. Ind.*, vol. 78, pp. 96–107, 2016.
[23] N. Qazi and B. L. W. Wong, "An interactive human centered data science approach towards crime pattern analysis," *Inf. Process. Manag.*, vol. 56, no. 6, p. 102066, 2019.
[24] N. Zamin and A. Ghani, "A Hybrid Approach for Malay Text Summarizer," in *Proc. Int. Multi-Conference Eng. Technol. Innov*, 2010.
[25] S. Alias, M. G. K. H. Siti Khaotijah, and T. Tan Ping, "A Malay Text Corpus Analysis for Sentence Compression using Patterngrowth Method," *J. Teknol.*, vol. 78, no. 8, pp. 197–206, 2016.
[26] A. Alsaffar and N. Omar, "Integrating a Lexicon Based Approach and K Nearest Neighbour for Malay Sentiment Analysis," *J. Comput. Sci.*, vol. 11, no. 4, 2015.
[27] T. Al-Moslmi, S. Gaber, A. Al-Shabi, M. Albared, and N. Omar, "Feature Selection Methods Effects on Machine Learning Approaches in Malay Sentiment Analysis," in *The Proceeding of 1st ICRIL-International Conference on Innovation in Science and Technology (IICIST 2015)*, 2015, pp. 444–447.
[28] N. I. Zabha, Z. Ayop, S. Anawar, E. Hamid, and Z. Z. Abidin, "Developing Cross-lingual Sentiment Analysis of Malay Twitter Data Using Lexicon-based Approach," *Int. J. Adv. Comput. Sci. Appl.*, vol. 10, no. 1, pp. 346–351, 2019.
[29] M. A. Saloot, N. Idris, and R. Mahmud, "An architecture for Malay Tweet normalization," *Inf. Process. Manag.*, vol. 50, no. 5, pp. 621–633, 2014.
[30] H. M. Hanum, Z. A. Bakar, N. A. Rahman, M. M. Rosli, and N. Musa, "Using Topic Analysis for Querying Halal Information on Malay Documents," *Procedia - Soc. Behav. Sci.*, vol. 121, pp. 214–222, 2014.

IOP Conf. Series: Materials Science and Engineering **769** (2020) 012016 doi:10.1088/1757-899X/769/1/012016

A Survey on Crowdsourcing Awareness In Indonesia Micro Small Medium Enterprises

[1] S.Wulandari [2] M.Rahmah

Faculty of Computing , College of Computing and Applied Sciences
Universiti Malaysia Pahang, Lebuhraya Tun Razak, Kuantan
[1] yani@dinamika.ac.id [2] drrahmah@ump.edu.my

Abstract. Crowdsourcing is one of the trends in e-business nowadays. It gives solutions for any distributed problem by making a call to crowds via the internet. Crowdsourcing process can solve the company resources issues mainly related to man, machine, money, method, material and marketing. In Indonesia, the best example of crowdsourcing is Gojek Company. The main problem of being faced by Micro Small Medium Enterprise (MSME's) is to obtain the business and employee. By implementing the concept of crowdsourcing, it will solve the stated problem. The objective of this research is to survey the MSMEs awareness of crowdsourcing technology as business solutions. Descriptive statistics approach was used as a research methodology. It based on meeting session and a survey using a questionnaire to 102 MSME East Java, Indonesia as the respondents. The results show 88% of MSMEs in Indonesia does not seem aware that crowdsourcing can helps them to solve their company administration problem. Only 10% aware of the existing of crowdsourcing and will help them in fulfilling human resources, business machines, materials, accessing the market and providing business method. In conclusions, crowdsourcing technology should be exposed to all MSMEs in Indonesia. One framework on how to embedded crowdsourcing into their company should be developed. This framework will help them to used crowdsourcing as one of the ways to enhance their company performance and marketing.

1. Introduction

Micro Small Medium Enterprises (MSMEs) in Indonesia, being identify by the experts have facing the problems in enhancing business competitiveness, such as fulfilling human resources, working, management and technological skill, limitations on access to relevant business information, difficulties in obtaining raw materials, high transportation costs, expansion of business and marketing networks[1]–[3]. According to[4], technological innovation in MSMEs is significant to improve their business performance. In modern society, e-business technology can help to foster a complex business task. Complex tasks and utilize business, and human life cannot be accomplished by an individual or a single computer. It needs to be done by collaborations from thousands of workers all around the world with the help of computers. Through online, the crowd can take part in the undertaking of the task in business[5]. Crowdsourcing is one of the trends in e-business nowadays. Crowdsourcing is one of the technological innovations that can help MSMEs to solve problems owned by MSMEs. It can be done through outsourcing resources to the crowds on internet [6], [7]. for Small business with limited capital, Crowdsourcing is the way out to solve a business problem about resources such as man, material, money, and method from the crowd on Internet [8]. Nowadays, more business applies crowdsourcing as a cost-effective alternative compared to employees or suppliers[9]. Crowdsourcing has become an essential aspect of various business models.

Crowdsourcing is an open innovation strategy, represents the actions of companies or organizations to take functions that have been carried out by employees and outsource them to crowd resources on the internet [10]. [11] defines crowdsourcing as a process of making calls on the internet network to explore creative ideas or to solve problems, or evaluate business problems, and allow anyone to send solutions. [12] used the term crowdsourcing as a combination of the words crowd on the internet and the outsourcing process to many people. Crowdsourcing is widely implemented in the process of developing new products [13]. [14] stated that crowdsourcing is the process that consists of making an open online call for a creative idea, or problem-solving, or evaluation or any other type of business issues, and to let anyone submit solutions[15]. Crowdsourcing presents several potential applications, open to future developments which have already implemented in companies or institution[16]. Crowdsourcing not only implement in huge Companies but also been implemented in MSMEs and has been identified as an innovation needed in MSMEs development[17]. Nonetheless, recently, very few studies have shown that open innovation strategies, such as crowdsourcing also exist in MSMEs [18]

This paper discusses the awareness of crowdsourcing technology as solutions among Micro Small Medium Enterprises in Indonesia. This paper aims to get information about the current condition in Indonesian MSMEs, crowdsourcing awareness in Indonesian MSMEs, crowdsourcing infrastructure in Indonesian MSMEs and obstacles adopting crowdsourcing technology to improve MSMEs performance.

2. Related Work

Based on the Law of the [10], Micro Small and Medium Enterprises or MSMEs can be interpreted as a productive business owned by individuals or individual business entities that meet the criteria of Micro Enterprises. Small-scale business is a productive economic enterprise that is independent. It carried out by individuals or business entities that are not subsidiaries or not branches of companies that are owned, controlled, or become part of either directly or indirectly from Medium or Large Businesses. The total net worth or annual sales of the Micro Businesses are shown in table 1:

Table 1: MSMEs SIZE

MSMEs Size	**Criteria**	
	Asset	Income
Micro	Max 50 Million Rupiahs	Max 300 Million Rupiahs
Small	> 50 Million – 500 Million Rupiahs	> 300 Million – 2,5 Billion Rupiahs
Medium	>500 Million – 10 Billion Rupiahs	> 2,5 – 50 Billion Rupiahs

Ministry of Cooperatives and MSMEs classified MSMEs in ten areas, as follows (i) Agriculture, (ii) Mining and Excavation, (iii) Processing Industry, (iv) Electricity, (v) Gas and Water, (vi) Construction, (vii) Hotel, Restaurant and Trade, (viii) Transportation, (ix)Finance, and (x)Services

Crowdsourcing concept can be used in MSMEs to enhance the marketing and problem of the employee. It has been used for various corporate and institutional activities ([11], [12] and has been identified as an innovation needed in MSMEs development. Crowdsourcing is a new method that refers to outsourcing activities carried out by companies to a community or online association where the company offers a gift to the community if it completes several tasks determined by the company [13]. Crowdsourcing as an online, distributed problem-solving and production model that leverages the collective intelligence of online communities to serve specific organizational goals[14]. Online communities, also called crowds, are allowed to respond to crowdsourcing activities promoted by the organization, and they are motivated to respond for a variety of reasons[15]. Figure 1 shows the crowdsourcing process. Crowdsourcing process starts at an organization that has a task that needs to be performed. Then an involvement of community (crowd) that is willing to complete the job voluntarily. This process continues in an online environment that allows the work to take place and the community to interact with the organization, and the last process is the mutual benefit for the organization and the community.

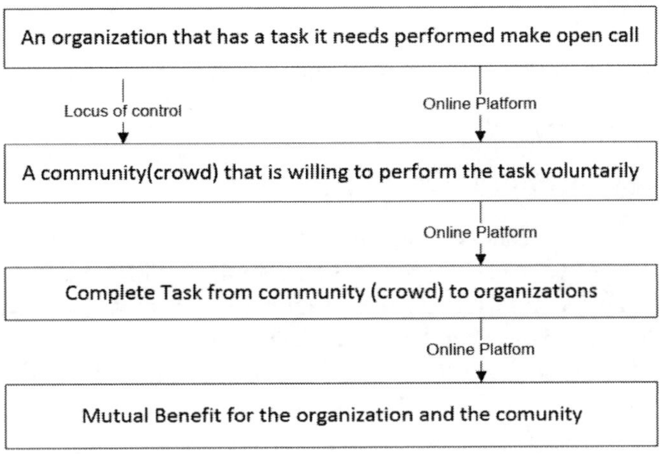

Figure 1. *Crowdsourcing* Model [9]

The crowdsourcing process can be embedded in company management[16]. The Indonesia MSMEs may advise their business to implement and producing new thoughts for authoritative advancement through those stages in the crowdsourcing process. The MSMEs can enhance their company by doing the exercise by bundling, creation or assembling, web improvement, showcasing, item fixing, site advancement, coordination's, preparing, new item advancement, item configuration, publicizing and advancement, thoughts age, critical thinking and more[17][18]. Those exercises are increasingly conservative and cost sparing at whatever point redistribute to outer groups instead of performing inside. Besides, MSMEs additionally could upgrade the imagination and association advancement to get new thoughts for new item advancement or item enhancement through publicly supporting. These may cultivate hierarchical development by tapping outer skill's information and their inventiveness[19]

Besides, MSMEs may offer their items, administrations, aptitudes, skill or even the errands or tasks from their company to the groups in the network. The MSMEs may contract low maintenance crowd workers that perhaps among alumni, ability or a gathering of individuals from networks to finish the assignments. It is because crowd workers in the publicly supporting stage comprise of different novices from low capacity to particular aptitudes. The MSMEs likewise may publicize or tell their company or administrations through web-based social networking or publicly supporting stage to the enormous association or groups to play out the errands and activities. [20]found that the generation procedure of MSMEs are typically work escalated, and such company generally fill in as providers for the substantial assembling company. In this manner, the publicly supporting practice is generally huge to the MSMEs business. It is set as opposed to redistributing to the characterized experts or expert bodies that would be caused additional working expenses to them. In Indonesia, crowdsourcing technology and implementation still low due to limited understanding among MSMEs.

3. Research Methodology

This paper focuses on to find out an awareness of crowdsourcing technology as an e-business strategy in MSMEs. The research was conducted in East Java, Indonesia. East Java Province was chosen because it is the best driver of cooperatives and MSMEs at the national level in 2017. It can be a barometer of innovation and competitiveness of MSMEs in Indonesia. The analytical method used in this study is a descriptive statistical data method. Primary data was obtained through interviews and questionnaire to 102 respondents. The secondary data in this study was obtained from MSMEs Data in

IOP Conf. Series: Materials Science and Engineering **769** (2020) 012016 doi:10.1088/1757-899X/769/1/012016

East Java province, purposely to get a better understanding of East Java MSMEs, taken from [21]. Questionnaire has been constructed based on the crowdsourcing model, which must contain:

i. Company (MSMEs) that have a task to be performed

ii. The crowd willing to performed the task

iii. Mutual benefit for both company and the crowd

iv. Infrastructure to make an open call

The questionnaire's constructs were then are tested to 20 MSMEs owner or managers, and being refined to ensure relevance and comprehension in the MSMEs environment. The questionnaire gathered responses using 5-point Likert scales. A numerical score was associated with each response, and this reflected the degree of attitudinal favorableness, with 'strongly disagree' associated with number '1' on the scale and 'strongly agree' associated with number '5'. The survey also consisted of a combination of open-ended and closed questions. The small amount of open-ended questions invited free comments where it was not always possible to predict the range of responses to a particular [22]The use of closed questions served to generate and gather information quickly by the researcher[23]The following research questions were addressed in this study:

RQ 1: Do the MSMEs have a task or problem to be performed through an open call?

RQ 2: What Kind of MSMEs task to be performed?

RQ 3: Do MSMEs need cheap and effective solutions from internet to solve their problem?

RQ 4: Do MSMEs aware about crowdsourcing?

RQ 5: Do MSMEs have infrastructure to make an open call via the internet?

4. Result

The survey provided 102 MSMEs respondents which have established in minimum three years. They are located in 38 cities or regency in East Java province, Indonesia. The details of the MSMEs category are stated in Figure 2 which, 32% are micro-MMSMEs that have assets below 50 million rupiahs, 51% are small-scale MMSMEs that have assets of 500 million rupiahs and 17 % small MMSMEs that have assets below 5 Billion Rupiah.

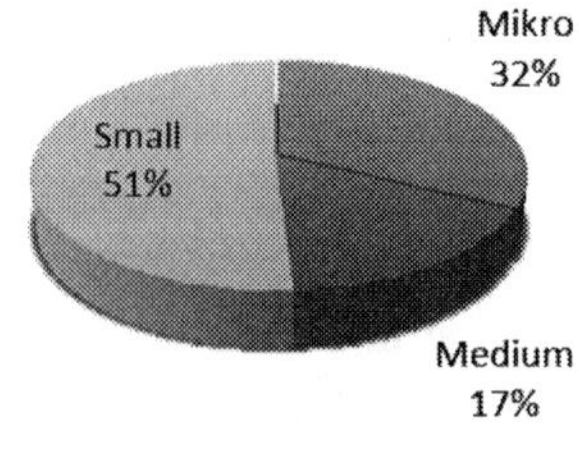

Figure 2 : Respondent Profile by MSMEs Size

Crowdsourcing implementation from the view of MSMEs Owner or Chief Executive Officers (CEO) is shown in Figure 3. General Manager or managers approve that, the publicly of supporting practices and its advantages. With a total of 75% MSMEs, 24% by the manager and the last 1% by the general manager. It can be concluded that this questionnaire was filled out by people who knew the actual condition of MSMEs. MSMEs business sectors that were surveyed were traded with a total of 61% MSMEs, restaurants including food and beverage with 33% and Hotels as number 6%.

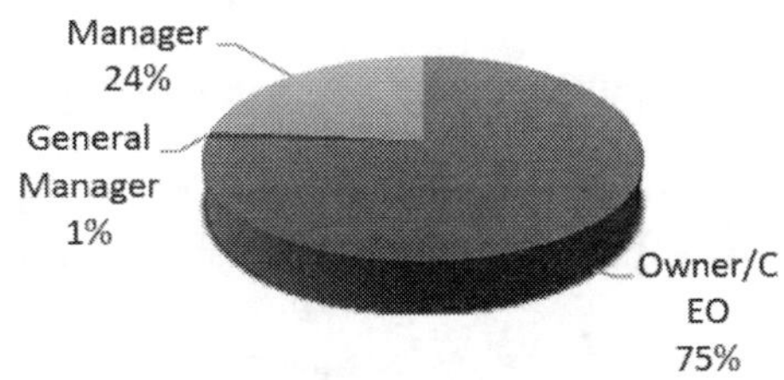

Figure 3 : Respondent Profile by Positions

Base on Figure 4, 51% MSMEs still run their businesses in real or require physical businesses such as houses, shop houses, workshops, outlets, showrooms, offices, hotels, etc. In contrast to businesses that are run offline, only 3% MSMEs run their businesses online. This number is still minimal compared to the number of internet users in East Java which reached 28 million people. While 46% of MSMEs of businesses runs offline, but also have online as a complement to social media, marketplace, and websites.

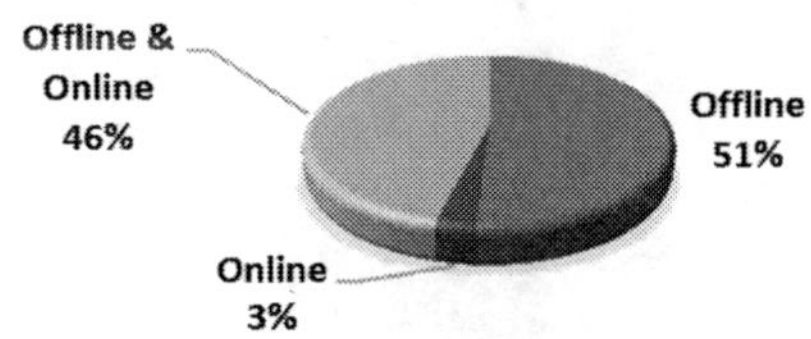

Figure 4 How MSMEs Doing Their Business

Out of 102 MSMEs surveyed, only 2% of MSMEs that had international coverage, 14% of Indonesia coverage is, 21% of East Java provinces, a local city 28% and near their resident is 35%. Current conditions of MSMEs' profit in the current year showed 71% in stable condition, while those with sales conditions rose up by 26%. MSMEs with sales conditions fell by 22% MSMEs while 2 MSMEs 2% did not have a sales record, so they did not know the status of their business was stable, up or down. These results show that Indonesia still far away from other countries of using e-business in MSMEs.

4.1 RQ 1: Do the MSMEs have a task or problem to be performed through an open call?

Based on the survey to 102 MSMEs, the result shows that 100% MSMEs' answer that the company have a task to be performed. Based on[13], we can conclude that MSMEs company need varying knowledge, heterogeneity and number of the task of complexity and modularity to be completed. In order to be called as crowdsources.

4.2 RQ 2: What Kind of MSMEs task to be performed?

The answer from the survey also listed tasks to be performed in MSMEs, as shown in table 2.

Table 2: Survey on SME's Experiences Hindrance Performing Task 2018

No	Task	Result	Current Solution
1.	Fulfilling Human Resources	100 %	Word of mouth Job Vacancies
2.	Fulfilling Materials/Goods/Services Production Materials	97 %	Buy Raw Materials/Goods/Services Production Materials

3.	Fulfilling Machines/Transportation	11 %	Buy devices / business machines
4.	Fulfilling Business Capital	100%	Bank Loan, Own Assets
5.	Acessing the Market	100%	Word of mouth marketing, Social Media, Internet
6.	Fulfilling Providing Solutions/Method	95%	Looking for expert help

The table shows that Human Resources is very crucial tasks to MSMEs. These problem can be solved by implementing crowdsourcing to fulfil the lack of manpower of their company.

4.3 RQ 3: Do MSMEs need cheap and effectiveness solutions from internet to solve their problem?
Based on the results, the data shows that 99% of MSMEs need cheap and effective solutions from internet to solve their problem. They also want to expand the Crowdsourcing Platform to improve their businesses and only 1% do not want it. Crowdsourcing able to get cheap and practical solutions because the crowd voluntarily bring their resources to answer crowdsources call.

4.4 RQ 4: Do MSMEs aware about crowdsourcing?
The survey shows that 88% of the respondent does not know about crowdsourcing and only 5% aware of the existing of crowdsourcing as seen in Figure 5.

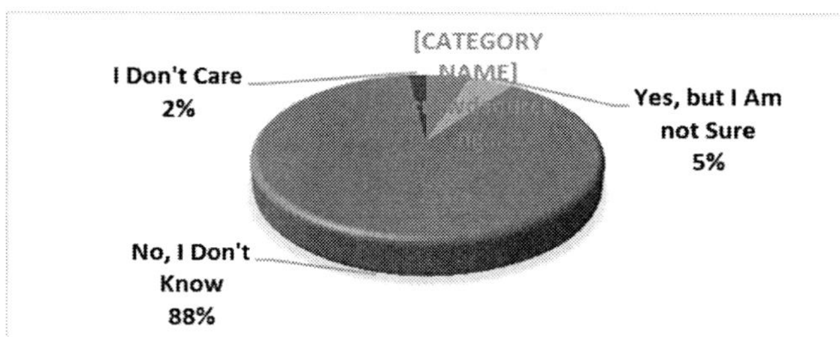

Figure 5 Crowdsourcing Awareness

These result show that the awareness of crowdsourcing is still low among Indonesian MSMEs.

Table 3 shows the percentage of crowdsourcing usage experience based on their company tasks.

Table 3: Survey on MSMEs' who have experience using crowdsourcing platform 2018

No	Task	Result
1.	Experiences using crowdsourcing platform to solve human resources task	20 %
2.	Experiences using crowdsourcing platform to solve Materials/Goods/Services Production Materials task	20 %
3.	Experiences using crowdsourcing platform to solve devices/business machines/Transportation task	33 %
4.	Experiences using crowdsourcing platform to solve capital	5 %
5.	Experiences using crowdsourcing platform to solve Accessing the Market	66 %
6.	Experiences using crowdsourcing platform to collect creative solutions	0%

From 12 companies that had an experience using crowdsourcing, 66% has implemented it for marketing purposes. These shows that the entire crowdsourcing platform is not being used in their company.

4.5 RQ 5: Do MSMEs have the infrastructure to make an open call via the internet?
The last crowdsourcing initiative is an infrastructure to make an open call (Brabham, 2003). So, the survey has monitored the penetration of computer and internet technology in MSMEs as a basic need

for adopting Crowdsourcing in the future. The result shows 92.36% of MSMEs already have IT devices in the form of smartphones, laptops, computers, and tablets, and 91% connected to the internet, as shown in figures 6 and 7.

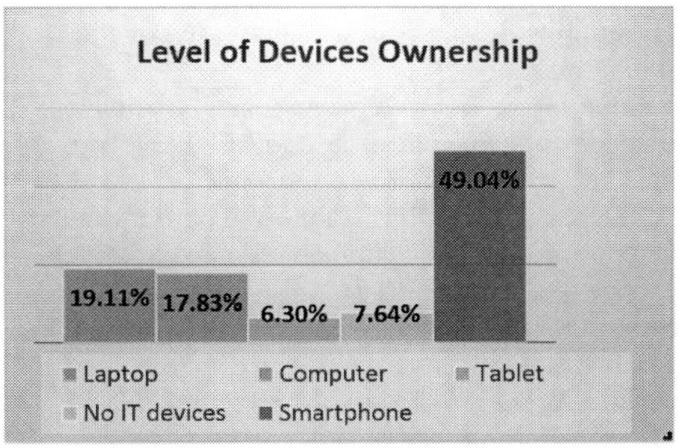

Figures 6 : Level of devices ownership

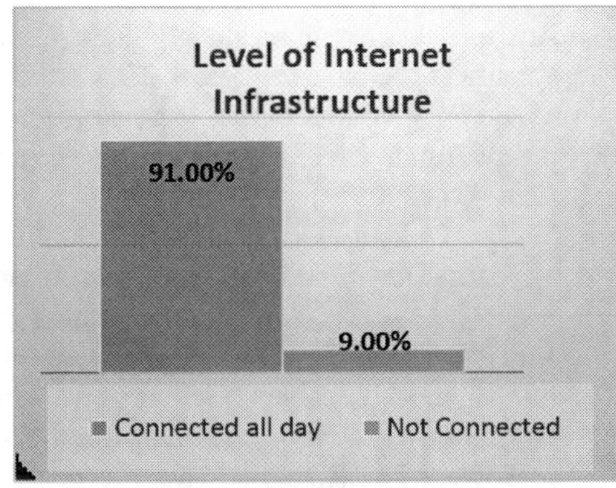

Figure 7 Level of internet infrastructures

These shows that their infrastructure is ready to implement the crowdsourcing tasks in the future.

5. Discussion and Conclusions

Our study indicates that there is low awareness of Crowdsourcing among Micro Small Medium Enterprise (MSMEs) in Indonesia. These is due to no exposure to them about the usage of Crowdsourcing and minimal knowledge about e-business technology. Although existing device and internet infrastructure can support the crowdsourcing implementation. Unfortunately it only for inadequate company administration and the usage of crowdsourcing by the experience MSMEs is limited to marketing purposes.

As for recommendations, before MSMEs be migrate to the e-business Crowdsourcing model, a more in-depth investigation needs to be carried out to gain more information for combining business model and crowdsourcing model. In future, one framework on how to embedded crowdsourcing into MSMEs company will be developed. This framework will helps them to used crowdsourcing as one of the way to solve their company problem and enhancing their performance and marketing.

Acknowledgement
This research work is supported by Universiti Malaysia Pahang (UMP) under FRGS/1/2018/ICT04/UMP/02/1 and RDU190112 grant. The author would like to express gratitude and heartiest thanks to the Faculty of Computing and all of MSMEs involves in this research.

References

[1] Asian Development Bank, "Asian Development Outlook 2018: How Technology Affects Jobs," 2018.
[2] T. H. Capital, "Survey of Entrepreneurs and Msmes in," 2018.
[3] D. C. Lantu, M. S. Triady, A. F. Utami, and A. Ghazali, "Pengembangan Model Peningkatan Daya Saing UMKM di Indonesia: Validasi Kuantitatif Model," *J. Manaj. Teknol.*, vol. 15, no. 1, pp. 77–93, 2016.
[4] M. A. Bek, N. N. Bek, M. Y. Sheresheva, and W. J. Johnston, "Perspectives of SME

innovation clusters development in Russia," *J. Bus. Ind. Mark.*, vol. 28, no. 3, pp. 240–259, 2013.

[5] J. Wortman Vaughan, "Making Better Use of the Crowd: How Crowdsourcing Can Advance Machine Learning Research," *J. Mach. Learn. Res.*, vol. 18, pp. 1–46, 2018.

[6] Roslina, AH & Shahirah, MF. (2019). Implementing white hat search engine technique in ebusiness website, 311–314, 10.1145/3306500.3306533

[7] M. de R. Van de Vrande, Vareska., Jeroen P. J. De Jong., Wim Vanhaverbeke., "Open innovation in SMEs: Trends, motives and management challenges. Technovation, 29," pp. 423–437, 2009.

[8] S. Qin, D. Van Der Velde, E. Chatzakis, T. McStea, and N. Smith, "Exploring barriers and opportunities in adopting crowdsourcing based new product development in manufacturing SMEs," *Chinese J. Mech. Eng.*, vol. 29, no. 6, pp. 1052–1066, 2016.

[9] J. Prpić, "John Prpić -Next Generation Crowdsourcing for Collective Intelligence Collective Intelligence 2016 Next Generation Crowdsourcing for Collective Intelligence," 2016.

[10] R. Indonesia, *Undang Undang Republik Indonesia Nomor 20*, no. 1. 2008.

[11] J. Howe, "The Rise of Crowdsourcing," no. 14, pp. 1–5, Jun. 2006.

[12] L. & N. S. Huston, "Implementing Open Innovation," *Technol. Manag.*, pp. 50:2, 21–25, 2015.

[13] D. C. Brabham, *Crowdsourcing*, vol. 6, no. 2. Masachusetts: MIT Press, 2013.

[14] D. C. Brabham, "Crowdsourcing As a Model for Problem Solving: Leveraging the Collective Intelligence of Online Communities for Public Good," *Int. J. Res. into New Media Technol.*, vol. 14, no. 1, pp. 75–90, 2008.

[15] M. Vuković, "Crowdsourcing for enterprises," *Serv. 2009 - 5th 2009 World Congr. Serv.*, no. PART 1, pp. 686–692, 2009.

[16] M. Weiss, "Crowdsourcing Literature Reviews in New Domains," *Technol. Innov. Manag. Rev.*, vol. 6, no. 2, pp. 5–14, 2016.

[17] M. F. Mansor, H. A. Halim, and N. H. Ahmad, "Leveraging crowdsourcing practices in Small and Medium enterprises (SMEs)," vol. 21, no. 4, pp. 1–11, 2018.

[18] X. J. Niu, S. F. Qin, J. Vines, R. Wong, and H. Lu, "Key Crowdsourcing Technologies for Product Design and Development," *Int. J. Autom. Comput.*, vol. 16, no. 1, pp. 1–15, 2019.

[19] J. Eke Lawrence, "the Use of the Internet in Small To Medium-Sized Enterprises," no. June, 2002.

[20] M. Hossain, "A review of literature on open innovation in small and medium-sized enterprises," *J. Glob. Entrep. Res.*, vol. 5, no. 1, p. 6, 2015.

[21] D. K. dan U. K. M. P. J. Timur, "Data UKM," *2018*, 2018. [Online]. Available: http//www.diskopukm.jatimprov.go.id/. [Accessed: 30-May-2018].

[22] R. B. Frary, "Hints for designing effective questionnaires, Practical Assessment, Research & Evaluation, 5, (3), 1-6. Gartner," 1996.

[23] T. Boynton, P.M and Greenhalgh, "Selecting, designing, and developing your Questionnaire," 2004.

A Model Of Biodiversity Taxonomy Database (iTaxo) For Managing And Analysing Kenyir Landscape Biodiversity Data

Salwana Mohamad @ Asmara[1], Mustafa Man[2], Mohd Tajuddin Abdullah[3]

[1]Faculty of Computing, Universiti Malaysia Pahang, Pahang, Malaysia
[2]School of Informatics and Applied Mathematics Universiti Malaysia Terengganu, Terengganu, Malaysia
[3]Institute of Tropical Biodiversity and Sustainable Development, Universiti Malaysia Terengganu, Terengganu, Malaysia

Email: salwanamohamad@ump.edu.my

Abstract. Biodiversity data management and analysis is a complex task that must be handled by biodiversity researchers. The large amount of data and complicated interaction of the data are the main problem in the biodiversity field. Data accessibility also becomes the main issue to create a data sharing environment among other biodiversity researchers. This is important element since the data is related to each other. A standardized form of data model for biodiversity data is not well implemented in real world. This study discussed about the database model for biodiversity taxonomy data (iTaxo) in order to manage the biodiversity data. We implement the Kenyir landscape which is located at the state of Terengganu in Malaysia as our case study data since the landscape is rich with biodiversity species and ecosystem. As a result, the model can be applied to manage, analyse and evaluate the biodiversity status at the landscape for conservation effort.

1. Introduction

Biodiversity is a composition of species and genetic variability of plants, animals, and microorganisms [1], [2]. The composition includes entities identification at the level of ecosystem, species and genetic diversity. The interactions between them are known as ecosystems and ecological processes. This means that biodiversity data can be assessed at three different levels which are ecosystem, species and genetic diversity. The data of biodiversity can be categorised into three groups which are taxonomy, geo-spatial and temporal data [3]. The effort to analyse the biodiversity in the context of ecosystem and functional aspects is the main challenge of this study area [4]. In order to perform the biodiversity analysis, the diversity of data must be mapped appropriately by the database system. The data should be analysed and represented in historiographical form in order to contribute the valuable information and understanding of policy making in biodiversity [5].

There are two main factors that lead to difficulty for researchers in biodiversity study to manage the biodiversity data. The factors are the complicated relationship of the biodiversity data and large amounts of the data. The complicated relationship in biodiversity data is caused by the lack of standardised and acceptable relationship of names, taxonomic concept, clades and globally unique identifiers (GUIDs) to define and build the relationships of these entities [6]. In terms of large amount of data, Porter [7] suggests the database architecture for taxonomic dataset to be designed with practicality for 'wide' database instead of 'deep' database. This is because of taxonomic dataset contain many different kinds of data and observation which is typically do not have tools for analysis and the tools function mostly on metadata content.

Besides the difficulty to manage the biodiversity data, the accessibility of the data also become the main issue in biodiversity research. The data accessibility effected the data sharing environment and decision making [8]. Data sharing is the ability to share the same data resource to others and reuse the data for analysis [9]. By sharing the data, every researcher should aware what data that have been discovered by other researchers in order to enable the data comparison and prevent the redundant research work. The data storage for the sharing mechanism must be persistent and sustainable for long term use.

The standard of data model for the storage and management of the biodiversity data is hard to find as it is not well implemented as a practical database [5]. This is because of the incompatibility between the logical view of data model design and the application view from the real scenario of end user. The complexity of the logical design has led to the impracticality of database model for the biodiversity research [10].

This work describes about the database model of iTaxo with reference to a case study in the Kenyir landscape. The landscape situated at the state of Terengganu in Malaysia. Tasik Kenyir covered by more than 340 islands spread out in water catchments area of 38000 hectares. Therefore, Tasik Kenyir is rich with biodiversity sources that cover the terrestrial and aquatic area. The objective of this database model is to create a platform of an integrated directory for all known species that will be updated by all biodiversity researchers especially in Tasik Kenyir landscape. While other biological taxonomy databases have focused their content on collecting and store the data, this proposed database model aims to emphasis on providing the analysis of biodiversity and geospatial data. The geospatial data management of the biodiversity data by implementing in-field data collection may contribute to the species distribution mapping, data analysis and modelling. It may become as a resource for studies on biodiversity that will contribute to the decision making in conservation effort.

2. Related Works

There are many projects that applied methods to manage and analyse biodiversity data worldwide. Malaysia Biological Diversity Database (MyCHM) **[11]** that has been developed by Forest Research Institute Malaysia (FRIM) is a mechanism for exchanging and integrating information on biodiversity. In 2015, MyCHM has been improved by Malaysia Biodiversity Information System (MyBIS) that purposes to be a centralized repository for biodiversity information in Malaysia. Various agencies from the area of environmental, forestry, wildlife and public institutions are the data provider for this system. However, MyBIS does not specify the class of species in the analysis of the status of biodiversity. In analysis map menu, MyBIS has provided the visual map of protected area such as forest reserves and sanctuaries in Malaysia but it does not emphasized the analysis of geospatial data of found species.

Global Biodiversity Information Facility (GBIF) [12] is an international open data infrastructure that allows the public to access data about all types of life on Earth. It provides a centralized access to an enormous number of biodiversity data. The project is also assisting institutions to publish data according to common standards. It also provides a map Application Programming Interface (API). The mapping API is a web map tile service to visualize GBIF data on interactive maps and overlay data from other sources in a simple way.

Integrated Taxonomic Information System (ITIS) [13] is an open access database with information on species names. Every species is described by hierarchical classification with standardized nomenclature. It contains documented taxonomic information of flora and fauna from both aquatic and terrestrial habitats. The system is merely functioning to store and manage the biodiversity data that will allow biological information sharing among researchers and related agencies. Taxonomic Database Working Group (TDWG) [14] are efforts to establish standards and infrastructure for integration and interoperability of data from biological collections. It is also an option of biodiversity applications deals with the management of taxonomic information and geographic distribution of species.

A database system for United Kingdom (UK) biodiversity data has been developed as National Biodiversity Network's Gateway (NBN) [15]. The system is a platform for researchers and public to learn and understand the UK biodiversity data by allowing users to record, share, explore and use the data. The accuracy of species identification is important to maintain the quality of the biological data. Thus, NBN has carried out a verification and validation of the data by expert volunteer verifiers. Users also can explore distributions of species records, whole datasets, protected sites and habitats using the interactive map tool. Despite the complete information that is available in the system, it is lack of statistical analysis for species assessment and the trend of the biodiversity status.

Most of the projects are objectively developed to collect, store and manage biodiversity data. The data standards have been applied to enable the accessibility and distribution of information through the web. iTaxo may have the same objective of assisting biodiversity researchers in managing their data about the found species. However, its main contribution is focusing on Kenyir landscape by detailing the hierarchy of the species taxonomy and integrating several analysis and visualization element in the database. Also, we allowed different access of users depend on the user roles that can be applied for Kenyir landscape researchers, student and public.

3. Database Model for iTaxo

This data model aims to describe the representation of the biodiversity data. The database model is designed by considering the taxonomic classification as in Figure.1.

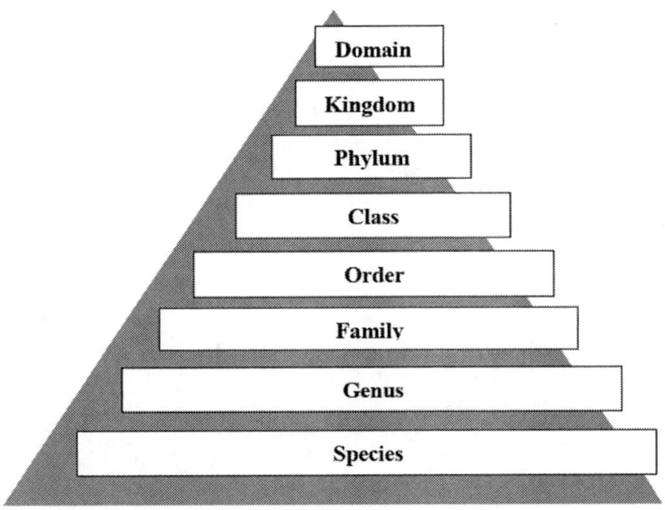

Figure 1. Biological taxonomy

The hierarchical biology taxonomy presented within this research is the basis for the relational database model. Biological taxonomy is the science of classification according to a pre-determined system of the biodiversity [16]. It includes the taxonomic group of Domain, Kingdom, Phylum, Class, Order, Family, Genus and Species. A taxon or a group of organisms are given name and a rank of taxonomy hierarchy. With the science of taxonomy, taxonomist will defines what belong or does not belong to a taxonomic group. In biological taxonomy, informatics play an important role to develop databases as the repositories for biological information [17]. This effort can integrates the numerous of biological data that is distributed across the various disciplines of biology.

3.1. Data Structure of iTaxo

The data structure of database model for iTaxo is defined by applying the notation of set theory which may contain repeating elements. It is a basic representation of data structure to develop list theoretic relational databases and operations on them. The data structure contains twelve sets of data entities which are Species, Genus, Family, Order, Class, Phylum, Kingdom, Domain, Wildlife_Status, Type_of_Forest, IUCN_Status and Location. The sets and the elements are notated in Table 1 and Table 2:

Table 1. Sets notation

Entities	Set Notation
Species	A
Genus	B
Family	C
Order	D
Class	E
Phylum	F
Kingdom	G
Domain	H
Wildlife_Status	I
Type_of_Forest	J
IUCN_Status	K
Location	L

Table 2. Sets elements

Set Elements	Notation
ID_Species	a_1
Name_Species	a_2
references	a_3
notes	a_4
photo	a_5
ID_Genus	b_1
Name_Genus	b_2
ID_Family	c_1
Name_Family	c_2
ID_Order	d_1
Name_Order	d_2
ID_Class	e_1
Name_Class	e_2
ID_Phylum	f_1
Name_Phylum	f_2
ID_Kingdom	g_1
Name_Kingdom	g_2
ID_Domain	h_1
Name_Domain	h_2
ID_Location	i_1
Location	i_2
GPS_Coordinate	i_3
Year	i_4
ID_Wildlife	j_1

Wildlife_Status	j_2
ID_Type	k_1
Name_Type_of_For est	k_2
ID_IUCN	l_1
IUCN_Status	l_2

The intersection and union among the sets of data entities represent the relationship of the database model. Every set share some of the same elements to relate each other is denoted as:

i. $A \cap B = \{b_1\}$
ii. $B \cap C = \{c_1\}$
iii. $C \cap D = \{d_1\}$
iv. $D \cap E = \{e_1\}$
v. $E \cap F = \{f_1\}$
vi. $G \cap H = \{h_1\}$
vii. $I \cap A = \{j_1\}$
viii. $J \cap A = \{k_1\}$
ix. $K \cap A = \{l_1\}$
x. $L \cap A = \{i_1\}$

The set of Species share the same elements of ID_Genus, ID_Type, ID_Wildlife, ID_IUCN, ID_Location from sets of Genus, Type of Forest, Wildlife Status, IUCN Status and Location is denoted as:

$$A \cup B \cup L \cup I \cup J \cup K = \{b_1, i_1, j_1, k_1, l_1\}$$

Every shared element is an identifier of the values from other set or entities that will describe the overall elements or attributes of the related set.

3.2. Entity-relationship diagram of iTaxo

The entities and relationships represented in the taxonomy must be evaluated to design the database model. Figure 2 shows the entity-relationship diagram of the taxonomy database.

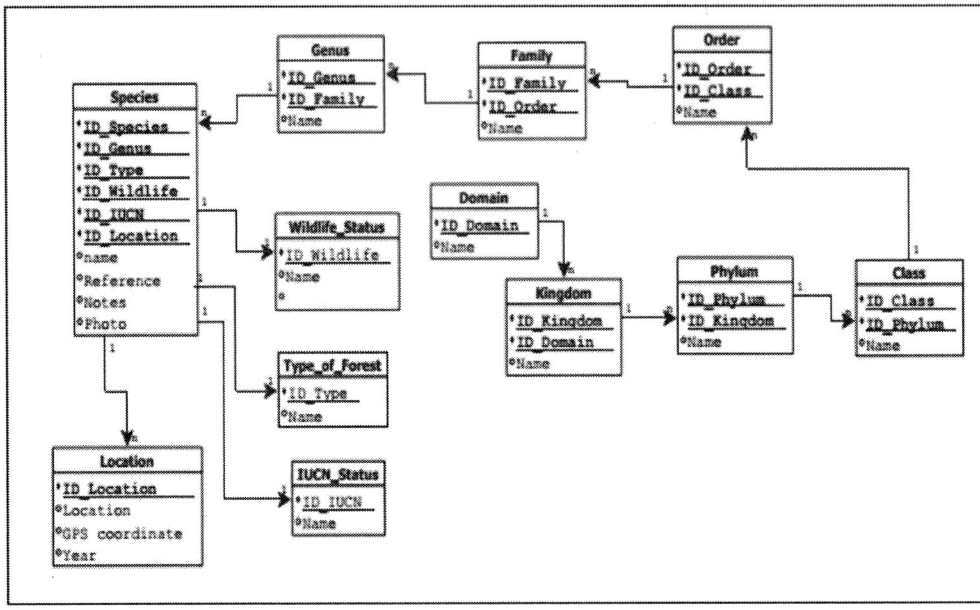

Figure 2. Entity-relationship diagram of taxonomy database

The main entities are defined based on the biological taxonomy hierarchy which also create the main data tables which are Domain, Kingdom, Phylum, Class, Order, Family, Genus and Species. Entities of Wildlife_Status, Type_of_Forest and IUCN_Status are defined to describe the species status and the habitat. An entity of Location is created to link the geospatial information to every species found.

The entities relationship of the taxonomy hierarchy has been designed by defining the cardinality of the relationship as one to many (1:n). For example, one type of Genus can have many type of Species. The hierarchical structure of the taxonomy identified the parent child relationship amongst the entities. The eight domains and subcategories in the taxonomy will function as parent and child entities. Every entities of the taxonomy hierarchy are linked by defining the foreign key (FK) of the child entities that comes from the primary key of the parent entities. For example, every Species are related to their Genus by referring to ID_Genus which is a primary key (PK) of Genus as shows in Table 3 and Table 4. The ID_Genus is defined as the foreign key for Species entity.

Table 3. Fields of species data table

Column	Data type	Description
ID_Species (PK)	Int	Primary key
ID_Genus (FK)	Int	Link to Genus type
ID_Type (FK)	Int	Link to type of forest
ID_Wildlife (FK)	Int	Link to Wildlife Act status
ID_IUCN (FK)	Int	Link to IUCN Redlist status

IOP Conf. Series: Materials Science and Engineering **769** (2020) 012017 doi:10.1088/1757-899X/769/1/012017

ID_Location (FK)	Int	Link to location information
Name	Varchar	Full name of the species
Reference	Varchar	Reference of the species
Note	Varchar	Additional note of the species

Table 4. Fields of genus data table

Column	Data type	Description
ID_Genus	Int	Primary key
ID_Family	Int	Link to Family type
Name	Varchar	Full name of the genus

Beside the taxonomy hierarchy, the foreign key concept is also used to link the species information to other entities such as Wildlife_Status, Type_of_Forest, IUCN_Status and Location. For example, Table 5 and Table 6 show the data table of Wildlife_Status and Location.

Table 5. Field of wildlife_status data table

Column	Data type	Description
ID_Type	Int	Primary key
Name	Varchar	Full name of the genus

Table 6. Field of location data table

Column	Data type	Description
ID_Type	Int	Primary key
Location	Varchar	Full name of the location
GPS coordinate	Varchar	GPS coordinate of the found species
Year	Int	Discovered year of the species

4. Result and Discussion

4.1 Taxonomy Information

In the taxonomy information interface, user enable to insert, delete, update and view the taxonomy information of a specimen. Based on the database structure, the taxonomy hierarchy have been defined as a separated entities. The foreign key concept in data attributes is used to relate each of the hierarchy by sharing the unique ID of primary key. Every species may share the same hierarchy of taxonomy with other species. Therefore, we have designed a separate interface to create the flexible environment for storing and managing the taxonomy information. Figure 3 shows the insert new kingdom taxonomy interface.

Figure 3. Insert new kingdom taxonomy interface

4.2 Species Information

The species information can be inserted through this interface by allowing the user to choose the type of taxonomy hierarchy. Every taxonomy hierarchy that have been created in the taxonomy information interface will be displayed on drop down list as shows in Figure 4.

IOP Conf. Series: Materials Science and Engineering **769** (2020) 012017 doi:10.1088/1757-899X/769/1/012017

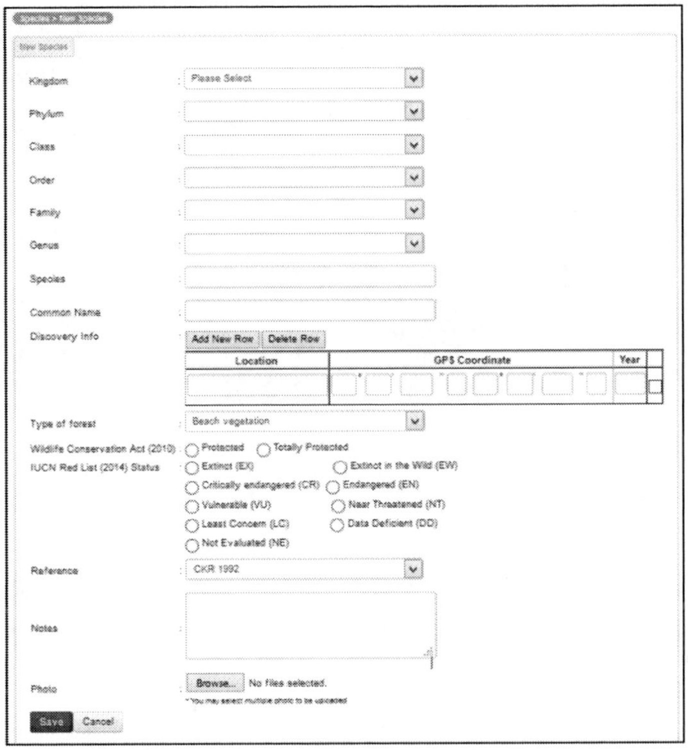

Figure 4. Species Information Interface

The discovery information will allow user to put the location, GPS coordinate and year of the found species. The habitat of the species can be specified by choosing the type forest. The status of the species are categorised by using two types of standard which are Wildlife Conservation Act (2010) status and IUCN Red List status. Photos of the species also can be uploaded in this interface.

The data will be displayed by categorised it into five types of information of the species which are taxonomy and nomenclature, taxonomy hierarchy, discovery information, status and others. Figure 5 shows the data view of taxonomic biodiversity data.

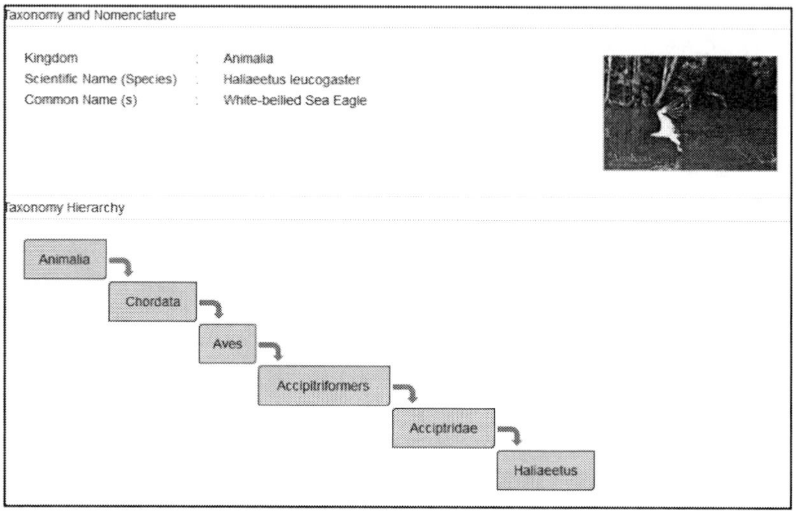

Figure 5. Data view of taxonomy biodiversity data

5. Conclusion

The biological taxonomy developed in this study has been implemented in a database designed for taxonomy biodiversity data of Kenyir landscape. We present a relational database model of iTaxo that combine the field data of taxonomic nomenclature, spatial and temporal data. The primary use of iTaxo is to provide a data repository for users in biodiversity research and public as well. Every user has different authorization of the system access based on their status as a researcher, system administrator and public. The database will have to be updated as new data is validated. The taxonomy and database from this study can promote standardization in the design of taxonomies across a range of biodiversity research.

The outcome of iTaxo is the accessibility of the integrated biodiversity data that will enable data sharing among the researchers. Data retrieval for following analyses can be done by organising primary datasets on biodiversity in a consistent and flexible database. Based on the analyses, it may also supporting decision making among the related users in terms of conservation effort of nature and policy making. Also, the informative knowledge of the database will educate the public about the status of current biodiversity sources that can promote the awareness of preserving our nature ecosystem. The main part of assessing and understanding the status of biodiversity in Kenyir landscape can be done by combining the large amount of biodiversity data in the context of the status of new found species, species threatened and species distribution.

Acknowledgment

Appreciation conveyed to Universiti Malaysia Pahang RDU180363 grant and Kenyir Geopark Development Fund grant.

References

[1] S. Pagiola, J. Kellenberg, L. Vidaeus, and J. Srivastava, "Mainstreaming biodiversity in agricultural development," *Financ. Dev.*, vol. 35, pp. 38–41, 1998.

[2] C. R. Margules, R. L. Pressey, and P. H. Williams, "Representing biodiversity: data and procedures for identifying priority areas for conservation," *J. Biosci.*, vol. 27, no. 4, pp. 309–326, 2002.

[3] A. Z. M. Kamruzzaman, H. Selamat, and M. T. Wahid, "Conceptual Design of Biodiversity Data Model (BiDaM) using Object Relational and Event Based Approach," in *Proceedings of the Postgraduate Annual Research Seminar*, 2005, pp. 77–81.

[4] T. Lotz, J. Nieschulze, J. Bendix, M. Dobbermann, and B. König-Ries, "Diverse or uniform? — Intercomparison of two major German project databases for interdisciplinary collaborative functional biodiversity research," *Ecol. Inform.*, vol. 8, pp. 10–19, Mar. 2012.
[5] G. C. Bowker, "Biodiversity datadiversity," *Soc. Stud. Sci.*, vol. 30, no. 5, pp. 643–683, 2000.
[6] C. S. Parr, R. Guralnick, N. Cellinese, and R. D. M. Page, "Evolutionary informatics: unifying knowledge about the diversity of life," *Trends Ecol. Evol.*, vol. 27, no. 2, pp. 94–103, 2012.
[7] J. H. Porter, "Scientific Databases," in *Ecological Data: Design, Management and Processing*, no. 1, W. K. Michener and J. W. Brunt, Eds. Oxford: Blackwell Science, 2000, pp. 48–69.
[8] S. Mohamad, "Generating visualisation for crime scene investigation based on probability result of knowledge-based system," *Int. J. Comput. Appl. Technol.*, vol. 56, no. 1, pp. 18–26, 2017.
[9] N. Enke, A. Thessen, K. Bach, J. Bendix, B. Seeger, and B. Gemeinholzer, "The user's view on biodiversity data sharing — Investigating facts of acceptance and requirements to realize a sustainable use of research data —," *Ecol. Inform.*, vol. 11, pp. 25–33, 2012.
[10] N. F. Johnson, "Biodiversity informatics," vol. 52, pp. 421–438, 2006.
[11] MyCHM, "Malaysia Biological Diversity Database," 2008. .
[12] GBIF, "Global Biodiversity Information Facility," 2004. [Online]. Available: http://www.gbif.org. [Accessed: 01-Jun-2016].
[13] ITIS, "Integrated Taxonomic Information System," 1990. [Online]. Available: http://www.itis.gov/. [Accessed: 10-Apr-2016].
[14] TDWG, "Taxonomic Database Working Group," 1994. .
[15] N. B. N. Gateway, "NBN Gateway," 2012. [Online]. Available: https://data.nbn.org.uk/. [Accessed: 31-Oct-2016].
[16] M. Farquhar, "Definitions of quality of life: a taxonomy," *J. Adv. Nurs.*, vol. 22, no. 3, pp. 502–508, 1995.
[17] S. F. Mawatari and H. Okada, "Taxonomy : A Discipline Central to Integrative Biology," in *Proceedings of International Symposium on "Dawn of a New Natural History - Integration of Geoscience and Biodiversity Studies*, 2004, pp. 29–39.

Predicting Graduate Employability based on Program Learning Outcomes

Wan Nor Afiqah Wan Othman, Aziman Abdullah, and Awanis Romli

Faculty of Computing (FKOM), College of Computing and Applied Sciences, Universiti Malaysia Pahang, Kuantan, Pahang, Malaysia

Email: wnor_afiqah93@yahoo.com, aziman@ump.edu.my

Abstract. Many studies based on the literature and adopted approach by Ministry of Higher Education regarding graduate employability are using survey. This approach is lack with on-demand analytical capability for impactful decision making. There is a lack of study that predicts the duration of graduate to get employed based on quantitative analysis. Since all institutions of higher education are compulsory to adopt and implement outcomes-based education (OBE), this study aims to develop a predictive model on GE based on program learning outcomes (PLO) data. There are two data sources used in this study, institutional academic database and online feedback from graduate. This study used simple linear regression to measure the degree of relationship between the category of PLO with the duration of graduate to get employed. This study received 47 responses from 216 with a response rate of 22%. PLO1 and PLO6 which are '*knowledge*' and '*problem solving and scientific skills*' respectively show high significance values on the duration of graduate to get employed. The linear models developed based on PLO1 and PLO6 were validated with error rate analysis and evaluated with error rate frequency analysis. The results show the model has potential value to be used to predict graduate employability performance within the time frame (6 months) as determined by Ministry of Higher Education. With prediction capacity from the developed model, more intervention program can be strategically planned to assure graduate can be employed in time and in-field.

1. Introduction

Graduates' employability is issues that have been rise to many policy implications for higher education in many developing countries such as Germany, United Kingdom, China, Vietnam as well as Malaysia [1–9] especially in securing in-field employment [10]. Several factors are contributing to the graduates' employability issue [3]. The main is an individual factor regarding to the skills gap, between what has acquired and what is been expected. It refers to the graduate's talent aspect in promoting their skills not only knowledge-discipline skills but also include soft-skills as well. This issue becomes more critical where World Economic Forum claims that "demand for various skills in 2020, more than one third (36%) of all jobs across all industries require complex problem-solving as one of the core skills, rather than 1 in 20 jobs (4%) for physical abilities" [11]. The employability includes all the aspects of individuals proactive behavior, personal management attitude and career management aptitude in order to get a necessary job [12].

Tracer study is a current approach as an assessment tool to collect and trace graduate's data through standardize survey. This approach aims are to gain experimental and observational data that give the impacts to the education programme and improve the transition of graduates from education to the labour market [13–18]. In Malaysia, Ministry of Education through Department of Higher Learning has conducted a yearly National Graduate Tracer Study. It is a web-based survey system collaborate with all higher education institutions (HEIs) upon graduation to improve the graduates' employability since 2006 [19]. Tracer study should be a tool that benefits stakeholders. However, there are several drawbacks which involved with key problems of data quality and lack of insight to the right audience. This data is also limited and restricted the institutions to provide practical improvements on the impact of universities' courses to the related field in industries [17], [18], [20], [21]. Hence, the lack in real time of decision making, comprehensive data and inclusive graduate data respectively of tracer study should be emphasized.

This study aims to develop a predictive model for graduates' employability duration to get employed using Simple Linear Regression (SLR) method. The key research questions are stated as below:

RQ1: Which PLO's that highly influence the duration of graduate to get employed?

RQ2: Is it feasible to predict the duration of graduate employability?

2. Literature review

2.1. Malaysia Qualification Framework (MQF)

The Malaysian Qualifications Framework (MQF) which was approved under the Malaysian Qualifications Agency (MQA) Act, 2007, is the instrument which sets national classifications of qualifications, levels of learning achievements based on learning outcomes, prescribed academic load at each level and is associated with the title of a named qualification. All academic programme in Malaysia higher education will be accredited only when it is compliance with the framework. The framework defines there are at least eight standard program learning outcomes (PLO) that any program must comply with. Table 1 shows the attributes of eight domains in PLO

Table 1. The eight standard program learning outcomes (PLO)

Program Learning Outcomes (PLOs)	Attributes
PLO-1	Knowledge
PLO-2	Practical Skills
PLO-3	Social Skills and Responsibilities
PLO-4	Values, Attitudes and Professionalism
PLO-5	Communication, Leadership and Team skills
PLO-6	Problem Solving and Scientific Skills
PLO-7	Information Management and Lifelong Learning Skills
PLO-8	Managerial and Entrepreneurial Skills

The purpose of the framework is to assure the quality of higher education. One of the quality issues in higher education is graduate employability. It is expected to offer much better insight for various stakeholders as been mentioned by Ministry of Higher Education (MoHE) in Malaysia Education Blueprint 2015-2025 to address the issue on graduates' employability.

2.2. Prediction on Graduate Employability

Zwick & Sklar [22] applied a linear regression model to predict first year college grades (FGPA) as well as degree completion using previous school grades and SAT scores. They concluded that the result in evaluating the contributions of SAT score and GPA is statistically significant in the prediction of graduation. Another study [23] also adopted linear regression analysis to investigate the predictors of graduates' employability based on eight graduates' skills. From the hierarchical regression results, it shows the eight predictors explained 37% variance of employment status. The study concludes that value-related graduates' skills are vital for employability. Based on the related works, this study also adopts simple linear regression method in predicting graduate employability based on the program learning outcomes data.

3. Methodology
Figure 1 shows the research model adopted in this study. Predictive Analytics Process adapted from the actual version (refer [24]) was used.

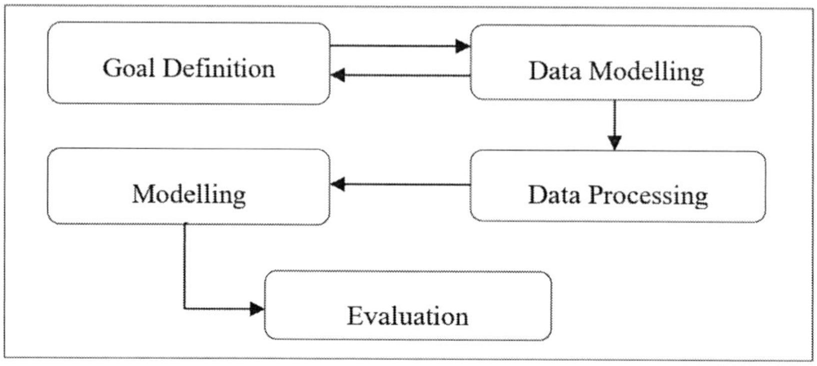

Figure 1. Predictive analytic process – modified version from Poornima & Pushpalatha (2018)

3.1. Phase 1: Goal definition
The goal defined for this study is to identify which competency that affects the duration of graduate to employed. This part is very important that will determine the required data model useful for the predictive model.

3.2. Phase 2: Data modelling
There are two data sources used in this study which are an institutional academic database (PTMK UMP) and online feedback from graduate. Table 2 illustrate the summary of the data model adopted in this study.

Table 2. Data model

Data	Attributes	Source
Program Learning Outcomes Attainment	**Student ID**, Enrolled programmed, Semester, Faculty and all eight program learning outcomes attainment scores.	Institutional Academic Database (permission granted).
Graduate Employability	**Student ID** and dates of first employment after graduation.	Online feedback.

3.3. Phase 3: Data processing
Time duration is a data value that dynamically changes over time itself. We only asked the graduate to response simply the first date that they are get employed for the first time after graduation. Based on the given date, we performed data processing to find the duration by calculating the number of days between two known dates based on equation (1).

$$Duration = Date\ of\ First\ Job - Date\ of\ Senate\ (graduation) \quad (1)$$

Table 3 shows the structure of the data how duration of employed graduate is been produced. This study adopts a common spreadsheet application with existing formula in calculating the number of days between two dates.

Table 3. Data discretion of employment within six months attributes

Student ID	Date of Senate	Date of First Job	Duration (Number of Days)
Student 1	2018-07-25	2018-08-01	7
Student 2	2018-07-25	2018-01-07	166

Lastly, data normalization process involved where it classifies values of data by using minimum and maximum steps. For example, within scales 0.0 to 1.0.

3.4. Phase 4: Modelling

Modelling phase can be done by training and testing data into statistical method or algorithms to create the model. The model of candidates, selection and validation are included in this phase. In this study, we select simple linear regression (SLR) as a method that will be used to build the model. The reason why SLR been used is because it is simple, direct, effective and easy to understand [25].

Model of candidates is a process to derives the correlation coefficient (r) and coefficient of determination (R^2) for the relationship between y - dependent variable (duration of employment within six months after graduation) and x – independent variable (eight PLO) of each candidate. Pearson Product Moment Correlation (r) [26] is used. The significance testing (p-value approach) is used to make conclusions of hypothesis where p-value are compared to a significance level (α). To summarize:

p-value $< \alpha$ => reject H_0 => accept H_a

p-value $\geq \alpha$ => fail to reject H_0

where H_a is expecting the graduate being employed in their related field of study within six months of their graduation. While H_0 is considered as vice versa from H_a.

Model selection process produces the best SLR model to be choose among the eight PLOs. In order to get the best model to use for the testing and forecasts of upcoming data sets, the statistical method need to be carried out, then the steps given need to be followed:

Step 1: Determine the R^2 and r. If positive value, r will be rejected.

Step 2: Determine the p-value where the result of attributes is significant relationship.

Model validation process is implemented when the selected model is tested using same dataset to identify the prediction accuracy performance. It is to prove the attributes used in this study able to build the correct models. The accuracy performance of model is validated based on the error rate of the predicted values. The error rate of the model is defined as:

$$\text{Error Rate (\%)} = (|\,\text{Predict-Actual}\,| \,/\, \text{6 Months}) \times 100\% \quad (2)$$

where,

Predict = Prediction duration using SLR model

Actual = Actual duration from raw data

6 Months = Approximately 182 days

3.5. Phase 5: Evaluation

The evaluation phase is to measure the effectiveness of model in UMP as a case study. The model was evaluated with frequency analysis of error rate.

4. Results and discussions

This section provides the relevant results for data analysis of prediction on graduate employment within six months after graduation. Discussions of the results obtained were provided in the form of diagram and tables based on the significant relationship between the duration of employment within six months and eight PLO using SLR method.

4.1. Statistical properties of SLR model

Table 4 provides the results of correlation coefficient for the relationship between both variables using Pearson Product Moment Correlation (r). This r value aims to determine the strength of relationships among both variables either positive or negative correlation.

As shown in table 4, the correlation result of duration against PLO 7 in Pre-LI group achieved the highest r values at 0.2587 compared to others. There are weak negative and positive correlation was analyzed. In this study, the positive correlation referred to the result of duration and PLOs variables are both increased. It means students took a long time to get a job after graduation even though they got high-grade points for their PLO. This happens due to the dataset of which almost 47 candidates get high-grade points or there are some zero values in Pre-LI and LI groups of each PLOs. In this research, the expected relationship is a negative correlation where better PLO attainments, the less time for fresh graduate to secure a job offer.

Table 4. Pearson product moment correlation of the relationship between duration and PLO 1 to 8 for Pre-LI and LI

PLO	Pre-LI	LI
PLO 1	-0.2144	-0.2027
PLO 2	-0.0776	0.0524
PLO 3	-0.1698	0.0718
PLO 4	-0.1465	-0.0107
PLO 5	-0.0858	0.04941
PLO 6	-0.2515	-0.2022
PLO 7	**0.2587**	-[a]
PLO 8	-0.0700	-0.0669

[a] '-' No data were found for each student in PLO 7 (LI group).

The results of coefficient of determination (R^2) is provided in table 5. The highest R^2 value is also on PLO 7. The $R^2 = 0.0669$ means that around 7% variation in duration can be explained by PLO 7 in Pre-LI group. This means that the model generated from the relationship between these two variables is more precise to predict than the other models.

Table 5. Coefficient of determination of the relationship between duration and PLO 1 to 8 for Pre-LI and LI groups

PLO	Pre-LI	LI
PLO 1	0.046	0.0411
PLO 2	0.006	0.0027
PLO 3	0.0288	0.0052
PLO 4	0.0215	0.0001
PLO 5	0.0074	0.0024
PLO 6	0.0632	0.0409
PLO 7	**0.0669**	-[b]
PLO 8	0.0049	0.0045

[b] '-' No data were found for each student in PLO 7 (LI group).

The significance testing by using p-value is implemented to check the significance of the independent or predictors in the model. Table 6 shows the significant values of both variables with a cut-off value for the p-value is $\alpha = 0.25$ (25%). This case is based on the mission of National Graduate Employability Blueprint 2012-2017 [27]. The null hypothesis, in this case, is H_0: $\beta_1 = 0$, i.e., that there is no significant linear relationship between the duration and PLO 1 to 8 for Pre-LI and LI groups.

Table 6. Significant testing of relationship between duration and PLO 1 to 8 for Pre-LI and LI groups

PLO	Pre-LI	LI
PLO 1	**0.1479**[*c]	**0.1717**[*c]
PLO 2	0.6043	0.7265
PLO 3	0.2539	0.6313
PLO 4	0.3259	0.9433
PLO 5	0.5662	0.7415
PLO 6	**0.0882**[*c]	**0.1730**[*c]
PLO 7	**0.0791**[*c]	-[d]
PLO 8	0.6403	0.6551

[c] * Significant.
[d] '-' No data were found for each student in PLO 7 (LI group).

As shown in table 6, the significant values ($p < 0.25$) with n = 47 for the relationship between duration and PLO 1 to 8 refers to the bold values. However, there is insufficient evidence to conclude that PLO 2 until 5 and PLO 8 for both groups are significant linear relationship to the duration. It because of the correlation coefficient is not significantly different from zero. Therefore, do not reject the null hypothesis, H_0: $\beta_1 = 0$.

4.2. Model validation and evaluation

The results from model validation with error rate analysis show variation of the degree of errors from the dataset that been tested with the model. Based on the error rate, we evaluate the performance of the model by performing frequency analysis using a histogram chart as shown in figure 2 and figure 3.

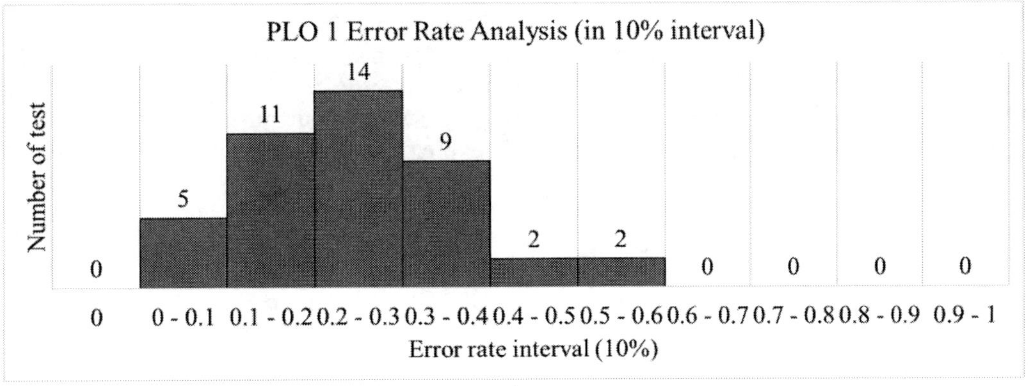

Figure 2. The PLO 1 error rate analysis (in 10% interval)

Figure 2 shows the distribution of error rate between actual and prediction result in 10% interval range. As the error rate of PLO 1 shown in figure 2 increases at the range of 0.2 - 0.3, this model minimally predicts the correctness of the actual duration.

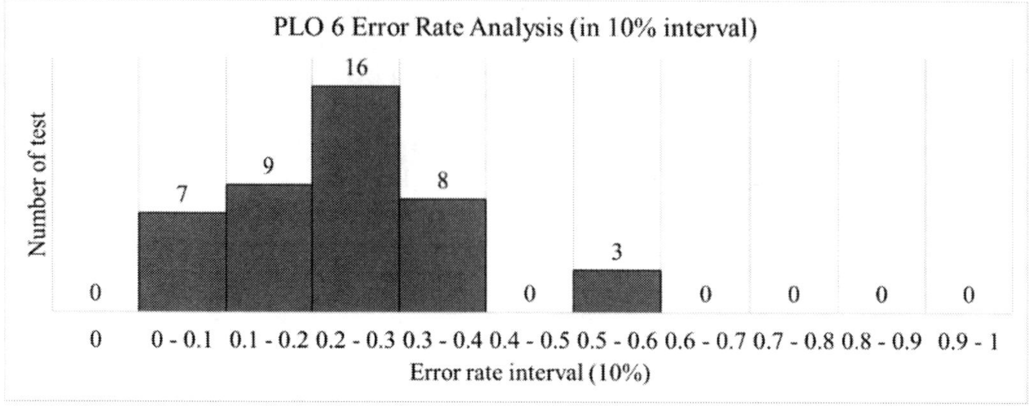

Figure 3. The PLO 6 error rate analysis (in 10% interval)

Figure 3 shows the PLO 6 pattern of error rate analysis is approximately the same as PLO 1. For both analyses, it found that up to 0.6 out of 1% have the errors. In this case study, it considered as better forecasting models where the error rate does not exceed 1%. This also means it accurately predicts the duration to get employed before six months after graduation.

5. Conclusion

As a conclusion, there is high feasibility that the duration of graduate gets employed after finishing their study can be predicted based on the learning outcomes attainment data. According to the findings, it can be summarized that PLO1 and PLO6 show highly significant value on the duration of the graduate to get employed. The negative correlation between both PLO's against duration to get employed indicates that the more competent graduate on those skills, the more likely they can get employed earlier. Furthermore, the result from model validation and evaluation support this conclusion.

By identifying this competency that highly affect graduate employability, more student intervention and enrichment program can be done for student before they go for an industrial internship. Hence, it ensures that students are more ready for the internship compares to before. This model, however, may only applicable for the academic program that has an industrial internship at the last semester of study.

For future research, this study is required to develop comprehensive models able to predict the graduates' employability for all academic programs. Besides, a small dataset was found which need to increase the number of respondents in order to get high response rate. Future work also should carry out the deployment phase. This phase is a generating report of the outcomes gets through the models. It produced for university reference as this is the valuable information that can give a better decision making.

6. Acknowledgement

This work was supported by Universiti Malaysia Pahang from grant PGRS1903108.

References

[1] Department of Statistics M 2018 *Labour force in Malaysia*
[2] Bai L 2015 Graduate unemployment : dilemmas and challenges in China's move to mass higher education *China Q.* **185** 128–44
[3] Yusof N and Jamaluddin Z 2015 Graduate employability and preparedness : a case study of University of Malaysia Perlis (UNIMAP), Malaysia *Geogr. J. Soc. Sp. 11* 129–43
[4] Sumanasiri E G T, Yajid M S A and Khatibi A 2015 Review of literature on graduate employability *J. Stud. Educ.* **5** 75
[5] Tran T T 2015 Is graduate employability the 'whole-of-higher-education-issue'? *J. Educ.*

Work **28** 207–27
[6] Zheng J, Tang J and Gao J 2012 Construction of graduate employment service system based on public information platform *AASRI Procedia* **1** 410–5
[7] Leo M 2016 *What you didn't know about fresh graduate unemployment in Malaysia [infographic]*
[8] ILO 2013 *Global employment trends for youth 2013* vol 1
[9] Shao H R 2011 *A study of the employability-based curriculum development in american universities : focus on business schools*
[10] Futch K W 2014 *Factors that predict time to in-field employment of associate degree graduates: a study of one college in the technical college system of georgia*
[11] World Economic Forum 2016 The future of jobs employment, skills and workforce strategy for the fourth industrial revolution *Growth Strateg.* 2–3
[12] Clarke M 2008 Understanding and managing employability in changing career contexts *J. Eur. Ind. Train.* **32** 258–84
[13] Aina L O and Moahi K 1999 Tracer study of the Botswana library school graduates *Educ. Inf.* **17** 215–44
[14] Kyaw U K 1992 *A tracer study of recent graduates : implications for education and manpower planning*
[15] Manda M and Matidza I 2016 A tracer study of 2011-2015 graduates from the land management programme at Mzuzu University *J. Educ. Soc. South. Africa* **3** 45–67
[16] Damotharan L V 2012 *Tracer study on AIMST University students using data mining*
[17] European Training Foundation 2017 *Tracer studies evaluating the impact of training programmes*
[18] Schomburg H 2016 *Carrying out tracer studies guide to anticipating and matching skills and jobs* vol 6
[19] Latif L A and Bahroom R 2010 OUM's tracer study: a testimony to a quality open and distance education *ASEAN J. Open Distance Learn.* **2** 35–47
[20] Rogan M and (NALSU) N A L S U 2016 *Tracing graduates into the labour market*
[21] Sanchez M P R and Diamante V A M 2017 Graduate tracer study of the college of nursing *Malaysian J. Nurs.* **8** 41–7
[22] Zwick R and Sklar J G 2005 Predicting college grades and degree completion using high school grades and SAT scores: the role of student ethnicity and first language *Am. Educ. Res. J.* **42** 439–64
[23] Mahazan A M, Siti Nubailah M Y, Ummi Salwa A B and Wan Mohd Fazrul Azdi W R 2015 Predictors of graduate employability: mediating roles of leadership, ethics, and religiosity *Int. Acad. Res. J. Bus. Technol.* **1** 126–36
[24] Poornima S and Pushpalatha M 2018 A survey of predictive analytics using big data with data mining *Int. J. Bioinform. Res. Appl.* **14** 269
[25] König R 2009 *Predictive techniques and methods for decision support in situations with poor data quality*
[26] Pearson K. 1900 On the criterion that a given system of deviations from the probable in the case of a correlated system of varibales is such that it can be reasonablly supposed to have arisen from random sampling *Phylosophical Mag.* **50** 157–75
[27] Ministry of Higher Education Malaysia 2012 *The national graduate employability blueprint 2012-2017*

IOP Conf. Series: Materials Science and Engineering **769** (2020) 012019 doi:10.1088/1757-899X/769/1/012019

Student Learning Progress as Predictor for Graduate Employability Performance

Wan Nor Afiqah Wan Othman and Aziman Abdullah

Faculty of Computing (FKOM), College of Computing and Applied Sciences, Universiti Malaysia Pahang, Kuantan, Pahang, Malaysia

Email: wnor_afiqah93@yahoo.com, aziman@ump.edu.my

Abstract. Graduate employability is a major concern for higher education industry. There is a lack of research on the use of program learning outcomes (PLO) data to predict graduate employability performance especially on the duration they get employed. Therefore, our motivation in this study is to investigate how PLO data can be used to predict graduate employability performance. This study adopted quantitative analysis as a research method by using Simple Linear Regression to measure the highest correlation and significance values between learning progress and duration graduate to get employed. The PLO data from all semesters were segmented into four-time segments: 1st SEM, MID SEM, Pre-LI and LI. The slope value of linear model from time series analysis of four-time segments is used as a value to determine the performance of student learning progress. 47 responses (22% response rate) from 216 graduates who completed their study from Faculty of Computing, Universiti Malaysia Pahang in 2018 has been received as a case study. We found that learning progress from PLO 3 and PLO 6 which are '*Social Skills and Responsibilities*' and '*Problem Solving and Scientific Skills*' respectively, show significant values on the duration to get employed. This study highlights student learning progress is potential to be used as a predictor for graduate employability performance.

1. Introduction

In the era of Industrial Revolution 4.0, there is more and more higher education institution putting their concern on the issue of graduate employability. Graduate employability is very much influence with individual competencies in the field as well as generic competencies such as problem solving, communication and other skills most researchers refer as soft skills.

Many literature studies on student performance using data mining techniques. Although the techniques are capable to predict the academic performance, it is, however, depending on what data or attributes are been used. Most study using attributes like CGPA and psychometric data to predict academic performance [1]. These attributes mostly are not much helpful for the institution to predict graduate employability based on their skills or competencies level. Therefore, our motivation in this study is to investigate how learning outcomes data can be used to predict graduate employability performance. This prediction is important for higher education institution to improvise teaching and learning strategy through interventions approach. This is well supported by EDUCAUSE study [2] where most institution (96%) cited that improving student outcomes was a goal for their student success studies.

This study aims to identify the feasibility of student learning progress based on learning outcomes to predict graduate employability performance. To facilitate this study, the key research questions are:

RQ1: How to develop a predictive model for student learning progress?
RQ2: Does learning progress significance or influence graduate employability performance?

2. Literature review

2.1. Learning outcomes

Learning outcomes can be referred to as a student's competencies that are shown after the completion of their studies [3–8]. The international trends of learning outcomes in higher education are based on 'student-centred' approach [9]. This approach points out the outputs related to student competencies. This learning outcomes stated in the program and course information are part of the important component in outcomes-based education (OBE). The program learning outcomes (PLO) is a foundational requirement in the OBE system [10]. In Malaysia Qualification Framework (MQF) version 1.0, there are eight standards of program learning outcomes (PLO) that any institution in Malaysia minimally must comply in order to be accredited. Table 1 shows the eight stipulated domains in PLO.

Table 1. The eight domains of program learning outcomes (PLO)

Program Learning Outcomes (PLOs)	Attributes
PLO-1	Knowledge
PLO-2	Practical Skills
PLO-3	Social Skills and Responsibilities
PLO-4	Values, Attitudes and Professionalism
PLO-5	Communication, Leadership and Team skills
PLO-6	Problem Solving and Scientific Skills
PLO-7	Information Management and Lifelong Learning Skills
PLO-8	Managerial and Entrepreneurial Skills

2.2. Learning progress

A potential way to visualize the students regulate effort is by using learning progress. The progress illustrates how much the student has learned a specific skill during their learning over time [11]. The continuous feedback to students is important as they can monitor and control their learning progress. Winne & Hadwin highlighted the importance of data that influence the learning progress: *without reliable, revealing and relevant data that support learners to make valid inferences about how they control and monitor their learning, learners will be handicapped* [12].

Molenaar, Horvers & Baker applied moment-by-moment learning progress to investigate the students' self-regulated learning [13]. This study concluded that learning progress has a relationship with students' accuracy and learning, but not with effort. This finding is consistent with another study [11], where the effort is not showed directly, but by visualizing the student recent progress can make them connect the progress with their recent effort. Another literature [14] used data mining to predict and solve the problem on student performance estimation, student progress, and student potentials. They used student progress indicators and attribute causal relationship predictor where it shows up the factor that affects the student performance. The results indicate the proposed tools give correct and accurate results as well as perform a better understanding on student progress. The study, however, did not address the graduate employability performance.

2.3. Prediction on Graduate Employability

Study by [15] adopted linear regression to look into the predictors of graduates' employability based on graduates' competencies. The study used a questionnaire for data collection and found that value-related graduates' competencies are important for employability. However, the study did not predict the duration of graduate to be employed and does not use learning outcomes data. Based on the related works, this study also adopts the simple linear regression method in predicting graduate employability through student learning progress based on the program learning outcomes data.

2.4. Research Gap

Although there are many literature studies on learning outcomes, studies on learning outcomes that addressing graduate employability performance are still lacking particularly in predicting the duration to get employed. It is found that there are only 38 literatures on graduate employability (title) that associate with learning outcomes (keyword) based on our simple search (1:05 pm; 14/9/2019) from google scholar database. Figure 1 shows the relative comparison in Venn diagram the total number of literatures related to '*learning outcome*' (LO) and '*graduate employability*' (GE). In this study, those articles will not be reviewed since our study are based on the case study in Universiti Malaysia Pahang (UMP) which we can confirm is the first attempt that UMP conducting a study to predict graduate duration based on learning outcomes data.

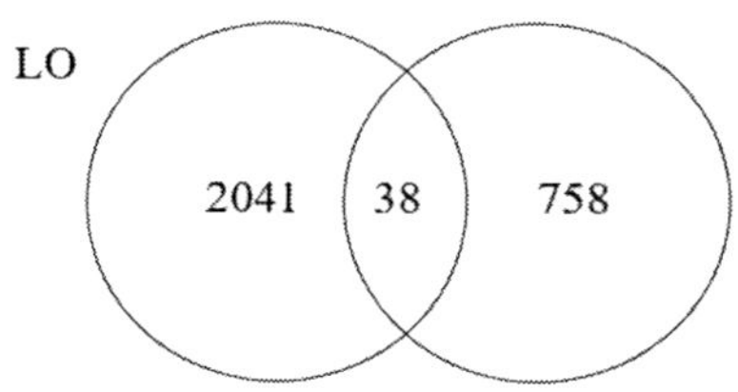

Figure 1. Total number of articles based on the keywords from googles scholar

3. Methodology

Figure 2 shows the research model adopted in this study. Predictive Analytics Process adapted from the actual version (refer [16]) was used.

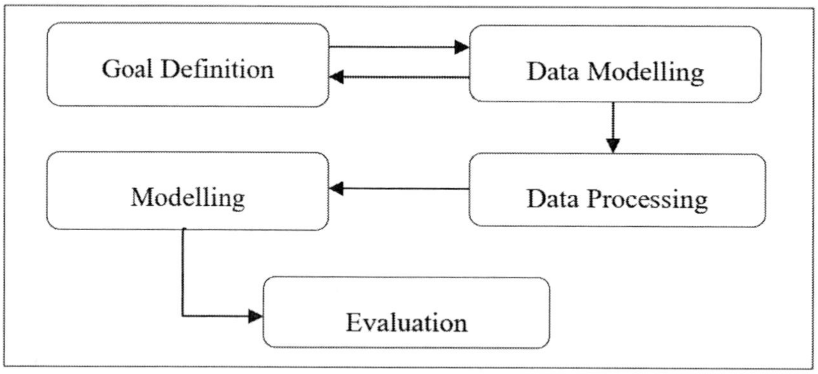

Figure 2. Predictive analytic process – modified version from Poornima & Pushpalatha (2018)

3.1. Phase 1: Goal definition

The goal defined for this study is to identify which competency through learning progress that affects the duration of graduate to employed. This part is very important that will determine the required data model useful for the predictive model. In this study, the goal for the predictive model is based on the mission of National Graduate Employability Blueprint 2012-2017 [17] which to ensure 75% of the graduate be employed within six months in their field of study.

3.2. Phase 2: Data modelling

There are two data sources used in this study which are an institutional academic database (PTMK UMP) and online feedback from the graduate. Table 2 illustrates a summary of the data model adopted in this study.

Table 2. Data model

Data	Attributes	Source
Program Learning Outcomes Attainment	**Student ID**, Enrolled programmed, Semester, Faculty and all eight program learning outcomes attainment scores.	Institutional Academic Database (permission granted).
Graduate Employability	**Student ID** and dates of first employment after graduate.	Online feedback.

3.3. Phase 3: Data processing

Time duration is a data value that dynamically changes over the time itself. We only asked the graduate to response simply the first date that they are get employed for the first time after graduate. Based on the given date, we performed data processing to find the duration by calculating the number of days between two known dates based on equation (1).

$$Duration\ (day) = date\ of\ employment - date\ of\ graduation \quad (1)$$

Table 3 shows the structure of the data how duration of employed graduate is been produced. This study adopts a common spreadsheet application with the existing formula in calculating the number of days between two dates.

Table 3. Data discretion of employment within six months attributes

Student ID	Date of graduation	Date of Employment	Duration (Number of Days)
Student 1	2018-07-25	2018-08-06	12
Student 2	2018-07-25	2018-11-26	124

Another data used to develop our model is student learning outcomes data. There are variations of the total mark for each PLO been assessed in each semester. We transform the data by producing the PLO attainment score by calculating the ratio of student score relative to total mark within scales 0.0 to 1.0 based on equation (2). For example, if a student got a score of 80 marks over 100 total marks for PLO1, then the ratio will be 0.8.

$$PLO\ Attaintment\ Score = \frac{Total\ Student\ Mark}{Total\ Full\ Mark} \quad (2)$$

The achievement of learning outcomes may not sufficient to reflect the capability of a student in learning since the learning process itself requires a progressing approach [18]. To measure the learning progress, this study used the slope value of linear equation based on learning outcomes data projected with time-series analysis. To produce time-series analysis, the data of graduate learning outcomes have been segmented into four-time segments: 1st SEM, MID SEM, Pre-LI and LI as shown in figure 3. The 1st SEM is referring to the first semester when student admitted or enrolled in the program. The MID SEM is the maximum score of PLO from the second semester and the last two semesters before they went to internship. While Pre-LI and LI are referring to the last semester graduate or student studied in university and the semester when they were doing industrial internship. The reason for this segmentation is to standardize the data model since there are variations in term of the total number of semester graduates completed their studies.

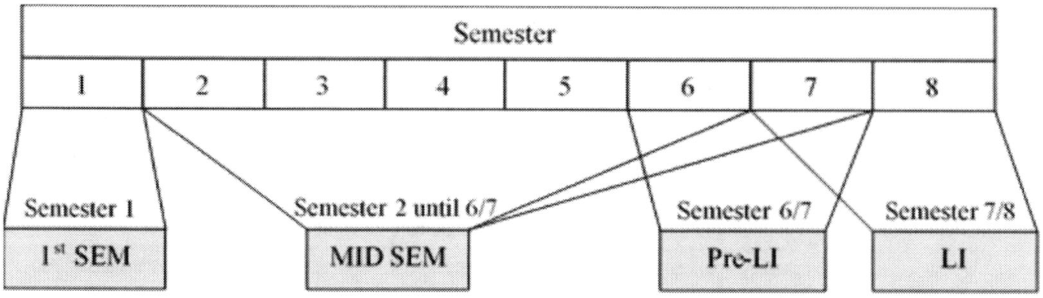

Figure 3. Data discretion of semester 1 until 8 into four groups

Once all the data for each graduate been segmented into four-time segments, then time series analysis is been conducted as illustrated in figure 4. With this dataset, we calculate the value of slope from the linear equation model as a value that represents student learning progress of each PLO for each student.

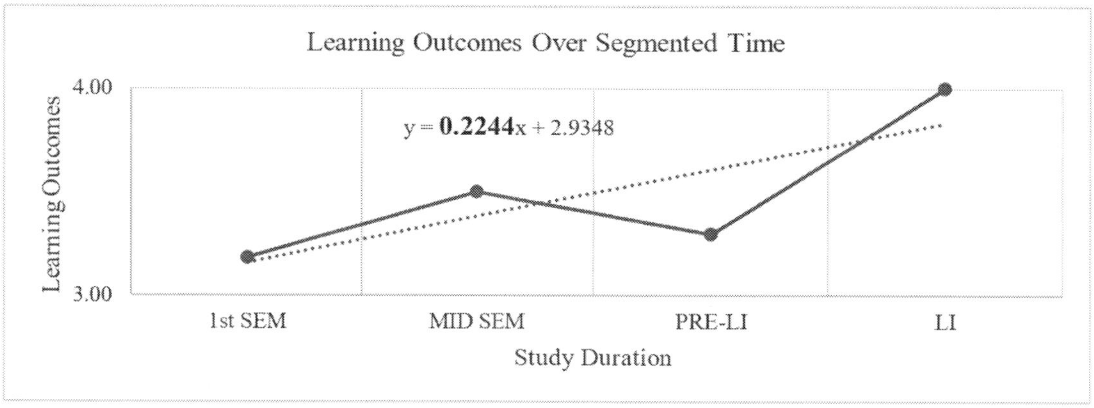

Figure 4. Learning progress model for one student

3.4. Phase 4: Modelling

Modelling phase in which training and testing data into statistical methods will be performed to create the model. The model of candidates, selection and validation are included in this phase. In this study, a simple linear regression (SLR) method will be used to build the model. The SLR is used because it is simple, direct, effective and easy to understand [19].

Model of candidates is a process to determine the correlation coefficient (r) and coefficient of determination (R^2) for the relationship between y - dependent variable (duration of employment within six months after graduation) and x – independent variable (learning progress) of each candidate. Pearson Product Moment Correlation (r) [20] is used. The significance testing (p-value approach) is used to make conclusions of the hypothesis whether the data support or rejects the null hypothesis (H_0). The p-value is compared to a significance level (α). In short, there are:

p-value $< \alpha$ => reject H_0 => accept H_a

p-value $\geq \alpha$ => fail to reject H_0

where H_a is expecting the graduate being employed in their related field of study within six months upon graduation. While H_0 is considered as vice versa from H_a.

Model selection process produces the best SLR model to be select among the eight PLOs. To obtain the best model to use for the testing and predicting upcoming data sets, the statistical method should be carried out, then the steps given need to be followed:

Step 1: Determine the R^2 and r. If positive value, r will be rejected.

Step 2: Determine the p-value where the result of attributes is significant relationship.

The process of model validation aims to identify prediction accuracy performance. It is implemented when the selected model is tested using the same dataset. The model validation is conducted in order to prove the attributes used in this study able to build the correct models. The accuracy performance of the model is validated based on the difference between the predicted value and actual value relative to the acceptable range of employed duration (6 months). The error rate of the model is defined as:

$$Error\ rate\ (\%) = \frac{Predict - Actual}{Goal} \times 100 \quad (3)$$

where,
Predict = Predicted duration using SLR model
Actual = Actual duration from raw data
Goal = 6 months or approximately 182 days

3.5. Phase 5: Evaluation
The evaluation phase is to measure the performance of linear model in UMP as a case study. The frequency analysis of error rate will be conducted to evaluate the model.

4. Results and discussions
This section provides the relevant results for data analysis of prediction on graduate employment within six months after graduation. Discussions of the results obtained were provided in the form of diagram and tables based on the significant relationship between the duration of employment within six months and learning progress based on eight PLOs using SLR method. We received 47 responses from 216 graduate who completes their study in 2018. The response rate of 22% is acceptable for us to proceed for further analysis.

4.1. Statistical properties of SLR model
Table 4 provides the results of correlation coefficient using Pearson Product Moment Correlation (r), coefficient of determination (R^2), and significant testing (p-value) for the relationship between both duration of graduate to be employed and learning progress.

Table 4. The statistical analysis of employability duration and learning progress

PLO	Correlation, r	Coefficient of determination (R^2)	p-value
PLO 1	-0.0713	0.0051	0.6338
PLO 2	0.0773	0.006	0.6057
PLO 3	0.3197	0.1022	**0.0285** *[a]
PLO 4	0.0848	0.0072	0.5709
PLO 5	-0.0194	0.0004	0.8971
PLO 6	-0.2250	0.0506	**0.1284** *[a]
PLO 7	-0.0750	0.0056	0.6165
PLO 8	0.0168	0.0003	0.9109

[a] *Significant.

For the relation depicted in table 5, the highest r and R^2 values, as well as the significant values ($p < 0.25$) with $n = 47$, are at PLO 3 and PLO 6. The cut-off value for the p-value is $\alpha = 0.25$ (25%) is based on the mission of National Graduate Employability Blueprint 2012-2017 [17]. Hence, the model generated from both PLO 3 and PLO 6 is more precise to use for prediction purpose than the other models.

4.2. Model validation and evaluation

The model validation results by using error rate analysis indicate a variation of the degree of errors from the dataset that already tested with the model. Based on the error rate of the predicted values, the frequency analysis as shown in figure 5 and figure 6 to evaluate the model performance.

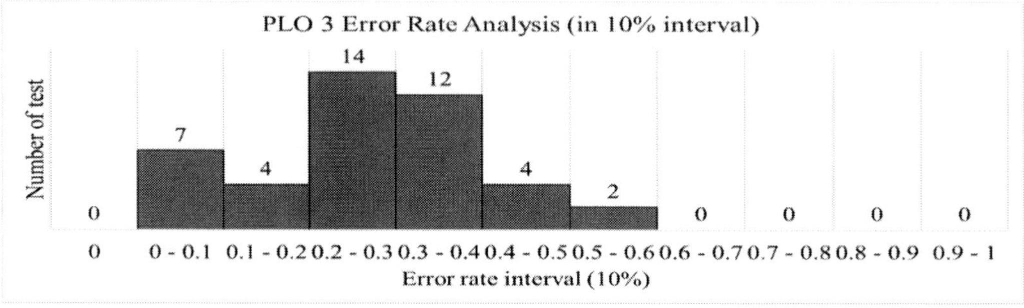

Figure 5. The PLO 3 error rate analysis (in 10% interval)

The distribution of error rate between actual and prediction result in 10% interval range is presented in figure 5. It shows that between the range of 0.2 – 0.3 and 0.3 – 0.4 are the highest error detected in this model. Means that, this model is less correctness in prediction and actual duration.

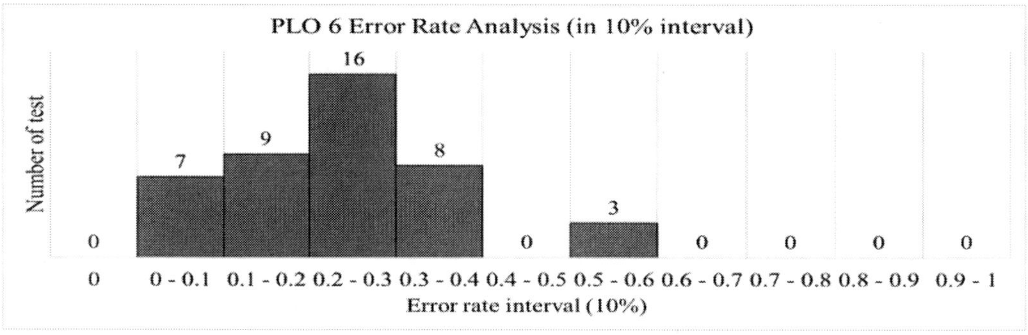

Figure 6. The PLO 6 error rate analysis (in 10% interval)

Figure 6 illustrates the result of PLO 6 model where the error rate is increased at the range of 0.2 – 0.3. This model also shows the same meaning as result in PLO 3. For both models, it analyses the error rate is within 0.6 out of 1%. It can be summarized that this prediction model shows an acceptable or tolerable error rate since it does not exceed 1%. The predicted result as compared to the actual indicates the error is within the targeted timeframe of graduate to get employed.

5. Conclusion

By using simple linear regression (SLR), a predictive model can be developed based on learning progress to predict employability duration to get employed. This method clarifies our first research question (RQ1). As for RQ2, the results show that learning progress of PLO 3 which is '*Social Skills and Responsibilities*' and PLO 6 where '*Problem Solving and Scientific Skills*' shows significant impact on graduate employability performance. The predictive model based on PLO 3 and PLO 6 shows an acceptable error rate that fulfil the mission of National Graduate Employability Blueprint 2012-2017. This suggests that there is high feasibility that the learning progress can be used as predictor for graduate employability performance. Since student learning progress can be obtained from student learning outcomes data, therefore more effective strategies can be plan in improving the quality of curriculum that gives impact on graduate employability performance while they still in study duration.

Acknowledgement
This work was supported by Universiti Malaysia Pahang from grant PGRS1903108.

References

[1] Shahiri A M, Husain W and Rashid N A 2015 A review on predicting student's performance using data mining techniques *Procedia Comput. Sci.* **72** 414–22
[2] Parnell A, Jones D, Wesaw A and Brooks D C 2018 *Institutions' use of data and analytics for student success*
[3] Fitzpatrick J J, Byrne E P and Kennedy D 2009 Making programme learning outcomes explicit for students of process and chemical engineering *Educ. Chem. Eng.* **4** 21–8
[4] Emily Greenleaf, Megan Burnett P G 2008 *Developing learning outcomes: a guide for University of Toronto faculty*
[5] Adam S 2004 *A consideration of the nature, role, application and implications for European education of employing 'learning outcomes' at the local, national and international levels*
[6] Accreditation C for H E 2003 *Statement of mutual responsibilities for student learning outcomes: accreditation, institutions, and programs*
[7] Wastiau-Schluter P 2004 *Focus on the structure of higher education in Europe 2004/05*
[8] O'Brien D J 2009 *Writing learning outcomes*
[9] Kennedy D 2006 *Writing and using learning outcomes: a practical guide*
[10] H. Taib, S M Salleh, B A Md Zain, M.A. Azlan, S. Mahzan, Hafeez Z A, Ong, P., S. Ahmad, M. N.A Rahman, N.F. Nasir, M. Azham Azmi. H.A Rahman Z N 2017 Programme learning outcomes assessment and continuous quality improvement in Faculty of Mechanical and Manufacturing, UTHM *IOP Conf. Ser. Mater. Sci. Eng.* 165
[11] Baker R S, Hershkovitz A, Rossi L M, Goldstein A B and Gowda S M 2013 Predicting robust learning with the visual form of the moment-by-moment learning curve *J. Learn. Sci.* **22** 639–66
[12] Winne P H and Hadwin A F 2013 nStudy: tracing and supporting self-regulated learning in the internet *International Handbook of Metacognition and Learning Technologies. Springer International Handbooks of Education.* pp 293–308
[13] Molenaar I, Horvers A and Baker R S 2019 What can moment-by-moment learning curves tell about students' self-regulated learning? *Learn. Instr.* 101206
[14] Yang F and Li F W B 2018 Study on student performance estimation, student progress analysis, and student potential prediction based on data mining *Comput. Educ.* **123** 97–108
[15] Mahazan A M, Siti Nubailah M Y, Ummi Salwa A B and Wan Mohd Fazrul Azdi W R 2015 Predictors of graduate employability: mediating roles of leadership, ethics, and religiosity *Int. Acad. Res. J. Bus. Technol.* **1** 126–36
[16] Poornima S and Pushpalatha M 2018 A survey of predictive analytics using big data with data mining *Int. J. Bioinform. Res. Appl.* **14** 269
[17] Ministry of Higher Education Malaysia 2012 *The national graduate employability blueprint 2012-2017*
[18] Harden R M 2007 Learning outcomes as a tool to assess progression *Med. Teach.* **29** 678–82
[19] König R 2009 *Predictive techniques and methods for decision support in situations with poor data quality*
[20] Pearson K. 1900 On the criterion that a given system of deviations from the probable in the case of a correlated system of varibales is such that it can be reasonablly supposed to have arisen from random sampling *Phylosophical Mag.* **50** 157–75

IOP Conf. Series: Materials Science and Engineering **769** (2020) 012020 doi:10.1088/1757-899X/769/1/012020

A Conceptual IR Chatbot Framework with Automated Keywords-based Vector Representation Generation

A S Lokman[1,2], M A Ameedeen[1,2] and N A Ghani[1,2]

[1] IBM Centre of Excellence, Cybercentre, Pahang Technology Park, Lebuhraya Tun Razak, 26300 Gambang, Kuantan, Pahang
[2] Faculty of Computing, Universiti Malaysia Pahang, Lebuhraya Tun Razak, 26300 Gambang, Kuantan, Pahang

abbas@ump.edu.my

Abstract. This paper proposes a conceptual remodel of Information Retrieval (IR) chatbot framework designed to eliminate the need for large Question-Answer (QA) pair dataset in chatbot's machine learning training and knowledge base development. Within ten proposed framework's components, we describe Ans2Q: a Neural Network model for question type approximation, and HR6: an IR score ranking calculation based on Ans2Q output. Fundamentally, these two components are the variance in which the proposed framework differs from others. Together with process flow explanation, we also provide several related formulas that hopefully can be used to implement this framework. Our general aim with this framework is to provide a tool that can be used to develop close domain chatbot with small knowledge and no readily available QA pair datasets.

1. Introduction

Development of Information Retrieval (IR) chatbot usually requires a large Question-Answer (QA) pair dataset. Current trends use Neural Network (NN) to predict the most similar Q to user's input query x, then retrieve the paired Answer (A) to the Question (Q) as chatbot's output/response. Major drawback for this model is the need for QA dataset. For general domain, there are lots of widely made available QA datasets [1][2][4], but for specific domains, such datasets might be unavailable due to small/non-specific domain scope [3]. As an example, new business venture might have all needed information to answer customer question regarding their product but do not yet have customer service conversational data. For matured organization, such conversational data is what is being used to create the QA pair dataset [5]. To create QA pair without Q data is a tedious process. For traditional rule-based chatbot, multiple keywords sets need to be created for each A response. For modern NN chatbot, Q for each A needs to be manually created by human. With motivation to provide non-tedious process of preparing QA pair for such datasets, we propose a new conceptual framework for IR chatbot that can automatically generate Q representation for each given A data. Our framework uses the traditional approach of keywords sets, while also implementing modern NN for question type classification and automated Keywords-Based Vector (K-VR) representation generation.

IOP Conf. Series: Materials Science and Engineering **769** (2020) 012020 doi:10.1088/1757-899X/769/1/012020

In essence, our paper makes following contributions:

- We propose a remodeled IR chatbot framework that is built upon K-VR similarity process.
- We propose a new score calculation method named HR6 for scoring IR candidate in ranking process based on weighted Kipling Method's word.
- We propose a new NN model for approximating question type based on sentences' parameters and its vector representation.

The rest of this paper is structured as follows: Next section will present related works that directly reflect various components in our proposed framework. Next subsequent section will discuss our propose framework from overall architecture to details on each components. This paper ends with future work in regards to our intended implementation and conclusion.

2. Related Work

Developing a chatbot with pre-developed framework has become a norm nowadays [6][7][8]. Although the framework design is mostly centered towards certain chatbot type, many argue that such design is needed in order to achieve full end-to-end usability in chatbot development. Modern chatbot design can generally be categorized into several base elements which are: 1) knowledge (open or close domain), 2) response generation (retrieval or generative), 3) text processing (vector embedding or latin alphabet), and 4) Machine Learning (ML) model (usually using neural network).

2.1. Knowledge Domain

A chatbot can be categorized as open or close domain. Domain in this sense is the knowledge that the chatbot is covering. Open domain means the chatbot must know general knowledge covering from recent news topic, entertainment, etc., as well as basic human knowledge. If the chatbot knowledge is specific to certain area such as customer service, psychology, etc., the system will be classified as close domain chatbot. Researchers have concluded that close domain chatbot is easier to build and currently producing good results [9][5][10] while open domain chatbot is still hard to build and have been producing a fair amount of false positive results [1][4].

2.2. Response Generation

Two basic methods for chatbots to generate/produce a response are retrieval and generative. A lot of variations from these two methods have emerged but the base process largely remains the same with retrieval being a process of selecting best output from shortlisted candidates, and generative being flexible output generation based on input sequence and trained classifiers. Among others, Seq2Seq [11] is the most famous model for both retrieval and generative methods. Originally designed for Neural Machine Translation (NMT) system, Seq2Seq architecture is based on encoder and decoder framework whereby encoder will receive input and decoder will produce output that is relevant to the input (like translating one language to another). Abbreviated from Sequence-to-Sequence phrase, Seq2Seq encoder takes input sentence as a sequence of words, analyzing each word (one-by-one in sequence) while generating its representation (one word at a time), and finally generating the whole input representation to be passed on to decoder module. Decoder then deciphers the input representation and generates an output (one word at a time) until all encoded representations are deciphered. With such capability and demonstrated good results in finding/calculating similarities (similarity of encoded representation between input and candidate output), researcher/developer has implemented Seq2Seq in ranking most similar output/response towards the input (retrieval) and also generate a totally flexible output in word or character based that is sequentially reflected the input (generative).

2.3. Text Processing
Word embedding (WE) or vector representation of word are real numbers in vector space that can denote a semantic relationship (by distributional hypothesis) between words within specific vocabulary. The relationship is defined based on distributional hypothesis that postulate "Words that occur in the same contexts tend to have similar meanings" [12]. Because it is a real number, statistical as well as arithmetical calculation towards WE is highly possible thus making it feasible to be used in statistical ML model. Concerning chatbot architectural design, most system that uses ML model (for training or real-time implementation) seem to also utilizes WE in their respective text processing modules.

2.4. Machine Learning Model
Machine learning nowadays is everywhere. Anything with big data repository is being transformed into ML powered systems, and any system with less data is being restructured to collect more data for future ML/AI roadmap. In modern chatbot architecture, ML (in particular Neural Network (NN)) can be seen as a core all-around technology where it is being used in input preparation and processing, as well as output processing and production.

The use of ML in chatbot's input preparation stage is mainly on generating WE. As discussed before, modern chatbot text processing module utilizes WE due to heavy use of statistical ML methods. Overall, there are two ways to generate vector representation of words which are count-method and predictive-method (ML method). While count-method is actually counting word co-occurrence based on specified context, predictive-method just predicts it. Although it seems unlikely (actual vs prediction), [13] systematic comparison has shown that predictive-method is far superior than count-method, thus becoming method of choice among researcher.

3. Framework
This paper proposes a remodel IR chatbot framework that is structured upon two main components: 1) Keywords-based vector representation (K-VR) and 2) Kipling representation (Kp). Both components will be auto-generated and become the representation of chatbot answer/response. In other words, K-VR and Kp is a Q pair for chatbot A data. For system to generate K-VR and Kp, A data must be segmented into sentence/s specific for answering one scope of question. For example, answer regarding phone number and address (contact information) should not be mixed with answer regarding product specification and type (product information). Following Figure 1 visualized our proposed framework in ten components within three interconnected layers.

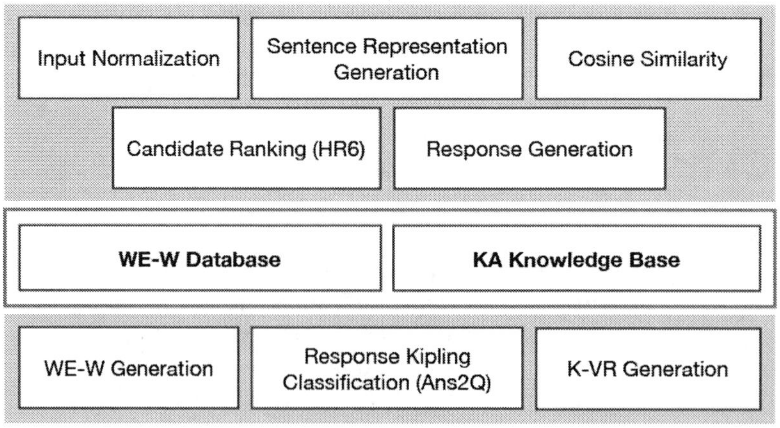

Figure 1. IR chatbot framework with K-VR and Kp (using Ans2Q) generation.

Referring to Figure 1, our proposed framework contains ten components grouped into three separate but interconnected layers. The top layer (Front-end layer) contains five components: 1) Input Normalization, 2) Sentence Representation Generation, 3) Cosine Similarity, 4) Candidate Ranking (HR6) and 5) Response Generation. The middle layer is a data layer that comprises of two databases: 1) WE-W Database and 2) KA Knowledge Base. The bottom layer (Back-end layer) contains three components: 1) WE-W Generation, 2) Response Kipling Classification (Ans2Q) and 3) K-VR Generation. Following are description for each layer:

Front-end layer - This layer will handle all real-time processes during chatbot interaction with user, from getting user's input until generating chatbot's output. Following Figure 2 depicted overall data flow for this front-end layer.

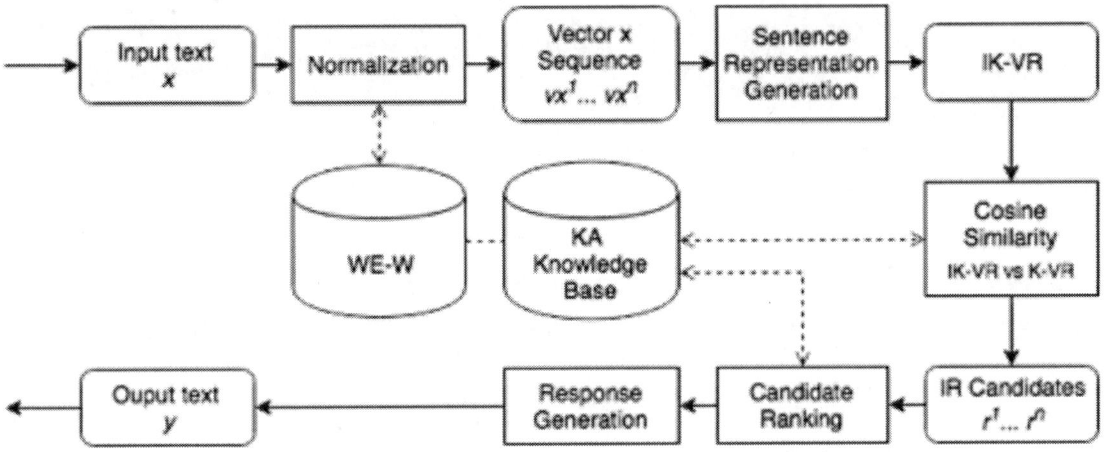

Figure 2. Data flow for front-end layer.

Referring to Figure 2, input x from user will go through Normalization process where each input word will be converted from latin alphabet to vector value based on WE-W (Word Embedding - Weight) database (WE-W will be further explained in framework component 6). Normalization will produce *Vector x Sequence* value that later be used in Sentence Representation Generation process (further explanation is in framework component 2). Next process is Cosine Similarity calculation between representation of user's query (IK-VR) and stored chatbot responses (K-VR). Further explanation on Cosine Similarity is in component 3 while K-VR is in component 10). Output from collective Cosine Similarity processes are top IR/response candidates that later will be scored using our proposed ranking function HR6 (further explanation is in component 4). Based on HR6 result, highest ranked answer will be generated by chatbot as an output y to user's input x.

Middle layer - This layer bridges a connection between front-end and back-end layer. Population of both databases in this layer is done through back-end layer components. After initial population, both databases will be ready for front-end layer utilization. Continuous KA Knowledge Base database population for new data will be done by back-end layer during implementation.

Back-end layer - This layer will handle all pre-processing details before front-end layer can utilizes the middle layer. After going live (implementation), this layer will be used for updating middle layer with new data (if any). In usual circumstances, only Ans2Q and K-VR Generation components will be used during implementation phase.

As briefly described above, connection between all three layers are through the middle layer such that front-end and back-end layer are not directly connected. With regards to given description, next sub sections will explain in detail each component functionalities together with its underlying logic calculation.

3.1. Input Normalization

This first component of front-end layer will take user's raw text data as input x, tokenize it into each individual word/symbol, retrieve each token's embedding from WE-W database, and create the vector x sequence $(vx^1 \ldots vx^n)$ where vx^i is the vector embedding and weight for each respective token. Token that did not have its embedding data in WE-W will be represent as zero vector $(0^{\rightarrow})$ with zero weight value.

3.2. Sentence Representation Generation

This second component of front-end layer will take vx sequence and produce IK-VR (*Input Keywords - Vector Representation*) value for every vx weight that is more than threshold t. Formula for IK-VR calculation is as follows:

$$\mathrm{IKVR} = \frac{1}{n}\sum_{i=1}^{n}(\mathrm{WE} \cdot \mathrm{tfidf})_i \text{ for tfidf} > t \quad (1)$$

3.3. Cosine Similarity

This third component of front-end layer will take IK-VR value and compare it with K-VR value from KA Knowledge Base using cosine similarity measurement. Calculation formula is as follows:

$$\cos(\theta) = \frac{\sum_{i=1}^{n} A_i B_i}{\sqrt{\sum_{i=1}^{n} A_i^2}\sqrt{\sum_{i=1}^{n} B_i^2}} \quad (2)$$

Given equation calculates the cosine similarity between two vectors where A_i and B_i are components of IK-VR and K-VR respectively. Output from this component is top ten list of IR candidates $(r^1 \ldots r^{10})$ that have the highest cosine similarity score.

3.4. Candidate Ranking (HR6)

The forth component of front-end layer will ranked IR candidates by calculating its HR6 score. HR6 (abbreviated Highest Ranking Six) is proposed to substitute the usual IR ranking function such as Okapi BM25 [14] by incorporating Kipling representation value Kp produced by our Ans2Q model. To generalize our function, Kipling representation Softmax KpS will be used instead of Kp one-hot representation. While one-hot representation is needed for supervised Ans2Q training (as discrete label class), HR6 needed more flexible value to cater for possible appearance of all six Kipling question words (Who, What, When, Where, Why and How) in user's input query. As such, we propose HR6 to be calculated using following equation (3). Given IK-VR from user input query Q and KpS value linked to candidate answer R from KA Knowledge Base, the HR6 score of R is:

$$\mathrm{score}(R, Q, KpS) = -\cos(\theta) + \left(\overrightarrow{KpS} \cdot \overrightarrow{IKp}\right), \quad (3)$$

where $\cos(\theta)$ equals cosine similarity between IK-VR and K-VR as calculated in *Cosine Similarity* equation, $KpS \rightarrow$ equals vector value of KpS (Kipling representation Softmax) associated with candidate R and $IKp \rightarrow$ equals vector value of IKp (Input Kipling representation) associated with input query Q. *KpS* is a six-dimensional vector with sum to one value, generated from Ans2Q Softmax layer. *KpS* is an original Softmax value before being transformed into one-hot encoding of *Kp* value (further explanation is in section 3.9. Response Kipling Clasification (Ans2Q)). *IKp* is a six-dimensional one-hot encoding value generated by matching each word in input query to six Kipling question words. We defined the sequence of Kipling words to be as follows: Who, What, When, Where, Why and How. As such, if the word "When" is presents in input query, *IKp* value would be [0,0,1,0,0,0]. We expect the input query to contain no more than one Kipling word but in the case where multiple words are presence, HR6 formula should relatively work the same as it calculates the dot product of two vectors (*KpS* and *IKp*). In essence, candidate R will get higher HR6 score when: 1) Cosine similarity value is low, 2) *IKp* has one *true* value (not all *false* or *zero* value), and 3) *IKp true* value index is same with *KpS* highest value index.

After all HR6 scores for IR candidates have been calculated, this component will rank candidates ($r^1 ... r^{10}$) with highest HR6 score on top to lowest HR6 score on bottom (highest score means the best matching candidate).

3.5. Response Generation
The fifth or last component of front-end layer will generate response (output y) back to user based on candidate ranking result. The highest ranked answer *A* linked to candidate K-VR and *Kp* value will be generated as output y provided that its HR6 score is above certain threshold. If the highest HR6 score is below the set threshold, this component will classify user's query as not-understandable by chatbot thus generating generic response as to get more/new input data.

3.6. WE-W Database
The first component of middle layer is Word Embedding - Weight (WE-W) database. WE-W contains two separate values for every word in general and specific dictionary. Those values are Word Embedding *WE* and Weight *W*. To populate WE-W as to become cumulative embedding database, two sets of dictionaries are required. One is general dictionary (base embedding) and the other is specific dictionary (complement embedding). General dictionary can be from any available public corpora (depending on chosen language), while specific dictionary must be from intended close domain corpora. Regarding this framework, WE-W database will be populated using component *8. WE-W Generation*.

3.7. KA Knowledge Base
The second component of middle layer is Keywords-Answer (KA) Knowledge Base. Generally, this database contains prewritten response/answer for chatbot to generate. Specifically, this database comprises of three separate values for every response item. Those values are: 1) *K* or K-VR that is a single representation of keywords for each *A*, generated using component *10. K-VR Generation* of this framework, 2) *A* that is a chatbot response/answer directly retrieved from specific/close-domain document, and 3) *Kp* that is a Kipling representation generated using Ans2Q model (component *9. Response Kipling Classification (Ans2Q)* of this framework). To get *K-VR* and *Kp*, WE-W must first be produced. Regarding this framework, K will be generated using component 10. K-VR Generation while Kp will be generated using component 9. Response Kipling Classification (Ans2Q).

3.8. WE-W Generation
The first component of back-end layer is a module to generate WE-W value for component 6. WE-W database. As mention before, WE-W is a group of two separate values which are WE and W. To generate WE, one can use any WE methods available nowadays such as GloVe [15] and Word2Vec [16]. Same for W, any word weighting methods can be used to calculate W. One of most used method in IR and chatbot field is TF-IDF. Term Frequency – Inverse Document Frequency (TF-IDF) is a numerical statistical method (a product of two statistics: TF and IDF) that calculates the importance of term (a word or phrase) based on its occurrence frequency in multiple documents (or in defined corpus space). The TF-IDF value increases proportionally to the number of times a term appears in the document (TF) [17] and is counterbalance by the number of times a term appears in the whole corpus (IDF) [18]. In other word, TF-IDF value is high when the term appears a lot in certain documents, but is low when the term appears a lot in a whole dictionary.

3.9. Response Kipling Classification (Ans2Q)
The second component of back-end layer is a module to generate *Kp* value for component *7. KA Knowledge Base*. To generate Kp, we propose a new NN model named Ans2Q (abbr. Answer-to-Question). To put thing into perspective as to why this model is proposed, consider following two questions: "How to do mileage claim?" and "What is mileage claim?". With IK-VR generation, every input word will be assigned a weight value that represents its semantic significance towards the whole

dictionary corpus. As such, the words "how", "to", "do", "what" and "is" will be regard as insignificance while the words "mileage" and "claim" will become the input's keywords. For that reason, both questions will have same interpretation by the system while in true context its represent different question type that requires different answer. Ans2Q intent to solve this issue by creating another tier in Answer representation that is a probability of which question type is most likely to be use in order to get the Answer. Following Figure 3 illustrates our proposed neural network model.

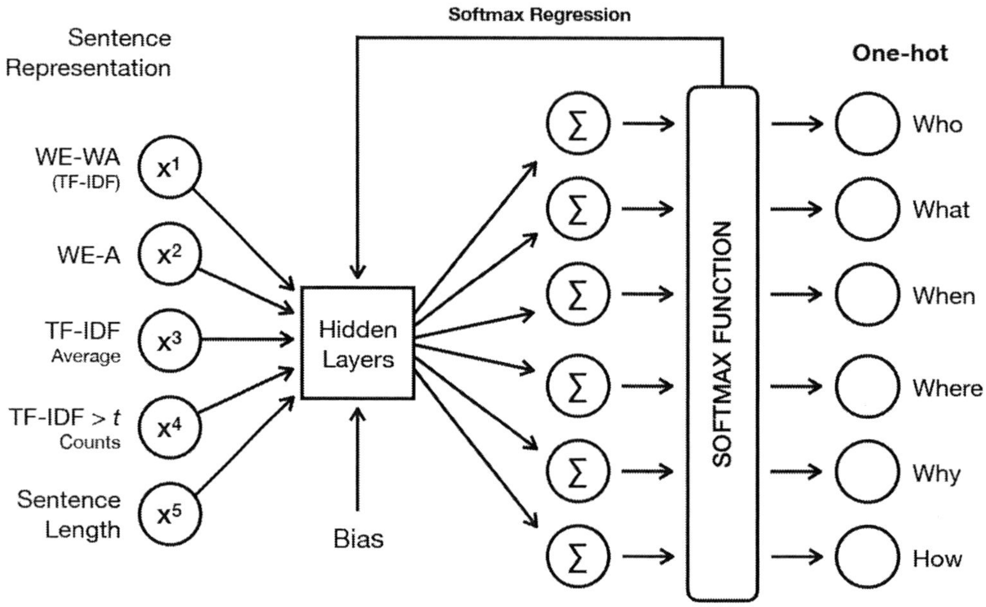

Figure 3. Neural network model for multinomial classification of Kipling question type.

Proposed model is intended to predict one-hot value of Kipling representation *Kp* for each Answer *A* data based on several independent variables of sentence/Answer representation. One-hot value is produced from Softmax function that takes the final value from Softmax Regression process that is done towards the model's hidden layers. Independent variables (input features) for Ans2Q are as follows:

- WE-WA (TF-IDF): Word Embedding - Weighted Average using TF-IDF value as weight for all word in the sentence.
- WE-A: Word Embedding - Average of all word in the sentence.
- TF-IDF Average: Average value of all words' TF-IDF.
- TF-IDF > t Counts: Total number of word with TF-IDF more than threshold.
- Sentence Length: Total number of word/token in the sentence.

All independent variables are Real Number value that need to pre-calculate prior to entering this model. As for dependent variables of one-hot value (label classification), it is a six-dimensional vector with binary matrix value (e.g [1,0,0,0,0,0] for Who and [0,1,0,0,0,0] for What). This vector will be determined based on Softmax function output where highest probability will be valued 1, while others will be 0 (this is the Kp value). Despite one-hot value, all original Softmax values for all classes will also be stored in database for later used (this is a complimented Kp value denoted as Kipling representation Softmax or KpS). In general, our proposed multinomial distribution probability can be represented using following Softmax equation that is an extension of logistic regression formula (sigmoid function) towards multiple classes problem.

$$p(y = j|\mathrm{x}) = \frac{e^{(\mathrm{w}_j^T \mathrm{x} + b_j)}}{\sum_{k \in K} e^{(\mathrm{w}_k^T \mathrm{x} + b_k)}} \quad (4)$$

Given Softmax equation calculates the probability of *Kp* class with index *j* based on given value x where w is weight, x is feature value, *b* is bias and *k* is the class label. *T* in the formula refers to applied layer (the Softmax Function layer) where in our case is represented by (6,1) vector; a six-dimensional vector with sum to one value.

3.10. K-VR Generation

The third component of back-end layer is a module to generate K-VR value for component *7. KA Knowledge Base*. K value is a vector representation (in real number) of keywords, designate for each *A* data in KA Knowledge Base. In our work, we denote *K* as K-VR that is an abbreviation of *Keywords - Vector Representation*. K-VR is calculate using following equation where *n* equals total number of word in *A* data with TF-IDF value is more than *t* threshold. The implementation of *t* is to limit *K* value as to only represents significant word for each document in specific dictionary, hence the term "keywords" (to be noted that K-VR equation is the same as IK-VR equation described in section 3.2. Sentence Representation Generation).

$$\mathrm{KVR} = \frac{1}{n} \sum_{i=1}^{n} (\mathrm{WE} \cdot \mathrm{tfidf})_i \ \text{ for tfidf} > t \quad (5)$$

4. Future Work

As with other conceptual proposal, proof of concept implementation is crucial. For this proposed framework, first task is to perform data preparation for middle layer database storage. We intend to pursue this by using specific banking institution (Bank X) Standard Operating Procedure (SOP) as our small scope domain data. Our case study is to develop a chatbot that can answer Bank X staffs' questions regarding their organizational SOP. Raw SOP documents will be normalized as to become group of sentence that represent specific topic in question. With normalized SOP data, back-end layer components can be prepared and trained to generate needed QA pairs for front-end layer components and its real-time processes.

5. Conclusion

This paper proposed a conceptual framework for IR chatbot that focuses on automated Q (question) representation generation. Our framework's main contributions are Ans2Q; a NN model for question type approximation and HR6; an IR score ranking calculation based on Ans2Q output. Although only conceptual, we provided detailed explanation on functionalities as well as calculation formula such that it is clear for anyone to implement this framework. Same with other NN training, biggest implementation challenge would be to fine tune Ans2Q so that Kp approximation is as good as manual human process. Sentence/answer representation parameters/features that goes into Ans2Q also need to be reevaluated after training as those proposed are only as an initial hypothesis.

References

[1] Serban, I. V., Sankar, C., Germain, M., Zhang, S., Lin, Z., Subramanian, S., Kim, T., Pieper, M., Chandar, S., Ke, N. R., Rajeshwar, S., Brebisson, A., Sotelo, J. M. R., Suhubdy, D., Michalski, V., Nguyen, A., Pineau, J., Bengio, Y. (2017). A deep reinforcement learning chatbot. *arXiv preprint arXiv:1709.02349*.

[2] Lokman, A. S., & Ameedeen, M. A. (2018, November). Modern Chatbot Systems: A Technical Review. In *Proceedings of the Future Technologies Conference* (pp. 1012-1023). Springer, Cham.

[3] Lokman, A. S., & Jasni M. Z. (2010). Chatbot enhanced algorithms: a case study on implementation in Bahasa Malaysia human language. In *International Conference on Networked Digital Technologies*, pp. 31-44. Springer, Berlin, Heidelberg.
[4] Liu, H., Lin, T., Sun, H., Lin, W., Chang, C. W., Zhong, T., & Rudnicky, A. (2017). RubyStar: A Non-Task-Oriented Mixture Model Dialog System. *arXiv preprint arXiv:1711.02781*.
[5] Qiu, M., Li, F. L., Wang, S., Gao, X., Chen, Y., Zhao, W., Chen, H., Huang, J., Chu, W. (2017). Alime chat: A sequence to sequence and rerank based chatbot engine. *In Proceedings of the 55th Annual Meeting of the Association for Computational Linguistics (Volume 2: Short Papers)*, (Vol. 2, pp. 498-503).
[6] Wei, C., Yu, Z., & Fong, S. (2018, February). How to Build a Chatbot: Chatbot Framework and its Capabilities. In *Proceedings of the 2018 10th International Conference on Machine Learning and Computing* (pp. 369-373). ACM.
[7] Lommatzsch, A. (2018, June). A Next Generation Chatbot-Framework for the Public Administration. In *International Conference on Innovations for Community Services* (pp. 127-141). Springer, Cham.
[8] Ma, S. P., & Ho, C. T. (2018). Modularized and Flow-Based Approach to Chatbot Design and Deployment. *Journal of Information Science & Engineering*, *34*(5).
[9] Yin, Z., Chang, K. H., & Zhang, R. (2017, August). DeepProbe: Information Directed Sequence Understanding and Chatbot Design via Recurrent Neural Networks. *In Proceedings of the 23rd ACM SIGKDD International Conference on Knowledge Discovery and Data Mining* (pp. 2131-2139). ACM.
[10] Cui, L., Huang, S., Wei, F., Tan, C., Duan, C., & Zhou, M. (2017). SuperAgent: A Customer Service Chatbot for E-commerce Websites. *Proceedings of ACL 2017, System Demonstrations*, 97-102.
[11] Sutskever, I., Vinyals, O., & Le, Q. V. (2014). Sequence to sequence learning with neural networks. *In Advances in neural information processing systems* (pp. 3104-3112).
[12] Harris, Z. S. (1954). Distributional structure. *Word*, *10*(2-3), 146-162.
[13] Baroni, M., Dinu, G., & Kruszewski, G. (2014). Don't count, predict! A systematic comparison of context-counting vs. context-predicting semantic vectors. In *Proceedings of the 52nd Annual Meeting of the Association for Computational Linguistics (Volume 1: Long Papers)* (Vol. 1, pp. 238-247).
[14] Robertson, S., & Zaragoza, H. (2009). The probabilistic relevance framework: BM25 and beyond. *Foundations and Trends® in Information Retrieval*, 3(4), 333-389.
[15] Pennington, J., Socher, R., & Manning, C. (2014). Glove: Global vectors for word representation. In *Proceedings of the 2014 conference on empirical methods in natural language processing (EMNLP)* (pp. 1532-1543).
[16] Mikolov T., Chen K., Corrado G., & Dean J. (2013). Efficient estimation of word representations in vector space. *arXiv preprint arXiv:1301.3781*.
[17] Luhn, H. P. (1957). 1 A Statistical Approach to Mechanized Encoding. *IBM journal*.
[18] Sparck Jones, K. (1972). A statistical interpretation of term specificity and its application in retrieval. *Journal of documentation*, *28*(1), 11-21.

Acknowledgments
This work was supported in part by Department of Higher Education, Ministry of Education Malaysia under the Fundamental Research Grant Scheme (FRGS), through Universiti Malaysia Pahang (Ref: FRGS/1/2018/ICT02/UMP/02/12).

IOP Conf. Series: Materials Science and Engineering **769** (2020) 012021 doi:10.1088/1757-899X/769/1/012021

Pathfinding Algorithms in Game Development

Abdul Rafiq[1], Tuty Asmawaty Abdul Kadir[2], Siti Normaziah Ihsan[3]

[1,2,3] Soft Computing and Intelligent System Research Group (SPINT)
Faculty of Computing, College of Computing and Applied Sciences
Universiti Malaysia Pahang, Kuantan, Pahang, Malaysia

E-mail: tuty@ump.edu.my

Abstract. This review paper provides an overview of a pathfinding algorithm for game development which focuses on the algorithms and their contribution to game development. The algorithms were categorised based on their search performance. The aim of this paper is to investigate and provide insights into pathfinding algorithms for game development in the last 10 years. We summarise all pathfinding algorithms and describe their result in terms of performance (time and memory). The result of this paper is metaheuristic techniques have better performance in terms of time and memory compared to heuristic techniques as a pathfinding algorithm.

1. Introduction

The rising global popularity of video games was the trigger to the increasing research interest in solving many AI issues related to video games such as decision-making, movement, strategy, and pathfinding. Commonly, pathfinding for a non-player character uses up a lot of CPU power and memory. This is a problem that has attracted constant attention from researchers. Pathfinding depends on the game environment where the obstacle may be static or dynamic. Video games mostly consist of three components which are player character, non-player character and others. Player character and non-player character are important roles in a video game. Player character is the main character controlled by a user while non-player character or known as NPC is a game object not controlled by a player in a video game [1]. Pathfinding is used for a non-player character (NPC) to find a path between the origin point to the goal point.

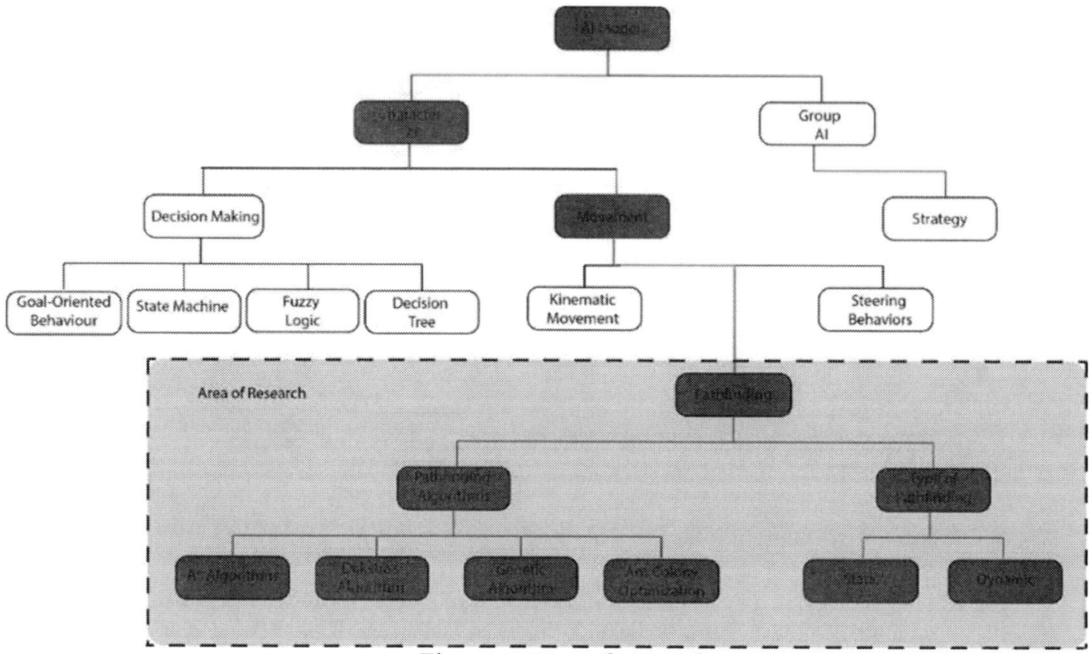

Figure 1. Area of research

Figure 1 illustrates an AI model for a video game. The AI model splits into three sections: Movement, Decision Making and Strategy. Decision-making refers to the character deciding what to do next. Movement refers to the character being able to move anywhere. Strategy refers to group strategy required to coordinate a team. The Movement and Decision-making sections contain algorithms that work for player characters or non-player characters. Pathfinding is part of movement besides kinematic movement and steering behaviours. The techniques or methods used to determine the path in the video game are called the pathfinding algorithms. Common pathfinding algorithms were developed to solve pathfinding problems such as Dijkstra [2], A* algorithm [3], genetic algorithms [4], and ant colony optimisation [5]. There are two types of pathfinding for a non-player character in a video game. The first is static pathfinding where the target node as the player does not move. The second is dynamic or real-time pathfinding where the target node as the player moves freely and randomly.

2. Pathfinding in Games

Pathfinding is a plotting node to find the shortest or minimum path between two points, which is from source to destination by a computer application. Pathfinding is a major component of many important applications in the fields of video games, robotics [5], crowd simulation [6], and GPS [7]. This paper focuses on pathfinding algorithms in video game development. Pathfinding algorithms are used for the agent or non-player character to find a path between the origin point to the goal point. Pathfinding is one of the requirements to create a realistic non-player character in a video game. A video game can be fun and entertaining especially when the non-player character is realistic enough. However, the main problem for a video game is the need for an optimal pathfinding for non-player characters. Garham has supported this by stating that the common problem in a video game is to find suitable pathfinding for agent movement. Pathfinding is implemented in any condition such as static, dynamic and real-time environments. The techniques or methods used to determine the path in a video game are called pathfinding algorithms. Pathfinding algorithms were developed to solve pathfinding problems such as Dijkstra, A* algorithm, genetic algorithms, and ant colony optimisation. Pathfinding algorithms are used to solve the shortest path problem and the optimal path. Usually, A* and Dijkstra's algorithms are used as a solution method to find the shortest path. Ant colony optimisation (ACO) and genetic algorithm are used as a search technique to find the minimum cost path in a graph.

2.1 Heuristic techniques

Heuristic is used to solve the problem in a faster and efficient way by optimal solution, accuracy, and precision [6]. Heuristic algorithms aim to find a good solution to a specific problem like pathfinding in a reasonable amount of computation time but with no guarantee of efficiency. 'Heuristic' means to find in Greek. Based on previous studies, many algorithms were developed to solve pathfinding problems such as the A* algorithm, Dijkstra's algorithm, and Depth First Search. The A* algorithm and Dijkstra's algorithm are the most popular techniques. All these techniques can be categorised as heuristics.

2.1.1 A algorithms*

The A* algorithm is one of the popular techniques used for pathfinding due to its accuracy and performance. It is used to find the shortest path between two nodes. The A* algorithm has been applied in several video game genres such as real-time strategy games [9], role-playing games [10], racing games [11] and turn-based strategy games. The A* algorithm was introduced by Hart, Nilsson, and Raphael in 1967 [12] to solve many problems, pathfinding in a video game being one of them. In the Non-Player (NPC) context, pathfinding is used to guide between two node points in order to capture the player character.

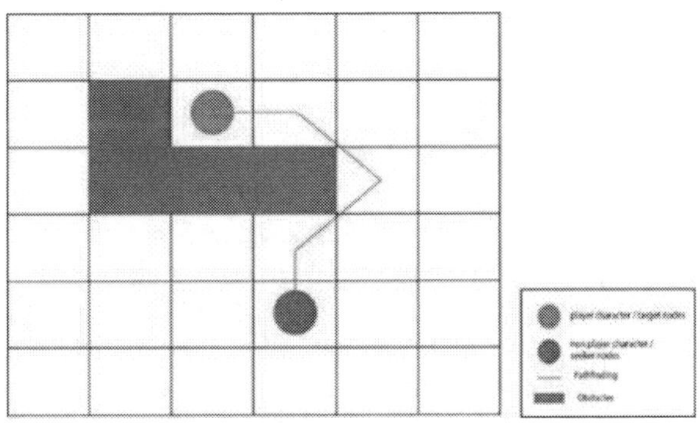

Figure 2. A* algorithm

The A* algorithm always tries to carry pathfinding by exploring the minimum value or lowest path to give the best minimum solution. The A* algorithm implements a heuristic function to evaluate that lowest path. The heuristic function allows the algorithm to quickly and accurately estimate the path. The advantage of the A* algorithm is it is very easy to understand the flow and logic. Due to its simplicity, the A* algorithm has always been chosen by programmers to solve pathfinding problems. This is because the A* algorithm finds the minimum solution by finding the shortest path. A* uses a heuristic function $f_{(n)}$ to determine the node. The value of the function $f_{(n)}$ is:

$$f_{(n)} = g_{(n)} + h_{(n)} \quad (1)$$

$g_{(n)}$ is the cost that is required to reach a target node from the starting node. $g_{(n)}$ will calculate the cost so far to reach the target node. $h_{(n)}$stands for heuristic value, where estimate form node to target node. If the grid has obstacles, $f_{(n)}$ will estimate and pick the lowest cost to give a good result. The A* algorithm becomes Dijkstra's algorithm when $h_{(n)}$ is zero which is guaranteed to find the shortest path.

IOP Conf. Series: Materials Science and Engineering **769** (2020) 012021 doi:10.1088/1757-899X/769/1/012021

2.1.2 Dijkstra's algorithm

Dijkstra's was introduced in 1970 by Holland as one of the best path algorithms [13]. Dijkstra's algorithm is a classic graph search algorithm which can find the shortest path between two points on the graph. It has been applied in many areas such as network routing [14], public transportation [15], and logistics [16]. Dijkstra's algorithm was the only choice for a long time in pathfinding until the A* algorithm became a popular method for pathfinding in video games [17]. We will discuss how Dijkstra's algorithm works as a pathfinding algorithm.

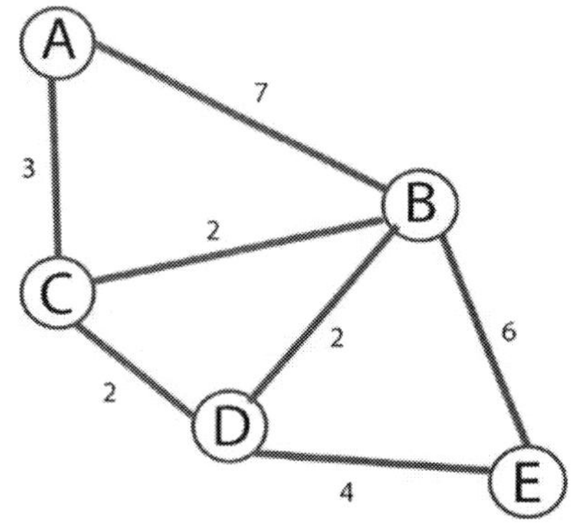

Figure 3. Dijkstra's algorithm

Figure 3 shows a graph with five nodes (A, B, C, D, E) and seven paths. There are many possible paths that will allow us to reach the E node from the A node. Dijkstra's algorithm will help us find the shortest path between the two nodes. From the starting node which is the A node, it visits the path with the smallest value known as the distance. Once it has moved to the smallest node, it will check each of its neighbouring nodes. For each neighbour node, they will calculate the distance from the starting node. If the distance or current path is less than another path, it will update the shortest path.

Table 1. The path node

Node	The path from a node	Shortest path
A	0	-
B	7 & 5	5
C	3	3
D	5,9,11	5
E	9,14,15 & 13	9

Table 1 shows the shortest path from the A node to each node. There are two pathways from A to B and the shortest path is 5. The only path from A to C has a value of 3. We want the shortest path from the A to E nodes. After calculating all the possible paths, we finally find the shortest path by visiting all the related nodes. The shortest distance from the A to E nodes is 9. Dijkstra's algorithm is a simple searching algorithm and easy to understand. It will explore each node to find the shortest path. The advantage of Dijkstra's algorithm is the coding is very easy to implement. Although Dijkstra's algorithm is easy to understand, it is not the best method to solve pathfinding problems.

2.2 Meta-Heuristic techniques

Metaheuristics are general high-level strategies that combine lower-level techniques for exploration and exploitation of the search space. Metaheuristic is a higher level heuristic. Generally, it performs better than heuristic. Metaheuristics can reduce search time and look good enough to solve complex pathfinding for a video game. Based on the study, metaheuristic algorithms such as genetic algorithm and ant colony optimisation were used to solve the pathfinding problem in a video game. Metaheuristics are based on some natural phenomenon. The most successful metaheuristic algorithms have been inspired by natural systems. For example, ant colony optimisation (ACO) and the bee algorithm were developed based on the behaviour of animals.

2.2.1 Genetic Algorithm

The genetic algorithm is one of the popular techniques used for searching in computer science. It was used to find a heuristic solution for optimisation and search problems in a large area of space. The genetic algorithm was introduced by John Holland [7]. It has been applied in many areas such as biomimetic invention [8], automotive design [9], engineering design [9] and robotics [10]. The genetic algorithm uses biological methods such as mutation and inheritance. GA converts the decision variables of the search problem into strings. The strings act as candidate solutions and the search problem is referred to as chromosomes. The alphabets act as genes and the values of genes are called alleles. For example, a salesman has a travelling problem and the route represents chromosomes while the city is a gene. In video games, the genetic algorithm has been applied as a pathfinding algorithm. The genetic algorithm and best-first search were used to optimise the search for the path between two points [11]. The result shows that GA has a better performance on a map with obstacles compared to best-first search. Other than that, the experiment to optimise pathfinding using a techniques genetic algorithm A* algorithm. The result shows that RTP-GA has a better performance in term of searching time on maps with obstacles compared to the A*algorithm [4]. Another result shows GAMMAs find paths quicker than the A* algorithm in terms of time [10].

2.2.2 Ant Colony Optimisation

Ant Colony Optimisation (ACO) is a technique inspired by the behaviour of ants when navigating a path from their nest to food sources. It was first described in 1997 by Gambardella Dorigo [12]. The Ant Colony Optimisation (ACO) technique is based on the ants' ability to find the shortest path between their nest and food sources. Firstly, each of the ants starts moving randomly. Each member of the ant colony tries to find a food source. After finding food, they will communicate with each other via pheromone trails to navigate the shortest path. Pheromones are a type of chemical substance released into the environment by ants. The connection line will be created as pathfinding between the starting point (nest) to the target point (food).

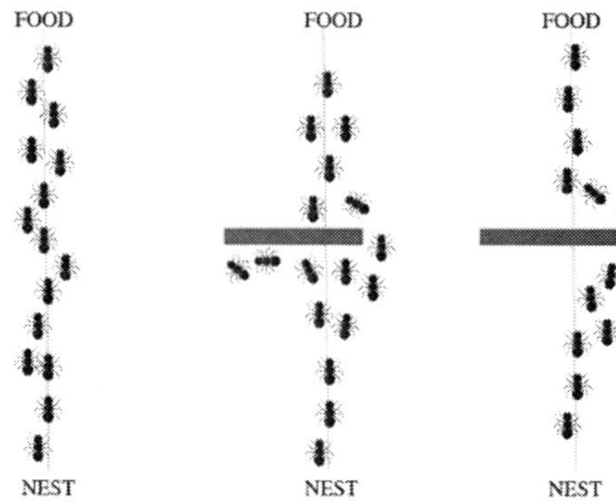

Figure 4. Ant Colony Optimisation

Figure 4 shows that the main idea of Ant Colony Optimisation (ACO) is mimicking real ant behaviour. In the first picture, each ant moves in a line from its nest to the food. In the middle picture, an obstacle is placed in the line between their nests to the food. Each ant moves randomly left and right to avoid the obstacle. The ants go around the obstacle and choose to turn right because it will reach the food quicker, whereas by turning left it will take a longer time and path. Pheromones are a chemical substance released into the environment by ants. Finally, the pheromones show a shorter and faster path around the obstacle.

The main objective of Ant Colony Optimisation (ACO) is as a search technique to find a minimum cost path in a graph. Ant Colony Optimisation (ACO) is used in many applications in the fields of video games [13], robotics [14], submarine [15] and so on. In the field of robotics, this paper proposes Ant Colony Optimisation (ACO) to improve path planning for robots. Ant Colony Optimisation (ACO) is applied in pathfinding for a robot to find an optimal path from a source point to a goal point while avoiding any obstacles in the configuration space. Other than that, Ant Colony Optimisation (ACO) is used for vehicles such as submarines. This paper is a study of a modified Ant Colony Optimisation (ACO) technique to improve pathfinding for a submarine in the ocean [15]. The Ant Colony Optimisation (ACO) technique has an advantage over the A* algorithm. The ant colony can adapt to changes in real time. According to V. Selvi, the ant colony optimisation (ACO) technique can be used in a dynamic application because it is able to adapt to changes such as new distances [16]. Due to its advantage, many researchers have applied and improved the ant colony technique to solve a certain complex environment.

3. Findings

Table 2 shows the summary of pathfinding algorithms used for game development. The papers are selected from those published between 2007 and 2018. Only 10 papers are related to the study. The summary is categorised into 5 sections; Year, Authors, Pathfinding Algorithms, Time, and Memory.

Table 2. Summary of pathfinding algorithms

Year	Authors	Pathfinding Algorithms	Time	Memory
2007	Leigh [17]	Genetic algorithm, A* algorithm	GAMMAs find paths quicker than A* algorithm	-
2011	Machado [4]	Genetic algorithm, A* algorithm	RTP-GA has a better performance in term of searching time on maps with obstacles compared to A*algorithm	-
2012	Santos [11]	Genetic algorithm	PPGA has a better performance on the map with obstacles compared to Best-First Search	-
2012	Recio [18]	Ant Colony Optimization (ACO)	-	-
2016	Hunkeler [5]	Ant Colony Optimization (ACO), Genetic algorithm	Ant Colony Optimization has a better performance compared to genetic algorithm	-
2016	Zikky [2]	Dijkstra's algorithm, A* algorithm	A* algorithm better than Dijkstra's algorithm in solving the shortest pathfinding problem	-
2017	Firmansyah [19]	A* and improved A* algorithm	Improved A* get the fastest path, efficient and short time compared to basic A* algorithm	-
2017	Primanita [20]	Iterative Deepening A* (IDA*), A* algorithm	If the map is without any obstacle IDA* better than A* in terms of time usage but IDA* worse than A* if the map has obstacles.	If the map is without any obstacle, IDA* better than A* in terms of CPU memory
2018	Sabri [3]	A* algorithm, Bee algorithm	A* algorithm faster than Bee in a simple map	Bee better than A* in a complex map
2018	Sazaki [21]	Hybrid A* Algorithms	Hybrid A* get better results compared to A* algorithm on an empty track	Hybrid A* get better results compared to A* algorithm on an empty track

4. Results

In this section we summarise and conclude all the pathfinding algorithms such as the A* algorithm, Dijkstra's algorithm, Genetic algorithm, and Ant Colony Optimisation used for game development. We start with the common pathfinding algorithms used for game development. Then, we summarise and conclude the result based on the performance of time and memory.

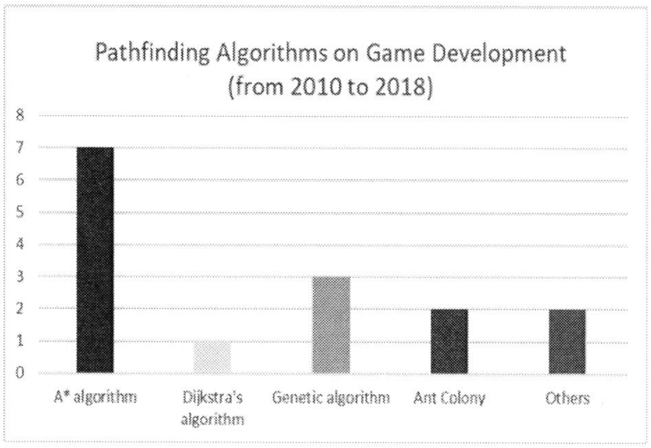

Figure 5. Result of Pathfinding Algorithms

The bar graph illustrates the pathfinding algorithms (Dijkstra's algorithm, A* algorithm, Genetic Algorithm, Ant Colony Optimisation, and Others) used for game development from 2010 until 2018. Overall, The A* algorithm is the most popular technique applied to pathfinding in game development. Both Dijkstra's algorithm and Best First Search (Others) are lesser used techniques applied on pathfinding for game development. Furthermore, metaheuristic techniques such as genetic algorithm and ant colony optimisation have seen a constant increase in usage as pathfinding algorithms from 2010-2019.

In terms of heuristic, the A* algorithm is the most commonly used technique for pathfinding compared to others such as Dijkstra's algorithm and best-first search. The advantage of the A* algorithm is its flow and logic are very easy to understand. Furthermore, the A* algorithm uses a heuristic function which allows the algorithm to quickly and accurately estimate a path. Due to its advantages, the A* algorithm has always been chosen by game programmers as the pathfinding algorithm in game development.

There are many available metaheuristic techniques, but we have found that the genetic algorithm and ant colony optimisation are most commonly used in pathfinding for games. Both algorithms are compared with the A* algorithm. Based on the study, both algorithms perform better than the A* algorithm in terms of time and memory usage. Metaheuristic is a higher level heuristic and generally performs better.

4.1 Time

In this section, we gathered all the pathfinding algorithms based on the result of time. The papers are selected from those published between 2007 and 2018. Only nine papers are related to the study. All the pathfinding algorithms were used for game development. We summarise and conclude the pathfinding algorithms in terms of time performance.

Table 3. Result of time

Years	Authors	Result (Time)
2007	[17]	GA better than A*
2011	[4]	GA better than A*
2012	[11]	GA better than First Search
2016	[5]	ACO better than GA
2016	[2]	A* better than Dijkstra
2017	[19]	Improved A* better than A*
2017	[20]	IDA* better than A*
2018	[3]	A* better than Bee (in the simple map)
2018	[21]	Hybrid A* better than A*

Table 3 shows the results of pathfinding algorithms used for game development based on the performance of time. For heuristic techniques, the A* algorithm is faster than Dijkstra for calculating and searching a path [2]. However, a few researchers improved the A* algorithm. In 2017, Firmansyah [19], Primanita [20] and Sazaki [21] improved the A* algorithm and the result was much better than the basic A* algorithm. Overall, for heuristic techniques, the A* algorithm has a good result compared to others, but it can be bettered by the improved A* algorithm.

Metaheuristic techniques are better than heuristic techniques. From the results in Table 4, the Genetic algorithm is faster than the A* algorithm [17], [4]. The Genetic algorithm is also better than another heuristic technique like Best First Search [11]. Metaheuristic is a higher level of heuristic. Generally, it performs better than heuristic. In 2016, the experiment by Hunkeler showed that Ant Colony Optimisation is faster than the Genetic Algorithm [5]. Overall, we can conclude that metaheuristic techniques such as GA, ACO can be a pathfinding algorithm and the results show that they are better than heuristic techniques.

4.2 Memory

In this section we gather all the pathfinding algorithms based on the result of memory. The papers are selected from those published between 2017 and 2018. Only three papers are related to the study. All the pathfinding algorithms were used for game development. We summarise and conclude the pathfinding algorithms in terms of memory usage performance in this section.

Table 4. Result of memory

Year	Author	Result(Memory)
2017	[20]	IDA* better than A* on an empty map
2018	[3]	Bee algorithm better than A* on a complex map
2018	[21]	Hybrid A* gets better results compared to A* algorithm on an empty track

Table 4 shows the results of pathfinding algorithms used for game development based on the performance of memory usage. Three of 10 papers investigated the performance of pathfinding algorithms in terms of memory. For heuristic techniques, the improved A* algorithms like IDA* [10] and Hybrid A* [31] have better results compared to the basic A* algorithm. From the result, we can conclude that improving the algorithm results in reduced memory usage. For a complex map, metaheuristic techniques are very good compared to heuristic techniques. For example, the Bee algorithm shows a better result in a complex map compared to the A* algorithm [2]. Overall, we can conclude that metaheuristic techniques such as the Bee algorithm can be a pathfinding algorithm and the result shows that it is better than heuristic techniques like the A* algorithm.

5. Summary and Future Trend

This paper provided a brief description of the common pathfinding algorithms (A* algorithm, Dijkstra's algorithm, Genetic algorithm, and Ant Colony) used for game development. In this review paper, we summarised all the pathfinding algorithms and described the result based on their performance in terms of time taken and memory usage. Based on the result, we found that the A* algorithms are the most popular techniques among the other techniques. The improved A* algorithm gets the fastest path and shortest time. However, it takes too much memory to calculate the path in a dynamic environment. Overall, metaheuristic techniques have better performance in terms of time and memory compared to heuristic techniques. In the future, we need good pathfinding algorithms that work for new technology such as Virtual Reality (VR), Augmented Reality (AR), and Hologram.

Acknowledgements

The authors would like to thank Universiti Malaysia Pahang (UMP). This work is supported by Universiti Malaysia Pahang (UMP) and funded by Ministry Education Malaysia under FRGS Grant FRGS/1/2016/ICT01/UMP/02/2 and Master Research Scheme (MRS).

References

[1] C. Foudil, D. Noureddine, C. Sanza, and Y. Duthen, "Path Finding and Collision Avoidance in Crowd Simulation," *J. Comput. Inf. Technol.*, vol. 17, no. 3, p. 217, 2009.

[2] M. Zikky, "Review of A* (A Star) Navigation Mesh Pathfinding as the Alternative of Artificial Intelligent for Ghosts Agent on the Pacman Game," *Emit. Int. J. Eng. Technol.*, vol. 4, no. 1, pp. 141–149, 2016.

[3] A. N. Sabri, N. H. M. Radzi, and A. A. Samah, "A study on Bee algorithm and A∗ algorithm for pathfinding in games," *ISCAIE 2018 - 2018 IEEE Symp. Comput. Appl. Ind. Electron.*, pp. 224–229, 2018.

[4] A. F. D. V. Machado *et al.*, "Real time pathfinding with genetic algorithm," *Brazilian Symp. Games Digit. Entertain. SBGAMES*, pp. 215–221, 2011.

[5] I. Hunkeler, F. Schar, R. Dornberger, and T. Hanne, "FairGhosts - Ant colony controlled ghosts for Ms. Pac-Man," *2016 IEEE Congr. Evol. Comput. CEC 2016*, pp. 4214–4220, 2016.

[6] L. A. Wolsey, "Heuristic Algorithms," *Integer Program.*, no. January, p. 17, 1998.

[7] J. Car, "An Introduction to Genetic Algorithms.," *Artif. Life*, vol. 3, no. 1, pp. 63–65, 2014.

[8] A. Cenys, D. Gibavicius, N. Goranin, and L. Marozas, "Genetic algorithm based palm recognition method for biometric authentication systems," *Elektron. ir Elektrotechnika*, vol. 19, no. 2, pp. 69–74, 2013.

[9] H. Yu and N. Yu, "Application of Genetic Algorithms To Vehicle Suspension Design," pp. 1–9, 2003.

[10] T. W. Manikas, K. Ashenayi, and R. L. Wainwright, "Genetic algorithms for autonomous robot navigation," *IEEE Instrum. Meas. Mag.*, vol. 10, no. 6, pp. 26–31, 2007.

[11] U. O. Santos, A. F. V Machado, and E. W. G. Clua, "Pathfinding Based on Pattern Detection Using Genetic Algorithms," *XI Brazilian Symp. Games Digit. Entertain.*, pp. 64–72, 2012.

[12] M. Dorigo and C. Blum, "Ant colony optimization theory: A survey," *Theor. Comput. Sci.*, vol. 344, no. 2–3, pp. 243–278, 2005.

[13] I. Sabuncu, "Work-In-Progress: Solving Sudoku Puzzles Using Hybrid Ant Colony Optimization Algorithm," *2015 1st Int. Conf. Ind. Networks Intell. Syst.*, pp. 181–184, 2015.

[14] H. Wang, Z. Wang, L. Yu, X. Wang, and C. Liu, "Ant Colony Optimization with Improved

Potential Field Heuristic for Robot Path Planning," *Chinese Control Conf. CCC*, vol. 2018-July, pp. 5317–5321, 2018.

[15] Y. Shan, "Study on submarine path planning based on modified ant colony optimization algorithm," *Proc. 2018 IEEE Int. Conf. Mechatronics Autom. ICMA 2018*, pp. 288–292, 2018.

[16] Vs. DrRUmarani Lecturer and A. professor, "Comparative Analysis of Ant Colony and Particle Swarm Optimization Techniques," *Int. J. Comput. Appl.*, vol. 5, no. 4, pp. 975–8887, 2010.

[17] R. Leigh, S. J. Louis, and C. Miles, "Using a genetic algorithm to explore A*-like pathfinding algorithms," *Proc. 2007 IEEE Symp. Comput. Intell. Games, CIG 2007*, pp. 72–79, 2007.

[18] G. Recio, E. Martin, C. Estebanez, and Y. Saez, "AntBot: Ant colonies for video games," *IEEE Trans. Comput. Intell. AI Games*, vol. 4, no. 4, pp. 295–308, 2012.

[19] E. R. Firmansyah, S. U. Masruroh, and F. Fahrianto, "Comparative analysis Of A∗ and basic theta∗ algorithm in android-based pathfmding games," *Proc. - 6th Int. Conf. Inf. Commun. Technol. Muslim World, ICT4M 2016*, pp. 275–280, 2017.

[20] A. Primanita, R. Effendi, and W. Hidayat, "Comparison of A∗ and Iterative Deepening A∗ algorithms for non-player character in Role Playing Game," *ICECOS 2017 - Proceeding 2017 Int. Conf. Electr. Eng. Comput. Sci. Sustain. Cult. Herit. Towar. Smart Environ. Better Futur.*, pp. 202–205, 2017.

[21] Y. Sazaki, A. Primanita, and M. Syahroyni, "Pathfinding car racing game using dynamic pathfinding algorithm and algorithm A∗," *Proc. - ICWT 2017 3rd Int. Conf. Wirel. Telemat. 2017*, vol. 2017-July, pp. 164–169, 2018.

IOP Conf. Series: Materials Science and Engineering **769** (2020) 012022 doi:10.1088/1757-899X/769/1/012022

Text-based emotion prediction system using machine learning approach

[1]Ahmad Fakhri Ab. Nasir, [1]Eng Seok Nee, [1]Chun Sern Choong, [1]Ahmad Shahrizan Abdul Ghani, [1]Anwar P. P. Abdul Majeed, [1]Asrul Adam and [2]Mhd. Furqan

[1]Faculty of Manufacturing and Mechatronic Engineering Technology, Universiti Malaysia Pahang, Pekan, Malaysia
[2]Department of Computer Science, Universitas Islam Negeri Sumatera Utara Medan

Abstract. Text-based input becomes a common channel for humans in sharing their opinions/emotions to the product or service through online social media, shopping platform etc. Humans are easy to make errors in interpreting emotions, especially the emotion that derived from text based. The main aim of this study is to develop text-based emotion recognition and prediction system. Several market challenges facing in the advancement of emotion analysis with accuracy being the main issue. Therefore, four supervised machine learning classification algorithms such as Multinomial Naïve Bayes, Support Vector Machine, Decision Trees, and *k*-Nearest Neighbors were investigated. The model was developed based on Ekman's six basic emotions which are anger, fear, disgust, joy, guilt and sadness. Data pre-processing techniques such as stemming, stop-words, digits and punctuation marks removal, spelling correction, and tokenization were implemented. A benchmark of ISEAR (International Survey on Emotion Antecedents and Reactions) dataset was used to test all models. Multinomial Naïve Bayes classifier resulted the best performance with an average accuracy of 64.08%. Finally, the best model was integrated to graphical user interface using Python Tkinter library to complete the whole system development. Besides, the detailed performance of the best model such as *tf-idf* and count vectorizer, confusion matrix, precision-recall rate, as well as ROC (Receiver Operating Characteristic) score were also discussed. Text-based emotion prediction system to interpret and understand human emotions was successfully developed.

1. Introduction

Human emotions can be expressed across two channels which is via verbal (expressed emotion in words, sounds or speech) and non-verbal (includes expression through facial, gesture or body movement). The way of human-computer interaction which allow computers to interpret and understand human's emotional and attentional expressions is necessary for the computer in several applications [1]. Emotion recognition (also known as Artificial emotional intelligence) is an affective computing whereby the development of systems utilised in the process of detecting, interpreting, and predicting human emotional state such as anger, happiness, sadness, etc. Affective computing is a device for computer interacts naturally with human where "affect" refers to emotion expression and the computing that measure, identifies with, emerge from and intentionally emotion affect [2]. This kind of system may have plenty helpful applications, for example it can use to measure how pleasant the residents of the country are. Since the late 20th century, most of the government and association in this globe concerned about the happiness economics which is the study of measurement on residents' happiness advance around health, life quality, sociology and economics field. This system is also useful in suicide prevention which initially interpreting user suicidal thoughts that shared in online platforms then some further action is taken to save the user with heavy suicidal thought [3]. Besides, emotion recognition system also applied in analysing and understanding users or customer's satisfaction toward product or services based on customer review [4]. This indirectly advantages to a company in improving company's profit.

Nowadays, the World Wide Web website has emerged into Web2.0 which allow humans to communicate with each other through social media sites (such as Facebook, Twitter, and Instagram), video sharing sites (such as YouTube) or blog. Human widely interact with computer via texts while multimodal human-computer interaction denoted to be appealing. It cannot be denied that most of the people like to share their opinion or express their emotions via online social media or comment [17-18]. Thus, expressing emotion in texts or words become common, a huge amount of data especially textual data are generated and detecting the text-based emotion plays the biggest challenge for either human or machine. Publishing negative or positive complaints on highly visible websites increases the likelihood that the public will be conscious of the customer's complaint. For example, if a person with many "followers" or "friends" published his/her opinions in a simple manner of expression on social media (such as Facebook, Instagram and Twitter), it may go "viral". Yet, a negative comment might reflect a big influence to the company reputation. For this reason, answering the complaints made by customers become a quite challenging process for company. They need to train their staff to be as possible on how to handle customer complaints, especially dealing with social media channels. In such a context, it is important to analyse and predict early stage customer's emotion of a service conversation.

The rest of the paper is organized as follows: Section 2: Related works; presented the background knowledge of textual based emotion recognition. Section 3: Methodology; portrayed the process flow, the strategy, and the components of research, including materials and methods used. Section 4: Results and discussions; explained and discussed all of the results obtained from the experiment. Section 5: Conclusions; provided the research conclusions and the recommendations for future research works.

2. Related works

Textual based emotion recognition and detection is a recent field of study which is familiar to sentiment analysis. Previous studies summarised that analysis on emotion detection can be divided into two types: (i) sentiment analysis and (ii) emotion analysis (or affective computing). Usually, sentiment analysis and emotion analysis are applied interchangeably. Based on the comparison between sentiment and emotion analysis done by [5] was clearly distinguished the concept for both analysis. Sentiment analysis is used to calculate and to detect the emotional polarity (positive, negative or neutral) from user input in textual based. Conversely, emotion analysis used to recognize and interpret the state of emotion expression from user input texts depends on several emotion categories. According to psychological models, human's emotion can be categorised into six basic emotions (Ekman's model) such as anger, disgust, fear, happiness, sadness and surprise [6]. Based on the collection information of previous studies on emotion recognition, Ekman emotion categories has been utilised and popular in most of the affective computing research studies and in various systems that develop to detect emotional state from text-based data [7-11].

There are several approaches that can be utilised in emotion detection: (i) lexicon-based approaches and (ii) machine learning based approaches (including supervised and unsupervised learning). Lexicon based approach (is defined as a word-dictionary) is a technique that just utilised single or multiple lexical resources to detect human emotion. Lexicon-based approaches considered as an extension work of keyword-based approaches that referred certain predefined keyword to detect the emotion in text [12-14]. Conversely, machine learning based techniques is achieved with the construction and development of algorithms to make the computer able to "learn" from data and making data driven predictions without being explicitly compute. There are two types of machine learning: (i) supervised and (ii) unsupervised.

Supervised machine learning techniques is depending on a human-labelled training dataset. It is completed via interpreted a set of labelled training data then inferred a function that mapping a new user input to output according to sample input-output pair or training dataset [15]. A vast and structured collection of words of a labelled corpus is required annotated with emotional tags. Handling with annotation task is time-consuming and unproductive and make the approach less efficient. However, there is some of the related studies on textual based emotion recognition in Twitter with micro-blog post where the collected corpus is labelled and annotated automatically by hashtags and existed emoticon [16-20]. Besides, some studies introduced supervised machine learning approaches in emotion detection by using Naïve Bayes classifier [21-22]. Supervised machine learning approaches can be performed by developing an emotion classifier with collected corpus from micro-blogging platform based on multi-class SVM (Support Vector Machine) kernels and analyse the sentiments followed the Ekman basic emotions model [23-24]. Other studies utilised SVM technique in analysis emotion classification, such as [9, 18, 25-26]. Moreover, there is another related work on textual based emotion recognition using supervised machine learning presented in [17, 27, 29]. On the other hand, unsupervised learning is a machine learning technique which trying to make a decision by analysing the hidden pattern in unlabelled training data in order to develop emotion classification models [11, 28]. Unlike supervised machine learning approach, input and output data are unknown and unlabelled in unsupervised learning algorithms.

Research has demonstrated the performance of supervised learning using difference types of machine learning algorithm. However, a limited or no attempts have thus far devoted to measure the performance for several machine algorithms on this field. Therefore, the current investigation also aims to evaluate the performance of four different text classifiers such as Support Vector Machine (SVM), Naïve Bayes, *k*-Nearest Neighbour (*k*-NN) and Decision Tree.

3. Methodology

This section shows the methodology of this work and the overall research framework is illustrated in Figure 1.

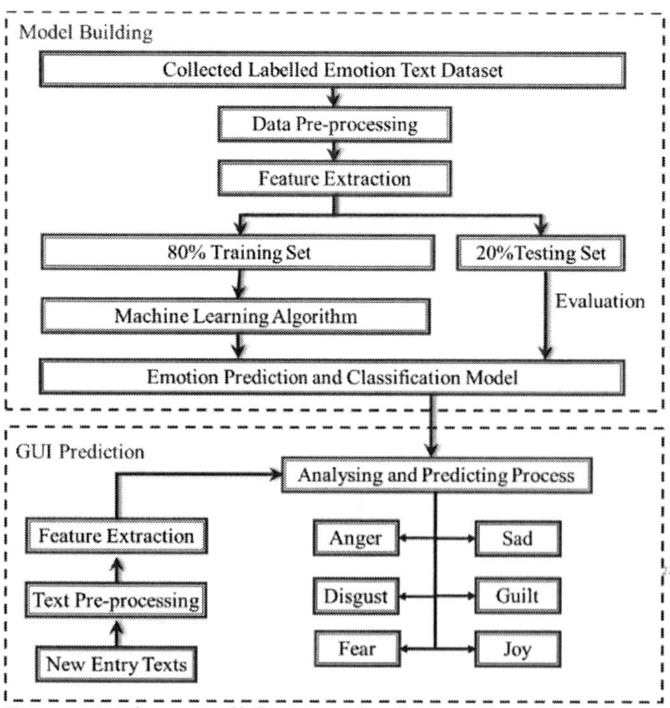

Figure 1. The overall framework of research methodology.

3.1. Model of Emotions
The emotion model created was tailored to fit with customer service applications. The emotions selected were categorized into two main sections, namely negative and positive. The negative emotions will consist of five basic emotions labelled as anger, sadness, fear, disgust and guilt and the positive emotions is happiness.

3.2. Data Source
Several resources were used in this project such as ISEAR (The International Survey on Emotion Antecedents and Reactions) dataset and NLTK (Natural Language Toolkits) corpus. ISEAR dataset is an emotion-labelled-datasets where the data has been cleaned and slightly normalized. It contains seven major emotions: joy, fear, anger, sadness, disgust, shame and guilt. Since ISEAR is a cleaned and established database, it was being tested for finding the best classification algorithms. Then, ISEAR also being further evaluated. Meanwhile, NLTK (Natural Language Toolkits) corpus is used as text processing libraries since in a real life, the input text is usually contains stop-words, wrong spelling, etc. Therefore, these unwanted text is needed to be removed. All data source utilised in this study is illustrated in table 1.

Table 1. Information of data source

Emotion class	ISEAR
Anger	1096
Disgust	1096
Sadness	1096
Fear	1095
Joy	1094
Guilt	1093
Total number	6570

3.3. Programming Language and Tools
In this project, Python programming language is fully utilised in developing text-based emotion prediction system. It is a free, open source software which consists of modules for designing Graphical User Interface (GUI) and connecting to relational database. Moreover, it contains numerous tools and libraries which suitable to use in this project such as natural language toolkit that helped in text processing and data analysis it was frequently utilised in most of the artificial intelligent project such as building neural network or machine learning model with the help of built in libraries. The list of libraries and toolkits in Python that required in this project: (i) Pandas, (ii) Scikit-learn (sklearn), (iii) Natural Language Toolkit (NLTK), (iv) Numpy, (iv) Matplotlib, (v) TextBlob toolkits, and (vi) Tkinter (GUI toolkit)

3.4. Model Development
First and foremost, text dataset in csv file with the collected labelled emotion was imported and loaded using Pandas dependency. Before splitting the dataset into training and testing, the labelled emotion text undergoes data pre-processing and feature extracting processes.

3.4.1. Data pre-processing. (i) Stemming - every single or plural English word can be a noun, a verb, and an adjective. For example: the single word of "programmer" and plural word of "programmers" are as noun form, "program" as verb form and "programming" as adjective form which are return to stem word of "program" after stemming process. In short, stemming process is a process of removing the prefix or suffix and returned word into originate word which might not be an actual word in dictionary such as word "angry" will become "angri". Stemming is helping in minimizing database size and improving retrieval performance. (ii) Removing stop-words - the purpose of data cleaning is to remove unwanted word that does not contain any significant emotion message in application. Stop-word is an unwanted frequent occurring word which hold needless space in database and expand the processing period. The sample of text is "you're", "you've", "you'll", "couldn't", "didn't", "doesn't", etc. (iii) Remove all the digits (0-9) and punctuation marks (such as [#"$!%&'^_` ()./:;<>=*+,-?@[]{}|\~]:). (iv) Spelling correction - the input text which contain misspelling is converted into correct word based. For example: the typing error of word "speling" undergoes correcting process and will result as "spelling". (v) Tokenization - it is a process used to split the input data or texts into pieces of single word named tokens. For example, the input text, "Hello world! This is Python programming" is changed to ["Hello", "world", "!", "This", "is", "Python", "programming", "."].

3.4.2. Feature extracting. Feature extraction is a process of transforming data into features which is capable used for machine learning model. Typically, machine learning algorithms programmed with numeric. Hence, the text or word is then mapped into numerical vectors feature (a series of numbers). This process is called text vectorization or "Bag of Words" representation. The features can be extract by using (i) count vectorizer - consider a straightforward method in feature extracting as it just counts the number of times a word/token appears in given document, or (ii) term frequency- inverse document frequency (tf-idf) term weighting technique in Scikit-learn machine learning library. It represents a statistical measurement method of evaluating the word significant in given document. Term frequency (*tf*) refer to how frequent a term appears in a document against total number of words in the document.

$$tf = \frac{number\ of\ occurence\ of\ a\ term\ in\ document}{total\ number\ of\ terms\ in\ the\ document} \tag{1}$$

Inverse document frequency *(idf)* is the measurement of the selected term's weight in the document. It is also explained as the log of the total number of given documents divided by the number of documents in which the selected term exists.

$$idf = \log\left(\frac{total\ number\ of\ given\ documents}{number\ of\ document\ with\ existing\ selected\ word}\right) \tag{2}$$

Both count vectorizer and *tf-idf* vectorizer consist adjustable *n*-gram size feature and *n*-gram is a consecutive word. Thus, 1-gram is just one word, also called unigram, while 2-gram is called bigram, 3-gram is called trigram and so on. While working with n-gram, all n-grams with a degree less than or equal to n are generated.

3.4.3. Classification algorithms.
(i) Multinomial Naïve Bayes classifier is a Naïve Bayes classifier for multinomial models which is suitable for classification with discrete features. It is a classification algorithm based on the Bayes theorem with independence assumption among predictors. This classifier learns through a document classification algorithm, and is based on a simple usage of the Bayes' rule:

$$P(c|d) = \frac{P(d|c)P(c)}{P(d)} \tag{3}$$

wherein,
c is a class,
d is a document,
P(c) is a class probability,
P(d) is the probability of *a* document,
P(d|c) is conditional probability of the class for the given document *d*,
P(c|d) is conditional probability that document *d* belongs to class *c*.

(ii) Support Vector Machine (SVM) represents a supervised machining learning algorithm in analysing data used for classification and regression analysis. SVM is defined as a discriminative classifier by using separating hyperplane. This means that SVM execute an optimal hyperplane that maximises the margin between two classes and classifies new input data with a set of given labelled training dataset. (iii) Decision trees approach is a flow chart likely tree structure which applied the if-then-else statement technique and can be used in solving classification and regression problems. A decision tree describes as a tree where each node denoted as a feature, each branch denoted as a decision or rule and each leaf denoted as class labels or distribution. (iv) *k*-Nearest Neighbor represents a nonparametric supervised machine learning technique which able to use in classification and regression analysis. It is widely applied in the field of object or pattern recognition, text classification, data mining, and others [García-Laencina et al., 2009]. The algorithm categories an unknown input data by comparing the unknown input data with the training data to which it is high feature similarity. Then, the distance between the input features and all training data is calculated using Euclidean distance which is calculated by the square root of the sum of squared distance between two points. Euclidean distance (*d*) is calculated using equation 4, where the distance between two vectors, xA and xB, is calculated.

$$d = ((xA_0 - xB_0)^2 + (xA_1 - xB_1)^2)^{1/2} \quad (4)$$

3.4.4. Performance measures. The documents were tested for face detection and calculated on the basis of the confusion matrix shown in table 2 [29, 30] which contains the classifier's decisions in the rows, and the actual decision about classification in the class in the columns. The four fields of the table contain number of true positive (*TP*), true negative (*TN*), false positive (*FP*) and false negative (*FN*) classified documents.

Table 2. Confusion matrix

		actual	
		YES	NO
predicted	YES	TP	FP
	NO	FN	TN

(i) Accuracy is expressed as the proportion of correctly classified cases over all cases and is calculated according to formula:

$$accuracy = \frac{(TP+TN)}{(TP+TN+FP+FN)} \quad (5)$$

(ii) Precision is expressed as the proportion of positive cases that are correctly recognized as positive over all cases classified as positive and is calculated according to the formula:

$$precision = \frac{TP}{(TP+FP)} \quad (6)$$

(iii) Recall is expressed as the proportion of positive cases that are correctly recognized as positive over all actual positive cases and is calculated according to the formula:

$$\text{recall} = \frac{TP}{(TP+FN)} \quad (7)$$

(iv) A measure that combines precision and recall is called the F_1-measure and represents their weighted harmonic mean. It is calculated according to the formula:

$$F_1 = \frac{(2 \times precision \times recall)}{(precision+recall)} \quad (8)$$

(v) The last evaluation measure to be mentioned here is the ROC (Receiver Operating Characteristic) curve. It is a graphical representation of the binary classifier performance on which the curve represents a compromise between true positive and false positive cases.

4. Results and discussions

4.1. Four classification algorithms analysis

In first experiment, four classification algorithms were trained. Training score was collected by splitting dataset to 80% for training set and 20% for testing set. Meanwhile, cross-validation (CV) score was obtained using 5-CV. All setup for classification algorithms are based on Python sklearn library default parameters. The recognition accuracy results of four classification algorithm is as shown in Figure 2. Theoretically, the score in learning curve is used to describe how good the model is. If the training score is higher, the model is performed better. Based on Figure 2, the learning curve of SVM showed the lowest and unstable training score compare to others algorithms which is around 18.8% only with 500 training examples. By increasing the number of training examples, the accuracy of training score decreases. Same goes to the *k*-NN which showed a low training score (around 0.65 score). Thus, SVM and *k*-NN algorithms can be said not suitable in this textual emotion prediction project as both classifiers represented low performance and executed a bad learning curve with high bias, poor fit and poor generation in training a textual emotion prediction model.

The learning curve of Multinomial Naïve Bayes and Decision Tree classifier showed training score at maximum regardless of training examples. Besides, the CV score (marked with green line) increases with increasing of the training examples. The learning curve of Decision Tree showed bigger gap between CV score and training score compared to Multinomial Naïve Bayes, therefore it indicates that Multinomial Naïve Bayes classifier is proven to have a better performance. However, the training score is slightly dropped on increasing of training examples may cause due to mislabelled examples or duplicate example with different label. To conclude the analysis of the classification algorithms selection, an average mean score is calculated based on results from Figure 2. Multinomial Naïve Bayes classifier illustrated a better performance with highest accuracy mean score (64.08%) compared to SVM (15.37%), Decision Tree classifiers (52.36%) and k-NN (47.95%). Thus, Multinomial Naïve Bayes classifier is selected as the machine learning algorithm in this project.

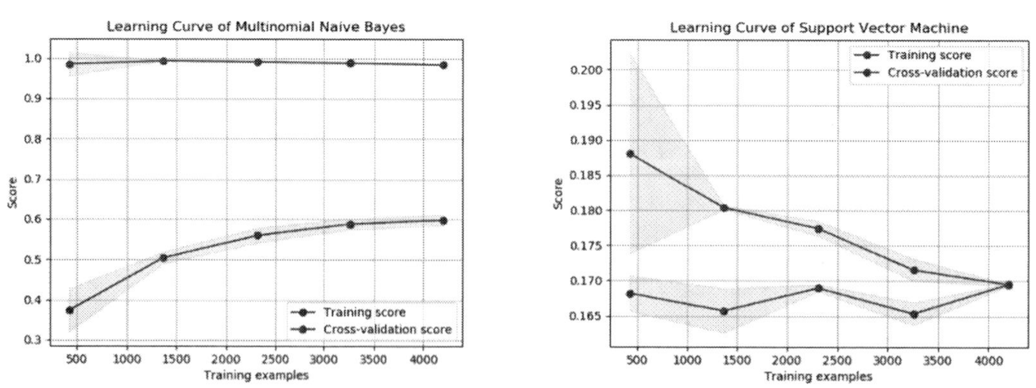

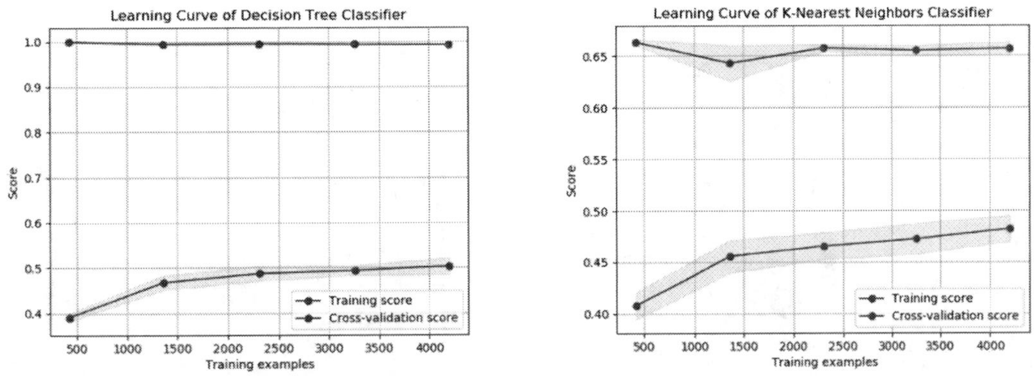

Figure 2. Learning curve for four classification algorithms.

4.2. The effect of feature extraction method on recognition accuracy

Figure 3 shows the comparison of recognition accuracy for two feature extraction methods on Multinomial Naïve Bayes algorithm. The accuracy score line of *tf-idf* vectorizer (blue line) is slightly above the accuracy score line of count vectorizer (red line). This indicates that the *tf-idf* vectorizer helps in improving the performance of the model by executing a higher recognition accuracy compared to count vectorizer. Besides that, the n-gram size feature can influence the result of recognition accuracy. The features extracted with trigram (n=3) from the document showed the highest recognition accuracy achieved is >64%. Hence, *tf-idf* vectorizer is selected as feature extraction method for this implementation.

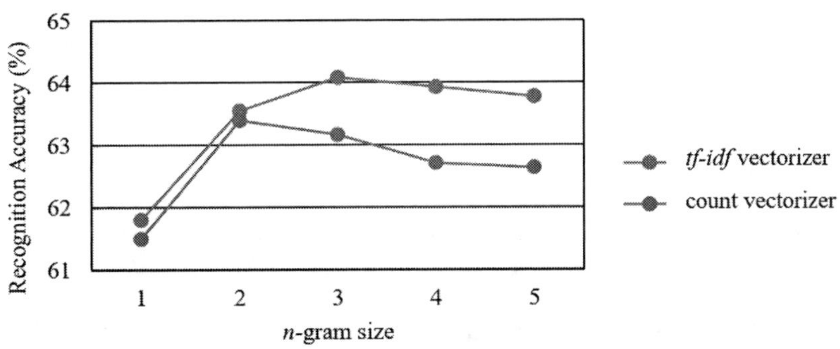

Figure 3. Recognition accuracy score comparison between *tf-idf* vectorizer (blue marked) and count vectorizer (red marked) with different size of n-gram.

4.3. The best model detailed analysis

Figure 4 shows the detailed performance of the developed model using confusion matrix, precision-recall rate, and ROC score. As previously mentioned, the dataset with 6570 sentences were split into 80% for training set and 20% for testing set. Thus, there is total 1314 testing data were picked-up randomly to measure the performance of the final model.

Precision is a measure of a result relevance while recall is a measure of how many relevant results are returned. The area under precision-recall curve (denoted as area in Figure 4) represent an average precision and it summarised the skill of a model across thresholds. Based on precision-recall results, that class 'fear', 'joy', and 'sadness' showed up more than 70% large of area under curve compared to 'anger', 'disgust', and 'guilt' classes. This indicates that 'anger', 'disgust' and 'guilt' have higher misclassification rate and overlapped each other. This is supported by confusion matrix table whereby the percentage of misclassification between 'anger', 'disgust' and 'guilt'.

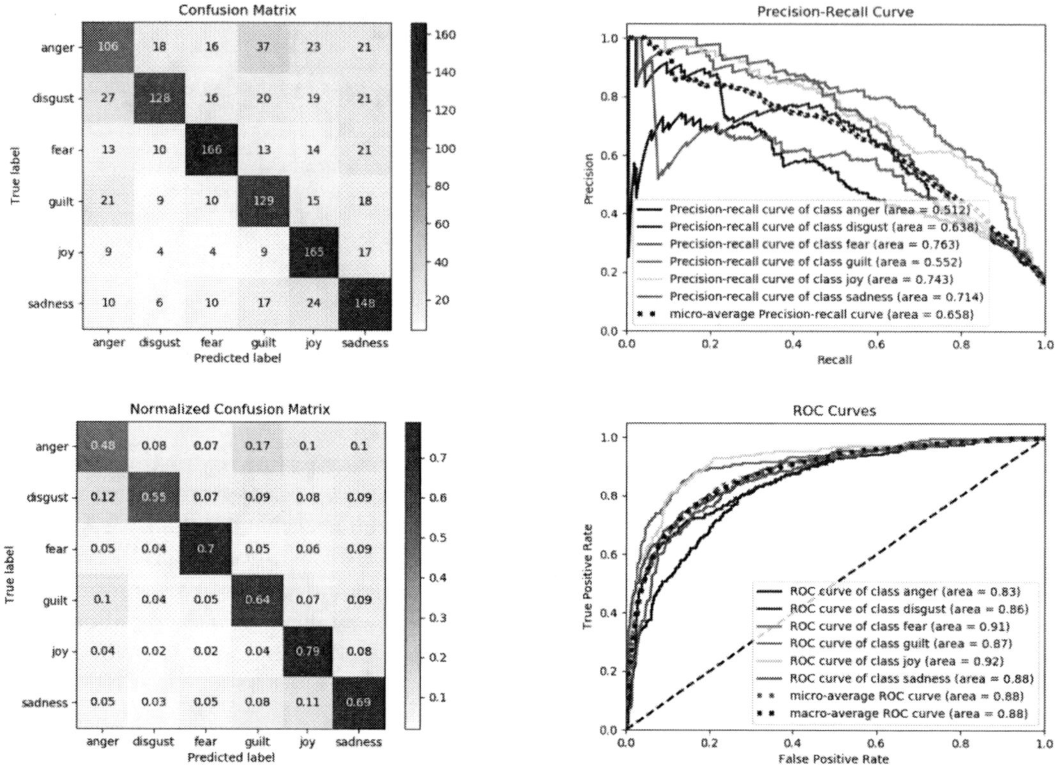

Figure 4. The detailed performance evaluations on the final model based on confusion matrix, precision-recall rate, and ROC score.

4.4. The Graphical User Interface

The design of the GUI is as shown in Figure 5. In GUI, user is allowed to insert English based message in the message entry box (beside "Customer Messages" text). In this example, the entry message is "Predict comes in good packaging. A new and solid plastic rack. Will buy again". Then, "Analyse" button is clicked to start the prediction process. The input texts undergo text pre-processing and feature extracting process, then it fitted into the final predictive model to predict the text's emotion classes by calculating the probability. Each probability of emotion classes is calculated and shown by using emoji stickers. The probabilities of emotion classes are converted into percentage for easy comparison. An emotion with a highest percentage is predicted as the estimate customer's emotion and the final emotion is displayed in the label box below (refer to Figure 5). After analysing process, the message is predicted as 'JOY' as the percentage of emotion joy is showed the highest (96.71%) compare to others (anger = 0.6%, sad = 2.62%, fear = 0.02%, disgust = 0.04% and guilt = 0.0%).

IOP Conf. Series: Materials Science and Engineering **769** (2020) 012022 doi:10.1088/1757-899X/769/1/012022

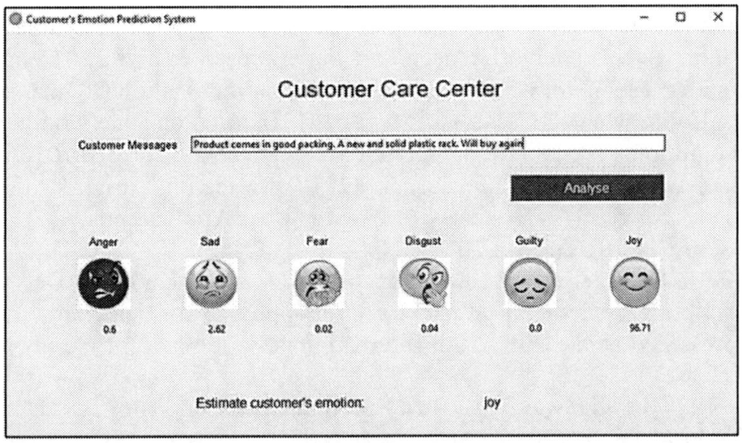

Figure 5. GUI layout for text-based emotion prediction system.

5. Conclusions

This paper presents the comparison of classifier's performance among four machine learning algorithms, Multinomial Naïve Bayes, Support Vector Machine, Decision Tree Classifier and *k*-Nearest Neighbors Classifier. Multinomial Naïve Bayes classifier results in a better performance and thus it was selected and proposed to develop a textual emotion prediction model. Besides, the investigation on feature extraction method revealed that *tf-idf* with 3-gram size is better for ISEAR dataset. An English text-based emotion prediction system with a simple graphical user interface for customer's emotion prediction from text system was successfully developed. The system is mainly focusing on analysing the Ekman's six basic emotions (anger, fear, disgust, joy, guilt and sadness). Although this is just only a base model, however the developed prediction system has a prospective future and its accuracy relies upon the type of classification algorithm applied, correct labelled emotional text database, data cleaning, and feature extraction method. Emotion texts might appear in complex sentence or called mix-emotion (examples: angry-sad, happy-angry) elements. To solve this problem, it is suggested to add some features or rules-based approaches in future work which might help in detecting emotions from text more correctly and perform better than the present result.

6. References

[1] Sebe, N., Cohen, I., Gevers, T., & Huang, T. S. (2005). Multimodal approaches for emotion recognition: a survey. Paper presented at the Internet Imaging VI

[2] Picard, R. W. (2003). Affective computing: challenges. Int. J. Hum.- Comput. Stud., 59(1-2), 55-64. doi:10.1016/s1071-5819(03)00052-1

[3] Desmet, B., & Hoste, V. (2013). Emotion detection in suicide notes. Expert Systems with Applications, 40(16), 6351-6358. doi:10.1016/j.eswa.2013.05.050

[4] Bafna, K., & Toshniwal, D. (2013). Feature based Summarization of Customers' Reviews of Online Products. Procedia Computer Science, 22, 142-151. doi:10.1016/j.procs.2013.09.090

[5] Munezero, M. D., Montero, C. S., Sutinen, E., & Pajunen, J. (2014). Are They Different? Affect, Feeling, Emotion, Sentiment, and Opinion Detection in Text. IEEE Transactions on Affective Computing, 5(2), 101-111. doi:10.1109/TAFFC.2014.2317187

[6] Ekman, P. (1992). An Argument For Basic Emotions (Vol. 6)

[7] Agrawal, A., & An, A. (2012, 4-7 Dec. 2012). Unsupervised Emotion Detection from Text Using Semantic and Syntactic Relations. Paper presented at the 2012 IEEE/WIC/ACM International Conferences on Web Intelligence and Intelligent Agent Technology

[8] Chen, Y.-L., Chang, C.-L., & Yeh, C.-S. (2017). Emotion classification of YouTube videos. Decision Support Systems, 101, 40-50. doi:10.1016/j.dss.2017.05.014

[9] Li, W., & Xu, H. (2014). Text-based emotion classification using emotion cause extraction. Expert Systems with Applications, 41(4), 1742-1749. doi:10.1016/j.eswa.2013.08.073

[10] Muljono, Winarsih, N. A. S., & Supriyanto, C. (2016, 5-6 Aug. 2016). Evaluation of classification methods for Indonesian text emotion detection. Paper presented at the 2016 International Seminar on Application for Technology of Information and Communication (ISemantic)
[11] Yasmina, D., Hajar, M., & Hassan, A. M. (2016). Using YouTube Comments for Text-based Emotion Recognition. Procedia Computer Science, 83, 292-299. doi:10.1016/j.procs.2016.04.128
[12] Chatzakou, D., Vakali, A., & Kafetsios, K. (2017). Detecting variation of emotions in online activities. Expert Systems with Applications, 89, 318-332. doi:https://doi.org/10.1016/j.eswa.2017.07.044
[13] Rabeya, T., Ferdous, S., Ali, H. S., & Chakraborty, N. R. (2017, 22-24 Dec. 2017). A survey on emotion detection: A lexicon based backtracking approach for detecting emotion from Bengali text. Paper presented at the 2017 20th International Conference of Computer and Information Technology (ICCIT)
[14] Shivhare, S. N., & Sri Khetwat, S. (2014). Emotion Detection From Text Documents (Vol. 4)
[15] Russell, S. J., & Norvig, P. (2016). Artificial intelligence: a modern approach: Malaysia; Pearson Education Limited
[16] Bandhakavi, A., Wiratunga, N., Deepak, P., & Massie, S. (2014). Generating a word-emotion lexicon from# emotional tweets. Paper presented at the Proceedings of the Third Joint Conference on Lexical and Computational Semantics (* SEM 2014)
[17] Hasan, M., Rundensteiner, E., & Agu, E. (2014). EMOTEX: Detecting Emotions in Twitter Messages
[18] Roberts, K., A Roach, M., Johnson, J., Guthrie, J., & M Harabagiu, S. (2012). EmpaTweet: Annotating and Detecting Emotions on Twitter
[19] Suttles, J., & Ide, N. (2013). Distant Supervision for Emotion Classification with Discrete Binary Values (Vol. 7817)
[20] Wang, W., Chen, L., Thirunarayan, K., & Sheth, A. P. (2012, 3-5 Sept. 2012). Harnessing Twitter "Big Data" for Automatic Emotion Identification. Paper presented at the 2012 International Conference on Privacy, Security, Risk and Trust and 2012 International Conference on Social Computing
[21] Krishnan, H., Elayidom, M. S., & Santhanakrishnan, T. (2017). Emotion Detection of Tweets using Naïve Bayes Classifier. Emotion.
[22] Strapparava, C., & Mihalcea, R. (2008). Learning to identify emotions in text.
[23] Cbalabantaray, R., mohd, M., & Sharma, N. (2012). Multi-Class Twitter Emotion Classification: A New Approach (Vol. 4)
[24] Li, J., Xu, Y., Xiong, H., & Wang, Y. (2010, 16-17 Aug. 2010). Chinese text emotion classification based on emotion dictionary. Paper presented at the 2010 IEEE 2nd Symposium on Web Society
[25] Luyckx, K., Vaassen, F., Peersman, C., & Daelemans, W. (2012). Fine-grained emotion detection in suicide notes: a thresholding approach to multi-label classification. Biomedical informatics insights, 5(Suppl. 1), 61-69. doi:10.4137/BII.S8966
[26] Drucker, H., Burges, C. J. C., Kaufman, L., Smola, A., & Vapnik, V. (1997). Support vector regression machines. Support Vector Regression Machines(7)
[27] Perikos, I., & Hatzilygeroudis, I. (2016). Recognizing emotions in text using ensemble of classifiers. Engineering Applications of Artificial Intelligence, 51, 191-201. doi:10.1016/j.engappai.2016.01.012
[28] Mohri, M., Rostamizadeh, A., & Talwalkar, A. (2012). Foundations of Machine Learning: The MIT Press.
[29] Karnalim, O. (2014). Language-Agnostic Source Code Retrieval Using Keyboard & identifier Lexical Pattern. International Journal of Software Engineering and Computer Systems, 4(1), 29-47

Acknowledgments

The authors acknowledge the internal grant from Universiti Malaysia Pahang (RDU180383).

IOP Conf. Series: Materials Science and Engineering **769** (2020) 012023 doi:10.1088/1757-899X/769/1/012023

Generic Code Clone Detection Model for Java Applications

Al-Fahim Mubarak-Ali[1], Shahida Sulaiman[2]

Faculty of Computing, 26300 Gambang, Pahang, Malaysia.[1]
School of Computing, Faculty of Engineering, Universiti Teknologi Malaysia, 81310 UTM Johor Bahru, Johor, Malaysia.[2]

fahim@ump.edu.my[1],shahidasulaiman@utm.my[2]

Abstract. Code clone is a common term used for codes that are repeated multiple times in a program. There are Type 1, Type 2, Type 3 and Type 4 code clones. Various code clone detection approaches and models have been used to detect a code clone. However, a major challenge faced in detecting code clone using these models is the lack of generality in detecting all clone types. To address this problem, Generic Code Clone Detection (GCCD) model that consists of five processes which are Preprocessing, Transformation, Parameterization, Categorization and Match Detection process is proposed. Initially, a pre-processing process produces source units through the application of five combinatorial rules. This is followed by the transformation process to produce transformed source units based on the letter to number substitution concept. Next, a parameterization process produces parameters used in categorization and match detection process. Next, a categorization process groups the source units into pools. Finally, a match detection process uses a hybrid exact matching with Euclidean distance to detect the clones. Based on these processes, a prototype of the GCCD was developed using Netbeans 8.0. The model was compared with the Generic Pipeline Model (GPM). The comparisons showed that the GCCD was able to detect clone pairs of Type-1 until Type-4 while the GPM was able to detect clone pair for Type-1 only. Furthermore, the GCCD prototype was empirically tested with Bellons benchmark data and it was able to detect clones in Java applications with up to 203,000 line of codes. As a conclusion, the GCCD model is able to overcome the lack of generality in detecting all code clone types by detecting Type 1, Type 2, Type 3 and Type 4 clones.

1. Introduction

Java is a programming language that is used to develop software and it was released in 1995. It is an open source language that is widely used in Netbeans or Eclipse to develop software. The original intention of the Java language was intended to let application developers write once, run anywhere (WORA). This means that the compiled Java source codes can run on all platforms that support Java without the need for recompiling the code again [1]. A lot of applications are developed using Java language. This is due to it being open source [1] and has more development resource compared to other programming language. Therefore, this makes Java language is more preferable among developers in developing applications. An experiment was conducted to see the occurrence of code clones in Java applications. Several Java applications totaling up to 512 000 lines of codes used in this experiment [1]. A total of 6% of the 512 000 lines of codes or 30 720 lines of codes from the tested Java applications contains clones. The analysis of this experiment concluded that large and early year developed Java applications contains more code clones compared to the smaller and later years built Java applications. There are two reasons contributing to this situation [1]. The first is because of the intense pressure of preparing the application for the stakeholders in limited time. The second reason is the absence of generic modules in Java in which would have prevented a lot of clones if existed. Therefore, it is essential that the code clone or also known as duplicated code is also an issue in Java applications.

Code clones can be detected through models. Three most notable code clone detection models available are Generic Clone Model [2], Generic Pipeline Model [3] and Unified Clone Model [4]. Generic Clone Model is a model that defines clone exists in a program [2]. The main use of this model is to describe clones. The model allows separation of concerns between clone detection, description and management using layers. This model has a clear separation of clone detection process definition using layers. Furthermore, the model is more focused on management of code clone that was driven by the operational aspects of code clon detection and removal. Generic Pipeline Model is a code clone detection model that has a combination of five processes to detect code clone. There are five processes involved in this generic pipeline model [3]. The first process is parsing process and it transforms source code into source units. The second process is pre-processing process and the purpose is to normalize the source units that was obtained from the previous process. The third process is pooling process and the purpose of this process is to group pre-processed source units into groups or pools based on user defined criteria. The fourth process is comparing process and the purpose of this process recursively compares source units in all pools using a divide and conquer strategy. The fifth process the filtering process and the purpose of this process is to remove irrelevant clone candidate sets from the result set. This process is utilized in removing non relevant candidate sets out of the result set. Unified Clone Model is an attempt in having a generic model that can represent all the results of all code clone tools [4]. It was designed through the different clone representations of existing tools. The outcome of the analysis has been divided into four groups which are detection for clone triage and management, integration of additional data from other sources, replication of scientific studies, and benchmarking of clone detection techniques.

Model is an attempt to have a unified process in detecting all code clone types. The attempt could be seen through the Unified Clone Model although this model is still in the design phase. The most advanced code clone detection model to date is the Generic Pipeline Model that detects on exact and near exact clones in Java applications. Exact clones refer to Type-1 clones while near exact clones refer to Type-2 clones. As a whole, Java is a programming language that is used to develop software. The absence of generic modules in Java causes the occurrence of code clone. In order to detect code clones that bring negative impacts to the software, code clone detection approaches and models have been proposed to detect the code clones. Models consist a combination of processes that pre-processes, transforms, detect and display code clone detection results. The models apply the existing code clone detection approaches as part of its process. Although code clone detection models are relatively new and few to have as a prototype, yet the existing code clone detection models have been a forward attempt in having a unified process that detects code clone regardless to the various code clone terminologies. Therefore, this research focuses on proposing a code clone detection model that detects all types of code clones in Java applications. Section 2 shows the proposed work while Section 3 shows the evaluation of the proposed work. Section 4 shows the findings while Section 5 concludes this paper.

2. Generic Code Clone Detection Model for Java Applications

Generic Code Clone Detection is a code clone detection model that is aimed to detecting all code clone types in Java applications. It consists of five processes which are pre-processing, transformation, parameterization, categorization and match detection process. The five combined processes form the proposed which is the GCCD. This model is aimed at detecting code clone Type 1, Type 2, Type 3 and Type 4. Figure 1 shows the high level view of Generic Code Clone Detection Model.

2.1 Pre-Processing Process

Source code refers to the codes that written in a source file of an application. There are five combined rules used to achieve the aim of this process. The first rule is PR-1: Remove package and import statements. The purpose of this rule is to remove package and imports in the source code. Package name usually shows the name of the package or project while imports refer to the classes that are imported into the package or project. Although the imports are classes, these imports are just merely imported statements. Therefore, this rule is designed to remove the import statements and package names from the source file.

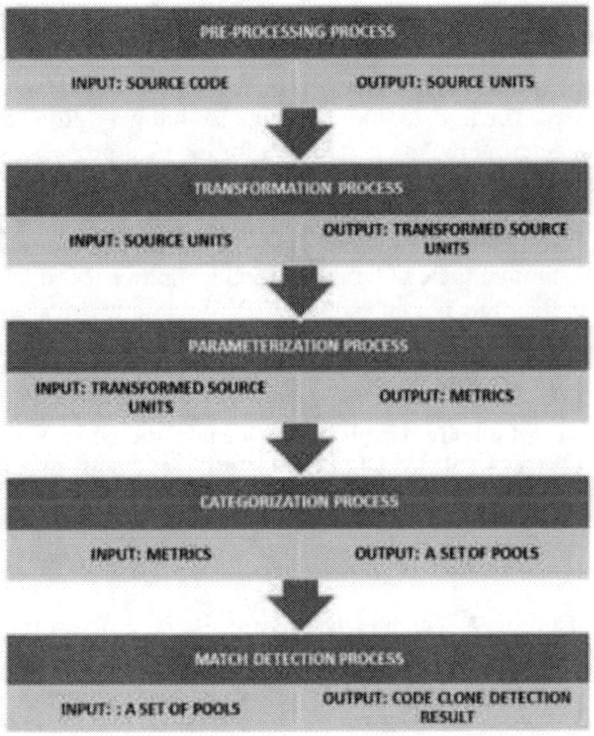

Figure 1. Generic Code Clone Detection Model

The second rule of this process is PR-2: Remove comments. Comment is an important component in programming as it serves as notes of a written source code for software developers and maintainers. Comments are usually in new lines or between source code lines and its occurrence is essential. Comment styles vary due programmers' style of programming and due to the programming language itself. Therefore, this rule attempts to remove comment lines occur in the source codes. The third rule of this process is PR-3: Remove empty lines. Empty lines refer to lines that have no content. The empty lines occur in most of the developed application and in nature happen due to programmers not cleaning up their code after developing or maintaining the code. Therefore, it is not harmful in removing these lines as it does not serve any importance in code clone detection result. The fourth rule in this process is PR-4: Regularize function access keyword to public. There are three function accesses in most of the programming languages which are public, private and protected. This rule regularizes all the function accesses into a single function access; which is public. Unlike CCFinder [5] that removes the function accessibility keywords completely, these try to preserve the code based on the function granularity by not removing these keywords but generalizing it to the same keyword. Furthermore, the keyword will serve as the constant value for ratio calculation. The fifth and final rule of this process is PR-5: Regularize source codes to lowercase. Source codes are written in different ways by programmers and therefore produce a different style of source code writing. Variables can be easily varied due to lowercase and uppercase letters; hence, causing a difference in the code clone detection results. As an example, the variable stringbuffer can be written as Stringbuffer, stringBuffer or even StringBuffer. Although it is the same variable, but code clone detection approach such as exact or string based approach will consider these variables as different variables hence affecting the code clone detection result. Therefore, it is important to regularize the variable in the same manner. This rule regularizes these variables into lowercase form.

The output refers to the output produced by this process. The output of this process is normalized source codes or better known as source units. Source unit is still in the form of source code. Each source unit represents a function of the source code.

2.2 Transformation Process

This is the second process in the GCCD. The aim of this model is to transform the source units into measurable units or numbers. The measurable units are also known as transformed source units is then are used as the medium in determining the parameters for the next process. The input of this process is normalized source codes or better known as source units that were produced from the previous process. The source units are still in the form of source code and each source unit represents a function of the source code. This process uses a letter to number substitution concept to transform the source code into measurable units. The substitution is done based on the location of the alphabet. For instance, a is the first alphabet in vocabulary therefore it is substituted to 01. The same concept applies for other alphabets in the vocabulary.

The output of this process is transformed source unit are in numerical form. The transformed source units are then divided into two groups which are *header (h)* and body (b). Header refers to the start of a transformed source unit prior the body part of a transformed source unit while body is the body of a transformed source unit. As an example to elaborate the header and body of a function, assume a function called Function A with the content of:

public void setAbbrev (String abbrev) this.abbrev.setText (abbrev) ;

After going through the first process which is the pre-processing process, the source unit of Function A appear as:

public void setabbrev string abbrev thisabbrevsettext abbrev

Therefore, the header and body of a function of Function A:

header (h) : *public void setabbrev string abbrev* and body (b) : *thisabbrevsettext abbrev*

2.3 Parameterization Process

This is the third process in the GCCD. The aim of this process is to create parameters or metrics that will be used for categorization and match detection process. The input of this process is the transformed source units that are obtained from the previous process. As mentioned before, the transformed source units are in the form of numerical. The parameter that is used for detection is header average ratio and body average ratio. In order to obtain the header average ratio and body average ratio, four metrics need to be obtained from the transformed source units. Table 1 shows the metrics that are extracted from the transformed source units.

Table 1. Metric extracted from the transformed source units

Metrics	Description
header code count	Amount of code in header
body code count	Amount of code in body
header ratio	Ratio of header
body ratio	Ratio of body
average header ratio	Average ratio of header
average body ratio	Average ratio of body

In order to obtain an average ratio, the ratio must be obtained first. As mentioned in the previous process, the access function of all the function has been transformed into the public. Therefore, the transformed source unit contains the same value of access function after going through the transformation process. By using this as the benchmark value, each source unit is divided with this value to obtain the ratio value of each code in header and body of the source unit. As an example to elaborate the detail calculation of average ratio for each transformed source unit, lets assume a transformed source unit has header, *TSUha*, and body, *TSUhb* . Therefore the ratio of the transformed source unit is:

$$R_A = \frac{(A_1 A_2 A_3 \ldots A_n)}{P_1} \quad (1)$$

$$R_B = \frac{(B_1 B_2 B_3 \ldots B_n)}{P_1} \quad (2)$$

where;

P_1 is value of access function that starts with public
R_A is ratio for each transformed source unit in header
R_B is ratio for each transformed source unit in body
$A_1A_2\, A_3..An$ is value in transformed source units in header
$B_1B_2\, B_3..\, Bn$ is value in transformed source units in body

Once the ratio of header and body in each function has been obtained the next step is to calculate the average ratio of header and body in each source unit. The average ratio of header and body in each source unit is:

$$AVR_A = \frac{(R_A)}{C_A} \quad (3)$$

$$AVR_B = \frac{(R_B)}{C_B} \quad (4)$$

AVR_A is average ratio for transformed source unit in header AVR_B is average ratio for transformed source unit in body CA is code count for each transformed source unit in header CB is code count for each transformed source unit in body. The output of this process is the aforementioned metrics that will be used for in the next phase which are categorization process.

2.4 Categorization Process

This is the fourth process in the GCCD. The aim of this process is to categorize and group these source units into a group of clone based on the same average ratio value of header and body between functions. The input of this process is the set of metrics obtained from the previous process. The categorization is done based on the average ratio value of header and body between transformed source units. The transformed source units are grouped into three pools based on the same average ratio value of header and the same average ratio value of body between functions. As an example, if transformed source unit A has the same average ratio value of header with transformed source unit B, therefore these two transformed source units is grouped into the same group or better known as the first pool. This process will be continued until all the transformed source units that have the same average value of header is grouped together in the same pool. As for the remaining transformed source units, these transformed source units is then grouped in the same group based on the same average value of body or better known as the second pool. It is done until all the remaining functions have been grouped together. If there are remaining functions after the categorization based these two groups, the remaining functions will be grouped into another group or better known as the third pool.

The output of this process is a set of pools that has been grouped based on the same average ratio value of header and the same average ratio value of body between transformed source units. The output of this process serves as the input of the final process which is the match detection process.

2.5 Match Detection Process

This is the fifth and final process in the GCCD. The aim of this process is to detect the code clone. The input of this process is the pairs and groups of source units based on three categories. The match detection process uses a hybrid detection technique of exact matching with Euclidean distance. As mentioned before, there are three pools obtained from the previous process. The match detection starts by finding the exact clone or better known as Type-1 clone; and near exact clones or better known as Type-2 clone. Therefore, there is two stage of exact matching being used to detect Type-1 and Type-2 clones. The first stage is the exact matching technique is used in detecting the same average ratio value of header and body between source units in the first pool. The compared functions that have the same average ratio value of header and body are detected as Type-1. The second stage is the exact matching technique is used in detecting the same average ratio header value but different body value and the same average ratio body value but different header value between functions in the first and second pool. The clones that are detected through this stage are known as Type-2. The remaining average ratio header and body value from the first and second pool is combined into the third pool for the next step of this process. As for the remaining average ratio header and body value, Euclidean distance is applied. Assume there are two source units which are A and B. Therefore, the Euclidean distance, ED, between A and B calculated as:

$$ED_{AB} = \sqrt{(headerA - headerB)^2 + (bodyA - bodyB)^2} \quad (5)$$

where;

ED_{AB} is Euclidean distance of Function A and Function B
headerA is the average ratio header of A
bodyA is the average ratio body of A
headerB is the average ratio header of B
bodyB is the average ratio body of B

The calculation of the Euclidean distance is applied to the remaining average ratio header and body values in the third pool. Once the calculation is done, it is the function is then grouped to Type-3 and Type-4 based on the distance obtained. Type-3 clone are taken from the range of 85% - 100% [7] while the remaining is defined as Type-4.

3. Evaluation

Table 2 shows the overall result of the detected clone pairs using the GCCD for Java applications in Bellons benchmark data (Bellon et al., 2007). There are total 7281 clone pairs of Type-1, Type-2, Type-3 and Type-4 detected in J2sdk1.4.0-javax-swing using the GCCD. There are 877 clone pairs or 12% of the total clone pairs detected for Type-1 while 3697 clone pairs or 50.8% of the total clone pairs detected for Type-2. As for Type-3, there are 1710 clone pairs or 23.5% of the total clone pairs detected and there are 997 clone pairs or 13.7% of the total clone pairs detected for Type-4. As for Eclipse-jdtcore, there are total 11268 clone pairs of Type-1, Type-2, Type-3 and Type-4 detected using the GCCD. There are 626 clone pairs or 5.6% of the total clone pairs detected for Type-1 while 2886 clone pairs or 25.6% of the total clone pairs detected for Type-2. As for Type-3, there are 4265 clone pairs or 37.9% of the total clone pairs detected and there are 3491 clone pairs or 30.9% of the total clone pairs detected for Type-4. There are total 2685 clone pairs of Type-1, Type-2, Type-3 and Type-4 detected using the GCCD in

Eclipse-ant. There are 185 clone pairs or 6.9% of the total clone pairs detected for Type-1 while 552 clone pairs or 20.5% of the total clone pairs detected for Type-2. As for Type-3, there are 581 clone pairs or 21.6% of the total clone pairs detected and there are 1370 clone pairs or 51% of the total clone pairs detected for Type-4. As for Netbeans-javadoc, there are total 595 clone pairs of Type-1, Type-2, Type-3 and Type-4 detected using the GCCD. There are 99 clone pairs or 16.6% of the total clone pairs detected for Type-1 while 341 clone pairs or 57.3% of the total clone pairs detected for Type-2. As for Type-3, there are 102 clone pairs or 17.1% of the total clone pairs detected and there are 53 clone pairs or 9% of the total clone pairs detected for Type-4.

Table 2. Detection result based Bellon benchmark data

Clone Type	**Bellons Benchmark Data (Java)**			
	J2sdk1.4.0-javax-swing	**Eclipse-jdtcore**	**Eclipse-ant**	**Netbeans-javadoc**
Type-1	877	626	185	99
Type-2	3697	2886	552	341
Type-3	1710	4265	581	102
Type-4	997	3491	1370	53
TOTAL	7281	11268	2688	595

4. Evaluation

The aim of producing a generic source unit representation is essential as it helps in processing and transforming the source codes in a more unified form that can be used in detecting code clone. The first three rules in the pre-processing process are essentially needed as it is the noise or uninformative section of the source code; thus does not harm the information or structure of the source codes. The next two rules are to regularize the source code in a form that is acceptable at the function level and reduce the dissimilarity due the lowercase and uppercase of the source codes. The second process in the Generic Code Clone Detection which is the transformation process is aimed to transform the source units into measurable units or numbers. The measurable units are also known as transformed source units act as an intermediate representation of the source code that will be used for clone detection purposes. By adopting the substitution cipher concept, the code to numerical transformation allows the application of the heuristic method to be applied to the source codes. Regarding on the runtime performance of these two processes, the combination of these two processes produces the highest runtime performance for large datasets compared to the small sized data sets. The third process which is the parameterization process is the process that produces the metrics that is used for categorization and match detection process. The aim of the categorization process is to categorize and group these source units into a group of clone based on the same average ratio value of header and body between functions. Apart from the source unit representation, the match detection technique plays an important role in detecting code clones accurately. The applied match detection approach in the fifth process which is the match detection process is a heuristic approach that uses elements from the source units itself in determining the clones. The combination ratio and distance is a combination of basic metrics but yet produces results in detecting for all the code clone types.

As seen in the empirical evaluation between the GCCD and Generic Pipeline Model, the different definition between the similarity pair and clone pair highly influences the code clone detection result. Similarity definition that is used in the Generic Pipeline Model; is a loose definition for clones as such that the computed pairs would not fit any common notion of a clone but could still be useful for a certain application of pair definition. This means a certain level of the allowance is given to near miss clones or non-clones to be claimed as clones. As for the GCCD that adopts clone pair definition, it is more rigid in having exact or near exact match pair of clones. Therefore, the final output which the code clone detection differs due the different definition of the clone or similarity pairs definition. A different definition of clone pairs yields different code clone detection result. Therefore, a more loosened clone pair definition might influence the code clone detection result.

5. Conclusion

This work shows that having a generic model for code clone detection is possible with the combination of essential processes of detecting code clone. Possible future work that can improve the code clone detection process and performance can be done through improvement of the pre-processing process in supporting code clone detection in other structural and procedural programming language. There are various programming languages exist in the current programming world; therefore it is essential the existence of clones on the applications of these programming languages. With the enhancement of the rules in the pre-processing process, the GCCD can support code clone detection in other programming languages tool. For future enhancement, this research can be extended through application of Artificial Neural Network [8].

Acknowledgement

The authors are grateful to Universiti Malaysia Pahang (UMP) in partially supporting this work through UMP Internal Grant (Grant ID: RDU180308).

References

[1] Dagenais, M., Patenaude, J.-F. F., Merlo, E. and Lagu¨e, B. (2002). Advances in Software Engineering. pp. 95–110.

[2] Giesecke, S. (2006). Generic modelling of code clones. In Koschke, R., Merlo, E. and Walenstein, A. (Eds.) Duplication, Redundancy, and Similarity in Software, Dagstuhl Seminar Proceedings, vol. 06301. Internationales Begegnungs- und Forschungszentrum fuer Informatik (IBFI), Schloss Dagstuhl, Germany.

[3] Biegel, B. and Diehl, S. (2010). Highly Configurable and Extensible Code Clone Detection. In 2010 17th Working Conference on Reverse Engineering. 237–241.

[4] Kapser, C. J., Harder, J. and Baxter, I. (2012). A Common Conceptual Model for Clone Detection Results. In Proceedings of the 6th International Workshop on Software Clones. IWSC '12. Piscataway, NJ, USA: IEEE Press, 72–73.

[5] Biegel, B. and Diehl, S. (2010). Highly Configurable and Extensible Code Clone Detection. In 2010 17th Working Conference on Reverse Engineering. 237–241.

[6] Kamiya, T., Kusumoto, S. and Inoue, K. (2002). CCFinder: a multilinguistic tokenbased code clone detection system for large scale source code. IEEE Transactions on Software Engineering. 28(7), 654–670.

[7] Kodhai, E. and Kanmani, S. (2014). Method-level code clone detection through LWH (Light Weight Hybrid) approach. *Journal of Software Engineering Research and Development*, 2(1), 1-29.

[8] Fakharudin A. S., Mohamad, M. A., Johan, M. U. (2009). Newspaper vendor sales prediction using artificial neural networks. *International Conference on Education Technology and Computer 2019*. 339-343.

Template Matching Analysis using Neural Network for Mobile Iris Recognition System

Anis Farihan Mat Raffei, Siti Zulaikha Dzulkifli and Nur Shamsiah Abdul Rahman

Multimedia & Computer Vision Research group, Faculty of Computing, College of Computing and Applied Sciences, Universiti Malaysia Pahang, Malaysia

Email: anisfarihan@ump.edu.my

Abstract. Today, the use of mobile phone among individuals with an advanced technology, acquisition and computation resources is crucial in securing personal data and services as well as in identifying a person. Furthermore, there has been growing interest of researchers using mobile as an acquisition device to capture an eye image in a non-cooperative environment (static motion and at different distances) for iris recognition. However, the use of mobile as an acquisition device still facing some noise factors because the quality of image captured in this environmental condition is low compared to a camera device. Besides, low awareness of user in handling the mobile devices and lack technical experiences in capturing iris images with this environmental condition is generally uncontrolled contribute to producing unpredictable low quality of eye images. Due to these issues, it led to incorrect segmentation of iris boundaries and subsequently lowered the ability to match the iris features. Hence, this condition contributes to performance degradation of iris recognition system. Therefore, to improve the ability to match the iris features in mobile iris recognition system, a combination with neural network is proposed. The iris database used to test against these methods is CASIA-Iris-M1-S2. The proposed method, a combination of Hamming distance and neural network has achieved a better result where it increases the existing method in term of accuracy for mobile iris recognition under non-cooperative environment.

Keywords: Mobile iris, Template Matching, Neural Network, Non-Cooperative Environment.

1. Introduction

Latest technology has been widely used in electronic devices especially in mobile [1]. Besides, mobile devices are widely known in term of portability and flexibility for different tasks [2]. However, many issues related to security flaw have been identified in mobile devices [3]. The demand of secure data and services led to the development of biometric recognition systems on mobile devices. One of the largest deployed application of biometric technologies is fingerprint [4].

Although the touch-based sensing interface in fingerprint biometric is very convenient, but the hygienic problems on the sensor surface causes users to avoid use of this technology. Besides, the existence of foreign elements on the sensor surface and the small sized sensors has degraded the performance of fingerprint biometric on mobile devices. Moreover, according to [5], the fingerprint recognition system was vulnerable to fake fingerprints in mobile device. The reliability of iris recognition system has increased the interest to integrate this technology in mobile devices because this technology uses a touchless sensor and non-intrusive to the fake iris [6]. Besides, this technology is reliable to visual recognition of a person's when imaging is captured at distances of less than a meter [7,8]. However, the integration of this technology with mobile device can be degraded due to low central processing unit in mobile device, user's gazing point and existence of noises such as motion-blurred, occlusion of eyelids and specular reflection. Hence, to improve the ability to match the iris in mobile iris recognition system, a method of neural network is suggested.

2. Literature Review

Template matching is a continuous step of iris recognition after feature extraction. During template matching process, the templates obtained from feature encoding requires a matching metric which it measures a similarity between two iris templates. The metric will produce a range of value when two iris templates from the same eye image known as intra-class while another range of value generated from two different irises is called as inter-class. From these templates, distinct and separate value should obtain in order to make decision either the two templates are from two different irises or same iris.

Various template matching methods such as Hamming distance, weighted Euclidean distance (WED) and normalised correlation has been proposed to measure the eye template. Hamming distance provides a degree of dissimilarity bits between two equal lengths of binary patterns. The integrodifferential operator (IDO) locates inner and outer circle of iris and use Hamming distance techniques to calculate the matching of two templates [9]. Past researchers had addressed that HD provides better performances in iris recognition system and it is a fast and simple technique [10]. Besides, it also adaptable as many image data types has been tested by using HD such as twin image, noisy image and non-cooperative image. A new technique has been proposed in template matching which is hybrid of fragile bit distance with HD and measures the accuracy to the HD technique alone [11]. It used in twin iris data and the proposed techniques gives promising results.

The WED derived from old Pythagoras theorem which is defined as linear distance between two points in the dimensional space. It used mostly to compare two different templates especially if the templates are composed of integer values. WED also is the commonly technique has been used when its related dealing with distance as it can have from one up to n-dimensional spaces. Two researchers have addressed the advantages of WED performance in iris recognition system. According to [12], the WED is very compatible to be embedded with classification technique to produce higher accuracy and WED is the simple techniques since the equation of this technique is only repeated Pythagoras theorem formula. This method was used by [13] with wavelet transform and multi-channel Gabor filtering to represent the iris texture. This system proves by using that method, accuracy rate reach 93.8% [11].

However, computational involved in determining distance is very complex and according to [14], it requires a lot of calculation process and requires massive computation process. Wildes has been used normalized correlation between the acquired and database representation for goodness of match. This technique has advantageous over standard correlation as it able to account the local in image intensity that corrupt the standard correlation calculation.

3. Mobile Iris Dataset: CASIA-Iris-M1-S2

This database is divides into three different distances which are 20cm, 25cm and 30cm. For each distance, 10 images were captured so that each person contributes to 30 images. Table 1 presents eye images in this dataset.

IOP Conf. Series: Materials Science and Engineering **769** (2020) 012024 doi:10.1088/1757-899X/769/1/012024

Table 1. Mobile Iris Dataset: CASIA-Iris-M1-S2.

Distance	Eye Images		
20 cm		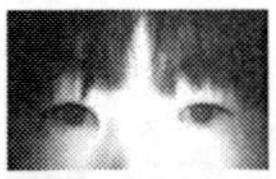	
25cm		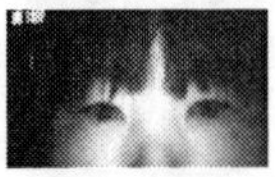	
30cm			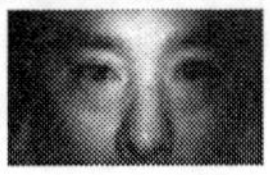

4. Methodology

Figure 1 presents the research framework. Circular Hough transform is a common algorithm that used to detect and determine simple geometric shape of an objects such as lines, circles, curves which present in the image. Circular HT is used to infer radius and centre coordinates of the pupils and iris regions [1]. Pre-processing has been done before the iris segmentation process. Degradation in segmentation accuracy normally caused due to low contrast of between pupil and iris [15]. Thus, before proceed to iris segmentation process, contrast enhancement is required to overcome this issue. Segmentation process started with limbic and continuously followed by pupillary boundaries. Firstly, calculate derivatives of intensity values of an eye image by generated an edge map, then thresholding the result of it. Binary edge map is then analysed on HT space to determine three parameters of one circle which is (h_o, k_o, Rad). Circular HT formula is shown below:

$$HT(h_o, k_o, Rad) = \sum_{n=1}^{j}(hn, kn, h_o, y_o, Rad) \quad (1)$$

where (h_n,k_n) is a line pixel and HT(h_n,k_n,h_o,k_o,Rad) equal to 1 if (h_o,k_o) is on circle and 0 if otherwise. Location of (h_o,k_o,Rad) with maximum value of HT(h_o,k_o,Rad) has been chosen as a parameter vector for strongest circular boundary while j equal to 1 to *n*. Process of segmentation ends with eyelashes and eyelid removal of an iris image by using linear HT and thresholding algorithm.

IOP Conf. Series: Materials Science and Engineering **769** (2020) 012024 doi:10.1088/1757-899X/769/1/012024

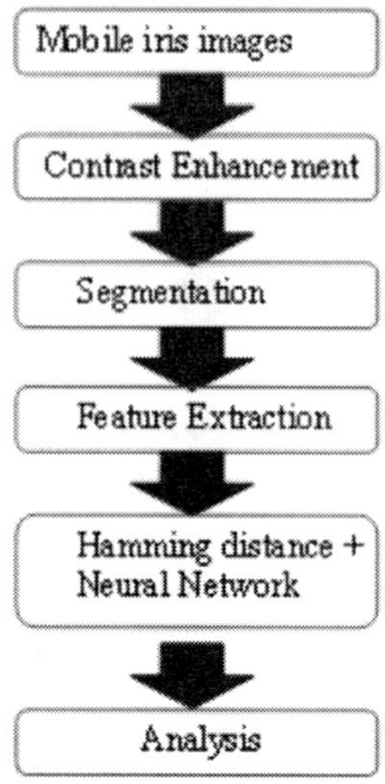

Figure 1. Research framework.

Next process after segmentation has been done is normalization. Normalization of an eye image is needed in order to transform the iris region to a fixed dimensions for allowing comparisons to happen. This process will produce iris regions that having same constant dimensions so that two eyes images will of same iris will having characteristics features at same spatial location. For normalization method, Daugman's Rubber Sheet (DRS) techniques was chosen. This techniques remaps each point to a pair of polar coordinates (r, θ) as r is on the interval $[0,1]$ and θ is an angle $[\theta, 2\pi]$. Remapping of iris region from (x, y) on Cartesian coordinates to the normalised a non-concentric polar representation is modelled as follow;

$$I(x(r,\theta), y(r,\theta)) \rightarrow I(r,\theta) \quad (2)$$

with

$$s(r,\theta) = (1-r)x_p(\theta),$$

$$y(r,\theta) = (1-r)y_p(\theta) + ry_l(\theta)$$

where $I(x,y)$ is the iris region image, (x,y) is the original Cartesian coordinates, (r,θ) is the corresponding normalised polar coordinates and x_p, y_p and x_l, x_l refer to coordinates of the pupil and iris boundaries along the θ direction.

Feature extraction provide accuracy to iris recognition process when the feature of iris were extracted. For this research study, one dimensional of log Gabor filter has been used. The formula of log Gabor filter is shown as follow;

$$LGB(h) = exp((-(log(h/h_0))^2/2(log(\sigma h_0)))^2, \quad (3)$$

where h_0 is the centre frequency and σ is the bandwidth of the filter.

In order to determine the uniqueness of iris patterns, the value of Hamming Distance (HD) need to be obtained by comparing the templates of two different eyes. Thus, the result of distribution HD were examined. HD acts as a matching algorithm where two of different templates is compared to get the number of bits. By using HD, capability to determine either the two patterns of iris templates were generated by the same iris or from two different irises. Hamming distance is described as the total of exclusive-OR between X and Y divided by N, where N is the total number of bits in bit pattern. The formula of Hamming Distance as follow:

$$HD = \frac{1}{N}\sum_{j=1}^{N} Xj\ (XOR)\ Yj \quad (4)$$

Each individual iris region has contained feature with high degrees of freedom, it will produce a bit pattern that is independent with another iris region. This means, it will highly correlated if the two iris codes produced from the same iris. Hamming Distance between two patterns is equal to 0.5 if the iris template was generated by two different irises. While if the two templates are from the same iris, the value of Hamming Distance should be nearer to 0.0 as they are highly correlated and the bits between two iris code should be agree.

Designing the neural network model is the important step in the implementation phase. In this step, elements such as input nodes and output nodes are determined. The hamming distance was set as an input while targeted output was set as an output for this model. Ideal output calculated from the neural network will be used to measure the error rate and mean square and back-propagation is used to train the model.

5. Analysis and Results

By using HD, capability to determine either the two patterns of iris templates were generated by the same iris or from two different irises. Each individual iris region has contained feature with high degrees of freedom, it will produce a bit pattern that is independent with another iris region. This means, it will highly correlated if the two iris codes produced from the same iris. Hamming Distance between two patterns is equal to 0.5 if the iris template was generated by two different irises. While if the two templates are from the same iris, the value of Hamming Distance should be nearer to 0.0 as they are highly correlated and the bits between two iris code should be agree.

In this research, the output data get from the ANN was used to calculate FRR, FAR and the accuracy of mobile iris recognition. FRR is defined as measures the probability the enrolled individual not been identifies while FAR defined as the measure of the probability in the biometric system is not correct which mean the system can successfully accept attempt by an authorized user. Below shows the formula of FRR, FAR and the accuracy of mobile iris recognition.

5.1. Accuracy of Mobile Iris Recognition For Left and Right Sides

Table 2 show the accuracy of mobile iris recognition for left and right eye sides. For left eye side, at threshold value of 0.05, the accuracy of mobile iris recognition is 99.63% where it is the highest accuracy among all and at threshold value of 0.15, accuracy of mobile iris recognition decreases to 99.59%. When threshold value is 0.25, the accuracy of mobile iris recognition is 99.48% while at threshold value of 0.35, the value of accuracy is 99.31%. Accuracy of mobile iris recognition is 99.11% where it is the lowest among all where the threshold value is set to 0.45.

For right eye side, at threshold value of 0.05, the accuracy of mobile iris recognition is 99.75% where it is the highest accuracy among all and at threshold value of 0.15, accuracy of mobile iris recognition decreases to 99.69%. When threshold value is 0.25, the accuracy of mobile iris recognition is 99.56% while at threshold value of 0.35, the value of accuracy is 99.36%. Accuracy of mobile iris recognition is 99.11% where it is the lowest among all where the threshold value is set to 0.45.

Based on the calculation result, it can be concluded that as the threshold value nearer to 0.5, the accuracy of mobile iris recognition was decreased while as the threshold nearer to 0, the accuracy of mobile iris recognition increased. This is due to the two templates were generated from the different irises when the value of threshold nearer to 0.5 while when it is nearer to 0, it is mean that the two templates were generated from the same iris.

Table 2. Accuracy of mobile iris recognition for left eye side.

Threshold	Left Eye			Right Eye		
	FAR (%)	FRR (%)	Accuracy (%)	FAR (%)	FRR (%)	Accuracy (%)
0.05	0.0264	0.7167	99.63	0.0306	0.4704	99.75
0.15	0.0247	0.8049	99.59	0.0279	0.6016	99.69
0.25	0.0200	0.8600	99.56	0.0240	0.8590	99.56
0.35	0.0100	1.2600	99.36	0.0160	1.2590	99.36
0.45	0.0005	1.7826	99.11	0.0043	1.7784	99.11

Table 3 shows the accuracy of mobile iris recognition of the proposed method. At threshold value of 0.05, the accuracy of mobile iris recognition is 99.36% where it is the highest accuracy among all and at threshold value of 0.15, accuracy of mobile iris recognition decreases to 99.00%. When threshold value is 0.25, the accuracy of mobile iris recognition is 98.88% while at threshold value of 0.35, the value of accuracy is 98.53%. Accuracy of mobile iris recognition is 98.05% where it is the lowest among all where the threshold value is set to 0.45.

Table 3. Accuracy of mobile iris recognition of the proposed method.

Threshold	FAR (%)	FRR (%)	Accuracy (%)
0.05	0.1955	1.0900	99.36
0.15	0.1580	1.8400	99.00
0.25	0.1455	2.0900	98.88
0.35	0.1080	2.8400	98.53
0.45	0.0043	3.8400	98.05

According to [10], the best technique result is when low FRR, low FAR and high accuracy was obtained from the comparisons of techniques. From the result obtained to compare Hamming distance and Euclidean distance, value of FRR for Hamming distance is 2.5% while Euclidean distance is 8.75%. For value of FAR, Hamming distance is 8.75% while Euclidean is 12.5%. Table 4 shows the summarize accuracy of the existing and the proposed methods. From the table, it can be conclude where the proposed method combination of Hamming distance and Artificial neural network has the lowest value of FAR which is 0.12% and the lowest value of FRR which is 2.34% which lead to the highest accuracy among the other techniques with value of 98.77%.

Table 4. Summarization accuracy of the existing and the proposed methods.

Methods	FAR (%)	FRR (%)	Accuracy (%)
Hamming Distance	8.75	2.50	96.48
Euclidean Distance	12.50	8.75	93.35
Hamming Distance + Artificial Neural Network (This Study)	**0.12**	**2.34**	**98.77**

6. Conclusion

The proposed method which is combination of Hamming distance and Artificial neural network has provided good results to match unique iris feature of eyes com-pared to existing method such as Hamming distance and Euclidean distance method. This is because of the value of FAR and FRR of proposed method gave the lowest value and thus increased the accuracy of mobile iris recognition. Analysis from the graph of equal error rate between Hamming distance and combination of Hamming distance and Artificial neural network showed that there is no intersection between them. This is due to the value of FAR and FAR for both methods did not equal. As a conclusion, from the experimental result of those two methods it proofed that the proposed method of this research study has gave an enhancement in term of accuracy for mobile iris recognition by using a dataset under non-cooperative environment.

7. Acknowledgement

The authors would like to express their appreciation to the Institute of Automation, Chinese Academy of Sciences which provided CASIA-Iris-M1-S2 dataset. This research was funded by the UMP Research Grant Scheme (RDU180378).

References

[1] Zhang Q, Li H, Sun Z and Tan T 2018 Deep feature fusion for iris and periocular biometrics on mobile devices *IEEE Transactions On Information Forensics and Security* vol 13 p 2897-2912.

[2] Gragnaniello D, Sansone C and Verdoliva L 2015 Iris liveness detection for mobile devices based on local descriptors *Pattern Recognition Letters* vol 57 p 81-87.

[3] Thavalengal S and Corcoran P 2016 User authentication on smartphones: Focusing on iris biometrics *IEEE Consumer Electronics Magazine* vol 5 p 87-93.

[4] Sharma L and Manish M 2018 Mobile banking transaction suing fingerprint authentication *Proceedings of the International Conference on Inventive Systems and Control*.

[5] Yildirim N and Varol A 2015 Android based mobile application development for web login authentication using fingerprint recognition feature *Proceedings of the Signal Processing and Communications Applications Conference*.

[6] Aginako N, Santana M C-, Navarro J L-, Otzeta J M M- and Sierra B 2017 Periocular and iris local descriptors for identity verification in mobile applications *Pattern Recognition Letters* vol 91 p 52-59.

[7] Galdi C and Dugelay J- L 2016 Fusing iris colour and texture information for fast iris recognition on mobile Devices *Proceedings of the International Conference on Pattern Recognition*.

[8] Reddy N, Rattani A and Derakhshani R 2016 A robust scheme for iris segmentation in mobile environment *IEEE Symposium on Technologies for Homeland Security*.

[9] Tania U T, Motakabber S M A and Ibrahimy M I 2014 Template matching techniques for iris recognition system *Proceedings of the International Conference on Computer and Communication Engineering: Emerging Technologies via Comunication Convergence* pp 9-11.

[10] Patil S, Gudasalamani S and Iyer N C 2016 A survey on iris recognition system *Proceedings of the International Conference on Electrical, Electronics, and Optimization Techniques*.

[11] Hollingsworth K P, Bowyer K W and Flynn P J 2011 Improved iris recognition through fusion of hamming distance and fragile bit distance *IEEE Transactions on Pattern Analysis and Machine Intelligence* vol 33(12) p 2465-2476.

[12] Kasim S, Hassan R, Mohd N S, Ramlan R, Mahdin H and Fudzee M 2017 A comparative study of different template matching techniques for twin iris recognition *International Journal on Advanced Science, Engineering and Information Technology*, vol 7(4-2 Special) p 1581-1588.

[13] Ukpai C O, Dlay S S and Woo W L 2015 Iris feature extraction using principally rotated complex wavelet filters (PR-CWF) *Proceedings of the International Conference on Computer Vision and Image Analysis Applications*.

[14] Podder P, Khan T Z, Khan M H, Rahman M M, Ahmed R and Rahman M S 2015 An efficient iris

segmentation model based on eyelids and eyelashes detection in iris recognition system *Proceedings of the International Conference on Computer Communication and Informatics*.

[15] Radman A, Zainal N and Suandi S A 2017 Automated segmentation of iris images acquired in an unconstrained environment using HOG-SVM and GrowCut *Digital Signal Processing: A Review Journal* vol 64 p 60-70.

IOP Conf. Series: Materials Science and Engineering **769** (2020) 012025 doi:10.1088/1757-899X/769/1/012025

A Review of Divorce Information Management System

Aziman Abdullah[1], Asar AK[2], Solehah Shamsuddin[2] and Mohammad Affendy Omardin[3]

[1]Faculty of Computing, Universiti Malaysia Pahang, Malaysia.
[2]Centre of Human Sciences, Universiti Malaysia Pahang, Malaysia.
[3]Faculty of Engineering Technology, Universiti Malaysia Pahang, Malaysia.

E-mail: aziman@ump.edu.my

Abstract. The social issue associate with divorce is not on the divorce process itself but rather than the impact to the individuals who experience divorce. Since divorce already become a stigma in society, most people perceived more on negative view instead of seeing it as a process to solve issues in marriage. With the high penetration of Internet technology globally especially the use of social network, it becomes a factor how people can cheat in marriage that lead to divorce. However, studies on the use of Internet or information technology in managing divorce is still lack. This paper aims to analyse the existing literature in Scopus database on divorce information management system and identify targeted stakeholders of the system. Our findings confirmed that there is lack of study on divorce information management system.

1. Introduction

According to Department of Statistics Malaysia[1], the crude divorce rate (CDR) has decreased from 2.1 (2016) to 2.0 (2017) per 1,000 Muslim population. Although the trend shows a declining one, the real issue needs to be focus on divorce is the impact to the respective individuals during and after the divorce. With more women's involvement in employment, there is an economy consequence of divorce on a country[2]. Children, the most precious national treasure for survival of the nation are the most vulnerable group affected by divorce.

Recent study[3] claims that there is a high risk of women with parental divorce to develop mental disorders, chronic stress, loneliness, avoidance and anxiety. Another study[4] claims that there is significant association of multiple divorces with severe coronary artery disease in women but not in men. Children are highly significant to experience mental health problems based on the existing literatures[5–11]. Due to divorce itself has become a stigma in society[12], many people perceived it as a bad thing and may overlook the way how the affected individuals especially woman and children who should properly be supported in coping with the situation. Perhaps getting divorce might be the best solution for the case of abusive spouse[13] and there are positive side[14] of it instead of being seeing in absolute negative. With more people get exposed with Internet particularly on the social network[15,16], there is high possibility of increase number of divorces. In China, broadband internet subscription is positively affect the number of divorces in areas with lower education levels as well as areas with a higher income growth rate[17]. However, studies on using Internet specifically information system as a strategy for mediation and coping on divorce and post-divorce process are still in infancy stage.

Therefore, our study addresses the effort done in existing studies focusing on divorce management system as the subject of research on adopting Internet technology. This paper aims to identify the key gaps of divorce management with the focus on information system.

2. Method

Systematic literature review is the research method that differs from traditional or conventional review. With more research works been published and indexed digitally, harnessing this advantage offer better and smarter approach in research. A systematic literature review was conducted to synthesize and better understand (i) how divorce information management system is currently conceptualized in the scientific literature, (ii) how existing studies on divorce are associated with information management and (iii) what gaps and research needs exist. The following questions guided the analysis:

1. How information system has been adopted in divorce management?
2. What is the research impact in term of citation and publication over the time?
3. Which key gaps exist in understanding, characterizing and related stakeholders through information system associated with divorce?

Table 1 illustrates how information technology has been adopted in various context. The results based on Scopus indexed database using a combination of key words, the 1st layer meant for us to find total literatures that used the word "*divorce*" in the title of the article. While the 2nd layer was used to filter or scoping the results from the 1st layer based on "*manage**" and "*info**" keywords, inclusion and exclusion criteria.

Table 1: Literature selection

Database	**Search**	
	1st Layer	**2nd Layer**
Scopus	*"Divorce"* (Title)	AND "*manage**" (All Fields) AND "*info**" (All Fields)
*Inclusion criteria	• Peer-reviewed articles from 2009 to 2019 • English literature • Articles conducting experiment or empirical work on usage of Internet, information system or related theories in divorce management.	
*Exclusion criteria	• Review articles, opinion pieces, non-peer-reviewed literature, articles in press and citations. • Articles that not written in English • Research that published more than 10 years	

Note: The inclusion and exclusion only apply for the 2nd layer search.

3. Results and Discussion

Based on the literature review conducted; the researchers found some main findings to be highlighted as follow:

Table 2: Number of articles initially identified in the review of the Information System Adoption

Source	Scopus
Retrieval Date	15 August 2019
1st Layer	1,805
2nd Layer	34
Final	7

It seems that the adoption of Internet technology still lacking in the world. There is none of the existing works that coming from developing countries including the Muslim world. While in the developed countries, particularly the West; the divorce online management consists only a few. UK and Netherland seem to be the only governments which have adopted online services for the divorce management as shown in the following table. This finding indicates that there is potential benefit or value proposition of Internet technology can offer for divorce management system.

Table 3: Review of the related literatures

Paper ID	Concept	Context	Stakeholder	Country
P1	Information theory, data privacy, legal	Guides for psychologists to manage confidentiality and the releasing of information[18].	Mediator	USA
P2	Information usage, health	Perception of adolescents of custodial parents through negative information about the other parent[19].	Divorcee, Children	USA
P3	Information System, mediation support	Children Information Management System for mediation on families[20].	Children, Mediator	USA
P4	Information System, mediation support	implement online service for divorce mediation[21].	Divorcee, Mediator, Regulator	UK
P5	Information System, mediation support	Usefulness of participatory visual methodology in gathering information of complex thoughts and feelings from adolescents regarding their experiences with parental divorce[22].	Children	USA
P6	Information System, mediation support	support and information by community of practice. Sharing experience to facilitate others in coping with divorce[23].	Divorcee, Mediator	UK
P7	Information System, mediation support	divorce mediation support system through online [24].	Divorcee, Mediator	Netherlands

3.1. *Research Impact*

To further understand the research impact of the reviewed articles, publication data such as year of publication and citation were captured and presented in Table 4.

Table 4: Citation of Research Articles

Articles ID	Publication Year	Scopus	Google Scholar
P1	2009	8	23
P2	2010	43	52
P3	2010	7	16
P4	2010	10	15
P5	2011	1	3
P6	2012	1	3
P7	2014	4	11

The citation number indicates that research on divorce information management system is still lacking considering the total citation in other indexing database (Google scholar) of the oldest literature in 2009. The highest citation is from articles P2 where it is related with mental health or health science. It is common that articles related with health science are highly cited. While literature in the latest year published in 2014 seems does not make much impact although UK government (applicable to England and Wales) already implements online information system to facilitate the divorce process[25].

3.2. Research Gap

Based on the previous discussion, most of the study on adopting information system in managing divorce are from developed countries. There is an absence of study from Islamic countries or from Asia region. Perhaps due to limited source in Scopus and language used for articles limits the visibility of the work. Nevertheless, 7 articles out of 1,805 articles related on divorce associated with information system which equivalent to 0.4% distribution is significant enough to highlight the research gap in this area.

Table 3 shows a clear and obvious research gap in the divorce information management system although developed country like United Kingdom already adopting online services to facilitate the process of managing divorce. To better extract and highlight the research gap, we cluster the context of stakeholders and produce a matrix shown by Table 5.

Table 5: Stakeholders Matrix

Paper ID	Divorcee	Children	Mediator	Regulator
P1			Yes	
P2	Yes	Yes		
P3		Yes	Yes	
P4	Yes		Yes	Yes
P5		Yes		
P6	Yes		Yes	
P7	Yes		Yes	

Table 5 shows the synthesis of the existing literature on divorce information management system is lack of integrated multi-stakeholder approach. Based on the existing literatures, there is no a single study that integrate multi-stakeholder in adopting information system in managing divorce.

4. Conclusion and Recommendation

This paper aim to address the effort done in existing literatures on divorce management system as the subject of research on adopting information system. There are limited studies on the usage of information system in managing divorce which dominated from developed countries. The number of citations also confirm that there is not much attention has been put by researchers on this matter perhaps due to lack of multi-disciplinary approach in addressing the issue.

As for recommendation, there is critical needs on how information system should be used and designed in smart features in managing divorce. This issue can be effectively been studied through addressing multi-stakeholder needs in multi-disciplinary approach. Divorce should be perceived as a process to move on in life and the right of people who are affected especially children can be protected in the legal system with smarter divorce information system. Adoption of artificial intelligence (AI) or computer simulation such as to estimate the cost of life expanses of the children that divorce parents need to be accountable can open a new promising solution. Thus, future research addressing the framework is critical that integrate multistakeholderism[26], data analytics[27] and society 5.0[28–30] concept towards meeting the United Nation sustainable development goals.

5. Acknowledgement

This work was supported by Universiti Malaysia Pahang [grant number RDU190309].

References

[1] Mahidin M U 2018 Marriage And Divorce Statistics, Malaysia, 2018 *Dep. Stat. Malaysia,*
[2] De Vaus D, Gray M, Qu L and Stanton D 2017 The economic consequences of divorce in six OECD countries *Aust. J. Soc. Issues* **52** 180–99
[3] Schaan V K, Schulz A, Schächinger H and Vögele C 2019 Parental divorce is associated with an increased risk to develop mental disorders in women *J. Affect. Disord.* **257** 91–9
[4] Daoulah A, Al-kaabi S, Lotfi A, Al-Murayeh M, Nasseri S A, Ahmed W, Al-Otaibi S N, Alama M N, Elkhateeb O E, Plotkin A J, Malak M M, Alshali K, Hamzi M, Khunein S Al, Abufayyah M and Alsheikh-Ali A A 2017 Inter-ethnic marriages and severity of coronary artery disease: A multicenter study of Arabian Gulf States *World J. Cardiol.* **9** 371
[5] Clarke-Stewart K A, Vandell D L, McCartney K, Owen M T and Booth C 2000 Effects of parental separation and divorce on very young children *J. Fam. Psychol.* **14** 304–26
[6] Strohschein L 2012 Parental Divorce and Child Mental Health: Accounting for Predisruption Differences *J. Divorce Remarriage* **53** 489–502
[7] Taylor R and Andrews B 2009 Parental depression in the context of divorce and the impact on children *J. Divorce Remarriage* **50** 472–80
[8] Haimi M and Lerner A 2016 The Impact of Parental Separation and Divorce on the Health Status of Children, and the Ways to Improve it *J. Clin. Med. Genomics* **4**
[9] Uphold-Carrier H and Utz R 2012 Parental Divorce Among Young and Adult Children: A Long-Term Quantitative Analysis of Mental Health and Family Solidarity *J. Divorce Remarriage* **53** 247–66
[10] Chase-Lansdale P L, Cherlin A J and Kiernan K E 1995 The Long-Term Effects of Parental Divorce on the Mental Health of Young Adults: A Developmental Perspective *Child Dev.* **66** 1614–34
[11] Strohschein L 2005 Parental divorce and child mental health trajectories *J. Marriage Fam.* **67** 1286–300
[12] Konstam V, Karwin S, Curran T, Lyons M and Celen-Demirtas S 2016 Stigma and Divorce: A Relevant Lens for Emerging and Young Adult Women? *J. Divorce Remarriage* **57** 173–94
[13] Dekeseredy W S, Dragiewicz M and Schwartz M 2017 *Abusive endings: Separation and divorce violence against women* (University of California Press)
[14] Halligan C, Chang I J and Knox D 2014 Positive Effects of Parental Divorce on Undergraduates *J. Divorce Remarriage* **55** 557–67
[15] Valenzuela S, Halpern D and Katz J E 2014 Social network sites, marriage well-being and divorce: Survey and state-level evidence from the United States *Comput. Human Behav.* **36** 94–101
[16] Clayton R B, Nagurney A and Smith J R 2013 Cheating, Breakup, and Divorce: Is Facebook Use to Blame? *Cyberpsychology, Behav. Soc. Netw.* **16** 717–20
[17] Zheng S, Duan Y and Ward M R 2019 The effect of broadband internet on divorce in China *Technol. Forecast. Soc. Change* **139** 99–114
[18] Zimmerman J, Hess A K, McGarrah N A, Benjamin G A H, Ally G A, Gollan J K and Kaser-Boyd N 2009 Ethical and Professional Considerations in Divorce and Child Custody Cases *Prof. Psychol. Res. Pract.* **40** 539–49
[19] Afifi T D and McManus T 2010 Divorce disclosures and adolescents' physical and mental health and parental relationship quality *J. Divorce Remarriage* **51** 83–107
[20] Holtzworth-Munroe A, Applegate A, D'Onofrio B and Bates J 2010 Child Informed Mediation Study (CIMS): Incorporating the children's perspective into divorce mediation in an American pilot study *J. Fam. Stud.* **16** 116–29
[21] Dingwall R 2010 Divorce mediation: Should we change our mind? *J. Soc. Welf. Fam. Law* **32** 107–17
[22] Hartman L R 2011 The way they see it: How do adolescents experience life following parental divorce? *Int. J. Interdiscip. Soc. Sci.* **6** 93–9

[23] Paechter C 2012 Online learning in divorce: How the formation of a community of practice in a divorce support website enables and inhibits learning *E-Learning Digit. Media* **9** 392–405
[24] Bollen K N L, Verbeke A L and Euwema M C 2014 Computers work for women: Gender differences in e-supported divorce mediation *Comput. Human Behav.* **30** 230–7
[25] Government U 2019 Get a divorce
[26] Raymond M and DeNardis L 2015 Multistakeholderism: anatomy of an inchoate global institution *Int. Theory* **7** 572–616
[27] Silva B N, Diyan M and Han K 2019 Big Data Analytics *SpringerBriefs in Computer Science* pp 13–30
[28] Mayumi Fukuyama 2018 Society 5.0: Aiming for a New Human-centered Society *Japan SPOTLIGHT* 8–13
[29] Salgues B 2018 *Society 5.0* (Hoboken, NJ, USA: John Wiley & Sons, Inc.)
[30] Shiroishi Y, Uchiyama K and Suzuki N 2018 Society 5.0: For Human Security and Well-Being *Computer (Long. Beach. Calif).* **51** 91–5

IOP Conf. Series: Materials Science and Engineering **769** (2020) 012026 doi:10.1088/1757-899X/769/1/012026

A Novel Learning Engagement Data Model (LEDM) for Online Attendance System

Aziman Abdullah[1], Asar AK[2] and Nur Alnisa' Anis Alanna Binti Ruzelan[1]

[1]Faculty of Computing, Universiti Malaysia Pahang, Malaysia.
[2]Centre of Human Sciences, Universiti Malaysia Pahang, Malaysia.

E-mail: aziman@ump.edu.my

Abstract. Student engagement is a very interesting subject in higher education. While many studies assess student engagement through survey, this approach claimed in literatures is lack of contextual analysis for decision making. Our motivation in this study is to integrate a simple way to assess student engagement of face-to-face session in blended learning approach within the online attendance system by identifying the data model supporting insightful analytics. This study aims to propose a new learning engagement data model incorporating behaviour, emotional and cognitive engagement for online attendance system. We found an interesting insight which there is a relationship of student engagements with the learning outcomes attainment. Initial findings in this study show potential values how our proposal may benefit higher education in adopting smarter way to measure student engagement while taking student attendance during face-to-face session in blended learning implementation.

1. Introduction

Student engagement is a very important indicator in measuring student success in their study. With digital technology advancement, particularly Internet, many higher education institutions start to explore and adopt blended learning approach as it is claimed[1] can engage student in learning. Since the approach still require physical presence of both teacher and student[2], student attendance system has potential to be used as a medium to understand student engagement. There are claims [3,4] that class attendance has strong relationships with academic performance. The claims used attendance data to indicate student physical presence in class. However, it did not indicate a complete student engagement during the class either student did achieve the learning outcomes during the class session. A complete student engagement should include all the dimension of student engagement theory which are cognitive, emotional and behaviour engagement.

Existing attendance system especially the one used manual signature disturbed student's attention since student need to write down their initial by queuing. Technology based attendance systems mostly are designed with the goal or emphasize on process efficiency in recording the data but not on learning engagement aspect. Therefore, this study aims to propose a novel engagement data model that act as a foundation towards measuring complete student engagement in class with analytics. This study is been facilitated with the following research questions: -

RQ1: How behaviour, emotional and cognitive engagement affect student learning outcomes attainment?

RQ2: How online attendance system can support learning engagement?

2. Literature Review

2.1. Student Engagement

Student engagement can be conceptualized as integration of three interrelated learning dimensions or domains which are cognitive engagement, emotional engagement and behavioural engagement. According to literature[5], these three dimensions can be described as:-

- behavioral engagement: students' participation in education, including the academic, social and extracurricular activities of the school
- emotional engagement: students' emotional reactions in the classroom and in the school (a sense of belonging or connectedness to the school)
- cognitive engagement: students' investment in their learning (motivation and self-regulation).

There is a claim[6] that face-to-face or in class activities comes with "*more engagement*" and "*immediate feedback*" which are foundation for student to progress in learning. But measurement and statistical methodologies are yet another challenges highlighted in literature[7] regarding student engagement especially on the most complex one, emotional engagement. Many research and implementation to measure or assess student engagement adopt a survey approach[8–10] at the end of semester which lack of insight for effective decision making. To facilitate better teaching practice, a contextual analysis is required for instructor to understand a complete dimension of student engagement and how it affects the attainment of the learning outcomes.

2.2. Online Attendance System

In general, student attendance or absence is a very practical indicator in assessing student engagement[11], managing student at risk[12] and predicting academic performance[13]. Student attendance is useful as a direct representation of student behaviour engagement in the class. However, conventional student attendance system only captured student identity either student's signature or the instructor's notes based on his observation, which is not productive nor smart when it comes to large number of students. Student attendance system that integrate biometrics[14], Quick Response or QR code[15], Radio Frequency Identification (RFID)[16] and voice recognition [17] mostly focusing on the technological aspect to acquire identity data of student. It however never about measuring student engagement. With more online services offered through cloud technology, there is a promising opportunity how a simple and smart attendance system that integrate student engagement can be developed using cloud technology.

2.3. Blended Learning

Blended learning can be easily understood as combination of face-to-face and online learning. It means, physical presence of both teacher and student is still required in blended learning strategy[18]. Blended learning is claimed to has an impact on student engagement[19], academic performance[20] and student learning outcomes[21]. One example of blended learning approach is flipped classroom[22] which is helpful for instructor and learner in making the class (face-to-face) session enrich with enquiries and promote student with high order thinking[23]. Blended learning implementation can be instrumental for institution strategic planning in transforming institution towards the concept of smart university[24].

3. Method

3.1. Participants and Procedure

The data were collected from the students who enrolled in a course Semester 2 in 2018/2019 session. Table 3.1 indicates the profile of the target population and selected course.

Table 3.1. Participants of study

Course Name	Web Engineering (BCS2243)
Student Size	38
Enrolment duration	28 Jan 2019 – 26 May 2019
Number of week	14 weeks
Number of lecture session	1 session per week (2 hours)
Number of laboratory session	1 session per week (2 hours)
Total face-to-face session for student	28 sessions

3.2. *Data Model*

This study adopted quantitative method for data collection. A simple online attendance system which include all dimension of student engagement during the face-to-face class. The data model designed in this study that been integrated in the system is shown in figure 3.1.

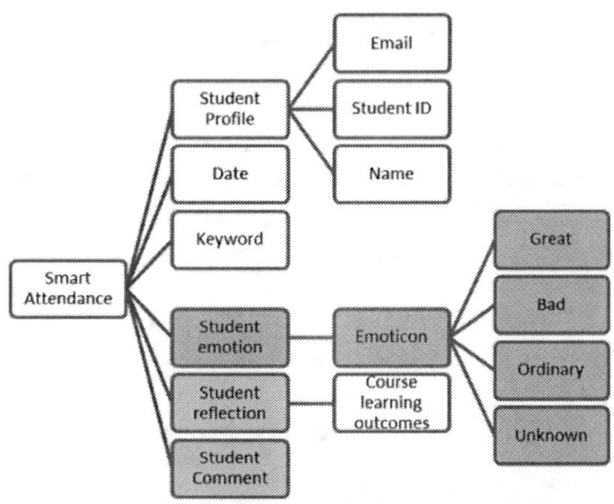

Figure 3.1: Proposed Learning Engagement Data Model (LEDM)

Based on figure 3.1, student profile, date and keywords are data that categorized for behavior engagement. The 'keyword' is used as a field for student to fill based on the instructor's random keywords mentioned during class. It is used to allow only students who are attend physically to the class can fill the field and submit their attendance form. This should minimize the attempt of student who do not attend to fill the online attendance form since it is accessible on Internet. The calculation model to assess behavior engagement is by calculating the total number of attendances relative to total face-to-face session student should attend in a semester as shown in equation (1).

$$B = \frac{\sum Attended}{total\ session\ per\ semester} \quad (1)$$

'Student emotion' category represent the emotional engagement which represented in emotion icon or also known as emoji in the online attendance system. Table 3.2 shows the category of emoji used as emotional identifier in this study.

IOP Conf. Series: Materials Science and Engineering **769** (2020) 012026 doi:10.1088/1757-899X/769/1/012026

Table 3.2 Emotional Engagement Scale

Emoji	Emotion description
 Great	Student feel great with the class learning experience.
 Just so-so	Student feel ordinary with the class learning experience.
 Not Telling	Student choose to not inform his/her feeling.
 Got a bad day	Student feel bad with the class learning experience.

In order to measure emotional engagement level, we develop our own emotional engagement index based on the frequency of each emotion relative to the total number of class attended by student. In other words, we try to measure student's feeling during the class session. The calculation model of the index is shown in equation (2).

$$E = \frac{\sum Emoji}{total\ class\ attended} \quad (2)$$

For cognitive engagement, we integrate a self-rating mechanism where student need to do quick reflection of the lesson oriented to the course learning outcomes that clustered to cognitive domain. The scale and cognitive engagement item in online attendance form is shown in figure 3.2. The calculation model for cognitive engagement is formulated in equation (3). It calculates the ratio of total score of student attendance based on the level of learning experience against the maximum score student could get based on highest level of learning experience. In this case we used 3 level of attainment measured in online attendance system as shown in figure 3.2.

$$C = \frac{\sum(level \times attendance)}{total\ class\ attended \times 3} \quad (3)$$

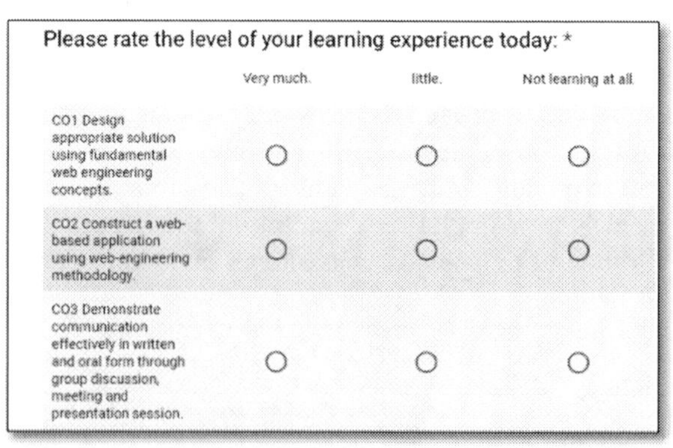

Figure 3.2: Cognitive Engagement Scale

3.3. Learning Impact Analysis

In order to justify the impact and value of our proposed model, we analyse the engagement level for each dimension with the summative assessment in the final exam (controlled environment) represents course learning outcomes in cognitive category. Scatter plot has been selected for clustering and visualize the relationship between student engagement with the course learning outcomes attainment (cognitive) assessed in summative assessment (final exam score).

4. Results and Discussion

4.1. Learning Engagement vs Learning Outcomes Attainment

Our first research question is to understand how behaviour, emotional and cognitive engagement affect student learning outcomes attainment? The following subsections explain how each dimension of student engagement are affecting student learning outcomes attainment.

4.1.1. Behaviour Engagement vs Learning Outcomes Attainment

There is weak negative correlation with $r = 0.15$ between behaviour engagement and cognitive attainment in the summative assessment (final exam) as shown in figure 4.1. This finding suggest that the attendance may not significance to certain student based on two arguments (1) there are student who got good score in final exam with poor number of attendances and (2) there are student who got poor score (<50 points) but attend more than half of the session. The potential implication from this initial finding is the institutional policy in providing more flexible learning experience. Common administrative issues like scheduling in course offering due to lack of resources (instructor or location) affect significantly the opportunity for student to enrol the courses that been offered. As a result, student may extend their semester to complete the required course enrolment and this directly affect their graduation duration.

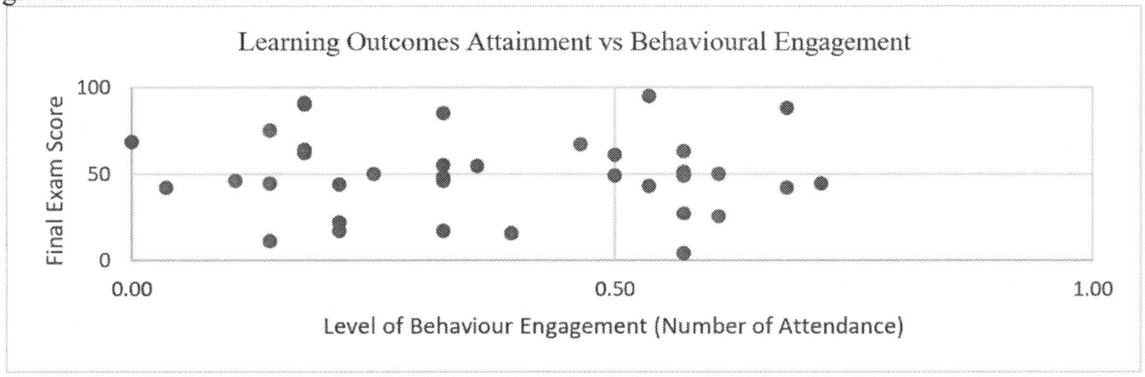

Figure 4.1: Behaviour Engagement Analysis

4.1.2. Emotional Engagement vs Learning Outcomes Attainment

There are four different emotion used in our study. Before we analyse them, we perform a simple test to identify outliers in order to produce better analysis. Figure 4.2 shows the outliers for the data captured in four type emoji used in the system. Based on the figure, we suggests that the emotion with label '*Got a bad day*' which represent negative emotion during the class is not significant to be analysed due to lack of data value variation in dataset and most student do not experience that feeling during class. Another data that will be ignore for analysis is the emotion with label '*Not telling*' due to it does not indicate any feeling. The rationale to include this type of emotion is to give an option to student for not informing their feeling to instructor during online attendance taking. In this section, we only analyse the positive emotion with label '*Great!!!*' and how it relates with learning outcome attainment.

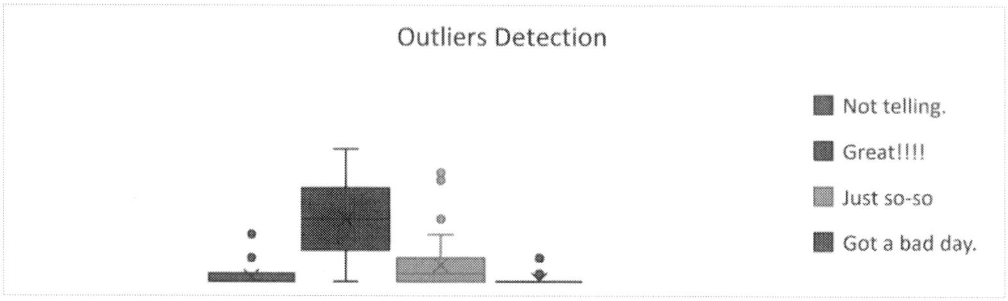

Figure 4.2: Outliers Analysis

Our analyses on relationship between emotion engagement with course learning outcomes attainment are shown in figure 4.3. Based on the visual analysis, there are more student who did not feel positive and excitement during the face-to-face session that manage to get a good score (>50) in final exam. Nevertheless, there are also student who feel positive and excitement but did not doing well during the final exam (score <50).

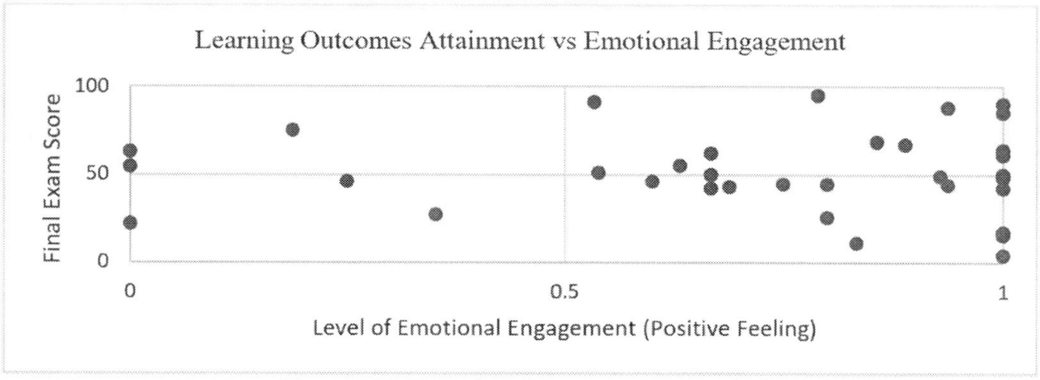

Figure 4.3: Emotional Engagement Analysis

4.1.3. Cognitive Engagement vs Learning Outcomes Attainment

Cognitive engagement is measured based on the self-reflection activity where student need to rate their perceived level of learning experience for each face-to-face session in the online attendance form. Figure 4.4 shows all student have high cognitive engagement during the class session. The findings indicate that there are several students who have very high level of cognitive engagement (perceived score =1) but their score in final exam still in a poor attainment. There are high variations of these two indicators which suggest that more research can be done to uncover the insight from this data.

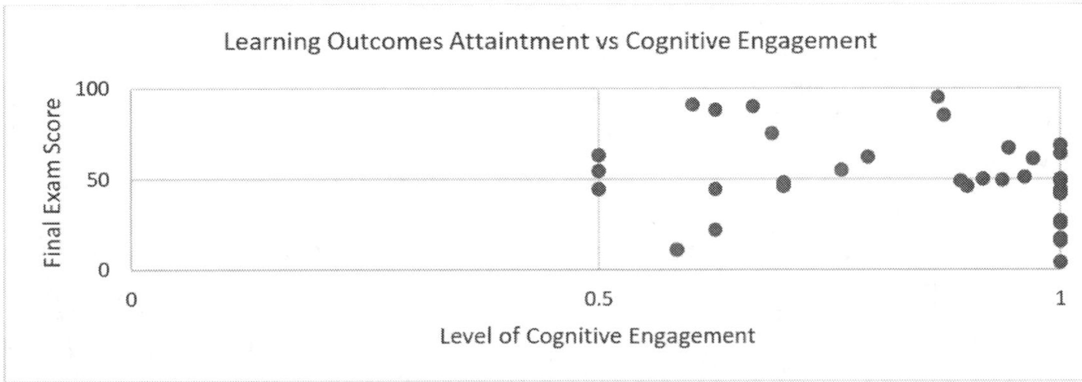

Figure 4.4: Cognitive Engagement Analysis

4.2. Online Attendance, Learning Engagement and Blended Learning

Based on the findings in figure 4.1, figure 4.3 and figure 4.4, we can confirm that our proposed data model integrated in online attendance can support learning engagement analytics. The results answer our second research question on how online attendance system can support learning engagement analytics.

5. Conclusion and Future Works

This paper aims to propose a novel contextual engagement data model that capable to measure student engagement in class using attendance analytics. Based on the findings, the proposed data model (figure 3.1) for online attendance system allowing learning analytics to be implement targeting on all dimension of student engagement. This data model in measuring student engagement is the first to be reported in research.

As for recommendation, the proposed data model can be integrated in existing student online attendance system for campus wide implementation. With the integration, more insight on engagement can be produced particularly on the aspect of curriculum improvement and how institution can further understand and manage student engagement in effective and predictive manner. In the future, we are planning to integrate this data model in online attendance system for the general courses enrolled by large number of students to study either the size of the class and location in the class do affect level of student engagement. Issue on holistic student development associated with the quality of welfare among student who live off-campus accommodation also potential can be studied based on our proposed learning engagement data model. We also plan to adopt emotional engagement elements in our existing research study regarding online divorce management system to study how engagement play a role in the quality of decision making.

6. Acknowledgement

This work was supported by Universiti Malaysia Pahang [grant number RDU190309].

References

[1] Holley D and Dobson C 2008 Encouraging student engagement in a blended learning environment: The use of contemporary learning spaces *Learn. Media Technol.* **33** 139–50

[2] Owston R 2013 Internet and Higher Education Blended learning policy and implementation : Introduction to the special issue *Internet High. Educ.* **18** 1–3

[3] Bijsmans P and Schakel A H 2018 The impact of attendance on first-year study success in problem-based learning *High. Educ.* **76** 865–81

[4] Benyo B, Sodor B, Doktor T and Fordos G 2012 Student attendance monitoring at the university using NFC *Wirel. Telecommun. Symp.*

[5] Fredricks J A, Blumenfeld P C and Paris A H 2004 School engagement: Potential of the concept, state of the evidence *Rev. Educ. Res.* **74** 59–109

[6] Kemp N and Grieve R 2014 Face-to-face or face-to-screen? Undergraduates' opinions and test performance in classroom vs. online learning *Front. Psychol.* **5**

[7] Fredricks J A, Filsecker M and Lawson M A 2016 Student engagement, context, and adjustment: Addressing definitional, measurement, and methodological issues *Learn. Instr.* **43** 1–4

[8] Wang M Te, Fredricks J A, Ye F, Hofkens T L and Linn J S 2016 The Math and Science Engagement Scales: Scale development, Validation, And psychometric properties *Learn. Instr.* **43** 16–26

[9] Fredricks J A, Wang M Te, Schall Linn J, Hofkens T L, Sung H, Parr A and Allerton J 2016 Using qualitative methods to develop a survey measure of math and science engagement *Learn. Instr.* **43** 5–15

[10] Leong F 2014 National Survey of Student Engagement *Encyclopedia of Counseling*

[11] Lee J, Song H-D and Hong A 2019 Exploring Factors, and Indicators for Measuring Students' Sustainable Engagement in e-Learning *Sustainability* **11** 985

[12] Bicard D F, Lott V, Mills J, Bicard S and Baylot-Casey L 2012 Effects of Text Messaged Self-Monitoring on Class Attendance and Punctuality of At-Risk College Student Athletes ed J Tiger *J. Appl. Behav. Anal.* **45** 205–10

[13] Lukkarinen A, Koivukangas P and Seppälä T 2016 Relationship between Class Attendance and Student Performance *Procedia - Soc. Behav. Sci.* **228** 341–7

[14] Charity A, Okokpujie K and Etinosa N-O 2017 A bimodal biometrie student attendance system *2017 IEEE 3rd International Conference on Electro-Technology for National Development (NIGERCON)* vol 2018-Janua (IEEE) pp 464–71

[15] Baban M H M 2014 Attendance Checking System Using Quick Response Code For Students At The University Of Sulaimaniyah *J. Math. Comput. Sci.* **10** 189–98

[16] Mahesh Sutar, Mahesh Patil S W 2016 Smart Attendance System Using RFID In IOT *Int. J. Adv. Res. Comput. Eng. Technol.* **5** 2278–1323

[17] Uddin N, Rashid M M, Mostafa M G, Belayet H, Salam S M, Nithe N A, Rahman M W and Halder S 2016 Development of Voice Recognition for Student Attendance *Glob. J. Hum. Soc. Sci. G Linguist. Educ.* **16** 1–8

[18] Moskal P, Dziuban C and Hartman J 2013 Blended learning: A dangerous idea? *Internet High. Educ.* **18** 15–23

[19] Holley D and Oliver M 2010 Student engagement and blended learning: Portraits of risk *Comput. Educ.* **54** 693–700

[20] Broadbent J 2017 Comparing online and blended learner's self-regulated learning strategies and academic performance *Internet High. Educ.* **33** 24–32

[21] López-Pérez M V, Pérez-López M C and Rodríguez-Ariza L 2011 Blended learning in higher education: Students' perceptions and their relation to outcomes *Comput. Educ.* **56** 818–26

[22] McNally B, Chipperfield J, Dorsett P, Del Fabbro L, Frommolt V, Goetz S, Lewohl J, Molineux M, Pearson A, Reddan G, Roiko A and Rung A 2017 Flipped classroom experiences: student preferences and flip strategy in a higher education context *High. Educ.* **73** 281–98

[23] Styers M L, Van Zandt P A and Hayden K L 2018 Active learning in flipped life science courses promotes development of critical thinking skills ed E L Dolan *CBE Life Sci. Educ.* **17** ar39

[24] Uskov V L, Bakken J P, Karri S, Uskov A V., Heinemann C and Rachakonda R 2018 Smart university: Conceptual modeling and systems' design *Smart Innovation, Systems and Technologies* vol 70 pp 49–86

The prediction of undergraduate student performance in chemistry course using multilayer perceptron

Che Akmal Che Yahaya, Che Yahaya Yaakub, Ahmad Firdaus Zainal Abidin, Mohd Faizal Ab Razak, Nuresa Fatin Hasbullah and Mohamad Fadli Zolkipli

Faculty of Computing, College of Computing and Applied Sciences, Universiti Malaysia Pahang, 26300 Kuantan Pahang Malaysia

E-mail: pcn18003@stdmail.ump.edu.my, aqemal.yahaya@gmail.com

Abstract. Chemical industry are key elements for changing crude materials to our ordinary objective merchandise. This has achieved an immense move in how things work. The disappointment pace of an understudy in a science course is additionally expanding with regards to requesting the compound specialists. Understudies who enlist the science course regularly bomb in the first or consequent semesters. Moreover, understudies are likewise unfit to comprehend in the event that they can adjust and graduate effectively in this program. The objective of this exploration is to foresee the future utilization of improved science for understudies to fizzled or graduate by upgraded Multilayer Perceptron (MLP) arrangement with Adaboost. The exactness of the outcomes is 92.23% percent.

Keywords: chemistry, undergraduate, education, machine learning, multilayer perceptron, neural network

1. Introduction

The chemistry course is an important education that teaches a student the interaction between matter and energy surrounding us [1]. It involves the world around us, such as medicine, cooking, cleaning, and environmental issues and also material that involves electronic devices. Many decades of studies in chemistry have resulted in the growth of several thousand molecular descriptors describing a variety of possibly any compound characteristics. Therefore, chemistry one important subject that can serve as one of the knowledge that can be implemented in routine activities. Among another subject, this research chooses chemistry subject because numerous times of science research has prompted the improvement of a few thousand atomic descriptors that depict a scope of properties of possibly any compound in our daily life [2].

However, for academic purposes these subjects tend more students to fail at their school level. Among 97,095 SPM candidates, 3.9 percent of them were failed [3]. In addition, many students unable to discover that they have the talent in chemistry courses and able to pass in flying colors. Hence, it is important to predict a student either graduates or fail before they register in this course.

By predicting the student performance (fail or graduate), it will highlight the students that have suitable in the chemistry course which will consume for 3 years. On the other hand, this prediction also will identify the students that attempt to fail and therefore the student may change other courses or to remind the student to give more concentration in a chemistry course. Accordingly, these review adopts machine learning to predict a student either suitable in the first place before entering the course. For conducted the situation, machine learning Multilayer Perceptron (MLP) has been chosen because it surpasses all the classifiers in the experiment [4]. While Adaboost for well-known boosting arrangement calculation, and it one of the learning type classifiers. It can be classified grouping calculation that improves the expectation execution during the preparation information, which blunders to improve the MLP [4]. Multilayer Perceptron is a part of artificial intelligence consciousness. It applies to a variety of methods for assessing dataset-based capacities. Such capacities, called Multilayer Perceptron (MLP), would then be able to be utilized to foresee future results. In this unique situation, a calculation is a predefined arrangement of steps that accepts a lot of information as sources of info and changes it through scientific tasks. MLP of different types uses models and calculations as structure hinders for making ecological forecasts and surmisings. The MLP algorithm will discover designs in the information, at that point anticipate the result of something that has never been considered.

This study proposes Adaboost-multilayer perceptron (MLP) to predict the performance of a student in a chemistry course. The contributions from this research as follows:

a) In order to develop outstanding performance based on machine learning predictive, implemented the real datasets from Universiti Malaysia Pahang (UMP) students in Pahang, Malaysia. With the real dataset and true situation, is able to provide great input for the machine learning training.
b) To investigate the multiple features based on student entry qualifications (matriculation, STPM, and diploma), MUET results, gender (male or female), Malaysian citizenship (Bumiputera), states in Malaysia and status (graduate or fail).
c) Implement Adaboost as an addition to MLP as a strong learner machine learning for efficient results.

The structure of the research is as follows. Section 2 reviews the related works and compare with the proposed study. Section 3 explains the methodology in the experiment. Section 4 provides results derived from experiment. Finally, section 5 delivers a conclusion from the results.

2. Literature reviews

Each paragraph discusses the factors that have an effect on the student's success in the chemical course. In addition, the machine learning methodology used will be clarified briefly and addressed in relation to the previous work for this study.

There are several factors that impact student performance in gaining the results:

a) *The lack of interest in Chemistry course*. Chemistry is a subject of great significance to science. Many students are afraid to learn chemistry because it is difficult to understand. This situation makes students uninterested in studying and completing the course. Students tend, however, to perform better if they are interested in the subject.

b) *Problems with language and communications*. Some of the words used in the science subject vary from those used in other subjects and in ordinary communication. In contrast, the subject of science is mostly in English. This makes it even more difficult for students to learn and understand that English is poor. It causes problems for students to understand the material of the chemical topic well [2]. It includes the ambiguity of the form of the sentence and the language used by lecturers and educators.

c) *Mismatch Technique by Lecturers/Teachers*. Different lecturers and instructors have their own style of teaching. Some of them favored practical instruction and some are not. They might be happy with their way, but not with the students. Teachers or lecturers must be able to approach students in a constructive and comfortable way so that they can research and learn better.

2.1. Machine learning
Machine Learning is a class of calculation that empowers applications to be progressively exact in anticipating results without being unequivocally redone. The fundamental explanation of ML is to make computations that can get input data and use quantifiable assessment to foresee execution while reviving yields as new data winds up available.

Machine learning algorithms is regularly named supervised or not supervised. The supervised algorithm requires an information researcher or information investigator with ML skill to give both information and yield, notwithstanding giving criticism on the precision of forecasts during calculation training [5]. Information researchers choose what factors or highlights of the model ought to be assessed and used to make expectations. When learning is done, the calculation must apply what has been figured out how to new information.

Unsupervised algorithms do not need to be equipped with the required result information. Alternatively, they use an iterative approach called in-depth training to analyse information and come to conclusions. Unsupervised learning algorithms, also known as neural networks, are used for more complex computing tasks than supervised learning systems, including image recognition, speech-to-text, and natural language generation. These neural networks work by millions of examples of training data and automatically finding sometimes small comparisons between many variables [6]. When equipped, the algorithm can use its association datasets to interpret new data. Such algorithms have only become possible in the age of big data because they require a massive amount of training data.

2.2. Types of machine learning
Just as there are almost infinite applications for machine learning, there are unlimited of machine learning algorithms. They range from relatively easy to highly complex. Here are some of the most widely used versions:

This class of machine learning algorithm involves identifying a correlation -- generally between two variables -- and using that correlation to make predictions about future data points.

a) *Decision trees*: these models use perceptions of specific activities and decide the best way to land at the ideal result [7].
b) *K-means clustering*: This model gatherings a given number of datasets that focus on a particular number of bunches dependent on comparative attributes.
c) *Neural networks*: Such profound learning models utilize a lot of preparing information to set up a relationship between numerous factors to figure out how to process approaching information later on.
d) *Reinforcement learning*: This territory of profound learning includes models that go through numerous endeavors to finish a system. Steps that produce good results will be remunerated and steps that produce undesired results will be punished until the calculation learns the ideal procedure.

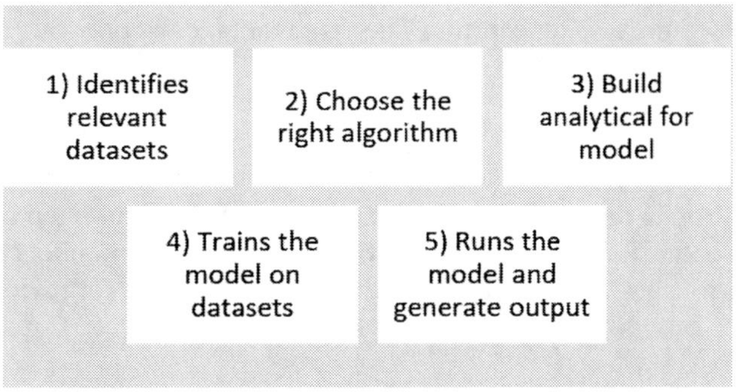

Figure 1. The machine learning methodology

IOP Conf. Series: Materials Science and Engineering **769** (2020) 012027 doi:10.1088/1757-899X/769/1/012027

2.3. Artificial neural network

A neural network system (ANN) is one of the most ordinarily utilized calculations. It is a PC model that has been utilized as a PC program. Complex connections among info and yield information groupings can be demonstrated utilizing ANNs. It was likewise used to recognize shrouded designs in informational indexes when the quantitative methodology, which is an express model dependent on the multifaceted nature of the issue, is tedious or isn't as of now conceivable. ANNs supplant these obscure practical associations with evaluated (roughly) adaptively assembled capacities. [4].

Neural systems are a lot of calculations, displayed freely after the human mind, which is intended to perceive designs. They translate tactile information through a sort of machine discernment, marking or grouping crude info. The examples they perceive are numerical, contained in vectors, into which all certifiable information, be it pictures, sound, content or time arrangement, can be deciphered

Neural networks help us cluster and classify the categories. It helps to group unlabelled data according to similarities among the example inputs, and they classify data when they have a labelled dataset to train on. Neural networks can also extract features that are fed to other algorithms for clustering and classification; thus neural networks as components of larger machine-learning applications involving algorithms for reinforcement learning, classification, and regression [5].

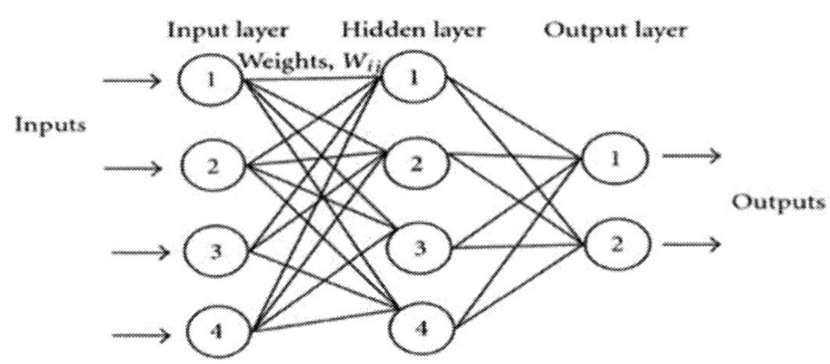

Figure 2. Schematic represent layer ANN

2.4. Boosting

Some random learning calculation precision can be improved by utilizing a general technique that is boosting. Adaboost (Adaptive Boosting) was distributed in 1995 by Freund and Schapire. A large number of the common sense troubles of the prior boosting calculations have been survived. The more solid and predictable outcome is acquired by learning the powerless calculation more than once in a progression of rounds that are given by the promoter calculation. After each time the basic learning calculation is named, it delivers another feeble forecast law. Likewise, Adaboost is utilized to test the improved classifier for the recognizable proof of understudy execution in the concoction course. There are a few focal points and detriments of boosting. Improves the consequences of the examination and upgrades the product for improved execution. The disadvantages of boosting are touchy to clamours and hard to actualize on the ongoing system.

2.5. Multilayer perceptron

The multilayer perceptron is one of the neural system framework's directed learning calculations. The multilayer perceptron comprises of a system of straightforward interconnected neurons or hubs and is completely connected to every hub associated with every hub in the following and past layers [8].

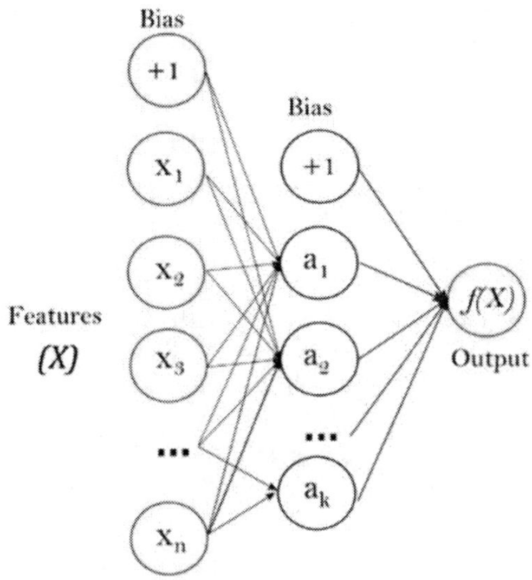

Figure 3: Multilayer perceptron

Advantages of using Multilayer Perceptron that can train non-linear models and learn models in real-time using partial fit and MLP drawbacks if hidden layers have a non-convex loss function where more than one local minimum occurs. As a consequence, different random weight initializations can lead to different accuracy of validation. Including MLP, it involves tuning a range of hyper parameters, such as the number of hidden neurons, layers, and iterations, and it is difficult to implement and interpret. The students ' information is based on their research history and the list of studies will be explained [9].

2.6. *Related works*

On the examination of the related work, the discoveries of past investigations, which likewise utilized explicit ML classifiers, contrasted from the test data[10]. For the examination, the Multilayer Perceptron (MLP) has a superior result, 92.5 percent, than different classifiers, including gullible Bayes and Forests, to assess the number. More investigation utilizing five separate classifiers uncovers that utilitarian trees (FT) are of the most elevated dependability (95 percent) and genuine positive worth (TRP) (96.7 percent).

In the past examination utilizing a similar Fisher test and the wellness test, the potential danger of utilizing the test is that the information utilized ought to be the recurrence or tally as opposed to the next kind. Meanwhile, the multilayer perceptron is equipped for perusing and testing information regardless of whether the data isn't in recurrence or checks. In the other case presented by the NVivo program, it is basic and proficient to utilize. It likewise improves the exactness of subjective investigations [11].

The cross-validation test is reasonable for the technique for identifying the exhibition of the understudies utilizing the Multilayer Perceptron (MLP) in the Adaboost algorithm. Cross-validation overlays the information multiple times and prepares the information to accomplish precise exactness. In light of the preparation dataset check, the exactness of the Adaboost calculation is 92.23%. It utilized to address the issue of over-fitting and to make forecasts increasingly broad [12].

Table 1. Comparison with previous paper and propose studies

Related works	Research titles	Proposed works	Methods
[1]	A longitudinal study of engineering student performance and Retention. I. Success and failure in the introductory course.	Detect the performance of students whether success or failure.	Used Fisher's exact test for independence between two categorical variables.
[2]	The inaccuracy of self-evaluation variable of student risk of failing the first year in chemistry.	Predict the risk of failing for the first-year students in a chemistry course.	Goodness-of-Fit test of Hosmer and Lemeshow (1991) using SPSS Statistics 17.0.
[3]	Root-exploit Malware Detection using Static Analysis and Machine Learning	Detect root-exploit malware in Android OS.	Identify the root-malware using three different machine learning classifiers.
[4]	Discovering optimal features using static analysis and a genetic search-based method for Android malware detection	To search for the best and smallest number of features to enhance accuracy and reduce the amount of complexity, noise, and irrelevant data in detecting malware.	Used five machine learning classifiers, Naïve Bayes (NB), functional trees (FT), J48, random forest (RF), and multilayer perceptron (MLP).
Current work	To Predict the Future of Undergraduate Student's Performance in Chemistry by Using ML	Detect the performance of students in the chemical course	Used Multilayer Perceptron in to detect performance (fail or success)

3. Methodology

The strategy tended to concern the structure of the test look into and the use of the techniques. The points of this work were to be utilized as a source of perspective for exploratory examination. It is to guarantee that the objectives of this exploration are accomplished. The machine learning programming called WEKA variant 3.8 is utilized for the investigation of the usage of characterization systems. It will likewise be talked about the stages forecast students ' success in the chemical course. Then we're showing the machine learning process and method used.

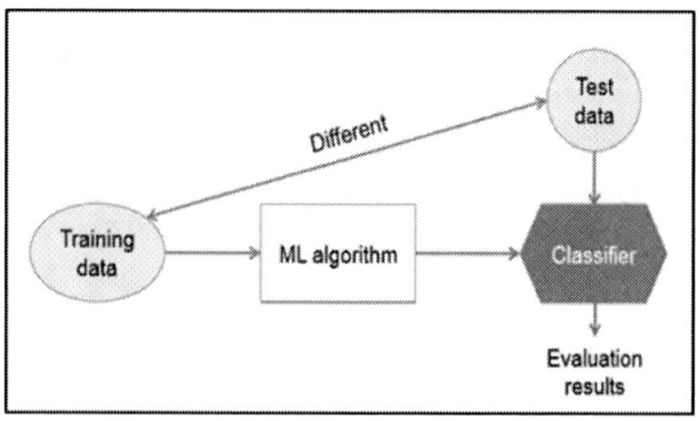

Figure 4. Machine learning process

So as to do an evaluation dependent on Weka tools as a software, it is essential to foresee the nature of understudies in the chemistry course whether to fall flat. Preparing information and test information are being set up to foresee understudy execution. Prescient preparing information will be utilized to get familiar with the ML algorithm and will at that point be founded on the preparation information of the framework. Next, the assessment result is gotten by assessing the information produced utilizing the MLP classifier strategy.

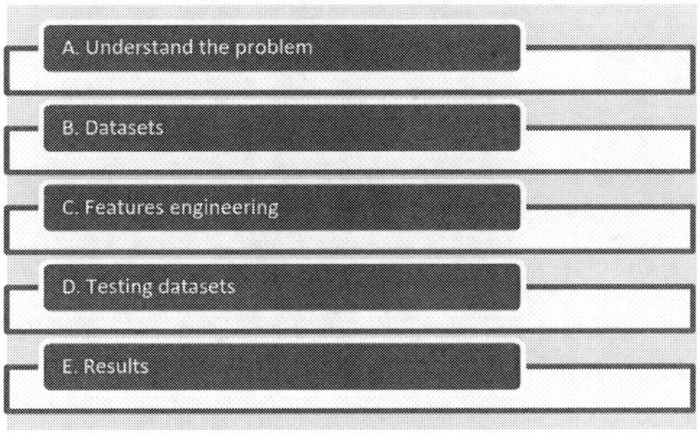

Figure 5. Methodology used

3.1. Understand the problems

The problems occur when the situation become hard to solve. For this situation, students who have initiatives can use other varieties way to solve the problems.

The problem statements from the previous section derived various questions. Hence, the research question derived as follows:

1) What are the features that determining the students will excellence or fail in a chemistry course? For instance: previous MUET results, entry qualification from matriculation, STPM, Diploma?
2) How to discover the students will fail in chemistry in the future?
3) How to discover the students will achieve excellent results in chemistry in the future?

3.2 Dataset

To predict the performance in a chemistry course, therefore, there is a need to utilize a decent dataset from students as a real situation. It collected datasets from the academic center of University Malaysia Pahang (UMP) which counted years from 2002 until 2015. It is important to note that this research is the first experimental data that scrutinizes this dataset.

3.3 Features

In features engineering, this research is in a raw form. The cleansing step data involves ambiguous, incomplete, and missing. Once it does, we were able to extract the features and provides many discoveries in student status (graduate and fail).

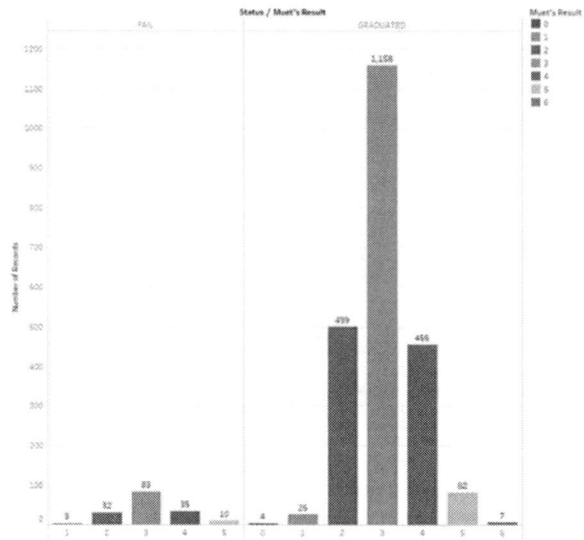

Figure 6. Features analysis by entry qualification, MUET

Figure 6 demonstrates the status of male and female students, whether to fail or succeed on the basis of their enrolment qualifications and MUET tests. Malaysia University English Test (MUET) is an English skill test administered by the Malaysian Examination Council (MEC) [15]. In this test, band 1 is the score for the student that hardly able to use the English language, however, the band 6 indicates that the students were very fluently in practicing the English language. Based on Student qualification before registering a chemistry course

In order to register the undergraduate course (bachelor degree) in Malaysia, there are three types of qualifications: (1) Diploma; (2) STPM; and (3) Matriculation.

A diploma is one of the qualification one step below an undergraduate degree, in Malaysia Diploma courses, are often taken up by students looking to enter the workforce at the earliest possible time, by equipping them with practical skills and knowledge for their specific industry. Entry into a Diploma course only requires students to have completed their high-school education, making it a very popular option amongst students in Malaysia. If students decide to change their minds at a later point, they also have the option to transfer to Year 2 of their chosen Undergraduate Degree as well, since Diploma courses typically range between 24 to 36 months [11].

Sijil Tinggi Persekolahan Malaysia (STPM, English: Malaysian Higher School Certificate) is a pre-university examination carried out by students in Malaysia. Formerly recognized as the Higher School Certificate (HSC). The HSC was a predecessor to the GCE A-Level in the United Kingdom and is still the name of the pre-university exam in some Australian states.

Since 1982, the STPM has been administered by the Malaysian Examinations Council (MEC), which has also been running the Malaysian University English Test (MUET) since 1999. However, national exams such as the Ujian Penilaian Sekolah Rendah and Sijil Pelajaran Malaysia are all set and reviewed by the Malaysian Examinations Syndicate (MES). Nevertheless, both the MEC and the MES are under the Ministry of Education [12].

STPM is one of the two main pre-university systems for admission to public universities in Malaysia. The other is a one-year enrolment program run by the Ministry of Education. STPM is not the only credential approved in addition to the Master Program and the Malaysian Higher Islamic Religious Certification (STAM). Technically, applicants can apply for admission to degree-level courses with a variety of pre-university examinations considered to be comparable to STPM, including A-level. Nevertheless, all those applying for university must have approved the MUET. STPM is internationally recognized by many universities, especially those in the Commonwealth of Nations, the United States and the Republic of Ireland. Many universities find STPM results to be equal to GCE A-Level results [13].

The Malaysian Matriculation Program (in Malay language i.e. Program Matrikulasi Malaysia or all the more regularly known as just Matrikulasi) is a 1 to the 2-year pre-college course that enables understudies to seek after a degree upon effective consummation. Matrikulasi is one of the most looked for after pre-college programs among Malaysians as it is an incredibly practical course into tertiary instruction. Understudies just need to pay a little enrollment charge, and the remainder of the expenses are borne by the Malaysian government. Registration understudies additionally get a stipend each semester for their everyday costs. Ordinarily, understudies who picked a possibility for Matriculation will keep on pursuing their degree with neighborhood open establishments. So, Malaysian Matriculation is likewise perceived by a few remote colleges, for example, those from Australia, United Kingdom, Indonesia, New Zealand, and a few other Commonwealth countries [14]. The chart demonstrates that the vast majority of the understudies who got Band 3 in MUET (unobtrusive English client) had the option to effectively graduate in science. On the other hand, the lowest number of MUET results (band 6–a highly qualified user) was the lowest student graduate in chemistry. It suggests that students are able to graduate in chemistry even though they were moderate users of English. This indicates which Band 3 enrolment students in MUET take the most of the courses. The students of Band 3 in MUET is the Standard English that the students have adept at. For most of the universities in Malaysia, the prerequisite for a student to pursue his or her studies is to get Band 3 into their MUET performance. The students who got the 5th band and 6 the course often takes place as they are knowledgeable in English, which is the main language used in universities. This is seen as their strength because it will help them learn and understand English quickly.

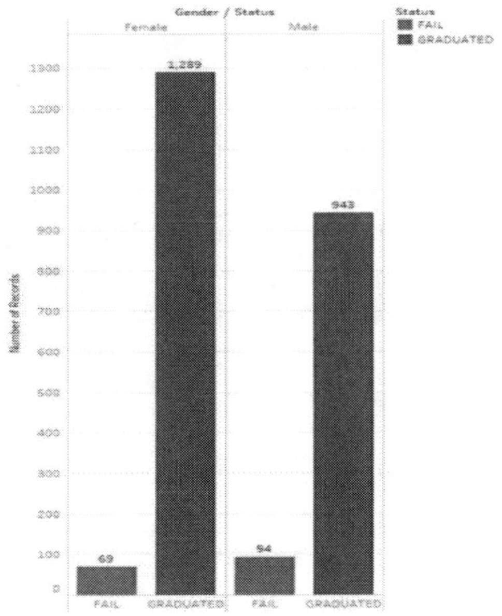

Figure 7. Entry qualification based on gender

Figure 7 shows the status of the student's status based on their gender. Female students are more advance to pass the course compare to male students.

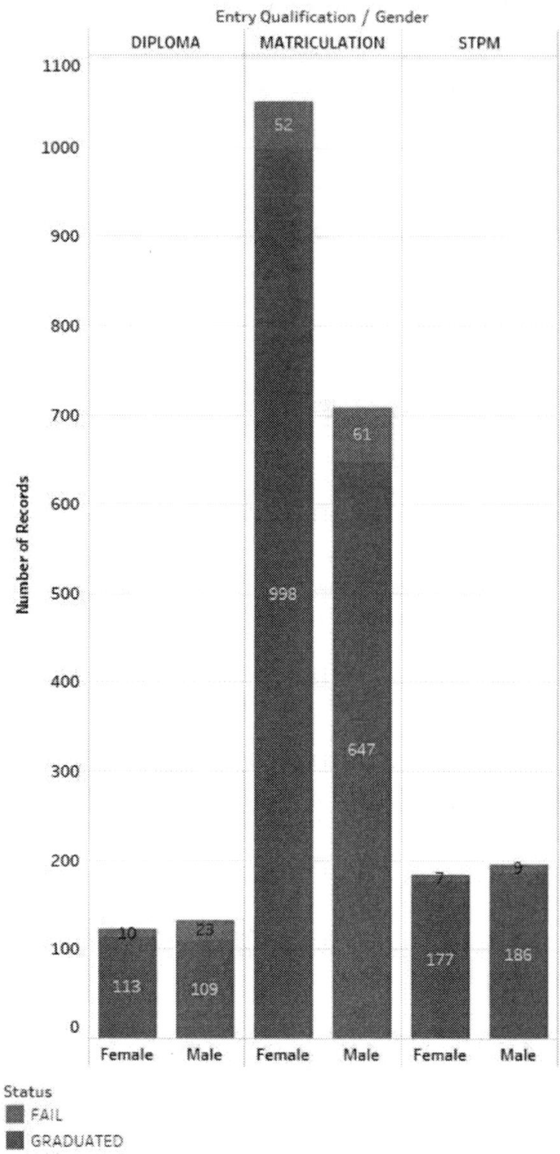

Figure 8. Result of student's performance by entry qualification and gender

Figure 8 shows the status of the student's status based on their entry qualification and gender. Female students from matriculation have the highest value in passing the course rather than students from the diploma.

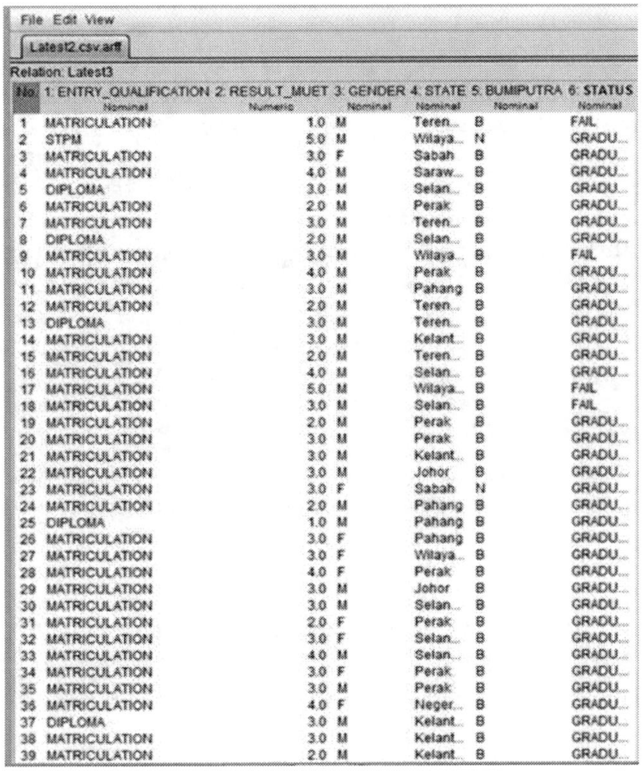
File Edit View
Latest2.csv.arff
Relation: Latest3

No.	1: ENTRY_QUALIFICATION Nominal	2: RESULT_MUET Numeric	3: GENDER Nominal	4: STATE Nominal	5: BUMIPUTRA Nominal	6: STATUS Nominal
1	MATRICULATION	1.0	M	Teren...	B	FAIL
2	STPM	5.0	M	Wilaya...	N	GRADU...
3	MATRICULATION	3.0	F	Sabah	B	GRADU...
4	MATRICULATION	4.0	M	Saraw...	B	GRADU...
5	DIPLOMA	3.0	M	Selan...	B	GRADU...
6	MATRICULATION	2.0	M	Perak	B	GRADU...
7	MATRICULATION	3.0	M	Teren...	B	GRADU...
8	DIPLOMA	2.0	M	Selan...	B	GRADU...
9	MATRICULATION	3.0	M	Wilaya...	B	FAIL
10	MATRICULATION	4.0	M	Perak	B	GRADU...
11	MATRICULATION	3.0	M	Pahang	B	GRADU...
12	MATRICULATION	2.0	M	Teren...	B	GRADU...
13	DIPLOMA	3.0	M	Teren...	B	GRADU...
14	MATRICULATION	3.0	M	Kelant...	B	GRADU...
15	MATRICULATION	2.0	M	Teren...	B	GRADU...
16	MATRICULATION	4.0	M	Selan...	B	GRADU...
17	MATRICULATION	5.0	M	Wilaya...	B	FAIL
18	MATRICULATION	3.0	M	Selan...	B	FAIL
19	MATRICULATION	2.0	M	Perak	B	GRADU...
20	MATRICULATION	3.0	M	Perak	B	GRADU...
21	MATRICULATION	3.0	M	Kelant...	B	GRADU...
22	MATRICULATION	3.0	M	Johor	B	GRADU...
23	MATRICULATION	3.0	F	Sabah	N	GRADU...
24	MATRICULATION	2.0	M	Pahang	B	GRADU...
25	DIPLOMA	1.0	M	Pahang	B	GRADU...
26	MATRICULATION	3.0	F	Pahang	B	GRADU...
27	MATRICULATION	3.0	F	Wilaya...	B	GRADU...
28	MATRICULATION	4.0	F	Perak	B	GRADU...
29	MATRICULATION	3.0	M	Johor	B	GRADU...
30	MATRICULATION	3.0	M	Selan...	B	GRADU...
31	MATRICULATION	2.0	F	Perak	B	GRADU...
32	MATRICULATION	3.0	F	Selan...	B	GRADU...
33	MATRICULATION	4.0	M	Selan...	B	GRADU...
34	MATRICULATION	3.0	F	Perak	B	GRADU...
35	MATRICULATION	3.0	M	Perak	B	GRADU...
36	MATRICULATION	4.0	F	Neger...	B	GRADU...
37	DIPLOMA	3.0	M	Kelant...	B	GRADU...
38	MATRICULATION	3.0	M	Kelant...	B	GRADU...
39	MATRICULATION	2.0	M	Kelant...	B	GRADU...

Figure 9. Show the features of data for each student

The data that is used for the training set filter before it evaluates using Cross-Validation test. In order to evaluate the accuracy of the performance of the students, test data need to be prepared. The test data are based on student's features such as entry qualification, result MUET, gender, state, *bumiputra*, and status. The data is prepared in .arff file name and then will be evaluated using the algorithm model. Figure 9 shows an example of the data.

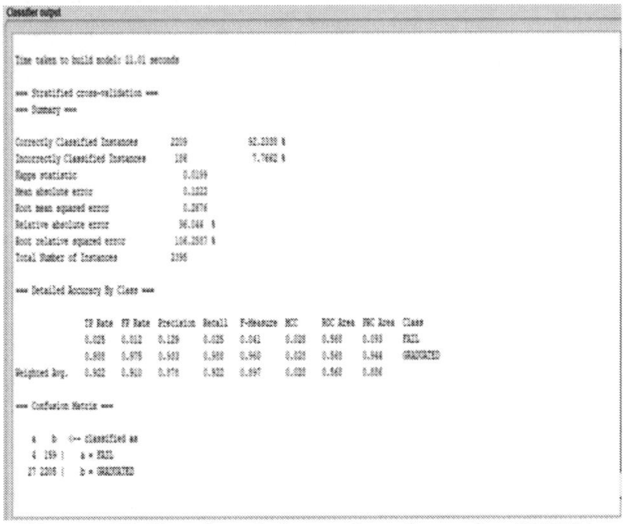

Figure 10. Result training and testing using WEKA

3.4 Testing datasets
Figure 10 indicates the likelihood calculation of the student's success or failure that depends on the characteristics of each student. It is documented after the data has been educated. The outcomes of the training data were analyzed using cross-validation, the training set given, the training set used, the percentage split. Implementation is used cross-validation testing as the training and testing set are different parts and the information in the test part is omitted from the training set. Cross-validation is utilized to beat the issue of over-fitting and to make expectations increasingly succinct.

Datasets will be outfitted with the Adaboost-MLP calculation to discover how precise the prescient model is. Adaboost classifier consolidates low classifier calculations to enable a solid classifier. By entering two classifiers, the reliability of better results will be increased [14]. Predictive train information is generalized to make the classifier more broadly available. Cross-validation is folded 10 times and train data to achieve correct accuracy.

3.5 Evaluation
To evaluate the Adaboost-MLP in predicting the student accuracy benchmark, which shows the percentage of the correctness in classifying the performance either graduate or fail.

4. Result
This section defines the results of the prediction of the performance of the students by using the data being collected from UMP. The Adaboost-MLP technique predicted 92.23 percent accuracy in classifying the performance of students before they enroll in the chemistry course. Hence, with this Adaboost – MLP model, before register chemistry course, this study able to predict either the student graduates or fail in this course in the near future.

5. Conclusion
This section offers an end on anticipating the accomplishment of understudies in a science course utilizing machine learning. Eventually, the goal is to test and apply an MLP calculation to anticipate understudy disappointment or graduate execution dependent on highlights (ENTRY QUALIFICATION, RESULT, GENDER, STATE, BUMIPUTRA STATUS). In validating the forecasting model, cross-validation method has been implemented. This work shows that Adaboost-MLP is applicable to predict student performance and the student status who will fail or graduate the course. If the prediction result is failed, the students are encouraged to focus more during the class sessions if they are still interested in chemistry course. On the other hand, the students are suggested to enroll for other courses that interest them. If the prediction result is graduated, this means that the student may have a talent in chemistry and be able to pass through in flying colors. Nevertheless, this estimation also depends on the student's ability and commitment to research the length of the chemistry course.

Acknowledgment
This work supported by Universiti Malaysia Pahang, under research grant by Faculty of Computing, Universiti Malaysia Pahang (RDU180360). The authors would like to thank the Academic Management Division (BPA), Gambang campus for providing them the dataset.

References
[1] A. M. Helmenstine, "Why study chemistry?" 2019.
[2] G. B. Goh, N. O. Hodas, and A. Vishnu, "Deep learning for computational chemistry," J. Comput. Chem., vol. 38, no. 16, pp. 1291–1307, 2017.
[3] Berita Harian, "Pendidikan Negara pada landasan betul," 16-Mar-2018.
[4] M. F. A. Razak, N. B. Anuar, F. Othman, A. Firdaus, F. Afifi, and R. Salleh, "Bio-inspired for Features Optimization and Malware Detection," Arab. J. Sci. Eng., 2017.
[5] "Mobile Malware Detection by Analyzing Official Market Metadata," Comput. Secur.
[6] A. Firdaus, N. B. Anuar, M. F. A. Razak, and A. K. Sangaiah, "Bio-inspired computational

paradigm for feature investigation and malware detection: interactive analytics," Multimed. Tools Appl., vol. 76, no. 280, pp. 1–37, 2017.

[7] B. Sanz, I. Santos, C. Laorden, X. Ugarte-Pedrero, P. G. Bringas, and G. Alvarez, "PUMA: Permission Usage to detect Malware in Android," in Advances in Intelligent Systems and Computing, 2013, pp. 289–298.

[8] P. P. K. Chan and W. K. Song, "Static detection of Android malware by using permissions and API calls," in International Conference on Machine Learning and Cybernetics, Lanzhou, China, 2015, vol. 1, pp. 82–87.

[9] S. Arshad, M. A. Shah, A. Wahid, A. Mehmood, H. Song, and H. Yu, "SAMADroid: A Novel 3-Level Hybrid Malware Detection Model for Android Operating System," IEEE Access, vol. 6, pp. 4321–4339, 2018.

[10] N. Pehlivan, U.; Baltaci, N. ; Acarturk, C. ; Baykal, "The analysis of feature selection methods and classification algorithms in permission based Android malware detection," Comput. Intell. Cyber Secur., pp. 1–8, 2014.

[11] L. Ye, C. Shuniak, R. Oueini, J. Robert, and S. Lewis, "Can they succeed? Exploring at-risk students' study habits in college general chemistry," Chem. Educ. Res. Pract., vol. 17, no. 4, pp. 878–892, 2016.

[12] H. Fröhlich, O. Chapelle, and B. Schölkopf, "Feature Selection for Support Vector Machines by Means of Genetic Algorithms," in Proceedings of the 15th IEEE International Conference on Tools with Artificial Intelligence (ICTAI'03), Sacramento, California, USA, 2003, pp. 142–148.

[13] EduAdvisor, "Everything You Need To Know About MUET in Malaysia," 2018.

[14] S. Sheen, R. Anitha, and V. Natarajan, "Android based malware detection using a multifeature collaborative decision fusion approach," Neurocomputing, vol. 151, pp. 905–912, 2015.

[15] Eduadvisor",Matriculation 2018.[Online]. Available: https://eduadvisor.my/matriculation/ [Accessed:21-Jan-2019]

[16] MPM, "Malaysians Examination Council," 2018. [Online]. Available: http://portal.mpm.edu.my/en/muet1. [Accessed: 21-Jan-2019].

Improving Green IT Assessment Tool Using Case-Based Reasoning Concept

Chua Xiao Ying[1], Mazlina Abdul Majid[1], Steve Anthony Lojuntin[2]

[1] Faculty of Computing, College of Computing and Applied Sciences, Universiti Malaysia Pahang, Kuantan, Pahang Malaysia
[2] Technical Development and Facilitation Division, Sustainable Energy Development Authority (SEDA) Malaysia

Email: xiaoying.chua@hotmail.com, mazlina@ump.edu.my

Abstract. In order to know how to measure their level of green practices, green assessment tool is there to act as a guide or as some might called it as a bench marker. Although, there are existing green assessment tools but there is inadequate existing green assessment tool with a recommendation feature. Recommendation feature is meant to aid users especially new user with their green practices. In this paper, case-based reasoning concept is chosen to produce this recommendation features. Besides that, Rapid Application Development Model is used in this research. The expected result of this research is to improve an existing green assessment tool to provide recommendation to user. With this paper, it can help raise user's level of green awareness and improve their practices.

Keywords: *GiTaT, Case-Based Reasoning Concept, Green Assessment Tool*

1. Introduction

Sustainability is the process of maintaining change in a balanced fashion, in which the exploitation of resources, the direction of investment, the orientation of technological development and institutional change are all in harmony and enhance both current and future potential to meet human needs and aspirations [1]. Sustainable development goals have been identified by the 2005 World Summit on Social Development [1]. In an effort to reduce the impact of advancement on the environment, green concept is popularized. Examples of the green concept that is introduced are such as green building, green computing, green information system (green IS), green technologies and others more. The goals of green computing are such as reduce the use of hazardous materials, maximize energy efficiency during the product's lifetime, and the recyclability of products. Therefore, tools to measure green computing/ICT is needed. Such tool is already in existence and Green IT Assessment Tool, GiTaT is one of the latest tools that exist.

It not only used as a benchmark to measure green in green computing/ICT, but it also suggests the best practices to its user. GiTaT is easily accessible as it is a web-based tools. It was built to be a smart and user-friendly tool by performing evaluation, and able to give rating, reports of the user level of green and able to give user the best practice suggestions. GiTaT measures current Green IT/IS practices implemented in organizations with consideration to environment, people and profit. Besides, as a web-based platform, it promotes sustainability by organizations towards evaluating, benchmarking and rating their current Green IT/IS practices. GiTaT offers inexpensive and practical guidelines to implementing Green IT/IS. Another thing about GiTaT is that it replaces the manual paper self-assessment methods based on checklists previously used in organizations. Electronic certification on Green IT/IS assessment taken is generated autonomously. Decision making recommendations is provided for organizations in implementing Green IT/IS best practices towards achieving sustainability.

Therefore, this paper proposes on developing a more friendly-user Green Assessment Tools using cased-based reasoning concept [2]. The idea of this research is to ensure the information related to Green IT/IS are always up to date. There are two problem which are determine that leads to this research which are the level of awareness of public towards green computing are low and so is the level of their practices of green computing in their daily life. In order to increase their level of awareness and practices, they need to be able to keep up to the latest information on green practices. The research question to be answered in this research is "How to ensure the green practices information is always the latest update?" The aim of this research is to improve the existing green assessment tool (GiTaT) [3] that can store new information gathered from users input and filter information according to user needs [4]. In order to accomplish this aim, several measurable objectives must be achieved which are such as to design the case-based reasoning architecture for the new feature in the Green IT Assessment Tool (GiTaT), to develop the case-based reasoning algorithm [5] into the Green IT Assessment Tool (GiTaT) and to validate the performance of case-based reasoning concept implemented into the Green IT Assessment Tools (GiTaT).

For this paper, there are scope of user and scope of function. The scope of user are public and private companies. Meanwhile the scope of function is to store new information on green practices from users and also for users to retrieve information according to their needs on green practices. This system will be able to store new information on green practices making the tool an advanced system with an everlasting lifecycle.

2. Materials & Methods

The methodology used in this research is rapid application development model. The phases that will be involved in this project are analysis phase, design phase, prototype phase, testing phase and lastly thesis writing phase. Each of the phase is done as aligned to the objectives of this research project as shown in Figure 1 below. This is to determine that this research project fulfils the objectives of this research project.

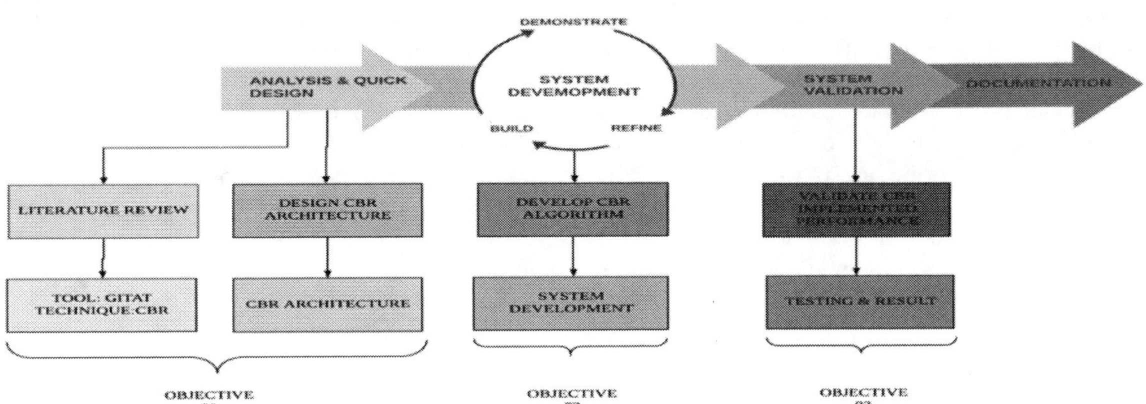

Figure 1. Research operation flow

Analysis phase is where the direction of this research project is identified. The direction of this research project is identified by gathering requirements especially. In this research-based project, all the information on this research can be found in literature review [6]. In literature review, information related to this research is covered. This information are past researches which are used as a baseline or guideline for this research. Discussion on green practices for sustainable development were also done in literature review. The discussion was on the existing green assessment tools and followed by the recommendation technique that is used in this research-based project. Lastly, concludes with type of methodology model that will be used in this research-based project.

Based on the analysis made in literature review, it was decided that this research will be using Green IT Assessment Tool (GiTaT) as the existing green assessment tools that will be improve on. Following that, the improvement that this research-based project will do is to create a suggestion and filter search system for new information in GiTaT. Case-based reasoning technique is the chosen concept after an analysis was done in recommendation techniques in literature review. This phase is the beginning step to achieve our Objective 1 which is to design the case-based reasoning architecture for the new feature in the Green IT Assessment Tool (GiTaT). Design phase means designing based on the user requirements and detailed analysis of the new system. Detail description of this research-based project architecture is explained in Main Results. This phase is completely aligned with our Objective 1 which is to design the case-based reasoning architecture for the new feature in Green IT Assessment Tool. In the search interface, users need to enter a keyword and pick the condition to filter the search. Detail description of system development can be found in Main results. This phase, System Development phase is to fulfil Objective 2 which is to develop the case-based reasoning algorithm into the Green IT Assessment Tool (GiTaT). Testing and implementation phase are to determine the success or failure of building the prototype. It is to determine whether the aim and objectives of the research have been achieved. If it is not achieved, some modification to the prototype part will be made to ensure the success of the project. Detailed explanation of this phase can be found in main results. This phase is aligned to fulfil Objective 3, to validate the performance of case-based reasoning implementation into the Green IT Assessment Tool (GiTaT) as a recommendation feature.

The hardware used in this research are computer and modem. Computer is the main hardware of the project implementation. Meanwhile modem is to connect to the internet to run the system to do this research. As for the software used are web browser, Visual Studio Code and XAMP application. Web browser such as google chrome and Microsoft Edge is used to run php file while Visual Studio Code is used to code or write the php file. XAMP application is used to connect to database.

3. Main Results

As mentioned in the Introduction, the aim of this project is to improve the existing green assessment tools by adding a new feature in it to increase the awareness and level of green practice. In order to do that, the case-based reasoning architecture for the assessment identification in Green IT Assessment is designed. After that, the case-based reasoning algorithm is developed into GiTaT. Lastly, validating the performance of Green IT Assessment Tools (GiTaT) with the implementation of case-based reasoning concept in it. In order to ensure the research question is answered, the objectives will be used as a guideline to ensure the aim of this research project is achieved.

The primary procedure is to collect the dataset which is usually done by collecting it from the Internet or an existing project that have been done by an individual. The dataset used in this project is created by an individual that created Green IT Assessment Tool to measure IT practitioners and IT managers current Green IT/IS practise adoption and implementation in CEs. The dataset consists of components listed in each Green IT/IS category as shown in Table 1.

Table 1. Dataset of this research-based project; green it/is categories & its components

Green IT/IS Categories	Components
Green Creation	Concerned about the energy consumption of cooling and lighting in our enterprise.
	Concerned about the efficiency of powering our enterprise hardware infrastructure
	Considers environmental factors in the design of enterprise infrastructure and IT infrastructure (servers, storage and network).
	Relocate our enterprise's data centre near clean sources of energy or renewable energy.
	Use electricity from Green energy providers in our enterprise.
	Enforces power management in our enterprise.
Green Distribution	Install software to make production more environmentally friendly in our enterprise.
	Retire energy inefficient systems in our enterprise.
	Analyses software induced energy bill separately from overall corporate.
	Engaging the service of a professional service provider to maintain our data centre.
	Install more energy efficient lightings in our enterprise.
	Upgrades to efficient transformers and UPS in our enterprise.
	Auditing the power efficiency of systems and technologies for lesser power consumption.
	Eliminates and de-commission unused services and systems.
Green Sourcing	Install software to make material sourcing more environmentally friendly.
	Buys recycled hardware equipment for enterprise use.
	Makes preference to hardware suppliers that have a Green track record.
	Gives weight to environmental considerations in enterprise equipment procurement [7].
	Deploy environment-friendly software procurement policy.
	Purchase hardware equipment from vendors that offers take back option.
Green Usage	Environmental consideration in planning IT usage and operations.
	Applied power management features of IT equipment regularly used.
	Turn off associated data centre system when not in use to saves energy.
	Print report on both side of a paper to reduce paper wastage.
	Utilizes IT equipment that has functions to monitor workloads and shut down when unused.
	Uses free cooling within the enterprise to reduce incurred energy cost.
End of Life	Recycle consumable equipment such as batteries, ink cartridges, and paper.
	Disposes of hardware equipment in an environmentally friendly manner.
	Carryout policy on managing electronic waste.
	Reuse hardware equipment.
	Refurbish old, out-dated and obsolete hardware equipment.

In this paper, architecture means the process and the product of planning, designing and constructing structures [8]. Below is the flowchart of the case-based reasoning architecture that is designed and used for this research-based project [9].

As shown as the Figure 2, the architecture of the case-based reasoning for this research project is the flow of the system design to aid users. It is a continuous cycle. It starts with user searches for explicit knowledge of Green IT/IS Best Practices and this can be considered as a new case coming in by searching for more knowledge to find a solution. Searches is made by using the case IT/IS case description which consist of case id, case category, case title, case description, and case content. After that, case retrieval is made. Existing cases from the database with similar word or phrase that is related to the current new case is retrieved to determine if those existing case's solution is a match to current new case. If the existing case is a match, then the new case problem is solved, and the new problem is determined to be case reuse.

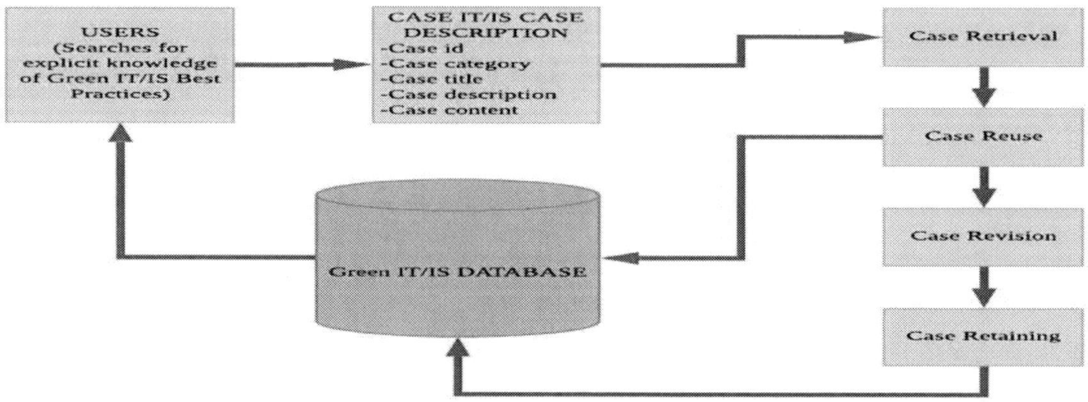

Figure 2. Case-Based Reasoning Architecture of this Research Project

Case reuse is using existing case solution to solved new case by adapting the solution to the new problem/case and which later will be recorded into the database. But if the existing case could not be used to adapt to solve the new problem case, case revision will be made. Case revision is a new solution after being evaluated by the satisfaction of the solution's function, the solution will be revised. After that is followed by case retaining, this is where decision is made whether the new problem case that was revised was retained. Objective 1 which is to design the case-based reasoning architecture for the new feature for the Green IT Assessment Tool (GiTaT).

The database architecture of this research is a 1-tier architecture. All the required components for a software application or technology involves on a single server or platform. This architecture concept is such as that in one place is where everything is stored such as all the elements of an application, including the interface, middleware and back-end data.

Coding in this research-based project is to develop the new feature system into the existing tool which in this case is Green IT Assessment Tool (GiTaT). For this research-based project, the coding language used was php and css. The tools used to create this system are XAMP and Visual Studio Code. Upon completion of this phase, Objective 2 which is to develop the case-based reasoning algorithm into the Green IT Assessment Tool (GiTaT) has successfully been achieved.

Interfaces is the end product or also known as the output of the development of this research. Interfaces is the medium for user to communicate with the computer. Besides that, it is easy to use that even computer illiterate is able to use it too as interfaces uses human language like English instead of machine language.

In the suggestion interface, user is required to select the categories and input the title, content and keyword in respective boxes. Then hit the submit button so that the new green practice information they keyed in is recorded into one of the tables named "storage" in the database which is named "information" for this research-based project.

Figure 3. Suggestion Interface & Search Interface

In the search interface, click the "Search" tab on the top menu bar to enter into search interface. There is a box for user to input value in it which is word that they like to search on. The search system/process only starts after user input and hit the search button right beside the search input box. Below that is a container that display all the information that is currently stored in the table "storage". In this container. It is title as front page and followed by displayed of information from the table "storage" which was input by many of this tool's user for the new green practices they learn about or attained.

Figure 4. Search Page Interface (storage) & Best Practices Interface

In search (Storage) interface, user will be redirected to search page interface as shown as in the figure above. The content of this page is the display of number of results with the search word input and the content with the keyword. For example, the input search word was recycle in the search (Storage) interface. Hence the result displayed in search page interface is related with the search word recycle. In this case we only have data with the word recycle from the database table named "storage" Hence, Search Page interface content displayed "There are 1 result!" and followed by the content with keyword which are "recycle old computer" and "recycle" respectively.

In best practices interface, user need to click Best Practices tab at the top menu bar. Just like the search box and button in search interface, it is used to search but instead it is used to search in a different table in the same database. For this search, it searches in table "bestpractices". Information stored in "bestpractices" table are information that is backed with literature review. Once the input in "storage" table have literature review that matches it, admin of this tool will add the new practice information into "bestpractices" table. Beside the search button is a link named advanced CBR search to filter the search in more detailed manner. The difference between these two searches, is that this search, searches according to the input entered. As long as the title, content, category and keyword have the search word, it will appear in the result page. Whereas the case-based reasoning filter search is searches with conditions. After the search box row, the row below is titled as Front Page is the displayed of all existing information in "bestpractices" table. Every existing category, title and description exist in "bestpractices" table is displayed at font page.

Green IT/IS Best Practice Case Retrieval interface, users need to select Advanced CBR Search link beside the search button at Best Practices interface. User can input the case word or case phrase and later tick the appropriated tick box. Follow by selecting to filter the search by using case retrieval method or green practice category. In the case retrieval method, selection can be made between retrieve using case title and case content or using retrieve using case title only or retrieve using case content only at the dropdown menu. While green practice category, its dropdown menu consists of Green Design, Green Production, Green Procurement, Green Operations, Green Disposal, Water Management, Green Creation, Green Distribution, Green Sourcing, Green Usage, and End of Life that can be chosen from. After selected the conditions or filter, user needs to hit or click n the Retrieve Similar Cases button.

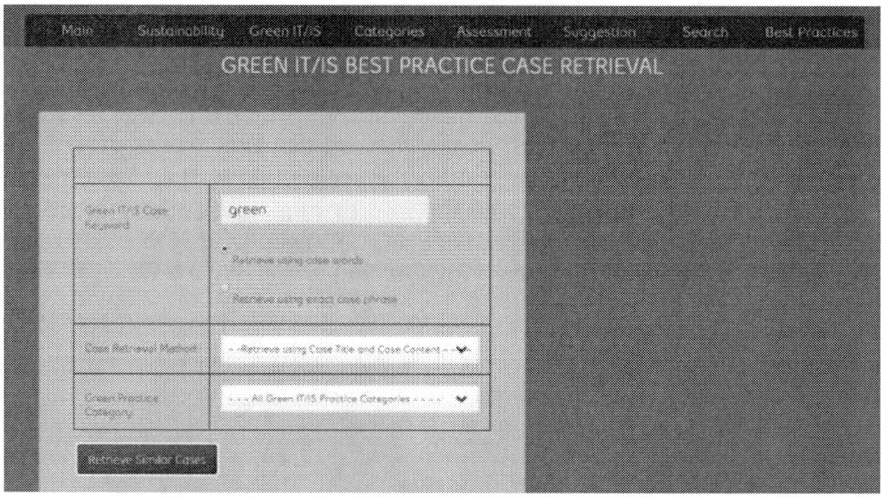

Figure 5. Green IT/IS Best Practice Case Retrieval

To validate the performance of case-based reasoning implemented into the Green IT Assessment Tool (GiTaT) is the final and last objective, Objective 3. In order to validate the performance of this research-based project's system, a few random people were invited to do the testing of the project. After testing the project, they are asked to fill up a feedback form consisting of two questions. The two questions that were being asked are an open-ended question which is "What is your opinion on GiTaT towards enhancing the enterprise environmental performance for sustainable development?" and a rating question which is "Will you recommend GiTaT to be used in the company?" [10]. The feedback forms have received many positive reviews. Hence, Objective 3, to validate the performance of case-based reasoning into the Green IT Assessment Tool (GiTaT) is accomplished.

4. Conclusion

It has come to attention how unaware and low the practice of green is among the public especially in green IT/IS and computing. As a solution to that, an existing green assessment tool which is Green IT Assessment Tool (GiTaT) [11] is used to improvised. It is improvised to allow user to search for related information that they wish to expand their knowledge on by applying the idea or concept of case-based reasoning in it and also to ensure the green practices information are always up to date.

Acknowledgement

This research is financially supported by Ministry of Education and Universiti Malaysia Pahang UMP Fundamental Research Grant Scheme (FRGS) (FRGS/1/2018/ICT04/UMP/02/4) and (RDU190167).

IOP Conf. Series: Materials Science and Engineering **769** (2020) 012028 doi:10.1088/1757-899X/769/1/012028

References

[1] World Summit Outcome. 2005. Retrieved from http://data.unaids.org/topics/universalaccess/worldsummitoutcome_resolution_24oct2005_en.pdf

[2] Alrawhani, E. M., Basirona, H., & Sa'ayaa, Z. Real Estate Recommender System Using Case-Based Reasoning Approach. Retrieved from http://journal.utem.edu.my/index.php/jtec/article/viewFile/979/569

[3] Bokolo, A.J., Majid, M.A. and Romli, A., 2018. A proposed model for green practice adoption and implementation in information technology based organizations. Problemy Ekorozwoju, 13(1).

[4] Hankel, A., Heimeriks, G., & Lago, P. 2017. Green ICT Assessment for Organisations. Journal of ICT Standardization, 4(2), 87–110. https://doi.org/10.13052/jicts2245-800X.421

[5] Kolodner, J. L. 1992. An Introduction to Case-Based Reasoning*. Artificial Intelligence Review (Vol. 6). Retrieved from http://alumni.media.mit.edu/~jorkin/generals/papers/Kolodner_case_based_reasoning.pdf

[6] Masood, M., & Mokmin, N. A. M. 2017. Learning material recommendation based on case-based reasoning similarity scores. In AIP Conference Proceedings (Vol. 1891, p. 020088). AIP Publishing LLC. https://doi.org/10.1063/1.5005421

[7] Morelli, J. 2011. Environmental Sustainability: A Definition for Environmental Professionals. Journal of Environmental Sustainability, (Volume 1, Issue 1). Retrieved from http://paperity.org/p/84319497/environmental-sustainability-a-definition-for-environmental-professionals

[8] Rapid Application Development Model - RAD Model. Retrieved December 3, 2018, from https://www.testingexcellence.com/rapid-application-development-rad/

[9] Shepperd, M. Case-based Reasoning and Software Engineering. Retrieved from https://pdfs.semanticscholar.org/63d4/ce202444c882bbe624147a1fdba9f185fde9.pdf

[10] Strange, T., & Bayley, A. 2008. Insights Sustainable Development. Retrieved from www.oecd.org/insights

[11] Anthony, B., Majid, M.A. and Romli, A., 2018. Green IS diffusion in organizations: a model and empirical results from Malaysia. Environment, Development and Sustainability, pp.1-42.

IOP Conf. Series: Materials Science and Engineering **769** (2020) 012029 doi:10.1088/1757-899X/769/1/012029

Detection of Proliferative Diabetic Retinopathy in Fundus Images Using Convolution Neural Network

Hasliza Abu Hassan[1,*], Marzuqi Yaakob[1], Sasni Ismail[1], Juwairiyyah Abd Rahman[1], Izyani Mat Rusni[1], Azlee Zabidi[2], Ihsan Mohd Yassin[3], Nooritawati Md. Tahir[3] and Suraiya Mohamad Shafie[4]

[1] Department of Engineering, Faculty of Engineering and Life Sciences, Universiti Selangor, Malaysia
[2] Faculty of Computing (FKOM), College of Computing and Applied Sciences, Universiti Malaysia Pahang
[3] Faculty of Electrical Engineering, Universiti Teknologi Mara, Shah Alam
[4] Ophthalmology Clinic, Prince Court Medical Centre, Kuala Lumpur, Malaysia.

E-mail: *ellyza90210@yahoo.com

Abstract. Convolution Neural Network (CNN) is one of the techniques under Artificial Neural Network (ANN) used to develop a Deep Learning Neural Network (DLNN) algorithm for detection of Proliferative Diabetic Retinopathy (PDR) on the fundus images. About 116 PDR and 150 Non-Proliferative Diabetic Retinopathy (NPDR) of fundus images retrieved from the publicly available MESSIDOR database applied in this research. This study consisted three objectives that included the execution of two pre-processing techniques on the data-set which were resizing and normalizing the fundus images, developed deep learning operational Artificial Intelligence (AI) network of feature extraction algorithm for detection of PDR on the fundus images and determined the output classification of the network encompassing the accuracy, sensitivity and specificity. There were five different parameters carried out along this research. Here, Parameter 5 showed the best performance among the five parameters based on the value of accuracy, sensitivity, and specificity that was 73.81%, 76%, and 69% respectively.

1. Introduction

Medical topographies related to the Diabetes Mellitus (DM) has reported since three millennial years ago in the ancient Egyptians manuscript[1]. The phrase 'Diabetes' was first introduced by a venerable Greek Physician, Aretaeus of Cappodacia [2]–[4]. Later, in the year of 1675, Thomas Willis a Britain Physician introduced the word 'Mellitus' after discovered extra sugar in blood and urine of patients [5]. By the year of 1776, Dobson who is also a Britain Physician proved the cause of DM due to the presence of excess sugar[5]. While in the year of 1857, a France physician, Claude Bernard comes out with term of excess glucose [5], which is referring to the simplest form of sugar found in blood and urine. Then, two Austria physician discovered the organ in the human body, which is the pancreases [5] that responsible for controlling the level of glucose in blood and urine at the year of 1889. After that, in the year of 1921, two physicians from Canada found the hormone called insulin that in charge of restraint of the glucose level in blood and urine [5], [6].

Today, the statistics have shown that about one over hundred thousand of all over the world people are suffering DM and the number will be twice in the future[1], [2], [6], [7]. According to[8], [7], based on the study carried by the World Health Organization (WHO) from the year 2005 until 2011, the number of diabetic patient worldwide estimated to be increased to 366 million by the year of 2030 which is the number is double from ten years ago. Anyhow, the number of diabetic patients were overreached where the latest statistics reported about 383 million diabetic patients in year 2013 and prognosis to be 592 million in year 2035 [8]. In Malaysia, statistics on the year 2015 reported about 3.3 million cases of diabetes[6]. Mention in The Star, July 2018[9], a study showed by the National Diabetic Institute (NADI), Malaysia is at the top-notch in Asia and among the highest rank of DM cases which about 2.5 million of the adult, those are 18 years old and above are diagnosed with DM. From the data showed that the increment is very terrifying.

The NADI Chairman, Dr. Mustaffa Embong also said DM has become a silent killer in this nation due to asymptomatic condition led to others heroic health cases like heart attack, stroke, and failure of the kidney as well as diabetes retinopathy (DR), which can cause blindness[9]. As the advance state of DM, the development of DR is very progressive[10] illness. Study showed by[11], DR is the resultant of retinal vasculature destruction which affecting the loss of vision and blindness. This terrify cases not only affecting the underdeveloped and developing country but also haunting the developed country like the United State (US). The research done by Shuang Yu[12] mention that, the American Academic of Ophthalmology (AAO) did numerous study regarding to DR and had discovered two stages of DR which are the Non Proliferative Diabetic Retinopathy (NPDR) and Proliferative Diabetic Retinopathy (PDR).

According to Deepashree [13], a patient with NPDR may suffer three phases of retinal abnormalities with are Mild NPDR, Moderate NPDR and Severe NPDR. The mild NPDR is the first phase of NPDR with an existence of at least a unit of micro-aneurysm. While the moderate NPDR is an existence of more than one unit of micro-aneurysm, hemorrhages and exudates. For the severe NPDR is a condition where the presence of large quantity of the microaneurysm, lots of hemorrhages, and many microvascular disorders.

Whilst the PDR is the advanced state or the phase after the NPDR where a new formation of blood vessel occurs. According to Shuang Yu[12], any patient diagnosed with PDR is at the most complexed conditional of a blood vessel in fundus which the occurrence of revascularization. It is parallel to Deepashreel [13] and H. Aguirre[14], PDR is the late stage where the retina entered the state of neovascularization or in other words growth of new blood vessel was being triggered inside the retina. Same as Snehal[15], the PDR is the circumstances of the formation of new frangible and tenuous blood vessels which highly exposed to the occurrence of haemorrhages.

2. Literature review

The time frame of the evolution of the industry is shown in Figure 2.1. According to [17], AI is the backbone of IR 4.0. The IR 4.0 consists of five main pillars as shown in Figure 2.1. The goals of 4.0 are to bring recent technology into a new era that has a higher level of operational efficiency and productivity as well as better automatization.

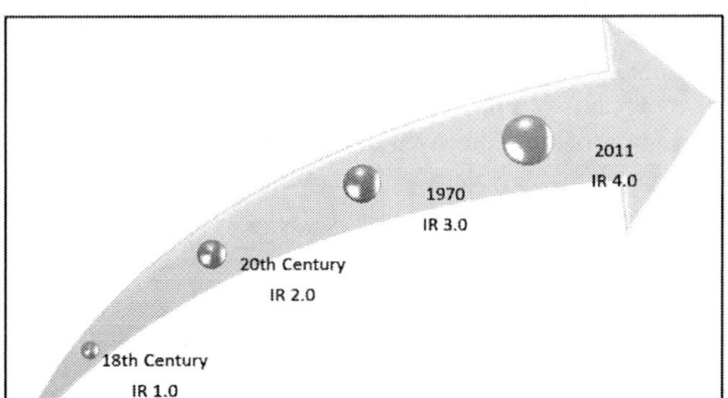

Figure 2.1: The Revolution of Industry Time Frame[16]

Whereas, in Figure 2.2 shows the five main pillars of IR 4.0. The first pillar referred to employment of the internet of thing (IoT). The IoT is the interconnection via the internet of computing devices embedded in everyday objects, enabling them to deliver and receive information [16]. The second pillar is the development of a cyber-physical system (CPS). The CPS is a mechanism that is controlled or monitored through a computer-based algorithm, firmly structured with the internet and its users. The third pillar is regarding the enhancing of digitalization approach through an enterprise architecture (EA) for the vast sector. For example in the transformation in the sector of transportation industry where an integrated digital system capable to generate robust big data analytics (BDA) that improve the decision-making and allow the intention crosswise disciplines and stakeholders[18]. The fourth pillar related to enterprise integration (EI). The IR 4.0 is urging companies to re-examine the way they do business. Hence, smart manufacturing, smart maintenance, and an endless scale of automation and integration in an overall spec of enterprise processes will change current purchase, productivity, manufacturing, sales, and maintenance processes [17]. Lastly, the fifth pillar is associated with empowering information and communication technologies (ICT). The rise of ICT in the recent era is has a strong correlation with the other pillars to cultivate the smart factory [19].

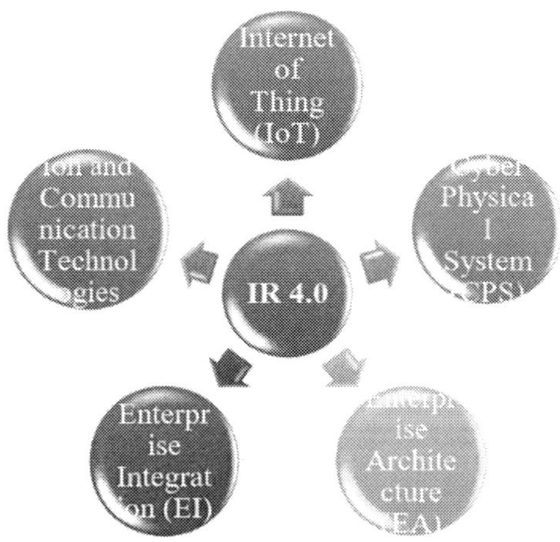

Figure 2.2: The Five Pillars of IR 4.0[17]

Nowadays, an AI used in immeasurable sectors. Especially in the manufacturing industry to operate the machinery and robotics [14]. Others, in image processing field AI plays important roles in the developments recent electronic devices like detection of DR in fundus images [17]. In the 11th Malaysia Plan, the study of AI is important to bring Malaysia close to the era of Industry 4.0[15].

AI is the knowledge and invention of computer systems to allow any mechanism performed tasks that normally demand human intelligence such as visual perception, speech recognition, decision-making, and translation between languages[20]. The development of AI is significant to the growth of recent technology. In figure 2.3 shows how AI has interrelation with the development of Industry 4.0 and some techniques of AI that commonly heard. The Artificial Neural Network are computational standards that can function similar to human nervous system and there are few types of ANN [22]. In Table 2.1 shows some prior paper that implements this method in their study.

Table 2.1: The employment of CNN technique in the prior works

Authors	Focus
[23]	Training data sampling for the detection of haemorrhages in color fundus images

[24]	To diagnose the presence of DR from digital fundus images and identify the accuracy classifying its severity.
[25]	The detection of blood vessels in fundus color images
[26]	For segmentation, detection, and classification of DR in fundus images
[27]	To evaluate the accuracy of the AI network model in the detection of DR through a screening programme of fundus images

CNN also known as Conv. Net [28] is one of the powerful methods in ANN that commonly used in images recognition. As listed in Table 2.0, all prior study that had been implementing this technique. Synonym with its name, convolution represents a multiple or numbers of layer build in a network. Figure 2.4 shows the example of a CNN architecture.

3. Research methodology

A total of 226 fundus images retrieved from the MESSIDOR database were used for training and testing the proposed network. The fundus images consist of two classes; Proliferative Diabetic Retinopathy (PDR) and Non- Proliferative Diabetic Retinopathy (NPDR) where each class contains 116 od dataset and 150 of dataset respectively.

There are two figures; Figure 3.1 shows the block diagram of the operational of the network and Figure 3.2 shows the flowchart of the proposed network.

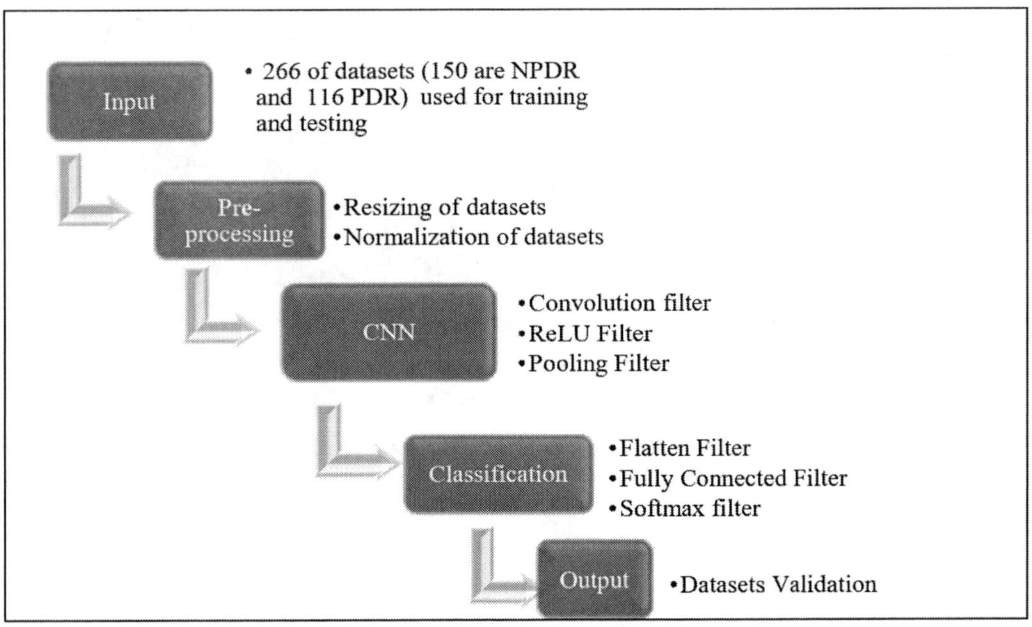

Figure 3.1: Block Diagram on the Operational of the proposed network

The Block diagram in Figure 3.1 explained the proposed method on the principle of the network operations. The first block indicated as the input for the proposed network, defined the total number of datasets used for training and testing. 150 of the datasets were diagnosed with NPDR while the other 116 datasets were diagnosed with PDR. These datasets retrieved from the MESSIDOR database.
The following block named as the pre-processing consists of two operations. The first operation was the resized of the input fundus images and the second operation was the normalization of the dataset. The convolution filter used to scale down the input dataset by 0.1 percent of the original dimensional (2240-unit width x 14880-unit height) to a new dimensional (224-unit width x 149-unit height).
The third block is the training and testing session for the proposed network named as CNN. In this block contains of three types of filters that carry out different operation on the dataset. The first filter named as convolution filter. Whereas, the second filter named as ReLU filter. And the third filter named as max-pooling filter.
The fourth block called as a classification that consists of few filters. The measurement of the network performances been carried out.
The fifth block is the output validation where analysis on the classification outputs of the network were carried out. The classification outputs for the proposed method such as sensitivity (SN), specificity (SP), and ROC were shown in the results.
Comparison on the classification outputs of the networks were carried out based on five cases. In every case, the ratio of datasets used for training and testing were different. The performance of the networks also being compared to the prior works.
Figure 3.2 shows a flowchart on the operation of the proposed method. In the beginning, 266 of fundus images that include 116 of PDR dataset and 150 of NPDR dataset. All the dataset was retrieved from the MESSIDOR database.
Then, all the dataset undergoes a pre-processing procedure called 'Resizing'. During this procedure, the original size of all dataset was reduced from 2240-unit width x 1488-unit height into 224-unit width x 1488-unit height by the scale of 0.1.
After scaling down, the fundus images were normalized. The dataset with different background colour will be standardized in this procedure. For the next procedure, the PDR and NPDR dataset were randomly selected by the network to be trained. The number of training datasets for five parameters were varied starting from 50%, 55%, 60%, 65% and 70% for both PDR and NPDR dataset.
The selected datasets were used by the network for training. Whereas, for the unselected datasets were used by the network for testing. After that, the classification outcomes such as accuracy, sensitivity and specificity for five parameters were classified using CNNs technique.

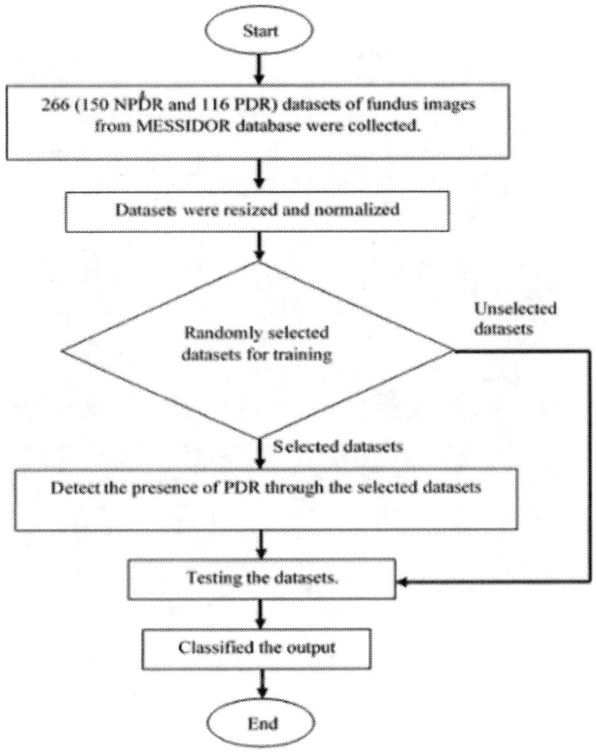

Figure 3.2: Flowchart of the Proposed Method

3.1. Input Database
The dataset of fundus images retrieved from the MESSIDOR database. This database is a research program funded by the French Ministry of Research and Defense for over 15 years, which owned numbers collection of fundus images. Furthermore, this database is the most popular used by the scientists to facilitate theirs studies on computer-assisted diagnoses of diabetic retinopathy (DR).

3.2. Input Dataset
The datasets refer to the fundus images consist of variety background color and fundus condition; NPDR and PDR fundus images. Not all of fundus images can be used as the input datasets for training and testing in this network. A few factors need to be considered in selecting the fundus images. Clear fundus images vision is the most significant factor to allow the network to learn the fundus images condition. Otherwise, the poor vision of the fundus images will reduce the performance of the networks and lead to defective output. Next is the condition of the fundus images where NPDR and PDR fundus images were selected for training. Figure 3.3 shows some samples of fundus images after being selected manually as the input datasets for training and testing.

IOP Conf. Series: Materials Science and Engineering **769** (2020) 012029 doi:10.1088/1757-899X/769/1/012029

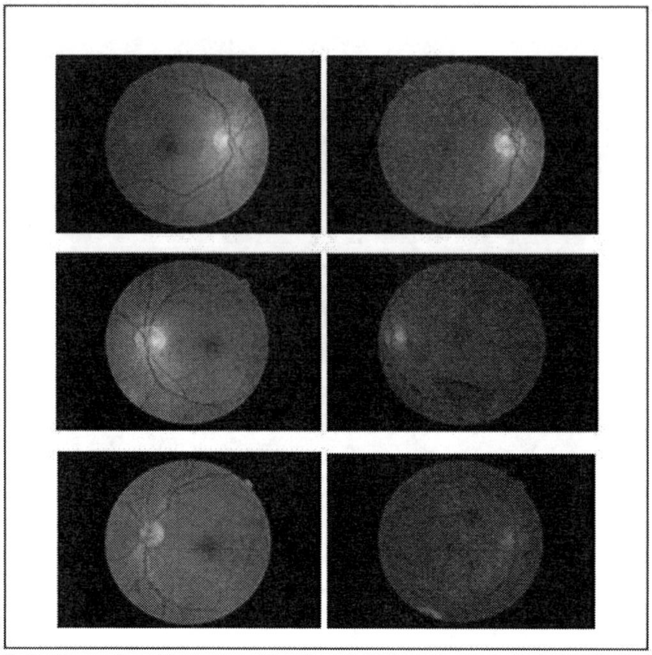

Figure 3.3: Sample of various fundus images from MESSIDOR database

3.3. Pre-Processing

This section consists of two parts; resizing and normalization. Both were crucial to generate precise output of the networks. The resizing of the input datasets in this procedure referring to change the size of the fundus images while the normalization of the datasets referring to standardize the background red green blue (RGB) color of the fundus images.

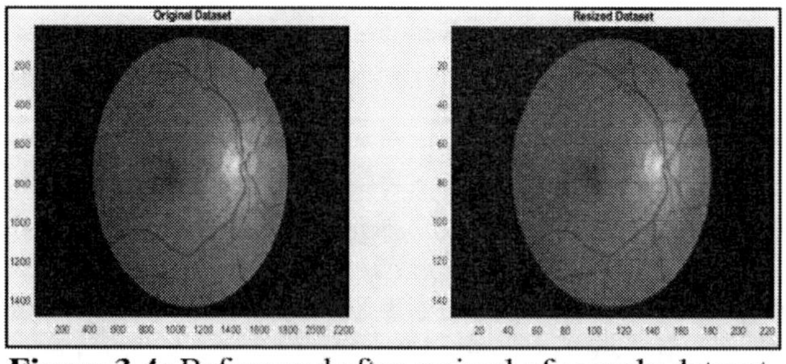

Figure 3.4: Before and after resized of sample dataset.

The dataset from the MESSIDOR came with variation sizes. Then, all the fundus images were resized into a standard scale. This was due to variation scale of dataset that may affecting the networks to operate. A dataset with a large scale may increase the processing time and requires higher performance of central processing unit (CPU) to perform the desired method.

However, if the scale of the dataset was resized to a very small size, it may reduce the information and resolution of the fundus image. Also, the compatibility of the CPU performance needs to be identified to allow the networks to operate at optimum range. Figure 3.4 shows a sample of a dataset before and after being resized. Here, the original size of the dataset was 1488 unit wide and 2240 unit height and after resized on scale of 0.1%, its became 149 unit wide and 224 unit height.

Next, since the fundus images that retrieved from the MESSIDOR database were variety in color background as shown in Figure 3.4 as the RGB color of each datasets defer to one another, normalization was done to standardize the color of the fundus images.

3.4. Classification using CNNs Techniques

Convolution Neural Networks (CNNs) also known as Conv.Net or comp.net is an artificial neural network that is so far been most popularly used for analyzing images. Although image analysis has been the widespread use of CNNs, they can also be used for other data analysis or classification problems as well. Most generally, CNNs can be illustrated as an ANN that has some types of specialization for being able to pick out or detect patterns and make sense of them.

This pattern detection is what makes CNNs so useful for image analysis. CNNs has hidden layers called convolutional layers and these layers are precisely differentiates CNNs with a standard multilayer perceptron (MLP). These convolutional layers do just like any other layer, a convolutional layer receives input then transform the input in some way and then outputs the transform input to the next layer. With a convolutional layer operation;

i. Function of Convolution Layer in Feature Learning

Convolution layer is the main layer in feature learning that does most of the computational dense stimulating. The convolution layer consists of a cluster of learnable filters. Every filter is small spatially with three-dimensional which is the width and height as well as the depth (w x h x d).

In the proposed network, three set of convolution layer with a different size were designed. The first set has a dimensional of 3 x 16 x 1. The second set has a dimensional of 3 x 32 x 1 and the third set has a dimensional of 3 x 64 x1.

ii. Function of ReLU Layer in Feature Learning

The ReLU layer applied an elementwise activation function. For example the max(0,x) thresholding at zero. Commonly, this layer present at every set of convolution layer and pooling layer. This layer ensure the size and the volume of the dataset unchanged after being filter.

iii. Function of Pooling Layer in Feature Learning

The next layer present in feature learning is the pooling. Its function is to reduce the spatial size of the fundus images representation to decrease the amount of parameters and computation in the network. A pooling layer operates on each feature map independently. There are two kinds of pooling layer types, which are the max pooling, and the sum or mean pooling.

3.5. Classification Outcomes

Basically, a classification outcome consists of few categories;

i. Accuracy;

Accuracy = No. of Predicted/Total no dataset [13].

ii. Sensitivity, Sn ;

Sensitivity defined as the proportion of datasets with a positive index test among all the datasets that is have the PDR;

Sn = (TP) / (TP + FN) [13]

iii. Specificity, Sp ;

Specification, Sp defined as of the proportion of all datasets index test among the datasets that do not have the PDR;

Sp = (TN) / (TN + FP) [13]

4. Result and discussions

Experimental analysis and testing are running on a Toshiba Satellite laptop Intel Core i7 Central Processing Unit (CPU), 2.1 (GHz) with 4 GB of Random-Access Memory (RAM). Window 7 was installed as the operating system.

4.1. Summary of Training Progress Graph for All Parameters

The training progress graph provided three important information during the network undergo training session. Firstly, it provides the percentages of training accuracy validation of the network carry out deep learning on the dataset. Second, it shows the pattern on data loss over iteration and epoch. Epoch referred to one forward pass and one backward pass of all training dataset while iteration is one forward and one backward pass of each batch size of dataset [25]. The third significant information that can gain from this graph is the duration time of the network used to complete the training session.
For this study, the number of iteration and epoch was remained the same for five different parameters. Table 4.1 shows the summary for the testing accuracy for every parameter. From Figure 4.1, the configuration indicated by the blue line showed an increased pattern form. However, due to some feature information on the dataset that had been loss during the resized session, the pattern did not rise rapidly. Therefore, mean value for every increment were counted, as shown by the dotted line.
Then, the rate amount of data loss during the training was shown by the red line. The pattern of the rate loss showed decreasing because of the epoch and number of iterations allowed the network to retrain the dataset repeatedly. Noticed that, the faint color of blue and red line at these graphs represented the actual progress of the network while the brighter color was after smoothed by the three filters; soft filter layer, flatten filter layer and fully connected filter layer. As the result, the line with the brighter color almost closed to the dotted line.
It was noticed that, as the percentage of information loss reduced, the percentage of accuracy increased. It means the network able to capture features information on the dataset and filtered out noise on the dataset during training session. Therefore, the more the percentage of information loss, the higher the accuracy of the network.

Table 4.1: Summary of Training Progress for All Parameters

Parameters	**No of dataset (batch size of dataset)**		**Testing Accuracy (%)**	**Training Time (s)**
	Training	**Testing**		
1	50 PDR: 50 NPDR Total = 100	66 PDR: 100 NPDR Total = 166	71.08	30
2	55 PDR : 55 NPDR Total = 110	61 PDR : 95 NPDR Total = 156	71.15	19
3	60 PDR : 60 NPDR Total = 120	56 PDR : 90 NPDR Total = 146	71.23	13
4	65 PDR : 65 NPDR Total = 130	51 PDR : 85 NPDR Total = 136	71.32	15
5	70 PDR : 70 NPDR Total = 140	45 PDR : 80 NPDR Total = 126	73.81	15

4.2. Summary of Confusion Matrix for All Parameters

The output class represented the actual number of the dataset while the target class represented the predicted number of the dataset. Both output class and target class have two classes that are denoted as '0' represented the NPDR dataset and '1' represented PDR dataset.

Table 4.2 shows the summary for all parameters that consists of number of dataset used for both training and testing as well as the number of dataset in each class that named as True Positive (TP), True Negative (TN), False Positive (FP), and False Negative (FN).

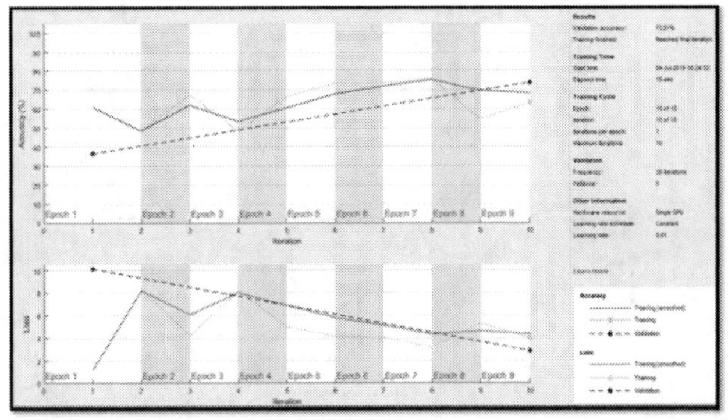

Figure 4.1: Sample of Training Progress Graph for Parameter 5

Table 4.2: Summary of Classes Based on the Confusion Matrix of All Parameters

Parameters	No of dataset		Classes			
	Training	**Testing**	**TP**	**TN**	**FP**	**FN**
1	50 PDR : 50 NPDR	66 PDR : 100 NPDR	100	18	0	48
	Total = 100	**Total = 166**	**Total = 166**			
2	55 PDR : 55 NPDR	61 PDR : 95 NPDR	95	16	0	45
	Total = 110	**Total = 156**	**Total = 156**			
3	60 PDR : 60 NPDR	56 PDR : 90 NPDR	76	28	14	28
	Total = 120	**Total = 146**	**Total = 146**			
4	65 PDR : 65 NPDR	51 PDR : 85 NPDR	80	17	5	34
	Total = 130	**Total = 136**	**Total = 136**			
5	70 PDR : 70 NPDR	45 PDR : 80 NPDR	68	25	12	21
	Total = 140	**Total = 126**	**Total = 126**			

4.3. Summary of Receiver Operating Characteristic (ROC) for All Parameters

Table 4.3 shows the summary of information from all ROC graph for all parameters. To get a better performance, the value of the Specificity and Sensitivity should closer to 100%. The ROC graphs for Parameter 1 until Parameter 5 were shown in Figure 4.3(i), Figure 4.3(ii), Figure 4.3(iii), Figure 4.3(iv), and Figure 4.3(v) respectively.

The y-axis of the ROC graph represented the True Positive Rate (TPR) that also indicated the sensitivity (Sn) of the network during testing session. Whereas, on the x-axis of the ROC graph represented the False Positive Rate (FPR) that also used to indicate the value of specificity of the network but the value of the FPR need to deduct by value of 1.

For Figure 4.3 (i), the value of the specificity is 100%, whereas the value for sensitivity is 67% (1-0.33). This showed large gap percentage values between sensitivity and specificity. The possible reason that lead to this are the poor resolution of the dataset used for testing. Others possible reason like under train also may led to this performance.

Then, Figure 4.3 (iii) shows the value of specificity 67% whereas the value for sensitivity is 73% (1-0.27)

Table 4.3: Summary of Information of All ROC Graph

Parameters	Specificity, S_p (%) (S_p=1-FPR)	Sensitivity, S_n (%) (S_n=TPR)
1	100	67.00
2	100	68.00
3	68.00	72.00
4	78.00	70.00
5	69.00	76.00

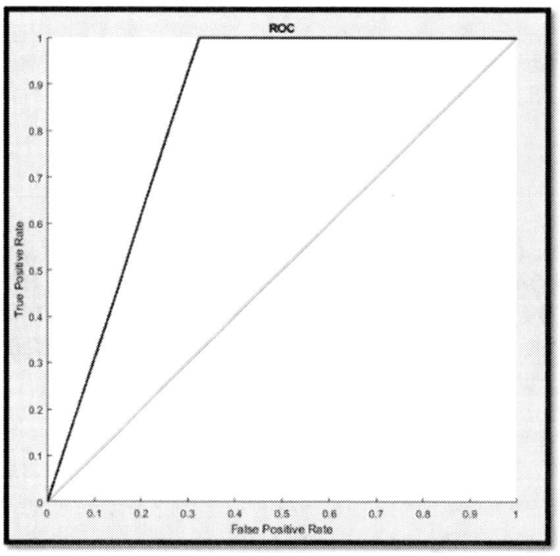

Figure 4.3(i): Receiver Operating Characteristic (ROC) Graph for Parameter 1

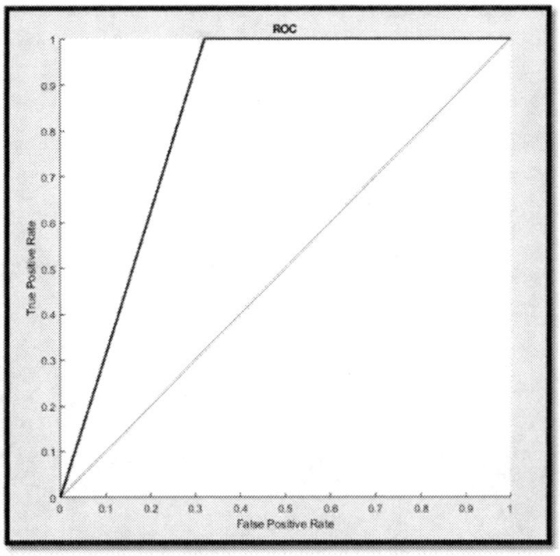

Figure 4.3(ii): Receiver Operating Characteristic (ROC) Graph for Parameter 2

Like Parameter 1, Figure 4.3(ii) shows the value for specificity is 100% whereas the value for sensitivity is 68.0% (1-0.32).

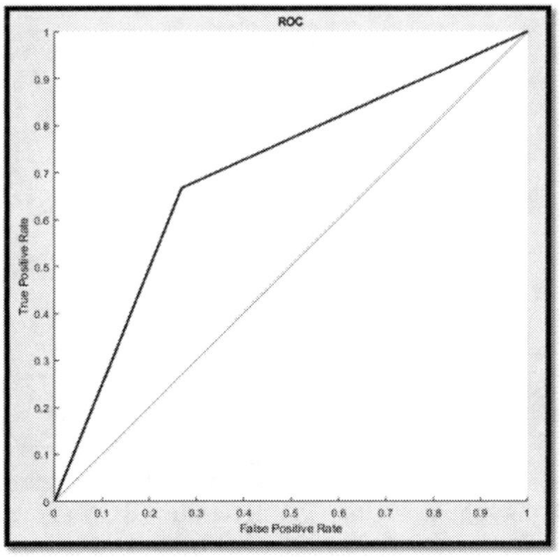

Figure 4.3(iii): Receiver Operating Characteristic (ROC) Graph for Parameter 3

This showed that, during the testing session, the proposed network only able to clarify the dataset with PDR with value of 67% correct. The reason that led to this output due to the poor resolution of dataset used for tested.

Based on Figure 4.3(iv), the value of specificity is 77% whereas the value for sensitivity is 70 %. This value was higher than the Parameter 3 but lower than Parameter 1 and Parameter 2.

Meanwhile, Parameter 5 in Figure 4.3(v) shows the value of specificity closer to Parameter 4 which is 68% and the value of sensitivity is 76%.

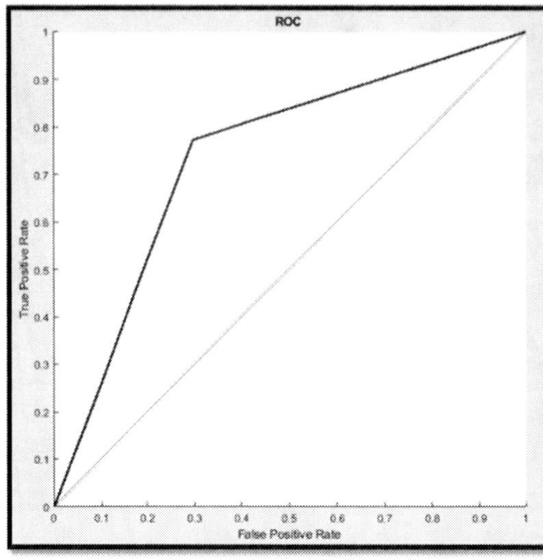

Figure 4.3(iv): Receiver Operating Characteristic (ROC) Graph for Parameter 4

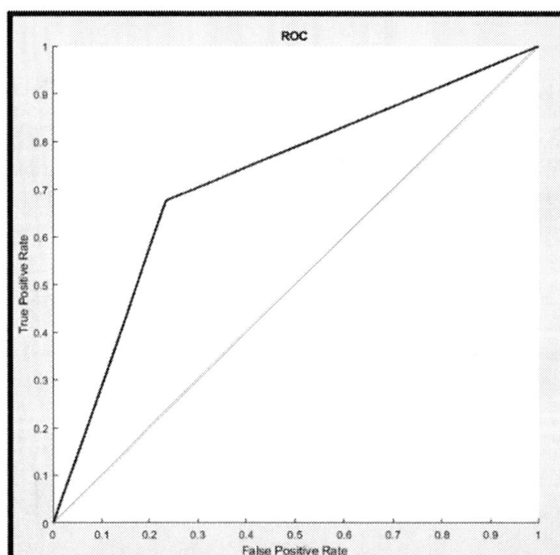

Figure 4.3(v): Receiver Operating Characteristic (ROC) Graph for Parameter 5

5. Conclusions

Based on this research, all the objectives which mention earlier were achieved. As a conclusion, Parameter 5 showed the best performance among five parameters by having the highest accuracy, which was 73.81%. In addition, the percentage value of sensitivity and specificity for parameter 5 were 76% and 69% respectively. The values of parameter 5 from this study were not as good as the prior work due to several constraints that were faced during this research. Firstly, it was due to the low of specification of the computer that used to run the proposed network where only limited numbers of dataset could be used for training and testing. Other than that, the low resolution of dataset image also had influenced the network to detect the presence of PDR. This is because, during the resizing procedure, some information on the original dataset may diminished and made the network unable to detect that information which will affect the percentage of output value. As for future study, higher specification of computer is highly recommended to allow greatest numbers of dataset to be used for training and testing as well as increase the performance of the network to detect the presence of PDR in the dataset.

6. References

[1] H. A. Hassan, N. M. Tahir, I. Yassin, C. H. C. Yahaya, and S. M. Shafie,, 2015 ,"Visualisation of exudates in fundus images using radar chart and color auto correlogram technique," Proc. - Int. Conf. Comput. Vis. Image Anal. Appl. ICCVIA 2015.

[2] H. A. Hassan, N. M. Tahir, A. Zabidi, I. M. Yassin, and M. Karbasi, 2017, "Classification Of Visualization Exudates Fundus Images Results Using Support Vector Machine Classification Of Visualization Exudates Fundus Images Results Using Support Vector Machine Classification Of Visualization Exudates Fundus Images Results," J Fundam Appl Sci, no. 4S, pp. 19–44.

[3] H. A. Hassan, A. I. Yassin, and A. Zabidi, 2016 ,"Discrimination Of Exudates And Non Exudates Pixels In Fundus Images And Classification Of Color Autocorrelogram Features Using Multilayer Perceptron And Support Vector Machine," vol. 11, no. 20, pp. 11930–11943.

[4] H. A. Hassan, N. M. Tahir, I. Yassin, A. Zabidi, C. H. C. Yahaya, and S. M. Shafie, 2013 ,"Automated optic disc removal in fundus images using iterative heuristics and morphological operations," Proc. - 2013 IEEE Conf. Syst. Process Control. ICSPC 2013, no. December, pp. 230–233.

[5] A. M. Ahmed, 2002 ,"History of diabetes mellitus.," Saudi Med. J., vol. 23, no. 4, pp. 373–8.

[6] H. A. Hassan, N. M. Tahir, A. I. Yassin, and A. Zabidi, 2016 ,"Discrimination of exudates and non exudates pixels in fundus images and classification of color autocorrelogram features using multilayer perceptron and support vector machine," ARPN J. Eng. Appl. Sci.

[7] H. A. Hassan, N. M. Tahir, A. Zabidi, and I. M. Yassin, 2018 ,"Visualization Of Exudates Fundus Images With Graphical User Interface For Early Detection Of Diabetic Retinopathy Special Issue," J Fundam Appl Sci, vol. 10, no. 6S, pp. 920–933.

[8] L. Kovács, B. Kurtán, G. Eigner, I. Rudas, and C. Chee Kong, 2015, "Analysis of a Novel Time-Delay Diabetes Model," in European Control Conference (ECC) , pp. 15–17.

[9] TheStarOnline, 2018 "Malaysia has highest rate of diabetes in Asia, says Nadi chairman," Star Online, p. 2018–2019.

[10] G. Gupta, S. Kulasekaran, K. Ram, N. Joshi, M. Sivaprakasam, and R. Gandhi, 2017 ,"Local characterization of neovascularization and identification of proliferative diabetic retinopathy in retinal fundus images," Comput. Med. Imaging Graph., vol. 55, pp. 124–132.

[11] R. A. Welikala et al. 2014, "Automated detection of proliferative diabetic retinopathy using a modified line operator and dual classification," Comput. Methods Programs Biomed., vol. 114, no. 3, pp. 247–261.

[12] S. Yu, D. Xiao, and Y. Kanagasingam, 2016 ,"Automatic detection of neovascularization on optic disk region with feature extraction and support vector machine," Proc. Annu. Int. Conf. IEEE Eng. Med. Biol. Soc. EMBS, vol. 2016-Octob, pp. 1324–1327.

[13] D. Devaraj and S. C. P. Kumar, 2014 ,"Blood vessels segmentation with GUI in digital fundus images for automated detection of diabetic retinopathy," Proc. 2014 Int. Conf. Contemp. Comput. Informatics, IC3I 2014, no. c, pp. 915–920.

[14] H. Aguirre-ramos, J. G. Avina-cervantes, I. Cruz-aceves, J. Ruiz-pinales, and S. Ledesma, 2018, "Blood vessel segmentation in retinal fundus images using Gabor filters , fractional derivatives , and Expectation Maximization," Appl. Math. Comput., vol. 339, pp. 568–587.

[15] S. D. Kasurde and S. N. Randive, 2016, "An automatic detection of Proliferative Diabetic Retinopathy," Int. Conf. Energy Syst. Appl. ICESA 2015, no. Icesa, pp. 86–90.

[16] Y. Lu, 2017, "Journal of Industrial Information Integration Industry 4 . 0 : A survey on technologies , applications and open research issues," J. Ind. Inf. Integr., vol. 6, pp. 1–10, 2017.

[17] I. Revolution, 2017, "Industrial revolution 4.0 04".

[18] M. Jayakrishnan, abdul karim Mohamad, and A. Abu, 2018, "Digitalization Approach Through An Enterprise Architecture For Malaysia Transportation Industry," vol. 9(13), no. disember, pp. 834–839.
[19] S. H. Bonilla, H. R. O. Silva, M. T. da Silva, R. F. Gonçalves, and J. B. Sacomano, 2018, "Industry 4.0 and sustainability implications: A scenario-based analysis of the impacts and challenges," Sustain., vol. 10, no. 10.
[20] M. Salman, A. W. Ahmed, O. A. Khan, B. Raza, and K. Latif , 2017 ,"Artificial Intelligence in Bio-Medical Domain An Overview of AI Based Innovations in Medical," vol. 8, no. 8, pp. 319–327.
[21] Economic Planning Unit, Eleventh Plan 2016-2020 Malaysia Anchoring Growth on People, 2016, vol. 31.
[22] A. Neural, N. Currently, and B. Used, 2019, "6 Types of Arti cial Neural Networks Currently Being Used in Machine Learning Feedforward Neural Network – Arti cial Neuron :," pp. 1–12.
[23] M. J. J. P. Van Grinsven, B. Van Ginneken, C. B. Hoyng, T. Theelen, and C. I. Sánchez, 2016, "Fast Convolutional Neural Network Training Using Selective Data Sampling: Application to Hemorrhage Detection in Color Fundus Images," IEEE Trans. Med. Imaging, vol. 35, no. 5, pp. 1273–1284.
[24] H. Pratt, F. Coenen, D. M. Broadbent, S. P. Harding, and Y. Zheng, 2016, "Convolutional Neural Networks for Diabetic Retinopathy," Procedia - Procedia Comput. Sci., vol. 90, no. July, pp. 200–205.
[25] D. Maji, A. Santara, P. Mitra, and D. Sheet, 2016, "Ensemble of Deep Convolutional Neural Networks for Learning to Detect Retinal Vessels in Fundus Images,".
[26] S. Wan, Y. Liang, and Y. Zhang, 2018, "Deep convolutional neural networks for diabetic retinopathy detection by image classification," Comput. Electr. Eng., vol. 72, pp. 274–282.
[27] V. Bellemo et al. , 2019, "Artificial intelligence using deep learning to screen for referable and vision-threatening diabetic retinopathy in Africa: a clinical validation study," Lancet Digit. Heal., vol. 1, no. 1, pp. e35–e44.
[28] R. Prabhu, 2018, "Understanding of Convolutional Neural Network (CNN) Deep Learning," Medium.Com, pp. 1–11.

Acknowledgments
This work was supported by Selangor State Research Grant No: R&D–GPNS/18/01-UNISEL056 awarded by Selangor State and the authors would like to thank family's members and members of Universiti Selangor (UNISEL) for their support in completing this research. Thank you too to any party that either directly or indirectly give moral support or provide useful sources during accomplished this research study.

IOP Conf. Series: Materials Science and Engineering **769** (2020) 012030 doi:10.1088/1757-899X/769/1/012030

Swarm intelligence algorithms' solutions to the travelling salesman's problem

J B. Odili[1], A. Noraziah[2,3], Roslina Mohd Sidek[2]
[1]*Department of Mathematical Sciences, Anchor University Lagos*, Nigeria
[2]*Faculty of Computing,Universiti Malaysia Pahang, Kuantan 26300, Malaysia*
[3]*IBM Centre of Excellence, Universiti Malaysia Pahang, Kuantan 26300, Malaysia.*

E-mail: jodili@aul.edu.ng, noraziah@ump.edu.my, roslinams@ump.edu.my

Abstract. This paper presents research findings on the application of swarm intelligence techniques in computational intelligence to solve the travelling salesman's problem. The travelling salesman's problem finds real-life application in post office mail delivery, school bus routing, delivery of food to home-bound people etc. After a number of experimental procedures, the study concludes that all the comparative algorithms are very efficient in providing solutions to the benchmark travelling salesman's problems considered, though the Discrete Cuckoo Search and the African Buffalo Optimization have a slight edge in performance over the other comparative algorithms. In all, the study agrees with earlier studies in reaching the conclusion that swarm-based optimization techniques are not only effective but also are very efficient in providing solutions to the travelling salesman's problems.

1. Introduction

Since the beginning of the 20th century, man has deliberately been seeking ways to enhance their quality of living leading to several successful research investigations. One of the prominent scientific developments has been the travelling salesman's problem (TSP) developed in the 1930s [1]. Since its development, the TSP has attracted the attention of several researchers in Mathematics, Operations Research, Computational Intelligence and different aspects of engineering, thus leading to the concentration of several research procedures [2]. These research procedures have resulted in several outputs that have been of immense benefits to humanity today. Some notable real-life application domains of the TSP includes transportation (by air, land or sea) [3], job scheduling [4], vehicle routing [5], collision avoidance in robotics engineering [6], drilling of holes in motherboards [7] , logistics planning [8], microchips manufacturing [9], DNA sequencing [10] and, most recently, in autonomous vehicles technology [11, 12].

In Computer Science, particularly, TSP has become, consciously or unconsciously, a veritable yardstick for validating search capacity of optimization algorithms [13]. To this effect, several optimization algorithms ranging from those developed late in the last century such as the Ant Colony Optimization, Particle Swarm Optimization, Genetic Algorithm, etc. the 21st century optimization algorithms such as the Cuckoo Search, Bat Algorithm, Firefly Algorithm, African Buffalo Optimization, Artificial Bee Colony etc. have all investigated the TSP.

This study is a comparative investigation of search capabilities of different metaheuristics in arriving at solutions to the TSP with the aim of assisting the research community in making informed choices about a particular swarm intelligence metaheuristic whenever they are confronted with a problem that requires the application of the TSP.

The rest of this paper is organized thus: section two examines the concept of the TSP, section three discusses the comparative algorithms; section four highlights the experimental setting; section five deals with the experimental procedures cum discussion of results and section six draws conclusions on the study.

2. The travelling salesman's problem

The TSP is a simulation the of a particular anonymoussalesman who has customers spread across different locations of a large city or in a number of cities/towns in a given location. The duty of the salesman is to visit each of his customers, and then return to his starting point [13]. Travelling Salesman's Problems could either be symmetric or asymmetric. Symmetric TSP chronicles such problems where the travelling cost/distance between any two given locations in the graph is same in either to or fro directions. On the other hand, asymmetric TSP describes situations where the cost /distance between, at least two locations in the graph is not the same. The asymmetric situation finds practical application in one-way traffic or other civil engineering or commercial considerations [14].

In the simplest form, the TSP is described mathematically as: given a number of nodes/locations along with the traveling cost/weights between each pair, find the cheapest route that visits all of the nodes/locations and then returns to the starting node/location. In other words, mathematically, a TSP problem that has n nodes is described as a graph: $G = (N, A)$, where $N = \{1, 2 \ldots, n\}$represents a number of nodes, A =$\{i, j \in N\}$ represents a number of arcs. Each node represents a specific $n \in N$ and each arc ki, li serves as connector between nodes k and l. The traveling cost from node k to node l, utilizing a particular arc, is represented by ckl, meaning cost between node k to node l.

The benchmark TSP instances used in this study are available in TSPLIB95 [15]. The resultant output obtained from this experiment were used to compare those of other bio-inspired techniques such as the Bat Algorithm (BA), Firefly Algorithm (FFA), Cuckoo Search (CS), African Buffalo Optimization (ABO), Hybrid Algorithm, Ant Colony Optimization and the Artificial Bee Colony (ABC). The choice of the comparative algorithms is borne out of the fact that, aside the ACO, they are all recently developed algorithm, otherwise called the 21st century algorithms in addition to their posting some of the best results in literature. Moreover, the above-listed algorithms are very popular among researchers and have proven to be among the most effective swarm intelligence techniques ever developed [16].

3. Comparative algorithms

In view of the relevance of the TSP in terms of its practical applications in mathematics, engineering, science and technology, several algorithms have been applied to solve this problem. The algorithms of interest to us in this study are the 21st century algorithms since their results, generally, tend to be better than those of the 20th century algorithms such as the GA, ACO and PSO. The superiority of the 21st algorithms over its peers developed in the 20th century may be due to hardware improvements, the use of more sophisticated programming languages as well as improved programming skills. The comparative algorithms in this study are the Ant Colony Optimization [17], Artificial Bee Colony [18], Hybrid Algorithm [19], African Buffalo Optimization [20], Discrete Cuckoo Search [21], Firefly Algorithm [22] and the Bat Algorithm [23].

A closer look at our choice of algorithm indicates that, aside being population-based and recently-designed (except, of course the ACO, chosen for its exceptional search effectiveness), it incorporates a hybrid (HA) which is hybridization of the Ant Colony Optimization with the Artificial Bee Colony [19] as well as other optimization techniques that deploy different communication methods in organizing the herd/flock for their search enterprise. For instance, while the ACO employs ring topology that transmits from one ant to the other, the PSO employs the Von Neumann communication topology where a particle relates information to the immediate neighbours and the ABO and FFA employ the star topology where the search agents disseminate search information to the entire population [24]. Overall, these algorithms were chosen for this comparative investigation because of their effectiveness and popularity among researchers. A brief review of each of the comparative algorithms is hereby presented.

3.1. African buffalo optimization

African Buffalo Optimization (ABO) basically models the communal decision-making procedures communicative prowess, harnessing the collective intelligence of the entire herd and the extensive memory capacities of the African buffalos in their migration from one location to another in search of food in the vast African savannah and forests. Using two major vocalizations: attract */maaa/* and repulse */waaa/* calls, the buffalos are able to organize themselves in their search for solutions [25]. The */waaa/* calls is deployed to warn the buffalo herd of impending dangers arising from the presence of predators; express the lack of pastures in a particular location; scare an approaching inferior; assert dominance and, therefore, urge the herd to move on to better, safer or more rewarding locations (exploration). On the other hand, the */maaa/* call is used to reassure an inferior; express satisfaction with the security situation of a location and encourage the buffalos to further graze a field (exploitation) [26]. The ABO algorithm is presented in Figure 1:

1. **Begin**
2. Randomly initialize the buffalos to nodes at the search space;
3. **While** (not termination), do
4. For k=1: n (n= population), do
5. Evaluate the buffalos' exploitation fitness using: $m'_k = m_k + lp1(bg - w_k) + lp2(bp_k - w_k)$ and

$$w_k' = \frac{(w_k + m_k)}{\lambda}$$

6. Is bg_{max} updating? Yes, go to 11. If No in 10 iterations, go to 2
7. **End for**
8. **End while**
9. Output best solution.
10. **End**

Figure 1: ABO algorithm

So far, the ABO has been successfully applied to solve a few optimization problems. Some of ABO application areas includes parameter-tuning of PID Controllers for Automatic Voltage Regulators [27], energy and delay routing in Mobile Ad-hoc Network [28], numerical function optimization [29] etc. Because the application areas of this optimization technique are not yet widespread, it is difficult to pinpoint its weaknesses. Its strengths, however, includes fast speed due to the use of few parameters. Again, being a parameter-less optimization technique, it is user-friendly since the user does not have to bother with tedious parameter-tuning processes.

B. Cuckoo Search

The Cuckoo Search (CS) which is a simulation of the subtle attitude of the cuckoo bird was designed by X. Yang and S. Deb[30]. The cuckoo lays her eggs in the nests of other birds, sometimes of other species with the hope that those other birds will execute the maternal function of incubating those eggs. Whenever the host birds discover the cuckoo prank, it either abandons the nest or throw such strange eggs away. Otherwise, it goes ahead to incubate the eggs. The cuckoo bird, on its own perfects the act of subtlety by imitating the egg of its host to perpetuate its kind by such fraud. In CS, the host eggs in a nest represents an optimization solution while the cuckoo egg is a representation of a newer solution with the objective of using the newer solution to replace the existing one.

```
Begin
Objective function: f(x) x = (x1, x2 .... xj)
 Determine initial population of nests
While termination not reached
      Randomly generate a cuckoo by Levy flight using
          X_ij(t + 1) = X_ij(t) + α ⊕ Levy (λ)
        Evaluate the cuckoo fitness
        Randomly select a nest among available host nests
          If ( f_i (X_ij(t + 1)) > f_k(X_ij(t)) ) then
          Replace k by the new solution
          End if
       Abandon a fraction of worse nets and replace with new ones
       Keep the best solutions
      Rank the best solutions and obtain the current best
End while
Output the best result
End
```

Figure 2: Cuckoo Search Pseudocode

In CS, it is assumed that a cuckoo bird lays an egg at a time in any nest chosen randomly. The nest with the best quality/number of eggs carries on to the next generation. Moreover, there are a fixed number of nests and the cuckoo egg is discovered by the host bird with a given probability, usually between 0 and 1. The CS pseudocode [31] is presented in Figure 2. The CS is quite an effective algorithm obtaining results where other algorithms struggle [21]. In spite of seeming wide applications, the Cuckoo search has the problem of speed due to its use of several parameters and sometimes falling into local optima hence the emergence of several varieties of the algorithms such as discrete CS, Improved CS etc.

4. Experimental overview

In this study, the ABO was implemented using MATLAB on a desktop computer: Intel Duo Core ™ i7-3770 CPU, 3.40 GHz with 4GB RAM on a set of three symmetric TSP instances available in TSPLIB95. Also, please note that the statistical evaluations in this study were done using the IBM SPSS, version 22. The experimental parameters are presented in Table 1. These chosen parameter values are popular among researchers for their capacity to produce very good outcomes in solving TSP problems.

4.1 Experimental results and discussions

The first experiment involves eight benchmark TSP instances ranging from 52 to 225 cities. The experimental datasets are Berlin52, St70, Eil76, Pr76, KroA100, Eil101, Ch150, Tsp225. These TSP instances chosen in this study are not only complex but popular among researchers. The results obtained from the ABO was used to compare those from ABC, HA and ACO. The simulation result is presented in Table 2. Please note that in Table 2, the "Mean" refers to the mean fitness of each algorithm after 20 runs and the "PDB %" (Percentage Deviation from Best %) values are obtained using Equation 1

$$\left(\frac{\textit{Best Value}-\text{Optima}}{\text{Best value}}\right) \times 100 \qquad (1)$$

where 'Optima' simply refers to the benchmark optimum and 'Best Value' is the overall best outcome of each algorithm after 20 runs of each algorithm.

Table 1 reveals the comparatively good performances of the four algorithms under consideration with the ABO slightly outperforming the other algorithms in all test cases. For instance, the ABO managed to obtain the optimal outcome to two TSP instances: Berlin52 and Eil76. The other algorithms in spite of their good results could not obtain the optimal outcome to any of the problems. It must be emphasized that in as much as it is beneficial to obtain the optimum solution, metaheuristics, unlike their deterministic algorithms do not guarantee the optimum results.

Another clear observation is that the ABO obtained the nearest-optimal solution to the other six instances than its peers. Moreover, using the mean score as a yardstick of measurement, the ABO outcomes seem to be competitive. It posted the best average score in all instances. Of the three other algorithms, it is rather difficult to explain the below-par performance of the Hybrid Algorithm (HA) that uses a similar memory matrix like the ABO. The performance of the HA could be traceable to its use of several parameters. This is because the HA combines the ABC with ACO in its attempt at solutions. ABC and ACO algorithms deploy large number of parameters to obtain good solutions.

Table 1: ABO, ACO, ABC compared

Problem	No of Cities	Opt	Method	Best	Mean	PDB (%)
Berlin52			ABO	7542	7616	0.98
			ACO	7548.99	7659.31	1.52
	52	7542	ABC	9479.11	10,390.26	37.72
			HA	7544.37	7544.37	0.03
St70			ABO	676	678.33	0.01
			ACO	696.05	709.16	4.73
	70	675	ABC	1162.12	1230.49	81.73
			HA	687.24	700.58	3.47
Eil76			ABO	538	563.04	0
	76	538	ACO	554.46	561.98	3.04
			ABC	877.28	931.44	70.78
			HA	551.07	557.98	2.31
Pr76		108159	ABO	108167	108,396	0.01
	76		ACO	115,166.66	116,321.2	7.55
			ABC	195,198.9	205,119.6	89.65
			HA	113,798.5	115,072.	6.39
Kroa100		21282	ABO	21311	22163.8	0.001
	100		ACO	22,455.89	22,880.12	7.49
			ABC	49,519.51	53,840.03	152.94
			HA	22,122.75	22,435.3	5.40
Eil101		629	ABO	640	640	0.17
	101		ACO	678.04	693.42	7.96
			ABC	1237.31	1315.95	104.88
			HA	672.71	683.39	6.39
Ch150		6528	ABO	6532	6601	0.001
	150		ACO	6648.51	6702.87	2.61
			ABC	20,908.89	21,617.48	230.93
			HA	6641.69	6677.12	2.21
Tsp225		3916	ABO	3917	3982	0.0002
	225		ACO	4112.35	4176.08	8.22
			ABC	16,998.41	17,955.12	365.27
			HA	4090.54	4157.85	7.74

However, statistically, using the Wilcoxon Signed Rank Test, there is a significant difference between the best performances of the ABO with those of the optimal solution as can be seen from Table 3. Since ABO had the best result then there is no need to compare the outcome of the other comparative algorithms with the Optimal solution from the TSPLIB95. Rather our statistical comparison should be between the best results of the comparative algorithms. The comparative results are presented in Tables 2-5.

H_0: There is no significant difference between the results of ABO with those of ABC, ACO and HA

H_A: There is a significant difference between the results of ABO with those of ABC, ACO and HA

Table 2: Wilcoxon Signed Rank Test

Hypothesis Test Summary

	Null Hypothesis	Test	Sig.	Decision
1	The median of differences between Optima and ABObest equals 0.	Related-Samples Wilcoxon Signed Rank Test	.012	Reject the null hypothesis.

Asymptotic significances are displayed. The significance level is .05.

In comparing the best results of the ABO with those of the ACO, it was discovered that there was a significant difference in outcome in favour of the ABO (See Table 3). However, statistically there is no significant difference between the ABO best outcomes with those of HA best outcomes, hence we retain the null hypothesis (See Table 3)

Table 3: Wilcoxon Signed Rank Test

Hypothesis Test Summary

	Null Hypothesis	Test	Sig.	Decision
1	The median of differences between ABObest and ACObest equals 0.	Related-Samples Wilcoxon Signed Rank Test	.022	Reject the null hypothesis.

Asymptotic significances are displayed. The significance level is .05.

Table 4: Wilcoxon Signed Rank Test

Hypothesis Test Summary

	Null Hypothesis	Test	Sig.	Decision
1	The median of differences between ABObest and HAbest equals 0.	Related-Samples Wilcoxon Signed Rank Test	.068	Retain the null hypothesis.

Asymptotic significances are displayed. The significance level is .05.

Similarly, there is no significant difference between the best results of the ABO with those of the ABC as can be seen in Table 5. In any case a closer look at statistical Table 6 reveals that the Optima-ABO best has the best significant level of 0.094 compared to those of ABC (0.123), HA (0.256) and ACO (0.251). However, we reject the null hypothesis (H_0) and accept the alternative hypothesis (H_A)

Table 5: Wilcoxon Signed Rank Test

Hypothesis Test Summary

	Null Hypothesis	Test	Sig.	Decision
1	The median of differences between ABObest and ABCbest equals 0.	Related-Samples Wilcoxon Signed Rank Test	.317	Retain the null hypothesis.

Asymptotic significances are displayed. The significance level is .05.

IOP Conf. Series: Materials Science and Engineering **769** (2020) 012030 doi:10.1088/1757-899X/769/1/012030

4.2. ABO and cuckoo search on TSP

The second set of experiments on the symmetric TSP was performed involving some of the most recently-designed algorithms: the ABO and the Cuckoo Search (CS) algorithm. The experiment involved 35 symmetric TSP ranging from51 to 1379 cities. Table 6 presents the performance of two effective algorithms is solving TSP. The CS obtained 24 optimal solutions out of the 35.

Table 6: ABO and Cuckoo Search

TSP Cases	Opt	CS Best	Mean	SD	ABO Best	Mean	SD
eil51	426	426	426	0.00	426	427	30.05
berlin52	7542	7,542	7,542	0.00	7542	7659	36.04
st70	675	675	675	0.00	676	676	0.00
pr76	108,159	108,159	108,159	0.00	108167	108167	0.00
eil76	538	538	538.03	0.17	538	538	0.00
kroA100	21,282	21,282	21,282	0.00	21554	21554	0.00
kroB100	22,141	22,141	22,141.53	2.87	22160	22509	25.58
kroC100	20,749	20,749	20,749	0.00	20755	20755	0.00
kroD100	21,294	21,294	21,304.33	21.79	21347	21462	158.37
kroE100	22,068	22,068	2,281.26	18.50	22088	22088	0.00
eil101	629	629	630.43	1.14	640	640	0.00
lin105	14,379	14,379	14,379	0.00	14419	14453	254.12
pr107	44,303	44,303	44,307.06	12.90	44407	44407	0.00
pr124	59,030	59,030	59,030	0.00	59058	59058	0.00
bier127	118,282	118,282	118,359.63	12.73	118297	118863	577.79
ch130	6,110	6,110	6,135.96	21.24	6111	6111	0.00
pr136	96,772	96,790	97,009.26	134.43	96784	96784	0.00
pr144	58,537	58,537	58,537	0.00	58587	58587	0.00
ch150	6,528	6,528	6,549.9	20.51	6533	6601	62.41
kroB150	26,130	26,130	26,159.3	34.72	26169	26431	383.94
pr152	73,682	73,682	73,682	0.00	73730	73730	0.00
kroA200	29,368	29,382	29,446.66	95.68	29370	29370	0.00
kroB200	29,437	29,448	29,542.49	92.17	29487	29534	67.18
tsp225	3,916	3,916	3,958.76	20.73	3917	3917	0.00
gil262	2,378	2,382	2,394.5	9.56	2378	2378	0.00
a280	2,579	2,579	2,592.33	11.86	2579	2579	0.00
pr299	48,191	48,207	48,470.53	131.79	48211	48211	0.00
lin318	42,029	42,125	42,434.73	185.43	42101	42336	1543.43
rd400	15,281	15,447	15,533.73	60.56	15300	15300	0.00
fl417	11,861	11,873	11,910.53	20.45	11862	11862	0.00
pr439	107,217	107,447	107,960.5	438.15	107340	107340	0.00
rat575	6,773	6,896	6,956.73	35.74	6774	6810	7.67
rat783	8,806	9,043	9,109.26	38.09	8811	8881	0.00
pr1002	259,045	266,508	268,630.03	1,126.8	259132	259132	0.00
nrw1379	56,638	58,951	59,349.53	213.89	56653	56653	0.00

On the other hand, the ABO obtained six optimal results. Similarly, The CS obtained 10 optimal solutions in all 20 runs of the algorithm as could be seen in the Mean calculation of results. The runs that produced the accurate results in all runs of the CS are the eil51, berlin52, st70, pr76, kroA100, kroC100, lin105, pr124, pr144 and pr152. On this count, the ABO obtained three accurate solutions in all runs and these are in eil76, gil262 and a280. A closer examination, however, reveals that that the DCS obtained 'perfect' runs in TSP instances with cities ranging from 51-152.

In TSP instances with larger datasets running into hundreds, this effectiveness was halted. On the other hand, the ABO's 'good' runs are more widespread. Versatility, being one of the indices of a good algorithm [36], the ABO appears to be a more versatile algorithm in this case. Nonetheless, the CS proves to be a more effective algorithm than the ABO in terms of obtaining the optimal solutions to the TSP instances under investigation here. Furthermore, in terms of the standard deviation from the mean score, both algorithms performed excellently with DCS having 0.00 standard deviation in 10 instances (the eil51, berlin52, st70, pr76, kroA100, kroC100, lin105, pr124, pr144 and pr152). Here the ABO produced more consistent result with 0.00 standard deviation in 24 instances. This is a proof of good algorithm convergence: attributes of a good optimization method [36]. Statistically, as can be seen from Table 6, using the Wilcoxon Rank Sum Test, there exist significant differences in the results of both the ABO and the Optimal solutions as well as in those of CS and the Optimal results available in TSPLIB95 (See Tables 6a and Table 6b). However, statistically, there exist not significant difference in the performance of the CS and the ABO in solving the TSP under investigation (See Table 6c), resulting in the acceptance of the null hypothesis (H_O): H_O: There is no significant difference in the performance of the CS and the ABO in solving the TSP instances under investigation.
H_A: There is a significant difference in the performance of the CS and the ABO in solving the TSP instances under investigation.

Wilcoxon Signed Ranks Test

Table 6(a): Test Statistics

Hypothesis Test Summary

	Null Hypothesis	Test	Sig.	Decision
1	The median of differences between ABO and OPT equals 0.	Related-Samples Wilcoxon Signed Rank Test	.000	Reject the null hypothesis.

Asymptotic significances are displayed. The significance level is .05.

Table 6(b): Test Statistics

Hypothesis Test Summary

	Null Hypothesis	Test	Sig.	Decision
1	The median of differences between CS and ABO equals 0.	Related-Samples Wilcoxon Signed Rank Test	.638	Retain the null hypothesis.

Asymptotic significances are displayed. The significance level is .05.

Table 6(c): Test Statistics

Hypothesis Test Summary

	Null Hypothesis	Test	Sig.	Decision
1	The median of differences between ABO and OPT equals 0.	Related-Samples Wilcoxon Signed Rank Test	.000	Reject the null hypothesis.

Asymptotic significances are displayed. The significance level is .05.

IOP Conf. Series: Materials Science and Engineering **769** (2020) 012030 doi:10.1088/1757-899X/769/1/012030

With the .05 used as the degree of freedom, there exists no significant difference in CS and ABO to solving the TSP instances since the outcome of the ABO in relation to the optimal solutions (OPT) resulted in .638 significances. Since both outcomes are beyond .05, we retain the null hypothesis (H_O).

4.3. ABO with Bat Algorithm and Firefly Algorithm for TSP

The performance of three recently designed and excellent algorithms are examined in attempting solutions to the TSP, which are Bat Algorithm, Firefly Algorithm [22] and ABO as shown in Table 8.

Table 8: ABO with BA and FFA

TSP Cases	Optima	BA Best	ABO Best	FFA Best
Oliver30	420	420	425	420
Eil51	426	430	426	426
Berlin52	7542	7542	7542	7542
St70	675	675	676	675
Eil76	538	538	538	543
Kroa100	21,282	21,292	21554	21,282
Krob100	22,140	22,373	22160	22,183
Kroc100	20,749	20,802	20755	20,756
Krod100	21,294	21,727	21347	21,408
Kroe100	22,068	22,323	22088	22,079
Eil101	629	640	640	643
Pr107	44,303	44,618	44407	44,303
Pr124	59,030	59,030	59058	59,030
Pr136	96,772	100,485	96784	97,716
Pr144	58,537	58,588	58587	58,546
Pr152	73,682	74,172	73650	74,033

As is evident in Table 8, all the comparative algorithms performed excellently well with ABO obtaining four optimal solutions to BA's five and FFA's six. Comparatively, therefore, the FFA is more effective, followed by BA and ABO. It is rather interesting to note that BA and ABO obtained optimal results in datasets of less than 100 cities. It is only FA obtained optimal solution in two other TSP instances that involved more than 100 cities, specifically in pr107 and pr124. Similarly, it worthy of note that the performance of the ABO in the remaining instances is commendable, even when it fails to obtain the optimal result, its results are quite neat the optimal.

5. Conclusion

This paper examined the performance of six swarm-intelligence techniques in providing optimize solutions to the symmetric travelling salesman's problems. It is safe to assert optimize technique that need maximize output cum profits while minimizing the input [31]. It is our belief that the comparative performance of these 21st optimization would assist researchers in choice of swarm intelligence techniques whenever they are confronted with real-life. After a number of experimental procedures, this study concludes that though all the algorithms are very effective in solving the benchmark symmetric travelling salesman's problems under investigation. Even though, the statistical evaluations indicate that there is no significant difference in the performance of the comparative algorithms, however, a closer observation indicates the DCS and the ABO have slight edge over the other algorithms in providing solutions to the benchmark-travelling salesman's problem instances investigated.

Acknowledgment

Appreciation conveyed to Anchor University, Lagos; Ministry of Higher Education Malaysia for FRGS RDU190185; and University Malaysia Pahang Grant RDU170303 for financing this research.

References

[1] Odili, J.B.: 'Application of Ant Colony Optimization to Solving the Traveling Salesman's Problem', Science Journal of Electrical & Electronic Engineering, 2013, 2013

[2] Dorigo, M., and Gambardella, L.: 'Ant-Q: A reinforcement learning approach to the traveling salesman problem', in Editor (Ed.)^(Eds.): 'Book Ant-Q: A reinforcement learning approach to the traveling salesman problem' (2016, edn.), pp. 252-260

[3] Rodríguez, A., and Ruiz, R.: 'The Effect Of The Asymmetry Of Road Transportation Networks On The Traveling Salesman Problem', Computers & Operations Research, 2012, 39, (7), pp. 1566-1576

[4] Conway, R.W., Maxwell, W.L., and Miller, L.W.: 'Theory of Scheduling' (Courier Corporation, 2012. 2012)

[5] Hussein, I.I., and Stipanovic, D.M..: 'Effective Coverage Control for Mobile Sensor Networks with Guaranteed Collision Avoidance', IEEE Transactions on Control Systems Technology, 2007, 15, (1), pp. 642-657.

[6] Yang, H.C., Liu, K.J., and Hung, M. H.: 'Drill-Path Optimization with Time Limit and Thermal Protection", Advanced Materials Research, 2012, pp. 153-159.

[7] Rodríguez, A, and Ruiz, R.: 'The Effect of The Asymmetry of Road Transportation Networks on The Traveling Salesman Problem', Computers & Operations Research, 2012, 39, (1), pp. 1566-1576.

[8] Bochtis, D, and Sørensen, C.G.: 'The Vehicle Routing Problem in Field Logistics: Part II', Biosystems Engineering, 2010, 105, pp. 180-188.

[9] Hao, X.: 'Optimization Models and Heuristic Method Based on Simulated Annealing Strategy for Traveling Salesman Problem', Applied Mechanics and Materials, 2010, pp. 1180-1184.

[10] Hoffman, K.L., Padberg, M, and Rinaldi, G.: 'Traveling Salesman Problem', Encyclopedia of Operations Research and Management Science, ed: Springer, 2013, pp. 1573-1578.

[11] Savla, K., Frazzoli, E, and Bullo, F.: 'Traveling Salesperson Problems for The Dubins Vehicle', IEEE Transactions on Automatic Control, 2008, 53, (1), pp. 1378-1391.

[12] Waldrop, M.M.: 'No Drivers Required', Ed: Nature Publishing Group Macmillan Building, 4 Crinan St, London N1 9xw, England, 2015.

[13] Applegate, D.L., Bixby, R. E., Chvatal, V, and Cook, W. J.: 'The Traveling Salesman Problem: A Computational Study', Princeton University Press, 2011.

[14] Kefi, S, Rokbani, N, Krömer, P, and Alimi, A.M.:'A New Ant Supervised-PSO Variant Applied to Traveling Salesman Problem', Hybrid Intelligent Systems, ed: Springer, 2016, pp. 87-101.

[15] Reinelt, G.:'Tsplib95, 1995', URL http://comopt. ifi. uni-heidelberg. de/software/TSPLIB95, 2012.

[16] Chetty, S, and Adewumi, A.O.:' Comparison Study Of Swarm Intelligence Techniques For The Annual Crop Planning Problem', IEEE Transactions on Evolutionary Computation, 2014, 18, pp. 258-268.

[17] Dorigo, M, and Stützle, T: 'Ant Colony Optimization: Overview And Recent Advances', Handbook of Metaheuristics, ed: Springer, 2010, pp. 227-263.

[18] Karaboga, D, Gorkemli, B, Ozturk, C, and Karaboga, N.: 'A Comprehensive Survey: Artificial Bee Colony Algorithm And Applications', Artificial Intelligence Review, 2014, 42, pp. 21-57.

[19] UZ, M.G., Kiran, and E. Özceylan, E.: 'A Hierarchic Approach Based On Swarm Intelligence To Solve The Traveling Salesman Problem', Turkish Journal of Electrical Engineering & Computer Sciences, 2015, 23, pp. 103-117.

[20] Odili, J.B., Kahar, M.N., and Noraziah, A.: 'Solving Traveling Salesman's Problem using African Buffalo Optimization, Honey Bee Mating Optimization & Lin-Kerninghan Algorithms', World Applied Sciences Journal, 2016, 34, (7), pp. 911-916

[21] Ouaarab, A., Ahiod, B., and Yang, X.-S.: 'Discrete Cuckoo Search Algorithm for The Travelling Salesman Problem', Neural Computing and Applications, 2014, 24, (7-8), pp. 1659-1669

[22] Yang, X.-S., Hosseini, S.S.S., and Gandomi, A.H.: 'Firefly Algorithm for Solving Non-Convex Economic Dispatch Problems with Valve Loading Effect', Applied Soft Computing, 2012, 12, (3), pp. 1180-1186

[23] Yang, X.-S., and Hossein Gandomi, A.: 'Bat Algorithm: A Novel Approach For Global Engineering Optimization', Engineering Computations, 2012, 29, (5), pp. 464-483

[24] Sarathambekai, S., and Umamaheswari, K.: 'Task Scheduling in Distributed Systems using Heap Intelligent Discrete Particle Swarm Optimization', Computational Intelligence, 2017

[25] Odili, J.B., and Mohmad Kahar, M.N.: 'Solving The Traveling Salesman's Problem using The African Buffalo Optimization', Computational intelligence and neuroscience, 2016, 2016, pp. 1-12

[26] Odili, J.B., and Kahar, M.N.M.: 'African Buffalo Optimization (ABO): A New Meta-Heuristic Algorithm', Journal of Advanced & Applied Sciences 2015, 03, (03), pp. 101-106,

[27] Odili, J.B., Kahar, M.N., and Noraziah, A.: 'Parameters-tuning of PID Controller For Automatic Voltage Regulators using The African Buffalo Optimization', PLoS One, 2017, 12, (4), pp. 1-17

[28] Hassan, M.H., and Muniyandi, R.C.: 'An Improved Hybrid Technique for Energy and Delay Routing in Mobile Ad-Hoc Networks', International Journal of Applied Engineering Research, 2017, 12, (1), pp. 134-139

[29] Odili, J.B., and Kahar, M.N.M.: 'Numerical Function Optimization Solutions using The African Buffalo Optimization Algorithm (ABO)', British Journal of Mathematics & Computer Science, 2015, 10, (1), pp. 1-12

[30] Yang, X.-S., and Deb, S.: 'Cuckoo Search Via Lévy Flights', in Editor (Ed.)^(Eds.): 'Book Cuckoo search via Lévy flights' (IEEE, 2009, edn.), pp. 210-214

[31] Odili, J. B., & Kahar, M. N. M. 'African Buffalo Optimization', International Journal of Software Engineering and Computer Systems, 2016, 2, (1), pp. 28-50.

Dengue Outbreak Prediction Using an Improved Salp Swarm Algorithm.

Khairunnisa Amalina Mohd Rosli[1], Zuriani Mustaffa[2], Yuhanis Yusof[3] and Mohamad Farhan Mohamad Mohsin[4]

[1,2]Faculty of Computing, Universiti Malaysia Pahang, 26300, Kuantan, Pahang, Malaysia.

[3]School of Computing, Universiti Utara Malaysia, 06010, UUM Sintok, Kedah, Malaysia.

[4]UUM CAS Student Development and Alumni, School of Computing, Universiti Utara Malaysia, 06010, UUM Sintok, Kedah, Malaysia

khairunnisaamalina@gmail.com

Abstract: Dengue disease is the most common type of disease caused by mosquitoes. It is reported that dengue fever was first recognized in Thailand and Philippines in 1950. According to World Health Organization (WHO), dengue is a viral disease that spread in public environment where the number of dengue cases reported enlarge within 5 years by 1 million from 2.2 million in 2010 to 3.2 million in 2015. Until today, numerous studies by researchers to improve the prediction of dengue fever disease based on Computational Intelligence (CI) methods have been reported. The research includes study using Swarm Intelligence (SI) algorithm. This research study proposed an improved Salp Swarm Algorithm (iSSA) for dengue outbreak prediction. The original SSA will be enhanced by enriching the exploration and exploitation process for the sake of improving the accuracy of dengue outbreak prediction. This will be done by inducing a mutation based on Levy Flight. Later, the iSSA algorithm will be realized on dengue disease dataset. The proposed iSSA will be compared against the original SSA and another CI method known as Grey Wolf Optimization (GWO). Two evaluation indicators known Root Mean Square Error (RMSE) and as Mean Absolute Error (MAE) are proposed in this research study to evaluate the prediction model where the smaller the value obtained, more accurate the prediction model. The result demonstrated that the proposed model produces a better result compare two the other results where the value of MAE and RMSE of the proposed model is smaller compare to other two model.
Keywords— Dengue outbreak prediction, Levy Flight, Salp Swarm Algorithm

1. Introduction

In the last few decades, nature inspired metaheuristic algorithms becoming more popular and dominant in computational intelligence [1]. Computational intelligence and meta heuristic algorithm have evolve and become popular in computer science including machine learning, artificial intelligence and data mining [2]. Many of nature inspired algorithm have been proposed where these algorithms imitate the nature behaviour of animals and they are known as Swarm Intelligence (SI) algorithms [3]. For example, Artificial Bee Colony algorithm (ABC) inspired by the food foraging behaviour of honeybee colony [4]. Next, an algorithm that mimic an ant in nature to find the food searching path is called Ant Colony

Optimization (ACO) [5] and Particle Swarm Optimization (PSO) that simulates the social behaviour flocks of bird [6].

Recently, a relatively new SI algorithm has been proposed, namely Salp Swarm Algorithm (SSA) SSA is introduced in year 2017 by Mirjalili [7]. Exploration and exploitation characteristics are vital in determining the achievement of optimization algorithm where exploration behaviour will give guarantee that the algorithm can explore the search space precisely meanwhile exploitation has the ability to search around the current best solution and select the best solution [8]. SSA algorithm goal is to enhance the balance between exploration-exploitation behaviour where it focuses on particles motion [9][14].

Despite its excellent performance, SSA is still open for an improvement where it is reported that more exploitation than exploration leads to higher convergence speed but the risk of not reaching to the global solution. On the other hand, more exploration leads to slow convergence speed and solutions might not be improved well enough to find the global optimal. These demerits need critical attention as it would demote the prediction result. The main objective of this study is to minimize the objective function where lower the MSE value more promising the performance of the algorithm.

In literature review, there are numerous techniques has been proposed by researchers to improve the prediction of dengue outbreak. Various Computational Intelligence methods have been created to handle the issue of dengue fever disease. In year 2018, a research study to predict dengue outbreak has been proposed [10]. In this study LSSVM is hybrid with new optimization algorithm namely Flower Pollination Algorithm (FPA) is implemented for dengue outbreak prediction in Yogyakarta Indonesia. They highlighted to determine the optimal value of LSSVM hyper-parameter FPA is used as optimization tool. In 2015, a study in predicting Dengue Fever (DF) in Kuala Lumpur by implementing artificial neural network method has been introduced where they used Linear Regression Model [11]. This research study considered mean daily temperature, daily rainfall and EVI as their factors for the proposed model. Cubic Spline Interpolation method is crucial to find a meaningful data for training purpose. Prediction model is used as it can produces model with low variances and deal with irregular observable variables. This research study is using Mean Test Error (MTE) to evaluate the results. The result shows that more training data, it will provide more accurate result.

In addition, quite a number of improvements that involved the inducing of Levy Flight (LF) in SI algorithms have been proposed to enhance the exploration and exploitation. In [12] they proposed an improved PSO by implement Levy Flight (LF) techniques to update the particle's velocity and improve the capability of global search. A prediction model by implement LF into Cuckoo Search (CS) has been proposed in [13]. The modification of this study is focusing on the information exchange between two best eggs in algorithm or their best solution. Then CS is validated by using test function and compare the performances with GA and PSO. LF random walk is proposed to balance the exploration and exploitation characteristic of ABC [14]. Global optima of unimodal or multimodal functions are decreasing by implement LF random walk. They will find the step size in search space including numbers of scout bees is raised. PSO with Active Velocity Penalty (PSO-AVP) is proposed [15] to penalizes the particles velocities to ensure they are move within the search space to harmonize their exploration and exploitation behaviour.

Due to the efficiency of LF in improving the SI algorithm, this study proposed an improved SSA (iSSA) by embedding the LF function in exploitation and exploration process. Then iSSA later will be realized for dengue outbreak prediction.

This paper is presented as follows: section II will explain a description on SSA continue with the methodology that will be implemented in section III followed with experimental results in section IV and lastly the conclusions of the research study in section V.

2. Salp Swarm Algorithm

SSA algorithm comes from the family of Salpidae known as the family of salps. They have a unique transparent barrel-shaped body that similar to jelly fish including its movement where the water will be pumped through their body as propulsion to move forward. Main foundation of SSA algorithm is because of the swarming behaviour. Salp chain will be form in the deep ocean when a group of salp is combined. Researchers propose this algorithm to solve optimization problem. Leader and followers are the population of SSA where the salp at the front of the chain is known as the leader while the other salps that follow at the back is known as the followers. The followers will follow one another directly or indirectly led by their leader. Some remarks are listed to determine the ability of the proposed SSA algorithm in solving optimization problem [7].

• SSA algorithm will saves the best solution and assigns it to the food source variable to ensure it never gets lost though the whole population worsen.

• SSA algorithm updates the position of the leader salp parallel with the food source to ensure the leader salp can always explores and exploits the space around it.

• SSA algorithm will constantly updates the position of follower salps as they will slowly move towards the leader salp.

• Continuous movement of follower salps can prevent SSA algorithm from changing easily in local optima.

• $c1$ parameter is reduced adaptively over the course of iterations hence SSA algorithm will explores the search space and then exploits it.

• $c1$ is the main controlling parameter that will control the exploration and exploitation behaviour.

• SSA algorithm is simple and easy to implement.

3. Methodology

The methodology is divided into four (4) main sections namely data collection, iSSA Prediction Model, Salp Swarm algorithm with Levy Flight Technique and finally evaluation section. For data evaluation, it will discuss about data collection and also data pre-processing process, which is also known as data cleaning to remove noise data and revolved the data into understandable data. Next is prediction by using iSSA where iSSA will be applied into dengue disease raw dataset to demonstrate prediction performance. Then, Salp Swarm algorithm with Levy Flight Technique is introduced. Lastly, this section will explain the method used to evaluate the performance of iSSA in dengue outbreak prediction.

3.1. Data Collection

For this study, a real dataset of dengue in Malaysia the data is collected. Dengue dataset is provided by the researchers from Universiti Utara Malaysia. The collected data is from year 2003 until year 2008 There is a total of 11 observations of dengue dataset input. The output variable of this dataset is dengue cases from day 7 and onwards. Input and output variable are important to evaluate the output accuracy. In this study, there are 4 input variables involved for predicting dengue outbreak which are number of dengue cases, rainfall (mm), temperature (C) and humidity (%). These features will be used to assist the model in learning the relationship of the variable with dengue outbreak accuracy. For data analysis, MATLAB language was selected.

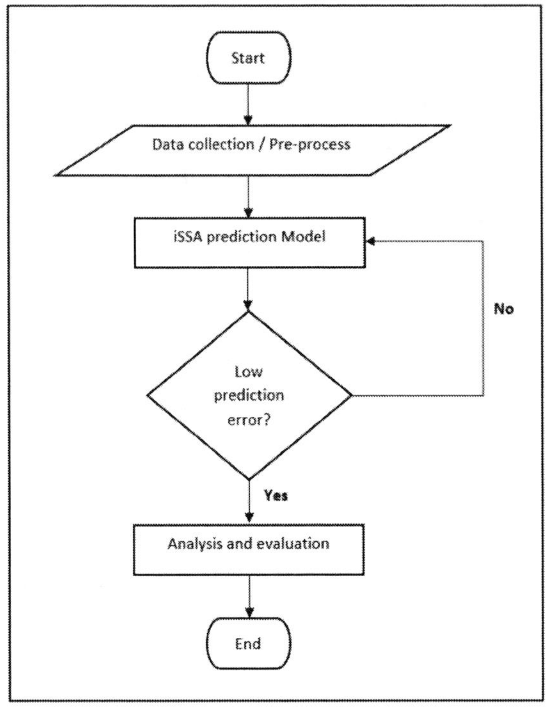

Figure 1. Methodology

Figure 1 explains the methodology of iSSA prediction Model. Firstly, data collection and data pre-process take place to remove meaningless information and redundant data before going to training stage. A good pre-processing technique will improve the efficiency of the result. Then the data will be analyzed by implement into iSSA model to determine whether they have low prediction error or vice versa. Then dengue dataset will be analyzed and undergo evaluation. The input and output of this research study is presented in Table 1.

Table 1. Dengue cases data

Input	Output
Dengue Case Humidity Rain Temperature	Dengue cases from day 7 and onwards

1) *Training, Validation and Testing:* Table 2 shows data proportion that has been set for simulation purposes.

Table 2. Data Proportion for Training, Validation and Testing

	Training	**Validation**	**Testing**
Data Proportion	70%	15%	15%

3.2 iSSA Prediction Model

SSA algorithm begins with approaching the global optimum. They will assign multiple salps with random positions. The fitness of each salp is calculated where the salp with the best fitness function will be elect as the best salp that will chase the food source represented by variable F. In the meantime, $c1$ parameter is updated. For each dimension, both position of the leader and follower are updated. If any salps goes beyond the search space, it will be brought back on the boundaries.

The above steps are executed repeatedly until the end criterion is fulfilled excluding initialization. It is highlighted that the food source will be updated during optimization as the salp chain will find a better solution by exploring and exploiting the search space. The above procedures show that the salp chain modelled can chase a moving food source. Hence, the salp chain can move towards the global optimum that changes over the course of iterations. Figure 2 below show the pseudocode of SSA algorithm:

```
Initialize the salp position χi(i = 1,2,......,n) considering ub and lb
While (end condition is not satisfied)
Calculate the fitness of search agent (salp)
F=the best search agent
Update c1 with equation (1)
        For each salp (χi)
        Update the position of leading salp with equation (2)
        Else
        Update the position of follower salp with equation (3)
End
End
Amends the salp based on the ub and lb
End
Return F
```

Figure 2. SSA pseudocode

$$x_j^1 = \begin{cases} F_j + c_1\left((ub_j - lb_j)c_2 + lb_j\right) & c_3 \geq 0 \\ F_j - c_1\left((ub_j - lb_j)c_2 + lb_j\right) & c_3 < 0 \end{cases} \tag{1}$$

$$c_1 = 2e^{-\left(\frac{4l}{L}\right)^2} \tag{2}$$

$$x_j^i = \frac{1}{2}(x_j^i + x_j^{i-1}) \tag{3}$$

Equation (1) is proposed to revise the position of leading salp where x_j^1 is the position of leader salp in the jth dimension while F_j is the food source in j^{th} dimension, ub_j is the upper bound meanwhile lb_j is the lower bound in j^{th} dimension and c_1, c_2, c_3 are random numbers between [0,1] interval.

In equation (2) c_1 are crucial parameter that will determine the exploration and exploitation criteria of the algorithm where l is the current iteration and L is the maximum iteration. Equation (3) indicates the position of the follower salp where x_j^i is the position of ith follower in j^{th} dimension. Random quantity that helps in exploration and exploitation of the search space play a crucial role in equation of SSA. In this study, Levy Flight (LF) will be implemented to stabilize the diversity and convergence in SSA. LF has local search and global search where it can be accomplished by coordinate the LF parameters and automatically adjust the step sizes [13]. This proposed technique able to improve the exploitation capability of SSA.

Exploration characteristic main objective is to determine the global optimum by identifying the unknown region of search space. Meanwhile exploitation is the ability to apply the knowledge of previous solution in order to find a better solution. In this study, the main purpose is to improve the performance and search space. It is highlighted to update the previous solutions step size plays an important role. The new solution may exceed the true solution if the step size is large, meanwhile if the step size is small, then the convergence rate may decrease that will result in poor performance.

```
Input function Min f(χ) and β
Select χ_i in swarm that will modify the position
Initialise τ = 1 and σ_h = 1
Compute σ_u with equation (4)
While (τ < ϵ)
        Compute step_size with equation (5)
        Generate new solution χ_i' with equation (6)
        Calculate f(χ_i')
        If f(χ_i') < f(χ_i) then
        χ_i = χ_i'
        End if
        τ = τ + 1
End
```

Figure 3. Levy Flight pseudocode

$$\sigma_u = \left\{ \frac{\Gamma(1+\beta)\sin(\pi\beta/2)}{\beta\Gamma[(1+\beta)/2]2^{(\beta-1)/2}} \right\} \tag{4}$$

$$step\_size(\tau) = 0.01 \times s(\tau) \tag{5}$$

$$\chi'_{ij}(\tau + 1) = \chi_{ij}(\tau) + step\_size(\tau) \times U(0,1) \tag{6}$$

The step size by using Levy distribution for exploitation of search area calculated as in equation (5). Factor of 0.01 is L/100 where L is typical length scale. Equation (6) where χ_{ij} is individual that will modify the position, U (0,1) is a random number between [0,1] range and $step\_size(\tau) \times U(0,1)$ is the random walk from levy distribution.

3.3 Salp Swarm Algorithm with Levy Flight Technique

The main objective for this research study is to enhance both exploration and exploitation criteria of SSA in order to accomplish better optimization performance. For that matter, Levy Flight technique is introduced to helps in exploration consequently of exploitation of the search space. A newly introduced parameter of Improved SSA will enhance its convergence speed and balance the ability of exploitation and exploration. In addition, this proposed algorithm will produce a better performance of the solution. The improved SSA is stated in the equation (7) including the pseudocode of iSSA show in Fig 4 below:

$$x_j^1 = \begin{cases} F_j + c_1((ub_j - lb_j)Levy + lb_j)c_3 \geq 0 \\ F_j + c_1((ub_j - lb_j)Levy + lb_j)c_3 < 0 \end{cases} \tag{7}$$

x_j^1 : the position of the first salp in j^{th} dimension.

F_j : the position of the food source in j^{th} dimension.

ub_j: is the upper bound.

lb_j : is the lower bound in j^{th}dimension.

c_1, $c_3$are random numbers in interval [0,1]

Begin

i. *Initialize the parameter of SSA (number of salps (s), iteration number (L) and salp position (x))*

ii. *Initialize a population of salp's position randomly*

iii. *Evaluate the fitness of each salp*

iv. *Set the number of iteration (l)*

v. *Update* c_1 *by equation* (1)

vi. *For each salp,*

If i==1 update the position of leading salp by equation (7)

Otherwise update the position of follower salp by using equation (3)

Evaluate the fitness of every salp

Update x^*

ix. *Increment l to 1*

x. *Repeat step (v,vi,ix)*

xi. *Return the best solution and fitness value*

End

Figure 4. iSSA pseudocode

4. Performance Evaluation Matrices

In this paper, Root Mean Square Error (RMSE) and Mean Absolute Error (MAE) are calculated to forecast the percentage accuracy of dengue outbreak cases. RMSE is the standard deviation of prediction error (residual). Residual are calculated to determine the robustness of the data around the line of best fit. MAE is a quantity used to measure the difference between two continuous variables. The equations are defined as follows:

$$RMSE = \sqrt{\frac{\sum_i^n (y_i - p_i)^2}{n}} \quad (8)$$

$$MAE = \frac{1}{n} \sum_{i=1}^{n} |y_i - p_i| z \quad (9)$$

Based on the above equation, yi is the actual value meanwhile pi is the prediction value of ith test data obtained from the proposed model and n is the number of data.

MAE and RMSE are important indicators of dengue prediction results where the smaller values obtained, more accurate the prediction model.

5. Result And Discussion

This section will present the result obtained from the proposed iSSA. The capability of iSSA is compared with the original SSA and another swarm algorithm known as Grey Wolf Optimization (GWO). The performance evaluation of iSSA, SSA and GWO are shown in Table 3.

Table 3. iSSA vs. SSA vs. GWO for Dengue Outbreak Prediction

	iSSA	**SSA**	**GWO**
MAE	1.84E-12	2.20E-12	3.022E-12
RMSE	1.357E-06	1.484E-06	1.4125E-06

From the table, it shows that iSSA recorded the lowest MAE and RMSE value of 1.84E-12 and 1.357E-06 respectively which is the lowest error compared with the original SSA that recorded 2.20E-12 of MAE value and 1.484E-06 of RMSE value. Meanwhile the value of MAE and RMSE value of GWO are 3.022E-12 and 1.4125E-06 respectively. The upper bound value is set to 1000 meanwhile 500 iteration is applied. Based on the result, it shows iSSA is reliable as both value of MAE and RMSE are smaller and has less error.

6. Conclusion

Main objective of this research study was to predict dengue outbreak cases based on the proposed algorithm namely iSSA. By using a real dataset with parameters of number of dengue cases, temperature (`C), rainfall values (mm), and humidity (%). The result achieved by the proposed iSSA was compared with the original SSA and another CI algorithm namely GWO. The performance of these algorithms is guided by MSE and RMSE where the lower the value more reliable the result. The finding of the study shows that the proposed algorithm has a promising result in comparison with the other two algorithms.

IOP Conf. Series: Materials Science and Engineering **769** (2020) 012031 doi:10.1088/1757-899X/769/1/012031

Acknoledgment

The author would like to thank to Faculty of Computing (FKOM), Universiti Malaysia Pahang.This study is funded under the UMP Research Grant. Grant No: RDU180366.

References

[1] Xin-She Yang, Su Fong Chien, Tiew On Ting, "Computational Intelligence and Metaheuristic Algorithms with Applications",The Scientific World Journal Volume 2014.

[2] X. S. Yang, Cuckoo Search and Firefly Algorithm: Theory and Applications, vol. 516 of Studies in Computational Intelligence, Springer, Heidelberg, Germany, 2014.

[3] H.-S. Yang, Engineering Optimisation an Introduction with Metaeuristic Application, 1sr es., John Wiley & Sons, New Jersey, 2010.

[4] D.T. Pham, A. Ghanbarzadeh, E. Koc, S.Otri, S.Rahim, M.Zaidi, The bees algorithm, Technical Note, Manufacturing Engineering Centre, Cardiff University, UK, 2005

[5] N.Zhang, J.Ji, C.Liu,N.Zhong, An ant colony optimization algorithm for learning classification rules in:Proceeding of 2006 IEEE/WIC, 2006, pp. 1034-1037.

[6] J.Kennedy, R.C Eberhart, Particle swarm optimization, in:Proceedings of IEEE International Joint Conference on Neural Networl, vol.4, 195, pp. 1942-1948

[7] S. Mirjalili, A. H. Gandomi, S. Z. Mirjalili, S. Saremi, H. Faris, and S. M. Mirjalili, "Salp Swarm Algorithm: A bio-inspired optimizer for engineering design problems," *Adv. Eng. Softw.*, vol. 114, pp. 163–191, 2017.

[8] C. Suvarna, A. Sali, and S. Salmani, "Efficient heart disease prediction system using optimization technique," *Proc. Int. Conf. Comput. Methodol. Commun. ICCMC 2017*, vol. 2018–Janua, no. Iccmc, pp. 374–379, 2018.

[9] C. Paper, "an Efficient Method To Predict Dengue," no. October, pp. 12–13, 2015.

[10] Z. Mustaffa, M. H. Sulaiman, F. Emawan, Y. Yusof, and M. F. M. Mohsin, "Dengue outbreak prediction: Hybrid meta-heuristic model," *Proc. - 2018 IEEE/ACIS 19th Int. Conf. Softw. Eng. Artif. Intell. Netw. Parallel/Distributed Comput. SNPD 2018*, pp. 271–274, 2018.

[11] C. Paper, "an Efficient Method To Predict Dengue," no. October, pp. 12–13, 2015.

[12] R.Jensi, G.Wiselin Jiji, An enhanced particle swarm optimization with levy flight for global optimization, Applied Soft Computing Elsevier, 2015

[13] Sangita Roy, Chaudhuri S.S, "Cuckoo Search Algorithm using Levy Flight: A Review", I.J Modern Education and Computer Science, 2013,12,10-15.

[14] Harish Sharma, Jagdish C.B, Arya K.V, Yang Xin-She. "Levy Flight artificial bee colony algorithm", International Journal of System Science, 2015

[15] Mounir Ben Ghalis, "Particle Swarm Optimization with an improved Exploration-Exploitation Balance", IEEE, 2008

IOP Conf. Series: Materials Science and Engineering **769** (2020) 012032 doi:10.1088/1757-899X/769/1/012032

Texture-based Feature using Multi-blocks Gray Level Co-occurrence Matrix for Ethnicity Identification

M Z Osman[1], **M A Maarof**[2], **M F Rohani**[2]

[1] Soft Computing and Intelligence System (SPINT), Faculty of Computing, Universiti Malaysia Pahang, Lebuhraya Tun Razak, 26300 Gambang, Kuantan, Pahang, Malaysia

[2] Information Assurance and Security Research Group (IASRG), School of Computing, Universiti Teknologi Malaysia, 81310, Skudai, Johor, Malaysia

Email: zamriosman@ump.edu.my

Abstract. Ethnicity identification for demographic information has been studied for soft biometric analysis, and it is essential for human identification and verification. Ethnicity identification remains popular and receives attention in a recent year especially in automatic demographic information. Unfortunately, ethnicity identification in a multi-class which consist of several ethnic classes may degrade the accuracy of the ethnic identification. Thus, this paper purposely analyses the accuracy of the texture-based ethnicity identification model from facial components under four-class ethnics. The proposed model involved several phases such as face detection, feature selection, and classification. The detected face then exploited by three proposed face block which are 1×1, 1×2 and 2×2. In the feature extraction process, a Grey Level Co-occurrence Matrix (GLCM) under different face blocks were employed. Then, final stage was undergone with several classification algorithms such as Naïve Bayes, BayesNet, *k*-Nearest Neighbour (*k*-NN), Random Forest, and Multilayer Perceptron (MLP). From the experimental result, we achieved a better result 2×2 face block feature compared to 1×1 and 2×2 feature representation under Random Forest algorithm.

1. Introduction

An ethnic group or ethnicity is a category of people which are identified based on the similarities of society, culture, ancestry, language or nation [1]. The ethnic group generally is formed based on the biological unit which made up from skin colour, head shape, hair, face shape and blood type.

Ethnicity identification can be referred as multi-class and inter-class identification problem. Multi-class problem involved several different ethnic groups and it was the most frequently studies by researchers [2]. Ethnic group exploited normally are Asian, African American, and Caucasian ethnics [3-5]. Meanwhile inter-class identification involved sub-ethnic group such as Chinese, Koreans, and Japanese. Several researchers adopted this dataset issue in intra-class ethnic problem such as [6, 7]. In multi-class ethnic, most of the researchers employed ethnicity identification based on two or three class ethnics. Two-class ethnic such as Asian and non-Asian [8], or Asian and Caucasian [9] are frequently employed. Three-class ethnic also showed a popularity where majority of researchers exploited three ethnics such as Asian, African American and Caucasian ethnics [10]. However, lack of ethnicity identification methods that exploited more than three ethnics since the performance of the methods reported to be dropped. The performance of two and three class problem reported with average accuracy of above 90%. Two-class ethnic revealed in superior accuracy with most of the results are above 94% until 99%. In addition to the ethnic class, Hispanic ethnic was only tested and reported by [11, 12]. According to [2], to have an accurate performance for ethnicity identification in uncontrolled environment is challenging. Feature representation also affect the performance of the ethnicity identification. Therefore, several searchers have been experimental with different feature representations such as skin tone feature [13, 14], and texture feature [15, 16]. Texture feature can be categorized into two: global and local features. Global feature known as holistic representation is most studied feature extraction used in ethnicity identification. It has the capability to preserve configural information which is the interrelations between facial regions. Meanwhile local feature also has gain popularity which capable in term of unconstrained ethnic identification [7]. One of example local texture is local binary pattern used by [15, 16]. Gabor feature is reported as a suitable feature for ethnicity identification. Unfortunately, Gabor feature produced high dimension of feature vector which might increasing the processing time.

The aim of this paper is to investigate the performance of ethnicity identification method by using different face blocks and GLCM feature extraction technique. This study also reveals the analysis on four-class ethnic groups which are Asian, African American, Caucasian and Hispanic.

2. Methodology

This section provides a brief on the methodology of the proposed methods. There are three main stages involved which are face detection, feature extraction, and classification. Figure 1 shows the flow of the proposed ethnicity identification method.

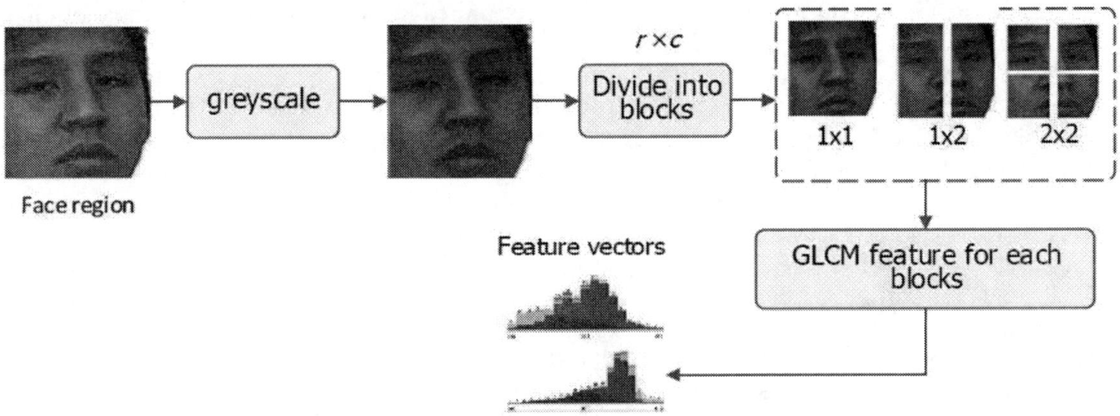

Figure 1. Texture feature using GLCM analysis and face blocks representation

2.1. FacePlace Database

FacePlace was created by the Tarrlab at Brown University. This database includes multiple image for over 200 individuals of many different races with consistent lighting, multiple views, real emotions, and disguises (and some participants returned for a second session several weeks later with a haircut, or new beard, etc). FacePlace is obtained from the Internet that contains four major ethnics such as Asian, Caucasian, African-American, and Hispanic groups. The images from this database were examined through the following stages.

2.2. Face Detection

A face detection is necessary to locate the existence of human in the image. For this purpose, a well-known Viola-Jones face detector was employed because of the face detection algorithm was not require color information to local the face region.

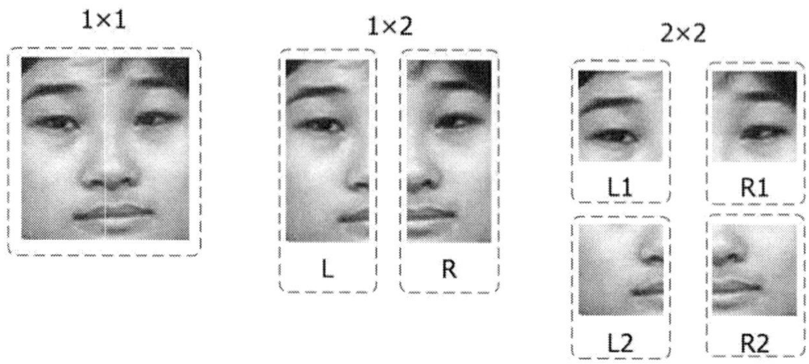

Figure 2. The design of multi-blocks face in three representations

2.3. Feature Extraction with GLCM

In this face, GLCM was employed to access the performance of ethnicity identification using textural features. GLCM also known as gray-level spatial dependence matrix. It calculates the frequency of pair pixels with specific values and in specified spatial relationship occur in the input image. Some of the textural features such as contrast, correlation, energy, and homogeneity are commonly derived to represent statistical texture information. Despite of these features, a total of 22 texture features employed to analyse the performance of the proposed method.

In this phase, three input representation which were applied to form the blocks. Example of blocks can be found in Figure 1 showing the flowchart of extracting texture feature. The detected face region was transformed into grayscale with fix scale for all detected faces. The gray-scale face region was then divided into r×c, where is the horizontal (row) and c is the vertical (column). The first block is the original size 1×1 representation. The second block divided the face region into two separated columns. The last block divided the face region into two row and two columns. Therefore, each sub-block in each blocks were calculated to extract the texture feature. The feature space for this study is the total of GLCM feature (22)×r×c. Labels in Figure 2 L1and R1is the first row, while L2 and R2 is the second row. Each of blocks represent different characteristic of ethnic groups. The 1×1 utilised the original gray-scale format as the global feature. Meanwhile 1×2 slicing the gray-scale face into two sub blocks. The 1×2 illustrated two sub blocks that slicing the left and right eye independently. The nose and mouth are separated each other. On the other hands, 2×2 slicing into more blocks which is four sub blocks. Each sub blocks in L1, R1, L2 and R2 have independently separated each other. L1andR1in the first row holds the eyes and eyebrow of left and right face view. L2 and R2 holds half of nose and mouth. Both nose and mouth are calculated independently. Therefore, each of three introduced blocks might characterised the ethnic group differently according to the texture appearance of sub blocks.

2.4. Classification Algorithms

During the experimental stage, there were five different classification algorithms were employed which are Naive Bayes, BayesNet, *k*-NN, Random Forest and MLP. Results of the study are presented in the following section.

3. Results and Discussion

This section described the findings and discussion on texture feature of human face using GLCM measures. There are three parts in this section where the results are based on the three types of input representation (face blocks) of 1×1, 1×2, and 2×2. The face blocks were designed to search for the best performance using GLCM measures. The results of texture-based feature are described in the following sections.

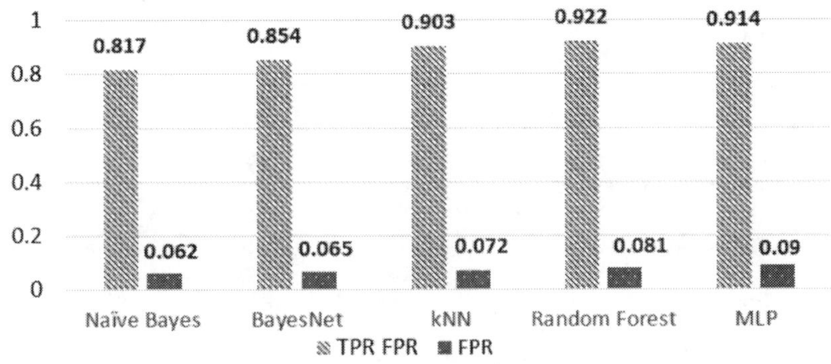

(a)

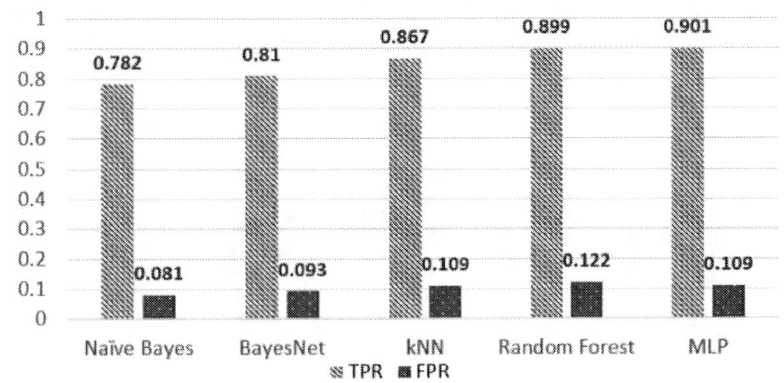

(b)

Figure 3. Comparison of TPR and FPR in texture-based 1×1 input face block using different classification algorithms in (a) TrainSet and (b) TestSet

3.1. GLCM Texture Feature with 1×1 Face Blocks

The texture feature for 1×1 face block was obtained from the original human face (the gray-scale input). The 1×1 face block produced a total of 22 texture features. Figure 3 present the train and test set results of ethnic identification based in the 1×1 face block using different classifiers.

In train set as shown in Figure 4, Random Forest presented the highest TPR with 0.922 and FPR 0.081. It is then followed by MLP with TPR 0.914, *k*-NN with TPR 0.603. In addition to the analysis, Figure 3(b) illustrates the comparison of testset which was also tested using different classifiers. Based on the Figure 5, MLP has the superiority with TPR 0.901 and FPR 0.109 compared to Random Forest which conquered in train set. According to the results collected in both train and test set, Random Forest and MLP showed a good performance with almost equal to TPR. Even though Random Forest and MLP produced the highest overall accuracy, they suffer in identifying Hispanic group successfully. Majority of the Hispanic ethnic was identified as Asian and Caucasian ethnic. Meanwhile, MLP identified all Hispanic as Caucasian ethnic. The findings show that the Hispanic could be having a similar characteristic as in Asian and Caucasian feature. Weak correlation in characterizing the Hispanic uniquely is the factor that Hispanic ethnic is challenging to identified.

3.2. GLCM Texture Feature with 1×2 Face Blocks

The texture feature further was investigated using 1×2 face blocks. This block was formed by dividing the human face region onto 1 row and 2 columns. Then, each subblocks was exploited to calculate the GLCM features. Hence, a total of 44 features were obtained using 1×2 face blocks. Figure 4 showed the results of TPR and FPR in train and test set data. According to Figure 4(a), the train set data achieved the highest TPR by Random Forest classifier with 0.930, followed by MLP with TPR 0.919, and k-NN with 0.914. Meanwhile, the test set in Figure 4(b), the Random Forest also revealed the superiority by producing TPR 0.899. Then, it was followed by MLP and k-NN respectively with TPR 0.869 and 0.857. The Naive Bayes and BayesNet produced TPR with less than 0.8.

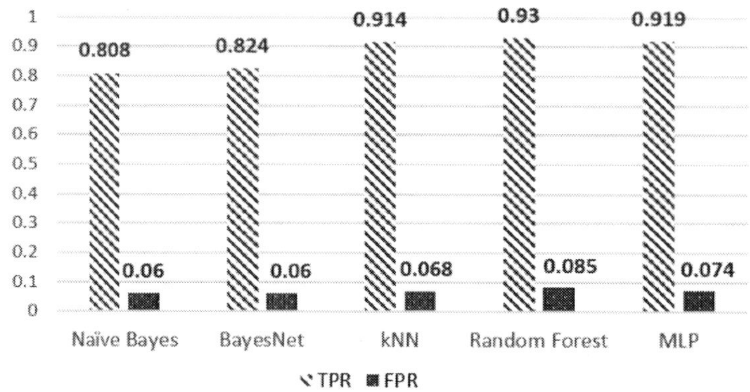

(a)

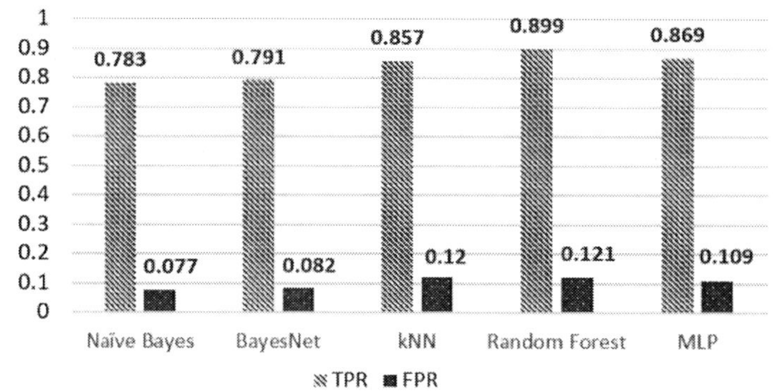

(b)

Figure 4. Comparison of TPR and FPR in texture-based 1×2 input face block using different classification algorithms in (a) TrainSet and (b) TestSet

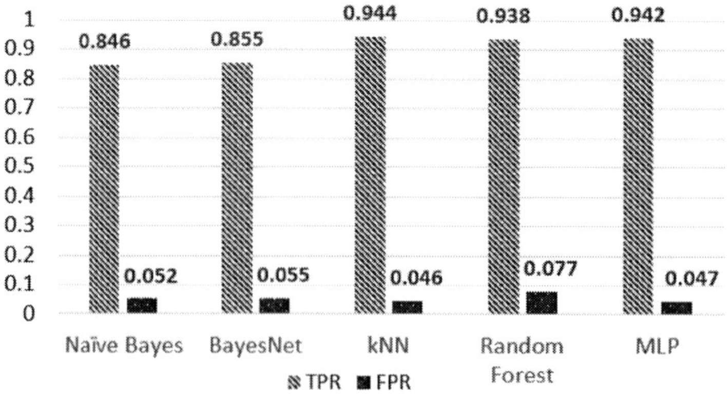

(a)

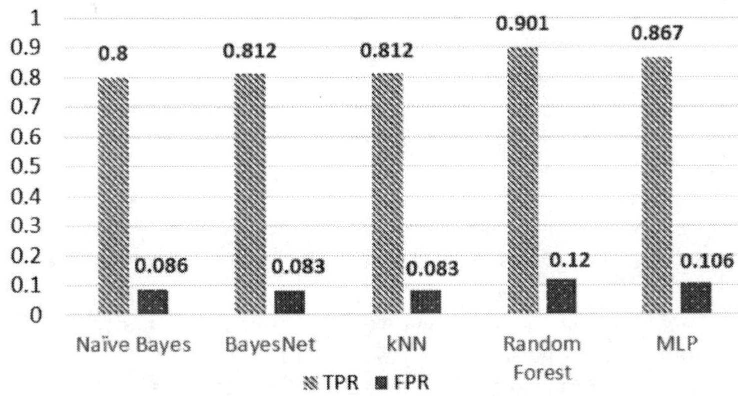

(b)

Figure 5. Comparison of TPR and FPR in texture-based 2×2 input face block using different classification algorithms in (a) TrainSet and (b) TestSet

	Asian	Caucasian	African	Hispanic
Asian	145	1	1	0
Caucasian	4	366	10	19
African	0	20	60	10
Hispanic	0	23	0	5

True Positive Rate for MLP classification in each ethnic.

Asian: 0.986
Caucasian: 0.917
African: 0.667
Hispanic: 0.179

Figure 6. Confusion matrix results texture-based 2×2 input face block using MLP classifier

3.3. GLCM Texture Feature with 2×2 Face Blocks

This block was formed by dividing the human face region onto 2 rows and 2 columns. Then, each sub blocks exploited to calculate the GLCM feature. Hence, a total of 88 features were obtained using 2×2 face blocks.

Figure 5 showed the results of TPR and FPR for train and test using different classification algorithms. In the train set as displayed in Figure 5(b), k-NN classifier has the highest TPR with 0.944, followed by the comparable TPR using MLP with 0.942, and Random Forest with TPR 0.938. The Random Forest classifier in Figure 5(b), surpass the MLP and k-NN classifiers. Despite TPR and FPR results, the confusion matrix shown in Figure 6 indicates MLP classifier fails to successfully to identify correct Hispanic group. However, the Hispanic group mostly was identified in Caucasian ethnic group.

4. Conclusion

As a conclusion, we have presented an experimental result using different classification algorithm under three face blocks and GLCM texture feature. From the experimental, we can see the accuracy of ethnicity identification was influenced by classifier and type of face block used. Four ethnics which are Asian, African American, Caucasian and Hispanic were used. Based on the findings, the proposed method performs well when employing 2×2 face block and using Random Forest classifier in the test set. However, the results also revealed that, the proposed method sometimes difficult to identify Hispanic class successfully. Hispanic mostly was wrongly identified as Caucasian ethnic group. This might due to close interrelation feature class between Hispanic and Caucasian ethnic group. Therefore, hybrid or fusion feature from relevant color or texture feature would help to reduce the gaps. For future enhancement, feature selection will be implemented in texture-based feature to reduce the dimension of feature vector. Other than that, deep learning method also one of potential advancement can be made for ethnicity identification.

Acknowledgement

We would like to show our gratitude to Universiti Malaysia Pahang (RDU vote numberRDU1803142) for supporting this research.

References

[1] Peoples, J. and G. Bailey, Humanity: An introduction to cultural anthropology. 2011: Cengage Learning.
[2] Anwar, I. and N.U. Islam, Learned features are better for ethnicity classification. Cybernetics and Information Technologies, 2017. 17(3): p. 152-164.
[3] Hosoi, S., E. Takikawa, and M. Kawade. Ethnicity estimation with facial images. in Automatic Face and Gesture Recognition, 2004. Proceedings. Sixth IEEE International Conference on. 2004. IEEE.
[4] Stavrianos, C., Facial Anatomy and Mapping Across Races' C. Stavrianos," C. Papadopoulos," O. Pantelidou,"I. Emmanouil and "N. Petalotis" Department of Endodology (Forensic Odontology),"Department of Removable Prosthodontics. 2012.
[5] Xiao, K., et al., Characterising the variations in ethnic skin colours: a new calibrated data base for human skin. Skin Research and Technology, 2017. 23(1): p. 21-29.
[6] Duan, X.-d., et al. Ethnic features extraction and recognition of human faces. in Advanced Computer Control (ICACC), 2010 2nd International Conference on. 2010. IEEE.
[7] Fu, S., H. He, and Z.-G. Hou, Learning race from face: A survey. IEEE transactions on pattern analysis and machine intelligence, 2014. 36(12): p. 2483-2509.
[8] Ding, H., et al. Facial ethnicity classification based on boosted local texture and shape descriptions. in Automatic Face and Gesture Recognition (FG), on 10th IEEE International Conference and Workshops. 2013. IEEE.
[9] Zarei, A. and D. Mou. Artificial neural network for prediction of ethnicity based on iris texture. 2012. IEEE.
[10] Bekhouche, S.E., et al., Pyramid multi-level features for facial demographic estimation. Expert Systems with Applications, 2017. 80: p. 297-310.
[11] Klare, B. and A.K. Jain. On a taxonomy of facial features. 2010 Fourth IEEE International Conference on Biometrics: Theory, Applications and Systems (BTAS). 2010. IEEE.
[12] Muhammad, G., et al. Race recognition from face images using Weber local descriptor. 2012. IEEE.
[13] Chen, H., et al., A Novel Race Classification Method Based on Periocular Features Fusion. International Journal of Pattern Recognition and Artificial Intelligence, 2017. 31(08): p. 1750026.
[14] Roomi, S.M.M., et al. Race classification based on facial features. in Computer Vision, Pattern Recognition, Image Processing and Graphics (NCVPRIPG), 2011 Third National Conference on. 2011. IEEE.
[15] Bougourzi, F., et al., A comparative study on textures descriptors in facial gender classification. 2017.
[16] Zighem, M., et al., Age Estimation Based on Color Facial Texture. 2017.

IOP Conf. Series: Materials Science and Engineering **769** (2020) 012033 doi:10.1088/1757-899X/769/1/012033

GbLN-PSO Algorithm for Indoor Localization in Wireless Sensor Network

M. Shahkhir Mozamir[1], Rohani Binti Abu Bakar[1], Wan Isni Soffiah Wan Din[2], Zalili Musa[1]

[1] Soft Computing and Intelligent System Research Group, Faculty of Computing, Universiti Malaysia Pahang, Kuantan, Pahang, Malaysia
[2] System Network and Security Research Group, Faculty of Computing, Universiti Malaysia Pahang, Kuantan, Pahang, Malaysia

Abstract. Localization is one of the issue in Wireless Sensor Network (WSN). The problem is the estimation error in the localization of the node. In this paper, we propose an indoor localization in WSN based on Global best Local Neighbourhood Particle Swarm Optimization (GbLN-PSO) algorithm. Three anchors are used as the distance measurement for each unknown node to be localized. A comparison of the performances of PSO and GbLN-PSO in terms of localization error (accuracy) and computation time is presenting using simulation in Matlab. The result shows that GbLN-PSO improved the accuracy approximately by 0.01% and reduced the computational time by 86.02% compared to PSO technique.

1. Introduction

Wireless Sensor Network raised quickly in distinctive area to monitor an application [1]. Two significant parts in WSN are unknown node and anchor. Unknown node has no information of their location, while anchor node having the information of location. To know the node position, each unknown node must have a process of estimation the location known as localization [2]. Localization is an important matter in area of WSN because the sensed data that has been detected by the node is purposeless if the information of the node position is unknown. Sensor nodes that have their location will conduct to increase the efficiency in power saving, routing and efficiency in collecting the source of the sensed information. The traditional technique used can solved the localization problem, but the estimated location is not accurate. Accuracy shows the reading how closed the truth node location and estimated node location. Knowing an unknown node position with high accuracy will lead the information to be delivered efficiency. The main problem in most current localization techniques is to achieve the high accuracy. The accuracy can be determined by calculating the percentage of localization error. The higher localization error gained; the lower accuracy will be achieved and vice versa. If the node position is wrongly estimated, the accuracy of the node will be reduced [3].

In the localization process, there are two types which are outdoor and indoor. The famous outdoor localization tool is Global Positioning System (GPS). GPS works by gaining minimum three satellites signals from the telecommunication station and through the process of localization. The outdoor localization has been used widely for the industrial applications of transportation, handling of the forest fires, logistical application and etc. [4]. Meanwhile the indoor localization is utilized to provide exact location of unknown node for indoor environment such as hospital, universities, shopping complex and etc. The vital challenges for indoor localization are need to face many obstacles that blocked the transfer of signals between receivers and emitters and a broad variety of shapes, sizes and materials that highly affect propagation of the signal more than in outdoor environment.

Tools of localization like an installation of GPS into sensor node, it may help to estimate the location. However, GPS still not a decent choice because the installation cost of the GPS at each of the node is super costly. Besides that, the node's lifetime will be shorted because GPS consume more power consumption from the battery to receives a signal and execute the localization process. In additional, GPS signal from satellites is powerless to pass through the buildings and it makes GPS tool is not a best choice for indoor localization. Because of that, different localization technique is required to take the place of the GPS for indoor localization [5].

Ranging phase is the approach that can be used in measuring the range between nodes. The ranging process will take the measurement angle or the calculation of the signal so that, it will produce a distance between two nodes. In the ranging phase, there are two versions which are range free and range based. There are four techniques that can be used in range based which are the Angle of Arrival (AOA), Time Distance of Arrival (TDOA), Time of Arrival (TOA) and Receive Signal Strength Indicator (RSSI) [6]. Besides that, there are five techniques in range free which are Approximate Point in Triangle (APIT), Hop-Count-Based localization, Multi-Hop, Centroid and Gradient [7]. The hybrid localization technique is determining as combination of two or more different techniques such as combine RSSI and AOA [8].

The rest of this paper are arranged as followed, the details about PSO and Variants PSO has been described in Section 2. Section 3 will be discussed about the proposed algorithm (GbLN-PSO), and Section 4 is the experiment setup. Section 5 is result and discussion, Section 6 will present the conclusion and potential area of improvement for research in future.

2. Related Works

At the initial of PSO study, many researchers are focussing on solving the optimization problems for a static environment. Wachowiak *et al.* address the constrained Economic Load Dispatch (ELD) problems and the author applied an Adaptive particle swarm optimization method to solve the problems [9]. Furthermore, Centripetal Accelerated Particle Swarm Optimization (CAPSO) has been introduced by Tareq *et al.* to increase the performance of particle in PSO algorithm by running the experiment of accuracy, convergence speed and global optimality searching algorithm [10]. This experiment was running by using real and binary search spaces are considered in the test space environment.

Nowadays, the researcher has made an improvement to the PSO algorithm in order to overcome the problems in dynamic environment. Stojkoska *et al.* introduced Multidimensional particle swarm optimization in dynamic environments to overcome premature particle convergence problem and speed up the convergence particle to the global optimum method by allowing each particle to converge at a different optimum [11].

Zhang *et al.,* has proposed the distance error from actual location and the estimation location of node, by employing the PSO as the optimization problem [8]. Viet Duc Le *et al.* has introduced PSO algorithm for distribution of localization in WSN to calculate the location of unknown nodes depends on the position data of anchors [12]. The PSO algorithm work by minimizing the localization error. It will enforce many communications and it can lead to take slowly, exhaustion and congestions of energy.

Different to Kulkarni *et al.*, recommended iterative distribution of node localization using PSO algorithm [13] where the set of Unknown node receives three or many anchors in their communication range and perform PSO algorithm with minimizing the localization error. This method makes each of the nodes unnecessary to transmit their position information to a central node in order to run the localization process. Moreover, all the node will be localized as long as the node in the range of their communication signal. However, using PSO algorithm for localization purposed need to consider the time complexity because higher computation time will influent higher computation energy.

The methods discussed above show that the localization process still have their weaknesses, especially the exploitation and exploration of particles in a search space. The particles are sometimes it does not work out the optimal solution, but its converging near to optimum peak. For that reason, this paper proposed the Global best Local Neighbourhood in Particle Swarm Optimization (GbLN-PSO) algorithm to overcome the problems of the unknown node localization.

3. GbLN-PSO algorithm

3.1. Global best Local Neighbourhood Particle Swarm Optimization

Various of applications using Computational Intelligence in WSN can be found such as Defence Advanced Research Projects Agency (DARPA) for military, Automated Local Evaluation in Real Time (ALERT) for environmental, and etc. PSO have been implement in various of filed such as object tracking, Wireless Network and detection problem.

In traditional PSO, each of the particles has the distance from the target location and it is represented as fitness or objective function for that problem. Every particle possesses its own *Pbest* (particle best as the best location itself), and in between all the particles, there will be a particle whose located closed to the target that called as *Gbest* (global best as the best among the particle location). In PSO algorithm, a set of particles will move quicker toward the way of the particle that have lowest value objective function. In each iteration, each particle's position and velocity are being updated and keep searching a new location. It is iterating continuously until the condition is satisfied.

$$d_i = \sqrt{(x-x_i)^2 + (y-y_i)^2} \tag{1}$$

$$f(x,y) = \frac{1}{M}\sum_{i=1}^{M}(\sqrt{(x-x_i)^2 - (y-y_i)^2} - \hat{d}_i)^2 \tag{2}$$

In Equation (1), *di* is the actual distance which is Euclidean distance. According to (1), the location of unknown nodes represents as (x, y). While (x_i, y_i) is represented as the position of ith anchor. The algorithm will find the coordinates of (x, y) by minimizing the fitness function. Where $f(x, y)$ represents as localization error such as in Equation (2) that used to calculate the error between estimated distance and actual distance of the nodes. After the estimation node are calculated, then the value will be used to generate the fitness function shown in Equation (3):

$$f_i(\hat{x},\hat{y}) = d_i - \sqrt{(x_i-\hat{x})^2 + (y_i-\hat{y})^2} \tag{3}$$

$(\hat{x},\hat{y})$ is the estimated coordinate of unknown node, (x_i, y_i) where $(i = 1,2,\ldots,M)$ is the actual coordinate of the ith anchor node. $(x_i-\hat{x})^2 + (y_i-\hat{y})^2$ is the distance between the ith anchor node and the unknown node, where $f_i(\hat{x},\hat{y})$ is the error value between measured and estimated distance of the ith anchor node and unknown node. d_i is measured distance between anchor and unknown node as shown in Equation (1). After the ranging phase is completed, the estimation phase will take place the initializing the location of the particles. Distribution of the particles in the area of the network size or search space.

IOP Conf. Series: Materials Science and Engineering **769** (2020) 012033 doi:10.1088/1757-899X/769/1/012033

The pseudocode of GbLN-PSO is given in Figure 1. In GbLN-PSO, the basic PSO algorithm has been added with another processes as shown in line 18 to 27 where every particle is compulsory to generate a small population of their Neighbourhoods as a virtual particle, and distributed along the location of the particle move to new location. The Neighbourhood particle generates using the Equation (4) as shown below:

$$l_i' = N(-a,a) + l_i \tag{4}$$

Where

$$a = x_i - x_{i-1}$$

$N(-a,a)$ is the random number to generate the neighbourhood particle. Then, the neighbourhood particle will calculate fitness and local best of every of particle. If the fitness of neighbourhood is better than local best particle $(fl_i' < fl_i)$, the fitness neighbourhood and the location particle will replace the local best and fitness value of local best, $fl_i = fl_i'$ and $l_i = l_i'$ respectively. Process of evaluating local best neighbourhood particles will carry out to evaluate global best. From Zalili *et al.* [21], this idea made particle more comprehensive in the searching process. The neighbourhood particle will make sure the particle not only focus on certain area. This process also controls the movement of particle to being converge early and let the particle to search along the same direction of global best result.

3.2. Communication Model

In this paper, Received Signal Strength Indication (RSSI) is implement to determine the distance from the anchor to unknown node by using as in Equation (5). The RSSI technique provides the mechanism to calculate the distance between sensor nodes. An unknown node calculates the distances from each of the anchors when it received signals from three anchors. From Equation (5), P is the power that received at reference distance, D_0, n is the path loss index, D is the nodes distance and $X\sigma$ is the noise parameter with zero mean Gaussian random variable.

$$RSSI = P - 10\eta \times (\frac{\log D}{D_0}) + X_\sigma \tag{5}$$

Algorithm 1 GbLN-PSO Algorithm	
	10. For $i = 1 : \mathrm{I}$
i = number of particles where $i = \{1,...,\mathrm{I}\}$	11. If $(f_{l_i} < f_i)$ then
	12. $f_{l_i} = f_i$
j = integer number where $j = \{1,...,\mathrm{J}\}$	13. $l_i = x_i$
x_i = location of particle i	14. End if
V_i = velocity of particle i	15. End loop
f_i =fitness value of particle i	16. End if
	17. For $i = 1 : \mathrm{I}$

g = global best
f_g =fitness value of g
l_i = local best of particle i
f_{l_i} = fitness value of l_i
l_i' =local best neighbourhood particle i
f_{l_i}' = fitness value of l_i'
w =weight
c =constant value
r =random value
S =search space
R =size of network

Step1: Initialization location of the particles

$x_i \sim N(0,1) * R$

Step 2: Evaluation of the population

For $n = 1 : N$
  Calculate $f_i \forall i$ in Equation (3)
  If $(t == 1)$
    For $i = 1 : \mathrm{I}$
      $l_i = x_i$
      $f_{l_i} = f_i$
    End loop
  Else

For $j = 1 : \mathrm{J}$
  $l_i' = N(-a, a) + l_i$
  Calculate f_{l_i}'
  If $(f_{l_i}' < f_{l_i})$ then
    $f_{l_i} = f_{l_i}'$
    $l_i = l_i'$
  End if
End loop
End loop
For $i = 1 : \mathrm{I}$
  If $(f_g < f_{l_i})$ then
    $f_g = f_{l_i}$
    $g = l_i$
  End if
End loop

Step 3: Updating of the particles

For $i = 1 : \mathrm{I}$
  $V_i = wV_i + c_1 r_1 (l_i - x_i) + c_2 r_2 (g - x_i)$
  $x_i = V_i + x_i$
End loop
End loop

Figure. 1 GbLN-PSO Algorithm

4. Experiment Setup

In order to evaluate the GbLN-PSO algorithm performance in WSN, we simulated the experiment using Matlab 2014 software and Intel(R) Core(TM) i7-2600 CPU @ 3.40 GHz. The parameters that we are applied for localization as shown in Table I. The parameters for PSO and GbLN-PSO is listed in Table II and the number of particle both PSO and GbLN-PSO show in Table III. In this experiment, the simulation has been run for 40 unknown nodes based on [28] and [27]. The weight and coefficient values is selected after the tuning parameter process is conducted.

Table 1. Simulation Parameter for Localization

Parameter	Values
Network Size	100×100 m^2
Anchors Node	3
Unknown Nodes	40

Table 2. Simulation Parameter for PSO and GbLN-PSO

Parameter	Values
Maximum Iteration	50
Weight	0.4
Coefficient	1.3
Random Number	[0,1]
Particle Position	Xmin=0, Xmax=100

Table 3. Number of Particle for PSO and GbLN-PSO

Algorithm	Number of Particle	Neighbor Particle	Total Particle
PSO	20	-	20
GbLN-PSO	5	3	20

The average of the localization error E ,calculated using Equation (6), which is delineate as the total error between truth node locations with the estimated, (x,y) location. Total error , $f(x,y)$ will be divided by total number of the unknown nodes, N.

$$E = \frac{f(x,y)}{N} \tag{6}$$

5. Result and Discussion

Figure 2 shows the comparison of required iterations number to converge on minimum value of unknown node. The PSO algorithms have applied 20 particles and GbLN-PSO algorithm applied 5 particles and 3 neighbourhoods respectively. In GbLN-PSO algorithm, less particle was applying because of existing of neighbourhood particles help the searching process. The convergence of the particle on PSO is fastest then GbLN-PSO. At the number iteration of 5th PSO have converge. It is different to GbLN-PSO, where the particles have been converging at iteration of 10th. GbLN-PSO able to avoid from being in the local optimum. The faster of convergence make the less exploitation of particle in the search area. Based on convergence result, the neighbourhood particles are distributed at the 6th of iteration. This is because at the 6th iteration, the neighbour particle will through the local search to avoid trapped into local optimal.

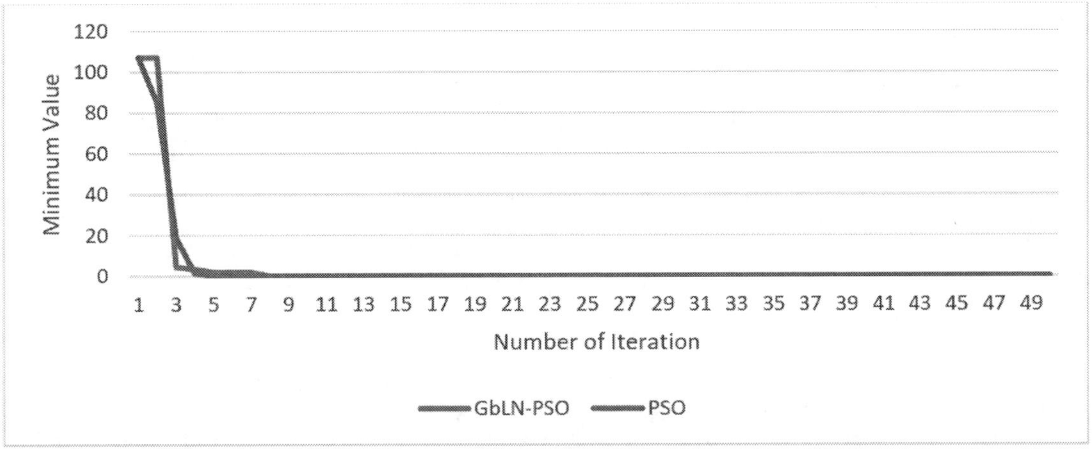

Figure. 2 Comparison numbers of iteration to converge on minimum value of unknown nodes

In this Figure 2 experiment, computational time is also evaluated to differentiation the attainment of the proposed algorithm. The time taken for computation is depends on how long particle take time to find the target, the number of particles. It is also can supposed that the longer computation time brings much energy consumption. Thus, we run the simulation and determine the value localization error and computation time for further analysis. The details of running experiment as shown in Table 4.

Table 4. Simulation Parameter for Localization

Algorithm	Average Localization Error, E_l (m)	Computational Time (s)
PSO	0.449675	213.3863
GbLN-PSO	0.444158	29.82545

From Table 4, PSO algorithm recorded high localization error at 0.449675m compared to GbLN-PSO algorithm is at 0.44960m. This is because GbLN-PSO algorithm has more exploration and exploitation of particle to find the optimum value. GbLN-PSO has an advantage that can avoid the particle from converge early compared to PSO. GbLN-PSO algorithm required less computational time at 29.82545 second compared to PSO which is take 213.3863 second. GbLN-PSO utilized neighbour particles that are scattered in the same iteration of the main particles. This neighbourhood particle features make the particle able to do the exploitation process and with less time taken to complete the process quickly.

In summary, GbLN-PSO able to localized the unknown node coordinates accurately, while PSO do so but with significant high localization error compare to GbLN-PSO. The impact of high localization error makes the accuracy of node position low. The simulation results show that the GbLN-PSO algorithm can minimize the localization error and the computational time also improved significantly compared to PSO. The GbLN-PSO algorithm has great result of the computational time as the time to localized node is decreased, it will save the power consumption. From the experimental result show that GbLN-PSO can achieved lower localization error and lesser computational time. This are two contributions in the unknown node localization in WSN.

6. Conclusion

As the conclusion this paper implemented Global best Local Neighbourhood Particle Swarm Optimization (GbLN-PSO) to solve indoor localization problem in Wireless Sensor Network (WSN). A set of unknown nodes perform localization by estimating the distance from different three anchors. The distance from anchor to unknown node is calculating using Receive Strength Signal Indicator (RSSI). Generally, PSO have converge earlier to find the optimal solution that make PSO trap into the local optimal problem. On the other hands, GbLN-PSO has improved the localization error and computational time. For further research, we suggest to improve the current algorithm (GbLN-PSO) into the mobile node.

Acknowledgments

The authors appreciate to the Ministry of Higher Education Malaysia and Universiti Malaysia Pahang (UMP) for supporting this research. This work is supported by the Postgraduate Research Grants Scheme (PGRS): Indoor Positioning Localization Based on RSSI Exploiting Intelligent Algorithm in Wireless Sensor Network, PGRS190384 Grant from UMP. This work is also financed by Universiti Malaysia Pahang under Master Research Scheme (MRS).

References

[1] Alomari A, Phillips W, Aslam N and Comeau F 2017 Dynamic Fuzzy-Logic Based Path Planning for Mobility-Assisted Localization in Wireless Sensor Networks *Sensors* **17** 1904
[2] Wei M, Chellali R, Yi Y, Wang T and Qin W 2015 Location Characterization in Noisy Range Network Localization 126–32
[3] Bekcibasi U and Tenruh M 2014 Increasing RSSI localization accuracy with distance reference anchor in wireless sensor networks *Acta Polytech. Hungarica* **11** 103–20
[4] Queiroz D V., Alencar M S, Gomes R D, Fonseca I E and Benavente-Peces C 2017 Survey and systematic mapping of industrial Wireless Sensor Networks *J. Netw. Comput. Appl.* **97** 96–125
[5] Sotenga P Z, Djouani K, Kurien A M and Mwila M 2018 Implementation of an indoor localisation algorithm for Internet of Things *Futur. Gener. Comput. Syst.*
[6] Kawahama Y and Katsuma R 2017 High-accuracy localization via measurements of RSSIs and LED light angles for low-cost WMSNs *2017 IEEE 7th Annu. Comput. Commun. Work. Conf. CCWC 2017* 1–7
[7] Kaur A, Kumar P and Gupta G P 2016 A weighted centroid localization algorithm for randomly deployed wireless sensor networks *J. King Saud Univ. - Comput. Inf. Sci.* 1–10
[8] Lin G H, Zhang J and Liu Z H 2018 Hybrid particle swarm optimization with differential evolution for numerical and engineering optimization *Int. J. Autom. Comput.* **15** 103–14
[9] Wachowiak M P, Timson M C and Duval D J 2017 Adaptive particle swarm optimization with heterogeneous multicore parallelism and GPU acceleration *IEEE Trans. Parallel Distrib. Syst.* **28** 2784–93
[10] Aote S S, Raghuwanshi Principal M M and Latesh Malik HOD R 2013 A Brief Review on Particle Swarm Optimization: Limitations & Future Directions *Int. J. Comput. Sci. Eng.* **2** 2319–7323
[11] Risteska Stojkoska B 2014 Nodes Localization in 3D Wireless Sensor Networks Based on Multidimensional Scaling Algorithm *Int. Sch. Res. Not.* **2014** 1–10
[12] Céspedes-mota A, Castañón G, Martínez-herrera A F and Cárdenas-barrón L E 2016 Optimization of the Distribution and Localization of Wireless Sensor Networks Based on Differential Evolution Approach **2016**
[13] Kulkarni R V. and Venayagamoorthy G K 2010 Bio-inspired algorithms for autonomous deployment and localization of sensor nodes *IEEE Trans. Syst. Man Cybern. Part C Appl. Rev.* **40** 663–75

A Comparison of Particle Swarm optimization and Global African Buffalo Optimization

Mohammed Adam Kunna Azrag[1*], Tuty Asmawaty Abdul Kadir[1], and Noorlin Mohd Ali[1]

[1] Faculty of Computing, College of Computing and Applied Sciences, Universiti Malaysia Pahang, 26300 Kuantan Pahang Malaysia

Mohammed87kunna@gmail.com

Abstract. The performance of Particle Swarm Optimization (PSO) brings attention to the field of algorithms when deals with different optimization problems. Due to her simple implementation, small consumption, and very effective in finding a solution in many problems, (PSO) becomes well known to the field of algorithms. In addition, the late proposed algorithms mostly are compared to the well-known algorithm such as PSO. Thus, the Global African Buffalo Optimization (GABO) was proposed lately and yet not been compared to the old well-known algorithms in terms of accuracy and time consumption. However, in this paper, a comparison between Particle Swarm Optimization (PSO) and Global African Buffalo Optimization (GABO) algorithms was performed. Five different nonlinear equations with their upper and lower boundaries values were selected as the test optimization functions problem in addition to PSO was applied to real case study. The experimental results illustrated the differences in the performances of both algorithms toward the optimum solution. At the end of the experiments, the PSO algorithm quickly convergence towards the optimum solution using a few particles and iterations rather than GABO. However, the experimental result showed that PSO achieved good results in all the test cases within a short time. In many cases, PSO and GABO are promising optimization methods.

Keywords: PSO; GABO; Convergence; Sphere Function; Rastrigrin function, Griewank Function; Rosenbrock Function; Shubert Function

1. Introduction

Optimization algorithms are applied in many applications such as business activities, engineering, and industrial designs with the aim of either minimizing or maximizing an objective function by methodologically selecting the input values [1]. Lately, nature-inspired optimization algorithms have increasingly been gaining popularity in the field of science and engineering [2]. This development has thrilled many researchers and they have deduced various reasons for it. Some researchers argue that these algorithms are successful because they were developed to replicate some of the most successful concepts in biological, physical, and chemical processes which occur naturally [3, 4]. With this situation, the issue of algorithmic choice always surfaces (since there exists so many to choose from) whenever the need for optimization arises [5]. There is a general understanding that the 'best' algorithm for solving any problem should mainly be selected based on the nature of the problem being faced. The 'No free lunch' optimization theorem reinforced this line of thought [6, 7]. In fact, there is no agreement on the

recommended principles guiding the choice of algorithms when faced with nonlinear large-scale optimization tasks [8].
In addition, the PSO was developed by Kennedy & Eberhart in 1995 with inspiration from the foraging behavior of birds. Different from evolutionary concept of Genetic algorithm (GA) which is based on Darwin's "survival of the fittest" concept, the PSO finds the optimal solution through collaboration between individuals. The study has found that the PSO is suitable for solving complex optimization problems [9,10]. The PSO algorithm has been greatly received and effectively applied to many function optimization and engineering technologies. It could be said that, through the analysis of biological communities, swarm intelligence of other complex behaviors such as their cooperation and competition can often produce some effective solutions to problems [9,10]. In addition, PSO algorithm has been successful in most cases than other algorithms such as GA, Simulating Annealing (SA), and Differential Evolution [11,12,13].
Thus, in this study, the PSO algorithm was studied based on the time consumption and convergence speed in order to reach global optima [9, 10]. Moreover, the Global African Buffalo Optimization (GABO) algorithm was proposed by [14] with inspiration from the global search behavior of the African Buffalo Optimization (ABO) algorithm [15]. The GABO algorithm showed better performances compared to the original ABO when studied with 2 optimization functions problem. However, this study focuses on the comparison between PSO and GABO algorithms by testing 3 global optimization functions based on the computational time and accuracy of each algorithm. The remaining part of this article is arranged as follows: Section two discussed the PSP and GABO methods while Section three explained the experimental process and the results of the study. Section four presented the conclusion of the study.

2. Methodology

2.1. *Particle Swarm Optimization*

The PSO was first presented by Kennedy & Eberhart [9, 10] as an optimization framework based on inspiration from the social life pattern and movement dynamics of birds, fish, and insects. In the PSO, the swarm is modelled by the particles and each particle has a velocity and position within a multidimensional space. It is used to find the best global position and its information on the best neighbour. When the particle in the swarm stochastically and independently search for their destination, they exchange information with each other, causing their movement towards the optimal point from different directions and paths. Consequently, they will be able to explore a wide search area, giving the best probability of establishing the global optimum [9, 10]. When compared to the other algorithms, one major advantage of PSO is that it is easy to implement because it does not require crossover, decoding, or encoding like the GA. Furthermore, its computational time is cost-efficient when compared to the other algorithms because it uses a small number of parameters. PSO is simple in both numerical and theoretic implementation; its simplicity, robustness, and stability made it suitable for application in several fields, such as electric motor, image processing, and automatic systems [16, 17, 18, 19]. Basically, the PSO is comprised of a swarm of "n" particles whose position individually represents a tentative solution to the fitness function within the dimensional search space. The updating of the particles' position is dependent on the following three factors, which are the particles' inertia weight ω, the local best position $p(t)$, and the global best position $G(t)$. These factors are calculated within the velocity and position of each particle using the following equations:

$$v_i(t+1) = \omega v_i + c_1 r_1 \big(p(t) - x_i(t)\big) + c_2 r_2 (G(t) - x_i(t))$$

$$x_i(t+1) = x_i(t) + v_i(t+1)$$

Where $v_i(t+1)$ is the velocity of particle i at iteration t, ω is the inertia weight equal to 1, $c_1 c_2$ are the acceleration constant, $r_1 r_2$ are a random number between (0,1), $p(t)$ is the local best position already found by particle i until iteration t, $G(t)$ is the global best position found in the entire particle i until iteration t, and $x_i(t)$ is the particle position i at iteration t.

2.2. *Global African Buffalo Optimization*

With the ability of the buffalos to recognize the global best position in a standard ABO algorithm at step 3 [13], the buffalos were made to perform a research of the best position found by the ABO buffalos to ensure that the global best position are the same and as close enough to the global optimum solution as possible. This can be seen in step 4. With this idea, some modifications of the standard ABO algorithm are proposed. The modified algorithm selected the best position which has already been found by the ABO buffalos and then set them as the best position to be tested by considering the lower and the upper values of the herds' best ($\boldsymbol{bgmax}$). After this, the algorithm then starts to search for the optimum global solution. In step 5 if the global optimum solution was found, then ABO move to step 6 if not, it repeats steps 2-5. But Step 6 in GABO is asking to ensure the global best is reached if not repeat step 2-6 (This is the primary difference between ABO and GABO). Step 7 outputs the optimum values which means the best minimum values. In this way, the buffalos can search for more domains. The Pseudocode of the proposed GABO is presented in Figure 1:

1. Initialize the buffalos randomly to nodes within the solution space;
2. Update the exploitation of the buffalos using:
   $mk' = mk + lp1(bg - wk) + lp2(bp.k - wk)$, Where mk and wk represents the exploitation and exploration moves, respectively of the kth buffalo (k=1,2,...,N), $lp1$ and $lp2$ are learning factors, bg is the herds best fitness, and bp is the individual best buffalos locations.
3. Update the exploration fitness of the buffalos using the equation below:
   $mk' = (wk + mk)$.
4. Set the best position bp to be the new dimension of the global best position $bgmax$.
5. Is the global best position found and $bgmax$ is updating? Yes, go to step 6. If no repeat step 1 to 5.
6. If the stopping condition is not reached, revert to step 2, else, proceed to step 6.
7. Result the best solution.

Figure 1: Pseudocode of GABO

The performance of GABO was greatly improved with the introduction of the search for the optimum global solution of the best position that has already been found by buffalos into the original version of ABO. This improvement was evident through the experimental study carried out on the benchmark problem of Sphere and Rosenbrock functions. To further illustrate the relevance of this study, the results of the experiments using the two non-linear tested functions were discussed and reported [9].

3. Experimental Setup

For comparison, five nonlinear functions and one case study was investigated.

3.1. *The Nonlinear functions*

The Sphere function is the first function as described by Eq. 1:

$$f(x) = \sum_{i=1}^{n} x^2 \tag{1}$$

Where $x = [x1, x2, \ldots, xn]$ is an n-dimensional real-valued vector.
The *Rosenbrock* function is the second function as described by Eq. 2:

$$f_1(x) = \sum_{i=1}^{n}(100\left(x_i - x_i^2\right)^2 + (x_i - 1)^2) \tag{2}$$

The generalized *Rastrigrin* function is the third function as described by Eq. 3:

$$f_2(X) = \sum_{i=1}^{n}(X_i^2 - 10\cos(2\pi X_i) + 10) \quad (3)$$

The fourth function is *Griewank* function described by equation (4):

$$f_3(X) = \frac{1}{4000}\sum_{i=1}^{n} X_i^2 - \prod_{i=1}^{n}\cos\left(\frac{X_i}{\sqrt{i}}\right) + 1 \quad (4)$$

The fifth function is *Shubert* function described by equation (5):

$$f_4(x) = (\sum_{i=1}^{5} i\cos((i+1)x_1 + i)(\sum_{i=1}^{5} i\cos((i+1)x_2 + i \quad (5)$$

As recommended by [9], the selection of the upper and lower as the original values was done to facilitate the benchmarking. The initialization of the lower and upper values of the two function is presented in Table 1.

Table 1: The lower and upper values

Function	The lower and upper values
$f(x)$	[-10, 10]
$f_1(x)$	[-5, 10]
$f_2(X)$	[-5.12, 5.12]
$f_3(X)$	[-100,100]
$f_4(X)$	[-10,10]

Studying the PSO and GABO algorithms, different population sizes were used for each function and these population sizes are 10, 20, and 30 particles and buffalos. The maximum number of iteration is set to be 30, 60, and 90 with the dimension is 5 in both algorithms. To get the mean global best position, each algorithm was evaluated 5 times.

3.2. *Case study*

Parameter estimation of the kinetics in metabolic model is difficult task due to this kinetics are reported from different laboratories in different condition, many pathways, and ordinary differential equations (ODE) consuming huge time. Thus, the main metabolic model of *Escherichia Coli* (*E. Coli*) formulated by [20, 22] was used as a case study due to its large scale kinetic parameters and the estimation may can be performed. This model contains 5 pathways with 172 kinetic parameters in addition to metabolites and enzymes.

4. Result

4.1. *The functions result*

The PSO global best position presented a perfect result when compared with the result of GABO. Moreover, as shown in Table 3, 4, 6, 7, 9, and 10, PSO presented a faster convergence speed towards the optimum values compared to GABO. However, PSO's convergence was quick in both functions. PSO required only 0.23 sec to achieve the global best position while 0.95 sec was required by GABO to achieve the same feat in *Sphere* function. In *Rosenbrock* function, PSO required only 0.35 sec to achieve the global best position while GABO required 1.03 sec. In *Rastrigrin* function, PSO required

IOP Conf. Series: Materials Science and Engineering **769** (2020) 012034 doi:10.1088/1757-899X/769/1/012034

0.41 sec to achieve the global best position while GABO required 0.92 sec. These comparisons are depicted below. In addition, the PSO algorithm searching wide area in all cases than GABO with 3100 and 2709 iteration respectively.
Table 2 presents the testing of the *Sphere* function using PSO & GABO on a 5-dimensional space with (10, 20, 30) particles and buffalos, and (30, 60, 90) iterations. Table 2 showed that PSO achieved a good global best position. In Table 3 & 4 below, the time consumption of *Sphere* function by PSO and GABO is presented. The PSO searching 3100 times in the *Sphere* function within 0.23s means that searching large space with a short time to ensure the optimum solution is given. While GABO searching 2709 times in the Sphere function within 2.42s.

Table 2: PSO & GABO results for *Sphere* function

Particles & buffalos size	Dimension	Iteration	PSO global best position	GABO global best position
10	5	30	2.2569e-5	1.2630e-4
		60	3.2074e-6	2.2053e-4
		90	2.7257e-7	2.0382e-5
20	5	30	5.2174e-7	2.2846e-4
		60	3.5328e-8	2.0258e-5
		90	2.0172e-9	3.0124e-3
30	5	30	3.2538e-9	2.5402e-5
		60	4.2860e-10	3.0245e-4
		90	4.2584e-11	3.5682e-6

Table 3: Time consumption of *Sphere* function by PSO

Function name	Calls	Total time	Self-time
PSO	1	1.169s	0.23s
Sphere	3100	1.146s	1.146s

Table 4: Time consumption of *Sphere* function by GABO

Function name	Calls	Total time	Self-time
GABO	1	3.37s	0.95s
Sphere	2709	2.42s	2.42s

Table 5 presented the testing of the *Rosenbrock* function using PSO & GABO. The parameters used are: 5-dimensional space with 10, 20, 30 particles & buffalos, and 30, 60, 90 iterations. There was improvement in the global best position of PSO compared to that of GABO (Table 5). Furthermore, PSO presented a faster convergence speed toward the optimum compared to GABO (Tables 6 & 7). However, it was observed that the PSO convergences quickly as it achieved the global best position within 0.035 sec while 1.03 sec was required for GABO to achieve a similar feat (Tables 6 & 7). Finally, the total time performance speed of the PSO and GABO algorithm is 1.607s and 2.59s respectively with 90 iterations.

Table 5: PSO & GABO results for *Rosenbrock* function

Particles & buffalos size	Dimension	Iteration	PSO global best position	GABO global best position
10	5	30	0	0.402
		60	0	0.375
		90	0	0.324
20	5	30	0	0.321
		60	0	0.284
		90	0	0.247
30	5	30	0	0.235
		60	0	0.215
		90	0	0.145

Table 6: Time consumption of *Rosenbrock* by PSO

Function name	Calls	Total time	Self-time
PSO	1	1.607s	0.035s
Rosenbrock	3100	1.572s	1.572s

Table 7: Time consumption of *Rosenbrock* function by GABO

Function name	Calls	Total time	Self-time
GABO	1	2.59s	1.03s
Rosenbrock	2709	1.56s	1.56s

In Table 8 below, the *Rastrigrin* function was tested using PSO and GABO. The parameters used are: 5-dimensional space with 10, 20, 30 particles & buffalos, and 30, 60, 90 iterations. It was discovered that the global best position of PSO improved better than the result of GABO as seen below in Table 5. Moreover, PSO's convergence towards the optimum was faster compared to GABO as presented in Tables 9 & 10. PSO took only 0.041 sec to reach the best global position while GABO required 0.92 sec to achieve the same feat (Tables 9 & 10). Finally, the total time performance speed of the PSO and GABO algorithm is 1.184s and 2.74s respectively with 90 iterations.

Table 8: PSO & GABO results for *Rastrigrin* function

Particles & buffalos size	Dimension	Iteration	PSO global best position	GABO global best position
10	5	30	2.3891e-2	1.0325
		60	2.2357e-4	1.0213
		90	3.6824e-2	1.0105
20	5	30	3.0217e-3	1.0023
		60	4.2017e-2	0.9982
		90	4.8170e-4	0.9407
30	5	30	1.0843e-5	2.5402e-5
		60	3.0281e-6	3.0245e-4
		90	4.9107e-7	3.5682e-6

Table 9: Time consumption of *Rastrigrin* by PSO

Function name	Calls	Total time	Self-time
PSO	1	1.184s	0.041s
Rastrigrin	3100	1.143s	1.143s

Table 10: Time consumption of *Rastrigrin* function by GABO

Function name	Calls	Total time	Self-time
GABO	1	2.74s	0.92s
Rastrigrin	2709	1.82s	1.82s

In Table 11 below, the *Griewank* function was tested using PSO and GABO. The parameters used are: 5-dimensional space with 10, 20, 30 particles & buffalos, and 30, 60, 90 iterations. It was discovered that the global best position of PSO improved better than the result of GABO as seen below in Table 5. PSO's convergence towards the optimum was faster compared to GABO (Tables 12 & 13). PSO required 0.051 sec to achieve the best global position while GABO took 0.71 sec to achieve the same feat (Tables 12 & 13). The total time performance speed of the PSO and GABO algorithm is 1.702 sec and 2.64 sec, respectively with 90 iterations.

Table 11: PSO & GABO results for *Griewank* function

Particles & buffalos size	Dimension	Iteration	PSO global best position	GABO global best position
10	5	30	0.00230	0.903
		60	0.00120	0.725
		90	0.00020	0.587
20	5	30	0.00040	0.504
		60	0.00051	0.358
		90	0.00064	0.179
30	5	30	1.4027e-3	0.165
		60	1.0004e-4	0.084
		90	2.2048e-5	0.067

Table 12: Time consumption of *Griewank* by PSO

Function name	Calls	Total time	Self-time
PSO	1	1.702s	0.051s
Griewank	3100	1.651s	1.651s

Table 13: Time consumption of *Griewank* function by GABO

Function name	Calls	Total time	Self-time
GABO	1	2.64s	0.71s
Griewank	2709	1.93s	1.93s

In Table 14 below, the *Shubert* function was tested using PSO and GABO. The parameters used are: 5-dimensional space with 10, 20, 30 particles & buffalos, and 30, 60, 90 iterations. It was discovered that the global best position of PSO improved better than the result of GABO as seen below in Table 5. Moreover, PSO's convergence towards the optimum was faster compared to GABO as presented in Tables 15 and 16. It took PSO 0.055 sec to achieve the global best position while GABO took 0.86 sec (Tables 15 & 16). The total time performance speed of the PSO and GABO algorithm is 1.326s and 2.79s respectively with 90 iterations.

Table 14: PSO & GABO results for *Shubert* function

Particles & buffalos size	Dimension	Iteration	PSO global best position	GABO global best position
10	5	30	0.0231	0.9237
		60	0.0095	0.8021
		90	0.0082	0.7292
20	5	30	0.0078	0.7023
		60	0.0064	0.6087
		90	0.0052	0.5742
30	5	30	0.0045	0.5247
		60	0.0031	0.4213
		90	0.0022	0.3214

Table 15: Time consumption of *Shubert* by PSO

Function name	Calls	Total time	Self-time
PSO	1	1.326s	0.055s
Shubert	3100	1.271s	1.271s

Table 16: Time consumption of *Shubert* function by GABO

Function name	Calls	Total time	Self-time
GABO	1	2.79s	0.86s
Shubert	2709	1.93s	1.93s

4.2. The estimation result

As it can be seen in the results function benchmark above, the PSO algorithm performed better than GABO algorithm in short time and searching wide area when testing five nonlinear functions. Thus, PSO was used to estimate 7 kinetic parameters of the main metabolic model of *E. coli* [20], were 8 metabolites of 12 experimental data [21] are optimized and described in Table 17 & 18 respectively. However, PSO algorithm achieved 29.36% distance minimization [22].

Table 17: Kinetic parameters estimation by PSO

Kinetics	**Original**	**lower**	**Upper**	**Kinetic estimation by PSO**
v_{max}^{pyk}	1.085	0.9	1.34	1.032
n_{pk}	3	2.5	3.25	2.647
icdh	24.421	23.9	24.6	24.306
k_{icdh}^{f}	289800	289799.4	289800.7	289799.65
$k_{icdhnadp}^{d}$	0.006	0.004	0.04	0.0372
$k_{icdhnadp}^{m}$	0.017	0.009	0.05	0.0482
v_{max}^{icl}	3.8315	3.3315	4.1	3.594

In Table 18 below, 8 metabolites of $(Glc, G6P, F6P, PEP, 6PG, Ru5P, Xu5P,$ and $E4P)$ were optimized well and the rest of the others metabolites are not optimized due to the nonlinearity of the model, the pathways, there is sum metabolites was lumped, and many kinetics reported or estimated from different laboratories in different conditions.

Table 18: The metabolites optimization

Metabolites	**Chassagnole**	**Kadir simulation**	**PSO Optimization**
Glc	0.0556	0.12203	0.1155
$G6P$	3.48	0.12989	0.21931
$F6P$	0.6	0.021457	0.022598
FDP	0.272	1.5186	2.5257
PEP	2.67	1.5076	1.9136
PYR	2.67	2.8279	3.1883
$6PG$	0.808	0.017854	0.01876
$Ru5P$	0.111	0.021398	0.022488
$Xu5P$	0.138	0.026516	0.027911
$S7P$	0.276	0.00473	0.0473
$R5P$	0.398	0.076388	0.027912
$E4P$	0.098	0.27837	0.003424
Distance minimization	0		29.36%

5. Conclusion

The performance of PSO and GABO algorithms were compared in this study based on selected test functions. The best position of the ABO served as the new search dimension for GABO when searching for the optimum values. PSO was studied to investigate the efficiency of the studied methods; five functions were employed, and one case study was investigated. The experimental outcome pointed towards PSO having a quicker convergence compared to the GABO. Because PSO works better than GABO, the kinetic parameters estimation was studied for PSO based on the main metabolic model of *E. coli*. Thus, PSO algorithm estimated 7 kinetic parameters and optimized 8 metabolites of 12 experimental data sets. Nevertheless, PSO was observed to perform accurately but requires more investigations on more complex multimodal, separable, and non-separable functions with GABO.

ACKNOWLEDGMENT
The authors would like to thank to Malaysia Ministry of Education for funding this work (grant number: FRGS/1/2018/ICT02/UMP/02/8: RDU 190181), all reviewers for the supportive comments and Universiti Malaysia Pahang for the support.

References

[1] Cheng, S., Shi, Y., & Qin, Q. (2015). Experimental Study on Boundary Constraints Handling in Particle Swarm Optimization from a Population Diversity Perspective. In Emerging Research on Swarm Intelligence and Algorithm Optimization (pp. 99-127). IGI Global.

[2] Rauff, J. (2015). Nature-Inspired Optimization Algorithms. Mathematics and Computer Education, 49(3), 208. Fusion, 57(5), 054007.

[3] Gandomi, A. H., & Alavi, A. H. (2012). Krill herd: a new bio-inspired optimization algorithm. Communications in Nonlinear Science and Numerical Simulation, 17(12), 4831-4845.

[4] Priami, C. (2009). Algorithmic systems biology. Communications of the ACM, 52(5), 80-88.

[5] Huang, H.-J., & Lam, W. H. (2002). Modeling and solving the dynamic user equilibrium route and departure time choice problem in network with queues. Transportation Research Part B: Methodological, 36(3), 253-273.

[6] Wolpert, D. H., & Macready, W. G. (1997). No free lunch theorems for optimization. IEEE transactions on evolutionary computation, 1(1), 67-82.

[7] Xu, H., Caramanis, C., & Mannor, S. (2012). Sparse algorithms are not stable: A no-freelunch theorem. IEEE transactions on pattern analysis and machine intelligence, 34(1), 187-193.

[8] Ellison, C. L., Finn, J., Qin, H., & Tang, W. M. (2015). Development of variational guiding center algorithms for parallel calculations in experimental magnetic equilibria. Plasma Physics and Controlled.

[9] Shi, Y., & Eberhart, R. C. (1999). Empirical study of particle swarm optimization. Paper presented at the Evolutionary Computation, 1999. CEC 99. Proceedings of the 1999 Congress on.

[10] Eberhart, Russell, and James Kennedy. "Particle swarm optimization." Proceedings of the IEEE international conference on neural networks. Vol. 4. 1995.

[11] Ou, C., & Lin, W. (2006, June). Comparison between PSO and GA for parameters optimization of PID controller. In *2006 International conference on mechatronics and automation* (pp. 2471-2475). IEEE.

[12] Wang, B., Wu, Z., & Zhao, Z. (2010). Performance comparison of GA, PSO, and DE approaches in estimating low atmospheric refractivity profiles. *Wuhan University Journal of Natural Sciences*, *15*(5), 433-439.

[13] Parwekar, P., Rodda, S., & Mounika, S. V. (2018). Comparison between Genetic Algorithm and PSO for Wireless Sensor Networks. In *Smart Computing and Informatics* (pp. 403-411). Springer, Singapore.

[14] Azrag, M. A. K., Kadir, T. A., Odili, J. B., & Essam, M. H. A. (2017). A Global African Buffalo Optimization. International Journal of Software Engineering & Computer Sciences (IJSECS), 3, 138-145.

[15] Odili, J. B., & Kahar, M. N. M. (2015a). African Buffalo Optimization (ABO): a New MetaHeuristic Algorithm. Journal of Advanced & Applied Sciences, 03(03), 101-106.

[16] Singh, Narinder, Sharandeep Singh, S. B. Singh, and Shelly Arora. "Half mean particle swarm optimization algorithm." Int. J. Sci. Eng. Res 3, no. 8 (2012): 1-9." International Journal of Scientific and Engineering Research (2012): 1-9.

[17] Mukherjee, V., & Ghoshal, S. P. "Intelligent particle swarm optimized fuzzy PID controller for AVR system." Electric Power Systems Research (2007): 1689-1698.

[18] Feng, D., Wenkang, S., Liangzhou, C., Yong, D., & Zhenfu, Z. "Infrared image segmentation with 2-D maximum entropy method based on particle swarm optimization (PSO)." Pattern Recognition Letters (2005): 597–603.

[19] Ghamisi, Pedram, Micael S. Couceiro, Fernando ML Martins, and Jon Atli Benediktsson. "Multilevel image segmentation based on fractional-order Darwinian particle swarm optimization." IEEE Transactions on Geoscience and Remote sensing. IEEE, 2014. 2382-2394.

[20] Kadir, T. A. A., Mannan, A. A., Kierzek, A. M., McFadden, J., & Shimizu, K. (2010). Modeling and simulation of the main metabolism in Escherichia coli and its several single-gene knockout mutants with experimental verification. *Microbial cell factories*, *9*(1), 88.

[21] Chassagnole, Christophe, Naruemol Noisommit-Rizzi, Joachim W. Schmid, Klaus Mauch, and Matthias Reuss. "Dynamic modeling of the central carbon metabolism of Escherichia coli." *Biotechnology and bioengineering* 79, no. 1 (2002): 53-73.

[22] Mohammed Adam Kunna Azrag, Tuty Asmawaty Abdul Kadir, Muhammad Nomani Kabir, and

Aqeel S. Jaber, "Large-Scale Kinetic Parameters Estimation of Metabolic Model of Escherichia Coli," *International Journal of Machine Learning and Computing* vol. 9, no. 2, pp. 160-167, 2019.

Applying Graph-based SLAM Algorithm in a Simulated Environment

Nazrul Azril Badioze Zaman[1], Shuzlina Abdul-Rahman[2], Sofianita Mutalib[3], Mohd Razif Shamsuddin[4]
[2,3,4]Research Initiative Group of Intelligent Systems,
[1,2,3,4]Faculty of Computer and Mathematical Sciences,
Universiti Teknologi MARA, 40450 Shah Alam, Selangor, Malaysia
azril.nazrul@gmail.com, {shuzlina, sofi, razif}@fskm.uitm.edu.my

Abstract. This paper explores the capabilities of a graph optimization-based Simultaneous Localization and Mapping (SLAM) algorithm known as Cartographer in a simulated environment. SLAM refers to the problem in which an agent attempts to determine its location in the immediate environment as well as constructing the map(s) of its environment. SLAM is one of the most important aspects in the implementation of autonomous vehicle. In this paper, we explore the capabilites of the Cartographer algorithm which is based on the newer graph optimization approach in improving SLAM problems. A series of experiments were tested in order to discover its Cartographer capabilities in tackling SLAM problems. Then, we compare the results of Cartographer with Hector SLAM, another graph-based SLAM algorithm. We present the results from the experiments which show some promising findings based on the amount of computer resources used and the quality of the map(s) produced.

Keywords: Autonomous Vehicle, Graph, Optimization, Simultaneous Localization and Mapping

1. Introduction

Autonomous Vehicle (AV) is not a new topic in today's technology nomenclature. Every now and then, news and articles that report on the topic of AV are published by a number of media streams. AVs, as the name implies, are vehicles that can perform driving or navigation tasks independently and move on its own with little to almost no input from a human driver. To achieve a full automation state, an AV must be able to process complex data with various attributes such as high dimensionality and unstructured data coming from various sources [1]. One of the most important criteria is for the vehicle to be able to receive data from its surroundings in order to scan the environment it is currently in. As an AV does not possess any senses such as vision, hearing, smell, taste and touch like humans do, most of the time, the problem with AV development is how the developers want to make the vehicle capable of sensing the environment it is currently situated in and its current position in real time [2]. In navigation, this problem is called Simultaneous Localization and Mapping (SLAM). As the name suggests, SLAM can be split into two parts namely the localization and mapping. Localization helps the agent 'know' and 'aware' of its own position in that particular environment it is currently in at any given time, while mapping helps the agent to construct and update the surrounding map(s) of the environment. Agent in this context includes robots, unmanned and autonomous vehicle which uses SLAM. Essentially, SLAM algorithms were used to help the agent make an estimation of the next step(s) it should take by employing and combining various types of calculations. Over the years, many algorithms have been developed in order to tackle this problem with each one of them varying in the results they are able to produce.

This research made full use of the Simulation Method in order to determine the capabilities of the SLAM algorithm in solving SLAM problems. Simulation method was chosen in order to cope with a number of limitations faced during the execution of this research such as not having a real, working AV or a mobile robot as an AV at our disposal and not having the expertise needed in order to carry out this experiment in the actual environment. Given these limitations, simulation helps in reducing the risks of any unforeseen incidents that may happen if this experiment is carried out in a actual environment [3]. Furthermore, it is also cost-effective, needing only computers to run the simulations instead of using a real car or a mobile robot that is costly, while still be able to deliver the same, if not a better result. This paper presents a case study based on a SLAM algorithm that was experimented in a simulated environment in order to access its capabilities in improving SLAM problems. The case study examines the capabilities of Cartographer algorithm and compared with Hector SLAM algorithm. The remainder of this paper is organized in the following order: Section 2 describes the terminologies that are used in this experiment that is SLAM and Cartographer. Section 3 presents the methodology and the experiments setup. Section 4 presents the results and findings from the experiments undertaken, and Section 5 concludes the paper.

2. Background Study

2.1. Simultaneous Localization and Mapping (SLAM)

Localization refers to the ability of an agent or vehicle to locate and keep track of its position in an unknown environment at any given time, while mapping refers to the process of constructing map(s) of the surroundings. SLAM refers to doing both of these tasks at the same time [4]. SLAM is important in autonomous navigation as it helps the agent to scan the environment it is currently in so that the agent can take the next steps in navigating such as the path-planning and decision-making [5][6]. As a result, SLAM enables the agent to move in an environment that it has no prior knowledge [7].

Over the years, a number of algorithms have been developed by researchers in order to solve SLAM problems based on the sensors that the agent used to scan the environment. At its core, SLAM can be categorized into two categories namely SLAM algorithm that uses particle filter and graph optimization. Particle usage has been proven essentially to solve various types of optimization, path finding and searching problems such as vehicle routing problem, vehicle assignment problem and others [8][9]. The use of particle filter for SLAM problems was first introduced in the 1980s. In the early days, SLAM algorithm that used particle filter estimated the position of an agent with landmarks by filtering out any noises in the environment [10]. With the advancement of technology, new methods were introduced to improve particle filter SLAM by adding features such as using the agent pose and data from the sensors to estimate the position of the agent [11][12]. Some examples of SLAM algorithm that is based on particle filter include Extended Kalman Filter (EKF), Unscented Kalman Filter (UKF), GMapping and FastSLAM [5][13].

Graph optimization is a newer approach to solve SLAM problems. First introduced in the 2000s, graph optimization consists of two processes namely graph construction and optimization [14]. The graph construction responsible for estimating the movement of an agent by constructing the image of the surroundings and detect changes in the image while the optimization process is used to optimize the agent's pose in the current environment. Graph optimization has been improved in recent times by adding sensors such as cameras and lasers to improve the quality of SLAM problems. Examples of graph optimization-based SLAM algorithms include algorithms such as Real-Time Appearance-Based (RTAB-MAP), Stereo Parallel Tracking and Mapping (S-PTAM), ORB SLAM, ORB SLAM2, Hector SLAM, GraphSLAM, LSD-SLAM and Cartographer [15][13]. Figure 1 summarizes an overview of SLAM taxonomy compiled and adapted from various research articles that can be categorized into two main categories namely filter-based and vision-based. Due to the page limitation, this paper is unable to describe this taxonomy in greater detail.

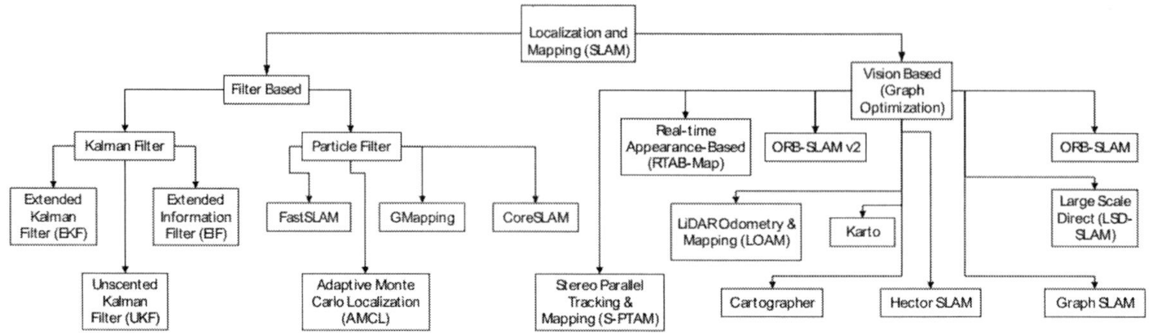

Figure 1. SLAM taxonomy adapted from various research papers

2.2. Cartographer Algorithm
Cartographer is an algorithm developed by Google used to tackle SLAM problems [16]. Cartographer utilizes laser sensors such as the Light Detection and Ranging (LiDAR) to recognize the environment instead of cameras. As stated by the developers, Cartographer does not use particle filters, which allows it to use less resources compared to other algorithms. In order to overcome error accumulation from the lack of any filtering, Cartographer uses pose optimization approach. In this approach, the laser scans the environment, the data of which in turn will be inserted into a submap. When a submap is completed, in which no new scans by the laser will be inserted, the submap will be used as a consideration for a loop closure. The completed submap will be compared with the agent's current pose. If there is a similar match between the agent's current pose in the submap, the submap will be used as a loop closure constraint to the optimization problem. As the optimization happens at such a fast rate, the loop closure happens every time an area is revisited. Hence, the loop closing process has to happen quickly before a new scan is registered by the laser. The efficiency of Cartographer algorithm during the loop closing process makes it stand out from the other algorithms [16].

3. Methodology
3.1. Establishing Workspace
This research, in its entirety, was conducted using the Ubuntu Operating System (UOS). The reason for the use of the system is because this research fully utilizes the use of the Robotic Operating System (ROS) in which it is only available on the Ubuntu OS at the time of this research was carried out. With that being said, the first major step in this research development was installing the Ubuntu OS into the laptop. This research used the Ubuntu version 16.04 to ensure that it was compatible with the version of ROS that was used. Ubuntu was installed in a dual boot method to accommodate the lack of resources. Hence, Ubuntu resided in the same laptop as the Windows OS. After successfully installing Ubuntu, the next step was to install the ROS on Ubuntu. The version of ROS that was used in this research is called 'Kinetic Kame'. Kinetic Kame is a version of ROS that received Long Term Support (LTS) from the ROS development team, which means it is much more stable and less buggy compared to other versions of ROS. Like any other applications on Ubuntu, ROS was installed by inserting the necessary commands into the Ubuntu terminal.

3.2. Installing Cartographer Algorithm
Cartographer was installed the same way as ROS, through the use of commands in the Ubuntu terminal. The tutorial for Cartographer installation can be found on the ROS wiki [17]. After successfully installing Cartographer, the Cartographer folder containing all the resources to run the algorithm should appear on the home directory of Ubuntu.

3.3. Installing Hector SLAM
The Hector SLAM package used in this project was retrieved from github instead of using the official package hosted on ROS wiki. The github version of the package has been modified to include integration with an agent that was used in the simulation without needing to install any additional dependencies. Hector SLAM was installed the same way as Cartographer, by inputting command into the terminal.

IOP Conf. Series: Materials Science and Engineering **769** (2020) 012035 doi:10.1088/1757-899X/769/1/012035

3.4. Running Simulation
After successfully installing the Cartographer algorithm, the next step taken was running the simulation to test the algorithm. Running these algorithms required calling a launch file, which contains all of the information of the processes that will be used during the simulation process.

4. Results and Findings
4.1. Computer Resource Usage
The computer resource usage was checked by using the system monitor program in the Ubuntu. Similar to the Task Manager in Windows OS, system monitor shows the processes of the programs and applications running. System monitor also shows the Central Processing Unit (CPU) usage and Random-Access Memory (RAM) usage of the processes running. In this experiment, the CPU and RAM usage were recorded while the Cartographer and Hector SLAM were simulated in the maze environment. No other processes were started during this experiment to get better readings on the computer resource usage of the Cartographer and Hector SLAM algorithms. This experiment was repeated ten times to get the average results for those two algorithms. Figures 2 and 3 show the results for both Cartographer and Hector SLAM on the first run.

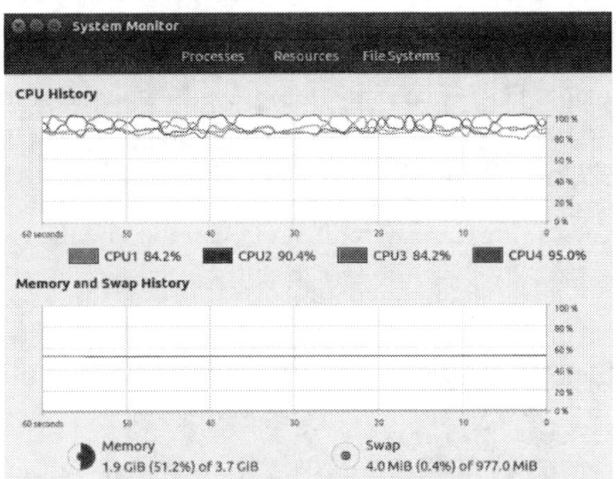

Figure 2. CPU and RAM usage while running Cartographer

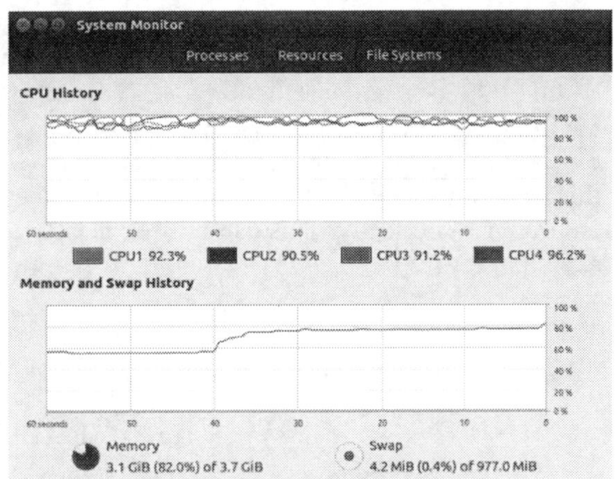

Figure 3. CPU and RAM usage while running Hector

Table 1 shows the overall results from the experiment. From the results table, Cartographer uses an average of 89.18% of CPU resources while Hector SLAM uses 91.82%. As for the RAM usage, Cartographer recorded an average of 55.2% while Hector SLAM uses 83.6%. Although there are no significant improvements between Cartographer and Hector SLAM in the CPU usage area i.e. a difference of 2.64%, it can be clearly seen that Cartographer uses a lot less RAM compared to Hector SLAM i.e. a difference of 28.4%. These findings directly correlate with the observation that Cartographer uses less resources overall as it utilizes pose graph optimization instead of particle filter.

Table 1. Results of computer usage experiment

Experiment	Cartographer		Hector	
	CPU (%)	RAM (%)	CPU (%)	RAM (%)
1	88.45	51.2	92.55	82.0
2	89.22	55.1	92.20	84.3

IOP Conf. Series: Materials Science and Engineering **769** (2020) 012035 doi:10.1088/1757-899X/769/1/012035

3	88.70	55.2	90.05	83.1
4	88.80	55.4	90.73	83.4
5	89.25	55.5	91.75	84.3
6	89.33	55.7	91.55	83.5
7	89.78	55.9	91.55	83.5
8	89.48	55.9	92.13	83.1
9	89.73	56.2	93.75	84.8
10	89.08	56.1	92.03	84.0
Average	89.18	55.2	91.82	83.6

4.2. Map Accuracy

The next experiment was conducted to test mapping capabilities of both algorithms. This experiment was conducted in a playpen environment instead of the maze and the suburb, where it is much easier to map the whole environment, which has less obstacles compared to the other two environments. For this research, the maps produced were displayed in RViz, a 3D visualization tool that comes together with a full ROS desktop installation. The generated maps can be seen by the greyed-out area of the environment in RViz. Figure 4, 5 and 6 shows the environment, layout of the environment as mapped by Cartographer algorithm and layout of the environment as mapped by Hector SLAM respectively. In this experiment, both algorithms show that they were able to capture and map the outline of the environment perfectly. This can be seen clearly by observing the environment in Gazebo and the map shown in RViz.

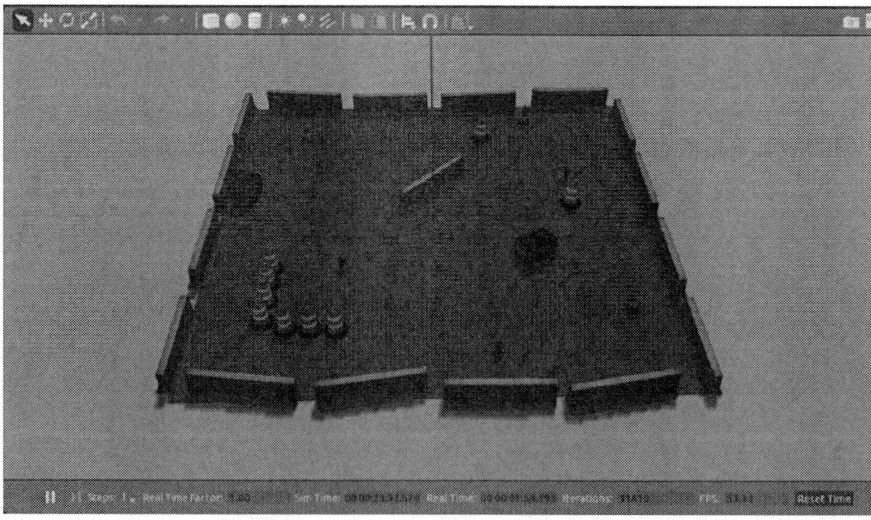

Figure 4. Environment used in this experiment

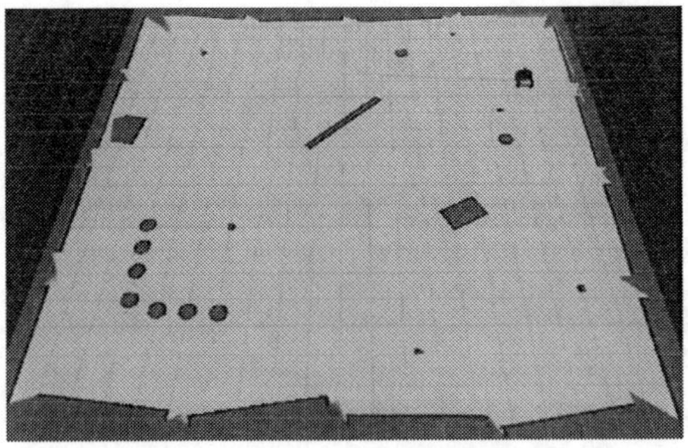

Figure 5. The environment mapped using Cartographer

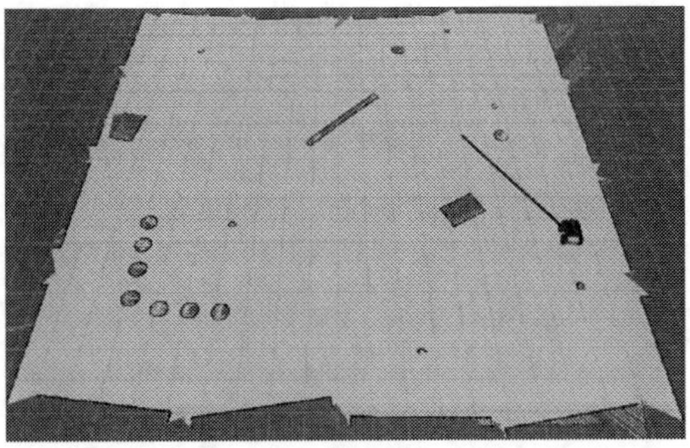

Figure 6. The environment mapped using Hector

It is worth to note that the map generated by Hector SLAM shows some anomalies in which the greyed-out area passes through what seems to be a static object and walls. In contrast, Cartographer did not encounter this problem and was able to produce a much sharper and clearer map of these two environments. As shown in Figure 7, the image on the left was adapted from the map produced by Cartographer, while the image on the right was from Hector SLAM. These problems (anomalies) may be caused by the lack of loop closure in Hector SLAM or various other reasons that are unidentifiable by the time this research was completed.

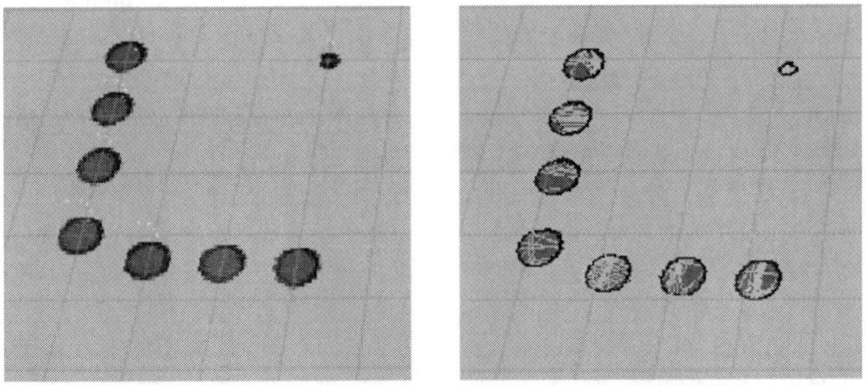

Figure 7. Comparison between maps produced by Cartographer (left) and Hector (right)

5. Conclusion

This paper shows the capabilities of graph optimization-based SLAM algorithm to work in a simulated environment. SLAM algorithms known as Cartographer and Hector were tested to uncover their capabilities in handling SLAM problems. Cartographer can be said to be more efficient when it comes to the resources used to run it. This can be observed during the first experiment where Cartographer recorded a significantly lesser computer resources used, mainly the CPU and RAM usage, compared to Hector SLAM. Cartographer was also able to produce a clearer and sharper map compared to its counterpart in this research the Hector SLAM. This is very important as the feature of SLAM as discussed earlier is the mapping part. In the future, a more in-depth work can be done with the Hector SLAM algorithm in order to identify what causes the anomalies during the mapping experiment. The Cartographer algorithm also can be compared with other SLAM algorithm with additional parameters in order to evaluate its capabilities in handling SLAM problem. Lastly, more research will be needed to experiment with other types of SLAM algorithms and to include more obstacles of various forms.

Acknowledgment
The authors would like to thank the Research Management Centre (RMC) under the grant of FRGS 600-IRMI/FRGS 5/3 (461/2019) of Universiti Teknologi MARA (UiTM) and the Faculty of Computer and Mathematical Sciences for all the given support.

References

[1] S. Abdul-Rahman, A. A. Bakar, and Z. A. Mohamed-Hussein, "An intelligent data pre-processing of complex datasets," *Intell. Data Anal.*, vol. 16, no. 2, pp. 305–325, 2012.

[2] T. T. O. Takleh, N. A. Bakar, S. A. Rahman, R. Hamzah, and Z. A. Aziz, "A brief survey on SLAM methods in autonomous vehicle," *Int. J. Eng. Technol.*, vol. 7, no. 4, pp. 38–43, 2018.

[3] S. Abdul-Rahman, M. S. Abd Razak, A. H. Mohd Mushin, R. Hamzah, N. Abu Bakar, and Z. Abd Aziz, "Simulation Of Simultaneous Localization And Mapping Using Point Cloud Data," *Indones. J. Electr. Eng. Comput. Sci.*, vol. 14, no. 2, pp. 1–10, 2019.

[4] O. Agunbiade and T. Zuva, "A Review: Simultaneous Localization And Mapping In Application To Autonomous Robot Agunbiade OY and Zuva T," no. May, 2018.

[5] Z. Kong and Q. Lu, "A brief review of simultaneous localization and mapping," *Proc. IECON 2017 - 43rd Annu. Conf. IEEE Ind. Electron. Soc.*, vol. 2017-Janua, pp. 5517–5522, 2017.

[6] C. Cadena *et al.*, "Past, present, and future of simultaneous localization and mapping: Toward the robust-perception age," *IEEE Trans. Robot.*, vol. 32, no. 6, pp. 1309–1332, 2016.

[7] A. R. Khairuddin, M. S. Talib, and H. Haron, "Review on simultaneous localization and mapping (SLAM)," *Proc. - 5th IEEE Int. Conf. Control Syst. Comput. Eng. ICCSCE 2015*, no. November, pp. 85–90, 2016.

[8] M. Yusoff, J. Ariffin, and A. Mohamed, "Solving vehicle assignment problem using evolutionary computation," *Lect. Notes Comput. Sci. (including Subser. Lect. Notes Artif. Intell. Lect. Notes Bioinformatics)*, vol. 6145 LNCS, no. PART 1, pp. 523–532, 2010.

[9] S. Masrom, S. Z. Z. Abidin, N. Omar, A. S. A. Rahman, and Z. I. Rizman, "Dynamic parameterizations of particle swarm optimization and genetic algorithm for facility layout problem," *ARPN J. Eng. Appl. Sci.*, vol. 12, no. 10, pp. 3195–3201, 2017.

[10] R. Smith, M. Self, and P. Cheeseman, "Estimating uncertain spatial relationships in robotics," in *Proceedings. 1987 IEEE International Conference on Robotics and Automation*, vol. 4, pp. 850–850.

[11] M. Montemerlo, S. Thrun, D. Koller, and B. Wegbreit, "FastSLAM: A Factored Solution to the Simultaneous Localization and Mapping Problem Michael," pp. 593–598, 2002.

[12] G. Grisetti, C. Stachniss, and W. Burgard, "Improved Techniques for Grid Mapping With Rao-Blackwellized Particle Filters," *IEEE Trans. Robot.*, vol. 23, no. 1, pp. 34–46, 2007.

[13] M. Filipenko and I. Afanasyev, "Comparison of Various SLAM Systems for Mobile Robot in an Indoor Environment," *2018 Int. Conf. Intell. Syst.*, no. September, pp. 400–407, 2018.

[14] N. Sünderhauf and P. Protzel, "Towards a robust back-end for pose graph SLAM," *Proc. - IEEE Int. Conf. Robot. Autom.*, pp. 1254–1261, 2012.

[15] B. M. F. Da Silva, R. S. Xavier, T. P. Do Nascimento, and L. M. G. Gonsalves, "Experimental evaluation of ROS compatible SLAM algorithms for RGB-D sensors," *Proc. - 2017 LARS 14th Lat. Am. Robot. Symp. 2017 5th SBR Brazilian Symp. Robot. LARS-SBR 2017 - Part Robot. Conf. 2017*, vol. 2017-Decem, pp. 1–6, 2017.

[16] Hess W, Kohler D, Rapp H, Andor D. Real-time loop closure in 2D LIDAR SLAM. In2016 IEEE International Conference on Robotics and Automation (ICRA) 2016 May 16 (pp. 1271-1278). IEEE.

A survey on technique for solving web page classification problem

Siti Hawa Apandi, Jamaludin Sallim and Rozlina Mohamed

Faculty of Computing, College of Computing and Applied Sciences, Universiti Malaysia Pahang, Gambang, Kuantan, Pahang, Malaysia

E-mail: sitihawa.apandi@gmail.com

Abstract. Nowadays, the number of web pages on the World Wide Web has been increasing due to the popularity of the Internet usage. The web page classification is needed in order to organize the increasing number of web pages. There are many web page classification techniques that have been proposed by the other researchers. However, there is no comprehensive survey on the performance of the techniques for the web page classification. In this paper, surveys of the different web page classification techniques with the result of the techniques achieved are presented. The existing works of web page classification are reviewed. Based on the survey, we found that the neural network technique namely Convolutional Neural Network (CNN) produce high F-measure value and meet the real-time requirement for classification compared to the other machine learning technique.

1. Introduction

Many people have uses Internet especially World Wide Web as a platform to search information. The Web rapidly flourished with the increasing number of web pages. A web page is a a web document that contains information in forms of textual content, images, audio and video [1]. The web browser is needed in order for the Internet users to view the web pages. The common web browsers used are Google Chrome and Mozilla Firefox.

The increasing number of web pages has caused some problems which it is difficult for the user to get relevant result when searching information in the search engine. Besides that, it is hard to maintained indices up to date with the increasing number of web pages in the web search systems. Thus to solve this problems, the web page classification is needed. The accurate and efficient classification of the web pages can make the search engines to retrieve the required information by the users quickly and efficiently even though there are huge information on the web [2-5]. The classification of the web page is also important for the development of web directories, discussion of specific topics Web and contextual advertising links on the analysis of the structure current site [5, 6].

There are two methods for web page classification which are traditional manual method and automatic method [2]. The traditional manual method of the web page classification use human effort which is done by the expert to manually classify the web page. This manual method takes a great deal of human effort and time consuming. It is impossible to be used for the web page classification because there is increasing number of web pages nowadays [2]. Thus, the suitable method to be used for the web page classification is the automatic method which uses the classifier in order to classify the web page.

There is an issue for the classifier which is huge scale dimensionality. This issue can be solved in the preprocessing phase of the web page classification process [2]. The preprocessing phase is very important process in the web page classification as it can provide better quality of training dataset. As a result, it can reduce the training and testing time. Besides, it can improve the performance result of the classifier [2, 7].

The other sections of this paper are structured as follows. In Section 2, the process of the web page classification is explained. This is followed by the review existing works of web page classification in Section 3. Lastly is the conclusion in Section 4.

2. Web page classification process

There is several process of web page classification as shown in Figure 1 which are collection of web pages, feature extraction, feature selection, build classifier, classify web pages and evaluate performance of classifier. The description of each process in web page classification is presented in the next subsection.

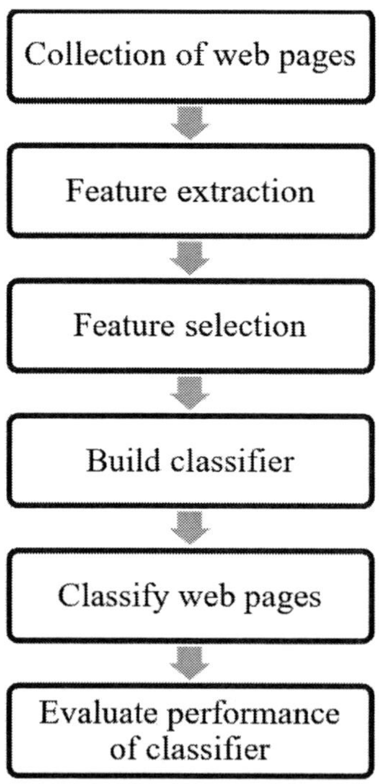

Figure 1. Process of web page classification.

2.1. *Collection of web pages*

Web pages are collected to be used as dataset in the web page classification. The dataset are split into two sets which are training and testing dataset. The training dataset is used to train the web page classifier, while the testing dataset is used to check the performance of the classifier [4].

2.2. *Feature extraction*

The feature extraction is one of the processes involved in the preprocessing phase that used to reduce the huge scale dimensionality issue. The irrelevant words and stop words that are found in the web pages is removed. The feature extraction process starts by extracting the raw content of the web pages with remove HTML tags and other WWW contents. Then, it continues with tokenization process which means breaking down text into small chunks known as token that can be phrase, word or symbols. Stemming or lemmatization is done next to reduce the tokens to their root words. For example, the root word "move" is selected to represent the other similar words such as "moves", "moving" and "moved". Then, the stop words such as propositions, conjunctions and words that have high frequency are filtering out [2, 3].

2.3. Feature selection

The feature selection is another one of the processes involved in the preprocessing phase that used to reduce the huge scale dimensionality issue. This is important in order to increase the accuracy and efficiency of the classifier. The purpose of feature selection is to select the best features that would represent the web page [6]. The feature selection can be categorized into three methods which are filter, wrapper and embedded. Table 1 presents the comparison of the three feature selection methods.

Table 1. Comparison of feature selection methods.

Feature selection method	Filter	Wrapper	Embedded
Technique	Statistical measures	Optimization algorithm	Combination of filter and wrapper method
Computational efficiency	Efficient	Inefficient	Inefficient
Computation time	Time efficient	Slow	Slow
Computational cost	Cheaper	Expensive	Expensive
Computational space	Less computational space	More computational space	More computational space
Complexity	Low	High	High
Generality	High	Less	Less
For high dimensional data	Suitable	Not suitable	Not suitable
Advantage		High classification accuracy	Reduce the computation time
Disadvantage	Does not deal with redundant features	Increased runtime	

2.4. Build classifier

The selected features set are used as the input data set to be fed to the classifier. In training phase, the classifier is build by using machine learning algorithm.

2.5. Classify web pages

In testing phase, the classifier uses the learned function to classify the web page and allocate the web page to particular categories [2, 3].

2.6. Evaluate performance of classifier

The metrics used to evaluate the performance of classifier are precision, recall, F-measure and accuracy. The confusion matrix as shown in Table 2 is a table that presents the actual label of category again the predicted label of category [2]. The parameters in the Table 2 are used in order to compute the value of all the performance evaluation metrics which are precision, recall, F-measure and accuracy.

Table 2. Confusion matrix.

	Predicted positive	Predictive negative
Actual positive	True Positive (TP)	False Negative (FN)
Actual negative	False Positive (FP)	True Negative (TN)

The description of each the performance evaluation metrics for the web page classification are as follows.

- Precision tells us that based on the results classified as positive, how many were actually positive. The precision can be calculated by the formula as in (1).

$$\text{Precision (P)} = TP / (TP + FP) \quad (1)$$

- Recall tells us how many were actually correctly classified as True Positive based on the all positive data. It can be calculated by the formula as in (2).

$$\text{Recall (R)} = TP / (TP + FN) \quad (2)$$

- F-measure is calculated based on the value of precision and recall. The F-measure can be calculated by the formula as in (3).

$$\text{F-measure} = (2 \times P \times R) / (P + R) \quad (3)$$

- Accuracy tells us how many were correctly classified data based on the total number of data. The formula in (4) is used to calculate the accuracy.

$$\text{Accuracy} = (TP + TN) / (TP + TN + FP + FN) \quad (4)$$

Of all the performance evaluation metrics discussed, it is better to use F-measure especially if there is an uneven class distribution which is the large number of actual negative. The accuracy is also useful to measure the performance but it is only when the values of False Positive and False Negative are almost same.

3. Existing works of web page classification

There are many techniques have been proposed by the other researchers for the web page classification. In this paper, we are going to review several existing works of web page classification. The Google Scholar is used to search for the papers related to the web page classification. The Google Scholar is used because there are many publication papers that can be found and it is easy to access. There are some criteria used to select the papers related to the web page classification that are going to be reviewed. Firstly, the paper published that is going to be reviewed need to discuss about the proposed technique to solve the web page classification. Then, the published paper selected is between the years of 2015 until 2018.

Table 3 presents a comparison of existing works of web page classification that has been arranged by descending order of year paper published. The comparison takes a look at the web page classification technique, the web page features used during the classification and the result achieved.

Table 3. Comparison of existing works of web page classification.

Author	Year	Title	Technique used	Feature of web page used	Performance evaluation metric used	Result
Shao, et al. [8]	2018	Effective Web-Page Classification Using Token-String CNN over URLs and Anchor Texts	• Token-string Convolutional Neural Network (CNN) based classifiers used to classify web pages	• URLs • Corresponding anchor texts	F-measure	Ranging from 0.93 to 0.99
					Classification time	Meets the real-time requirement. It take less than 3.2 seconds to classify 100K URLs
Wai, et al. [4]	2018	Ontology Based Web Page Classification System by Using Enhanced C4.5 and Naive Bayesian Classifiers	• The enhanced C4.5 decision tree and Naive Bayesian classifier used for classification of web pages • Use the ontology for semantic extraction about features	• Keyword • Semantic features	Accuracy	92.5%
Kiziloluk and Ozer [9]	2017	Web pages classification with Parliamentary Optimization Algorithm	• The Parliamentary Optimization Algorithm (POA) used to classify the web pages	• HTML tags	Accuracy	Ranging from 88.07% to 94.03%
Raj and Francis [10]	2017	Enhancements to Web Page Classification Based on Particle	• For feature selection, two optimization	• Terms and HTML tags	Precision	92.21
					Recall	87.69
					F-measures	89.89

		Swarm and Cuckoo Search Optimization	techniques namely Particle Swarm and Cuckoo Search used to refine and reduce feature sets • Naïve Bayes Classifier (NBC) used for classification of web pages		Accuracy	96.81
Bhatt, et al. [5]	2016	An Improved Optimized Web Page Classification using Firefly Algorithm with NB Classifier (WPCNB)	• The Firefly Algorithm (FA) used to optimize best feature selection • The Naive Bayes (NB) classifier is used for classification of web pages	• Term in HTML or XML tag	F-measure Accuracy	0.987 98.90%
Raj, et al. [11]	2016	Optimal web page classification technique based on informative content extraction and FA-NBC	• The Term Frequency -Inverse Document Frequency (TF-IDF) is used in the feature extraction phase • The Optimal Firefly Algorithm	• Terms and HTML tags	Precision Recall F-measure Accuracy	1.0 0.79 0.919 94.84

			(FA) based Naïve Bayes Classifier (FA-NBC) is used for classification of web pages			
Lee, et al. [3]	2015	Web page classification based on a Simplified Swarm Optimization	• The document frequency is used for feature selection • The Simplified Swarm Optimization (SSO) is used for web page classifier	• HTML tags and terms	Precision Recall F-measure CPU time	81.81% 83.46% 82.21% Require the greatest amount of the CPU time

Explanation about the feature selection methods as presented in Table 1 show that the filter method of feature selection is better than the other methods of feature selection. However based on the Table 3, we find that the technique involved in the existing works of web page classification mostly used optimization technique which is based on the wrapper method of feature selection [3, 5, 9-11]. This thing happens because the wrapper method of feature selection can produce high classification accuracy as stated in Table 1.

Based on the Table 3, we find that the commonly feature selected to be used in the existing works of web page classification is terms and HTML tags of web page [3, 10, 11].

From Table 3, we find the web page classifier that uses Convolutional Neural Network (CNN) which is one of the type of neural network, produce high value of F-measure [8]. This statement agreed by [7] that state the neural network work better than the other machine learning to be used as the web page classifier. It is also notice that the web page classification using CNN has advantage to meet the real-time requirement for the classification [8] compared to the work by Lee, et al. [3] which take greatest amount of CPU time.

4. Conclusion

This paper has discussed the basic idea of process and existing works in the web page classification. The process of web page classification includes collection of web pages, feature extraction, feature selection, build classifier, classify web pages and evaluate performance of classifier. The preprocessing process can be divided into two sub process which are feature extraction and feature selection. There are various techniques have been proposed for the web page classification. From the survey of the web page classification techniques, the Convolutional Neural Network (CNN) technique provides better result compared to other machine learning techniques. It achieves high F-measure value and meet real-time requirement for classification. This paper can be used as a guideline for the researchers who are interested to know the process and techniques involved in the web page classification. For the future work, we would like to explore more on the neural network for the web page classification.

5. Acknowledgment

This research work is supported by Universiti Malaysia Pahang Grant: RDU182207-2. This work also partially supported by Adnuri SMA Research Center (M) Sdn. Bhd.

References

[1] E. Suganya and D. S. Vijayarani, "Web Page Classification in Web Mining Research - A Survey " *International Journal of Innovative Research in Science, Engineering and Technology,* vol. 6, pp. 17472-17479, 2017.
[2] A. Osanyin, O. Oladipupo, and I. Afolabi, "A Review on Web Page Classification," *Covenant Journal of Informatics and Communication Technology,* vol. 6, pp. 11-32, 2018.
[3] J.-H. Lee, W.-C. Yeh, and M.-C. Chuang, "Web page classification based on a simplified swarm optimization," *Applied Mathematics and Computation,* vol. 270, pp. 13-24, 2015.
[4] H. P. M. Wai, P. P. Tar, and P. Thwe, "Ontology Based Web Page Classification System by Using Enhanced C4. 5 and Naïve Bayesian Classifiers," in *2018 International Conference on Intelligent Informatics and Biomedical Sciences (ICIIBMS)*, 2018, pp. 286-291.
[5] K. Bhatt, A. Singh, and D. Singh, "An Improved Optimized Web Page Classification using Firefly Algorithm with NB Classifier (WPCNB)," *International Journal of Computer Applications,* vol. 146, pp. 15-21, 2016.
[6] J. Alamelu Mangai, V. Santhosh Kumar, and V. Sugumaran, "Recent Research in Web Page Classification–A Review," *International Journal of Computer Engineering & Technology (IJCET),* vol. 1, pp. 112-122, 2010.
[7] L. Safae, B. El Habib, and T. Abderrahim, "A Review of Machine Learning Algorithms for Web Page Classification," in *2018 IEEE 5th International Congress on Information Science and Technology (CiSt)*, 2018, pp. 220-226.
[8] L. Shao, S. Yao, X. Zhou, J. Guo, and J. Wang, "Effective Web-Page Classification Using Token-String CNN over URLs and Anchor Texts," in *Proceedings on the International Conference on Artificial Intelligence (ICAI)*, 2018, pp. 105-111.
[9] S. Kiziloluk and A. B. Ozer, "Web pages classification with parliamentary optimization algorithm," *International Journal of Software Engineering and Knowledge Engineering,* vol. 27, pp. 499-513, 2017.
[10] A. M. J. Raj and F. S. Francis, "Enhancements to Web Page Classification Based on Particle Swarm and Cuckoo Search Optimization," *International Journal of Computer Science and Information Security (IJCSIS),* vol. 15, pp. 107-113, 2017.
[11] A. Raj, F. S. Francis, and P. J. Benadit, "Optimal web page classification technique based on informative content extraction and FA-NBC," *Computer Science and Engineering,* vol. 6, pp. 7-13, 2016.

IOP Conf. Series: Materials Science and Engineering **769** (2020) 012037 doi:10.1088/1757-899X/769/1/012037

Backpropagation Network Optimization Using One Step Secant (OSS) Algorithm

Solikhun[1] , Mochamad Wahyudi[1] , M. Safii[1] , Muhammad Zarlis [2]

[1]Doctoral Program, Faculty of Computer Science and Information Technology, Universitas Sumatera Utara, Sumatera Utara, Indonesia.
[2]Faculty of Computer Science and Information Technology, Universitas Sumatera Utara, Indonesia

Abstract. Education is one of the main indicators in national development efforts. The government in its efforts to realize national goals to educate the nation's life has made a policy for compulsory 9-year schools, namely elementary and junior high schools. To find out how many residents use the education facilities provided by the government can be seen through school enrollment rates (SPR). A high School Participation Rate (SPR) means showing greater opportunities to access education in general. This study aims to optimize artificial neural networks with the One Step Secant (OSS) algorithm. Artificial Neural Network (ANN) is part of the artificial intelligence system (Artificial Intelligence, AI) which is one of the artificial representations of the human brain that always tries to simulate the learning process in the human brain. The sample data used for optimization is SPR Indonesia data by province. Using 4 architectural models with 5 input variables, 1 shadow layer and 1 output. The best results obtained between architectures 5-4-1, 5-8-1, 5-16-1 and 5-32-1 are architectures 5-16-1. Obtained prediction accuracy comparisons using the One Step Secant (OSS) algorithm and standard algorithms namely 96.97% and 100%. The standard algorithm is superior in accuracy, the One Step Secant (OSS) algorithm is superior in terms of iterations.

1. Introduction

Education is one of the main indicators in national development efforts. According to Law No. 20 of 2003 concerning the National Education System, education is a conscious and planned effort to create a learning atmosphere and learning process so that students actively develop their potential to have religious spiritual strength, self-control, personality, intelligence, noble character, and skills needed by themselves , society, nation and state.

The government in its efforts to realize national goals to educate the nation's life has made a policy for compulsory 9-year schools, namely elementary and junior high schools. To find out how many residents use the education facilities provided by the government can be seen through School Participation Rate (EPR). A high School Participation Rate (SPR) means showing greater opportunities to access education in general. The importance of education in building human resources must be a mature calculation to manage it. One way to maximize business in the future is to know the picture that happened first. Accurate and accurate predictions can be a benchmark for seeing the future.

In computer science there is a technique that can be used to predict the future, namely Artificial Neural Networks (ANN) using the backpropagation method. This method is a very good method of dealing with the problem of recognizing complex patterns. But the standard backpropagation algorithm tends to be slow to reach convergence in getting maximum results. Therefore this algorithm can still be optimized to improve the results of accuracy. The One Step Secant (OSS) algorithm is an algorithm that can train any network during weight, the net input and activation functions have derivative functions. This algorithm will maximize time and increase the accuracy produced by standard backpropagation.

The object that will be predicted in this study is the school Participation rate (EPR) in education coverage aged 19 to 24 years, namely in higher education. To calculate school Participation numbers, a formula can be used:

$$EPR\ 19-24 = \frac{Number\ of\ residents\ aged\ 19\text{–}24\ years\ who\ are\ still\ in\ school}{Number\ of\ residents\ aged\ 19\text{–}24\ years} x\ 100\% \quad (1)$$

In a previous study, [1] [2] conducted a study to look at the factors that influence SPR, namely the teacher to student ratio, poverty rate and income per capita that have a significant effect on the SPR. This study produced a correlation between SPR and the social conditions of the community. The relationship formed is negative where if the poverty level increases, the SPR will decrease.

2. Method

2.1. Artificial Intelligence

Artificial intelligence is one area that is quite reliable in solving problems such as prediction (forecasting) [1]. AI is a very important discipline and it includes a number of well recognized and mature areas including Neural Network [2]–[4]. Artificial Intelligence (AI) is a general term that implies the use of a computer to model intelligent behavior with minimal human intervention. AI is generally accepted as having started with the invention of robots. The term derives from the Czech word robota, meaning biosynthetic machines used as forced labor [5].

2.2. Artificial Neural Network

Artificial Neural Network (ANN) is one of the studies of Artificial Intelligence and is a new computing technology in the field of computer science study. Neural networks mostly used for problem-solving in pattern recognition, data analysis, control and clustering [6][7]. Initially ANN were developed in the field of artificial intelligence and were first introduced for image recognition. The central concept was inspired by knowledge of the nervous system, especially the human brain with its closely connected neurons [8][9].

2.3. Backpropogation Neural Network

Backpropagation (BP) algorithm was used to develop the ANN model [10][11]. The typical topology of BPANN (Backpropagation Artificial Neural Network) involves three layers: input layer, where the data are introduced to the network; hidden layer, where the data are processed; and output layer, where the results of the given input are produced [12]–[14]. A backpropagation algorithm was used for training. It is a convenient and simple iterative algorithm that usually performs well, even with complex data. Unlike other learning algorithms (like Bayesian learning) it has good computational properties when dealing with largescale data [15][16].

2.4. Backpropagation Architecture

The back-propagation learning algorithm (BPLA) has become famous learning algorithms among ANNs. In the learning process, to reduce the inaccuracy of ANNs, BPLAs use the gradient-decent search method to adjust the connection weights. The structure of a back-propagation ANN is shown in Figure 2. The output of each neuron is the aggregation of the numbers of neurons of the 3 previous level multiplied by its corresponding weights. The input values are converted into output signals with the calculations of activation functions. Backpropagation ANNs have been widely and successfully applied in diverse applications, such as pattern recognition, location selection and performance evaluations [17][18].

2.5. Data Analysis

The data used in this study is the data of SPR ages 19-24 years by province obtained from the National Statistics Agency (bps.go.id). The data used is data from 2011 to 2017. The following is the data used in this study can be seen in table 1.

Table 1. Data on School Participation Rates Ages 19 – 24 years

The Province	2011	2012	2013	2014	2015	2016	2017
Aceh	27,68	28,55	29,18	32,93	33,07	33,94	34,28
Sumatera Utara	16,94	17,27	21,81	24,82	25,16	26,62	26,8
Sumatera Barat	23,95	27,55	30,66	32,89	33,13	34,71	35,45
Riau	15,34	15,81	22,04	24,48	24,85	26,18	27,28
Jambi	15,64	15,22	20,25	22,11	22,22	23,86	24,12
Sumatera Selatan	12,75	13,91	14,08	16,87	17	18,07	19,17
Bengkulu	17,02	19,64	24,12	28,14	28,37	28,93	29,9
Lampung	10,39	11,9	16,19	18,67	18,81	19,72	20,96
Bangka Belitung	8,63	9,3	9,46	12,22	12,73	13,81	14,99
Kep, Riau	9,67	10,14	14,85	17,4	17,69	18,58	19,13
Dki Jakarta	17,83	18,02	19,65	22,52	22,71	23,06	24,6
Jawa Barat	11,15	12,25	17,34	19,27	19,4	20,37	21,5
Jawa Tengah	11,51	11,83	17,42	20,48	20,57	21,59	22,13
Yogyakarta	44,17	44,69	45,86	49,08	49,17	49,95	51,33
Jawa Timur	12,69	14,59	19,49	21,84	21,95	22,67	23,34
Banten	13,56	15,97	18,08	19,61	19,68	20,74	21,33
Bali	18,93	18,99	19,84	23,59	23,75	25,36	26,56
Nusa Tenggara Barat	16,99	17,82	22,64	26,73	26,84	27,79	28,52
Nusa Tenggara Timur	17,4	17,92	22,88	26,22	26,54	26,75	27,8
Kalimantan Barat	11,94	14,17	19,27	23,18	23,32	24,75	25,8
Kalimantan Tengah	13,05	14,04	19,89	22,31	22,47	22,72	24,15
Kalimantan Selatan	13,62	16,48	16,95	20,36	20,53	21,89	23,53
Kalimantan Timur	16,92	20,33	25,04	27,34	27,55	28,88	30,04
Sulawesi Utara	15,16	16,12	16,36	20,91	21,31	22,82	24,22
Sulawesi Tengah	16,72	16,74	21,76	25,05	25,13	25,57	26,31
Sulawesi Selatan	21,46	23,17	27,8	30,23	30,64	31,48	32,16
Sulawesi Tenggara	21,48	23,62	24	28,78	28,89	29,31	30,03
Gorontalo	19,85	20,46	23,27	27,94	28,38	28,98	29,21
Sulawesi Barat	13,03	14,65	18,04	21,53	21,97	22,36	23,49
Maluku	26,71	28,98	33,8	36,44	36,6	37,51	38,2

Maluku Utara	19,33	21,79	26,42	30,85	31,25	31,75	32,1
Papua Barat	16,46	20,03	24,1	29,66	29,96	31,45	31,92
Papua	12,81	13,86	17,5	22,48	22,55	23,75	24,57

3. Results and discussion

3.1. Input and Target Data Transformation

The original data is pre-processed by artificial neural networks with the backpropagation method, in order to be understood, the data must be converted into numbers between 0 and 1 using the formula:

$$x' = \left(\frac{0{,}8*(x-x_{min})}{(x_{max}-x_{min})}\right) + 0{,}1 \qquad (2)$$

where :

- x' = Transformation Results
- x = Original Data
- x_{min} = Minimum Data
- x_{max} = Maximum Data

For training data used SPR data based on provinces with 5 input data, namely data from 2011 to 2015 with a target for 2016 while for testing data using 5 input data, namely data from 2012 to 2016 with a target for 2017, transformation of training and testing data is shown in Tables 2 and 3:

Table 2. Training Data

Data	X1	X2	X3	X4	X5	Target
Data 1	0,45691	0,47321	0,48501	0,55527	0,55789	0,57419
Data 2	0,25569	0,26187	0,34693	0,40333	0,40970	0,43705
Data 3	0,38703	0,45447	0,51274	0,55452	0,55902	0,58862
Data 4	0,22571	0,23452	0,35124	0,39696	0,40389	0,42881
Data 5	0,23133	0,22347	0,31770	0,35255	0,35461	0,38534
Data 6	0,17719	0,19892	0,20211	0,25438	0,25681	0,27686
Data 7	0,25719	0,30628	0,39021	0,46553	0,46984	0,48033
Data 8	0,13297	0,16126	0,24164	0,28810	0,29073	0,30778
Data 9	0,10000	0,11255	0,11555	0,16726	0,17681	0,19705
Data 10	0,11948	0,12829	0,21653	0,26431	0,26974	0,28642
Data 11	0,27237	0,27593	0,30646	0,36023	0,36379	0,37035
Data 12	0,14721	0,16782	0,26319	0,29934	0,30178	0,31995
Data 13	0,15396	0,15995	0,26468	0,32201	0,32370	0,34281
Data 14	0,76585	0,77560	0,79752	0,85785	0,85953	0,87415
Data 15	0,17607	0,21166	0,30347	0,34749	0,34956	0,36304
Data 16	0,19237	0,23752	0,27705	0,30571	0,30703	0,32689
Data 17	0,29297	0,29410	0,31002	0,38028	0,38328	0,41344
Data 18	0,25663	0,27218	0,36248	0,43911	0,44117	0,45897

Data 19	0,26431	0,27405	0,36698	0,42956	0,43555	0,43948
Data 20	0,16201	0,20379	0,29934	0,37260	0,37522	0,40201
Data 21	0,18281	0,20136	0,31096	0,35630	0,35930	0,36398
Data 22	0,19349	0,24707	0,25588	0,31977	0,32295	0,34843
Data 23	0,25532	0,31920	0,40745	0,45054	0,45447	0,47939
Data 24	0,22234	0,24033	0,24482	0,33007	0,33756	0,36585
Data 25	0,25157	0,25194	0,34600	0,40763	0,40913	0,41738
Data 26	0,34037	0,37241	0,45916	0,50468	0,51237	0,52810
Data 27	0,34075	0,38084	0,38796	0,47752	0,47958	0,48745
Data 28	0,31021	0,32164	0,37429	0,46178	0,47002	0,48126
Data 29	0,18244	0,21279	0,27630	0,34169	0,34993	0,35724
Data 30	0,43874	0,48126	0,57157	0,62103	0,62403	0,64108
Data 31	0,30047	0,34656	0,43330	0,51630	0,52379	0,53316
Data 32	0,24670	0,31358	0,38984	0,49400	0,49963	0,52754
Data 33	0,17831	0,19799	0,26618	0,35948	0,36080	0,38328

Table 3. Testing Data

Data	X1	X2	X3	X4	X5	Target
Data 1	0,47321	0,48501	0,55527	0,55789	0,57419	0,58056
Data 2	0,26187	0,34693	0,40333	0,40970	0,43705	0,44042
Data 3	0,45447	0,51274	0,55452	0,55902	0,58862	0,60248
Data 4	0,23452	0,35124	0,39696	0,40389	0,42881	0,44941
Data 5	0,22347	0,31770	0,35255	0,35461	0,38534	0,39021
Data 6	0,19892	0,20211	0,25438	0,25681	0,27686	0,29747
Data 7	0,30628	0,39021	0,46553	0,46984	0,48033	0,49850
Data 8	0,16126	0,24164	0,28810	0,29073	0,30778	0,33101
Data 9	0,11255	0,11555	0,16726	0,17681	0,19705	0,21916
Data 10	0,12829	0,21653	0,26431	0,26974	0,28642	0,29672
Data 11	0,27593	0,30646	0,36023	0,36379	0,37035	0,39920
Data 12	0,16782	0,26319	0,29934	0,30178	0,31995	0,34112
Data 13	0,15995	0,26468	0,32201	0,32370	0,34281	0,35293
Data 14	0,77560	0,79752	0,85785	0,85953	0,87415	0,90000
Data 15	0,21166	0,30347	0,34749	0,34956	0,36304	0,37560
Data 16	0,23752	0,27705	0,30571	0,30703	0,32689	0,33794
Data 17	0,29410	0,31002	0,38028	0,38328	0,41344	0,43593
Data 18	0,27218	0,36248	0,43911	0,44117	0,45897	0,47265
Data 19	0,27405	0,36698	0,42956	0,43555	0,43948	0,45916

Data 20	0,20379	0,29934	0,37260	0,37522	0,40201	0,42169
Data 21	0,20136	0,31096	0,35630	0,35930	0,36398	0,39077
Data 22	0,24707	0,25588	0,31977	0,32295	0,34843	0,37916
Data 23	0,31920	0,40745	0,45054	0,45447	0,47939	0,50112
Data 24	0,24033	0,24482	0,33007	0,33756	0,36585	0,39208
Data 25	0,25194	0,34600	0,40763	0,40913	0,41738	0,43124
Data 26	0,37241	0,45916	0,50468	0,51237	0,52810	0,54084
Data 27	0,38084	0,38796	0,47752	0,47958	0,48745	0,50094
Data 28	0,32164	0,37429	0,46178	0,47002	0,48126	0,48557
Data 29	0,21279	0,27630	0,34169	0,34993	0,35724	0,37841
Data 30	0,48126	0,57157	0,62103	0,62403	0,64108	0,65400
Data 31	0,34656	0,43330	0,51630	0,52379	0,53316	0,53972
Data 32	0,31358	0,38984	0,49400	0,49963	0,52754	0,53635
Data 33	0,19799	0,26618	0,35948	0,36080	0,38328	0,39864

3.2. *Defining Output*

The expected results at this defining stage are to look for patterns to determine the best value to predict. The test results are as follows:

a. The output of this prediction is the best architectural pattern to predict the amount of rice production by province by looking at the minimum error.

b. Categorization of training output (train) and testing (test)

The category for output is determined by the minimum error rate of the target, the categories are listed in Table 4.

Table 4. Categorization data

No	Explanation	Minimum Error
1	True	<= 0,05
2	False	> 0,05

3.3. *Result*

This study implements several architectures to obtain optimal results, the architecture (model) used can is summarized in Table 5:

Table 5. Architectural design

Characteristics	**Specification**
Architectural	1 *hidden layer*
Input Data	5
Hidden Layer	4, 8, 16, 32
Goal	0,01
Maximum Epochs	100000
Learning rate	0,01

Each of the architecture is processed using the standard backpropagation algorithm and the One Step Secant (OSS) algorithm. The best architecture can be seen from the accuracy of the truth, a little more epochs and the size of the MSE. The following is the accuracy data, the number of epochs and MSE from the tested model.

Table 6. Comparison of Standard Backpropagation with Backpropagation with One Step Secant (OSS)

Model	Criteria	Standard	OSS
5-4-1	Accuracy	93,94%	96,97%
	MSE	0,012766	0,004864
	Epochs	68	4
5-8-1	Accuracy	96,97%	96,97%
	MSE	0,009094	0,01385
	Epochs	69	9
5-16-1	Accuracy	100%	100%
	MSE	0,010414	0,008030
	Epochs	27	2
5-32-1	Accuracy	100%	96,97%
	MSE	0,008179	0,009440
	Epochs	39	5

From the results above it can be seen that the best model that can be used to predict is the 5-16-1 architectural model with 100% accuracy.

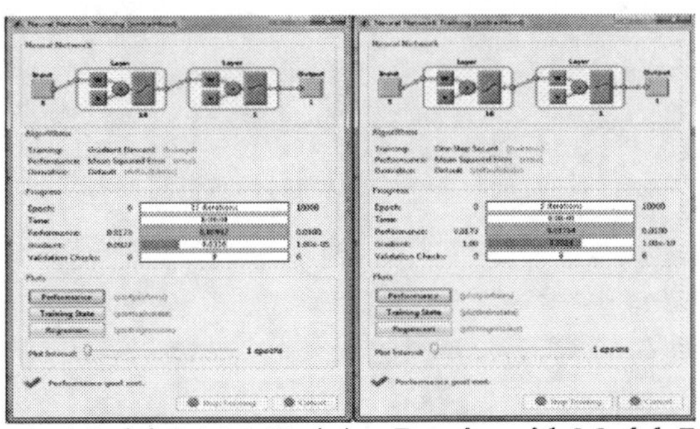

Figure 1. Model 5-16-1 Training Results with Matlab R2011A

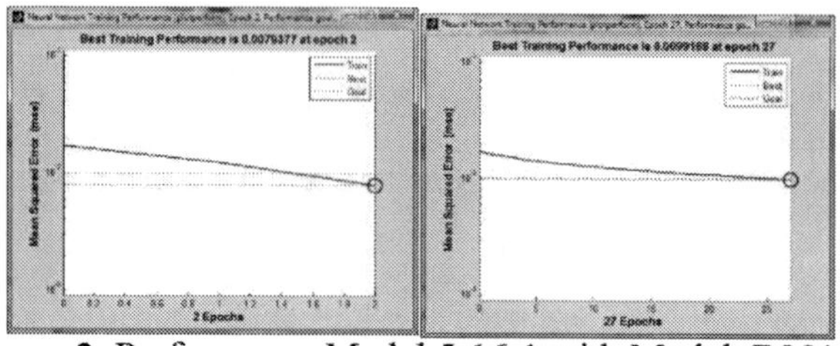

Figure 2. Performance Model 5-16-1 with Matlab R2011A

3.4. *Prediction of School Participation Rates*

By using the best architecture that has been obtained, the prediction of SPR is based on the province with the following results:

Table 7. Prediction with OSS algorithm

Backpropagation Algorithm One Step Secant						
No	The Province	Prediction	Target	Output	Error	Sse
1	Aceh	42,0635	0,72639	0,81571	-0,08932	0,00798
2	Sumatera Utara	33,3170	0,56252	0,67602	-0,11350	0,01288
3	Sumatera Barat	42,0184	0,72554	0,81987	-0,09433	0,00890
4	Riau	33,0966	0,55839	0,66876	-0,11037	0,01218
5	Jambi	31,3296	0,52529	0,63753	-0,11224	0,01260
6	Sumatera Selatan	18,1990	0,27928	0,41214	-0,13286	0,01765
7	Bengkulu	34,9601	0,59330	0,71355	-0,12025	0,01446
8	Lampung	20,9946	0,33166	0,51812	-0,18646	0,03477
9	Kep, Bangka Belitung	11,6416	0,15642	0,23226	-0,07584	0,00575
10	Kep, Riau	18,2126	0,27953	0,43566	-0,15613	0,02438
11	Dki Jakarta	31,6809	0,53187	0,65068	-0,11881	0,01412
12	Jawa Barat	23,6838	0,38204	0,55616	-0,17412	0,03032
13	Jawa Tengah	22,7233	0,36404	0,58578	-0,22174	0,04917
14	Di Yogyakarta	50,9297	0,89250	0,90668	-0,01418	0,00020
15	Jawa Timur	28,7599	0,47714	0,63751	-0,16037	0,02572
16	Banten	28,5603	0,47340	0,57365	-0,10025	0,01005
17	Bali	32,0137	0,53810	0,65621	-0,11811	0,01395
18	Nusa Tenggara Barat	33,0721	0,55793	0,69982	-0,14189	0,02013
19	Nusa Tenggara Timur	34,2631	0,58024	0,69833	-0,11809	0,01394
20	Kalimantan Barat	26,6172	0,43700	0,65034	-0,21334	0,04551
21	Kalimantan Tengah	28,5604	0,47340	0,65104	-0,17764	0,03155
22	Kalimantan Selatan	25,8532	0,42268	0,58185	-0,15917	0,02534
23	Kalimantan Timur	37,3613	0,63829	0,70193	-0,06364	0,00405
24	Sulawesi Utara	23,5823	0,38014	0,58174	-0,20160	0,04064

25	Sulawesi Tengah	32,5041	0,54729	0,68947	-0,14218	0,02022
26	Sulawesi Selatan	38,5247	0,66009	0,76787	-0,10778	0,01162
27	Sulawesi Tenggara	35,8477	0,60993	0,72314	-0,11321	0,01282
28	Gorontalo	34,1229	0,57762	0,71308	-0,13546	0,01835
29	Sulawesi Barat	25,2334	0,41107	0,61657	-0,20550	0,04223
30	Maluku	42,8217	0,74059	0,85322	-0,11263	0,01269
31	Maluku Utara	35,2826	0,59935	0,76311	-0,16376	0,02682
32	Papua Barat	32,3408	0,54423	0,71475	-0,17052	0,02908
33	Papua	22,6007	0,36175	0,62994	-0,26819	0,07193
Accuracy						96,97%
MSE						0,021878

Table 8. Prediction with standard algorithm

Algorithm Backpropagation Standar

No	**The Province**	**Prediction**	**Target**	**Output**	**Error**	**Sse**
1	Aceh	30,8526	0,51635	0,61903	-0,10268	0,01054
2	Sumatera Utara	31,4041	0,52668	0,61474	-0,08806	0,00775
3	Sumatera Barat	30,1669	0,50350	0,61973	-0,11623	0,01351
4	Riau	31,5960	0,53028	0,60797	-0,07769	0,00604
5	Jambi	32,7102	0,55115	0,63314	-0,08199	0,00672
6	Sumatera Selatan	23,7577	0,38342	0,51954	-0,13612	0,01853
7	Bengkulu	29,5445	0,49184	0,58951	-0,09767	0,00954
8	Lampung	25,9695	0,42486	0,59445	-0,16959	0,02876
9	Kep, Bangka Belitung	16,6250	0,24979	0,34991	-0,10012	0,01002
10	Kep, Riau	23,5082	0,37875	0,53794	-0,15919	0,02534
11	Dki Jakarta	33,1127	0,55869	0,64062	-0,08193	0,00671
12	Jawa Barat	28,3444	0,46936	0,61792	-0,14856	0,02207
13	Jawa Tengah	26,5437	0,43562	0,62673	-0,19111	0,03652
14	Di Yogyakarta	42,5008	0,73458	0,76966	-0,03508	0,00123
15	Jawa Timur	31,1146	0,52126	0,64590	-0,12464	0,01554
16	Banten	32,7348	0,55161	0,63027	-0,07866	0,00619
17	Bali	31,7920	0,53395	0,61437	-0,08042	0,00647
18	Nusa Tenggara Barat	29,4290	0,48968	0,60594	-0,11626	0,01352
19	Nusa Tenggara Timur	31,2334	0,52348	0,61693	-0,09345	0,00873
20	Kalimantan Barat	27,6106	0,45561	0,62074	-0,16513	0,02727
21	Kalimantan Tengah	30,6152	0,51190	0,64717	-0,13527	0,01830
22	Kalimantan Selatan	29,5870	0,49264	0,61355	-0,12091	0,01462

23	Kalimantan Timur	32,2305	0,54216	0,58414	-0,04198	0,00176
24	Sulawesi Utara	26,9006	0,44231	0,59950	-0,15719	0,02471
25	Sulawesi Tengah	31,0830	0,52066	0,63472	-0,11406	0,01301
26	Sulawesi Selatan	29,7949	0,49653	0,59996	-0,10343	0,01070
27	Sulawesi Tenggara	29,9012	0,49852	0,59142	-0,09290	0,00863
28	Gorontalo	28,9707	0,48109	0,59517	-0,11408	0,01301
29	Sulawesi Barat	28,2139	0,46691	0,62948	-0,16257	0,02643
30	Maluku	29,0290	0,48218	0,64385	-0,16167	0,02614
31	Maluku Utara	26,2997	0,43105	0,58470	-0,15365	0,02361
32	Papua Barat	24,9856	0,40643	0,54234	-0,13591	0,01847
33	Papua	24,8417	0,40373	0,62311	-0,21938	0,04813
Accuracy						100,00%
MSE						0,016016

4. Conclusion

The conclusion obtained from the research:

1. The iterations needed by the One Step Secant algorithm are fewer than the standard algorithms,
2. The more hidden layers not necessarily the results will be better.
3. The accuracy of the OSS algorithm and standards is not much different and even most standard algorithms have higher accuracy.

References

[1] A, Wanto, M, Zarlis, Sawaluddin, and D, Hartama, "Analysis of Artificial Neural Network Backpropagation Using Conjugate Gradient Fletcher Reeves in the Predicting Process," *Journal of Physics: Conference Series*, vol, 930, no, 1, pp, 1–7, 2017.

[2] H, Chiroma, S, Abdulkareem, A, I, Abubakar, and T, Herawan, "Kernel functions for the support vector machine: Comparing performances on crude oil price data," *Advances in Intelligent Systems and Computing*, vol, 287, pp, 273–281, 2014.

[3] R, B, F, Hakim, E, N, Sari, and T, Herawan, "Soft Solution of Soft Set Theory for Recommendation in Decision Making," *Advances in Intelligent Systems and Computing*, vol, 287, pp, 313–324, 2014.

[4] A, Lasisi, R, Ghazali, and T, Herawan, "Comparative Performance Analysis of Negative Selection Algorithm with Immune and Classification Algorithms," *Advances in Intelligent Systems and Computing*, vol, 287, pp, 441–452, 2014.

[5] P, Hamet and J, Tremblay, "Artificial Intelligence in Medicine," *Metabolism: Clinical and Experimental*, pp, 1–14, 2017.

[6] J, Adnan *et al.*, "Multilayer Perceptron Based Activation Function On Heart," *Journal of Fundamental and Applied Sciences*, vol, 9, no, 3, pp, 417–432, 2017.

[7] A, Wanto *et al.*, "Levenberg-Marquardt Algorithm Combined with Bipolar Sigmoid Function to Measure Open Unemployment Rate in Indonesia," 2018, pp, 1–7.

[8] A, Ehret, D, Hochstuhl, D, Gianola, and G, Thaller, "Application of Neural Networks with Nack-propagation to Genome-Enabled Prediction of Complex Traits in Holstein-Friesian and German Fleckvieh Cattle," *Genetics Selection Evolution*, vol, 47, no, 22, pp, 1–9, 2015.

[9] M, Fauzan *et al.*, "Epoch Analysis and Accuracy 3 ANN Algorithm Using Consumer Price Index Data in Indonesia," 2018, pp, 1–7.

[10] P, Antwi *et al.*, "Estimation of Biogas and Methane Yields in an UASB Treating Potato Starch Processing Wastewater With Backpropagation Artificial Neural Network," *Bioresource Technology*, vol, 228, pp, 106–115, 2017.

[11] S, P, Siregar and A, Wanto, "Analysis of Artificial Neural Network Accuracy Using Backpropagation Algorithm In Predicting Process (Forecasting)," *International Journal Of Information System & Technology*, vol, 1, no, 1, pp, 34–42, 2017.

[12] D, Huang and Z, Wu, "Forecasting outpatient visits using empirical mode decomposition coupled with backpropagation artificial neural networks optimized by particle swarm optimization," *PLoS ONE*, vol, 12, no, 2, pp, 1–17, 2017.

[13] A, Wanto, A, P, Windarto, D, Hartama, and I, Parlina, "Use of Binary Sigmoid Function And Linear Identity In Artificial Neural Networks For Forecasting Population Density," *International Journal Of Information System & Technology*, vol, 1, no, 1, pp, 43–54, 2017.

[14] B, Febriadi, Z, Zamzami, Y, Yunefri, and A, Wanto, "Bipolar function in backpropagation algorithm in predicting Indonesia's coal exports by major destination countries," *IOP Conference Series: Materials Science and Engineering*, vol, 420, no, 12089, pp, 1–9, 2018.

[15] A, Wanto *et al,*, "Analysis of Standard Gradient Descent with GD Momentum And Adaptive LR for SPR Prediction," 2018, pp, 1–9.

[16] N, Nasution, A, Zamsuri, L, Lisnawita, and A, Wanto, "Polak-Ribiere updates analysis with binary and linear function in determining coffee exports in Indonesia," *IOP Conference Series: Materials Science and Engineering*, vol, 420, no, 12089, pp, 1–9, 2018.

[17] S, Sumijan, A, Windarto, A, Muhammad, and B, Budiharjo, "Implementation of Neural Networks in Predicting the Understanding Level of Students Subject," *International Journal of Software Engineering and Its Applications*, vol, 10, no, 10, pp, 189–204, 2016.

[18] A, P, Windarto, L, S, Dewi, and D, Hartama, "Implementation of Artificial Intelligence in Predicting the Value of Indonesian Oil and Gas Exports With BP Algorithm," *International Journal of Recent Trends in Engineering & Research (IJRTER)*, vol, 3, no, 10, pp, 1–12, 2017.

Resilient Algorithm in Predicting Fertilizer Imports by Major Countries

Solikhun[1] , Mochamad Wahyudi[1] , M, Safii[1] , Muhammad Zarlis [2]

[1]Doctoral Program, Faculty of Computer Science and Information Technology, Universitas Sumatera Utara, Sumatera Utara, Indonesia,
[2]Faculty of Computer Science and Information Technology, Universitas Sumatera Utara, Indonesia

Abstract : In the last five years (2013-2017) Indonesia's fertilizer production experienced volatile growth, but overall tended to increase at a rate of 1,7% per year. The research aims to optimize artificial neural networks with a resilient backpropagation algorithm (Rprop), artificial neural networks are one of the artificial representations of the human brain that always tries to simulate the learning process in the human brain. Sample data used for optimization is fertilizer import data according to the main country of origin and uses 4 architectures, the best results are obtained between architectures 6-8-1, 6-12-1, 6-16-1, and 6-32-1 is architecture 6-32-1 with 100% accuracy.

1. Introduction

Fertilizers are different from supplements. Fertilizers contain raw materials that are needed for plant growth and development, while supplements such as plant hormones help smooth the metabolic process. Nevertheless, in fertilizers, especially artificial fertilizers, a number of supplementary ingredients can be added.

In the last five years (2013-2017) Indonesia's fertilizer production experienced volatile growth, but overall tended to increase at a rate of 1,7% per year. Production in 2013 was recorded at 11,37 million tons and in 2017 it increased to 12,10 million tons. There are 6 types of fertilizers produced domestically, and urea among them is the highest with production in 2013 of 6,70 million tons, then NPK 2,53 million tons and ZA / US amounting to 827 thousand tons. In 2017 urea production increased to 6,84 million tons, NPK of 3,28 million tons and ZA / US fell to 799 thousand tons.

Today there are many studies that utilize technology to facilitate human work. One of them is Artificial Neural Network (ANN), ANN is one branch of artificial intelligence that is modeled on neural networks in humans. One of the algorithms used to predict fertilizer is the resilient backpropagation (Rprop) algorithm. Resilient backpropagation (Rprop) algorithm is the result of the development of the backpropagation algorithm. Change the weight on backpropagation is influenced by the learning rate and depends on the slope of the error curve, the smaller the learning rate, the longer the learning process. While the greater the learning rate, the weight value will be far from the minimum weight. To overcome this, a new algorithm called Rprop was developed.

2. Method

2.1. *Artificial Intelligence*

Artificial Intelligence is one area that is quite reliable in solving problems such as prediction (forecasting) [1], AI is a very important discipline and it includes a number of well recognized and mature areas including Neural Network [2][4][5].

2.2. Artificial Neural Network
Artificial Neural Network (ANN) is one of the studies of Artificial Intelligence and is a new computing technology in the field of computer science stud. Neural networks mostly used for problem-solving in pattern recognition, data analysis, control and clustering [6][7][8][9].

2.3. Backpropagation Neural Network
Backpropagation (BP) algorithm was used to develop the ANN model [10][11]. The typical topology of BPANN (Backpropagation Artificial Neural Network) involves three layers: input layer, where the data are introduced to the network; hidden layer, where the data are processed; and output layer, where the results of the given input are produced [12][14][15][16].

2.4. Backpropagation Architecture
The structure of a back-propagation ANN is shown in Figure 2, The output of each neuron is the aggregation of the numbers of neurons of the 3 previous level multiplied by its corresponding weights. The input values are converted into output signals with the calculations of activation functions. Backpropagation ANNs have been widely and successfully applied in diverse applications, such as pattern recognition, location selection and performance evaluations [17][18].

2.5. Data Analysis
The data used in this study is data on fertilizer imports according to the main country of origin obtained from the National Statistics Agency (bps,go,id). The data used is data from 2010 to 2017. The following is the data used in this study can be seen in Table 1.

Table 1. Data on fertilizer import data according to the main country of origin

country of origin	2010	2011	2012	2013	2014	2015	2016	2017
China[1]	477,6	857,7	1 363,1	1 051,8	1 672,4	2 065,6	2 206,3	2 408,3
Canada	703,3	1 538,7	1 324,3	1 335,4	1 407,6	1 400,1	1 220,6	1 266,1
Russia	428,3	1 058,7	768,2	191,4	634,2	667,8	673,5	962,3
Malaysia	143,9	194,7	206,0	239,5	292,6	356,3	288,3	328,6
Norway	109,4	115,2	132,6	88,5	155,7	151,3	101,2	22,6
Australia	403,1	375,4	279,9	208,1	223,4	333,5	299,3	332,9
Germany	154,8	164,9	174,2	167,6	136,0	147,3	107,9	249,3
Jordan	104,2	197,4	268,7	359,2	170,3	107,1	131,9	257,9
South Korea	130,8	149,9	337,9	319,5	247,1	48,5	28,7	52,0
Japan	23,8	71,0	16,5	34,1	25,2	6,9	1,1	0,9
Thailand	52,0	61,1	43,4	18,2	2,8	0,5	5,0	1,4
Philippines	38,5	19,0	0,2	8,3	0,6	0,5	0,9	0,8

3. Results and discussion
The fertilizer import data will then be processed by ANN using the Resilient Backpropagation (Rprop) method. In order for data to be identified, the data must be transformed into a number between 0 and 1 (logsig). The data used includes data on fertilizer imports according to the main country of origin from 2010 to 2017 obtained by the website of the Central Statistics Agency (BPS). For training data, fertilizer import data are used according to the main country of origin with 6 input data, namely data from 2010 to 2015, while for the test data using 6 input data, namely data from 2011 to 2016 with the target for 2017. Transformation of data can be seen in Table 2 and 3.

IOP Conf. Series: Materials Science and Engineering **769** (2020) 012038 doi:10.1088/1757-899X/769/1/012038

Table 2. Training Data

Data	x1	x2	x3	x4	x5	x6	Target
Data 1	0,148177	0,186534	0,237536	0,206122	0,268749	0,308429	0,322628
Data 2	0,170953	0,255257	0,233621	0,234741	0,242027	0,241270	0,223156
Data 3	0,143202	0,206818	0,177502	0,119295	0,163980	0,167371	0,167946
Data 4	0,114501	0,119628	0,120768	0,124149	0,129507	0,135936	0,129073
Data 5	0,111020	0,111605	0,113361	0,108911	0,115692	0,115248	0,110192
Data 6	0,140658	0,137863	0,128226	0,120980	0,122524	0,133635	0,130184
Data 7	0,115601	0,116621	0,117559	0,116893	0,113704	0,114845	0,110868
Data 8	0,110495	0,119900	0,127096	0,136228	0,117166	0,110788	0,113290
Data 9	0,113179	0,115107	0,134079	0,132222	0,124916	0,104874	0,102876
Data 10	0,102382	0,107145	0,101645	0,103421	0,102523	0,100676	0,100091
Data 11	0,105227	0,106146	0,104360	0,101816	0,100262	0,100030	0,100484
Data 12	0,103865	0,101897	0,100000	0,100817	0,100040	0,100030	0,100071

Table 3. Testing Data

Data	X1	X2	X3	X4	X5	X6	Target
Data 1	0,186534	0,237536	0,206122	0,268749	0,308429	0,322628	0,343012
Data 2	0,255257	0,233621	0,234741	0,242027	0,241270	0,223156	0,227748
Data 3	0,206818	0,177502	0,119295	0,163980	0,167371	0,167946	0,197090
Data 4	0,119628	0,120768	0,124149	0,129507	0,135936	0,129073	0,133140
Data 5	0,111605	0,113361	0,108911	0,115692	0,115248	0,110192	0,102260
Data 6	0,137863	0,128226	0,120980	0,122524	0,133635	0,130184	0,133574
Data 7	0,116621	0,117559	0,116893	0,113704	0,114845	0,110868	0,125138
Data 8	0,119900	0,127096	0,136228	0,117166	0,110788	0,113290	0,126006
Data 9	0,115107	0,134079	0,132222	0,124916	0,104874	0,102876	0,105227
Data 10	0,107145	0,101645	0,103421	0,102523	0,100676	0,100091	0,100071
Data 11	0,106146	0,104360	0,101816	0,100262	0,100030	0,100484	0,100121
Data 12	0,101897	0,100000	0,100817	0,100040	0,100030	0,100071	0,100061

3.1. Defining Output

The results expected at this stage are defining to find patterns to determine the best value to predict. The test results are as follows:

a. The output of this prediction is the best architectural pattern to predict fertilizer import data according to the main country of origin with a minimum error.
b. Categorization of training output (train) and testing (test).

Table 4. Categorization Data

No	Explanation	Minimum Error
1	True	< = 0,05
2	False	> 0,05

3.2. Result

In this study several architectures were used to obtain optimal results, the architecture can be seen in Table 5.

Table 5. Architectural Design

Characteristics	Specification
Architecture	1 *Hidden Layer*
Data Input	6
Hidden Layer	8, 12, 16, 32
Goal	0,01
Maximum Epochs	100000
Learning Rate	0,01

Table 6.Training and Testing Standard Model 6-8-1

Training					Testing				
No	Target	Output	Error	SSE	No	Target	Output	Error	SSE
1	0,322628	0,04289	0,27974	0,078254	1	0,343012	0,26326	0,07975	0,006360
2	0,223156	0,30431	0,08115	0,006585	2	0,227748	0,97506	0,74731	0,558477
3	0,167946	0,26568	0,09773	0,009551	3	0,197090	0,91146	0,71437	0,510324
4	0,129073	0,05685	0,07222	0,005216	4	0,133140	0,05934	0,07380	0,005446
5	0,110192	0,05577	0,05442	0,002961	5	0,102260	0,05604	0,04622	0,002137
6	0,130184	0,07500	0,05519	0,003046	6	0,133574	0,07163	0,06194	0,003837
7	0,110868	0,05658	0,05428	0,002947	7	0,125138	0,05706	0,06808	0,004635
8	0,113290	0,05654	0,05675	0,003221	8	0,126006	0,05876	0,06725	0,004523
9	0,102876	0,05766	0,04522	0,002044	9	0,105227	0,05795	0,04728	0,002236
10	0,100091	0,05483	0,04527	0,002049	10	0,100071	0,05499	0,04508	0,002032
11	0,100484	0,05495	0,04553	0,002073	11	0,100121	0,05504	0,04508	0,002032
12	0,100071	0,05480	0,04527	0,002049	12	0,100061	0,05459	0,04547	0,002067
			Total	0,119998				Total	1,104104
			MSE	0,010000				MSE	0,092009
								Accuracy	83%

Table 7. Training and Testing Standard Model 6-12-1

Training					Testing				
No	Target	Output	Error	SSE	No	Target	Output	Error	SSE
1	0,322628	0,333684	0,01106	0,000122	1	0,343012	0,39833	0,05535	0,003064
2	0,223156	0,490995	0,26784	0,071738	2	0,227748	0,12072	0,10698	0,011444
3	0,167946	0,106104	0,06184	0,003824	3	0,197090	0,02506	0,17199	0,029582
4	0,129073	0,120543	0,00853	0,000073	4	0,133140	0,11292	0,02020	0,000408
5	0,110192	0,086001	0,02419	0,000585	5	0,102260	0,09347	0,00878	0,000077
6	0,130184	0,055631	0,07455	0,005558	6	0,133574	0,05130	0,08224	0,006764
7	0,110868	0,100671	0,01020	0,000104	7	0,125138	0,08522	0,03987	0,001589
8	0,113290	0,261737	0,14845	0,022037	8	0,126006	0,14528	0,01922	0,000369
9	0,102876	0,22735	0,12447	0,015494	9	0,105227	0,22545	0,12023	0,014455
10	0,100091	0,103548	0,00346	0,000012	10	0,100071	0,07865	0,02142	0,000459
11	0,100484	0,090371	0,01011	0,000102	11	0,100121	0,07724	0,02284	0,000522
12	0,100071	0,084585	0,01549	0,00020	12	0,100061	0,093176	0,00688	0,000047
			Total	0,11989				Total	0,068780
			MSE	0,00991				MSE	0,005732
								Accuracy	100%

Table 8.Training and Testing Standard Model 6-16-1

Training					Testing				
No	Target	Output	Eror	SSE	No	Target	Output	Eror	SSE
1	0,322628	0,10282	0,21981	0,048317	1	0,343012	0,74094	0,39792	0,158344
2	0,223156	0,33440	0,11124	0,012374	2	0,227748	0,88892	0,66117	0,437144
3	0,167946	0,18241	0,01446	0,000209	3	0,197090	0,74763	0,55054	0,303093
4	0,129073	0,13648	0,00741	0,000055	4	0,133140	0,12612	0,00702	0,000049
5	0,110192	0,12372	0,01352	0,000183	5	0,102260	0,11144	0,00918	0,000084
6	0,130184	0,36077	0,23059	0,053171	6	0,133574	0,31103	0,17746	0,031491
7	0,110868	0,13458	0,02371	0,000562	7	0,125138	0,13917	0,01403	0,000197
8	0,113290	0,06283	0,05046	0,002546	8	0,126006	0,14539	0,01939	0,000376
9	0,102876	0,05465	0,04822	0,002325	9	0,105227	0,10232	0,00291	0,000008
10	0,100091	0,09865	0,00144	0,000002	10	0,100071	0,11508	0,01501	0,000225
11	0,100484	0,11028	0,00980	0,000096	11	0,100121	0,12035	0,02023	0,000409
12	0,100071	0,10973	0,00966	0,000093	12	0,100061	0,10176	0,00169	0,000003
			Total	0,119934				Total	0,931425
			MSE	0,009994				MSE	0,077619
								Accuracy	75%

Table 9.Training and Testing Standard Model 6-32-1

Training					Testing				
No	Target	Output	Error	SSE	No	Target	Output	Error	SSE
1	0,322628	0,35285	0,03023	0,000914	1	0,343012	0,20459	0,13842	0,019160
2	0,223156	0,42085	0,19769	0,039082	2	0,227748	0,04181	0,18594	0,034574
3	0,167946	0,12804	0,03991	0,001593	3	0,197090	0,02471	0,17238	0,029714
4	0,129073	0,14345	0,01437	0,000207	4	0,133140	0,09851	0,03463	0,001199
5	0,110192	0,21169	0,10149	0,010301	5	0,102260	0,16947	0,06721	0,004517
6	0,130184	0,09340	0,03679	0,001353	6	0,133574	0,06175	0,07182	0,005159
7	0,110868	0,18224	0,07137	0,005094	7	0,125138	0,17542	0,05028	0,002528
8	0,113290	0,14516	0,03187	0,001016	8	0,126006	0,19851	0,07251	0,005257
9	0,102876	0,13524	0,03236	0,001047	9	0,105227	0,17450	0,06927	0,004798
10	0,100091	0,22999	0,12990	0,016874	10	0,100071	0,24336	0,14329	0,020532
11	0,100484	0,24692	0,14644	0,021444	11	0,100121	0,24390	0,14378	0,020672
12	0,100071	0,24196	0,14189	0,020131	12	0,100061	0,25109	0,15103	0,022811
			Total	0,119055				Total	0,170920
			MSE	0,009921				MSE	0,014243
								Accuracy	100%

Table 10. Training and Testing Resilient Model 6-8-1

Training					Testing				
No	Target	Output	Error	SSE	No	Target	Output	Error	SSE
1	0,322628	0,04181	0,28082	0,078858	1	0,343012	0,09659	0,24642	0,060724
2	0,223156	0,15896	0,06419	0,004121	2	0,227748	0,95293	-0,72519	0,525894
3	0,167946	0,16299	0,00496	0,000025	3	0,197090	0,77471	-0,57762	0,333644
4	0,129073	0,04767	0,08140	0,006626	4	0,133140	0,04942	0,08372	0,007009
5	0,110192	0,04694	0,06326	0,004001	5	0,102260	0,04713	0,05513	0,003039
6	0,130184	0,05913	0,07105	0,005048	6	0,133574	0,05671	0,07686	0,005908
7	0,110868	0,04757	0,06330	0,004007	7	0,125138	0,04785	0,07729	0,005973
8	0,113290	0,04783	0,06546	0,004285	8	0,126006	0,04935	0,07665	0,005876
9	0,102876	0,04865	0,05423	0,002941	9	0,105227	0,04909	0,05614	0,003152
10	0,100091	0,04633	0,05376	0,002890	10	0,100071	0,04643	0,05364	0,002877
11	0,100484	0,04642	0,05407	0,002923	11	0,100121	0,04645	0,05367	0,002881
12	0,100071	0,04630	0,05377	0,002891	12	0,100061	0,04617	0,05389	0,002904
			Total	0,118616				Total	0,959881
			MSE	0,009885				MSE	0,079990
								Accuracy	75%

Table 11. Training and Testing Resilient Model 6-12-1

Training					Testing				
No	Target	Output	Error	SSE	No	Target	Output	Error	SSE
1	0,322628	0,085276	0,23735	0,056336	1	0,343012	0,17389	0,17389	0,030239
2	0,223156	0,258470	0,03531	0,001247	2	0,227748	0,13728	0,13728	0,018847
3	0,167946	0,023932	0,14401	0,020740	3	0,197090	0,01303	0,01303	0,000170
4	0,129073	0,054688	0,07438	0,005533	4	0,133140	0,05259	0,05259	0,002766
5	0,110192	0,042358	0,06783	0,004601	5	0,102260	0,04678	0,04678	0,002188
6	0,130184	0,026821	0,10336	0,010684	6	0,133574	0,02639	0,02639	0,000696
7	0,110868	0,048712	0,06216	0,003863	7	0,125138	0,04238	0,04238	0,001796
8	0,113290	0,126583	0,01329	0,000177	8	0,126006	0,06641	0,06641	0,004410
9	0,102876	0,109511	0,00663	0,000044	9	0,105227	0,10971	0,10971	0,012035
10	0,100091	0,051821	0,04827	0,002330	10	0,100071	0,04046	0,04046	0,001637
11	0,100484	0,045729	0,05475	0,002998	11	0,100121	0,03983	0,03983	0,001586
12	0,100071	0,043132	0,05694	0,003242	12	0,100061	0,04678	0,04678	0,002188
			Total	0,111796				Total	0,078561
			MSE	0,009316				MSE	0,006547
								Accuracy	100%

Table 12. Training and Testing Resilient Model 6-16-1

Training					Testing				
No	Target	Output	Error	SSE	No	Target	Output	Error	SSE
1	0,322628	0,07769	0,24494	0,059995	1	0,343012	0,46926	0,46926	0,220205
2	0,223156	0,21597	0,00718	0,000052	2	0,227748	0,63869	0,63869	0,407919
3	0,167946	0,14952	0,01842	0,000339	3	0,197090	0,60278	0,60278	0,363340
4	0,129073	0,09126	0,03781	0,001430	4	0,133140	0,07857	0,07857	0,006174
5	0,110192	0,09168	0,01852	0,000343	5	0,102260	0,07751	0,07751	0,006008
6	0,130184	0,30257	0,17239	0,029717	6	0,133574	0,23517	0,23517	0,055304
7	0,110868	0,09204	0,01883	0,000355	7	0,125138	0,09514	0,09514	0,009052
8	0,113290	0,03991	0,07338	0,005384	8	0,126006	0,09674	0,09674	0,009359
9	0,102876	0,03535	0,06752	0,004559	9	0,105227	0,06472	0,06472	0,004188
10	0,100091	0,08217	0,01792	0,000321	10	0,100071	0,09751	0,09751	0,009508
11	0,100484	0,09226	0,00822	0,000068	11	0,100121	0,10285	0,10285	0,010579
12	0,100071	0,09619	0,00388	0,000015	12	0,100061	0,09211	0,09211	0,008484
			Total	0,102577				Total	1,110119
			MSE	0,008548				MSE	0,092510
								Accuracy 67%	

Table 13. Training and Testing Resilient Model 6-32-1

Training					Testing				
No	Target	Output	Error	SSE	No	Target	Output	Error	SSE
1	0,322628	0,25257	0,07005	0,004908	1	0,343012	0,19061	0,15240	0,023226
2	0,223156	0,25903	0,03587	0,001287	2	0,227748	0,06206	0,16568	0,027451
3	0,167946	0,02808	0,13987	0,019564	3	0,197090	0,03277	0,16432	0,027001
4	0,129073	0,00958	0,11949	0,014279	4	0,133140	0,00781	0,12533	0,015708
5	0,110192	0,01581	0,09438	0,008908	5	0,102260	0,01280	0,08946	0,008002
6	0,130184	0,01079	0,11940	0,014256	6	0,133574	0,00670	0,12687	0,016096
7	0,110868	0,01387	0,09700	0,009408	7	0,125138	0,01363	0,11151	0,012435
8	0,113290	0,01132	0,10197	0,010397	8	0,126006	0,01657	0,10943	0,011976

9	0,102876	0,01232	0,09056	0,008201	9	0,105227	0,01415	0,09108	0,008295
10	0,100091	0,01940	0,08069	0,006511	10	0,100071	0,02193	0,07814	0,006106
11	0,100484	0,02143	0,07905	0,006249	11	0,100121	0,02170	0,07842	0,006150
12	0,100071	0,02195	0,07812	0,006103	12	0,100061	0,02304	0,07702	0,005932
			Total	0,110069				Total	0,168379
			MSE	0,009172				MSE	0,014032
								Accuracy	100%

Each of these architectures is processed using the standard backpropagation algorithm and the Resilient Backpropagation (Rprop) algorithm. The best architecture can be seen from the accuracy of the truth, a little more epochs and the size of MSE, Data accuracy, Amount of Epochs and MSE from tested models.

Table 14. Comparison of Standard Backpropagation with Resilient Backpropagation algorithm (Rprop)

Model	**Criteria**	**Standard**	**Rprop**
6-8-1	MSE	0,084931	0,073837
	Epochs	3490	4
	Accuracy	83%	75%
6-12-1	MSE	0,005291	0,006043
	Epochs	636	3
	Accuracy	75%	67%
6-16-1	MSE	0,071648	0,085394
	Epochs	755	6
	Accuracy	100%	100%
6-32-1	MSE	0,013148	0,012952
	Epochs	170	2
	Accuracy	100 %	100 %

From the results above, it can be seen that the best model to predict is architecture 6-32-1 with 100% accuracy.

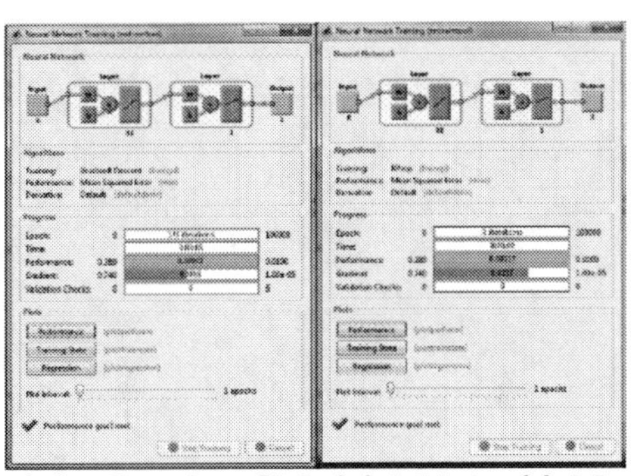

Figure 1. Training Results of Model 6-32-1 with Matlab R2011A

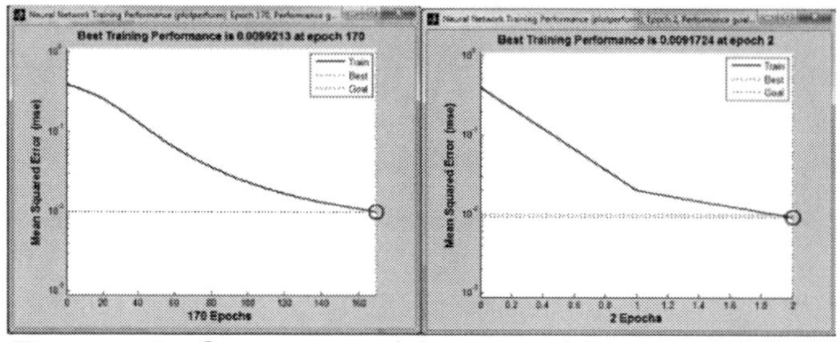

Figure 2. Performance Model 6-32-1 with Matlab R2011A

3.3. Prediction of School Participation Rates

By using the best architecture that has been obtained, the prediction of fertilizer imports according to country of origin is carried out and the results can be seen in Table 15 and 16 respectively.

Table 15. Prediction with the Resilient Backpropagation algorithm (Rprop)

Origin Country	Prediction	Target	output	Error	Sse
China[1]	882,894	0,343012	0,393242	-0,05747	0,003303
Canada	640,2384	0,227748	0,312629	0,103128	0,010635
Russia	-22,8691	0,19709	0,092336	-0,07029	0,00494
Malaysia	454,6694	0,13314	0,25098	-0,12832	0,016466
Norway	663,5241	0,10226	0,320364	-0,14025	0,019671
Australia	398,8438	0,133574	0,232434	-0,12223	0,014941
Germany	647,011	0,125138	0,314878	-0,11771	0,013856
Jordan	479,9736	0,126006	0,259387	-0,03219	0,001036
South Korea	659,2646	0,105227	0,318949	-0,12853	0,01652
Japan	730,9429	0,100071	0,342762	-0,09562	0,009144
Thailand	711,6193	0,100121	0,336342	-0,0903	0,008154
Philippines	722,4551	0,100061	0,339942	-0,09013	0,008123
Total					0,12679
MSE					0,010565
Accuracy					100%

Table 16. Prediction with standard algorithm

Origin Country	Prediction	target	output	Error	Sse
China[1]	624,2482	0,343012	0,307316	0,068996	0,004760
Canada	40,98268	0,227748	0,113548	0,210000	0,044100
Russia	-283,52	0,197090	0,005745	0,005660	0,000032
Malaysia	-268,472	0,133140	0,010744	0,004528	0,000021
Norway	-248,423	0,102260	0,017404	0,003084	0,000010
Australia	-270,222	0,133574	0,010162	0,003900	0,000015
Germany	-255,721	0,125138	0,01498	0,007585	0,000058
Jordan	-266,12	0,126006	0,011525	0,016104	0,000259
South Korea	-248,717	0,105227	0,017307	0,004299	0,000018
Japan	-239,105	0,100071	0,0205	0,010712	0,000115
Thailand	-238,978	0,100121	0,020542	0,009928	0,000099
Philippines	-237,839	0,100061	0,02092	0,010690	0,000114
Total					0,0496
MSE					0,004133
Accuracy					100%

4. Conclusion

From this study it can be concluded that:
a. The accuracy of the rprop algorithm and the accuracy standard are 100%.
b. The iteration needed by the resilient algorithm is smaller than the standard algorithm.
c. MSE resilient algorithm is smaller than standard algorithm.

References

[1] A, Wanto, M, Zarlis, Sawaluddin, and D, Hartama, "Analysis of Artificial Neural Network Backpropagation Using Conjugate Gradient Fletcher Reeves in the Predicting Process," *Journal of Physics: Conference Series*, vol, 930, no, 1, pp, 1–7, 2017.

[2] H, Chiroma, S, Abdulkareem, A, I, Abubakar, and T, Herawan, "Kernel functions for the support vector machine: Comparing performances on crude oil price data," *Advances in Intelligent Systems and Computing*, vol, 287, pp, 273–281, 2014.

[3] R, B, F, Hakim, E, N, Sari, and T, Herawan, "Soft Solution of Soft Set Theory for Recommendation in Decision Making," *Advances in Intelligent Systems and Computing*, vol, 287, pp, 313–324, 2014.

[4] A, Lasisi, R, Ghazali, and T, Herawan, "Comparative Performance Analysis of Negative Selection Algorithm with Immune and Classification Algorithms," *Advances in Intelligent Systems and Computing*, vol, 287, pp, 441–452, 2014.

[5] P, Hamet and J, Tremblay, "Artificial Intelligence in Medicine," *Metabolism: Clinical and Experimental*, pp, 1–14, 2017.

[6] J, Adnan *et al*,, "Multilayer Perceptron Based Activation Function On Heart," *Journal of Fundamental and Applied Sciences*, vol, 9, no, 3, pp, 417–432, 2017.

[7] A, Wanto *et al,*, "Levenberg-Marquardt Algorithm Combined with Bipolar Sigmoid Function to Measure Open Unemployment Rate in Indonesia," 2018, pp, 1–7.

[8] A, Ehret, D, Hochstuhl, D, Gianola, and G, Thaller, "Application of Neural Networks with Nack-propagation to Genome-Enabled Prediction of Complex Traits in Holstein-Friesian andGerman Fleckvieh Cattle," *Genetics Selection Evolution*, vol, 47, no, 22, pp, 1–9, 2015.

[9] M, Fauzan *et al,*, "Epoch Analysis and Accuracy 3 ANN Algorithm Using Consumer Price Index Data in Indonesia," 2018, pp, 1–7.

[10] P, Antwi *et al,*, "Estimation of Biogas and Methane Yields in an UASB Treating Potato Starch Processing Wastewater With Backpropagation Artificial Neural Network," *Bioresource Technology*, vol, 228, pp, 106–115, 2017.

[11] S, P, Siregar and A, Wanto, "Analysis of Artificial Neural Network Accuracy Using Backpropagation Algorithm In Predicting Process (Forecasting)," *International Journal Of Information System & Technology*, vol. 1, no, 1, pp, 34–42, 2017.

[12] D, Huang and Z, Wu, "Forecasting outpatient visits using empirical mode decomposition coupled with backpropagation artificial neural networks optimized by particle swarm optimization," *PLoS ONE*, vol. 12, no, 2, pp, 1–17, 2017.

[13] A, Wanto, A, P, Windarto, D, Hartama, and I, Parlina, "Use of Binary Sigmoid Function And Linear Identity In Artificial Neural Networks For Forecasting Population Density," *International Journal Of Information System & Technology*, vol. 1, no, 1, pp, 43–54, 2017.

[14] B, Febriadi, Z, Zamzami, Y, Yunefri, and A, Wanto, "Bipolar function in backpropagation algorithm in predicting Indonesia's coal exports by major destination countries," *IOP Conference Series: Materials Science and Engineering*, vol, 420, no, 12089, pp, 1–9, 2018.

[15] A, Wanto *et al,*, "Analysis of Standard Gradient Descent with GD Momentum And Adaptive LR for SPR Prediction," 2018, pp, 1–9.

[16] N, Nasution, A, Zamsuri, L, Lisnawita, and A, Wanto, "Polak-Ribiere updates analysis with binary and linear function in determining coffee exports in Indonesia," *IOP Conference Series: Materials Science and Engineering*, vol. 420, no, 12089, pp, 1–9, 2018.

[17] S, Sumijan, A, Windarto, A, Muhammad, and B, Budiharjo, "Implementation of Neural Networks in Predicting the Understanding Level of Students Subject," *International Journal of Software Engineering and Its Applications*, vol,.10, no, 10, pp, 189–204, 2016.

[18] A, P, Windarto, L, S, Dewi, and D, Hartama, "Implementation of Artificial Intelligence in Predicting the Value of Indonesian Oil and Gas Exports With BP Algorithm," *International Journal of Recent Trends in Engineering & Research (IJRTER)*, vol. 3, no, 10, pp, 1–12, 2017.

IOP Conf. Series: Materials Science and Engineering **769** (2020) 012039 doi:10.1088/1757-899X/769/1/012039

Implementation of Data Mining In Predicting the Study Period of Student Using the Naïve Bayes Algorithm

Sri Wahyuni[1], Murni Marbun[2]
[1] Departement of Teknik Computer Universitas Pembangunan Panca Budi Medan, Indonesia
[2] Department of Teknik Informatika STMIK Pelita Nusantara Medan, Indonesia

E-mail: sriwahyuni@dosen.pancabudi.ac.id

Abstract. The period of study of students is one of the parameters in the assessment of accreditation of university. Pembangunan Panca Budi University as one of the best institutions in North Sumatra needs to predict the period of study of students in order to know the strategies implemented so that students graduate on time. This study predicts the period of study using the Naïve Bayes Data Mining Algorithm method to students of the Pembangunan Panca Budi University of Computer System study program bachelor program with a study period of 4 years. Application designed using Naïve Bayes algorithm that works based on the shortest distance between two objects by determining the value of k. The value of k is a parameter to determine the closest distance between a new object and an old object. The data used as variables are student academic data, GPA for the first from 4 semesters, the average value of the National Examination during high school, and majors during high school. The results of the study showed that the data mining application that was designed could predict the period of study of students at the Pembangunan Panca Budi University.

1. Introduction

Data Mining is a semi-automatic process that uses statistical techniques, mathematics, artificial intelligence and machine learning to extract and identify potential and useful knowledge information stored in large databases. Having a very large amount of data, for example lecturers, staff, facilities and infrastructure in universities is currently required to have the ability to compete by utilizing all the resources they have. As mentioned earlier, the data mentioned is student data in the data set, for example data on potential students dropping out. Understanding the profile of potential drop out students can be done by revealing knowledge, also used to find out more about the profile of a student who is going to take the final examination whether it tends to fail or succeed. Knowledge possessed to understand, predict, and prevent failure is very important for the management of higher education, especially University of Pembangunan Panca Budi. Previous research was conducted using the Decision Tree C4.5 method where is it known that the most influential variables are from school origin [1]. Naïve Bayes algorithm works based on the shortest distance between two objects by determining the value of k. The k value is a parameter used in determining the closest distance between a new object and the previous object [2].

This knowledge can be used in helping lecturers to get to know the situation of their students, and can be used as early knowledge in the decision making process. Measures of student success or achievement can be seen from the Grade Point Average (GPA) reflecting all grades obtained by students until the current semester. GPA is obtained by summing all the quality values of all courses taken and dividing by the total credits (semester credit units).

By using data mining techniques, such as classification algorithms, it is possible to find characteristics of student achievement and use their characteristics to predict future achievements. The research was conducted at University of Pembangunan Panca Budi where student data obtained from the last five years of Pembangunan Panca Budi University are illustrated in table 1 below:

Table 1. Student Data Year 2010 until Year 2015

Year	Number of Student Entering	Number of Student Graduating
2010	1776 person	199 person
2011	1979 person	630 person
2012	2155 person	83 person
2013	1914 person	166 person
2014	3286 person	81 person
2015	3774 person	-

From the table above it can be clearly concluded that every year there is an increase in the number of students drop out. Then an application is needed that can predict the study period of students to reduce student numbers drop out. Data mining methods are often applied at the University level to analyze available data about academic data and extract new information and knowledge in support of decision making [3].

2. Research Methods

The stages of the research conducted in the Data Mining Implementation study in predicting the study period of students with the Naïve Bayes algorithm as follows:

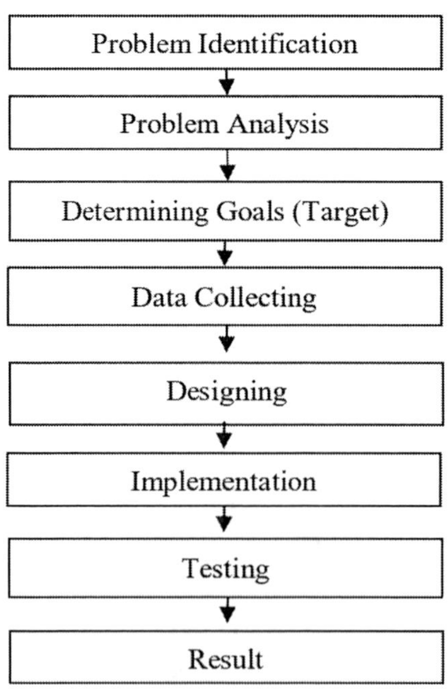

Figure 1. Research Framework

2.1 Data Calection

The data used in this study were obtained from University of Pembangunan Panca Budi. The data used as samples is 20 data. The data are 10 testing data and 10 data. Details of the data traning used can be seen in Table 2.

Table 2. Training Data for Student Study Period

Name	GPA Semester 1	GPA Semester 2	GPA Semester 3	GPA Semester 4	Average National Examina tion Score	High School Major	Studi Period
Andi	3.00	3.11	1.89	1.70	85	Science	> 4 Years
Anisa	2.72	2.32	1.89	1.60	78	Social	> 4 Years
Anggi	3.28	3.37	3.32	2.00	87	Science	<=4 Years
Farhan	3.61	3.16	2.89	3.11	82	Science	<=4 Years
Fatin	3.50	3.42	3.27	2.84	89	Social	<=4 Years
Siska	2.83	2.63	2.58	2.25	78	Science	> 4 Years
Fajar	1.50	1.89	1.93	1.44	81	Social	> 4 Years
Widya	3.06	3.89	3.53	3.55	89	Social	<=4 Years
Nikita	3.11	3.21	2.89	3.22	88	Science	<=4 Years
Edi	2.17	2.16	1.29	1.65	76	Science	> 4 Years

After getting the data to be used as training data, then test data or new data will be made which is seen in the table below and determined whether the new data is at which level.

Table 3. Testing Data

No	Name	GPA Semester 1	GPA Semester 2	GPA Semester 3	GPA Semester 4	Average National Examination Score	High School Major	Studi Period
1.	Agus	3,97	3.79	3.88	3.35	90	Science	-

Before searching the study period value for testing data, value limits will be determined first for 1-4 semester GPA and national exam scores as shown in table 4 below.

Table 4. Value Limit for 1 - 4 Semester GPA and Average National Examination Score

No	Attribute	Limit
1.	GPA Semester 1	<= 2.8 2.81-3.2 >=3.21
2.	GPA Semester 2	<= 2.8 2.81-3.2 >=3.21
3.	GPA Semester 3	<= 2.8 2.81-3.2 >=3.21
4.	GPA Semester 4	<= 2.8 2.81-3.2 >=3.21
5.	**Average National Examination Score**	<= 80 81-86 >=87

After carrying out the value threshold for criteria that has an integer value type then the classification process will be carried out using the naïve bayes method based on existing training data as will be displayed in the process below.

3. Discussion And Result

3.1 Manual Calculation

P (Ci)
P (Study Period > 4 years) = 5/10 = 0.5
P (Study Period <= 4 years) = 5/10 = 0.5
P (X|Ci)
Testing Data for Semester 1 is: 3.97, so the threshold taken is >= 3.21

1. GPA Semester 1
   P (GPA Semester 1 = ">= 3.21" |Study Period = "> 4 years ") = 0 / 5 = 0
   P (GPA Semester 1= ">= 3.21" | Study Period = "<= 4 years ") = 3 / 5 = 0.6
2. GPA Semester 2
   Testing Data for Semester 2 adalah: 3.79, so the threshold taken is >= 3.21
   P (GPA Semester 2 = ">= 3.21" | Study Period = "> 4 years ") = 0 / 5 = 0
   P (GPA Semester 2 = ">= 3.21" | Study Period = "<= 4 years ") = 4 / 5 = 0.8
3. GPA Semester 3
   Testing Data for Semester 3 adalah: 3.88, so the threshold taken is >= 3.21
   P (GPA Semester 3 = ">= 3.21" | Study Period = "> 4 years ") = 0 / 5 = 0
   P (GPA Semester 3 = ">= 3.21" | Study Period = "<= 4 years ") = 3 / 5 = 0.6
4. IP Semester 4
   Testing Data for Semester 4 adalah: 3.35, so the threshold taken is >= 3.21
   P (GPA Semester 4 = ">= 3.21" | Study Period = "> 4 years ") = 0 / 5 = 0
   P (GPA Semester 4 = ">= 3.21" | Study Period = "<= 4 years ") = 2 / 5 = 0.4
5. National Examnition
   Testing Data for the average national examination score is: 90, so the threshold taken is >= 87
   P (National Examination Score =" >= 87" | Study Period = "> 4 years ") = 0 / 5 = 0
   P (National Examination Score = ">= 87" | Study Period = "<= 4 years ") = 3 / 5 = 0.6
6. Major
   Testing Data for major is: Science
   P (major = "Science" | Study Period = "> 4 years ") = 3 / 5 = 0.6
   P (major = "Science" | Study Period = "<= 4 years ") = 4 / 5 = 0.8
   P(X | Study Period > 4 years) = 0 * 0 * 0 * 0 * 0 * 0.6 = 0
   P(X | Study Period <= 4 years) = 0.6 * 0.8 * 0.6 * 0.4 * 0.6 * 0.8 = 0.055296

P (X|Ci)*P (Ci)
P (Study Period > 4 years) * P(X | Study Period > 4 years) = 0 * 0.5 = 0
P (Study Period <= 4 years) * P(X | Study Period <= 4 years) = 0.055296 * 0.5 = 0.0276
So for existing data testing, the study period is included in the classification <= 4 years.

3.2 *Software Implementation*

For Input Training data, Training data inputting form, there are data that must be filled by an administrator, namely Student Identity Number, name of student, GPA semester 1, GPA semester 2, GPA semester 3, GPA semester 4, average national examination score, high school majors, and study periods.

Display testing data processing functions for the admin to display the testing data that has been inputted on the Testing form. The data displayed is Student Identity Number, name of student, GPA semester 1, GPA semester 2, GPA semester 3, GPA semester 4, average national examination score, high school majors, and study periods. Display output testing processing data can be seen in Figure 2.

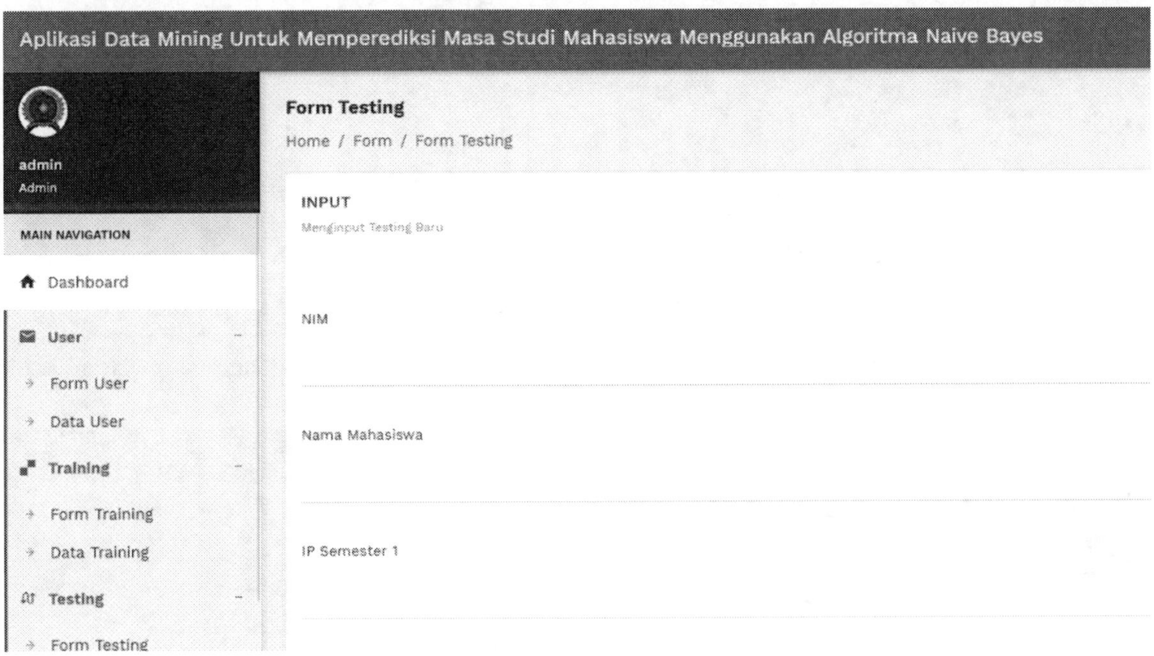

Figure 2. Training Data Processing

The Input Data Testing page is an image display for Input Data Testing. For the testing data inputting form, there are data that must be filled by an admin, namely Student Identity Number, name of student, GPA semester 1, GPA semester 2, GPA semester 3, GPA semester 4, average national examination score, high school majors. Input Data Testing page is an image display for input data testing. For the testing data inputting form, there are data that must be filled by an admin, namely Student Identity Number, name of student, GPA semester 1, GPA semester 2, GPA semester 3, GPA semester 4, average national examination score and high school majors.

Display testing data processing functions for the admin to display the Testing data that has been inputted on the Testing form. The data displayed is Student Identity Number, name of student, GPA semester 1, GPA semester 2, GPA semester 3, GPA semester 4, average national examination score, high school majors, and study periods. Display testing data processing output. The data testing processing form functions for the admin to display the Testing data that has been inputted on the Testing form. The data displayed is Student Identity Number, name of student, GPA semester 1, GPA semester

2, GPA semester 3, GPA semester 4, average national examination score, high school majors, and study periods.

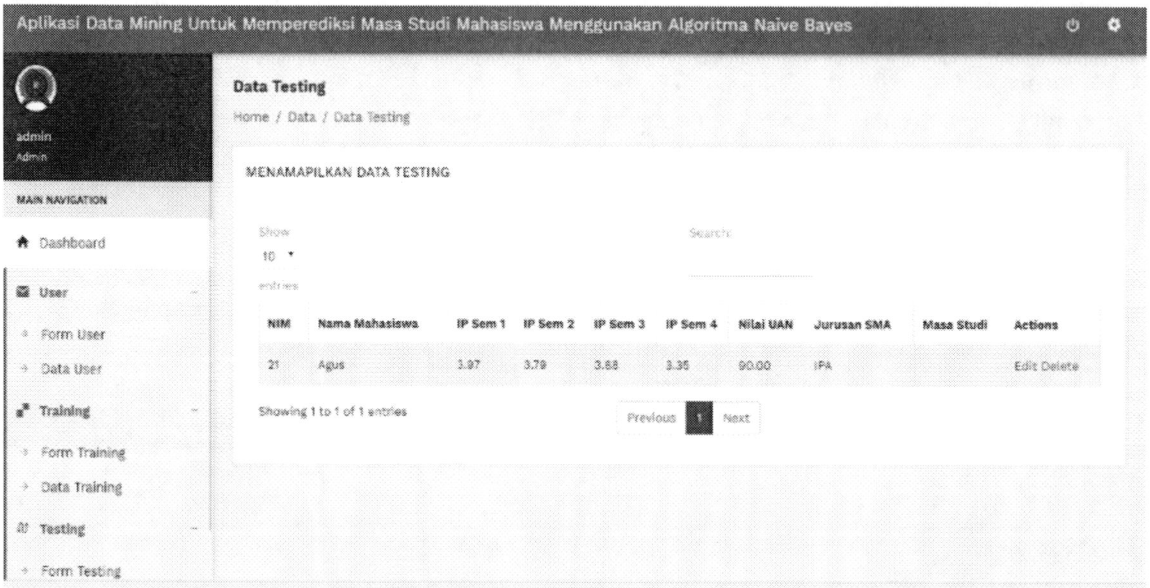

Figure 3. Testing Data Processing

This page displays the classification process using the Naive Bayes method where the table column displayed is student identity number, student name, probability of Naive Bayes probability for study period over 4 years, Naive Bayes probability value for the 4-year study period, and mass classification studies.

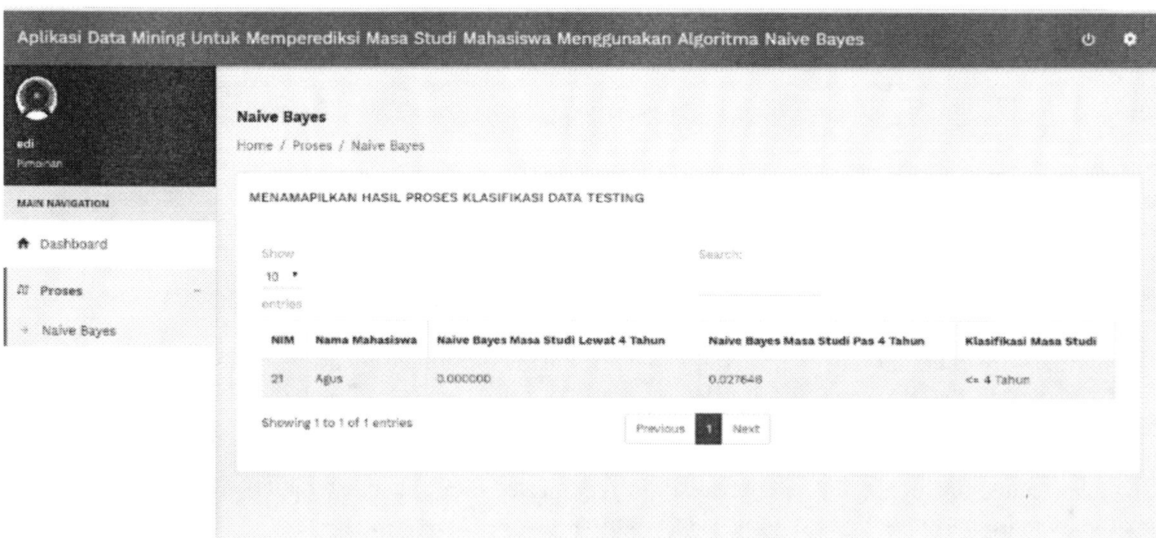

Figure 4. Mining Data Classification Process

4. Conclusion

Implementation of Data Mining with the Naïve Bayes Algorithm can predict the period of study the students of Computer Systems study program at the Pembangunan Panca Budi University. Based on a sample of 10 testing data with the variables used, the predicted results of the study period were exactly

4 years if the student came from the IPA department, the national examination scores 9 and IP scores above 3.0. The results of the implementation of a data mining system that was designed can be utilized to create a student graduation strategy on time and emphasize drop-out students.

References

[1] Wahyuni, S. 2018. *Implementasi Metode Decision Tree C4.5 Untuk Menganalisa Mahasiswa Drop Out*. ETHOS Jurnal Penelitian dan Pengabdian. Vol. 6, No.1:39-48, 2018
[2] Nur Sukma. 2017. *KLasifikasi Calon Pendonor Darah Dengan Metode Naive Bayes Classifier*. Malang: Universitas Brawijaya Malang
[3] D. Kabakchieva. 2013. *Predicting Student Performance by Using Data Mining Methods for Classification*. Sofia University
[4] Hariyanto, Eko and Rahim, Robbi. *Arnold's Cat Map Algorithm in Digital Image Encryption*. *International Journal of Science and Research (IJSR)*, Vol 5 Issue 10, 2016, pp 1363-65,
[5] Yeni, 2011, "*Algoritma Data Mining*". Penerbit Andi.
[6] Wahyuni, S. 2017. *Implementasi Rapidminer Dalam Menganalisa Data Mahasiswa Drop Out*. Medan: Jurnal Ilmiah Abdi Ilmu. Vol. 10, No.2:1899-1902

Optimizing Production Planning for Tempe Babe Within Goal Programming Method

Sulindawaty[1], Syahril Effendi[2], Muhammad Zarlis[3], Herman Mawengkang[4]
[1]Graduate Program of Computer, Department of Computer Science, Universitas Sumatera Utara, Medan, Indonesia
[2,3]Department of Computer Science, Universitas Sumatera Utara, Medan, Indonesia
[4]Department of Matematics, Universitas Sumatera Utara, Medan, Indonesia
Email: sulindawaty@gmail.com

Abstract. Production planning is the most important part of the production process in Tempe Babe. The material consideration in the production planning in this research is the optimizing of the amount of production in fulfilling the number of tempe demand, optimizing of the amount of profit Tempe obtained and the minimization of the amount of raw materials used for Tempe production. The implementation of the goal programming method can provide solutions for optimizing production planning on Tempe Babe, where the deviational variables in goal programming serve to demonstrate the possibility of negative and positive deviations from right side value of the destination function. The results of calculation analysis with goal programming, indicating the excess amount of product production will be added to the product that has the biggest advantage and the excess amount of production will be added to the product that has the number of requests Optimal profit from each product resulting in the amount of prediction of Tempe production. The results of this research provide the optimum solution for production planning on Tempe Babe every month. In addition, this study also resulted in a new mindset to find production planning models in the IoT environment as a continuation of this research.

Keywords: Production Planning, Tempe Babe, Goal Programming, optimization

1. Introduction

Production planning is part of operational planning within the company. Production planning is a plan about the number and type of products that the company will produce in a given period [3]. Some things to consider in drafting production planning are product optimization problems that can reduce production costs so that the low production cost levels can be achieved for production processes [4]. Product optimization In this case is maximising production volumes to meet demand, maximize profitability, and minimize the use of raw materials.

Tempe Babe is a business engaged in producing tempeh. When the demand for Tempe increases, Tempe Babe will also increase the amount of Tempe production as expected. Tempe Babe wants the amount of Tempe produced can fulfill all the demands, so that the gains gained also increase. The increasing number of tempeh is produced, resulting in increased raw materials used. The raw material supply to make tempeh owned by Tempe Babe is limited, so Tempe Babe wants a minimum use of raw materials but can still meet all consumer demand.

There are a number of destinations that Tempe Babe wants to achieve, which is maximizing profits, meeting all requests by minimizing the use of raw materials available. With this condition, it takes a mathematical model that can find its optimal solution. One of the mathematical models that can be used in production planning with some objectives is goal programming. This method of goal programming can be used for problems that have many goals. The goal programming method is able to provide an effective solution to the objectives that have been established [8]. This method is an extension of linear programming, where linear programming is only able to solve the problem with a single purpose while the goal programming is able to solve problems with multiple objectives [9].

2. The problem

The problem formulation in this research is how to optimize the limited resources (raw materials) of existing production in Tempe Babe using goal programming method.

3. Production Planning

Production planning is planning about what product and how many companies will be manufactured in a future period. Production planning is part of operational planning within the company[2]. Production planning is a very important part. Because the industry is often faced with complex problems in taking a decision to maximize the company's profit, so the company strives to maximize production volume in order to meet the number of requests Making optimal production planning [5].

In the preparation of production planning, the need to be considered is the production optimization so that it will be achieved the lowest cost level for the implementation of the production process [1]. This research makes the concept of optimization of production planning using goal programming method which is able to provide predictions on the production of Babe Tempe. Completion of goal programming using simplex method resulting in the production planning of Tempe Monthly.

4. Data Analysis

The Data used in this research is the amount of Tempe production, the number of Tempe requests, the profit amount of each Tempe, the amount of raw materials used to make tempeh and the amount of raw material availability.

Data collection is done by interview and direct observation. The interview was done by asking the owner Tempe Babe. The Data obtained from the interview are the amount of raw materials used to make tempeh and the price of each raw material used. To strengthen the results of interviews conducted direct observation of the amount of tempeh produced or the availability of raw materials used.

a. Analisis Goal Programming

Goal programming is applied to linear problems by incorporating various objectives in the model formulation. Each goal is expressed as a goal and is presented numerically. This numerically stated Goal is what it tries to accomplish [10]. Various goals are not always achievable at the same time because there can be deviation from the goal, so that in the formulation of goal programming, the goal in numerical for each goal must be set first. Then, the solution to be sought is to minimize the number of deviations of each goal against each of the goal [6].

Analysis of the goal programming method for the production planning of Tempe on Tempe Babe as follows:

1. Formulating the target constraints function
   In drafting a programing goal first must be specified a decision variable from the optimization model, then formulate the constraints function, and the function of the goal to be achieved.
   a. Specifying decision variables
      The decision variable in this issue is the production amount of each type of Tempe.
      Xi = Total Production of the product to-i

i = Product type is generated i=1,2,3,4.
$X1$ = Total production of Tempe 10x
$X2$ = Total production of Tempe 15x
$X3$ = Total production of Tempe 20x
$X4$ = Total production of Tempe 25x

b. Formulation of constraint function

In this study, there are several goals or objectives that want to be achieved to assist decision makers in making production planning, these objectives include:

Maximize production volume to meet the number of requests

The formulation of target constraints function maximizes production volumes to meet the number of requests is as follows:

With:

XI = number of products I produced
PI = Product Request Level I
DBI = deviation value under Pi
Dai = deviation value above pi

The formulation of target constraints function above can be used as follows:

$$X1 + DB1 - DA1 = 625$$
$$X2 + DB2 - DA2 = 547$$
$$X3 + DB3 - DA3 = 591$$
$$X4 + DB4 - DA4 = 585$$

In this case, the desired goal of Tempe Babe is to maximize the volume of production, hence the negative deviation (shortage of production amount) is zero so that the negative deviation should be

- Maximize profits

The base price for the manufacture of each Tempe varies because it uses a different amount of soybean depending on the size of Tempe. By taking into account the number of Tempe requests, the estimated amount of profit to be achieved can be calculated as follows:

Projection Advantage:

$$\sum UiXi \ 4 \ i=1$$

U = Profit for 1 seed Tempe
x = number of Tempe requests
i = Type Tempe

Then the projection gains:

Projected gains = (Rp.650x260)+(Rp.613x180)+(Rp.659x120)+(Rp.653 x20)
= Rp. 371,480

The equation of function for projection gains is:

$650X1 + 613X2 + 659X3 + 653X4 \geq 371{,}480$

In this case, the target of Tempe Babeis maximizing profit. Thus, it is expected that negative deviations (gains under the projected profit) are cultivated zero so that negative deviations (DB) must be minimized. Thus, the target constraint function for this function is: $650X1 + 613X2 + 659X3 + 653X4 + DB5 - DA5 = 371{,}480$

- Minimize raw material usage

The use and availability of raw materials as a function constraint is to see the relationship between usage and the availability of raw materials with the number of products produced. The amount of raw materials for each product must be smaller or equal to the availability of these materials. The formulations used are: $\sum \sum BlXi \leq BTl \ 4 \ i=1 \ 12 \ l=1$

In this case, the goal of Tempe Babe is minimizing the use of raw materials. So, it is expected that a positive deviation (excess raw material) is cultivated zero so that positive deviation (DA) must be mimised. Thus, the target constraint function for this function is:

$0.00429X1 + 0.00429X2 + 0.00375X3 + 0.00375X4 + DB6 - DA6 = 100$
$0X1 + 0X2 + 0.00435X3 + 0.00435X4+DB7 - DA7 = 100$
$0.001X1 + 0.001X2 + 0X3 + 0X4 + DB8 - DA8 = 30$
$0X1 + 0X2 + 0.00088X3 + 0.00088X4+DB9 - DA9 = 45$
$0.00975X1 + 0.01116X2 + 0.00280X3 + 0.00280X4+DB10 - DA10 = 75$
$0.00530X1 + 0.00530X2 + 0X3 + 0X4+DB11 - DA11 = 75$
$0.00008X1 + 0.00022X2 + 0.00217X3 + 0.00220X4+DB12 - DA12 = 30$
$0.00026X1 + 0.00048X2 + 0.000308X3 + 0.00030X4+DB13 - DA13 = 17$
$0X1 + 0.000035X2 + X3 + 0X4+DB14 - DA14 = 0.5$
$0X1 + 0X2 + 0.00024X3 + 0X4+DB15 - DA15 = 6,75$
$0X1 + 0X2 + 0X3 + 0.00006X4+DB16 - DA16 = 2$
$0X1 + 0X2 + 0X3 + 0.00035X4+DB17 - DA17 = 5$

In order for DA equation above the minimum, then the goal function equation becomes:
Min Z=$\sum DAi$ 17 i=6

2. Formulating Goal Programming formulation
   Based on the objectives-the goal that Tempe Babe wants to achieve is:
   - Maximize production Volume
   - Maximize profits
   - Minimize the use of raw materials and goal programming formulation

b. Solutions with the Simplex method

Production planning of tempeh on Tempe Babe using simplex method is as follows [7]:
1st iteration:
- The first step is to calculate the Zj value.
- The next step is to calculate the value of Cj-Zj. Once the Cj-Zj value is obtained, the next step is to see if the table is optimum. Table is optimal if the value of Cj-Zj ≥ 0 for the entire J. If the table is not optimal then the Cj-Zj value that has the largest negative value is selected as the key column. In this case, the X3 column is the key column.
- Once the key column is obtained, the next stage is selecting the key row. The key line is obtained from the row that has the smallest ratio value and the ratio value should be greater than zero.
- The next stage is to swap the base value on the key row with the decision variable on the key column.

2nd Iteration:
- The first step is to replace the DB3 base variable with the X1 decision variable. In the 1st Iteration simplex table The key element is a coefficient of X1 in line 1, so in the iteration table to 2 the key coefficient of the element should be 1, whereas for the coefficient of X1 in the other row, which is not the key element must be 0.
- In the second iteration and the next iteration of the step used is the same as the 1st iteration, which is calculating the Zj, Cj-Zj value, looking for the Cj-Zj value that has the largest negative value. Calculations are carried out until optimum table.

5. Result

Based on settlement using simplex method obtained the optimal solution because the whole Cj-Zj ≥ 0. Thus the optimum solution for production planning is Tempe Babe produces each month as follows:

- Production of Tempe X1 as much as 6000 Tempe
- Production of Tempe X2 as much as 5800 Tempe
- Production of Tempe X3 as much as 5700 Tempe
- Production of Tempe X4 as much as 5200 Tempe

This research resulted in a framework of thought for the development of the results of this research, where production optimization will be applied according to the needs in the IoT environment. The frame of thought is produced as follows:

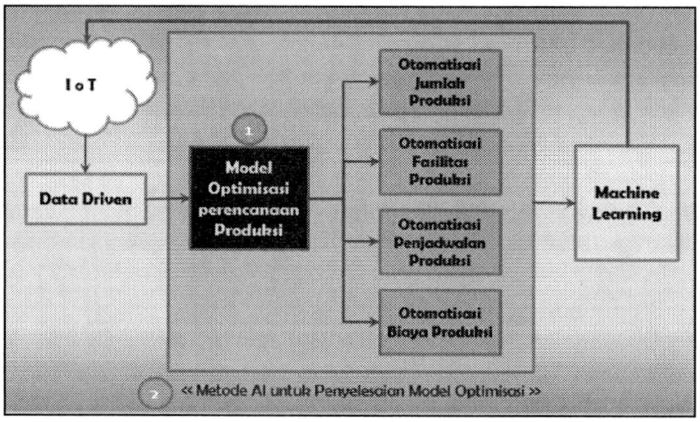

Figure 1: The Production planning optimisation Framework

Based on the image above this study will be developed with a big data scope that requires a more complex model. The research was developed to find a production planning model in the IoT environment by developing methods from previous research to find new methods for production planning model completion in the IoT environment. The use of IoT technology is expected to improve production planning, as it will be able to produce more directional and realtime planning.

6. Conclusions

The research of production planning optimization on Tempe Babe using the goal programming method can provide information to predict the amount of Tempe production each month. The results of this research provide the optimum solution for production planning on Tempe Babe every month. In addition, this study also resulted in a new mindset to find production planning models. The NEW mindset of this study will be developed in subsequent research by developing methods from previous research to find new methods for completion of production planning models in the IoT environment.

References

[1] Alarcón, F., Perez, D., & Boza, A. (2016). Using The Internet Of Things In A Production Planning Context. Brazilian Journal of Operations & Production Management. https://doi.org/10.14488/bjopm.2016.v13.n1.a8.

[2] Cardoso, S. R., Barbosa-Póvoa, A. P. F. D., & Relvas, S. (2013). Design and planning of supply chains with integration of reverse logistics activities under demand uncertainty. European Journal of Operational Research. https://doi.org/10.1016/j.ejor.2012.11.035

[3] Depuy, G. W., Usher, J. S., Walker, R. L., & Taylor, G. D. (2007). Production planning for remanufactured products. Production Planning and Control. https://doi.org/10.1080/09537280701542210

[4] Faustino Alarcóna; David Pereza; Andrés Bozaa,"using the internet of things in a production planning context", Brazilian Journal of Operations & Production Management13 (2016), pp 72-76.

[5] Krenczyk, D., Kalinowski, K., & Grabowik, C. (2012). Integration production planning and scheduling systems for determination of transitional phases in repetitive production. In Lecture Notes in Computer Science (including subseries Lecture Notes in Artificial Intelligence and Lecture Notes in Bioinformatics). https://doi.org/10.1007/978-3-642-28931-6_27

[6] Marpaung, J. 2009. Perencanaan Produksi Yang Optimal Dengan Pendekatan Goal Programming Di PT.Gold Coin Indonesia. Skripsi. Sumatera Utara: Universitas Sumatera Utara.

[7] Nafisah, L., Sutrisno, S., & H. Hutagaol, Y. E. (2018). Perencanaan Produksi Menggunakan Goal Programming (Studi Kasus di Bakpia Pathuk 75 Yogyakarta). SPEKTRUM INDUSTRI. https://doi.org/10.12928/si.v14i2.4913

[8] Papathanasiou, J., & Ploskas, N. (2018). Goal programming. In Springer Optimization and Its Applications. https://doi.org/10.1007/978-3-319-91648-4_6

[9] Siswanto, 2007. Operations Research Jilid 1. Jakarta: Erlangga.

[10] Zhang, Y., & Tao, F. (2016). Optimization of Manufacturing Systems Using the Internet of Things. Optimization of Manufacturing Systems Using the Internet of Things.

Enhanced Global and Local Curvature Properties for Corner Detection

Suraya Abu Bakar[1], Muhammad Suzuri Hitam[2], Wan Nural Jawahir Hj Wan Yussof[3] and Junaida Sulaiman[4]

[1,4]Faculty of Computing, Universiti Malaysia Pahang, Lebuhraya Tun Razak, 26300 Gambang, Kuantan, Pahang, Malaysia
[2,3] Faculty of Ocean Engineering Technology and Informatics, Universiti Malaysia Terengganu, 21030 Kuala Nerus, Terengganu, Malaysia

surayaab@ump.edu.my, suzuri@umt.edu.my, wannurwy@umt.edu.my, junaida@ump.edu.my

Abstract. Corner detection is basically a methods used to extract certain kind of features in images which could produce some information including the location or position of the corner points. Thus, in this paper an enhancement shape corner detection method is proposed to detect true corners of shape images. The overall performance of the proposed enhanced shape corner detector and six other existing shape detectors and descriptors including Harris, SUSAN, Harris-Laplace, CSS, SIFT and global and local curvature properties is presented. The experimental results of corner detection methods are tested using the benchmark binary image MPEG-7 Core Experiment Shape-1 Part B dataset. To measure the performance of corner detection evaluation, an appropriate number of true corners were determined.

1. Introduction

In the late 1970's, the first corner detector was developed and since that many corner detectors have been proposed. Figure 1 shows the timeline of corner detectors since late 1970 until 1999 that been discussed by Yadav and Yadav [1].

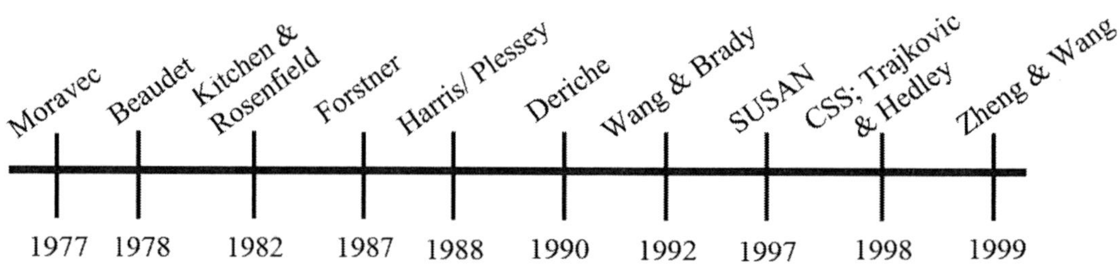

Figure 1. Timeline of corner detectors [1]

Corner detectors have been used in many applications for example in the field of image retrieval [2-4], object recognition [5] and image matching [6]. In computer vision area, corner detection is widely

IOP Conf. Series: Materials Science and Engineering **769** (2020) 012041 doi:10.1088/1757-899X/769/1/012041

used in pre-processing process for various applications including tracking for event cameras in real-time [7], motion detection [8] and mobile robots [9-10].

For robust image corners detection, Mokhtarian et. al. [11] proposed Curvature Scale Space (CSS). He and Yung [12] proposed an improved multi-scale corners detector with dynamic region of support based on CSS technique. The results indicated that it offers an effective and robust solution to images that contains extensively varying size features. Then, in the 2008 the work continued, where they [13] proposed a corners detector based on curvature where it detects fine and coarse features.

An enhanced corner detector has higher ability to detect high amount of true corners compared to the typical technique. Nevertheless, there are still a number of the true corners do not seem to be detected. During the analysis, the original methods of global and local curvature properties were modified to ensure only true corners identified in a single object binary shape dataset.

2. Global and Local Curvature Properties

The corner detector based on Global and Local Curvature Properties was proposed by He and Yung [13] and this work is the improved version from their previous works [12]. They proposed a corners detector based on curvature where it detects fine and coarse features precisely with low computational costs and their work consists of improved version of the original Curvature Scale Space (CSS). The aim of He and Yung [13] suggested steps is to employ global and local curvature properties and balance their effect once corners are extracted. Their proposed steps are described in two subsections as follows.

2.1. Initial list candidates for corners

In listing corner candidates, j th extracted corners are defined as in equation (1):

$$A^j = \{ P_1^j, P_2^j, \dots, P_N^j \} \tag{1}$$

where $P_i^j = (x_i^j, y_i^j)$ are pixels on the contour, N is the amount of pixels on the contour and x_i^j, y_i^j are the coordinates of the i th pixel on the j th contour.

2.2. Corner Evaluation

Region of Support (ROS) has been described by He and Yung [13] to employ global curvature characteristics of the neighbours which remove round corners but not the obtuse corners. Figure 2 shows the examples of round and obtuse corners. As can be observed in Figure 2(a) and 2(c), certain difference is not important though the curve around the corners is the biggest between its neighbours. However, the curvature of an obtuse corner as in Figure 2(b) may additionally have alike or maybe smaller than absolute maximum compared with round corner. From Figure 2(d), frequently the magnitude is expressive higher than its neighbours.

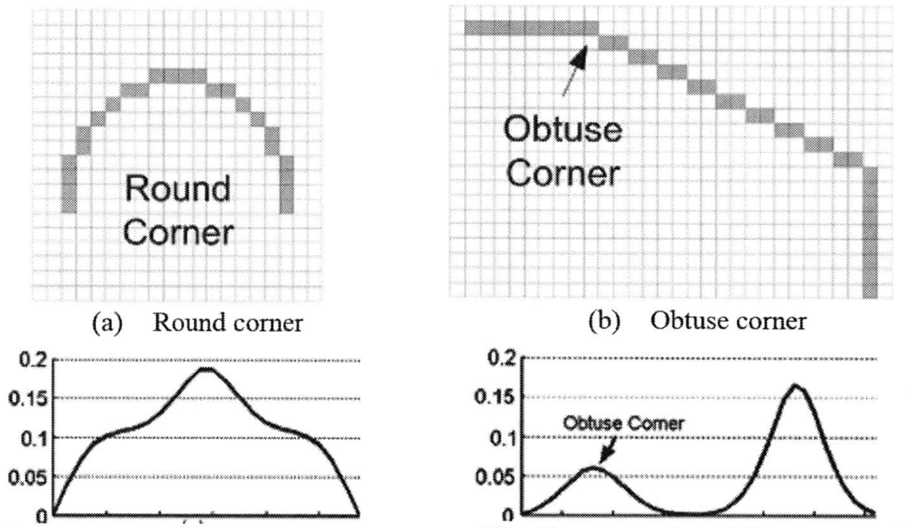

(a) Round corner

(b) Obtuse corner

(c) Curvature plot of round corner in (a)

(d) Curvature plot of obtuse corner in (b)

Figure 2. Types of round corners and obtuse corners [13]

The ROS of a corner is outlined by the segment of the contour, that is finite by the corners of two nearest curvature minima. The ROS of every corner is then accustomed calculate an area of local threshold where u is the position of the corner candidate on the contour, $L_1 + L_2$ is the size of the ROS targeted at u and R is the coefficient as defined in equation (2):

$$T(u) = R \times \overline{K} = R \times \frac{1}{L_1 + L_2 + 1} \sum_{i=u-L_2}^{u+L_1} |K(i)| \qquad (2)$$

where $\overline{K}$ is the mean curvature of the ROS. A true corner is declared if the curvature of the corner candidate is larger than $T(u)$ or otherwise it is eliminated from the list.

3. Enhanced Global and Local Curvature Properties

Ideally, the proposed enhanced shape corner detection is basically inspired from corner detection based on global and local curvature properties proposed by He and Yung [13]. An example process and images for enhanced corner detection can be found in our previous work [14]. To get clear or picture with the whole process, this sections presents the original steps of corner detection based on global and local curvature properties [13] and the proposed enhanced shape corner detection process as illustrated in Figure 3(a) and Figure 3(b), representatively. As can been seen in Figure 3(a), there are eight steps that have been suggested by them in detecting corners.

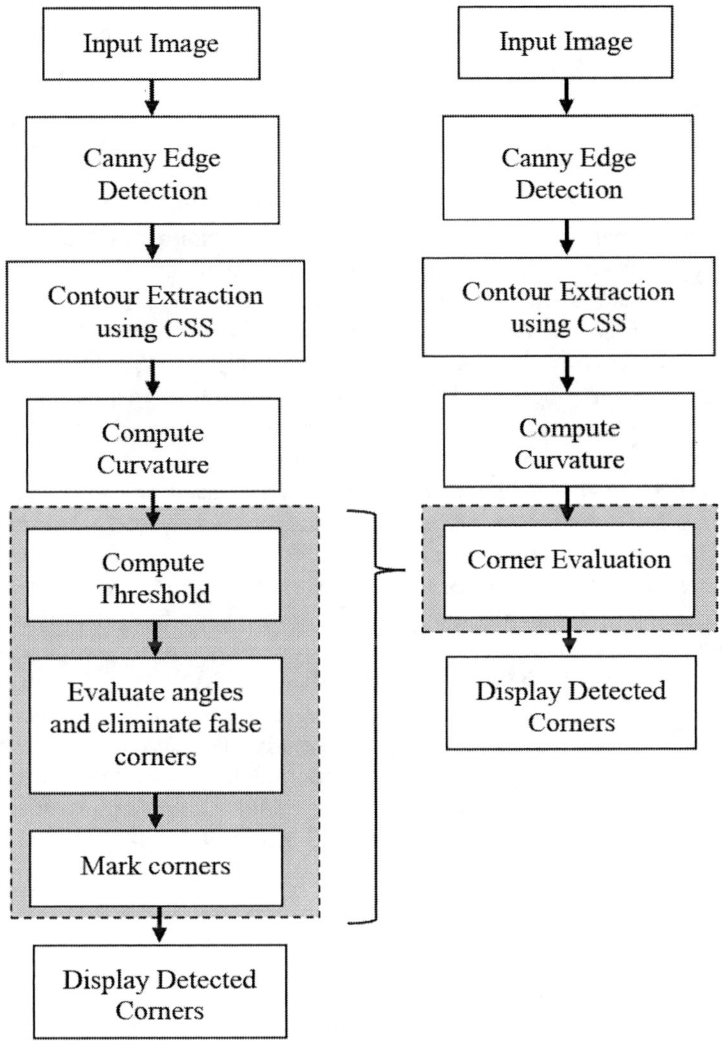

a. Corner detection process based on global and local curvature properties by He and Yung [13].

b. The proposed enhanced global and local curvature properties shape corner detection process.

Figure 3. The process flow of corner detector method

It begins with the input image and followed by the detection of edges by utilizing Canny edge technique. Then, the contour was extracted as within the CSS process. After that, the curvature was computed at a fixed low scale for every contour to preserve the true corners. It is followed by the process of threshold computation according to the mean curvature within a region of support (ROS). In this step, round corners will be removed by comparing the curvature of corner candidates with the adaptive threshold. Next, the end points of open contour are taken into consideration to mark it as corners unless it is very close to another corner. Finally, the detected corners will be displayed. The proposed enhanced shape corner detection process is presented in Figure 3(b).

As illustrated in Figure 3(b), the proposed enhanced shape corner detection has less steps process then the original global and local curvature properties. Given an input image, it first extracts the edges of the object shape using a Canny edge detector. After that, the same processes as in original global and local curvature properties in Figure 3(a) are applied until the threshold computation process. The process is complete until the corner detected results are displayed. Noticed that throughout the rules the used pairs of larger and equal to because the value acquired that later on will define as the corners.

4. Experimental Results

The experimental results of the proposed enhanced global and local curvature properties are presented and the numbers of true corner detected is compared with other existing methods. The experiments used the MPEG-7 dataset to assess the enhanced global and local curvature properties. The foremost common corner descriptors and detectors such as Harris, Harris Laplace, SUSAN, CSS, SIFT and Global and Local Curvature Properties has been assessed and compared to the enhanced global and local curvature properties concerning of true corners observation.

To show a clearer performance of different corner detection method, the overall number of recognized corners is calculated and the experimental results evaluation are summarized as shown in Table 1. The MPEG-7 Core Experiment Shape-1 Part B dataset were applied in this investigation. From Table 1, the number of corners points detected on MPEG-7 images under all of the six existing methods and the proposed enhanced methods are presented. As displayed in Table 1, Harris technique identified most noteworthy corners with the foremost corners identified under image category hammer, ray, bird and face images. Nevertheless, there are corners spotted on the straight line of the object. Furthermore, some identified corners are not suitable and not reliable for shape representation meanwhile some of them are either nearby to each other or overlooked the true corners of the shape. As a consequence, the entire shape of the object could not be fully represented.

A Harris Laplace and SUSAN methods almost detect the same number of corners except for ray, rat and bird images. Moreover, SIFT representing the second highest in detecting corners after Harris but as discussed in the previous sections, some of the corners detected are false corners or redundant. So that, most of the corners are not meaningful because of false corner does not give any significant information of the shape object. On the other hand, CSS produced the lowest total number of identified corners that unsuitable for shape representation since missing significant shape points as can be observed under all types of image category in Table 1.

Table 1. Corners Detection Evaluation Results using MPEG-7 Core Experiment Shape-1 Part B dataset.

Image with corner detected / Image Detector	Brick	Hammer	Rat	Ray	Bird	Face
Harris	40	17	71	120	61	50
Harris Laplace	21	14	28	33	18	31
SUSAN	21	15	58	82	43	34
SIFT	44	13	73	38	29	23
CSS	11	4	17	9	8	8
Global and Local Curvature Properties	14	7	26	21	14	10
Enhanced Global and Local Curvature	**15**	**8**	**28**	**22**	**15**	**12**

Corner detector using global and local curvature properties detected appropriate corner compared with CSS but there are seems some corners points cannot be detected. It can be highlighted that the corners detected by the proposed enhanced methods are more appropriate compare with original methods.

5. Conclusion

To summarize, the proposed methods detected good corners points where additional corner point could be found compared to the original approaches. Consequently, the detected corners are more accurate and every spotted corner points are significant in improving shape representation. Hence, it is appropriate to be applied in various applications such as computer vision and image matching. Detection appropriate corners of object shape is crucial so that all the features information of the significant points can be stored and used in retrieving the similar shape object.

Acknowledgements

This work was partly supported by Universiti Malaysia Pahang under grant number RDU190307.

References

[1] Yadav, A. and Yadav, P. 2009 *Digital Image Processing*. Published by USP/Laxmi Publications (P) Ltd., New Delhi.

[2] Kavitha, K., and Sudhamani, M. V. 2014 *Object Based Image Retrieval from Database Using Combined Features.* Paper presented at the Signal and Image Processing (ICSIP)

[3] Kuo-Lung, H., and Chieh-Hsien, L. 2012 *A Novel Image Retrieval Technique Based on Salient Image Features.* Innovations in Bio-Inspired Computing and Applications (IBICA)

[4] Bakar, S.A., Hitam, M.S. and Yussof, W.N.J.H.W 2019 *A Comparative Analysis of the Zernike Moments for Single Object Retrieval* Baghdad Science Journal vol. 16, pp.504-514

[5] Ben-Musa, A. S., Singh, S. K., and Agrawal, P. 2014 *Object detection and recognition in cluttered scene using Harris Corner Detection.* International Conference on Control, Instrumentation, Communication and Computational Technologies (ICCICCT).

[6] Xin, Z., Guojin, H., and Jiying, Y. 2009 *A Rotation Invariance Image Matching Method Based on Harris Corner Detection* International Congress on the Image and Signal Processing (CISP).

[7] Ignacio A. and Margarita C. 2018 *Asynchronous Corner Detection and Tracking for Event Cameras in Real Time* IEEE Robotics and Automation Letters Vol. 3, No. 4

[8] N. Algethami and S. Redfern, 2018 *Combining accumulated frame differencing and corner detection for motion detection* Proceedings of the Conference on Computer Graphics & Visual Computing (CGVC)

[9] Morgan Q., Kartik M., Shreyas S.S., Micheal W. and Yash M., 2019 *The Open Vision Computer : An Integrated Sensing and Compute System for Moile Robots* IEEE International Conference on Robotics and Automation (ICRA)

[10] D, Pebrianti, YH, Hao, NAS, Suarin, L, Bayuaji, Z, Musa, M, Syafrullah and I, Riyanto 2018 *Motion tracker based wheeled mobile robotoller system identification and controller design* Intelligent Manufacturing & Mechatronics

[11] Mokhtarian, F., S. Abbasi and J. Kittler, 1997 *A new approach to computation of curvature scale space image for shape similarity retrieval* Proceedingsof the 9th International Conference on Image Analysis and Processing.

[12] He, X. C., and Yung, N. H. C. 2004 *Curvature scale space corner detector with adaptive threshold and dynamic region of support* Proceedings of the 17th International Conference on the Pattern Recognition (ICPR).

[13] He., X. C., and Yung, N. H. C. 2008 *Corner Detection based on Global and Local Curvature Properties.* S P I E - International Society for Optical Engineering.

[14] Bakar, S.A., Hitam, M.S. and Yussof, W.N.J.H.W 2017 *Improved Global and Local Curvature Properties for Shape Corner Detection* Journal of Applied Sciences 458-566.

Coding mechanism and soft systems technique applied to integrate the fuzzy based Decision Support System with HRM factors in the SMEs of Bangladesh

Tarnima Warda Andalib[1]*, Hasliza Abdul Halim[1] and Muhammad Khairul Islam[2]

[1]School of Management, Universiti Sains Malaysia, Penang, Malaysia
[2]Faculty of Management and Accounting, University Tenaga Nasional Malaysia, Malaysia

E-mail: tarnima.warda.andalib@gmail.com, haslizahalim@usm.my

Abstract. In this article, the objective is to integrate fuzzy based decision support system (DSS) with human resource management framework (HRM) of the manufacturing SMEs in Bangladesh, where various decision-making styles regarding HRM factors are exhibited. In the decision support system, few decision-making styles are selected and sorted using fuzzy inference mechanism. Fuzzy based decision-support system is integrated with HRM framework by applying coding mechanism and soft systems technique. This fuzzy approach allows the researchers to examine the relations between decision-making styles and human resource management factors when there is uncertainty, vagueness, multiple elements of problems and solutions. Human resource management has various factors incorporating with hiring employees, career development and management, employees' performance management, handling management styles, building relationship between the managers and the employees. Well, decision support system is involved in each of this process. At present, in the competitive world using fuzzy logic approach in decision support system in human resources management has achieved particular stance and integrated decision support system in HRM assists the managers in making decisions meaningfully. The purpose is to provide results to the manufacturing SMEs that can assist the managers to exercise the most appropriate decision making style and take such a decision that is beneficial for all in the SMEs.

Key words: Decision support system, Decision making styles, Human resource management, factors, SMEs, Bangladesh

1. Introduction

Small and medium sized enterprises (SMEs) in Bangladesh has become widely important and crucial for the nation's job production, low cost local products and contribution in national GDP. Nevertheless, SMEs struggle themselves and face numerous obstacles from the initial stage. One of its core strength remain in its innovative manpower, no matter how many employees are working- if they are motivated and innovated enough, SMEs will run few miles ahead. Here, lies another difficult kind of scenario. Dealing with the employees, meeting their needs, protecting their rights and making the business run with low capital and lots of constraints become perilous. However, this is why researchers here has focused the DSS based HRM. If the SMEs' owners and managers from the very beginning emphasizes on the HRM and deploys a proper DSS to deal with various distinct factors of HRM then situation

IOP Conf. Series: Materials Science and Engineering **769** (2020) 012042 doi:10.1088/1757-899X/769/1/012042

becomes smooth. Researchers here have done multiple case studies using coding mechanism to find out the core factors of HRM framework and the core issues of decision making related to each of these factors. HRM frameworks usually provide HR managers' insights about employees' welfare and support in decision-making but most these decisions are based on human judgment and preference [3-5] [22]. Decision support system is a technological programmed application, which co-operates managers' decision making by using data and models to resolve semi-structured and unstructured issues [26]. Researchers here used fuzzy logic to make DSS applicable in the real time uncertain and crucial scenario and integrated this DSS with HRM of the SMEs by using soft systems techniques.

2. Background and Motivation of the study

Managers have various ways and styles to deal with the employees. But, the basic two styles are transactional and transformational styles. There are sub-styles in each of this main category. However, managers do suffer while making any decision regarding any process of Human resource management. Specially, when the managers have to keep in mind about employees' welfare and SMEs' goals at the same time. Therefore, in this study fuzzy based decision support system is integrated with the HRM factors, so that managers can get a support system while making their decisions regarding any HRM component [3-5].

3. Objective of the study

3.1 To define decision support system with fuzzy logic
3.2 To find out the HRM core processes where managers' decision making plays a role
3.3 To integrate fuzzy based decision support system with HRM processes by coding mechanism and soft systems techniques.

4. Literature

4.1. Human Resource Management (HRM)

HRM (Human resource management) framework has been the primary important department of any organization where it is a set of interrelated policies, factors, models and factors', where employees are the core resource to deal with certain philosophies by achieving certain goals of the organization [3-5, 19-21]. HRM has certain factors and activities like acquiring, training, appraising and compensating employees by attending to employees' relations, health, safety and rights and most of these are unstructured [3,4]. Understanding this important remit, various HRM processes developed throughout the past decades and still urging for more. Scholars have developed various HRM models and frameworks like Harvard Model, Commitment Model, Outcome Model, AMO framework and so on. All HRM frameworks have the core purpose and that is to assist the HR managers in dealing with human resources strategically. Decades provide evidences and incidents of failed managerial experiences when employees' are being treated inappropriately even though they are the most important asset in an organization mentioned by Khatri in 1999. Therefore, Andalib has proposed to include employee rights in the HRM framework in the companies of Bangladesh in 2018 [3-5, 25]. Therefore, in this study HRM integrated with international employees' rights platforms will be taken into account. UDHR (Universal Declaration of Human Rights) is one international standard, where Article 23: 3 mentioned that all employees have rights to a just and favourable remuneration also the workplace has to ensure social protection and human dignity and with ILO (International Labour Organization) protocols relating to employees' rights. This strategic dimension assists the managers not only to focus on the employees' welfare but also resolves various issues in the organization automatically, since employees stay more motivated at work and performs really well [2-8,11,21].

4.2. Decision Support System (DSS)

Decision Support Systems (DSS) is usually identified as a computer program based database with information that assists to decide the activities. A properly functional DSS is quite interactive and assists the managers to interpret the meaning of raw text, models, knowledge base and various documentations. Therefore, issue becomes transparent with properly taken decisions [1]. DSS also provides decisions in distinctive forms and formats since systems can perform while analyzing the semi-structured, un-structured and unstructured data. Initially DSS used to work with small data-subsets and these data-subsets were always model driven. Gradually, advancement in the technology improved this situation and later large databases were taken into consideration with complex data-sets and complicated analysis processes are involved [2,3]. The HRM core factors that should be present in the HRM framework are recruitment, career development, industrial relationship, employee rights, management and leadership styles and etc [3-5].

4.2.1. Types and Methods of Decision Support Systems. DSS has three types, which are active, passive and ubiquitous. Usually, passive DSS is a traditional one dealing with features those functions on the built-in knowledge-based having zero content and applied only for static user preference. Factors of passive DSS are rule based. Nevertheless, Active DSS is incorporated with learning capability, knowledge based system, expert and adaptive system. In this category, use of technology like machine and tool learning are involved. The third category ubiquitous DSS is a portable and mobile proactive system that initiates push and pull approaches [23]. DSS types depend on four factors, which are Model Management System, Communication and Advisory System (AS), Database Management System, Document Driven System, Knowledge Base System (KBS). These four models follow particular methods, which are model-driven, communication-driven, data driven, document-driven and knowledge-driven. Methods that emphasize on statistical, financial, optimization and simulation models are model driven. The numbers and values help to take the decisions and in this case less data-volume works perfectly. Decodes remains as open source of this method. Usually, communication-driven method works on the possibilities of tasks and group activities [4, 5]. In data –driven kind, huge volume of data like time-series data are used. In Document – Driven data mostly unstructured data is dealt with and shaped into electronic structure [6, 22]. Knowledge – driven data applies the prudent formulas, rules and procedures through the expert systems.

4.3. Fuzzy Logic

The Fuzzy method has a lot of good features that can be used in human resource management. Fuzzy logic looks at the problems and solutions in imprecise manner, pretty much in similar ways that human brain takes information in and reason with uncertainties, vagueness and judgments. Fuzzy logic can solve and deal with numerical data as well as linguistic knowledge, where mathematical description is not required. Fuzzy logic deals with more complex problems and absolutely differs from classical logic. Classical logic has only options which are either black or white or true or false objects in classical logic take on value of either zero or one [18, 22, 23]. Nevertheless, fuzzy logic deals with data and options in a wider range and objects can take any real value between 0 and 1. FL can make development and implementation much simpler. It needs no intricate mathematical models, only a practical understanding of the overall system behaviour. FL mechanisms can result to higher accuracy and smoother control as well. Zadeh observed in 1965 that conventional computer logic couldn't manipulate data that represented subjective or vague ideas, so he created fuzzy logic to allow computers to determine the distinctions among data with shades of gray, similar to the process of human reasoning [30]. FL recognizes not only clear-cut, black-and-white alternatives, but also the infinite gradations in between. FL differs from orthodox logic in that it is multivalued. Fuzzy deals with degrees of truth and degrees of membership. Fuzzy reasoning eliminates the vagueness by assigning specific numbers to those gradations. These numeric values are then used to derive exact solutions to problems. Fuzzy inference process: Fuzzification: Translate input into truth values Rule Evaluation: Compute output truth values Defuzzification: Transfer truth-values into output.

4.3.1. Fuzzy – method to design DSS. When decision-making is done using a multivalued system or multi logic system managers get more probabilities to make decision of one single problem and decision-making becomes easier and smooth. From mathematics the classic and traditional methods of management has been derived. In this case usually, two variables and multivalued logics are emphasized on each data. Here, the vagueness of data and expression of human feelings may remain (linguistic variables) since, there are lack of flexibility and lack of meticulousness in mathematical models. These days, applying fuzzy management knowledge using fuzzy theory seems like an innovative approach to deal with the uncertain responses and to start with the new approach. Theory of fuzzy systems using fuzzy logic theory and fuzzy measure theory are considered as imported variables and components in the models based on knowledge, experience, judgments, flexibility and decisions that eventually provides a gray area of every problems of the world 10,11]. Brain processes all information and these fuzzy systems. The processes that take place in the brain follows steps. Initially, recovering information and data is performed, then recognizing it, thinking and judging about it, evaluating it and finally deciding are the core steps. Therefore, fuzzy logic is applied in this study in the decision support system of an organization to make managers' take decision in a better manner. Applying the fuzzy method smoothens managers' decision-making processes and enhances the performances of recruitment, retention and management of the employees by the employers. Sometimes, components like loose and incomplete information, psychology and linguistics barriers reveal that decision-making process in today's world is very complex [18, 22, 30].

4.3.2. Fuzzy Inference Systems to design the decision set for DSS. Fuzzy inference is a process of articulating and manipulating the imported data and providing a format by applying fuzzy logic. Furthermore, the mapping exhibits the options and probabilities of the decisions those can be made based on what, why, how questions. This process also contains membership functionalities, operational values, rules of if-then, automatic controls, categorized data, analytical tool, vision of the programs and expert systems in the core and etc. Since, fuzzy inference is multi-disciplinary in nature, these systems can be called rule-based or set based on fuzzy rules, fuzzy modeling, fuzzy logic controllers, decision support systems, simple and vague fuzzy systems and etc [22]. The two basic types of fuzzy inference system, are 1) Mamdani and 2) Sugeno. Outputs of these inference systems differ from each other.

5. Methodology: Qualitative approach: Collection, Analysis and Validation

Since, this research integrates decision support system with HRM processes, qualitative approach is chosen as most appropriate because this narrows down the directly collected information from participants' into a central meaning by in-depth analysis procedure [16,17]. Here, ontological, axiological and methodological assumptions with post-positivism and constructivism Interpretivist paradigms are chosen because these reveals the procedures to develop this new HRM framework [3-5]. Multiple case studies (MCS) are used to establish a logical chain of evidence by eliminating the repetitive themes by using analysis, which are within-cases and cross-cases as per Yin in 2009,to find in-depth case descriptions through multiple sources of information (observations, interviews, audiovisual materials, documents and reports), to finalize the specific HRM themes and factors of the HRM framework as well as to find the crucial area where managers feel problem to make decisions [16,17,29]. Purposeful sampling is used to select these cases from a broad range of companies as per Miles, Huberman & Saldaña's concept and therefore, five (5) SMEs with 12 managers are chosen [1,24].

5.1. Step 1: Coding mechanism. Here, Coding technique has been applied as per Auerbach and Silverstein in 2003 to transform the found raw data into meaningful theoretical constructs, themes by eliminating the repeating themes [10]. Therefore, open codes, categorized codes, themes and main factors are finalized and eventually the new HRM framework is developed. NVIVO MAC tool is used to do in-depth coding, transcribing and organizing data. Research-data has been validated with conformability, credibility, justifiability, transferability, dependability and partial reflexivity, where data has been cross checked, eliminated by the participants' judgment, where scholars' evidences found, where list and chain of data found with consistency and where interpretations are done from chain of evidences [24, 27,30,31,32, 33].

5.2. Step 2: Fuzzy Logic applied in DSS of HRM. Mamdani Fuzzy inference method has been applied in the Decision Support System of HRM. In this study, to design the decision support system, Mamdani fuzzy inference system proposed by EbrahIm Mamdani in 1975 has been applied. Mamdani fuzzy inference system generates a fuzzy set theory as probable decision options for the managers. [22]. Fuzzy logic related to knowledge, experience, judgments and decisions of distinctive managers are incorporated in this study to design the Decision Support System. Active Decision Support system mainly has been observed in here to integrate with HRM, since active DSS is incorporated with learning capability, knowledge based system, expert and adaptive system. Nevertheless, Ubiquitous DSS has been tried to integrate in certain HRM component as well. DSS is based on basically data, model, communication and knowledge factors in this study [15,18].

5.3. Step 3: Soft Systems Techniques to integrate HRM and DSS. Checkland and Poulter's soft systems method (SSM) of 2010 is used to map these codes, themes and factors. In the findings, the desired factors are HRM Core factors, Managers' Decision making styles and the Decision support system related to the core processes [12].

6. Result and discussion

Researchers found the core HRM factors, which are important but also face challenges regarding decision making. Table 1, reveals these HRM factors. Nevertheless, from Table 2, managers' styles are found from the interviews with each SME manager and two distinct leadership styles have been found, which are transactional and transformational. In Table 3, DSS types are revealed with distinct fuzzy logic reasoning and probable sets of decisions that managers can take for certain HRM factors.

Table 1.Multiple case studies of the Bangladeshi SMEs'

SME Cases	Type and establishment year	Number of Managers (12 managers)	HRM component that creates difficulty in decision making
Case 1	Garments –Knitting, colouring, 2011	3	Employee rights, industrial relations, hiring
Case 2	Food Catering from Home, 2015	2	Hiring, compensation, employee rights
Case 3	Pastry Shop , 2016	2	Career progress, employee rights, leadership styles
Case 4	Garments– Men's wear, 2012	3	Grievance management, employee rights, industrial relation
Case 5	Processed Food Farm, 2010	2	Leadership styles, employee rights, hiring, medical

Table 2. Managers' Portfolio

Managers	Position in the company	Work Tenure (years)	Number of Subordinates	Managers' leadership style
P 1	HR Manager	7	23	Transactional
P 2	Administrative Manager	5	12	Transactional
P 3	Labour Manager	5	20	Transactional
P 4	Labour Manager	4	24	Transformational
P 5	SME Owner	8	60	Transactional
P6	Administrative Manager	4	7	Transformational
P7	SME Owner	5	11	Transformational
P8	SME Stakeholder	5	40	Transactional
P9	HR Manager	6	20	Transformational
P10	SME Owner	10	97	Transformational
P11	HR Manager	3	15	Transactional
P12	Senior Manager	3	8	Transactional

Table 3. Fuzzy based HRM

HRM Core Component	DSS Type	Fuzzy Logic Reasoning	Probable Decision/Solution
Hiring and Selecting (HS)	Active, +ve	Selection Criteria, Sorting Education Sorting, Sorting Experiences, Assessing Potentiality, Observing extra-curricular and interest area	1. Hire with good result in education. 2. Hire with more experienced years. 3. Hire with creative interests. 4. Hire with reasoning interests.
Remuneration & Benefits (RB)	Ubiquitous, ++ve	Sorting salary Numbers that inspire employees, Benefits and the factors, provident fund, gratuity and etc	1. Provide salary number with three/four zeros. 2. Provide gratuity and provident fund after 3 years / 5 years
Career Management (CM)	Ubiquitous, ++ve	Sorting number of experience years, score of performance appraisal, number of trainings, workshops, project management leads, entrepreneurial attitude	1. Providing training outside the country/inside the country/on job to all/to ones who show results. 2. Provide workshops and projects to entrepreneurial employees.
Leadership Style (LS)	Active, +ve	Mood assessment, Leaders' background check, leaders' previous experiences	1. Provide workshops to the leaders to improve their insights. 2. Provide education regarding humanitarian law and human rights guidelines to be a good leader. 3. Remove leaders with negative humane behaviour

Industrial Relation (IR)	Active, +ve	Events assessment where employees and managers have taken part together like SME programs of family, product show and etc, take inputs of each employees' scoring the managers and take input from performance appraisal	1. Create platforms to bring employees and management into discussions. 2. Give scores to employees and managers to assess
Employees' grievances and rights mechanism (EGRM)	Active, +ve	Observe turnover, complaints, observe and sort violations of rights,	1. Provide mental support to the victim employee. 2. Provide awareness program to the present employees

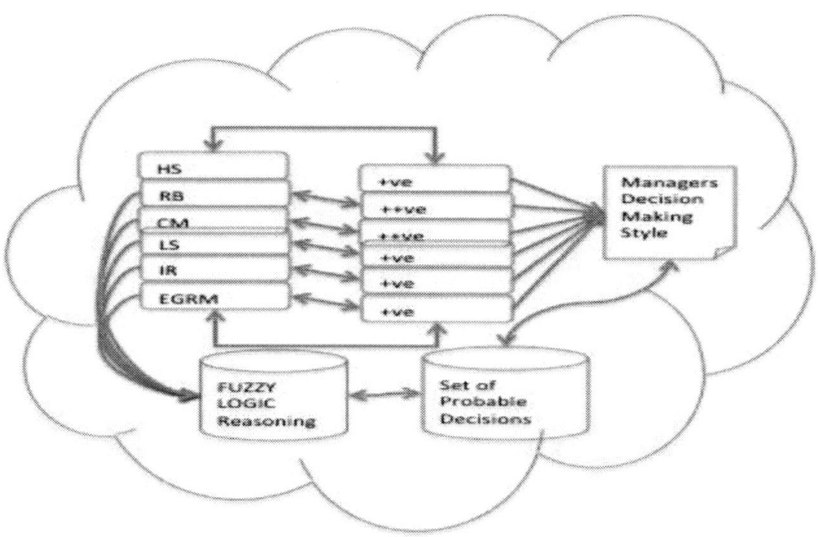

Figure 1: Conceptual framework of integrated HRM-DSS

Data that was found and saved in NVIVO tool are coded by following Auerbach & Silverstein's method. The found codes are of two kinds, 1) HRM factors and 2) DSS probable system. Checkland and Poulter's soft systems method (SSM) is to map these codes and themes. Researchers have incorporated and connected the fuzzy DSS with HRM of the SMEs' of Bangladesh [12]. Therefore, in figure 1, the conceptual framework has been disclosed.

7. Conclusion

HRM framework of the SMEs of Bangladesh has been incorporated with fuzzy based DSS in this study [11]. HRM is a complex process since human beings are employees with expectations and unpredictability. Nevertheless, there are sets of expectations which are predictable, therefore, each HRM component has been under scrutiny in this study. Through multiple case studies, the HRM factors are observed and the problems regarding the factors have been identified that makes it difficult for the managers to decide anything about that particular component. The fuzzy based DSS that has been developed has taken all these difficulties into account and identified numerous reasons for those problems and then have computed few probable solutions. Managers may use this set of probable solution and can choose one decision by himself or herself. Nevertheless, since DSS has been applied in the HRM arena, DSS is integrated with HRM factors by soft systems technique. The future direction of the work will be included in Internet of Things (IoT) concept with present proposed work [28-31].

References

[1] Akhtaruddin, M. *Corporate mandatory disclosure practices in Bangladesh*, The International Journal of Accounting, Elsevier, 2005, 40 (4), 399-422.

[2] Andalib, T. W., Darun, M. R. & Azizan, N. A. *East Asian trends of human resources management: Theories and practices*. International Journal of Human Resources Development and Management, 2019.

[3] Andalib, T. W., Darun, M. R., Halim, H. A, Sultana, U.S. An Innovated HRM Framework for Sustainability: Multiple Case Studies in Bangladesh, *The European Proceedings of Social & Behavioural Sciences EpSBS*, Olive Tree hotel, Penang, Malaysia, 17-19 October 2018, Future Academy, UK, 2019, 7, 62-73,8 Doi:https://doi.org/10.15405/epsbs.2019.08.7

[4] Andalib, T. W. & Darun, M. R. An HRM model for manufacturing companies of Bangladesh mapping employee rights' protocols and grievance management system. *Indian Journal of Science and Technology*, 2018, 11, 17

[5] Andalib, T. W. A study on convergence of international employee rights within human resources management practices in Bangladesh (Doctoral dissertation), scale 2, Pahang, Malaysia, Universiti Bangladesh Pahang, Nov 2018

[6] Andalib, T. W., Darun, M. R. & Azizan, N. A. Western Trends of Human Resources Management: Theories and Practices, *International Journal of Industrial Management (IJIM)*, (ISSN 2289-9286), 2015, 1, 1-9

[7] Andalib, T. W., Darun, M. R. & Azizan, N. A. A process of developing an HRM framework based on qualitative study in Technology companies in Bangladesh and Malaysia, *International Journal of Research in Business and Social Science* , (2147-4478), 2014, 3, 2, 148-163

[8] Andalib, T. W., Darun, M. R. & Azizan, N. A. Frustration of employees: Reasons, dimensions and resolving techniques, *WCIK E Journal of Integration Knowledge*, In: Proceeding of the World Conference on Integration of Knowledge (WCIK 2013), Langkawi, Kedah, 2013, 25-26 November, 1 (11), 15-25.

[9] Alauddin, MD., & Chowdhury, M. M. Small and Medium Scale Enterprise in Bangladesh prospects and challenges. *Global Journal of Management and Business Research: Finance,* 2015, 15(7), 1-9

[10] Auerbach, C. F., & Silverstein, L. B. *An introduction to coding and analysis8: Qualitative Data* New york and London: New York University Press, 2003

[11] Bangladesh Employees' Federation, BEF. *The Bangladesh Labour Act 2006,* 2009

[12] Checkland, Peter. *Soft systems methodology: a thirty-year retrospective*, Systems research and behavioral science, 2010, 17,1

[13] Chowdhury, M. S.; Islam, R.; & Alam, Z. Constraints to the development of small and medium sized enterprises in Bangladesh: An empirical investigation. *Australian Journal of Business and Applied Sciences*, 2013, 7(8): 690-696.

[14] Creswell, J. W. *Research Design Qualitative, Quantitative, and Mixed Methods Approaches.* London: SAGE Publications, 2013

[15] Denaï MA, Palis F, Zeghbib A. *Modeling and control of non-linear systems using soft computing techniques*. Applied Soft Computing, 2007, 1;7(3), 728-38.

[16] Eisenhardt, K. M., & Graebner, M. E. *Theory building from cases: opportunities and challenges*. Academy of Management Journal, 2007, 50(1), 25-32.

[17] Eisenhardt, K. M. Eisenhardt, k. *Building theories from case study research*. The Academy of Management Review, 1989,14(4), 523-550.

[18] Gorzalczany, M.B., & Piasta, Z. Neuro-fuzzy approach versus rough-set inspired methodology for intelligent decision support. Information Sciences, 1999,120 (1-4), 45-68.

[19] Halim, H. A., Ahmad, N. H., Ramayah, T.; Taghizadeh, S. K., Capturing the'pioneering Minds'via Human Capital: The Impact on Innovative Performance Of Malaysian Smes, *Asian Academy Of Management Journal*, 2016, 21.

[20] Hoque, A. S. M. M., Awang, Z., Jusoff, K., Salleh, F., & Muda, H. Social Business Efficiency: Instrument Development and Validation Procedure using Structural Equation Modelling, *International Business Management*, 2017, 11(1): 222-231.

[21] International Labour Organisation. *Handbook of procedures relating to international labour Conventions and Recommendations,* 2012.

[22] Jantan, Hamidah, Hamdan, A. R, Othman, Zulaiha Ali, *Intelligent Techniques for Decision Support System in Human Resource Management*, Decision Support Systems: Advances in, editor: Ger Elvin, 2010.

[23] Kwon, O., Yoo, K., Euiho S., UbiDSS*: A proactive intelligent decision support system as an expert system deploying ubiquitous computing technologies*, Expert Systems with Applications, 2005, 28(1), 149-161, DOI: 10.1016/j.eswa.2004.08.007

[24] Miles, M. B., Huberman, A. M., & Saldaña, J, *Qualitative Data Analysis. A Methods Sourcebook 3rd.* LosAngeles, London, New Delhi, Singapore, Washington DC: SAGE Publications, 2014.

[25] Muniapan, B. (2015*), Labour Relations and International Business: The Doctrine of Constructive Dismissal and Labour Relations in Bangladesh, In Beyond the UN Global Compact: Institutions and Regulations*, Advances in Sustainability and Environmental Justice, Emerald Group Publishing Limited, 17, 291-315.

[26] Qian, Z.,Huang,G.H., & Chan, C.W. Development of an intelligent decision support system for air pollution control at coal-fired power plants. Expert System with Applications, 2004, 26 (3), 335-356.

[27] Sitwala Imenda, Is There a Conceptual Difference between Theoretical and Conceptual Models?, Journal of Social Sciences, 2014, 38 (2), 185-195, DOI:10.1080/09718923.2014.11893249.

[28] Rahman, Md Arafatur, Muhammad Nomani Kabir, Saiful Azad and Jahan Ali, *On Mitigating Hop-to-Hop Congestion Problem in IoT Enabled Intra-Vehicular Communication,* The IEEE International Conference on Software Engineering & Computer Systems, Kuantan, August 19-21, 2015.

[29] Rahman, Md Arafatur, Ali, Jahan, Azad, Saiful, Kabir, Muhammad Nomani, *A Performance Investigation on IoT Enabled Intra-Vehicular Wireless Sensor Networks*, International Journal of Automotive and Mechanical Engineering (IJAME), 2017.

[30] Rahman, Md Arafatur, *Reliability Analysis of ZigBee based Intra-vehicle Wireless Sensor Networks,* Lecture Notes in Computer Science series (Vol. 8435), published by Springer, 2015

[31] Rahman, Md Arafatur, Anwar, Farhat, Azad, Saiful, *Integrated Metric-Ad Hoc On-Demand Distance Vector: A Routing Protocol Over Wireless Mesh Networks*, Journal of Computer Science, Science Publications, USA, 2009 Volume 05, ISSN: 1549-3636, pp 511-518 (SCOPUS)

[32] Stake, R, The art of case study research. Thousand Oaks, CA: Sage, 1995, 49-68.

[33] Yin, R. K. *Case study research design and methods*, Sage publications, 2009, 4, 240.

[34] Zadeh, A. *Fuzzy Sets, Information and Control*, 1965, 8(3), 338-353, doi.org/10.1016/S0019-9958 (65) 90241-X.

IOP Conf. Series: Materials Science and Engineering **769** (2020) 012043 doi:10.1088/1757-899X/769/1/012043

Artificial Intelligence (AI): A Review of its Uses in Language Teaching and Learning

Zuraina Ali
Department of Modern Languages, Universiti Malaysia Pahang (UMP), 26600 Pekan, Pahang Darul Makmur, Malaysia

zuraina@ump.edu.my

Abstract. The uses of Artificial Intelligence (AI) seems to be relevant in many fields nowadays due to its ability in providing a simulation of human intelligence processes that are handled by machines; in particular computer systems. This paper concerns with reviewing the uses of AI in language teaching and learning. In particular, it reviews the research on the uses of AI in its application in the learning and teaching of language. Qualitative research method; specifically content analysis, is employed as the technique to review the articles that are obtained from relevant databases. Findings from the study reveal that there are four (4) themes emerge in the uses of AI in relation to teaching and learning a language. The uses of AI for pedagogy, therefore, prove that its uses eases language teaching and learning.

1. Introduction

Artificial Intelligence (AI) is a system that will start to become visually invisible for multiple human tasks as it uses would become accustom like other technologies that are used in our everyday life. As a consequence, the dependency on AI, may be natural among human. It seems that in language learning and teaching, the term Computer Assisted Language Learning (CALL) gets a wider scope of meaning in its usage when AI is integrated in conducting classes [1]. In fact, with the relevant of AI in language teaching and learning, it may branch the term CALL to Intelligent CALL (ICALL) [2].

In terms of roles that can be explored by AI in language teaching and learning, one of it is it enables learners to speed up their learning of a particular language [1]. Learners are able to learn a language based on repetition of phrase patterns that show the relationships between words. An example of AI that is used for learning language is when a learner uses Siri; an operating system of iOS that is developed by Apple, to converse in English. This is especially relevant for learners of ESL (English as a Second Language) and EFL (English as a Foreign Language) [3]. More specifically, it helps the group of learners to improve their speaking; in particular pronunciation and listening skills.

Other AI technology that are commonly used nowadays are Google Now and Microsoft Cortana. The former is a search engine that is developed to listen and comprehend the command and queries of the users. The search engine opens up the most relevant webpage automatically following the voice command of the users. However, only users of Samsung Galaxy with its Android versions are able to use the AI [4]. On the other hand, Microsoft Cortana like the two (2) AIs previously discussed assist its users in making calls, sending messages, setting reminders, taking notes, recognizing music and many others [5]. Though Microsoft Cortana shares the same features of Apple's Siri and Google Now, its features in answering user queries are getting better in which it is able to track flights, add reminders for users upon their asking of directions – to name a few [6].

Despite that, the current review aims to identify the use of AI in the teaching and learning of language. Moreover, with the Fourth Industrial Revolution where technology is very much integrated in teaching and learning, AI is the technology that every educator needs to be exposed to. Using AI, it seems that the presence of a teacher nor the learning of language is necessary in spite of the fact that a machine is able to handle a class [7]. With its use, education can be more student-centered and, consequently, it leads to the empowerment of the learners.

2. Methodology

These sections explain the methodologies that are employed in the current review of articles. They discuss the research design in terms of sampling of the research articles, data collection procedures and data analysis procedures in synthesising the review of articles.

2.1. Research Design

The current study employs a qualitative research design; more specifically content analysis in finding the answer to the single research question that is formulated in the current study. Fraenkel, Wallen & Hyun [8] state that content analysis is the analysis of any written contents that are in textbooks, essays, newspapers, novels, magazine articles, cookbooks, songs, political speeches, advertisements and pictures. Having said that, the review of articles in this paper is derived from research articles as the researcher synthesises information relevant in these resources for the purpose of answering the question in point.

2.2. Sampling of research articles

The research articles that are gathered for the purpose of reviewing the uses of AI in teaching and learning of English; in particular discussions in 'Findings', are obtained from databases - Ebscohost, Science Direct and Scopus. Such is the eligibility criteria of the resources due to they are more reliable compared to using web search engines for instance Google and Yahoo. Moreover, the quality of the papers that are deposited in the databases are more guaranteed. Also, only research articles that are within five (5) years' time are selected as the samples of articles for the review. The reason being merely current research findings are discussed prior to understanding the uses of AI in language teaching and learning.

2.3. Data collection procedures

Four (4) steps are observed in collecting data for the current study. The first step is 'Determine relevant databases' in that selected databases namely Scopus, Ebscohost and Science Direct are chosen as to obtain the resources from. In the second step i.e. 'Scout relevant articles', articles that concern with the uses of AI in language teaching and learning are selected. Moreover, only articles that provide full papers are selected to be analysed. In step three, data synthesizing is carried out. Articles are read, comprehended and inferred to obtain the essence of contents that are written by researchers. Finally, in step four, themes are composed by which important and interesting data is identified [9]. The formulated themes are used to discuss the findings following the research question that is developed at the beginning of the writing this review paper.

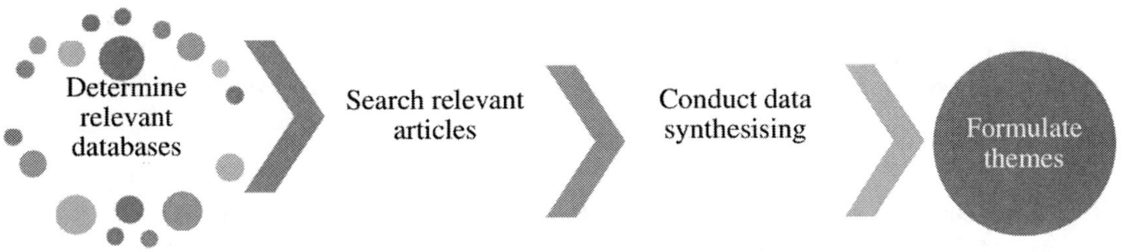

Figure 1. Steps in collecting data for the review of articles

2.4. Data analysis procedures

Five (5) stages are carried to analyze the research articles [8]. First, 'Specify Unit of Analysis' whereby the researcher determines phrases, words and sentences that may be typed in the database in her quest to find the relevant articles. Thesaurus is used to search the synonyms and definitions of the phrases, words and sentences. Second, 'Locate Relevant Data' in that relevant data from online databases is tracked down to find current research articles. Third, 'Develop a Concept or Rationale'. In this stage, skimming, scanning, analyzing and synthesizing are performed as the researcher read the contents that are printed in the articles. Fourth, in 'Develop a sampling plan', Step 3 is merged with Step 4 through which the researcher exercises a random sampling in finding appropriate information prior to skimming, scanning, analyzing and synthesizing the selected articles. In carrying out this stage, she tries to immerse herself; ideally being sensitive and observant taking her time to process the data in the selected articles [10]. Fifth, 'Formulate Coding Category' takes place as she makes a critical link between the research articles she is reading with that of creating codes and categories. This task that is called memoing [11] as it requires her to summarize the contents of the research articles. The question, 'What does the data says?' lingers in her mind as she concludes the significant and relevant findings that are manifested from the empirical studies written by scholars. Figure 2 shows the data analysis procedures that are carried out in the writing of this review article.

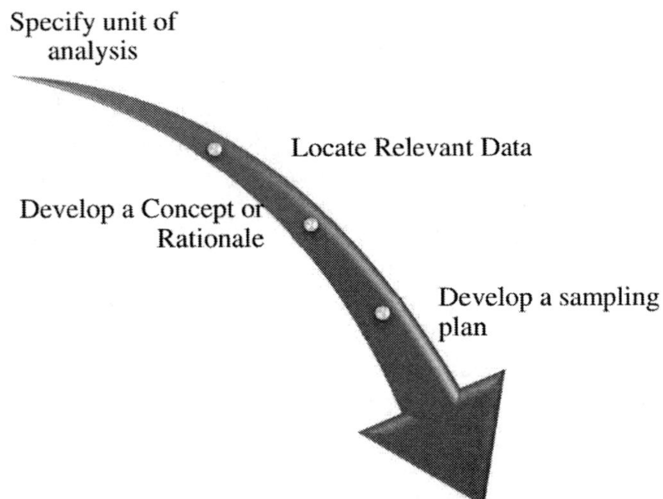

Figure 2. Data analysis phases for the review of articles

3. Research Findings

To recapitulate, the review seeks to find answer to the uses of AI in language teaching and learning. Four (4) themes are found in relation to the research question that is posted at the beginning of the review. These themes are:

3.1. AI is used to recognize natural human speech

The use of AI is able to assist researchers in developing natural language systems in that it helps to program more precise speech recognizing system [12]. In their study, AI helps to recognize oral communication that is demonstrated in people who are under stress. The experimental study employed small groups of subjects consisting of three (3) to 10 members. Their characteristics vary except that English is not their first language. Using AI in their study, it was found that the platform enables the researcher to identify the linguistics coping strategies that people develop when they are under stress due to physiological and psychological factors. Moreover, the use of AI in psycholinguistics is important as it can analyze the natural language processes in humans. Precisely AI in this regards can be programmed to become human-machine communication that is more functional, quicker and easier.

3.2. AI amalgamate Flipped Learning Classroom for effective learning

In the study that uses AI for the learning of English, its application is integrated with flipped learning approach [7]. Precisely, AI and flipped learning amalgamate in the blended classes and this combination provides many positive impact to the learning of the language. In terms of self-efficacy, it was found that students were more positive in learning English. This is true when students who were selected to be in the experimental group were more confident in their speaking than those in the control group. Likewise, students' listening skills in the experimental group outperformed those who were in the control group. Finally, it was found that students' extrinsic motivation was higher compared to intrinsic motivation in that the former factor i.e. determination to earn rewards, make students became more motivated to learn English rather than the motivation that was derived for the sake of merely performing the listening/ speaking activity.

3.3. AI promotes efficiency and effectiveness in teaching English

An analysis on the uses of AI found that English language teaching and learning can be improved with the use of the tool [13]. It helps practitioners in their teaching-related matter; more specifically in translation tasks. In the study, AI enabled practitioners of English to enhance listening activities as well as prepare students with native- like communication environment. In fact, it seemed that indirectly, the other two (2) language skills i.e. reading and writing can also be strengthen with the use of the tool. The researcher concluded that integrating AI in English teaching activities, enable students to interact and optimize their learning horizons. In fact, it may encourage positive development among the practitioners in their attempt to teach the language.

3.4. AI is used for assessment of speech

In a narrative writing concerning Web speech technology language learning application, it is found that AI is used in human conversation [14]. In other words, instead of using AI for human to talk to a computer, human now can talk through a computer. Previously, it was human-to-human communication when Computer-Mediated Communication is employed to communicate between one individual to another individual. Yet, with the use of AI, human are able to talk to machine akin to when they talk to a human. Daniels [14] further explains that Chatterbots or Artificial Conversational Ethics is one example of AI in that human are able to communicate through a machine. It holds intelligent conversation using a keyword matching technique. For instance, if a human asks the Chatterbots, 'What is your name?', the AI then will reply accordingly to the question based on the records of answers in its database. Hence, the assessment of speech can be made possible with the use of AI in today's world. In summary, Fig. 3 shows the overall themes that are developed after analyzing and synthesizing of articles are conducted.

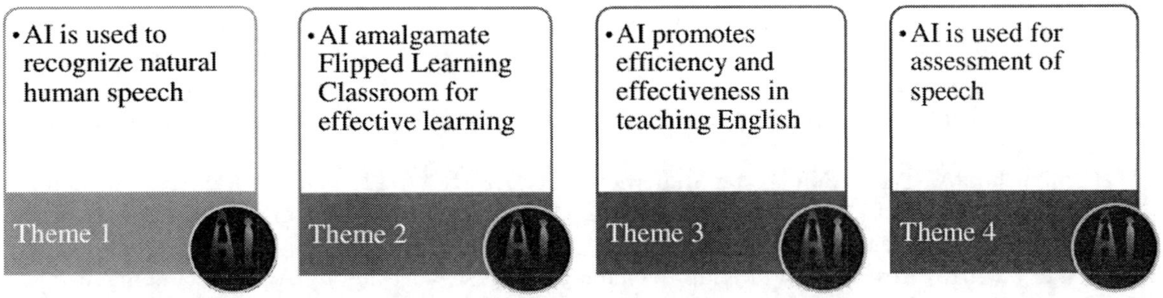

Figure 3. Formulated themes as findings for the review of articles

4. Conclusion

The purpose of the current reviews is to identify the uses of AI in language teaching and learning. This review has found that AI is a system that is used to understand natural human speech, able to incorporate flipped learning approach for teaching and learning of language, make learners become more competence and productive and able to assess human speech. Overall, the review strengthens the idea that AI is paramount in the language teaching and learning due to its various benefits. The findings will be of interest to educators who intend to integrate AI in their classrooms. Nonetheless, the study is limited by the lack of information on the uses of AI in other language skills namely reading and writing since the review of articles previously discussed concerns with its use in speaking and listening only. As such, a further review could assess the uses of AI for the two (2) skills to better understand its benefits in language teaching and learning as a whole.

Acknowledgement

The researcher would like thank the Ministry of Education, Malaysia who have granted the research scheme; Fundamental Research Grant Scheme (FRGS) under research ID RDU190188 to carry out the research works.

References

[1] Kite-Powell J. (2017). The Role of Artificial Intelligence and Language. Retrieved from https://www.forbes.com/sites/jenniferhicks/2017/12/30/the-role-of-artificial-intelligence-and- language/

[2] Kannan J, P Munday (2018) New Trends in Second Language Learning and Teaching through the lens of ICT, Networked Learning, and Artificial Intelligence. In: Fernández Juncal C., Hernández Muñoz N. (eds.) Vías de transformación en la enseñanza de lenguas con mediación tecnológica. Círculo de Lingüística Aplicada a la Comunicación 76, 13-30 http://dx.doi.org/10.5209/CLAC.62495

[3] Goksel-Canbek N, Mutlu M. E. (2016) in Errol Jayawardene (2015). On the track of Artificial Intelligence: Learning with Intelligent Personal Assistants. International Journal of Human Sciences, 13(1), 592-601. doi:10.14687/ijhs.v13i1.3549

[4] Vaidya T, Zhang Y, Sherr M, & Shields C. (2015, August). Cocaine noodles: exploiting the gap between human and machine speech recognition. In Proceedings of the 9th USENIX Conference on Offensive Technologies. 1-14. USENIX Association. Retrieved from https://www.usenix.org/system/files/conference/woot15/woot15-paper-vaidya.pdf

[5] Tariq K. (2014) Siri, Google Now, Cortana: The Ultimate Digital Voice Assistant? WCCFTech.com, Retrieved from http://wccftech.com/digital-voice-assistant/#ixzz3lcDQvCrq

[6] Elwany E, Shakeri S. (2014) Enhancing Cortana User Experience Using Machine Learning. Recall, 55(54.61), 24-24.

[7] Shin M. H. (2018) How to use artificial intelligence in the English language learning classroom. Indian Journal of Public Health Research & Development, 9(9), 557-561.

[8] Fraenkel J R, Wallen N E, & Hyun H H (2011) How to design and evaluate research in education. New York: McGraw-Hill.

[9] Maguire M, Delahunt B (2017) Doing a thematic analysis: A practical, step-by-step guide for learning and teaching scholars. AISHE-J: The All Ireland Journal of Teaching and Learning in Higher Education, 9 (3), 3351-33514

IOP Conf. Series: Materials Science and Engineering **769** (2020) 012043 doi:10.1088/1757-899X/769/1/012043

[10] Green J, Willis K, Hughes E, Small R, Welch N, Gibbs L, Daly J (2007) Generating best evidence from qualitative research: the role of data analysis. Australian and New Zealand journal of public health, 31(6), 545-550.
[11] Hesse-Biber S (2010) Analyzing qualitative data: With or without software. Boston College, Massachusetts, US
[12] Anikushina V, Taratukhin V, von Stutterheim C (2018) Natural Language Oral Communication in Humans Under Stress. Linguistic Cognitive Coping Strategies for Enrichment of Artificial Intelligence. Procedia computer science, 123, 24-28.
[13] Wei F (2018) Research on the Improvement of English Learning by Artificial Intelligence (AI). Journal of Advanced Oxidation Technologies, 21(2), 392-395
[14] Daniels P (2015) Using Web Speech Technology with Language Learning Applications. JALT CALL Journal, 11(2), 177-187.

IOP Conf. Series: Materials Science and Engineering **769** (2020) 012044 doi:10.1088/1757-899X/769/1/012044

Students' Continuance of Using E-Learning System: A Review of Conceptual Frameworks

N S A Rahman[1], A N Rosman[1] and N A Sahabudin[1]

[1]Faculty of Computing
Universiti Malaysia Pahang
26300 Kuantan, Pahang, Malaysia

shamsiah@ump.edu.my

Abstract. This research presented a review of conceptual framework regarded to find the factors of students' continued use of online learning system. The outcomes were reported according to a systematic review on the recent research articles (from 2013 until 2019) related with continued use of e-Learning system. From review of theoretical models and related theories we recommend to use Unified Theory of Acceptance and Use of Technology (UTAUT) to measure students' intention to continued use of e-learning system. The discussion of this research obtained revealed that the most common factors influence students' intention to continued use e-learning are performance expectancy, effort expectancy, social, self-efficacy, perceived usefulness and behavioral intention to continued use.

1. Introduction

E-learning is important in the educational process nowadays, that information and also communication technology is used to distribute learning content. For the management of studies, communication between students and teachers is important. Although students used E-learning to complete their task and for studying, they tend to face success and also the failure of understanding the online instruction. This may be related to student motivation, lecturer need to take part to stimulate the students to explain how the online environment may be used and study groups need to be built and always be aware of student's anxiety in using E-learning which might affect them in a negatively [1].

In universities nowadays, E-learning was found been used as an essence tool for learning management system. E-learning provides students an interactive web-based platform to perform exercises and other education activities [1]. The important thing to use E-learning in higher education is the level of satisfaction of the students towards online learning activity is paramount for efficient and for understanding its efficiency. The satisfaction of students is identified in terms of environment, it can be said that many factor affecting during face to face and online learning will determine the effect [2]. From previous research, E-learning was growing significantly due to the importance and implementation of E-learning at all universities all over the world. Despite the popularity of E-learning at universities level, there is a very low research paper on student continued usage interaction in E-learning in Malaysia [3]. Although many research on E-learning was done all over the world, the antecedents which affecting the use of E-learning in learning institution need to investigate due to the different way of implementation such as instructor teaching style and E-learning support [4, 5]. Some of the research considers whether countries work well on cross-border issues such as challenges and successfulness of

E-learning implementation [5]. The usage of E-learning had their own factors that affect students' success in using E-learning [6] and student level of satisfaction for efficient design [2].

Another study highlighted the needs of motivation to play as important role in ensuring effective use of E-learning. A student is motivated to strongly involve to explore and try a new system that will give them benefits [7]. In addition, the UTAUT model was applied to explain E-learning adoption which can assist in understanding behavioural intention to use E-learning. Preliminary works were conducted using survey on universities students to assess the application of UTAUT in E-learning [5]. Therefore, this research will discuss the detail of previous work done by other researchers to identify the factor of continued used E-learning. The following sections describes E-Learning, discussion of related theories and models, discussion of related work done by previous researchers and finally discussion of conclusion and future work.

2. E-Learning

E-learning in Malaysia is growing significantly from years to years due to the improvement in technologies. Universities in Malaysia implement E-learning in their education system for keeping up with technology development. E-learning is accepted due to the capability to connect students and teachers in education [5]. There are trends on implementation of E-learning in Malaysia which are learning management system (LMS), E-learning policy, E-learning governance, E-learning training and integration into teaching and learning [8]. From the recent research of E-learning in Malaysia at Universities Technology Malaysia (UTM), the Centre of Teaching and Learning (CTL) provide various educational supported based on courses in E-learning. Important feature of CTL is effectiveness than traditional computer-based [9]. Besides the implementation of E-learning in UTM, other universities in Malaysia also implement E-learning in their education whether IPTA or IPTS. There is research that show 20 public universities, seven private universities and three polytechnics that implement E-learning in their education [10]. University Malaysia Pahang (UMP) is one of 20 public universities that use E-learning in their education. UMP has implemented open source LMS in their education which is Moodle that is known and widely used because it is free and east to modify.

In the past two years, the use of E-learning was increased. E-learning also had been used as essential tool for learning management system through universities. Students used E-learning as a tool to conduct interactive web-based exercises [1]. This was supported by Embi [10] who found that most of the students' access E-learning based on course that they are taking daily and most of them can access their E-learning easily and everywhere either from computer laboratory, their hostel and their home. The important thing to use E-learning in higher education is the level of satisfaction of the students towards online learning activity is paramount for efficient and for understanding its efficiency. The satisfaction of students is identified in terms of environment, it can be said that many factor affecting during face to face and online learning will determine the effect [2].

3. Theoretical Model

The UTAUT is well-known as one of the common technology acceptance theories which derived bases of the Theory of Reasoned Action (TRA), the Technology Acceptance Model (TAM), Diffusion of Innovations (DOI) and Theory of Planned Behaviour (TPB) [11, 12]. The UTAUT model in Figure 1 explains the intent to continue usage behaviour of the user and use an Information system (IS). The theory has four key that direct determiners of usage intention and behaviour of the user. These include the performance expectancy, effort expectancy, social influence, and facilitating conditions [3, 11]. The UTAUT model also theorized experience, age, voluntariness and gender in the constructs list of the four key on the behavior of the user as moderators of the impact [12, 13, 14].

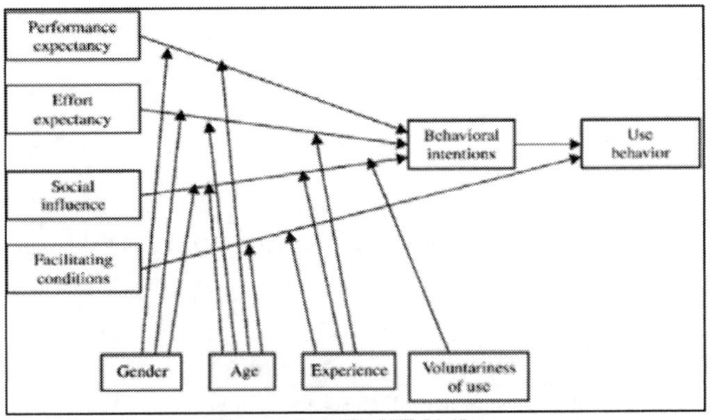

Figure 1. UTAUT [14]

Performance expectancy (PE) is defined where the system provides benefits in using of E-learning system to the user in performing their assignment or learning task. For effort expectancy (EE) is defined as a degree that easy to associate with the user in using E-learning. Social influence (SI) is defined as the extent which individual perceive for others to believe the importance of use in E-learning system [15].

Besides Technology Acceptance Model (TAM), UTAUT included social element which considered an important element in the learning environment. The advantages in today's society are the use of social networks and online communication that show significant social component [12, 16]. Although TAM acceptance theory in E-learning is used the most in other research, their result shows that literature in the field of E-learning and the acceptance theories other than TAM based on previous studies. The UTAUT model explains more on individual acceptance by using four moderators including age, gender, voluntariness and experience [11]. Some of the research paper used the UTAUT model as it seems reasonable to be used to study continued used of E-learning. These arguments lead us that UTAUT is a suitable and appropriate theory for this research [12, 17].

3.1. Related Prior Research

In this research, ten articles comprising journals and conference proceedings obtained from relevant databases between from 2013 until 2019 were used to review the previous frameworks used in prior research related to continued use intention in using e-learning. This section will discuss four frameworks related to the current research as follows:

Lin et al. [18] conducted a study towards using e-learning systems by extending of UTAUT. The study's goal was to centers around the level of correspondence between showing styles and perceived learning of students, which the aim and use of e-learning frameworks will legitimately direct together. Figure 2 represents the Education Behavioral Intention (EduBIM) model proposed by researchers of this study. The study enriches the determinants from UTAUT by coordinating intellectual individual contrasts to integrate the impacts of demographic moderators.

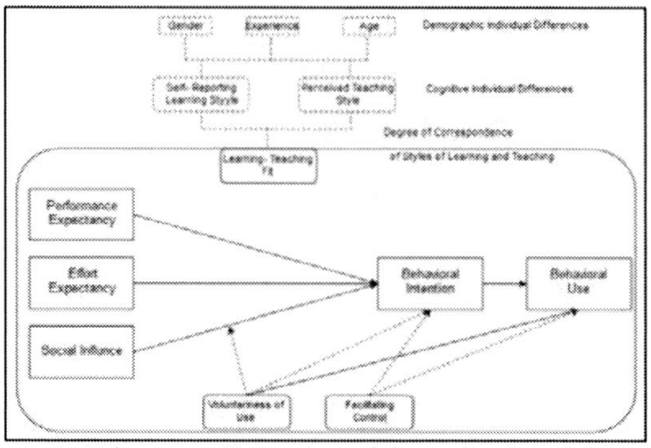

Figure 2. Education Behavioural Intention Model [18]

Zawaideh [3] in an article conducted a case study at one of Malaysian institution by proposing an acceptance model for e-learning services. The researcher recognized antecedents in clarifying the practice of e-learning situations among students' in that institution by using UTAUT model as shown in Figure 3. The investigation uncovered, performance expectancy, facilitating conditions, effort expectancy, social factors, cultural, behavioral intention, and user behavior as the variables that impact availability of students in Malaysia to utilize e-Learning regarding its degree.

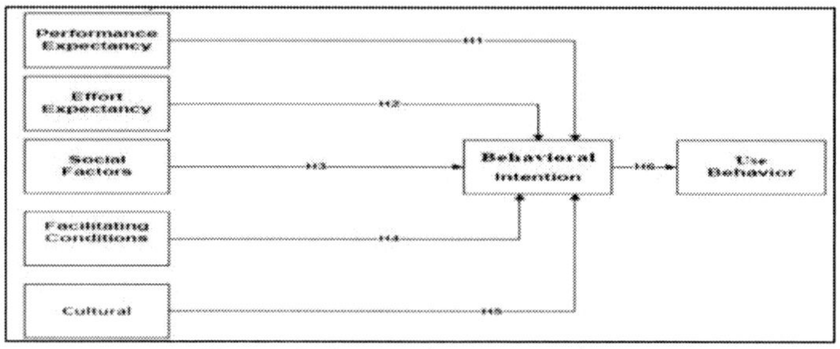

Figure 3. Research Framework [3]

Decman [12] carried out an investigation on the influences of previous education and gender by modeling the e-learning acceptance in higher education mandatory environments. The investigation's aimed to conduct a survey to assess the suitableness of UTAUT inside a particular required e-learning condition in higher education. A model was constructed (see Figure 4) comprising fundamental determinants from UTAUT including social influence, effort expectancy, performance expectancy and behavioral intention. The investigation also looks forward on the other moderating factors including students' past education and gender in the displayed structure.

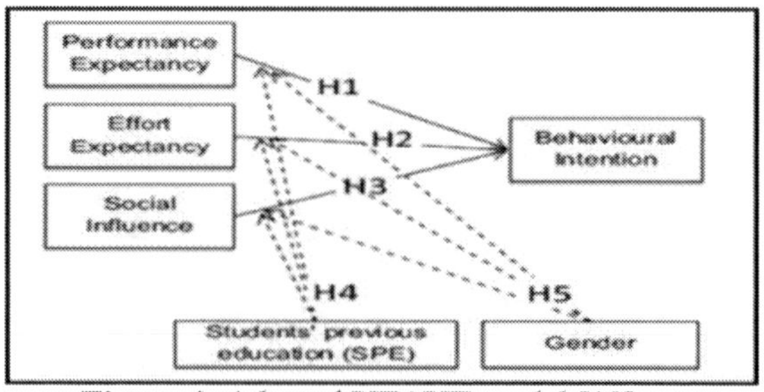

Figure 4. Adapted UTAUT model [12]

IOP Conf. Series: Materials Science and Engineering **769** (2020) 012044 doi:10.1088/1757-899X/769/1/012044

Bellaaj et al. [15] performed an empirical investigation on continuance using e-learning at University of Tabuk. The research built up a model (see Figure 5) in light of certain factors in the UTAUT. The discoveries uncover that performance expectancy, and effort expectancy decide the intention of continued use of e-learning system. Moreover, the examination likewise uncovered that with more grounded Internet experience, the impact of performance expectancy increases, and the influence of effort expectancy decreases. In the mean time, the impact of social influence on intention of continued use is seems to be more grounded for ladies than for men.

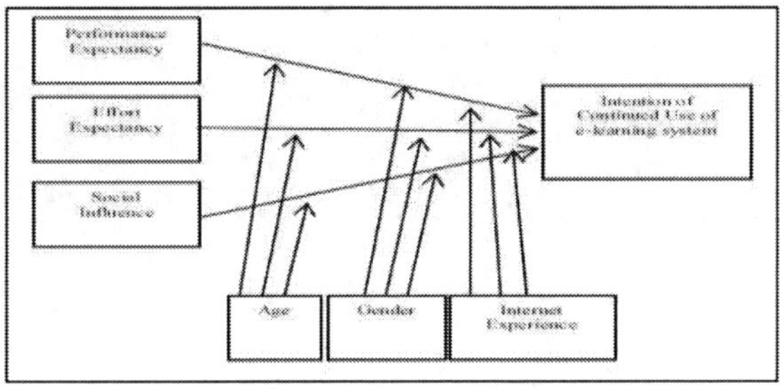

Figure 5. Research Framework [15]

4. Related Works

This section summarizes about the theory on frameworks details used in prior research works. From previous studies (see Table I), there are several theories were used such as TAM, UTAUT, CFS, ELRS SS and MSLQ. From the review, it was found that the numerous investigations were based on TAM and UTAUT. TAM appears to be heavily applied in technology acceptance research attributed to its simplicity and understandability [19] and credited to its power of explanatory for key factors which have significant effects on the acceptance of user and information system and technology adoption.

From Table 1, four out of ten research papers used the UTAUT model. The research of Zawaideh [3] focusing on some factors which are performance expectancy, social factor of cultural conditions that affect students to implement the ease of use E-Learning for education. According to Bellaaj et al. [15], the result shows that the critical factor is the effort expectancy and performance expectancy that determines the continued used of e-learning. The factor toward the social environment in e-learning implementation is stressing on performance expectancy [12]. For Ibrahim et al. [5], the study focuses that self-efficiency, perceived ease of use and intention use of e-learning is the critical factors that affect students' continuance of using e-learning for their education.

There are some common factors in continued used of E-learning that influence students. The most found to be key determinant factors are effort expectancy, social, performance expectancy [3, 8, 12, 15, 18], self-efficacy [2, 5, 8, 20], behavioural intention [12, 18, 20] and perceived usefulness [5, 8, 20]. In other word, these factors have significant positive effect towards student to use E-learning.

In this study, ten academic articles comprising of journal articles and conference proceeding articles were obtained from research databases related to this field, to assess the research thoroughly for the most influential factors on students' continuance of using e-Learning system. The result of such review led to a classification of identified variables about influential factors as illustrated in Table 2.

According to prior studies, Table 2 shows the most frequently referred influential factors that affect continued use. The most frequently introduced variables are performance expectancy [3, 12, 15, 18], social [3 12, 15, 18], self-efficacy [2, 5, 8, 20], effort expectancy [3, 12, 15, 18], perceived usefulness [5, 8, 20], behavioural intention [12, 18, 20, 22], perceived ease of use [5, 20], characteristics [5, 6] and motivation [2, 21].

Table 1. Related Work by Prior Research

No.	Author	Aim	Theory	Factors
1.	Alhabeeb & Rowley [6]	To compare the different of students and academic staff perspectives.	CSFs	Ease of access, characteristics, support and training, technology infrastructure and experience of the system
2.	Zawaideh [3]	To identify factors in interpreting e-learning environments among students in higher education.	UTAUT	Cultural, social, performance expectancy, effort expectancy and facilitating conditions.
3.	Ibrahim et al. [5]	To investigate the acceptance of e-learning among university students.	TAM	Characteristics, perceived usefulness, self-efficacy, perceived ease of use, course design and intention to use.
4.	Yilmaz [2]	To examine how far the level of students' readiness will impact students' motivation and satisfaction in using E-learning.	ELRS SS MSLQ	Motivation, learner control, self-direct learning and self-efficacy.
5.	Noesgaard & Orngreen [21]	To investigate the capabilities of E-learning by using an integrative review	Integrative Review	Interaction, motivation, experience and practice.
6.	Decman [12]	To evaluate and assess the appropriateness of UTAUT model	UTAUT	Social influence, performance expectancy, effort expectancy and behavioural intention.
7.	Bellaaj et al. [15]	To focus on virtual learning system for student continued use	UTAUT	Social influence, effort expectancy, performance expectancy and intention of continued use.
8.	Al-rahmi et al. [8]	To find the influential determinants that affecting E-learning in evaluating the effectiveness of E-learning.	TAM	Perceived usefulness, intention to use e-learning, satisfaction, effectiveness, self-efficacy, learner interface and learning community.
9.	Findik & Ozkan [20]	To understand higher education of bahavioral intentions instructors towards Learning Management Systems (LMS)	TAM	Subjective norm, perceived usefulness, technological complexity, perceived ease of use, self-efficacy and behavioural Intention
10.	Lin et al. [18]	To investigates explanatory models for adopting technologies	UTAUT	Behavioural intention, effort expectancy, performance expectancy, social influence and behavioural use.

Table 2. Summary of Most Frequently Referred Variables

Factors	Alhabeeb & Rowley [6]	Zawaideh [3]	Ibrahim et al. [5]	Yilmaz [2]	Noesgaard & Orngreen [21]	Dečman [12]	Bellaaj et al. [15]	Al-rahmi et al. [8]	Findik & Özkan [20]	Lin et al. [18]
Characteristic **(CH)**	√		√							
Ease of access **(EOA)**	√									
Support and training **(ST)**	√									
Technology infrastructure **(TI)**	√									
Experience of system **(EOS)**	√									
Performance expectancy **(PE)**		√				√	√			√
Social **(SC)**		√				√	√			√
Self-efficacy **(SE)**			√	√				√	√	
Facilitating Conditions **(FC)**		√								
Effort expectancy **(EE)**		√				√	√			√
Course Design **(CD)**			√							
Intention to use **(IU)**			√							
Perceived usefulness **(PU)**			√					√	√	
Perceived ease of use **(PEOU)**			√						√	
Self direct learning **(SDL)**				√						
Learner control **(LC)**				√						
Motivation **(M)**				√	√					
Interaction **(IN)**					√					
Experience **(EX)**					√					
Practice **(PR)**					√					
Behavioural Intention **(BI)**						√			√	√
Technological complexity **(TC)**									√	
Subjective Norm **(SN)**									√	

Factors	Alhabeeb & Rowley [6]	Zawaideh [3]	Ibrahim et al. [5]	Yilmaz [2]	Noesgaard & Orngreen [21]	Dečman [12]	Bellaaj et al. [15]	Al-rahmi et al. [8]	Findik & Özkan [20]	Lin et al. [18]
Behavioural use **(BU)**										√
Learner interface **(LI)**								√		
Learning community **(LCM)**								√		
Satisfaction **(S)**								√		
Effectiveness **(E)**								√		

5. Conclusion and Future Work

This paper critically examines the factors that influence students' continued use E-learning system by presenting the results on review of ten related articles retrieved from relevant databases between from 2013 until 2019. From the review, our results reveal that the most frequently referred influential factors that affect students' continuance of using E-learning system are performance expectancy, social, self-efficacy, effort expectancy, behavioural intention and perceived usefulness. Notwithstanding these limitations, it is suggested to have future exploration from other sources in investigating and identifying other factors. It is suggested, more research is needed to consider the use of theoretical and methodological approaches such as structural equation modeling in order to examine the students' continuance intention is investigated in future studies.

6. Acknowledgement

This research was fully funded by Universiti Malaysia Pahang, grant number RDU1803147.

References

[1] M Samir, A El-Seoud, I A T F Taj-Eddin, N Seddiek, M M El-Khouly, A Nosseir 2009 E-Learning and students' motivation: A research study on the effect of E-learning on higher education 20–26

[2] R Yilmaz 2017 Exploring the role of e-learning readiness on student satisfaction and motivation in flipped classroom Computers in Human Behavior **70** 251–260

[3] F H Zawaideh 2018 Acceptance Model for e-Learning Services: A Case Study at Al-Madinah International University in Malaysia *International Journal of Academic Research in Accounting, Finance and Management Sciences* **7** 14–20

[4] R Cheung, D Vogel 2013 Predicting user acceptance of collaborative technologies: An extension of the technology acceptance model for e-learning *Computers & Education* **53** 160-175

[5] R Ibrahim, N S Leng, R C M Yusoff, G N Samy, S Masrom, Z I Rizman 2017 E-Learning acceptance based on Technology Acceptance Model (TAM) *Journal of Fundamental and Applied Sciences* **9** 871-889

[6] A Alhabeeb, J Rowley 2018 E-learning critical success factors: Comparing perspectives from academic staff and students *Computers and Education* **127** 1–12

[7] S R Harandi 2015 Effects of e-learning on Students' Motivation *Procedia - Social and Behavioral Sciences* **181** 423–430

[8] W M Al-rahmi, M S Othman, L M Yusuf 2015 The Effectiveness of Using E-Learning in Malaysian Higher Education: A Case Study Universiti Teknologi Malaysia *Mediterranean Journal of Social Sciences* **6** 625–637

[9] W M Al-Rahmi, A A Saged, O Alfarraj, M S Othman, A I Alzahrani, N Alias, N S A Rahman 2018 Use of E-Learning by University Students in Malaysian Higher Educational Institutions: A Case in Universiti Teknologi Malaysia *IEEE Access* **6** 14268–14276

[10] M A Embi 2011 E-Learning in Malaysian Institutions of Higher Learning: Status, Trends and Challenges Mohamed Amin Embi Universiti Kebangsaan Malaysia (November) 14–15

[11] V Venkatesh, M G Morris, G B Davis, F D Davis 2003 User Acceptance of Information Technology: Toward a Unified View *MIS Quarterly* **27** 425-478
[12] M Decman 2015 Modeling the acceptance of e-learning in mandatory environments of higher education: The influence of previous education and gender *Computers in Human Behavior* **49** 272–281
[13] S Pahnila, M Siponen, X Zheng 2011 Integrating habit into UTAUT: The Chinese eBay case *Pacific Asia Journal of the Association for Information Systems* **3** 2
[14] V Venkatesh, J Y Thong, X Xu 2012 Consumer acceptance and use of information technology: extending the unified theory of acceptance and use of technology *MIS quarterly* **36** 157-178
[15] M Bellaaj, I Zekri, M Albugami 2015 The continued use of e-learning system: An empirical investigation using UTAUT model at the University of Tabuk *Journal of Theoretical and Applied Information Technology* **72** 464–474
[16] R P Bagozzi 2007 The legacy of the technology acceptance model and a proposal for a paradigm shift *Journal of the Association for Information Systems* **8** 4
[17] B Sumak, G Polancic, M Hericko 2010 An empirical study of virtual learning environment adoption using UTAUT *Second International Conference on Mobile, Hybrid, and On-Line Learning* 17-22
[18] P C Lin, H K Lu, S C Liu 2013 Towards an education behavioral intention model for e-learning systems: An extension of UTAUT *Journal of Theoretical and Applied Information Technology* **47** 1200–1207
[19] P Legris, J Ingham, P Collerette 2003 Why do people use information technology? A critical review of the technology acceptance model *Information & Management* **40** 191–204
[20] C D Findik, S Ozkan 2013 A model for instructors' adoption of learning management systems: Empirical Validation in higher education context *Turkish Online Journal of Educational Technology* **12** 13–25
[21] S S Noesgaard, R Orngreen 2015 The effectiveness of e-learning: An explorative and integrative review of the definitions, methodologies and factors that promote e-Learning effectiveness *Electronic Journal of E-Learning* **13** 278–290
[22] M A Al-Shara, M E Mufadhal, N A Sahabudin, R A Arshah 2019 Acceptance of online social networks as technology-based education tools among higher institution students: Structural equation modeling approach Scientia Iranica 26 136-144

A Hybrid Screen Size Independent Authentication Scheme for Smart Devices

Ahmad Faizuwan Bin Welhendri[1], **Saiful Azad**[1,2,*], **Md Sayfullah Al Noman Ranak**[1]

[1]Faculty of Computer Systems & Software Engineering, University Malaysia Pahang, 26300 Gambang, Kuantan, Pahang, Malaysia
[2]IBM CoE, UMP, 26300 Gambang, Kuantan, Pahang. Malaysia

E-mail: `saifulazad@ump.edu.my` (*corresponding author)

August 2019

Abstract. Although, several authentication schemes have already been proposed for smart devices; however, most of these schemes does not consider the fact that smart devices come in different sizes. Hence, they are not screen size independent — which is the point of interest in this paper. Again, alongside screen size independence, a secure scheme also must defend the aforementioned attacks. Taking these concerns into account, in this paper, a hybrid screen size independent authentication scheme is proposed for smart devices that integrates Vibration Code or VC and Press Touch Code or PTC using a juggling-based approach. Here, VC ensures resilience against the shoulder surfing attack since it is a sense based technique and hybridization of both these schemes contribute to attaining resilience against the smudge and brute force attacks up to some extent. In addition, the proposed scheme does not require a space more than the placement of a finger on the screen; and thus, it is screen size independent. The proposed scheme is evaluated in terms of security and functionality; and compared with other similar schemes where it outperforms the others.

Keywords: Graphical authentication, vibration code, press touch code, screen size independent scheme, hybrid authentication scheme.

1. Introduction

Due to offering an extensive variety of functionalities to users, smart devices are becoming a part of daily lives [1]. Therefore, for frequent access, many people prefer to store both sensitive and private data (e.g., passwords of other systems, sensitive documents, private images and so on) in these devices. However, due to lack of or insufficient protection, sometimes attackers capture these data and make them victimized. One of the easiest mechanism to protect contents in these devices is to utilize an access control mechanism. For that, authentication schemes are more preferred for these devices.

Till date, a considerable number of authentication schemes are already proposed. Most of these schemes could be broadly divided into three factors, namely *i*) knowledge or something

Hybrid Authentication

you know (e.g., PIN, textual password), *ii*) possession or something you have (e.g., token, smart card), and *iii*) inheritance or something you are (e.g., biometrics) [2]. Among them, knowledge-based schemes are utilized in most of the smart devices since they do not need any extra hardware (like inheritance) or any additional substance (like possession). Again, the knowledge-based schemes can be divided into several categories including textual schemes and graphical schemes. Among these categories, graphical schemes are more preferable to the general user due to their graphic oriented nature [3]. It has been established through several psychological studies that human brain can recall images more precisely than texts [4]. However, many of these schemes suffer from several attacks, such as smudge [5], shoulder surfing [6], and brute force [7] are the most prominent attacks. Over and above, most of these techniques are not screen size independent; whereas, screen of smart devices comes in different sizes. Due to this, sometimes it becomes difficult to provide credentials; specially on miniature smart devices. Therefore, in this paper, a hybrid scheme is proposed that is screen size independent and is resilient against the aforementioned three prominent attacks.

The subsequent sections of this paper are organized as follows. In Section 2, similar schemes are scrutinized to establish the necessity of proposing a new scheme. Afterwards, the proposed scheme is detailed in Section 3. Then, it is evaluated and compared with other similar schemes in Section 4. The concluding remarks of this paper and the future directions of this work are mentioned in Section 5.

2. Related Works

Since a specific scheme has specific defend mechanism that can tackle one or more attacks, a rational combination of two or more schemes would increase the level of security. In this case, two approaches are generally employed, namely *i*)multi-factor authentication and *ii*) hybrid authentication. In case of the former approach, as the name suggests, two or more factors (which are mentioned in Section 1, i.e., knowledge, possession, and inheritance) are combined to ensure higher level of security. A user would be granted access only after successfully presenting two or more pieces of evidences or factors. Conversely, in case of the latter approach, two or more schemes of a same factor are combined. In this paper, this approach is embraced since it takes considerably shorter time to authenticate over multi-factor approach. It is because, the selected schemes of a hybrid approach could be seamlessly combined, which is extremely unlikely in case of multi-factor approach.

At present, only a few hybrid authentication schemes are proposed. For instance, a such authentication scheme is proposed in [8], which utilizes a textual authentication scheme along with grid cells for generating credentials for Personal Digital Assistants (PDAs). However, it lacks adequate users' study to endorse their claim of defending attacks like brute force, shoulder surfing, and dictionary. In [9], another hybrid authentication scheme is proposed that is based on shape and text. Here, the shapes of strokes are provided as the elementary part of the credential followed by the text. One of the primary concerns of this scheme is its registration process where a user has to mention the original shape and strokes to the system. Any attacker who is observing the registration session would be able to capture the

IOP Conf. Series: Materials Science and Engineering **769** (2020) 012045 doi:10.1088/1757-899X/769/1/012045

Hybrid Authentication

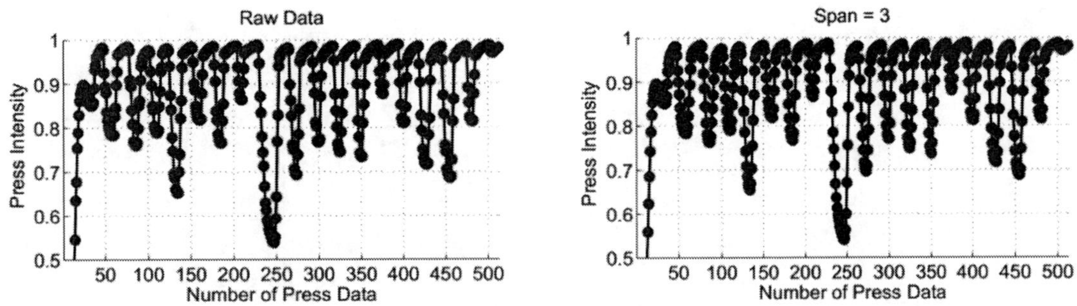

Figure 1. Raw data and Smoothed data for span = 3 are shown.

password. Again, because of the adaptability issue, usually users select considerably weak strokes, and thus, make this scheme vulnerable to several attacks. Again, another similar authentication scheme is proposed in [10] that combines image and text. The primary targets of this combination are to increase password space and memorability. However, this scheme is vulnerable to the shoulder surfing attack since both predecessors — image and text — are vulnerable to this attack. Therefore, designing a hybrid authentication scheme still remains an important issue to investigate.

3. Proposed Scheme

In this paper, a new hybrid authentication scheme is proposed that combines VC [11] and PTC [3] schemes. Here, the PTC is a special code, which is extracted from the pressure intensity values of a user and the VC is another special code, which is generated from the vibration. The justifications of combining these two schemes are that VC has the resilience against the shoulder surfing attack; however, it spends a long time during authentication. Conversely, PTC spends a short time for authentication; however, it experiences the shoulder surfing attack when attackers are relatively near. The details of these schemes and their combination mechanism are mentioned below.

3.1. PTC Scheme

The detail description of the PTC scheme is given in [3]. Therefore, a brief description is mentioned in this section for better understanding of the hybridization process. This scheme is based on the number of force presses provided by a user on a screen. For that, a Pressure Sensitive Screen (PSS) is obligatory, which can recognize the Press Intensity Value (PIV) at any certain time, τ_i. Generally, PIVs are recorded after every δ time interval. However, in our proposed scheme, PIVs are acquired through a system call, which is performed after every Δ time interval, where $\Delta > \delta$. Every system call returns a PIV, v_i which is later saved in a list, λ chronologically, i.e., v_i is acquired before v_{i+1} and after v_{i-1}. This way, the number of system calls as well as the number of PIVs in λ are reduced to facilitate data acquisition and data processing. This process starts with the placement of a finger on the screen and ends when the finger is taken off from the screen. In other words, data (i.e., PIV) acquisition starts

Hybrid Authentication

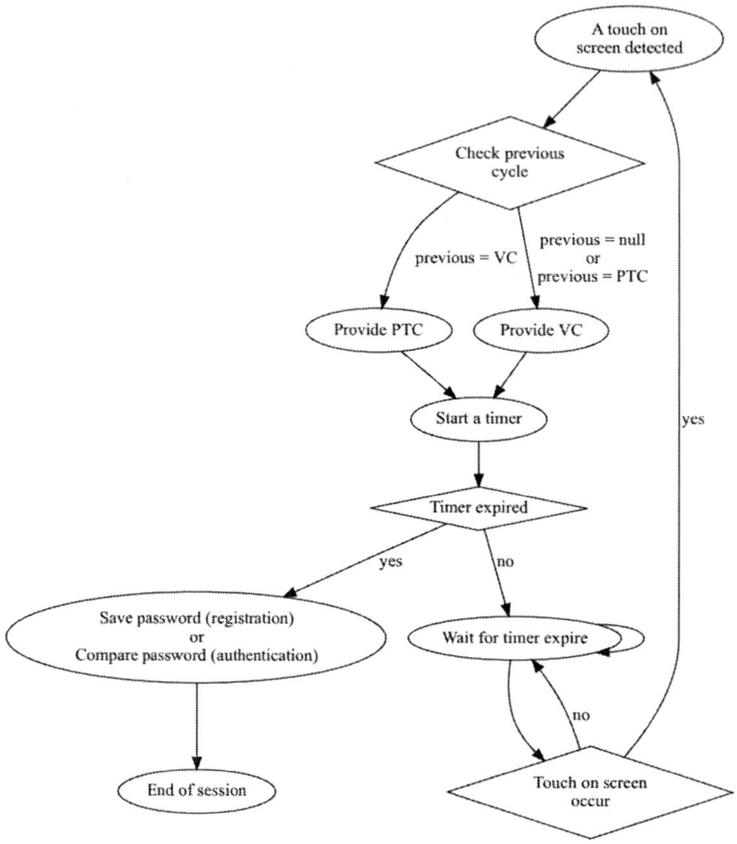

Figure 2. A high-level depiction of the proposed hybrid authentication scheme.

when a touched occur on the screen and ends when a detached occur from the screen.

After the completion of data acquisition phase, the subsequent phase is data cleaning. All the acquired data in λ are cleaned using the Moving Average Filtering Technique (MAFT) technique [12]. The following equation is employed in this case:

$$v_i' = \frac{1}{2N+1}\left(v_{(i+N)} + v_{(i+N-1)} + ... + v_{(i-N-1)} + v_{(i-N)}\right) \quad (1)$$

where v_i' is the PIV at i-th data point in λ after the smoothing process, N is the number of neighboring data points on either side of v_i, and $2N+1$ is the span. Among various spans like 3, 5, 7, 9, and others; span = 3 is selected taking the arguments in [3] into account. The justification of this choice is that higher span values flattens the peaks and hence, makes the presses undetected. Figure 1 demonstrates raw data and cleaned data for the span = 3.

Afterwards, the number of force presses is extracted from the cleaned PIVs in λ. As could be observed in Figure 1 is that force presses produce peaks. Therefore, a *1-D Peak* finding algorithm [13] is modified in [3], called *Press Touch Finding Algorithm (PTFA)*, which is also utilized in this paper for finding force presses or PTC. In brief, PTFA is a brute force based technique that selects a data point as a peak only if it is higher than its adjacent data points, and can be represented as below:

$$v_{i-1} < v_i > v_{i+1} \quad (2)$$

Hybrid Authentication

This equation identifies all local maximum peaks. The count of these local maximum peaks is the PTC. For instance, in Figure 1, PTC = 20 since there are 20 local maximum peaks. Afterwards, this value is either stored (in case of registration) or compared (in case of authentication).

3.2. VC Scheme

The detail description of the VC scheme is mentioned in [11]. However, for the better understanding of the proposed hybrid scheme, a brief description is given below. When a finger is placed on the screen, vibrations triggered with a specific time interval. It ends when the finger is lifted up from the screen alike the PTC. The number of vibrations that are sensed within this epoch is considered as the VC. Note that, for the VC scheme, vibrations need to be controlled (i.e., start, end, and interval between two vibrations) by the scheme. After the VC is counted, it is either stored for registration or compared for authentication.

3.3. Integration

For combining the aforementioned schemes, a juggle-based approach is considered, where the registration and/or the authentication commences with the VC, then the PTC, again the VC, and this sequence repeat until the session ending event occurs. Assuming every VC or PTC giving event as a cycle, the commencement of the next cycle must not exceed a fixed time, called interval threshold, τ_{max}. Conversely, it is considered as the end of the registration or authentication session. In other words, after completion of a cycle, the next cycle must commence before interval threshold expired. In case of ending the session, a user must not place finger within τ_{max} on the screen. A high-level pictorial description of the proposed scheme is given in Figure 2.

3.4. Registration and authentication

At first, a user has to register a credential for enabling authentication-based access control on a smart device. For that, the user need to place a finger on the screen. As mentioned in Section 3.3, the first cycle of the scheme is the VC. Once the VC is provided, the user has to lift the finger. Afterwards, before τ_{max} expired, s/he has to place the finger on the screen and need to provide the PTC; and then, the VC. The user can juggle between the VC and the PTC as long as s/he wants. For higher level of security, the users are suggested to repeat the cycles for a considerable number of times (for more details, see Section 4.1.3). The user can end the session by not touching the screen before τ_{max} timer expired. At the end of the session, the credential of the user is stored in the system for utilizing in authentication.

After the completion of the registration, every time during the access to the device, the user has to recall the credential. The process of authentication is similar like the registration. However, at the end, instead of storing the credential, the newly provided credential will be compared with the registered credential. The user will be given access only if both the credential matched.

Hybrid Authentication

4. Evaluation

The proposed scheme is implemented on the Android Operating System and is tested on a Huawei P9 Plus device (which is a PST enabled device). The performance of this scheme is analyzed below in terms of security and functionality.

4.1. Security Analysis

For analysing the security of the proposed scheme, its resilience against three most prominent attacks, namely shoulder surfing, smudge, and brute force are taken into account.

4.1.1. Shoulder Surfing Attack: Since the VC is a sense-based scheme, it is resilient against the shoulder surfing attack as only a user who touched the screen would be able to sense the vibration. Any other who is picking on the phone or capturing the authentication session using a camera would not be able to acquire the credential. In addition, the PTC also offers resilience against this attack up to a certain extent as demonstrated in [3] with an experimental evaluation. According to that, it offers: negligible or no defence when the distance is near (< 0.5 m), seldom recognize from a mid range distance (2 to 3 m), and hard to recognize from a long range distance (> 3 m). Hence, the proposed technique inherits shoulder surfing resiliency from its predecessors.

4.1.2. Smudge Attack: This attack occurs when oily residues or smudges remain on the screen of a device as a side effect of proving a credential. In [5], it has been exhibited that around 92% cases, it is possible to break credentials partially and 68% cases fully. However, in the proposed scheme, it is not possible to find the VC and/or the PTC using the oily residues or smudges; and hence, it is resilience against this attack.

4.1.3. Brute Force Attack: An authentication scheme generally attains resilience against the brute force attack when it offers a large password space. Note that, in this attack, all possible combinations are checked systematically to break a credential. However, when the password space is large, it becomes infeasible. This proposed scheme offers a comparable password space like the Personal Identification Number (PIN). It can be calculated as γ^{α}, where γ is the maximum number of allowable VC or PTC in a cycle and α is the number of cycles. When γ is 10 and α is 6, the password space of the proposed scheme is equivalent to a 6-digit PIN number, which is used in Automated Teller Machine (ATM) of banks in many countries and considered as a considerably strong password in terms of brute force attack when maximum three consecutive attempts are allowed within a certain time.

4.2. Functionality Evaluation:

This evaluation is most important since one of the objectives of this paper is to propose a screen size independent authentication scheme, which is a function of an authentication scheme.

Hybrid Authentication

Table 1. Comparison of the proposed technique with other similar techniques. Here, attacks are typed in the *italic font* and functions are typed in `typewriter font`.

Attack(s) / `Function(s)`	Authentication Schemes			
	Vibration Code	Press Touch Code	Knock Code	Proposed technique
Shoulder Surfing	H	M	M	H
Smudge	H	H	H	H
Brute Force	M	M	M	M
`Screensizeindependence`	Y	Y	Y	Y
`Short Authentication Time`	N	Y	Y	Y

Legends: L – Low, M – Medium, H – High, Y – Yes, N – No

4.2.1. Screen Size Independence Although, many authentication schemes are in operation at present; majority of them are not screen size independent. Hence, they are not appropriate for several smart devices, more specifically, for miniature devices where screen sizes are limited like smart watch, smart band and so on. For instance, most of the textual authentication schemes are not screen size independent since a full or partial keyboard will not fit in such devices. On the other hand, similar contention also could be raised for the graphical authentication schemes since they need a considerable screen size for displaying graphics. For the proposed scheme, a place for touching the screen is enough to provide both the VC and the PTC; and hence, it is a screen size independent authentication scheme.

4.2.2. Comparably low authentication time The proposed scheme spends relatively shorter registration or authentication time with respect to several similar techniques including VC and Vibration And Pattern (VAP) code [11]. Although, these schemes are resilient against the shoulder surfing and smudge attacks; its registration or authentication duration is relatively large. According to [3], between 4 to 10 seconds are spent for registration or authentication. Conversely, since the proposed scheme integrates the PTC which takes a shorter time for similar purposes; hence, the total duration would be reduced.

4.3. Comparison with other similar schemes

Among the existing authentication schemes, only a few can be considered as screen size independent authentication scheme, including VC, PTC, and Knock Code (KC) [14]. Their comparison with respect to security and functionality are mentioned in Table 1. As could be observed from the table is that PTC and KC are moderately resilient against the shoulder surfing attack; whereas, VC and the proposed technique offers the highest level of resistance against this attack. Again, all the compared schemes are highly resistant against the smudge attack and moderately resistant against the brute force attack.

In terms of functionality, all the proposed technique are screen size independent since they are chosen accordingly. Here, it is noteworthy to mention that all these compared schemes have variants that are not screen size independent. For instance, when these schemes are enhanced with grid cells, they are not screen size independent since a cell must be of a considerable size to provide VC or PTC or KC. Again, in case of short authentication time, all except VC spends a short duration for authentication. Generally, the VC takes a duration

Hybrid Authentication

between 4 to 10 seconds for authentication as mentioned earlier in Section 4.2.2.

From the above discussion, it can be observed that the proposed scheme offers either equal or more resilience against the attacks as well as the functionality. Therefore, we can conclude that the proposed scheme is superior over other similar schemes.

5. Conclusions & Future Works

In this paper, a new hybrid graphical authentication scheme is proposed that integrates both the VC and PTC using a juggling-based approach to ensure a high-level security in terms of attacks like shoulder surfing, smudge, and brute force until a certain extent. Moreover, the proposed scheme is screen size independent and can be implemented even in miniature smart devices including smart watch and smart band. The performance of the proposed scheme is compared with other similar screen size independent authentication schemes, and it is found that it performs superiorly over others.

In future, we would like to extend the proposed technique in a way that a user does not have to lift off the finger when one cycle is finished. For that, it is necessary to identify the switching conditions, i.e., when a user would like to provide the VC and when it is the PTC. Again, there would be no fixed scheme to start with. A user can start providing the PTC or the VC and then switched to the other.

Acknowledgement

This work has been partially supported by the RDU grant, RDU1703100, of University Malaysia Pahang, Malaysia.

Referencing

[1] Chang Y F, Chen C and Zhou H 2009 *Computer Standards & Interfaces* **31** 740–747
[2] Gibson D 2011 Understanding the Three Factors of Authentication URL `http://www.pearsonitcertification.com/articles/article.aspx?p=1718488`
[3] Ranak M S, Azad S, Nor N N H B M N and Zamli K Z 2017 *PLoS ONE* **12**
[4] Adebola O, Ithnin N, jali M Z and Akosu N 2013 *Journal of Theoretical and Applied Information Technology* **53** 124–130
[5] Aviv A J, Gibson K, Mossop E, Blaze M and Smith J M 2010 Smudge attacks on smartphone touch screens *Proc. of 4th USENIX Workshop on Offensive Technologies*
[6] Goucher W 2011 *Computer Fraud & Security* **2011** 17–20
[7] Calculating password complexity URL `https://thycotic.force.com/support/s/article/Calculating-Password-Complexity`
[8] Sreelatha M, Shashi M, Anirudh M, Ahamer M and Kumar V M 2011 *International Journal of Network Security & Its Applications* **3** 111–119
[9] Zhang Y, Monrose F and Reiter M K 2010 The security of modern password expiration: An algorithmic framework and empirical analysis *Proc. of 17th ACM conference on Computer and communications security* pp 176–186
[10] Mackie I and Yldrm M 2018 A novel hybrid password authentication scheme based on text and image *Proc. of IFIP Annual Conference on Data and Applications Security and Privacy*

Hybrid Authentication

[11] Azad S, Rahman M, Ranak M S A N, Ruhee B M F K, Nisa N N, Kabir N *et al.* 2017 *Computers & Electrical Engineering* **59** 99–110
[12] Ahmed N and Rao K 2012 *Orthogonal transforms for digital signal processing* 1st ed vol 1 (Springer Science & Business Media) ISBN 9783642454523
[13] Cormen T H 2009 *Introduction to algorithms* (MIT press)
[14] LG Corporation Korean Patent 10-1404234, 2014 Knock code

IOP Conf. Series: Materials Science and Engineering **769** (2020) 012046 doi:10.1088/1757-899X/769/1/012046

Stability Analysis of Relay Access Protocol (RAP) using Fuzzy Logic in Wireless Sensor Networks

Azamuddin Abdul Rahman[1], **Mohd Nizam Mohd Kahar**[2], **Wan Isni Sofiah Wan Din**[3]
[1,2,3]Faculty of Computing, Universiti Malaysia Pahang, Malaysia
[1]azamuddin@ump.edu.my; [2]mnizam@ump.edu.my; [3]sofiah@ump.edu.my

Abstract. Cluster Head (CHs) determination is critical whereas the combination of input parameters is expected to play an important role. However, each WSN's protocol uses a different combination of input parameters based on residual energy and several other input parameters, which produce different amount of energy consumption. Besides, most of the previous works on cluster heads selection do not take received signal strength indication (RSSI) into consideration. As a result, the sensor nodes with bad RSSI can be elected as CHs which lead to an increase in energy consumption. Moreover, the CHs use single-hop communication to forward their sensed data to the Base Station (BS). This results in high energy consumption and consequently, decrease the network lifetime. In this case, the inter-node distances may degrade the performance of the network due to long distance. Thus, this paper proposed a Relay Access Protocol (RAP) in order to reduce energy consumption in cluster-based wireless sensor networks (WSNs). The RAP protocol is a 2-tier protocol used for energy efficiency by selecting Cluster Heads (CHs) based on Fuzzy Logic (FL). The performance of the proposed protocol was evaluated under two different scenarios and compared with existing protocols through simulations. The experimental results show that the proposed scheme outperformed the existing protocols in terms of First Node Dead (FND) and Last Node Dead (LND) in all of the scenarios.

1. Introduction

The WSNs give focus on energy consumption and stability to provide longer network lifetime. This is due to the limited battery lifetime of the sensor nodes which cannot support long operational hours for some WSN's applications. Motivated by that, some research works have used fuzzy approach for CHs selection with the aim to prolong network lifetime and reduce energy consumption [1]. Fuzzy logic approach uses a combination of fuzzy logic input parameters enable efficient CHs selection and helps in reducing sensors battery power. Common works in the scope of CHs selection always consider the sensor nodes that having the highest number of neighbors and the highest remaining energy is elected as CH. However, the signal quality of the sensor nodes is negligible, hence it potentially drains sensor nodes energy even faster. By exploring previous researches in the area of energy consumption reduction, it reveals that more effort has been made towards CHs selection but a few works focus on noise and signal strength in a fuzzy descriptor perspective. For instance, MAP employs a fuzzy logical-based method to select CHs. Its fuzzy system considers three input parameters such as residual energy, centrality and communication cost. Nevertheless, it does not consider received signal strength indicator (RSSI) in which it may select CHs with badly received signal strength. In each sensor node, different sensor nodes receive different RSSI that affecting energy consumption besides data processing and control tasks. In particular, the effect of noise and interference in the WSNs compromises the data transmission due to numerous data transmission that leads to high energy consumption and low received signal strength [2],[3]. In the literature, the researchers rely only on the residual energy which is not suitable for CHs selection [4],[5]. This is due to the residual energy input parameters rely on other important attributes to reduce energy consumption. Consequently, CHs consume more energy as compared to non-CH nodes. In an inter-cluster transmission for single and multi-hop transmission, the problem of energy imbalance among sensor nodes could arise [6]. For single-hop transmission, CHs which are far away from BS drain out their energy faster due to long distance transmission. Also, for multi-hop transmission, CHs far away from the BS deplete their energy quickly because of the long distance transmission and also extra burden caused by traffic relaying. As a result, imbalance partitioning of the network occurs and thus affect the performance and energy consumption of the network. The work in [5] has demonstrated that if the relay nodes are distributed randomly and far away from the BS in the network, the sensor nodes die rather quick according to First Node Dead (FND) and Last Node Dead (LND) criteria.

In this paper, a protocol called Relay Access Protocol (RAP) is proposed based on MAP protocol. The MAP used a combination of residual energy, centrality and communication cost as input parameters for CHs selection using the Fuzzy approach [12]. The main idea of our proposed protocol was that CHs selection would be based on not only the centrality and node's residual energy as used in the MAP, but also the received signal strength factor. This signal strength indication (RSSI) was considered as one of the input parameters. Another key idea used in our protocol was during the selection procedures of relay nodes. The cost involved in relaying in terms of distance threshold was incorporated as the criteria for selecting the relay node instead of the random selection used in MAP [3]. Furthermore, the idea of extending the data transmission phase by clustering CHs with the nearest relay nodes was combined with the proposed relay nodes selection based on the K-Optimal approach. The clustering process of CHs with the relay nodes used the K-Means approach which can provide balanced energy to the relay nodes. Also, this proposed approach can reduce the transmission burden at the relay nodes, thereby minimizing energy consumption. The performance of our proposed enhancement protocol was compared with the existing protocols using FND, LND, and network lifetime as the performance metrics. The contributions of this paper are summarized as follows:

1) A new combination of fuzzy input parameters was proposed, which incorporates residual energy, centrality and RSSI.
2) The K-Optimal and K-Means techniques were used in a two-tier strategy network to efficiently deploy the relay nodes.
3) The effectiveness and stability of the proposed combination of input parameter were compared with other combination of input parameters on the basis of FND, LND under various scenarios.
4) The energy efficiency of RAP was evaluated and compared with existing protocols in terms of FND, LND and network lifetime under various scenarios.

The rest of the paper is organized as follows. In Section 2, we present related works. Section 3 presented the energy model. Then, the optimal relay selection technique is presented in Section 4. Next, Section 5 presents simulation settings. The simulation results of the proposed approach are presented in Section 6. Finally, Section 7 concluded this paper.

2. Related Works

The WSN is a collection of the connected sensors, embedding software and hardware through the base station to the Internet. The architecture of WSN consists of three main components known as sensor node, a base station (BS) or gateway, and user. It is commonly installed in the form of a cluster consist of a large number of sensor nodes to cover a wide area network. Compared to the current wireless standard network (i.e., IEEE 802.11a/b/g/ac), WSN aims to respond to the users where sensors collect and gather information to forward sensor/actuator data from/to the user via the base station in a small amount of data. The sensors are usually dependent upon non-rechargeable or replaceable power. Nevertheless, power consumption is a challenge due to the drawbacks of battery life.

CHs selection is one of the crucial aspects for prolonging network lifetime and reducing energy consumption. The right combinations of input to be used as fuzzy descriptor play a role to determine the CHs that can contribute to energy reduction. Previously, many WSNs protocols consider the conventional method using different algorithm approaches for CHs selection [4],[5],[6],[7]. The computational cost is high and complex to use traditional approaches whereas most of the works rely on the residual energy of the nodes to select the CHs.

The several numbers of works that consider the fuzzy logic approach has been explored to gain insights for obtaining the energy consumption efficiency in WSNs. Early work by [8] improved CHs selection in LEACH by adopting a fuzzy logic approach. Instead of considering the random number for CHs selection, three input parameter were used such as residual energy, concentration, and centrality. During the setup state, the fuzzy inference system estimates the residual energy and location information, and centrality for CHs selection. The remaining processes are the same with LEACH protocol such as advertisement message and join request. Similarly, work in [9] uses the same approach whereby the selection of the CHs based on residual energy and centrality. The authors claim that this proposed approach outperforms Fuzzy Gupta approach although their fuzzy approach does not consider concentration as a fuzzy descriptor. It means that the combination of residual energy and centrality has a strong correlation as opposed to the combination of residual energy, concentration, and centrality.

The work in [10] proposed a super-CH (SCH) that can send the information to the mobile BS. The CHs was selected based on residual energy, mobility of BS, and centrality. Mamdani fuzzy inference was used to select the SCH. The proposed clustering follows the LEACH in which each node generates a random number between 0 and 1. Also, the fuzzy K-Mean Algorithm (FKM) was used to partition data point into K clusters and the number of clusters were represented as a cluster center. Although the complexity is low, the mobility of the BS considerations is not really significant toward CHs selection. Another work in [11] proposed fuzzy-based unequal clustering which used residual energy, distance from the BS, node degree and competition radius as input parameters for CHs selection. This work attempt to overcome the overburden problem of CHs due to inefficient clustering. To mitigate such issue, the authors consider the distance from the BS to avoid hotspot problem. Two membership functions were used which is trapezoid and triangular. The trapezoid membership function is applied at the boundary variables and triangular membership function The drawback of the proposed protocol is that when the CHs are placed far away from the BS, those particular nodes might suffer from quick energy depletion due to long transmission range. The work in [12] proposed an adaptive fuzzy clustering protocol called as LEACH-SF which used a fuzzy c-mean algorithm for clustering. The selection of CHs was using the Sugeno fuzzy inference system based on residual energy, distance from the BS, and centrality. The novelty of this work is that the fuzzy rules table is defined by a bee colony algorithm instead specify it manually. However, the placements of the nodes were not discussed in detail where this aspect is important for energy balancing.

Recently, the work in [13] proposed a multi-clustering based on Fuzzy Logic (MCFL) by reusing the previous CHs as the current CHs. The election of the CHs based on residual energy and communication cost of CHs and its child nodes. Though the fuzzy approach was used, the proposed two combinations of input parameters commonly is not sufficient enough to produce a high number of CHs' output chances. Moreover, the distance between CHs and the BS is not considered for CHs selection. Later than that, the researchers attempt to utilize the multi-tier network for the scalability in the network. For example, two-tier Distributed Fuzzy Logic Based Protocol (TTDFP) is a multi-hop clustering protocol in which CHs is selected based on a combination of distance (CHs to the BS), residual energy and relative node connectivity [14]. Besides electing the CHs using such input parameters, the routing phase also considers a fuzzy logic approach to determine the type of route such as very low, low, moderately high, extra low, normal, extra high, moderately low, high and very high. It means that, if the route has high residual energy and close distance, then the chance value of the particular route is very high. In fact, this approach increases the complexity and computational cost due to the fuzzy system has to double the process in order to determine the CHs and the best route. The route can be simply determined by the shortest distance based on Euclidean distance calculation. MAP protocol fairly distributes all the sensor nodes according to the circle geometry calculation results in sensor nodes in 2-tier with three times more nodes [15]. The selection of the CHs in MAP based on residual energy, centrality, and communication cost. However, the RRSI was not considered for CHs selection and the relay nodes are randomly deployed in the network. Simulation results showed that this approach is more energy-efficient than LEACH and SEP.

3. Energy Model

Some performance metric in [14] have been used to evaluate the energy consumption. The energy consumption model which have been proposed in [15] rely on the transmitted packet with (k) bits over the distance (d). The energy consumption model is given by:

$$E_{TX}(k.d)=\begin{cases}k*E_{elect}+k*\varepsilon_{fs}*d^2, & d\leq d_o\\ k*E_{elect}+k*\varepsilon_{mp}*d^4, & d\geq d_o\end{cases} \tag{1}$$

where ε_{fs} is the amplifier's value per bit during data transmission if the distance $d \leq d_o$, E_{elect} is energy dissipated per (k) bit to run the transmitter, ε_{mp} is the amplifier's value per bit for data transmission if the distance $d \geq d_o$. The energy consumption of each node is calculated as:

$$E_{RX}=k*E_{elect} \tag{2}$$

As sensor nodes communicate with their respective CHs, the energy dissipation occur in CHs. The energy dissipation of the CHs can be obtained as follow:

$$E_{CH}=\left(\frac{N}{L}-1\right)*k*E_{elect}+\frac{N}{L}*k*E_{DA}+k*E_{elect}+k*\varepsilon_{fs}*d^2 \tag{3}$$

4. Network Model

The network model was based on 2-tier network model. The BS was placed in the centre of the network due to it has an effect on the energy consumption. The position of the BS can affect the energy consumption of the network. For that reason, the BS has been placed in the center of the network to reduce energy consumption [10].

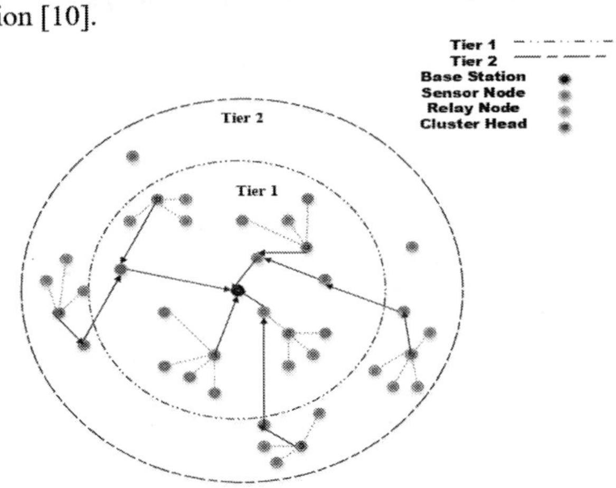

Figure 1. Data transmission of RAP

As shown in Figure 1, the circle area has been partitioned into 2 smaller areas. The diameter of each tier were 25 meters and 50 meters. The sensor nodes in 2-tier have three times more sensor nodes which is 75 according to the the circle geometry calculation. The sensor nodes obtained in 1-tier was 25. For data transmission, sensor nodes in 1-tier sent data to its respective CHS. Then, the aggregated data were sent to the relay nodes. In 2-tier, the relay nodes of 2-tier pass the data in 1-tier before reaching to the BS. The K-Optimal technique is used to obtain the relay nodes in respective tier. The optimal number of relay nodes is written as:

$$RN_{opt} = \left(\sqrt{\frac{U_i}{2\pi}}\right) * \sqrt{\left(\frac{\varepsilon_{fs}}{\varepsilon_{mp}}\right)} * \frac{U}{d^2} \tag{4}$$

The energy model takes ε_{fs} as the energy radio dissipates (d^2 power loss) and ε_{mp} is the amplifier energy of multi-path fading loss (d^4 power loss). U represents the total number of nodes and d represents the average distance between nodes and the BS. The value of ε_{fs} and ε_{mp} were considered as 10* 0.00000000001 which indicate as free space model and 0.0013 * 0.00000000001 represents as multi-path fading [39, 40,41]. The optimal number of relay nodes is obtained depending on average distance to the BS, the path of the nodes and total number of nodes deployed in the network. Thus, the number of relay nodes for 1-tier and 2-tier respective tier was obtained as:

In 1-tier,

$$RN_{opt}^{1-tier} = \left(\sqrt{\frac{U_{1-tier}}{2\pi}}\right) * \left(\sqrt{\frac{\varepsilon_{fs}}{\varepsilon_{mp}}}\right) * \frac{U}{d^2} \tag{5}$$

$$RN_{opt}^{1-tier} = \left(\sqrt{\frac{25}{2\pi}}\right) * \left(\sqrt{\frac{10*0.00000000001}{0.0013*0.00000000001}}\right) * \frac{100}{d^2} = 7 \tag{6}$$

In 2-tier,

$$RN_{opt}^{2-tier} = \left(\sqrt{\frac{U_{2-tier}}{2\pi}}\right) * \left(\sqrt{\frac{\varepsilon_{fs}}{\varepsilon_{mp}}}\right) * \frac{U}{d^2} \tag{7}$$

$$RN_{opt}^{2-tier} = \left(\sqrt{\frac{75}{2\pi}}\right) * \left(\sqrt{\frac{10*0.00000000001}{0.0013*0.00000000001}}\right) * \frac{100}{d^2} = 12 \tag{8}$$

According to Equation (5) and Equation (6), the number of relay nodes obtained was 7 and 12 at the initial state. Figure 2 shows the distribution of the nodes in 1-tier and 2-tier.

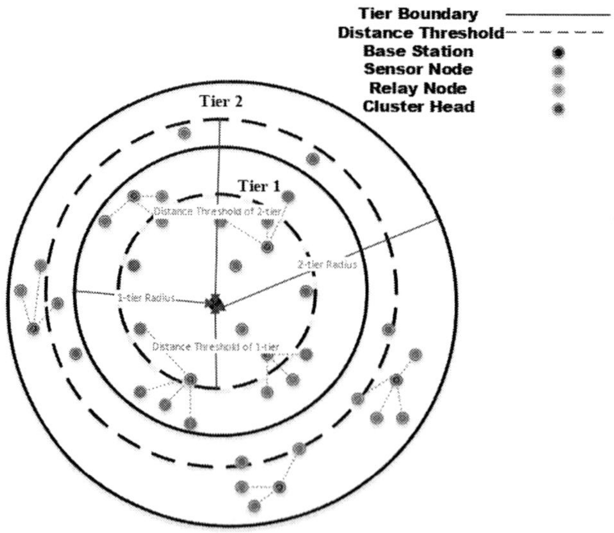

Figure 2. The position of optimal relay nodes

According to K-Means approach, the number of optimal relay nodes obtained can be used to determine the number of clusters that can be formed with the CHs. This is because the number of relay nodes can be used as a centre point to find the nearest CHs. There are five steps in K-Means clustering technique as follows:

1) Determine the optimal number of relay nodes.
2) The distinct data points which is the optimal number of relay nodes is obtained.
3) Calculate the distance between the CHs and relay nodes based on Euclidean distance calculation.
4) Assign the nearest CHs to the distinct point to form a cluster.
5) Calculate the mean and center of each cluster.

The distance threshold creates an upper bound so that the transmission distance between relay nodes and CHs can be minimized. Figure 2 shows the setting of distance threshold in the network. The distance threshold of 1-tier and 2-tier were set as $d_1 = 15m$ m and $d_2 = 30m$ m respectively. Consequently, the selected relay nodes form a cluster with the nearest CHs. The cluster formation in RAP can be divided into two types which is cluster formation of the CHs with the sensor nodes and cluster formation of the relay nodes with the CHs. For data transmission, in 1-tier, the data transmission journey start from child nodes in the cluster sent the data to its respective CH. Data aggregation is in a bottom-up manner from the child nodes. The sensed data from the child nodes are aggregated and compressed by the CHs. CHs collect the aggregated data from the child nodes to be sent to the relay nodes. Then, relay nodes forward the data to the BS. The transmission flow in 1-tier is shown in Figure 3.

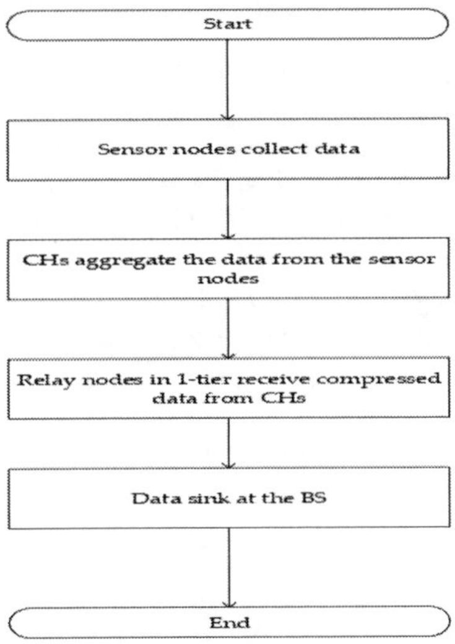

Figure 3. Data Communication in 1-tier

Similarly, in 2-tier, whereas the sensor nodes of 2-tier pass the data to its CHs. Then, CHs forward data to the nearest relay node in 1-tier. The transmission flow in 2-tier is shown in Figure 4.

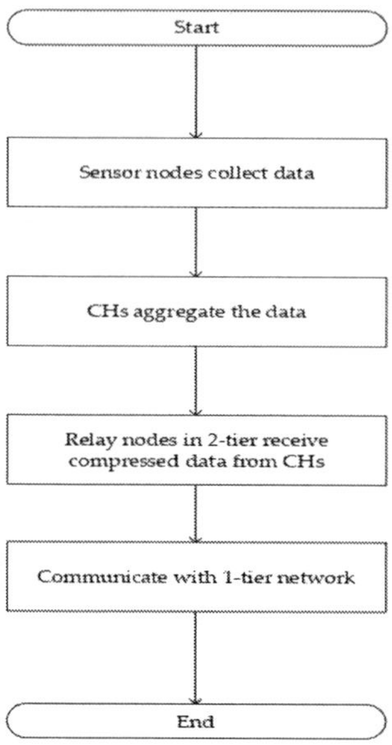

Figure 4. Data Communication in 2-tier

Since the network consists of 1-tier and 2-tier, the transmission occurs across these two tiers. In 2-tier, the CHs find the nearest relay node, *R2* to visit next. Then, *R2* forwards the data to the nearest relay nodes, *R1* in 1-tier (i.e., based on the certain criteria). Consequently, the *R1* forwards the data to the BS. Hence, establishing a connection from CHs to the relay nodes across 1-tier results in the larger hop (i.e., quad-hop) to the BS. The data aggregation is in a bottom-up manner from the child nodes in tier-1 and tier-2 and forwarded then to the BS for further processing.

5. Simulation Settings

The WSN network was simulated by employing MATLAB with the aim to evaluate its energy consumption. The proposed protocol known as Relay Access Protocol (RAP) protocol was compared with Stable Election Protocol (SEP) and MAP protocol in term of First Dead Node (FDN), Last Dead Node (LDN), and Total Dead Node (TDN). The simulation parameters setting in experiments are presented in Table 1.

Table 1. The simulation parameter settings

Parameter	Value
Number of Sensor Nodes	100 and 200
Network Area	100 M x 100 M
Base Station Coordinate	Center (50,50)
1-tier Diameter	25 M
2-tier Diameter	50 M
Energy (J)	1 J, 0.5 to 2 J
Number of bits	2000, 4000, 6000, and 8000

The experiments were divided into three sets, while the first and second sets observe the FDN and LDN that consists of a combination of input parameters i.e., ResCen, ResRSSI, and ResCenRSSI. This is to test the stability of RAP protocol, where the various number of bits were used as 2000, 4000, 6000 and 8000 bits. The initial energy (J) were carried to have 1 J and 0.5 to 2 J with 100 nodes which indicated as scenario 1 and scenario 2 respectively. Thirdly, performances of improved MAP 1 and improved MAP 2 were evaluated in terms of the number of dead nodes, FDN and LDN. For that, the proposed protocols are categorized into two types. For the first type, namely the improved MAP 1, the proposed protocol used the combination of input parameters (ResCenRSSI) without incorporating the relay selection approach. For the second type, namely improved MAP 2, the method of relay selection along with CHs fuzzy based selection technique was incorporated.

6. Simulation Results

6.1 The Stability of the RAP in Scenario 1

This section explains the FND, LND, and number of the dead node when the residual energy of the 100 sensor nodes is 1 J. There is a very strong correlation between FND and LND, where the longer the LND or network lifetime is, and the more slowly the FND is going to appear. Figure 5 and 6 show the FND of ResCenRSSI, ResCen and ResRSSI with 2000, 4000, 6000 and 8000 bits.

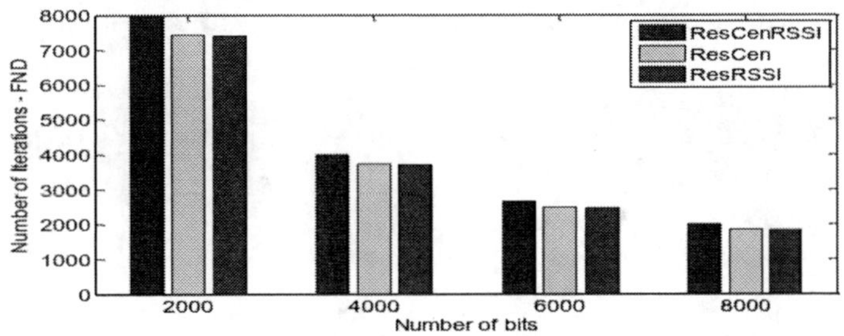

Figure 5. FND of improved MAP with 1 J energy of 100 sensor nodes

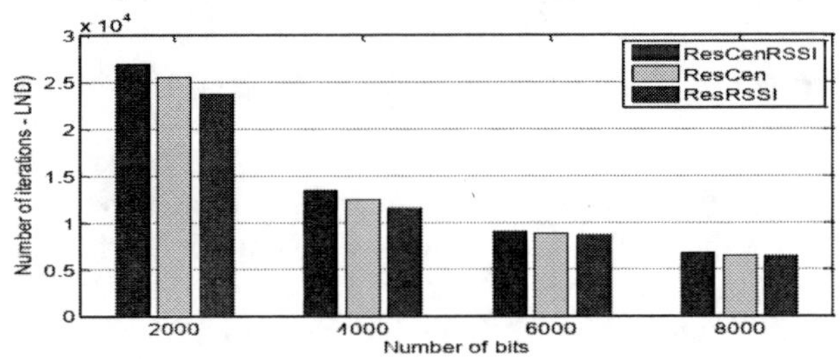

Figure 6. LND of improved MAP with 1 J energy of 100 sensor nodes

As shown in Figure 5 and 6, the results show that the FND appears with ResCenRSSI is much later than that in ResCen, and ResRSSI with all number of bits. The FND of ResCenRSSI was reported at 7997th, 3993th, 2877th, and 2438th iterations which is the highest value compared to the other input combinations for 2000, 4000, 6000, 8000 bits. ResCen and ResRSSI is following behind ResCenRSSI as 7722th, 3711th, 2474th, 1856 and 7712th, 3705th, 2465th, 1832th for 2000, 4000, 6000, and 8000 bits respectively, and ResRSSI has the lowest value of FND. Also, the LND of ResCenRSSI is much longer than ResCen and RSSI which indicate the longest network lifetime. Since inclusion of RSSI avoid low received signal strength to be elected as CHs, ResCenRSSI perform well with initial energy of 1 J. ResCen gives no considerations to the RSSI of sensor nodes, and their performance is much worse than that ResCenRSSI because low residual energy would deplete energy more quickly with bad RSSI as the number of bits are increased. On the other hand, ResRSSI does not consider centrality where this input parameter influences energy consumption as well.

6.2 The Stability of the RAP in Scenario 2

This sections explains the FND, LND and number of the dead node when the residual energy of the 100 sensor nodes is 0.5 to 2 J.

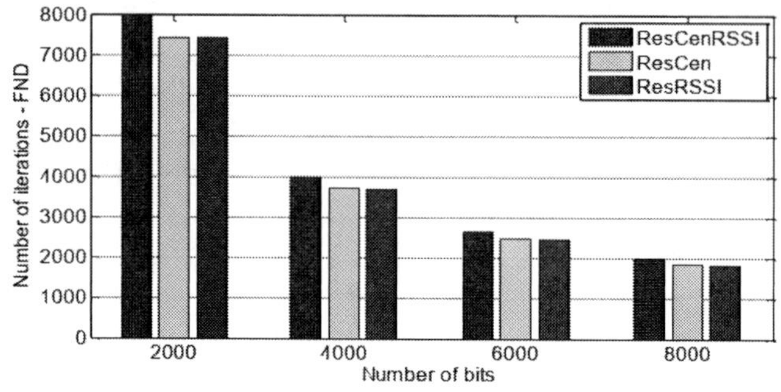

Figure 7. FND of improved MAP with 0.5 to 2 J energy of 100 sensor nodes

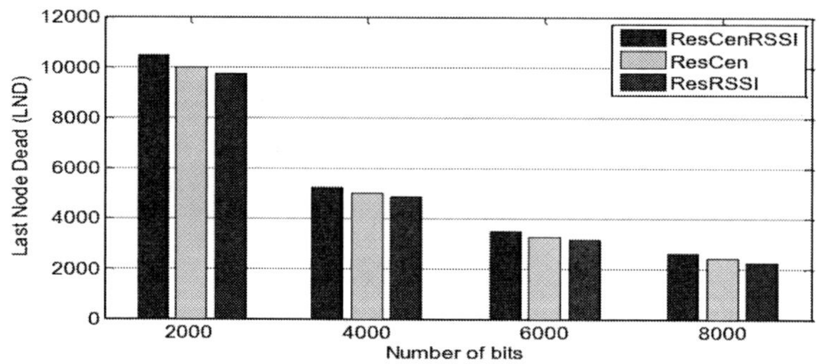

Figure 8. LND of improved MAP with 0.5 to 2 J energy of 100 sensor nodes

Figure 7 and 8 describe the FND and LND for various numbers of data: 2000 bits, 4000 bits, 6000 bits, 8000 bits, and 10000 bits for all combination of input parameters. As shown in the figures, in general, the iterations taken for the first and last node to die is more slowly as data increased from 2000 bits to 8000 bits compared in the first scenario (i.e., initial energy of 1 J). This is because the energy assigned to the nodes is much higher which is 2 J which lead to a longer network lifetime. The death time of the first node with ResCenRSSI is at the 7986th, 3993th, 2662th and 1997th iterations, whose performance increased by about 10% and 12 % than ResCen and ResRSSI for 2000 bits, 4000 bits, 6000 bits, and 8000 bits respectively. When the last node dead, the iteration of ResCenRSI for 2000 bits, 4000 bits, 6000 bits, and 8000 bits are increased about by 12% and 13% than ResCen and ResCenRSSI. Furthermore, the results show that the FND appears in ResCenRSSI is much later than in ResCen and ResRSSI, and the network lifetime is much longer. This is because the higher residual energy of the sensor nodes i.e., 2J, it will take longer iterations to die. Thus, it can be inferred that combination between residual energy, centrality, and RSSI is better energy consumption strategy with an initial energy of 2 J. The higher residual energy provided cause longer network lifetime. This situation corroborates that ResCenRSSI maintain its efficient operation fashion with the various number of bits.

6.3 Number of dead nodes evaluation

This section evaluates the energy consumption of SEP, MAP, improved MAP and improved MAP 2 in tern FND, LND and number of dead nodes. The improvement gained through improved MAP protocol is further laid out in each of the two scenarios.

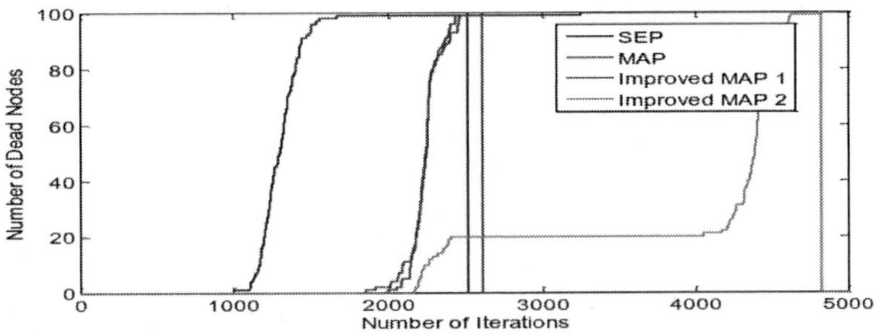

Figure 9. Number of Dead Nodes for Scenario 1

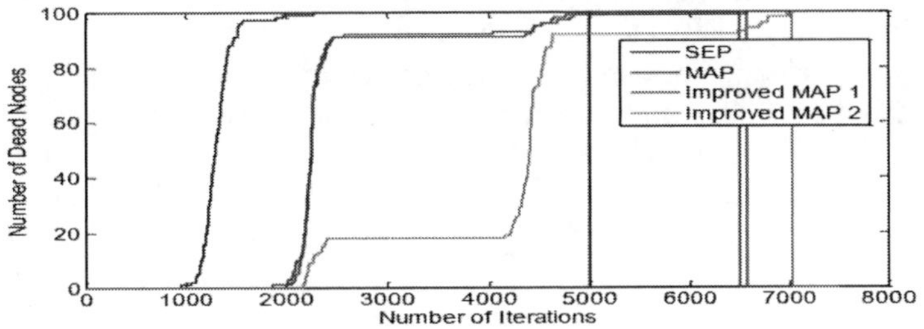

Figure 10. Number of Dead Nodes for Scenario 2

Figures 9 and 10 show the number of dead nodes with respect to iterations for Scenarios 1 and 2. As shown in Figure 9 and Figure 10, the FND of SEP protocol occurred at 1000th and 988th iterations and 60% and 40% nodes died at about the 1200th and 1100th iterations for Scenarios 1 and 2. The network lifetime of SEP was the poorest than other protocols. The reason for this was that the SEP protocol adopted the probabilistic CHs selection method [21]. Moreover, the placement of the BS in SEP was not centered, which may lead to high energy consumption and shorter network lifetime. In MAP protocol, the FDN occurred at about the 2401th and 6408th iterations and 80% and 90% nodes died at about 2200th and 3000th iterations for scenarios 1 and 2. The total dead nodes of improved MAP 1 was slightly reduced and also the FDN was slightly earlier which was at the 2511th and 6597th iterations compared with MAP for all scenarios.

The average improvement obtained in network lifetime in the improved MAP 2 protocol considering FND as the metric was about 26% for Scenario 1 and 24% in the case of Scenario 2 in comparison with MAP. With reference to LND as the network lifetime evaluation metric, the average improvement in Scenarios 1 and 2 was about 57% and 14%. The reason was that the improved MAP 1 method selects the CHs with high residual energy, high centrality, and with good signal strength. It can also be seen that the total dead nodes in improved MAP 2 dropped significantly during the 2500th and 2400thiterations for Scenarios 1 and 2 respectively. The BS in improved MAP 2 communicated with the relay nodes within the distance threshold, which caused the distance of data transmissions to be reduced and hence saved energy. Furthermore, the relay nodes efficiently formed clusters with the CHs based on the K-Means approach which provided load balancing, especially with the high number of the sensor nodes. As observed, the improved MAP protocol, i.e., MAP 2, achieved better energy efficiency and balancing than the MAP and SEP protocols. The reason was because the improved MAP protocol considered the impact of both intra-cluster and inter-clusters communication loads during operation. From the results, it can be concluded that the improved MAP was able to address the non-efficient CHs selection, heterogeneity of nodes, and quick energy depletion successfully.

7. Conclusions

In this paper, the Relay Access Protocol (RAP) has been proposed in order to improve the network energy efficiency. The FL-based CHs selection approach has been exploited in this work. The CHs selection used a combination of three input parameters which were residual energy, centrality, and RSSI. The CHs were assigned based on high residual energy, closeness to their child nodes, and good signal strength. Furthermore, the relay node selection was adopted to provide energy balancing and reduce the communication energy consumption of far-end CHs. The simulation results showed that the method of relay selection along with CHs fuzzy based selection technique namely as improved MAP 2 extend the lifetime of wireless sensor network 57% and 14% longer than the original MAP for all scenarios.

Acknowledgments

The authors would like to thank UMP PGRS Grant (190391) for providing financial support and facilities. In addition, they would like to acknowledge and send their gratitude to the Faculty of Computing and Systems Network & Security Research Group UMP for their support.

Reference

[1] M. Toloueiashtian and H. Motameni, "A new clustering approach in wireless sensor networks using fuzzy system," J. Supercomput., vol. 74, no. 2, pp. 717–737, Feb. 2018.

[2] W. I. S. W. Din, S. Yahya, R. Jailani, M. N. Taib, A. I. M. Yassin, and R. Razali, "Fuzzy logic for cluster head selection in wireless sensor network," vol. 050006, p. 050006, 2016.

[3] P. Rajeshwari, B. Shanthini, and M. Prince, "Hierarchical Energy Efficient Clustering Algorithm for WSN," Middle-East J. Sci. Res. Signal Process. Secur., vol. 23, pp. 108–117, 2015.

[4] H. Wu, Z. Zhong, and L. Hanzo, "A cluster-head selection and update algorithm for ad hoc networks," GLOBECOM - IEEE Glob. Telecommun. Conf., 2010.

[5] V. Pal, G. Singh, and R. P. Yadav, "Cluster head selection optimization based on genetic algorithm to prolong lifetime of wireless sensor networks," Procedia Comput. Sci., vol. 57, pp. 1417–1423, 2015.

[6] M. M. V. M. Kumar and A. Chaparala, "Dynamic energy efficient distance aware protocol for the cluster head selection in the wireless sensor networks," 2017 2nd IEEE Int. Conf. Recent Trends Electron. Inf. Commun. Technol., pp. 147–150, 2017.

[7] M. Mirzaie and S. M. Mazinani, "MCFL: An energy efficient multi-clustering algorithm using fuzzy logic in wireless sensor network," Wirel. Networks, vol. 24, no. 6, pp. 2251–2266, 2018.

[8] S. A. SERT, A. Alchihabi, and A. Yazici, "A Two-Tier Distributed Fuzzy Logic Based Protocol for Efficient Data Aggregation in Multi-Hop Wireless Sensor Networks," IEEE Trans. Fuzzy Syst., vol. 6706, no. c, pp. 1–15, 2018.

[9] W. I. S. W. Din, S. Yahya, M. N. Taib, A. I. M. Yassin, and R. Razali, "The combinations of selected parameters to prolong the network lifetime for cluster head selection in wireless sensor network," Proc. - Int. Conf. Intell. Syst. Model. Simulation, ISMS, vol. 2015-Septe, pp. 568–572, 2015.

[10] D. Agrawal and S. Pandey, "FUCA: Fuzzy-based unequal clustering algorithm to prolong the lifetime of wireless sensor networks," Int. J. Commun. Syst., vol. 31, p. e3448, 2017.

[11] B. Kan and L. Cai, "An A ccurate Energy Model for WSN Node and Its O ptimal Design," no. 2006, pp. 328–332.

[12] J. Peng, T. Liu, H. Li, and B. Guo, "Energy-efficient prediction clustering algorithm for multilevel heterogeneous wireless sensor networks," Int. J. Distrib. Sens. Networks, vol. 9, no. 2, p. 678214, 2013.

[13] S. Chao, R. WANG, H. HUANG, and L. SUN, "Energy efficient clustering algorithm for data aggregation in wireless sensor networks," J. China Univ. Posts Telecommun., vol. 17, pp. 104–122, 2010.

[14] Y. K. Meena, A. Singh, and A. S. Chandel, "Distributed Multi-Tier Energy-Efficient Clustering," Int. J. Comput. Theory Eng., vol. 4, no. 1, p. 1, 2012.

[15] G. Smaragdakis, I. Matta, and A. Bestavros, "SEP: A stable election protocol for clustered heterogeneous wireless sensor networks," Boston University Computer Science Department, 2004.

IOP Conf. Series: Materials Science and Engineering **769** (2020) 012047 doi:10.1088/1757-899X/769/1/012047

Performance analysis on round trip available bandwidth estimation tools for wireless mesh network

I E Kamarudin, M A Ameedeen, Z R M Azmi, M I U Ong, A S Lokman

Faculty of Computing, Universiti Malaysia Pahang, Malaysia

Email: edzereiq@ump.edu.my

Abstract. In wireless mesh network environment, available bandwidth (ABW) estimation is important to provide feedback on the quality of network infrastructure to end user. Recent study shows the emerging of Round Trip (RT) based ABW tools where it eliminates the need of having a receiver end. However, these tools are not fully tested in various network environment and benchmark with existing widely used ABW tools. In this paper, we perform an extensive performance analysis study on RT based tools with other widely used ABW tools by looking at the accuracy and consistency. In term of accuracy and consistency, RT based tools excel in these conditions.

1. Introduction

With the rapid increase of network applications and services available, available bandwidth (ABW) estimation plays an important role to provide feedback on the quality of the network infrastructure to support those services. For example, streaming services rely heavily on the available bandwidth to adjust its stream bitrate base on the current ABW result to further improve the streaming session.

A typical mistake by user is assuming that the actual maximum bandwidth capacity supported by any network in the path is based on the wireless network standard itself. For example, wireless mesh network (WMN) for IEEE802.11ac, the maximum bandwidth supported is 100Mbps. However, theoretically it can only support up to 97.80 Mbps [11]. This assumption is wrong because we need to factor in other consideration such as what kind of traffic going through the network. As a result, this will most probably lead to underestimating the total bandwidth requirement on given path.

Bandwidth capacity refer to the maximum throughput that can be transmitted across a path. The ABW refer to the unused capacity at the same path. As shown in equation 1, at any given time, a network path is either in idle state or in transmitting packets state, the definition of the ABW is by looking at the average unused bandwidth over some interval T. Therefore,

$$Ai(t,T) = \frac{1}{T}\int_{t}^{T+t} \big(Ci - \lambda i(t)\big)dt \qquad (1)$$

Where Ai (t; T) is the ABW at path i at time t, Ci is the path's bandwidth capacity, and λi is its network traffic. The ABW along the path is the minimum ABW of all traversed links.

To estimate the ABW, each tool uses either one of this technique, passive or active. Both techniques require a sender and a receiver end to complete the probing cycle. All tools will not work when either one end at the sender or receiver of the end-to-end path malfunction. Recent studies [1-3] shows that by introducing Round Trip (RT), it helps to eliminate the decency of having the need of a receiver. In addition, RT based tools also provide lower intrusion and better convergence when tested. However, this RT based tools were not tested extensively across different network settings such as varying different bandwidth. Furthermore, benchmarking testing was not done to compare with other widely used ABW tools such as Pathrate [4], Pathload [5] and IGI [6].

In this paper, we evaluate existing RT ABW and non RT ABW tools to look for accuracy and consistency. The evaluation is done by testing across different 802.11-based WMN, with various parameters such as physical data rate and by inducing external traffic in the path. Our work and contributions are highlighted as below:

- We select six ABW tools from RT and non RT probe type.
- We perform extensive testing in terms of different WMN physical rate and network conditions.
- We assess all tools base on accuracy and consistency. For accuracy, tools must be able to estimate ABW within the selected WMN bandwidth range. For consistency, tools must be able to provide the least fluctuation of over and underestimate the bandwidth.

2. Background

ABW measurement techniques are divided into two types of probing, passive based and active based probing. Congestion situation, packet loss, and delay performance are used to estimate the ABW in passive based probing. For active based probing, it sends probe-packets over a network path in order to estimate the ABW. Due to efficiency and reliability of estimations, active probe is preferable. Active-based ABW tool is implemented by inducing traffic congestion from sender, where probe packets are sent at increasing rates. At the recipient end, the probe packets delay is measured to determine the time or point at which they start to increase in a consistent basis. The ABW is then measured at the probe packet rate utilized by looking at turning point of the packets.

Active probing is classified into two model; Probe Gap Model (PGM) and Probe Rate Model (PRM). In PGM, estimation of ABW is done based on the readings at the cross-traffic rate in the path. Previous capacity reading of the network path is required to develop tools using this technique. The probing technique works when the sender transmits a pair of packets to the receiver end, where the pair packets transmit close enough to each other in time for packets to queue at the bottleneck path. The change in packet spacing can be determine by receiver end to make an estimation of the amount of cross traffic during the measurement time in the bottleneck path and then compute the ABW as the difference between the bottleneck path capacity and the cross-traffic rate. Spruce [7] and IGI [6,13] are examples of tools utilizing this approach. For PRM, it estimates the ABW based on the probe rate results between the sender and receiver. Along the path, when the probe traffic is sent at a rate lower compare to the available bandwidth, the arrival rate at the receiver end will match with the sender rate. On the opposite, when the probe traffic is higher compare to the available bandwidth, it will result in queuing and delay of transmitting probing packets. Based on this and by identifying at the turning point, PRM measures the available bandwidth, where the probe sending and receiving rates match. TOPP [8,14], Pathload [5,15] and NEXT [9] are examples of ABW tools that utilized this approach.

The main limitation of active based probing tools is that it needs to be deployed on both the sender and receiver end. They need to be installed and run at both end along the network path. To overcome this limitation, recent research [1-3] suggests that there is a need to deploy ABW tools at only one end, especially only at the sender, along the network path. This estimation of ABW is based on round-trip-time measurements with low intrusiveness in the network path and short convergence time to produce the estimation. RT-WBest is an example of tools utilizing this approach. It utilizes two-parts algorithm where at the first part, it uses the packet-pair dispersion technique to estimate path capacity. Once complete, at the second part, it sends a packet train to estimate available bandwidth.

3. Comparison of existing tools

In this section, we describe the testing environment setup and the validation methodology.

3.1 Tools and test environment.

We evaluate selected tools; RT-Best, Spruce, IGI, Pathchirp TOPP and Pathload in a testbed environment in Figure 1. Probe traffic for all tools is sent from ABW server to Node A. External traffic is generated by the traffic generator to Node B. The servers are connected to the access point 1 (AP1) with a wired 100Mbps LAN. The nodes are connected to the access point 2 (AP2). Table 1 shows the detail specifications of each devices.

Table 1. Testing devices specifications

Device	Specifications
AWB & Traffic generator server	Intel Core i7 3.0 GHz CPUs, 4 GB RAM, Ubuntu 18.04
Node A & B	Intel Core i5 2.6 GHz CPUs, 4GB RAM, Ubuntu 18.04
Access Point 1 & 2	D-Link AC1200 DAP1665

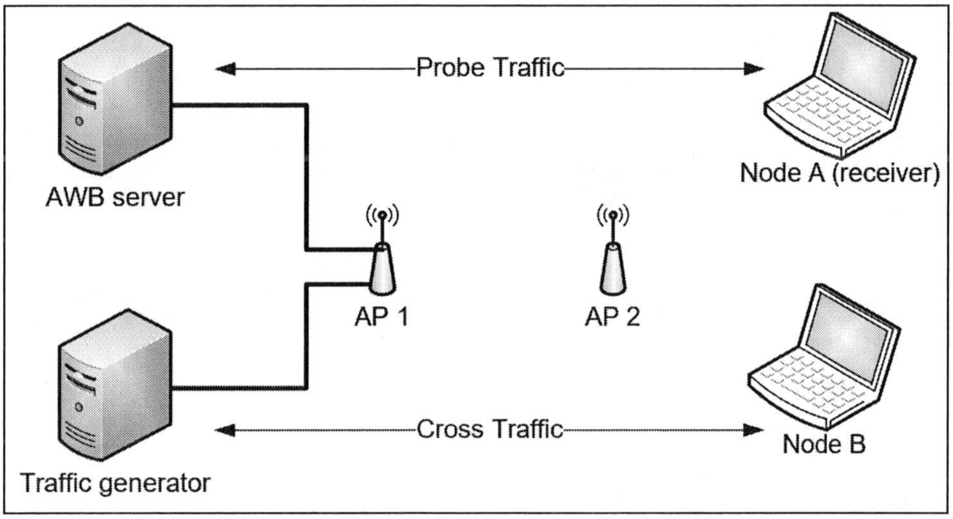

Figure 1: Experimental scenario

3.2 Testing method and evaluation
Probe traffic will be generated by the ABW server to Node A and then the Node A replies through the APs. To standardize the testing, we generate 150 packet pairs, with 70 as the length of the packet train, and 1500 bytes as the size of probe packet in IP layer [12]. Additional traffic will come from the Traffic generator server to the Node B, sharing the same path along both APs.

The testing is done in two conditions: 1) Probe test traffic is generated from ABW server to Node A without external traffic 2) Probe test traffic is generated from ABW server to Node A with external traffic from Traffic generator server to Node B. First test, ABW tools were tested by probing the network path where only traffic exists in the path was only from the tools itself. In the second test, external packets generated along the network path using TCP traffic, generated from traffic generator server, on top of the tools itself. This action causes both traffic to access the shared wireless channel and competing on the same path. Both conditions were then tested in three WMN environment 1) IEEE802.11b at maximum speed of 11Mbps across all link, 2) IEEE802.11g at maximum speed of 54Mbps across all link, 3) IEEE802.11ac at maximum speed of 100Mbps across all link.

For evaluation, each tool is tested for accuracy and consistency across the two types of network conditions, with three different network environments standard. In this test, we took 30 readings for each of the tools. The summary of all testing conducted as below:

- Probe test without external traffic: 1) IEEE 802.11b scenario at 11Mbps 2) IEEE 802.11g scenario at 54 Mbps 3) IEEE 802.11ac scenario at 100Mbps.
- Probe test with external traffic: 1) IEEE 802.11b scenario at 11Mbps 2) IEEE 802.11scenario at 54 Mbps 3) IEEE 802.11ac scenario at 100Mbps.

4. Experimental result

4.1 Accuracy

For accuracy, we take overestimation into consideration. Equation 2 shows the calculation for accuracy.

$$Accuracy[within\ range\ of\ selected\ WMN] = \frac{Number\ of\ times\ within\ range}{30} \times 100 \qquad [2]$$

For IEEE802.11b, 802.11g and 802.11ac, the theoretical value of the capacity throughout the network is 7.80 Mbps, 37.80 Mbps and 97.80 Mbps [10] respectively. Hence, in this testing, the benchmark accuracy reading of available bandwidth was set within each respective range. Only results that falls under this ranges will be considered as accurate in this testing. The measurement of RT-Best, Spruce, IGI, Pathchirp, TOPP and Pathload are shown in Figure 2 and 3 for IEEE 802.11b, Figure 4 and 5 for IEEE 802.11g and lastly Figure 6 and 7 for 802.11ac.

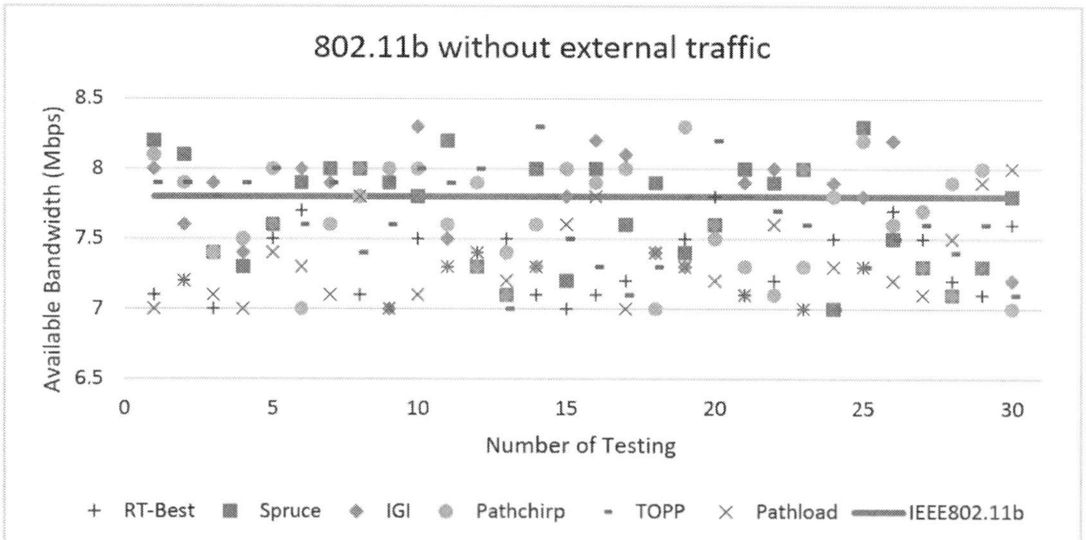

Figure 2: Estimation of ABW for IEEE 802.11b without external traffic

Based on this, RT-WBest was the only tool able to measure and produce result close to theoretical reading within the acceptable range of bandwidth for each WLAN, both for with and without external traffic scenario. When compare with other well know ABW tools, only RT-WBEST and Pathload were able to provide acceptable result. In contrast, Spruce, IGI, Partchip and TOPP provided few readings outside of the acceptable range of bandwidth for each WLAN. The abw reading was easily affected when tested with external traffic. IGI suffers the most in our case, almost 70% of the result are outside of the acceptable bandwidth range.

In summary when looking at the accuracy of each tools, RT-Wbest provide the most accurate result compare to the rest on all scenarios. It does not have significant impact during estimation like other tools, even when tested with external traffic.

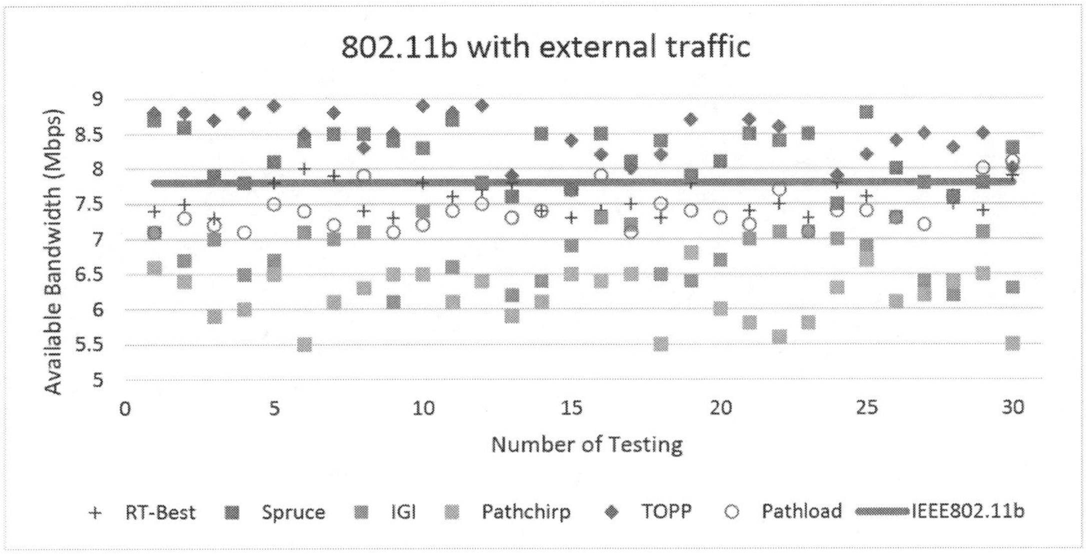

Figure 3: Estimation of ABW for IEEE 802.11b with external traffic

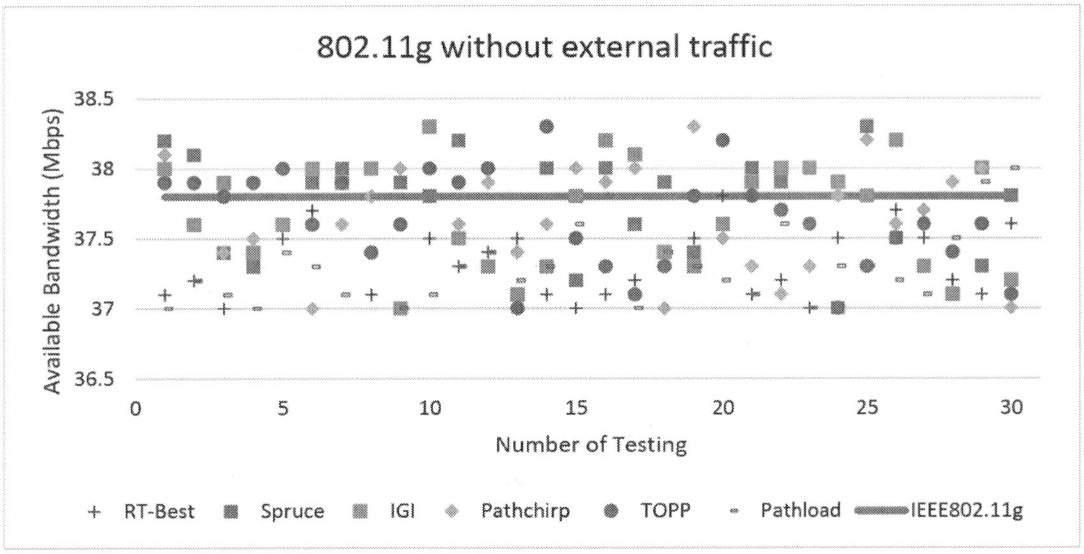

Figure 4: Estimation of ABW for IEEE 802.11g without external traffic

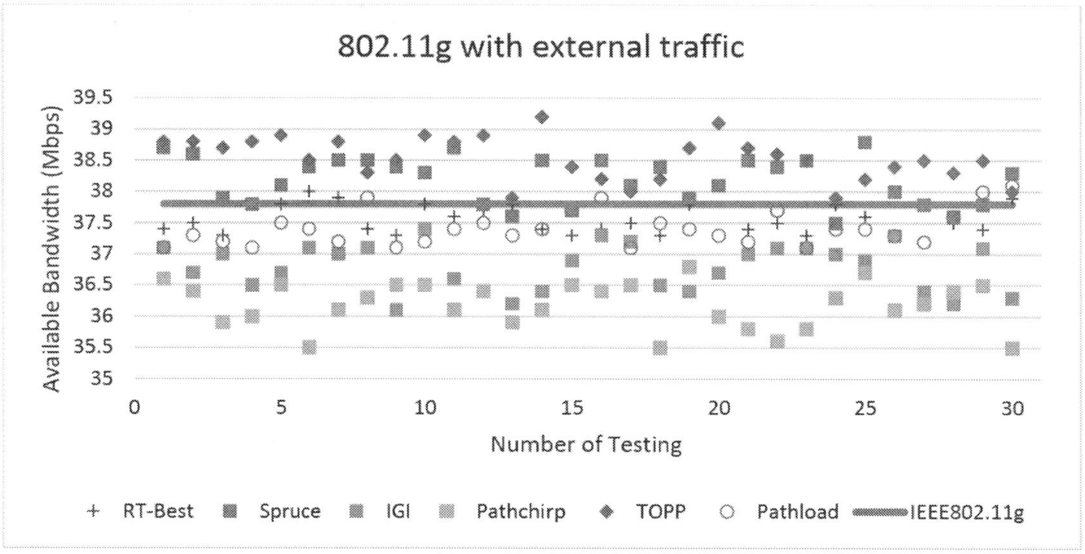

Figure 5: Estimation of ABW for IEEE 802.11g with external traffic

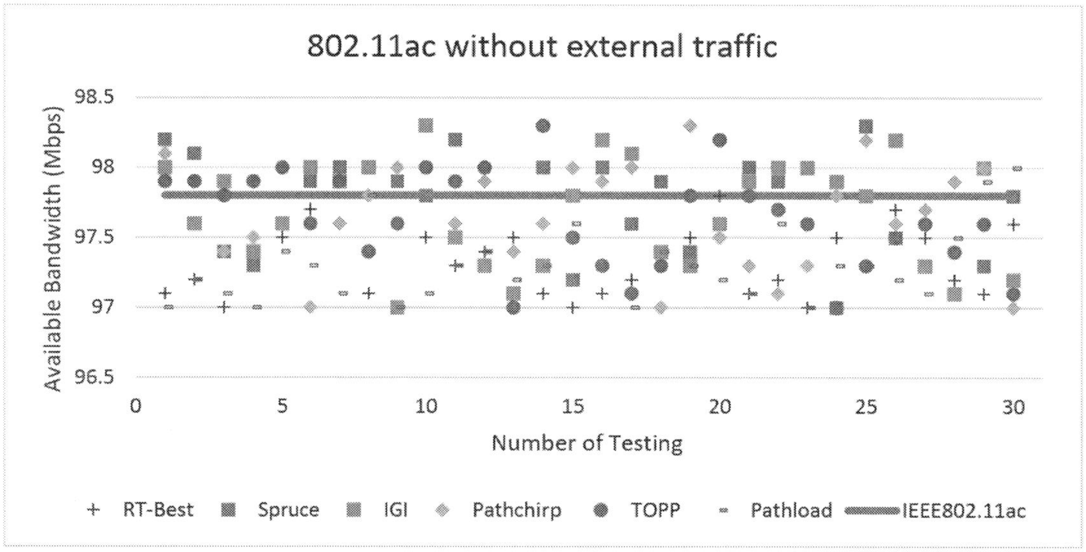

Figure 6: Estimation of ABW for IEEE 802.11ac without external traffic

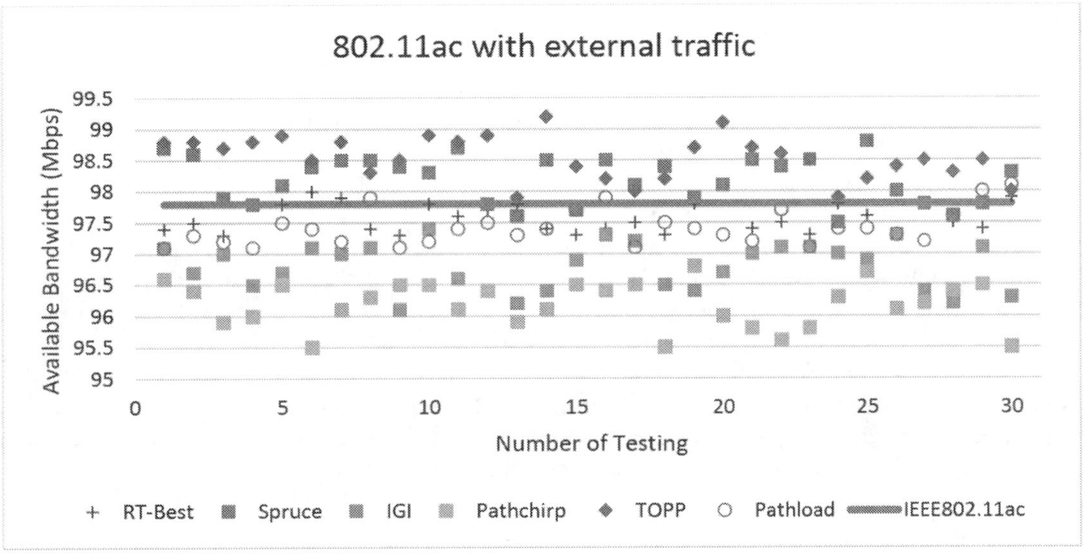

Figure 7: Estimation of ABW for IEEE 802.11ac with external traffic

4.2 Consistency

For consistency, we calculate the standard deviation for each tool base on equation 3. The lower the standard deviation value shows the accuracy of the tools in bandwidth estimation.

$$\sigma = \sqrt{\frac{\sum_{k-1}^{n}(X\,k - \mu)^2}{n}} \tag{3}$$

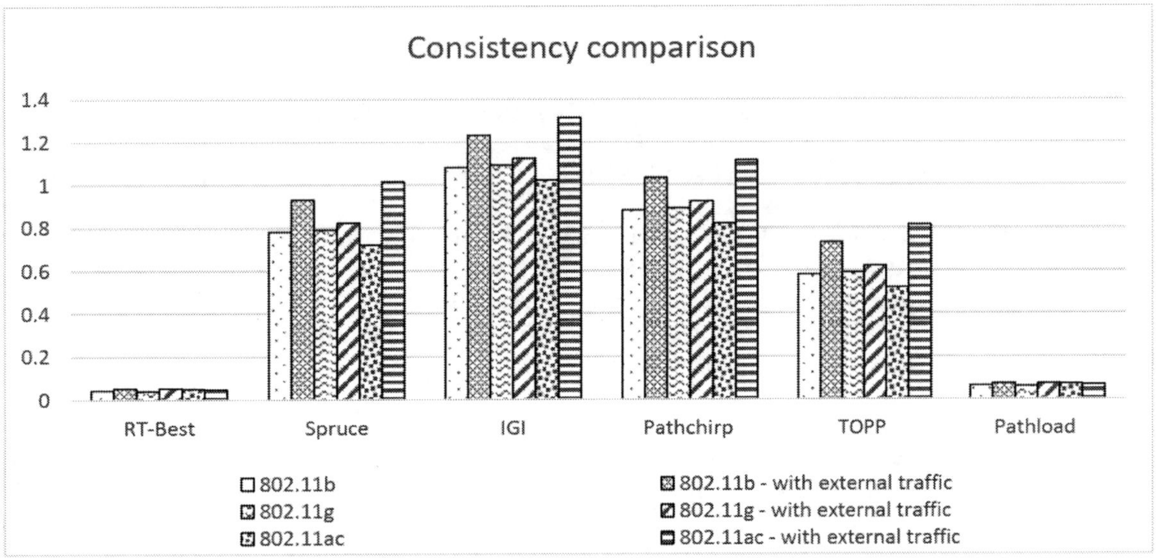

Figure 8: ABW tools consistency comparison

Figure 8 shows the standard deviation value for all tools in respective testing scenario. Based on our findings, RT-Wbest provided the highest consistency in estimation reading across all scenario and

WLAN. This finding is supported by the earlier accuracy result. It shows that by applying RT method, RT-WBest was able to provide a much consistent reading throughout the testing.

In summary, we conclude that RT- WABest provides more accurate results than Pathload, Spruce, IGI, TOPP and Pathchirp in all tested scenario. In contrast, IGI provided the least consistent reading among all, with similar result as the accuracy report.

5. Conclusion

In this paper, we performed a robust testing on six widely of ABW estimation tools; RT-WBest, Spruce, IGI, Partchip, TOPP and Pathload. All tools were tested in 802.11b, 802.11g and 802.11ac environment, with and without the existing of external traffic sharing the same path. In our testing, we look at the accuracy and consistency of each tools across the testing scenario.

The result shows that, no matter what the network condition, RT-Wbest provides more accurate and consistent reading in each evaluation case. It shows that by implementing RT, estimation of bandwidth can be further improved. IGI has the least accurate and consistent. In the future, we will implement more RT on existing ABW tools to improve their accuracy and consistency.

6. Acknowledgement

This research was conducted under the Systems Network and Security (SysNetS) research group, Faculty of Computing and supported by Universiti Malaysia Pahang Research Grant (RDU1703293)

7. References

[1] Salcedo, D., Guerrero, C. D., & Martinez, R. (2018). Available bandwidth estimation tools: Metrics, approach and performance. *International Journal of Communication Networks and Information Security*, *10*(3), 580–587.

[2] Abut, F., & Leischner, M. (2018). An Experimental Evaluation of Tools for Estimating Bandwidth-Related Metrics. *International Journal of Computer Network and Information Security*, *10*(7), 1–11.

[3] Yang, T., Jin, Y., Chen, Y., & Jin, Y. (2017). RT-WABest: A novel end-To-end bandwidth estimation tool in IEEE 802.11 wireless network. *International Journal of Distributed Sensor Networks*, *13*(2).

[4] Salcedo, D., Guerrero, J., & Guerrero, C. D. (2017). Overhead in available bandwidth estimation tools: Evaluation and analysis. *International Journal of Communication Networks and Information Security*, *9*(3), 393–404.

[5] Manish Jain, C. D., Jain, M., & Dovrolis, C. (2002). Pathload: A Measurement Tool for End-to-End Available Bandwidth. *In Proceedings of Passive and Active Measurements Workshop*, 14–25.

[6] Ishida, Y., Hanada, M., & Kanemitsu, H. (2018). Available Bandwidth Estimation Method using Delay Information, *22*(2), 329–332.

[7] Jakimoski, K., Arsenovski, S., Gorachinova, L., Chungurski, S., Iliev, O., Djinevski, L., & Kamcheva, E. (2016). Measurements of available bandwidth in computer networks. *International Journal of Grid and Distributed Computing*, *9*(4), 201–210.

[8] Khangura, S. K., & Fidler, M. (2018). Available bandwidth estimation from passive TCP measurements using the probe gap model. *2017 IFIP Networking Conference, IFIP Networking 2017 and Workshops*, *2018-January*, 1–9.

[9] Paul, A. K., Tachibana, A., & Hasegawa, T. (2016). NEXT-FIT: Available bandwidth measurement over 4G/LTE networks - A curve-fitting approach. *Proceedings - International Conference on Advanced Information Networking and Applications, AINA*, *2016-May*, 25–32.

[10] Delphinanto A, Koonen T, Zhang S, et al. Path capacity estimation in heterogeneous, best-effort, small-scale IP networks. In: *Proceedings of the 35th IEEE conference on local computer networks (LCN 2010)*, Denver, CO, 10–14 October 2010. New York: IEEE.

[11] Xiao, Y., & Rosdahl, J. (2002). Throughput and delay limits of IEEE 802.11. *IEEE Communications Letters*, *6*(8), 355–357. https://doi.org/10.1109/LCOMM.2002.802035

[12] Paul AK, Tachibana A and Hasegawa T. An enhanced available bandwidth estimation technique for an end-to- end network path. IEEE Trans Netw Serv Manag 2016; 13(4): 768–781.

[13] Kamarudin, I. E., Ameedeen, M. A., & Azmi, Z. R. M. (2015). Experimental Analysis on Available Bandwidth Estimation Tools for Wireless Mesh Network. In H. A. Sulaiman, M. A. Othman, M. F. I. Othman, Y. A. Rahim, & N. C. Pee (Eds.), *Advanced Computer and Communication Engineering Technology* (pp. 525–535). Cham: Springer International Publishing.
[14] S. Ekelin *et al.*, "Real-time measurement of end-to-end available band- width using Kalman filtering," in *Proc. 10th IEEE/IFIP Netw. Oper. Manag. Symp. (NOMS)*, Vancouver, BC, Canada, Apr. 2006, pp. 73–84.
[15] Zhong M, Hu P and Jadwiga I. Revisited: bandwidth estimation methods for mobile networks. In: Proceedings of the IEEE 15th international symposium on world of wire- less, mobile and multimedia networks (WoWMoM), Syd- ney, NSW, Australia, 16–19 June 2014. New York: IEEE.

IOP Conf. Series: Materials Science and Engineering **769** (2020) 012048 doi:10.1088/1757-899X/769/1/012048

Contextualizing The Framework For Knowledge Management System At The Organizational Level

[1]Mohammad Fakhrulnizam Mohammad, [2]Rusli Abdullah, [3]Marzanah Ab. Jabar & [4]Rozi Nor Haizan Nor
[1]Universiti Kuala Lumpur, Malaysian Institute of Aviation Technology
[2,3,4]Universiti Putra Malaysia, Faculty of Computer Science and Information Technology

Abstract. Formulation of knowledge management system (KMS) is significance to support knowledge management activities within organizations; moreover, information technology has become the driving factors for business strategies and directions. In the past studies, the focus of KMS establishment are diverse; ranging from its successful adoption, perceived usefulness of the users or on its supporting roles in organization strategy to improve their performance and smooth operations. Despite its benefits to the organization, however, gap has been identified on the lack of existence and definition of KMS framework at the organizational level studied by other researchers. Thus, the objective of this study is to formulate the KMS framework at the organizational level known as OKMS (organization-based knowledge management system). This study is performed by methodologically conducting a review and analysis of the past literature and related areas of concern on establishment of KMS. Outcome of this study will not only help to motivate for other future empirical and theoretical research for KMS at the organization level, but, it also provide a departing point for other KMS researches establishment for organizations.

1. Introduction

Knowledge is defined as an intellectual asset to organization [1]–[3] that drives the competitiveness of organization within robust environment. Knowledge can be defined as an information that are consolidated together with past background, experience, set of thinking and reflection [4]. During the age of knowledge economy (K-economy), the needs for acquiring, managing, transferring, sharing and conserving knowledge for organizations are more realized. Organizations have become more knowledge intensified whereby their sustainability in businesses relies heavily on their abilities to manage knowledge in and out of the organization. On that context, KMS is an information technology (IT) system that are meant to support management of knowledge activities in organization. Establishment of such system is imperative for organization to be more competitive [5] by systematically managing the knowledge. However, previous studies also found that different interpretation has been defined on the concept and definition of KMS at the organization level such as it is a system that produces, compile and reuse knowledge, approaches in organizing information or even managing resources and information discovery [6]. For instance, from the concept of system theories, systems can be categorized into several definitions as shown in Table 1.

Besides, an earlier case study conducted shows that one of the many factors resulting to the failure of KMS implementation in the organization is due to improper planning [8]. Thus, it is important to define a clear framework for the establishment of KMS at the organizational level precisely for knowledge intensive organization.

Table 1: Concept of System Theories. Source : [7]

Concept of System Theories	Description
General system theories	System can be classified into three types that are open system, closed system and isolated systems. Ensuring its operation ability, system needs to involve with a lot of interactions
Cybernetics	Define the capability of a system to adapt to robust challenges in the real environment as a single entity.
Organization	Organization can be known as open system and closed system that comprises of two major elements (can be defined as system) that are social component (human) and technical components (technology, equipment) or known as socio-technical systems.
Biology & Sociology	Entails the system capabilities in differentiating itself with the environment by emphasizes on the self-regulated system aspects. .
System Dynamics and Smart Systems	Describe the capability of a system to stand and operate on its own that support self and independence learning process that require input and feedback from the user.

2. Literature review

Although many studies have been conducted in the past on the establishment of KMS, however, there are still dearth of literature discussing on its development [9], [10]. Thus, indicates requirement for more studies especially in guiding for the establishment of KMS. As stated earlier, apart from various definition of KMS, but it is also concluded that many past KMS studies were more focuses on the adoption or receptance of KMS for individual. However, it is to note that the implementation of KMS are commonly relies on the direction as well as utilization at the organizational level, therefore the needs to emphasize a focus on the establishment of KMS at the organizational level. Moreover, since adoption of KMS in organization relies heavily on the adoption of its frameworks/model [11]–[17], therefore it is imperative for the establishment of KMS at the organizational level or known as OKMS framework. In addition to that, it is an emerging needs for organization to ensure a solid foundation have been developed for the establishment of any IT system because failure in implementing the system may occur to the increasing costs of implementation [18], [19]and resistance of users from using the system [20]. In-depth research have also suggested for the investigation on the roles of KMS in supporting organization to strategize their decision making process as well as implementing good governance in robust environments [21] in the future.

2.1 Generic idea of KMS framework

In studying the generic elements of KMS architecture model (KSAM), seven elements of KMS have been outlined comprises of people, strategy, source of knowledge, delivery interface, functionality, infrastructure and continual improvement [22]. Alongside with defining the elements for the model of KMS architecture, the same study also suggested external driving factors for the model. For instance, number of factors need to be considered for the source of knowledge such as experiences, network, information and events. This is due to the fact that, one of the reason for the failure of knowledge management (KM) implementation in organization is due to the non-existence of capacity to absorb external elements [23] Thus, capabilities of organization in acquiring and assimilate external knowledge into organizations drives them to be more competitive [24], [25].

KMS has also introduced as a concept of socio technical KMS (STKMS) that combines socio and technical approach as suggested in several studies by [26]–[29]. The basis of STKMS model comprises of five elements such as strategic leadership, organizational learning, organizational infrastructure, knowledge culture and technological infrastructure [29]. In addition to that, the roles of stakeholders of the system (actors) offer some improvement that are significant for effective KMS in organization [22]. Besides, the model had the details in emphasizing on what are needed by a KMS; which is the needs to derive the instances of delivery interface which does not highlighted earlier; nonetheless the STKMS model developed by [29] has provided a departing point for other studies in developing KMS model in various areas and field.

Alongside with the above, other studies have also defined KMS into several types of characteristics such as fragmented, content, process and capability based KMS [30], context wise, performance-led, enabler-savvy and supporting sustainability[31] and process based KMS [32]–[34]. Review of past studies conducted on KMS discovered that the emerging of KMS capabilities and features are one of the factors contributing to the issues and challenges of KMS implementation [35]. Hence, the differences in term of concepts and applied use of the system posits the needs for the establishment of the KMS framework at the organizational level. Therefore, this study posits that there is an emergent need to look at a standardized understanding on the foundation of KMS at the organizational level to address the scarcity of IT adoption studies conducted at the organization level; at the same time reducing the potential issues from happening that may arises from the diversities of features and capabilities.

2.2 Organization-based KMS

Objective of this study is to provide a framework and clear description of KMS framework known as OKMS; derives from the review of past studies to conceptualize the elements of KMS at the organizational level. However, it is discovered that there are still lack of studies focuses on the establishment of KMS at the organizational level. The only study discovered focuses on KMS at the organizational level has defined five components of OKMS [36] as shown in Figure 1. The model developed depicts the importance of combining technical approach with social approach in establishing KMS framework. In fact, framework of KMS cannot only relies on technical approach but also social approach [29], [36]. It is holds true because KMS is an IT system with the purpose to manage knowledge which involves interactions between two different types of knowledge; knowledge that are written (explicit) and knowledge that are known (tacit). As it being said, the process of knowledge in organization involves the conversion from tacit to explicit and vice versa [37]. Furthermore, capability of organization to absorb external knowledge (acquire, assimilate and transform) able to enhance the competitiveness of the organization [24], [25]. Thus, combination of social and technical approach justifies the KMS framework in the past studies and future.

In scoping the development of KMS, the main purpose of its development needs to be put in consideration. For instance, in ensuring a comprehensive development of KM model for banking sector, seven variables were suggested inclusive of (1) environment (2) people (3) technology (4) knowledge process (5) creation of knowledge(6) knowledge sharing and (7) knowledge retention [38]. In this study the researcher addressed on the importance of environment and people in making up the KM model for the banking sector. Environment refers to the surrounding or situation where all parties with the same interest and objective work together; which is also addressed previously on its importance in making up the elements of KMS framework [26], [29], [36], [39]. Meanwhile, elements related to knowledge activities implies on the importance of KM process in the framework. Moreover, the elements of knowledge process is also suggested in other areas/industries such as development of KMS in higher education [40]–[42] and South Korean based companies [43]. This study has addressed significance elements to be considered for KMS frameworks which are the needs for KM process, environment and people. Of which, these three elements has been highlighted in a later study [22]. In addition to that, it was highlighted that one of the significance factors to the adoption of KMS in organization is the knowledge sharing [44], which is one of the activities of knowledge.

From implementation aspects, it is suggested that the development of KMS in acquiring the knowledge from senior employee or known as master in their studied industry is by segregating into two stages which collecting and analysing data in the first stage and constructing the KMS based on manufacturing process [45]. As a result, the system does not only improve the skill and problem-solving capabilities of the junior level employee but also improving the quality of product in a whole. Although this study offer a golden rules on the implementation of KMS, but, since it will not be covered in this study, thus it will be a good points to ponder for other future researches emphasizes on the approach or effectiveness of KMS implementation.

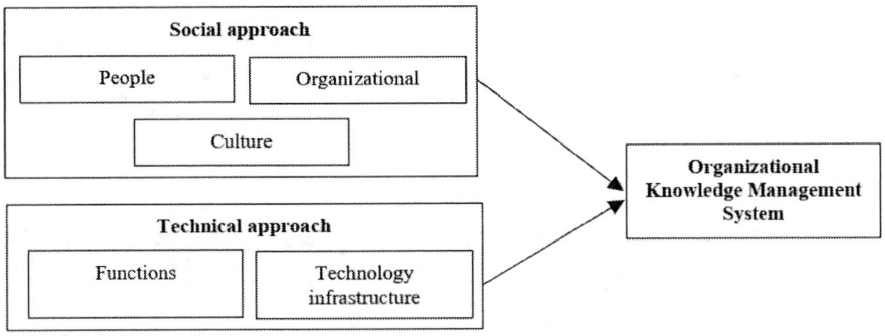

Figure 1: Organizational Knowledge Management Systems. Source : [36]

3. Methodology

In formulating KMS framework at the organizational level, there are three strategies adopted in this study. Firstly, past studies on KMS establishment of related concept were reviewed and analysed adopting the pragmatic approach. The concept of pragmatic approach combines and includes all possible elements that comprises of both social and technical aspect [46]. There is solid foundation in using and adopting the approach, since the similar approach were also performed in the past for integrating two models (TAM and TTF) by examining both underlying theories and analyse their similarities and differences in providing the integrated model [47] Furthermore, earlier review on empirical studies of IT adoption models conducted at the organization level suggested that combining more than one theoretical models in future would be able to support a more complex adoption of new IT system [48].

Secondly, the use of related theories such as systems theory is adopted in formulating the framework for KMS. From the past studies, the use of systems theory is yet to be seen in the development of KMS, but, since organization can be referred as one system that are supported by interrelated parts to achieve efficiency and effectivity [49], therefore this study decided to use systems theory in guiding the formulation of the OKMS framework. In fact, a review study discovered that systems itself can be segregated into five concepts which is one of them is cybernetics [7]. The concepts of cybernetics defined as capability of a system to adapt to robust challenges in the real environment as a single entity which resemble the implementation of KMS as an IT platform in managing knowledge in organization to support the challenges faced by organization. On the other hand, KM has been regarded as a process that has an input that will be processed and produced an output, whereby organization will be able to feed in their influence into the cycle to improve the strength of KM activities subsequently improve the performance of HEI [50]. Of which, the process as derived in that study carriers the four elements of a system theory which are input, process, output and feedback.

Thirdly, this study adopts the previous KMS framework/model developed in a study by [22], [29], [36] that offer more clearer overview of KMS framework architecture. Those three studies offer great contribution in providing the landscape of what architecture for KMS framework should be. Findings from the studies posits the same direction on the needs for combinations of social and technical approach in KMS framework.

4. Discussion and findings

Establishment of OKMS is essential to ensure effective operational framework in supporting the organizations' objective to remain competitive in a robust and competitive market. KMS is seen plays significance roles in enhancing and stimulating the innovation process in organization. Thus, motivation of this study is to provide a platform and reference for future studies interested in the application, adoption, implementation or development of KMS at the organizational level. One of the pertinent findings of this study is the future potential extent of its coverage for KMS establishment for organizations. Although many studies on KMS have been established in the past, however, there is still huge potential for the establishment of KMS and improvement especially at the organizational level.

Based on review of past studies on KMS development, various focus of studies has been indicated. As discussed earlier, the purpose of KMS establishment is to support and enable the management and implementation of KM effectively by using an IT platform. Thus, there are several areas that is significant to be addressed by the new OKMS framework that are not highlighted in the past study as follows:

4.1 Strategy

The roles of top management and leadership is important in ensuring the success and effectiveness of KM implementation. Thus, the elements of strategy leadership are essential in the proposed KMS framework for organization. One of the failure of KMS implementation in organization is due to the lack of top management commitment [51]–[54]. Besides, establishment of KMS in organization needs to be aligned with the vision and objective of the organization. Thus, the element of strategy does not only essential but also crucial. Since this component was not introduced in the earlier model, thus, it provides a great contribution of this study to fill up the gap for new KMS framework at the organizational level.

4.2 Knowledge management process

Knowledge processes involves the capturing, organizing, filtering and safekeeping of the knowledge [38], [42]. Past studies shows that the development of KMS in an organization of their industry emphasized on the elements of knowledge processes [55]. Each organization has its own processes that the implementation of any IT system needs to consider process elements in a system. Furthermore, the nature of knowledge activities in the organization is by allowing it to transfer and flow among and within people and system. For instance, in analysing the capabilities of KMS frameworks in supporting organizations' competitiveness, two elements are known as important - knowledge infrastructure and knowledge process capabilities [43] In fact, in facilitating the transformation process between human and systems, it is a concern to identify the knowledge required before adopting technology such as artificial intelligence [56].

4.3 Knowledge management (KM) tools

The purposes of KM tools are to allow easier contextualisation of information, allows easier transfer of information, facilitate the communication and networking as well as provide a human-computer interface between the systems and the user [57]. It also assists to facilitate the process of KM for interaction between internal and external. In the development and improving the design of knowledge database (KB) for KMS, it is suggested that organization to include content management [58] and document management [40]in the system. Content and document management are example of KM tools available for KMS. The use of KM tools can be segregated and defined by the different types of knowledge as shown in Table 2.

Table 2: KM tools and type of knowledge Source : [57]

Explicit knowledge	**Know how**	**Know who**	**Tacit knowledge**
System tools	Collaboration tools	CRM tools	Video conferencing
DBMS	E-mail	Social network analysis	Face-to-face facilitation
Data warehousing	Groupware	Knowledge portals	Other technologies

4.4 Delivery interface

Delivery interface is a functionality of KMS to provide an interface between the system and human (user of the system. The mode of interface can be in the form of push, pull or interactive mode [22]. As discussed in the same study, the mode of interface shall be determined by the needs of the culture and the user of the systems itself. For instance, the requirements for users with limited capabilities on visibility and hearing might be different with users with normal capabilities. Similarly, if the systems will be adopted in a critical environment, the notifications and alerts system might be useful to be considered.

4.5 Technical functions

In the earlier model developed in a study by [36], functions in the technical approach refers to the functions of operating and managing the knowledge such as using, finding, creating and packaging knowledge. However, in the new framework proposed, related knowledge activities are grouped within the KM process of social approach. In relation to that, this study posits the importance of adding technical functions such as security features, systems searching capabilities, decision support system and to some extent an intelligent agents functionality such as data mining. The needs for technical functions in KMS is not new as it has been highlighted earlier in a study by [40], [41], [59], [60].

5. Proposed framework

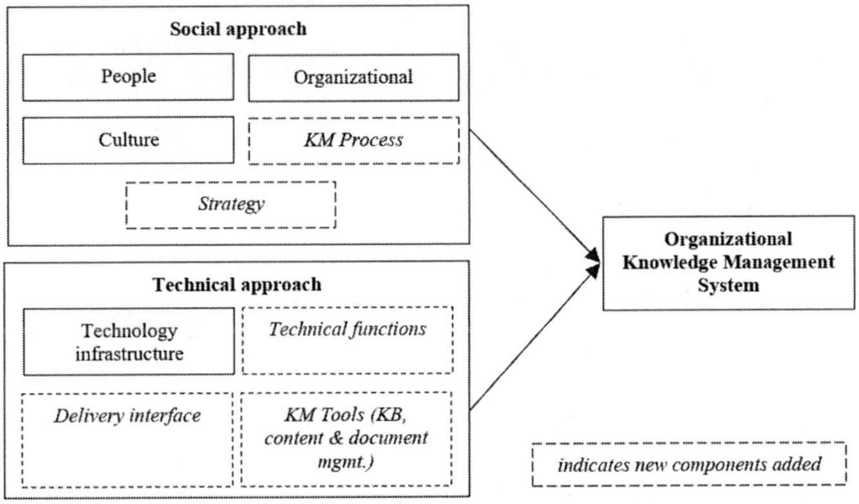

Figure 2: Proposed OKMS conceptual framework. Source: Drawn by researcher

Based on the intense and comprehensive discussion in Discussion and findings section, the proposed conceptual framework of OKMS is drawn as shown in Figure 2. As highlighted earlier, the conceptual framework is adopted from the earlier OKMS framework developed in a study by [36]. In referring to the newly proposed conceptual framework, there are five components of KMS proposed as a result of review of the past studies and KMS development in organization that are two components of social approach (KM process and Strategy) and three components for technical approach (technical functions, delivery interface and KM tools).

6. Conclusion

OKMS framework proposed in this study shall be able to provide a generic overview and architecture of what KMS framework for organization such as higher education institutions, training organization or organizations that emphasize on the importance of managing knowledge known as especially for knowledge intensive organization (KIO). Although previous studies on organizational KMS framework has successfully developed the framework at the organization level, but majority of the studies did not address the type of organization that might benefit from the implementation of KMS framework. However, as discussed earlier, the objective of KMS establishment is to ensure proper and systematic management of knowledge in organizations that relies on knowledge for sustainability or known as knowledge intensive organization (KIO). KIO is an organization that relies heavily on knowledge activities such as producing and reusing of knowledge [47].

Although earlier study has successfully offer broad overview of basic KMS architecture for organization which also adopted in this study, however, the previous framework studies conducted has failed to address on other technical needs of the systems such as technologies and tools needed to ensure the operability of the systems. At the same time, previous framework has also failed to emphasize in detailed the elements of KM process which should be the major focus of any establishment of KMS framework/system. In future, it is suggested that the framework be validated and tested empirically and scientifically. To ensure its practicality and usability, empirical and quantitative analysis will help to identify the applicability of the proposed elements. Prior to that, expert analysis is also needed in ensuring its operability from the perspectives of the experts and real situation implementation.

7. References

[1] M. Alavi and D. Leidner, "Knowledge Management Systems : Issues , Challenges , and Benefits," vol. 1, no. February, 1999.

[2] J. G. C. Y. Lee and M. E. Nissen, "Accelerating acculturation through tacit knowledge flows: refining a grounded theory model," *Vine*, vol. 40, no. 3/4, pp. 312–325, 2010.

[3] O. Novianto and D. Puspasari, "Knowledge Management System's Implementation in a Company with Different Generations: A Case Study," in *Procedia - Social and Behavioral Sciences*, 2012, vol. 65, pp. 942–947.

[4] T. H. Davenport, D. W. Delong, and M. D. Beers, "Successful knowledge management projects," *Sloan Manage. Rev.*, vol. 39, no. 2, pp. 43–57, 1998.

[5] W.-T. Wang and Y.-J. Lai, "Examining the adoption of KMS in organizations from an integrated perspective of technology, individual, and organization," *Comput. Human Behav.*, vol. 38, pp. 55–67, 2014.

[6] A. Alshehhi, "The Notion of Organizational Knowledge Systems," vol. 2, no. 3, 2016.

[7] C. Mele, J. Pels, and F. Polese, "A Brief Review of Systems Theories and Their Managerial Applications," *Serv. Sci.*, vol. 2, no. 1–2, pp. 126–135, 2010.

[8] P. Akhavan, M. Jafari, and M. Fathian, "Exploring the Failure Factors of Implementing Knowledge Management System in the Organizations," *J. Knowl. Manag. Pract.*, vol. 6, pp. 1–10, 2005.

[9] H. Jiang, C. Liu, and Z. Cui, "Research on Knowledge Management System in Enterprise," *IEEE*, 2009.

[10] A. Kaldi, A. Aghaie, and F. Khoshalhan, "KMS adoption in organizations," *2008 IEEE Int. Conf. Ind. Eng. Eng. Manag. IEEM 2008*, no. 17, pp. 37–41, 2008.

[11] M. A. Hameed, S. Counsell, and S. Swift, "A conceptual model for the process of IT innovation adoption in organizations," *J. Eng. Technol. Manag. - JET-M*, vol. 29, no. 3, pp. 358–390, 2012.

[12] S. Assegaff, A. R. C. Hussin, and H. M. Dahlan, "Knowledge Management System as Enabler for Knowledge Management Practices in Virtual Communities," *Int. J. Comput. Sci. Issues*, vol. 10, no. 1, pp. 685–688, 2013.

[13] H. Hussin, "A conceptual model of information technology innovation implementation effectiveness in higher education," *5th Int. Conf. Inf. Commun. Technol. Muslim World*, pp. 1–6, 2013.

[14] H. Lee and G. L. Roth, "A Conceptual Framework For Examining Knowledge Management In Higher Education Contexts," *New Horizons Adult Educ. Hum. Resour. Dev.*, vol. 23, no. 4, pp. 22–37, 2009.

[15] M. J. Manatos, C. S. Sarrico, and M. J. Rosa, "The integration of quality management in higher education institutions: a systematic literature review," *Total Qual. Manag. Bus. Excell.*, vol. 3363, no. July, pp. 1–17, 2015.

[16] K. Ooi, P. Teh, and A. Yee-Loong Chong, "Developing an integrated model of TQM and HRM on KM activities," *Manag. Res. News*, 2009.

[17] M. S. Owlia, "A framework for quality dimensions of knowledge management systems," *Total Qual. Manag. Bus. Excell.*, vol. 21, no. 11, pp. 1215–1228, 2010.

[18] P. Sinha, M. Arora, and N. Mishra, "Framework for a Knowledge Management Platform in Higher Education Institutions," *Int. J. Soft Comput. Eng.*, vol. 2, no. 4, pp. 96–100, 2012.

[19] Yu-Chung, H. ; Shi-Ming, H. Lin, Quo-Pin, and Mei-Ling-Tsai, "Critical factors in adopting a knowledge management system for the pharmaceutical industry," *Ind. Manag. Data Syst.*, vol.

105, no. 12, pp. 164–183, 2005.
[20] J. Li, M. Liu, and X. Liu, "Why do employees resist knowledge management systems? An empirical study from the status quo bias and inertia perspectives," *Comput. Human Behav.*, vol. 65, pp. 189–200, 2016.
[21] D. Blackman and M. Kennedy, "Knowledge management and effective university governance," *J. Knowl. Manag.*, vol. 13, no. 6, pp. 547–563, Oct. 2009.
[22] D. Williams, "Nuts and Bolts of a Knowledge Management System," *J. Inf. Knowl. Manag.*, vol. 14, no. 04, p. 1550035, 2015.
[23] J. S. Edwards, "Processes : Still the poor relation in the knowledge management family," in *Success and Failures of Knowledge Management*, Elsevier, 2016, pp. 59–69.
[24] W. M. Cohen and D. A. Levinthal, "Absorptive Capacity: A New Perspective on Learning and Innovation," *Adm. Sci. Q.*, vol. 35, no. 1, p. 128, 1990.
[25] S. a. Zahra and G. George, "Absorptive Capacity : A Review , Reconceptualization , and Extension," *Acad. Manag. Rev.*, vol. 27, no. 2, pp. 185–203, 2002.
[26] A. Ali, R. N. H. Nor, R. Abdullah, and M. A. A. Murad, "Developing Conceptual Governance Model for Collaborative Knowledge Management System in Public Sector Organisations," *J. ICT*, vol. 15, no. 2, pp. 171–191, 2016.
[27] S. Assegaff and A. R. C. Hussin, "Review of Knowledge Management Systems As Socio-Technical System," *Int. J. Comput. Sci. Issues*, vol. 5, no. 3, pp. 129–135, 2012.
[28] S. Lusa and D. I. Sensuse, "Study of socio-technical for implemetation of knowledge management system," *Int. J. Soft Comput. Softw. Eng.*, vol. 2, no. 1, pp. 14–23, 2012.
[29] S. Sajeva, "The analysis of key elements of socio-technical knowledge management system," *Econ. Manag.*, 2010.
[30] B. B. Nielsen and S. Michailova, "Knowledge Management Systems in Multinational Corporations : Typology and Transitional Dynamics," *Long Range Plann.*, vol. 40, pp. 314–340, 2007.
[31] M. Loon, "Knowledge management practice system: Theorising from an international meta-standard," *J. Bus. Res.*, 2017.
[32] R. J. Gil Herrera and M. J. Martin-Bautista, "A novel process-based KMS success framework empowered by ontology learning technology," *Eng. Appl. Artif. Intell.*, vol. 45, pp. 295–312, 2015.
[33] S. Sarnikar and A. V Deokar, "A design approach for process-based knowledge management systems," *J. Knowl. Manag.*, vol. 21, no. 4, 2017.
[34] R. Woitsch and D. Karagiannis, "Process-oriented knowledge management systems based on KM-services: the PROMOTE ® approach," *Intell. Syst. Accounting, Financ. Manag.*, vol. 11, no. 4, pp. 253–267, 2002.
[35] K. Iskandar, M. I. Jambak, R. Kosala, and H. Prabowo, "Current Issue on Knowledge Management System for future research: a Systematic Literature Review," in *Procedia Computer Science*, 2017, vol. 116, no. Iccsci, pp. 68–80.
[36] P. Meso and R. Smith, "A Resources-based View of Organizational Knowledge Management Systems," *J. Knowl. Manag.*, vol. 4, no. 3, pp. 224–234, 2000.
[37] I. Nonaka, R. Toyama, and N. Konno, "SECI, Ba and Leadership: A Unified Model of Dynamic Knowledge Creation," *Long Range Plann.*, vol. 33, no. 1, pp. 5–34, 2000.
[38] B. S. Tanaji, "Benefits of Knowledge Management System for Banking Sector," *Int. J. Comput. Sci. Commun.*, vol. 3, no. 1, pp. 133–137, 2012.
[39] R. Abdullah, M. H. Selamat, A. Jaafar, S. Abdullah, and S. Sura, "An Empirical Study of Knowledge Management System Implementation in Public Higher Learning Institution," *J. Comput. Sci.*, vol. 8, no. 1, pp. 281–290, 2008.
[40] R. Abdullah, S. Sahibuddin, R. A. Alias, and M. H. Selamat, "Knowledge Management System Architecture For Organizational Learning With Collaborative Environment," *Int. J. Comput. Sci. Netw. Secur.*, vol. 6, no. March 2006, pp. 237–246, 2006.

[41] L. Burita, "Knowledge management system for military universities cooperation," *IEEE*, pp. 1–4, 2014.
[42] J. A. Rah, S. Gul, and Z. Ashraf Wani, "University libraries: step towards a web based knowledge management system," *Vine*, vol. 40, no. 1, pp. 24–38, 2010.
[43] T. Cho and R. Korte, "Managing knowledge performance: testing the components of a knowledge management system on organizational performance," *Asia Pacific Educ. Rev.*, vol. 15, no. 2, pp. 313–327, Jun. 2014.
[44] F. D. Shrafat, "Examining the factors influencing knowledge management system (KMS) adoption in small and medium enterprises SMEs," *Bus. Process Manag. J.*, pp. 00–00, 2017.
[45] C. Y. Cheng, T. Y. Ou, T. L. Chen, and Y. Y. Chen, "Transferring cognitive apprenticeship to manufacturing process knowledge management system: A case study of small- and medium-sized coating industry," *Vine*, vol. 44, no. 3, pp. 420–444, 2014.
[46] L. Ngowi and N. H. Mvungi, "Socio-Technical Systems: Transforming Theory into Practice," *Int. J. Ind. Syst. Eng.*, vol. 12, no. 2, pp. 285–291, 2018.
[47] M. T. Dishaw and D. M. Strong, "Extending the technology acceptance model with task-technology fit constructs," *Inf. Manag.*, vol. 36, pp. 9–21, 1999.
[48] T. Oliveira, M. F. Martins, and U. N. De Lisboa, "Literature Review of Information Technology Adoption Models at Firm Level," *Electron. J. Inf. Syst. Eval.*, vol. 14, no. 1, pp. 110–121, 2011.
[49] C. Chikere and J. Nwoka, "The Systems Theory of Management in Modern Day Organizations - A Study of Aldgate Congress Resort Limited Port Harcout," *Int. J. Sci. Res. Publ.*, vol. 5, no. 9, pp. 1–7, 2015.
[50] K. Asma and M. Abdellatif, "A New Model for the Impact of Knowledge Management on University Performance," *J. Inf. Knowl. Manag.*, vol. 15, no. 04, p. 1650041, Dec. 2016.
[51] M. Avinash Sawant, O. Prakash Yadav, and C. Rokke, "A practical quality management system implementation framework for small-sized companies," *Int. J. Intell. Enterp.*, vol. 5, no. 2, pp. 173–193, 2018.
[52] R. Mohamad, "Knowledge management as innovation : organizational cultural factors affecting knowledge management practices in Malaysian Higher Educational administration," Victoria University, 2012.
[53] F. Savitri, D. Sahiraliani, and R. Yakhusna, "Knowledge Management Implementation Within the Higher Education Institutions in Bandung, Indonesia's City of Education," *South East Asia J. Contemp. Business, Econ. Law*, vol. 2, no. 1, pp. 59–63, 2013.
[54] Y. Wang and Y. Wang, "Determinants of firms' knowledge management system implementation : An empirical study," *Comput. Human Behav.*, vol. 64, 2016.
[55] J. Hahn and T. Wang, "Knowledge management systems and organizational knowledge processing challenges: A field experiment," *Decis. Support Syst.*, vol. 47, no. 4, pp. 332–342, 2009.
[56] A. Paszek and P. Wittbrodt, "The Method Of Building A Knowledge Management System In A Manufacturing Company," *Int. J. Mod. Manuf. Technol.*, vol. VII, no. 1, 2015.
[57] S. R. Ghani, "Knowledge Management: Tools and Techniques," *J. Libr. Inf. Technol.*, vol. 29, no. 6, pp. 33–38, 2009.
[58] Y. Kurniawan and S. E. Hiererra, "Analysis and Design of Knowledge Management Systems for School of Information System at XYZ University (A Case Study Approach)," *Int. J. Multimed. Ubiquitous Eng.*, vol. 9, no. 12, pp. 229–240, Dec. 2014.
[59] C. Chen and J. Huang, "How organizational climate and structure affect knowledge management — The social interaction perspective," *Int. J. Inf. Manage.*, vol. 27, pp. 104–118, 2007.
[60] W. Maseri, W. Mohd, and R. Abdullah, "Integrated Knowledge Management System Design : A Case Study in a Malaysian Higher Learning Institution," in *e-Proceedings of Knowledge Management International*, 2006, pp. 628–633.

IOP Conf. Series: Materials Science and Engineering **769** (2020) 012049 doi:10.1088/1757-899X/769/1/012049

Approaches in Creating Meta-requirement: A Systematic Literature Review

Muhamad Idaham Umar Ong[1,*], Mohamed Ariff Ameedeen[1], Imran Edzereiq Kamarudin[1]

[1]Faculty of Computing, College of Computing and Applied Sciences, University Malaysia Pahang, Gambang, 26300, Malaysia

* Correspondence: idaham@ump.edu.my

Abstract. Common development team will spend less resources on User Requirement Specification (URS) validation, in the case of development project that provides this URS in the phase of competing for tender. In order to overcome this, the use of meta-requirement in evaluation of requirements is being introduce. The purpose of this article is to present the finding of a systematic literature review of meta-requirements. A methodology that involve defining a systematic way of finding literature and presenting it in a meaningful manner has been adopted. Meta-requirement is an area that is station within the field of information system design theory which is contain in the design research body of knowledge. Based on the result of the analysis of the systematic literature review, it is shown that there is still advancement in the field of meta-requirement and there still improvement that can be done to further the knowledge and contribution in this area. It is suggested that further research that will define a highly comprehensive model of creating meta-requirement should be undergo and the issue of meta-requirement traceability against real-world requirements should be tackle.

1. Introduction

The term Software Engineering represents a knowledge area involving the activities of software development that should include requirements documentation, design principles, and other artifacts that will be the basis of the inner workings and deliverables functionality of a requested software product by the relevant stakeholders [1]. User requirements specification (USR) is one of the most critical artifacts in software development, representing the needs and the wishes of the stakeholder in a form of high-level abstract. Most common USR are being presented in the form of natural language such as use case descriptions, user stories, etc.[2]–[6].

Current issue and challenges in the intention of producing correct list of requirements through the process of requirements validation are; (1) Review process (2) Reviewer team (3) Requirements specification (4) Validation support tool (5) Organizational culture and (6) Governance [7]. The statistical data shown in a research done in 2011, major percentage of defect is originated from Requirements Engineering phase with a total of 56%, while design phase, coding phase and other phases are only concurred 27%, 7% and 10%. If the Requirements Engineering phase is being done correctly with the right method and technique, this will immediately affect the future phases by reducing the number of defects that will be discovered [8]. This is where the stated challenges will be overcome with the usage of meta-requirement.

The meaning of the term "meta-requirement" was described as classes of goals to which the theory (kernel theory) applies. The used of the term "meta-requirement" [9] was to simplify requirements in a sense where a design theory doesn't address a single problem but a class of problem. Another researcher defines meta-requirement is the source of today's system function. It is a minimum and indivisible requirement component. Requirement information may be a meta-requirement or a collection of meta-requirement [10]. Kaiya define meta-requirement as all requirements follow some laws when the requirements are proper. The author call such laws as meta-requirements [11].

The main objective of this paper is to summarize published literatures to build and present an overview of existing studies on meta-requirements in the context of meta-requirement creation. The second objective of this publication is to identify the major factors impacting and meta-requirement. Third is search for the unit of measurement(s) used to measure meta-requirement. And the fourth objective of this paper is to identify the limitations and future research in the field of meta-requirement creation in order to suggest areas for further research.

2. Meta-Requirement Literature

Based on a systematic literature search done throughout major scientific and academic databases, in the matter of meta-requirements there four names that are being considered as the founding fathers of meta-requirement. Their contribution towards meta-requirement are in the field of Design Research, more specific in Information System Design Theory (ISDT).

Meta-requirements as describe by researchers is the most basic form of requirement. Its main role in ISDT as the main artifact that describe the most basic function that a system must have, well in theory that is. The definition of ISDT is an initial draft of establish relationships between components of a system to achieve a specific result. Design theory must address the question of how to combine components and relationships to make subsystem and how to combine subsystems and relationships to make a systems [9].

In 1992, a work by J. Walls, Widmeyer, & El-Sawy [9] is the first to include the phrase "Meta-requirement" in his work and not just a buzzword with any relation to the field of ISDT. The future work that comes afterword in relation meta-requirement majorly in the field of ISDT, highly cited his work. Meta-requirement that was presented in his work is quite green, whereby future researchers has made improvement in the meta-requirement is being presented.

Käkölä, Koivulahti-Ojala, & Liimatainen [12]–[14] develops the product aspects of the ISDT for the class of Requirements and Release Management Systems (RRMS) with the involvement of meta-requirement. Advancement has been done thorough this work whereby the researcher introduces a different way of presenting meta-requirement. This is the first author to used 2-dimensional representation of meta-requirement that shows some quality type attributes to the meta-requirement itself.

3. Systematic Literature Review Methodology

This section will describe the method that will be used to conduct the systematic literature review. A detail explanation of Figure 1 will be made available in a journal publication. The research questions that will be answer at the end of the methodology are as follows:

- RQ1: What are the earliest and latest research works regarding meta-requirement?
- RQ2: What are the type and important factors in representing meta-requirement?
- RQ3: What are the limitations of existing work in the field of meta-requirement?

IOP Conf. Series: Materials Science and Engineering **769** (2020) 012049 doi:10.1088/1757-899X/769/1/012049

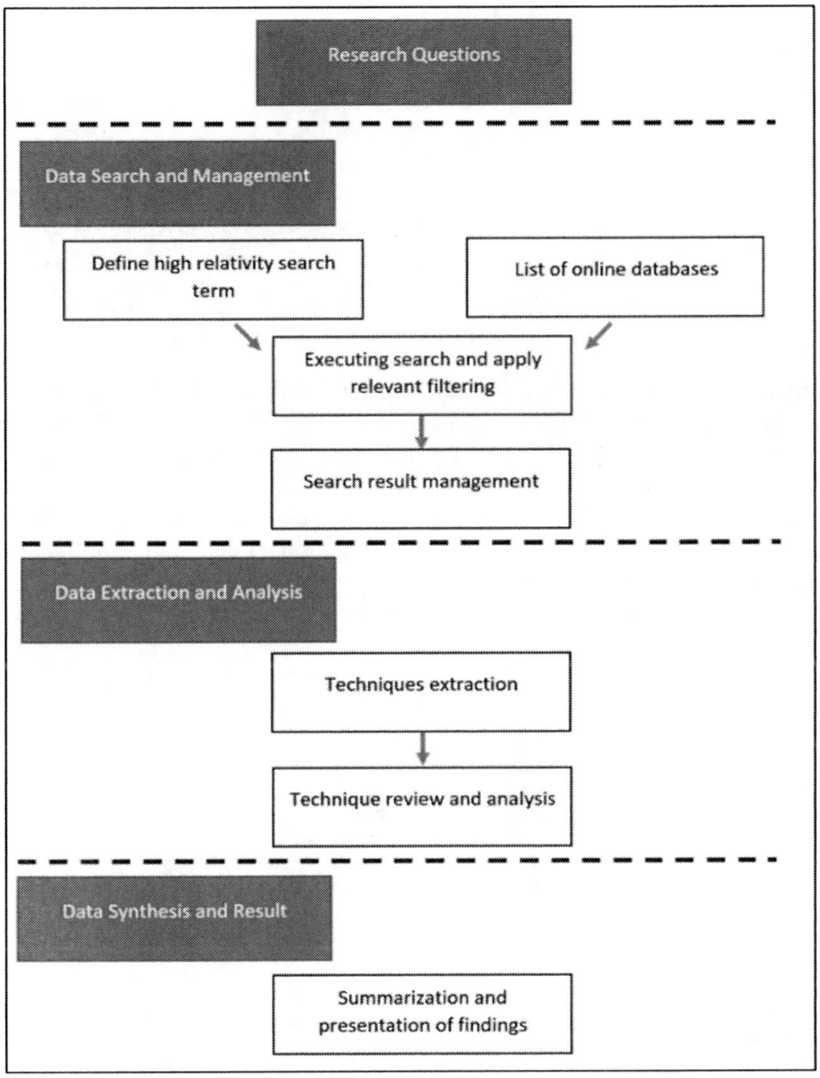

Figure 1. Overview of Steps Involved in Review Study

4. Result and Discussion

4.1. Earliest and Latest Research Works

Based on the result of the literature review, research on meta-requirement has been done way back since the early 90's and still continue until today. Even though the number of publications on meta-requirement are not popping up that frequent, it is also showing that there is still a lot that can be improve within the field of meta-requirement. Figure 2 shows the frequency of relevant publication based on year 2004 to 2018.

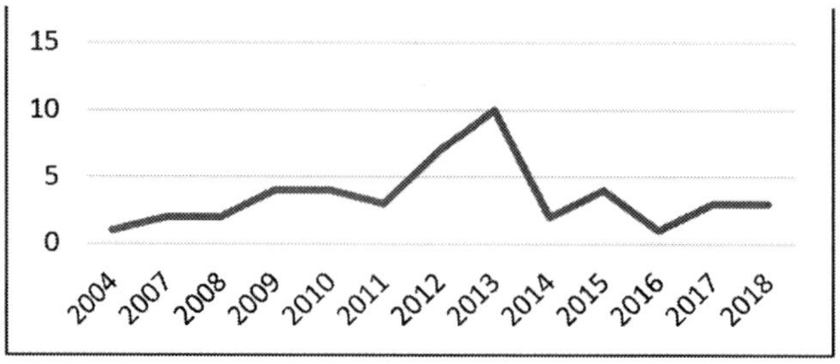

Figure 2. Number of papers by year of publication

There are multiple researches that utilizes the work of [9] such as [15], [16] in designing their needed solution based on the knowledge of design-science research. In the mist of these current researchers contribution and citation to [9], there are few researcher focus on the intervention that is needed in improving the knowledge of ISDT framework. Some researchers such as [12]–[14], [17]–[19] included a few improvement in the attention of contribution towards the presentation of meta-requirement in ISDT. But our focus will be more on the process of producing and validating in the process involve from kernel theory to meta-requirement in the ISDT. A major finding in the work of [20] stating that there is no explicit guideline existed in deriving meta-requirement from the selected kernel theory.

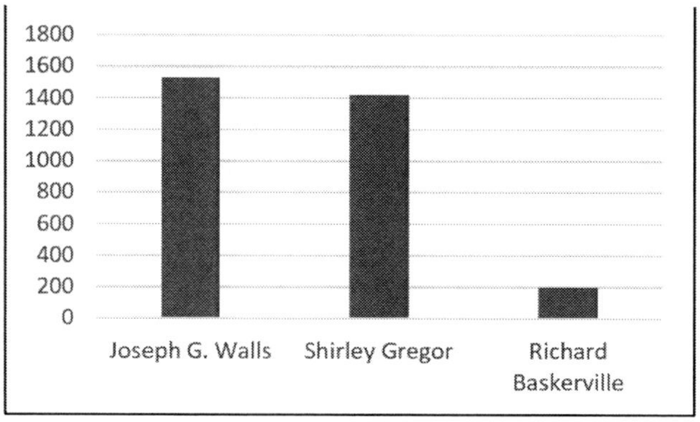

Figure 3. Number of Citations for Meta-Requirement Related Methods

Throughout the 45 selected studies, 3 different authors and their method of deriving meta-requirement are being applied by those researchers. First ever publication which give meaning to the word meta-requirement in the word of information system development was [9]. Meta-requirements was one of the product produce by Walls ISDT (refer to Figure 4) and was widely cited by researchers. The detail and in-depth information on the statistical number that were mention will be publish ion a separate journal.

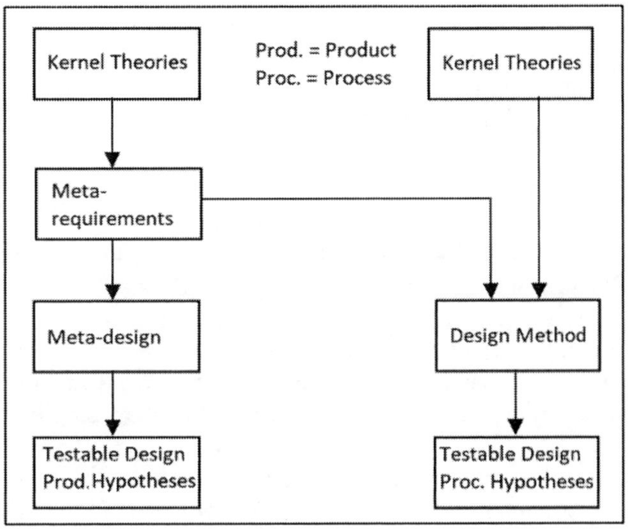

Figure 4. Component of Information System Design Theory [9]

4.2. Type and Important Factors in Representing Meta-Requirement

The important factors of meta-requirements is mainly base on the kernel theory that will be used in the method of deriving meta-requirement itself. For the case of [13], important factors are based on eSourcing Capability Model for Service Providers (eSCM-SP). If there are standard and important factor that can be translated to meta-requirement characteristics and attributes for the scope of Information System (IS) development, it is shown in publication by [21] and [12]. A full publication with information related to the applied kernel theory with regards to the scope of develop system are presented in **Error! Reference source not found.**

Based on the systematic literature review, it is concluded that most of meta-requirement related publication are presented in the form of 1-dimensional rather than 2-dimensional, this is shown based on Figure 5, that shows the number of publications that uses 1- and 2-dimensions type of meta-requirement. Further detail information is being shown on **Error! Reference source not found.**

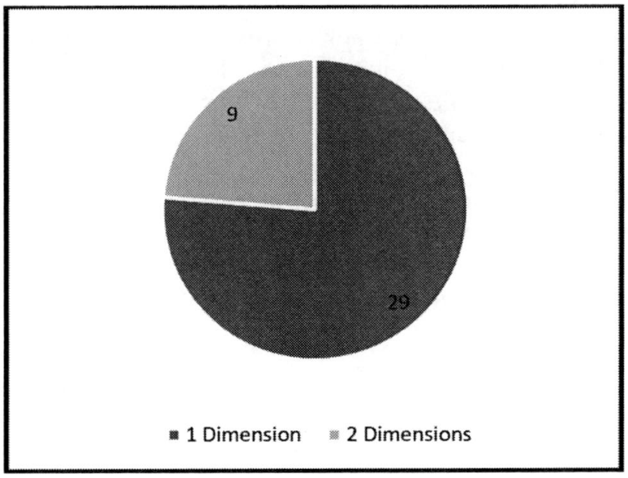

Figure 5. Meta-requirement Dimensional Presentation vs Quantity

IOP Conf. Series: Materials Science and Engineering **769** (2020) 012049 doi:10.1088/1757-899X/769/1/012049

4.3. Limitation of Existing Framework in the Field Meta-Requirement

Major contribution of the related publication is the creation of a design theory that can be use by future researchers in helping them to develop somewhat similar outcome. But this has also present limitation where throughout the course of this literature review. Based on the author knowledge, the following are the limitation discovered:

- There are no existing publication that shows the relationship of meta-requirement produced in ISDT or in part of the field of Design Science Research with the real-world artifacts, focusing on the relationship of meta-requirement and real-world requirement specification.
- Standard in depicting meta-requirements are not well established. Even though most of the researcher uses the similar syntactic define by [22], there is no clear research on the standard of writing meta-requirements. This was shown based on the findings of meta-requirement being presented in general language format or by following a syntactical format similar towards requirement engineering and the presentation of either single or 2-dimensional meta-requirement.
- Researchers [20] has commented that there is actually no explicit guideline in deriving meta-requirement. Even though it is clearly stated in the work of [9] the origin of meta-requirement, there is no instruction in the form of diagram or framework has been discovered throughout the progression of this paper.
- With the highly abstract nature of meta-requirement, there is no publication that shows other derivation of meta-requirement into other instance of an artefacts. For example, modelling meta-requirement seem not exist.
- No involvement of Object Management Group (OMG) in the context of meta-requirement in the field of Design Theory. This is a major indicator that meta-requirement is not included in OMG interest whereby most publication that are listed in this literature review are not references to OMG and meta-requirement is not define in any material produce by OMG. Even though [23]–[25] working on meta-requirement and also referencing towards OMG, it's not a direct referencing.
- Throughout the whole process of executing the systematic literature review, there no clear significant of meta-requirement being established between it and the main Software Engineering body of knowledge that is Object Management Group (OMG). There was [24]–[27] who cited OMG in his publication but there are a few argument that is raised. First, throughout the journal there was no proper definition of the term "meta-requirement" mentioned. Second, relationship of meta-requirement and OMG was not clearly stated. And finally, its publication was not related to the well-published "meta-requirement" definition in the context of Design Theory.
- Meta-requirement as one of the by-products of ISDT. The main star in ISDT is meta-design where it uses meta-requirement and testable hypotheses as a stepping stone in making there are artefacts for check and balance when considering the usefulness of the meta-design.
- No major relation of requirements engineering and meta-requirement. Through the literature that was reviewed, most of them are in context of design theory and only a few [23] was consider as using meta-requirement in the core of requirements engineering. But there are a few factors that being consider of these publications whereby the background of meta-requirement is not well established compared in the context of DT.

5. Conclusion and Future Works

There is still limited work in meta-requirement at present that provide guideline in producing meta-requirement. The process of making sure the traceability of each artifacts that are related to meta-requirement in its core field of Information System Design Theory is also another matter that needs to be taken seriously. From the artifacts of kernel theory, meta-design in both product and process and testable hypotheses, all of them are closely related to meta-requirement. Even though meta-requirement that is produce in ISDT is being consider as a theory, proving the effectiveness of the produced theory against its real-world counterpart is crucial.

Our future work will focus on formulating an algorithm that will function as a meta-requirement creation. The algorithm itself must be able to work in conjunction to ISDT environment taking into account every related artifact in the field. In order to achieve this vision, a systematic formulation and the right approach is needed to produce an algorithm that can work within the stated environment.

References

[1] I. Sommerville, *Software Engineering*. Pearson, 2011.
[2] I. Graham, *Requirements Engineering and Rapid Development*. Addison Wesley, 1998.
[3] S. Lauesen, *Software Requirements: Style and Techniques*, no. January 2002. Addison-Wesley, 2002.
[4] M. Luisa, F. Mariangela, and N. I. Pierluigi, "Market research for requirements analysis using linguistic tools," *Requir. Eng.*, vol. 9, no. 1, pp. 40–56, May 2004.
[5] A. S. Lokman and M. A. Ameedeen, "Modern Chatbot Systems: A Technical Review," Springer, Cham, 2019, pp. 1012–1023.
[6] Sutarman, M. B. A. Majid, J. B. M. Zain, and A. Hermawan, "Recognition of Malaysian Sign Language using skeleton data with Neural Network," in *2015 International Conference on Science in Information Technology (ICSITech)*, 2015, pp. 231–236.
[7] S. O. Mokhtar, R. Nordin, Z. A. Aziz, and R. M. Rawi, "Issues and Challenges of Requirements Review in the Industry," *Indian J. Sci. Technol.*, vol. 10, no. 3, 2017.
[8] K. Khan *et al.*, "Requirement development life cycle: The industry practices," *Proc. - 2011 9th Int. Conf. Softw. Eng. Res. Manag. Appl. SERA 2011*, pp. 12–16, 2011.
[9] J. G. Walls, G. R. Widmeyer, and O. A. El Sawy, "Building an Information System Design Theory for Vigilant EIS," *Inf. Syst. Res.*, vol. 3, no. 1, pp. 36–59, 1992.
[10] B. Chen, T. Yang, and S. Li, "Intelligent acquisition and integrated application technologies for manufacturing resources," in *Proceedings - 2008 Pacific-Asia Workshop on Computational Intelligence and Industrial Application, PACIIA 2008*, 2008, vol. 1, pp. 832–836.
[11] H. Kaiya, "Meta-Requirements for Information System Requirements: Lesson Learned from Software Ecosystem Researches," *Procedia Comput. Sci.*, vol. 126, pp. 1243–1252, 2018.
[12] A. Salo and T. K. Käkölä, "Groupware support for requirements management in new product development," *J. Organ. Comput. Electron. Commer.*, vol. 15, no. 4, pp. 253–284, 2005.
[13] Y. Lu and T. Käkölä, "An information system design product theory for the abstract class of integrated requirements and delivery management systems," *Proc. Annu. Hawaii Int. Conf. Syst. Sci.*, pp. 3677–3686, 2014.
[14] P. Forselius and T. Käkölä, "An information systems design product theory for software project estimation and measurement systems," *Proc. 42nd Annu. Hawaii Int. Conf. Syst. Sci. HICSS*, pp. 1–10, 2009.
[15] A. Van Looy, M. De Backer, G. Poels, and M. Snoeck, "Choosing the right business process maturity model," *Inf. Manag.*, vol. 50, no. 7, pp. 466–488, 2013.
[16] W. Maass and U. Varshney, "Design and evaluation of Ubiquitous Information Systems and use in healthcare," *Decis. Support Syst.*, vol. 54, no. 1, pp. 597–609, 2012.
[17] T. Käkölä, M. Koivulahti-Ojala, and J. Liimatainen, "An Information Systems Design Theory for Integrated Requirements and Release Management Systems," *2009 42nd Hawaii Int. Conf. Syst. Sci.*, pp. 1–10, 2009.
[18] R. Baskerville and J. Pries-Heje, "Explanatory Design Theory," *Bus. Inf. Syst. Eng.*, vol. 2, no. 5, pp. 271–282, 2010.
[19] Y. Lu and T. Käkölä, "An Information System Design Product Theory for integrated Order, Transportation and Warehouse Management Systems," *Proc. Annu. Hawaii Int. Conf. Syst. Sci.*, pp. 3717–3726, 2013.
[20] A. Amrollahi and B. Rowlands, "OSPM: A design methodology for open strategic planning,"

Inf. Manag., vol. 55, no. 6, pp. 667–685, Sep. 2018.

[21] Z. Tang and S. Liu, "The meta-requirements model study for trusted software," in *2010 6th International Conference on Wireless Communications, Networking and Mobile Computing, WiCOM 2010*, 2010, pp. 1–4.

[22] A. R. da Silva, "Linguistic Patterns and Linguistic Styles for Requirements Specification (I)," *Proc. 22nd Eur. Conf. Pattern Lang. Programs - Eur. '17*, no. I, pp. 1–27, 2017.

[23] A. Alebrahim, *Bridging the Gap between Requirements Engineering and Software Architecture*. 2017.

[24] A. S. Black, "Ehealth-As-a-Service: a Service-Based Design Approach for Large Scale Ehealth Architecture," 2018.

[25] M. Koivulahti-Ojala, "On UML modeling tool evaluation, use and training," *Jyväskylä Stud. Comput.*, p. 157, 2017.

[26] S. Sarnikar and A. Deokar, "Knowledge management systems for knowledge-intensive processes: Design approach and an illustrative example," *Proc. Annu. Hawaii Int. Conf. Syst. Sci.*, pp. 1–10, 2010.

[27] A. Alebrahim, "Framework for Identifying Meta-Requirements," in *Bridging the Gap between Requirements Engineering and Software Architecture*, Wiesbaden: Springer Fachmedien Wiesbaden, 2017, pp. 51–109.

Acknowledgments
This work was supported in part by University Malaysia Pahang under Grant RDU RDU191802-2.

IOP Conf. Series: Materials Science and Engineering **769** (2020) 012050 doi:10.1088/1757-899X/769/1/012050

User Requirement Validation: Challenge Exploration in Pre-project Execution

Muhamad Idaham Umar Ong[1],*, Mohamed Ariff Ameedeen[1]

[1]Faculty of Computing, College of Computing and Applied Sciences, University Malaysia Pahang, Gambang, 26300, Malaysia

* Correspondence: idaham@ump.edu.my

Abstract. The need for validation technique for user requirement or in other term Tender Specification at the Request for Tender (RFT) is vital. Without much resource for vendor to dispose of, the outcome of this research will proof useful. The purpose of this article is to determine the gap of initial and finish software product, method of eliminate or mitigate those gaps, list of OO artefacts and their relevancy towards the research. Finding of this article will be the basis of interconnecting user requirement specification and meta-requirement via UML. Based on the result, RE plays an important role in the survivability of a project, there are a few researches being done for solutions in the field of RFT specification validation, UML is the most relevant artefacts and usability of UML diagram ranked based on consistency is relevant toward connecting UML and meta-requirement. It is suggested that further research is to be carried out that will define the connection between user requirement specification in the form of UML to meta-requirement in order to achieve the overall objective of the research.

1. Introduction

There are many software development projects have failed to see the light of day. For project that were delivered to their respective owner, there is usually a problem where the developed product itself doesn't meet the initial expectation of the owners themselves. Delivered product has gone through more of a devolution rather than evolution from which the first day the initial prototype was presented. This give software developers a bad reputation in keeping their promises to develop the desired product.

Commonly, software developers on this day continues to produce monstrosity in the form of products that originated from (1) lack of understanding of the nature of the software itself, these include understanding from the aspect of software developer and stakeholders. (2) lack of knowledge management in young and in-experience software developers are struggling to maintain the standards that they predecessors has formally or informally established and (3) Problem in software quality assurance where not many projects willing to invest in latest testing technology and tools but accommodate cost effective testing that uses a lot of regression and unmanaged testing activities. [1]

A static published by [2], software development project failures occurs throughout the industry. It is noted that from the overall percentage of project involve in the study, less than 28% of software development project are being delivered successfully without facing any challenges. Table 1 summarizes the overall statistical result based on the study conduct by [2].

Table 1. Percentage of success/fail projects [2]

Project Outcome	**Size of Stakeholder's Organization**		
	Large	**Medium**	**Small**
Completed on-time and on-budget with all features initially specified	9%	16.2%	28%
Completed with burse time and budget, features were left-out	61.5%	46.7%	50.4%
Project cancelled	29.5%	37.1%	21.6%

It is also mentioned that some of the main reasons why project budget and timeline are overrun its related to the features that are initially specified during the earlier stages of the software development project. Developer having problem to ensure that the features is comply with the need of the system users. Things like; (1) Lack of user input, (2) incomplete requirements and specifications and (3) changing requirements and specifications [2] are the top 3 problem that are being faced by developer. And all those three problems are pointing to one specific area of knowledge in software engineering, and that is requirements engineering.

Software development project failures are not solely tied to the hardworking people of the Information Technology (IT) sector. Failures are wide spread and involves human. When looking into the area of Critical Success Factors (CSF) in software development project, some of the prominent factors are; (1) top management's commitment and involvement/support, (2) allocation of scarce resources, (3) communications among various stakeholders, (4) team configuration and structure and (5) social cohesion in the team and the complexity of the project and organizational culture. [3]

With earlier intervention by investing more resources and the correct tools and methodology, the problem in software development project could be overcome and projects will be delivered to their respective owner without much of an issue.

The purpose of this publication is to investigate and answer the following research questions present finding on; (1) Why software development fails in the scope of requirement engineering, (2) How developed methods are used to eliminate or mitigate those failures, and (3) Which artefacts are considered as the life-line of a software development project.

The remaining five sections are as follows. Section 2 presents the background of the study. Section 3 describe the adopted research methodology and threat to validity of study. Section 4 describe briefly the list of reviewed literatures in answering the research questions. Section 5 reports the results and discussion from the study. Finally, Section 6 concludes and outlines some possible future works.

2. Background of Study

This section will describe the list of background of the conducted study.

2.1. The starting point buyer and vendor

This section will describe the process of purchaser from an organization will looking to purchase a software solution from a software vendor. The process starts by a request is being put out internally that a need for a solution to a problem or opportunity is to be cater with the help of a software solution. The request will be advertised through either Express of Interest (EOI), Request for Information (RFI), Request for Proposal RFP), Request for Quotation (RFQ) or Request for Tender (RFT). Software vendors will start engaging with the organization. The RFT advertisement contain requirements, criteria and instructions that are to be abide by the vendors. The vendor will submit a reply either via tender or proposal specifying the service or solution that will be provided, buyers evaluate the vendors and project will be awarded to the winning vendor. Please be informed that the publication will not discuss in depth on the detail knowledge and processes of procurement, but more on how a purchaser's tender is being initiated and how software vendor approaches the buyers focusing the challenges faced by the software vendors.

2.2. From Request for Tender to Awarding Project

Purchaser is in need of a solution for a specific set of problems. Within the scope of solution with the utilization of software, buyers will look into different vendors which could provide the suitable solution to their problem. Some buyers will use different approach in order to gain more insight regarding vendors and suppliers for such request.

Purchaser will always try to balance the price and quality ratio of the solution. It is common that the best solution in the market doesn't come cheap and plus there is a saying "Nothing is perfect". In order to overcome the buyer's problem of finding the suitable solution, tenders are being advertise to encourages business to present their proposal on their take on how to solve the specified problem.

Project proposal is one of the medium widely used by software vendors as a way to communicate their intention and dealing with their potential buyer. The content of a project proposal is straight forward. It may contain information related to the decision maker but not limited to the following information; (1) Problem face by organization, (2) Vision of the project that will ensure that it is in line and closely related with the organizational strategies and vision, (3) Benefit of the project will deliver, (4) Deliverables in the scope of equipment, artefacts and tangible and intangible products, (5) Success criteria that will give confidence to stakeholders when project is executed. Usually referring to success criteria such as Specific, Measurable, Achievable, Realistic and Time bound (SMART) (6) Listing deadlines, overall development plan and approaches used such as the used of external vendors and the used of agile software development, (7) Cost and budget. From all this information, everything will be summarized in an executive summary. Some buyers will sometime be requested separate documentation of proposal are to be prepared, first proposal is related to the technical aspect of the project and second is financial matters.

Decision maker will make the decision to accept or reject the project proposal. First impression is critical at this moment, usually the sales people will try to make a big impact right at the start of this proposal presentation. Problem statement is key, it will explicitly show the current problem faced by the decision maker's organization which could be overcome had your proposed project has been in production. Also highlighting where opportunity was missed and where risk and cost were incurred that could have been prevented with the use of the proposed project.

2.3. The Challenge Faced by Software Vendors

Usually in the event of presenting software development project proposal for the decision makers, a prototype is being used to the express the expected outcome of the project and in a way increasing the decision maker's level of confidence towards the to-be awarded software vendor. Prototype are usually based on past successful project or it could an out of the wild combination of GUI and functions from a few different software projects. For a software vendor with strong background with similar past successful project, this decision of whether to take on the project or not would be a snap.

In a scenario where a detail list of thousands of specifications of the desired end-product is being listed out in the RFT, it is quite a daunting challenge where producing a highly convincing feedback that will meet and fulfill every single specification of the desired end-product. This is the task that will be led by Sales staff with the help and advice from a team [4] that consist but not limited to; (1) Sales staff, (2) Project Manager, (3) Technical specialists (Requirements Engineer), (4) Implementers, (5) Finance and commercial specialist, and (6) Legal experts. Some of the limitations [5], [6] that are imposed to these team are; (1) Time. This is a case by case scenario whereby the timeframe of between the advertisement of the RFT and project proposal preparation is quite short, (2) Resources. In current situation where the is no guarantee that the software vendor will be awarded with the project, not much resources will be invested in partaking this effort and (3) Competitive. Depending on the RFT itself, the level competitiveness will vary. It is worth mentioning that even large will invest their time in small projects.

Based on the limitation shown on the previous paragraph, the task of Requirements Engineer has somewhat seemed absurd. In the scenario stated, Sales staff and Project Manager will be relying on the input from the requirement engineer to ensure that the proposal will be up-to-par with the expectation of the decision maker. Based on [7], requirements engineer main target is to achieve the desired output of a software development project by balancing the three main aspect of a software project [7]; (1) Specification, deals with the degree of requirements understanding at a given time (2) Representation, copes with the different representations (informal and formal languages, graphics, sounds etc.) used for expressing knowledge about the system. Within RE there are three categories of representations. and (3) Agreement, deals with the degree of agreement reached on a specification.

With the current attention of a Requirement Engineering is briefly illustrated in Fig. 1, it is acknowledged that with the combination of the situation of proposal presentation of a tender with the standard of work that needed to achieve by Requirement Engineers is quite challenging. Hundreds or even thousands of specifications that are included in the RFT document are also representing the almost finalize user requirement. If the thousands of specifications that are included are being taken lightly, then software development work will projects suffer failure due to user requirements related problem such as changing requirements, unclear, ambiguous and unusable requirements and misunderstood user requirements and the failure to freeze requirements [8]–[14]. Cost and delivery overruns are resource or economic factors. Cost overruns and missed delivery can result in project termination. Publication by [15] includes increase of costs and timeline, actual project expenditures and delivery below the estimates and insufficient budget. These indirect factors may be the reason for the overrun. Failure can also be due to time and delivery below the estimates related to estimation issues in project management. Finally the depletion of funds can result in project termination.

2.4. The Desired Form of User Requirement

During the stage of proposal presentation based on the scenario setup of previous chapter, Requirements Engineer will be facing a conundrum where they will need to analyze the Tender specification that comes in the form of thousands of lines of user requirement to ensure that the content of those specifications are up-to-par with what is needed for development team to work with.

There are three type of user requirement quality that must be evaluated and decided by the project team; (1) correctness of requirements, (2) completeness of requirements, and (3) consistency of requirements. With the high expectation and standards that is defined for Requirements Engineer to work with and with great attention that the result will enable meaningful advice to the sales executive and project manager is seem to bleak. But will the proper mechanism that is developed by future researchers, the hope for a future where a meta-level analysis of tender specification in the form of User Requirement can be realized and will fulfil the needs of software development team.

3. Research Methodology

The research methodology is partially adopting the processes that was define by [16]–[18]. In order to answer the defined research question, the study is based on literature review of publications from publication journals article, conference proceeding and books. It is also taken into account publication that are source from websites and press articles. The information gathered on the stated topic of interest is considered as a mature area of interest and was viewed from a selection of sources that are considered from highly cited and well-established publisher such as SCOPUS indexed and ISI journals.

An initial search using the combination of the following terms: Systematic Literature Review, Systematic Mapping Study, Software Development, Software Engineering, Requirement Engineering, Challenges, Project Management, Tender, Failures, Gaps in Google Scholar. No publication dates were defined in used of restriction which was considered as irrelevant for the task of literature review. Each of the listed publications were reviewed and synthesize to highlight; (1) Failure and success factors of software development projects, (2) proposes method/tool of elimination or mitigating failures and (3) the most critical artefacts of RE.

Following initial analysis of the studies, papers were selected based on their relevance and provided insight specific to the interest. The literature review was structured around the analysis and synthesis of study findings to identify the relevant information and answering the defined research questions.

3.1. Threat to validity

Even though there is a possible method of producing an absolute literature search and result analysis through a rigorous and highly regressive search in every single source available in the world, there are always limitation that must be abide by your average researcher. In this section, threats to validity are being presented.

Consideration of abstraction of pure RE challenges in pre-project execution is not absolute possible with the close relation of the term computer science majorly in software engineering. Some of the findings presented may or may not be 100% aligned to the research questions but is made available in the study considering the relevancy of the information itself. This will be discussed further in section 5.

Another threat to the validity of this research is the humanity aspect of this research. There is no automated tool being used in this research and all information that is compiled for this research are being done through effort of normal human being. With the hope of being able to provide an absolute compilation of all existing meta-requirement creation method in the world, this research is also abiding to the constraints and limitation that is being put on the researchers themselves as normal human being.

4. Reviewed Literature

This section will describe the list of background of the conducted study. The following are the cited literature that are relevant in answering the defined research questions.

Title	Author	Year	Publisher	Related RQ
Perceived Causes of Software Project Failures - An Analysis of Their Relationships [19]	T. O. A. Lehtinen, M. V. Mäntylä, J. Vanhanen, J. Itkonen, and C. Lassenius	2014	Information and Software Technology	RQ1
Reducing Software Requirement Perception Gaps Through Coordination Mechanisms [20]	H. G. Chen, J. J. Jiang, G. Klein, and J. V. Chen	2009	Information and Software Technology	RQ1
Obsolete Software Requirements [21]	K. Wnuk, T. Gorschek, and S. Zahda	2013	Information and Software Technology	RQ1
Critical Requirements Engineering Errors Leads to Fails Software Project [22]	M. Talhe	2018	The Educational Review, USA	RQ1
A Study on the Software Requirements Elicitation Issues - Its Causes and Effects [23]	N. Kumari.s and A. S. Pillai	2013	3rd World Congress on Information and Communication Technologies	RQ1
A Quantitative Study to Identify Critical Requirement Engineering Challenges in the Context of Small and Medium Software Enterprise [24]	S. Besrour, L. B. A. Rahim, and P. D. D. Dominic	2016	3rd International Conference on Computer and Information Sciences	RQ1

A Total Benefits Strategy is a Valuable Approach in HR Outsourcing [25]	J. Miller	2008	Employment Relations Today	RQ1
Investigating requirement engineering techniques in the context of small and medium software enterprises [34]	S. Besrour, L. B. A. Rahim, and P. D. D. Dominic	2016	3rd International Conference in Computing and Information Science	RQ2
A Systematic Study on Requirement Engineering Processes and Practices in Mauritius [35]	G. Huzooree, V. Devi, and R. Lecturer	2015	International Journal of Advanced Research in Computer Science and Software Engineering	RQ2
A dynamic life-cycle model for the provisioning of software testing services [36]	Y. Lu and T. Käkölä	2014	Systems Science and Control Engineering	RQ2
Empirical studies of requirements validation techniques [37]	U. A. Raja	2009	2nd International Conference on Computer, Control and Communication	RQ2
A review on software requirements validation and consistency management [14]	M. Kamalrudin and S. Sidek	2015	International Journal of Software Engineering and its Applications	RQ3
A systematic identification of consistency rules for UML diagrams [38]	D. Torre, Y. Labiche, M. Genero, and M. Elaasar	2018	Journal of Systems and Software	RQ3

5. Result and Discussion

This section will showcase the result of this research. Each finding for respective research question will be presented in each different sub-heading.

5.1. RQ 1: Why software development fails in the scope of Requirement Engineering?

This section will discuss the findings based on the literature listed in section 4. From the view of [19], a software development project consists of people, tasks, methods and project environment where each of these must play its part in making sure everything working together in harmony. Fail to do so will cause the more prone to failure. The study also concluded that weakness identified in teamwork among project team, weak monitoring and management in task backlog, and finally small number of resources were invested in testing were the root causes of the failure.

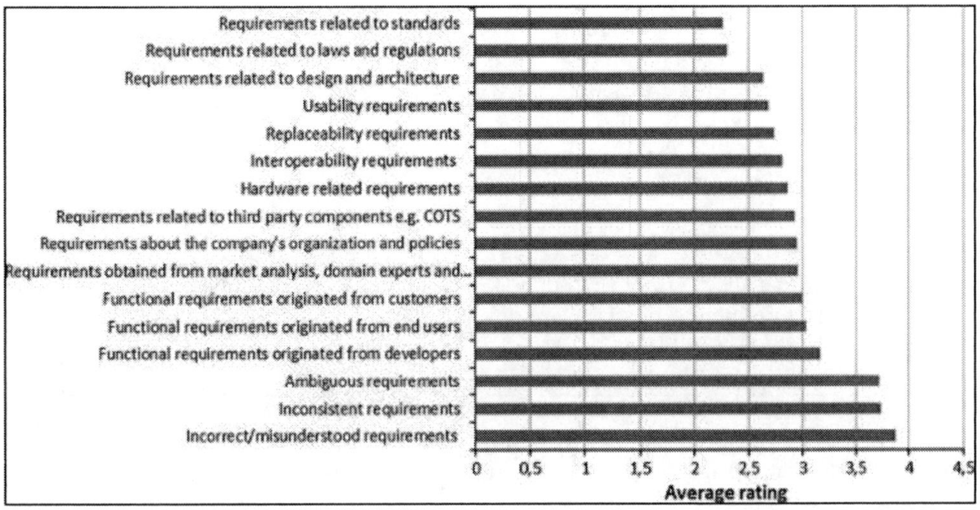

Figure 1. Type of OSRs Likely to Become Obsolete [21]

Also based on the same study, finding on the failures cause and it's causal are being illustrated. It gives an overview of the different kind of failure that could occur in a project and the inter-relationship between different aspect of a project. Even though the study focuses more on failures that are purely related to the area of software testing, failures that are related to the area of RE are also seems to be playing its major part as few of the reasons impacting the matters related to software project development project failure.

Communication and relationship between stakeholders and developers also gives critical impact in ensuring the project successfulness. A study by [20] confirms that communication among project team members, including stakeholders will surely give major impact in improving the overall performance of project through key area examination of relationships among pre-project association, horizontal coordination, vertical coordination and user-developer perception gaps.

On another study done by [21], establish relationship between the different type of RE problem with obsolete requirements. In other words, the impact of errors occurs in RE process will impact on the system functionality in future where function that are based on problematic user requirement will become irrelevant and most likely will not be used by the users. Figure 1 depict the type of Obsolete Software Requirements (OSR) rating.

Based on a study by [22], the author has listed errors that could occurs during the RE process. The different stages of RE process that was used in the study involves; (1) Elicitation and collection, (2) Analysis and investigation, (3) Specification and description, (4) Verification and (5) Management. Table 3 summarizes the publication findings.

Table 2. Type of Requirements Engineering Errors [22]

Type of Error	Description	Common Errors
Conception	Arise after the requirement is unwell defined, then the assuming thinking started	Unwell-defined requirement, lack of clarity, unneeded function, unneeded complexity, poor requirement quality, missing requirement, unprepared requirements engineer and inappropriate constraints
Specification	Requirement engineer can't able to describe or specify requirements that has been analysed	Requirements not traced, inadequate verification of requirements quality
Implementation	These errors get up when the idea	Inadequate requirements validation,

	was correct, the description was exact, but the requirement was not implemented correctly	inadequate requirements process, poor quality measurement
Visualization	These errors occur when proposed system area was represented by prototypes but not drawn against a result space area	Prototype Error
Requirement Management	Requirements stored in paper or spreadsheet rather than in requirement repository are difficult if not impossible to create, manipulate, and maintain	Excessive requirements instability including unmanaged scope creep, inadequate requirements management

The publication [23] by show the area of requirement engineering where to the concern of answer this publication, it details out the issue reported, the cause and the consequences against the different requirement elicitation issue factors. Table 4 summarizes the overall finding of the [23] findings. Each error that could happen in RE has vast consequences. With the role of RE consider as the main role in providing meaningful and highly sought inputs for a project, depending how good are those inputs will determine the successfulness of the project. As highlighted by the author the consequences of error occur in RE are the same errors occurs over and over, the cost of re-working project output, budget is being overruned, poor quality and progress monitoring systems, feeling of dissatisfaction among stakeholders towards the development progress, project development and process failure and tools that are not alignment with stakeholders environment, uncertainties, poor and ambiguine requirements, high costs of maintenance and operation, high changes frequency, conceptual inconsistency and flaws in resultant system.

Through the literature that was reviewed, some of the more critical challenges of RE are; (1) Vague and ambiguous requirements, (2) Incomplete Requirements, (3) inconsistent requirements, (4) Ownership and communicating requirement to owner and (5) Traceability of requirements [24]. It is also noted that publication in seven root causes related to project failure [25] suggested the following; (1) Incomplete Requirements, (2) User-related factors, (3) Incompetent project planning, (4) Weak support/involvement from management, (5) Lack of resources, (6) Weak business case and (7) Unclear development objectives. The most challenging activity in RE is change management.

Table 3. Requirements Elicitation Issues and Consequences [23]

Elicitation issue factors	**Issues**	**Consequences**
Changed	Management and political rules, acceptance criteria changes, unstable requirements, changes in nature of requirements overtime, user needs and understanding changes	Process overheads, re-work impacted project cost, quality, failure, requirements inconsistency, unusable
Communication	Articulation related, unaware of needs, mis-understanding amongst stakeholders, verbal and presentation skill, requirements-related, culture and perspective related, language barriers, change related	System failure, budget overruns, project failure, coordination issues, misunderstanding, poor definition of needs, broken communication links, imperfect specification, scope creep, abstract communication, low motivation, waste
Human	Conflicts, ambiguities amongst stakeholders, intra-group conflicts, communication, participation,	Withholding information, recognition failures, sabotage efforts

	cognition errors	
Knowledge	Understanding needs, domain related, problem analysis, knowledge sharing mechanisms	Project failure, low quality specifications, domain knowledge
Requirement	Issues related to documentation, knowledge, practice, prioritization, process, quality, requirements related, schedule, skills, technical, traceability, uncertainty, understanding	Repeated errors, re-work cost, budget overruns, poor quality systems, stakeholder dissatisfaction, project failure, process and tools mis-alignment, errors, uncertainties, poor requirements, high maintenance costs, frequent changes, conceptual inconsistency, incomplete domain knowledge, flaws in resultant system
Social and organization	Legal, policy and structure changes, complexity, cultural issues, time factor issues	Communication barriers, wait time, delays in GSD, legal consequences, affects trust
Scope	Scope ill-defined, over scoping	Requirement changes, quality issues, project delays and cancellations, customer expectations not met, communication gaps, wasted effort, requirements specifications not updated
Stakeholder	User-participation, stakeholder, staff	Poor specification correctness, completeness, consistency, risks, inefficiencies and duplication, communication problems, re-work, project delays, cost overruns, project failure
Tools, Techniques and Method	Tools, techniques, methods	Loss of information, requirements, delays in delivery, increased costs, decreased success rates, disorganized efforts, late discovery of requirements, lack of detailed approach, significant gap in RE theory, practice, requirements inconsistent and expectation mismatch

As a conclusion on the finding on RQ 1, it is undeniable that RE is very crucial in ensuring the success of a software development project. Overall finding can be summarize based on Figure 2.

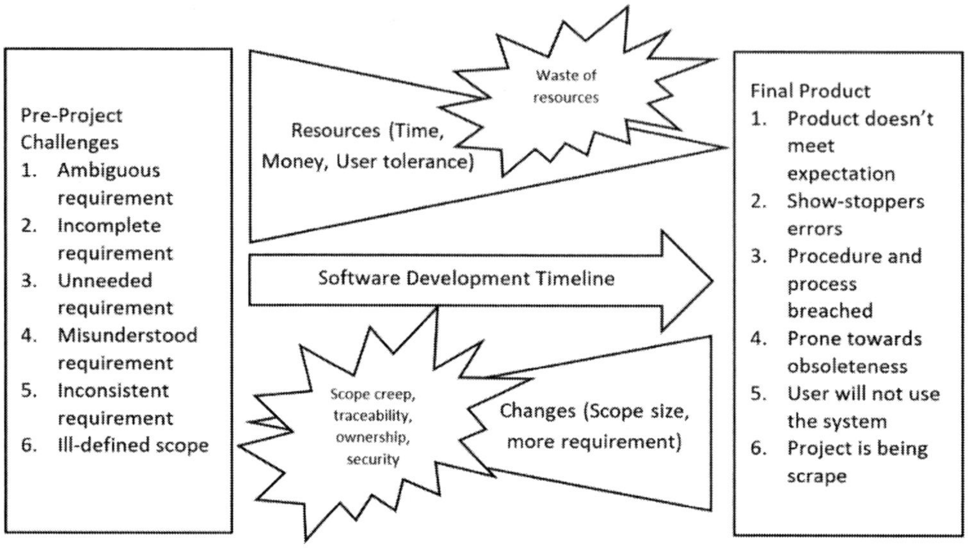

Figure 2. Summarization of RE Challenges and Impact in Pre-project vs End-product Expectation

5.2. RQ 2: How are RQ 1 problem are being eliminated or mitigated?

This section will be broken down into primarily two section, managerial solution and technical solution will be further broken down to requirement elicitation phase solution and requirements validation and verification phase solution. Managerial solution will be presenting solution for issues related to RE based on fixes on the project management level and technical will be based on the process and methods of RE.

This section will discuss the findings on the available solution on the managerial-level. Findings is presented based of proposed solution on both the Project Management and Requirements Engineering related activities.

This section will discuss more on the technical-based solution, majoring in RE. Firstly, lets discuss on how the main content of a software development RFT specification (user requirements) are being produce. Based on a study by [26] in the scope of Small and Medium Enterprise (SME), findings were presented on the ranking of most used requirement elicitation techniques based on the 15 different selection that are currently being used widely by the industry. It shows that most requirements are being collected based on Table 5.

Table 4. Score Summary of Identified RE Techniques [26]

Requirement Technique	**Level**
Interview	4.39
Joint Application Development (JAD)	4.35
ERD-Based Specification	4.23
Peer Review	4.17
Use-Case	4.17
Goal-Oriented	4.15
Check-List	4.11
Activity Diagram	4.09
Structured Natural Language (NL)	4.08
Misuse-Case	4.06
Software Requirement Specification	4.05
Brainstorming	4.02
Ethnography	2.18

Laddering	2.04
Observation	1.89

With the different method of requirement elicitation is being shown in the table above, Based on a study by [27], they have suggest multiple solution on the different issues that are faced by the four stages of RE; (1) Requirement Elicitation, (2) Requirement Analysis and Negotiation, (3) Requirement Specifications and (4) Requirement Validation.

Focusing on the research question, the next area of interest is the validation of requirements. It is crucial that in the effort of approaching a RFT, vendors must take the necessary action in preparing their winning presentation in order to convince the decision makers to pick them as the awarded vendor. Different method are being introduce by researchers and practitioners to give vendor the extra advantage when presenting their ideas to the decision maker.

For example, a method proposed by [28] suggested that LTesting can be utilize to generate a brief test plan that can be used in the vendor's proposal presentation. Other than that, it could also support decision-making for vendors to either compete or skip a RFP for providing service of software testing by producing an initial test report based on the specification provided. Based on the Figure 3 highlight the relevant proactive step in the stage of RFP. The detail description of the highlighted steps are; (1) The initiation step if where the RFT is being engage by the vendor, (2) LTesting will verify the test requirements (3) vendor will produce a brief test plan to be presented to the customer, (4) Contract is awarded, (5) A detail test plan is being produced and (6) Resource are being allocated to the newly awarded project.

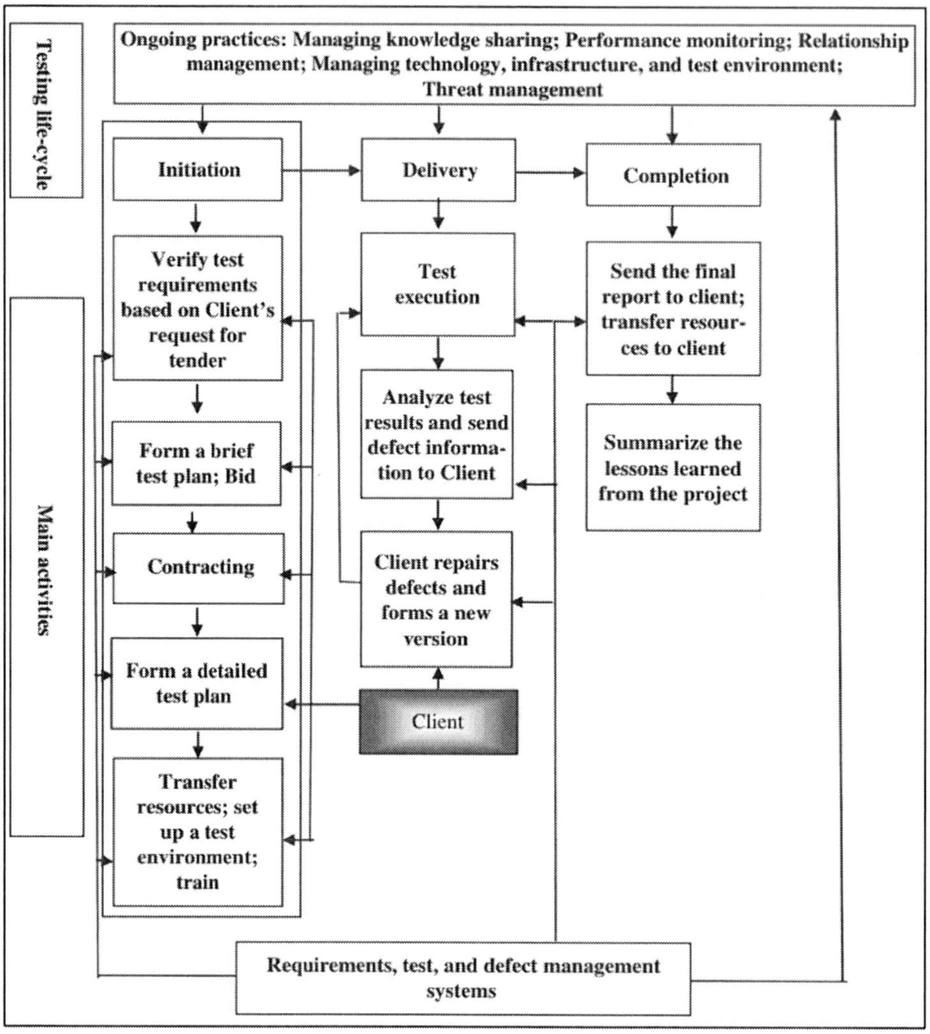

Figure 3. Generic, Dynamic eSourcing Life-cycle Model for the Provisioning of Testing Services [28]

With the assistance of the correct method or tool, vendor could shine brightly by being different and provide out-of-the-box information that other competing vendors can't. In the scope of this study, there are multiple techniques that are being proposed by researcher that could perform requirements validation but whether it can be utilize in the situation of RFT is still unknown. A study conduct by [29], their finding highlight which requirements validation technique that can be used in a low resource environment (refer to Table 10). It is suggested that prototyping is one of the most favorable tools that is suitable for the use in a RFT proposal presentation. But other vendors will also use prototyping and depending on the quality differences of those prototype will decide on the winning vendor.

Table 5. Comparison of Requirement Validation Techniques [29]

Resource Category	Requirements Inspections	Requirements Prototyping	Requirements Testing	Viewpoint Oriented Requirements Validation
Team Size	Large teams	Small teams	Large teams	Small teams

Cost	Costly	Less	Costly	Less
Company	Large companies	Small & Large companies	Large companies	Small & Large companies
Reuse	-	Yes	Yes	-
Customer Involvement	No	Yes	No	Yes

In conclusion, there research and solution being proposed by different researchers, but within the scope of solving this problem in RFT user specification validation is still green. To the authors best knowledge, the is no commercially available solution for the stated scope of problem.

5.3. RQ 3: Which OO artefacts considered as highly critical?

This section will present the finding related to RQ 3, which of the artefacts that are available in OO is considered as the most critical. A few studies has explicit shows the result in answer RQ 3. Based on a finding by [30], the most critical diagram or modelling tool in OO is Use Case and Class diagram in the scope of requirements validation. More detail in the Consistency aspect of requirements. Refer to Figure 4 for more detail on the publication findings.

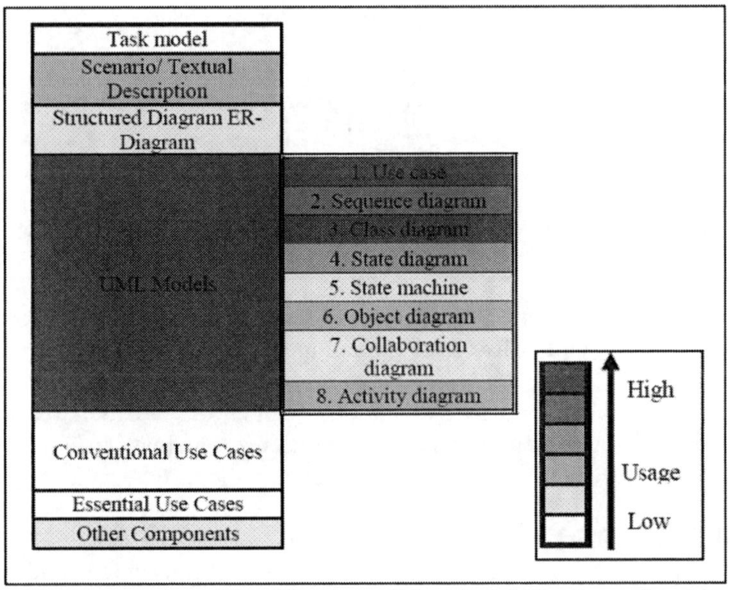

Figure 4. Heat Map Representation: Classification of the Model Used as a Semi-Formal Specification Approaches [30]

Another research has also exhibit detail finding on the issue of requirement consistency validation in OO. A study done by [31] summarize the following diagrams (ranked) are being widely use in requirement consistency check; (1) Class Diagram, (2) Interaction Diagram, (3) State Machine Diagram, (4) Use Case Diagram, (5) Activity Diagram, (6) Composite Structure Diagram and (7) Object Diagram as shown in Figure 5.

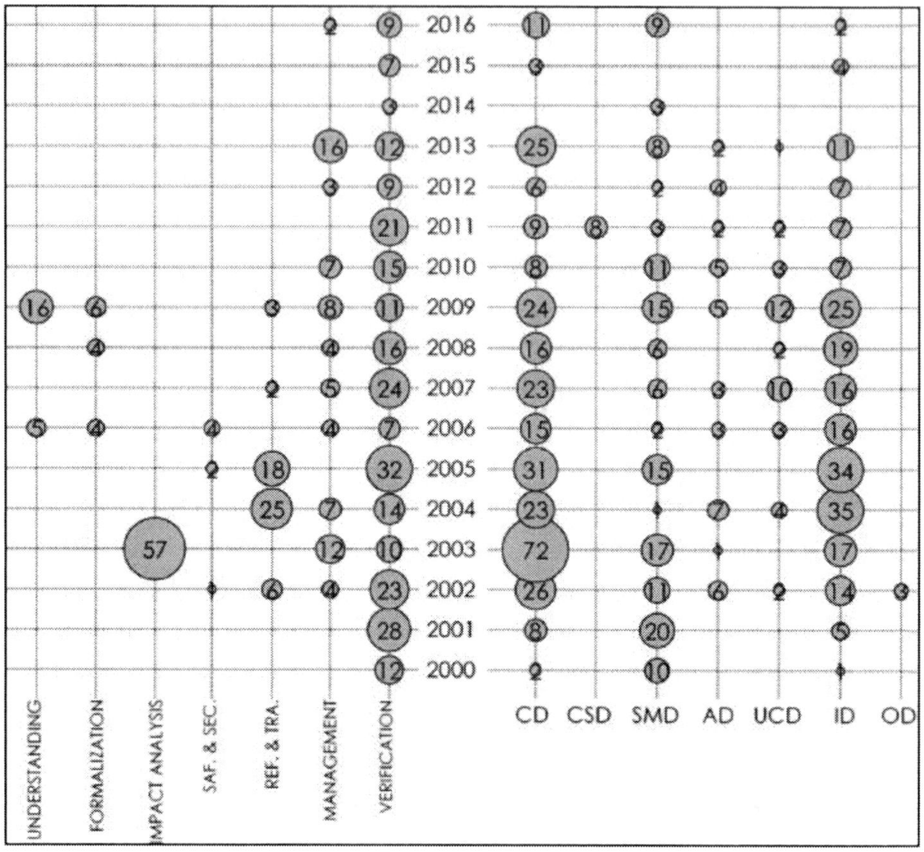

Figure 5. Summary of rules between UML diagrams presented over the years, coupled with diagrams and Software Engineering activities. [31]

From the findings shown, it is noted that the stated diagrams are being considered as relevant toward the continuation of this research. They will be used to evaluate in term of their relevancy in bridging the gap between user requirements specification and meta-requirements.

6. Conclusion and Future Work

There is still limited work in bridging the gap between user requirement and meta-requirement at present. The process of making sure the relationship of each artifacts that are related to UML and meta-requirement in its core field of software engineering is also another matter that needs to be taken seriously.

Our future work will focus on formulating an algorithm that will function as closing the gap between user requirements specification and meta-requirement. The algorithm itself must be able to work in conjunction to ISDT and OO UML environment taking into account every related artifact in the field. In order to achieve this vision, a systematic formulation and the right approach is needed to produce an algorithm that can work within the stated environment.

References

[1] P. Kapłański, "Ontology-Aided Software Engineering," pp. 1–161, 2012.
[2] The Standish Group, "Chaos Summary 2014," 2014.
[3] S. McCarthy, P. O'Raghallaigh, C. Fitzgerald, and F. Adam, "Social complexity and team cohesion in multiparty information systems development projects," *J. Decis. Syst.*, vol. 27, no. sup1, pp. 18–31, May 2018.

[4] D. Nickson, *Bids, Proposals and Tenders : Succeeding with effective writing.* British Informatics Society, 2012.

[5] Y. Lu and T. Käkölä, "An Information System Design Product Theory for integrated Order, Transportation and Warehouse Management Systems," *Proc. Annu. Hawaii Int. Conf. Syst. Sci.*, pp. 3717–3726, 2013.

[6] T. Käkölä, M. Koivulahti-Ojala, and J. Liimatainen, "An Information Systems Design Theory for Integrated Requirements and Release Management Systems," 2009.

[7] K. Pohl, Klaus, Pohl, and Klaus, "The three dimensions of requirements engineering: A framework and its applications," *Inf. Syst.*, vol. 19, no. 3, pp. 243–258, Apr. 1994.

[8] L. Wallace, M. Keil, and A. Rai, "Understanding software project risk: a cluster analysis," *Inf. Manag.*, vol. 42, no. 1, pp. 115–125, Dec. 2004.

[9] Y. K. Dwivedi *et al.*, "IS/IT project failures: A review of the extant literature for deriving a taxonomy of failure factors," in *IFIP Advances in Information and Communication Technology*, 2013, vol. 402, pp. 73–88.

[10] A. A. Alshazly, A. M. Elfatatry, and M. S. Abougabal, "Detecting defects in software requirements specification," *Alexandria Eng. J.*, vol. 53, no. 3, pp. 513–527, 2014.

[11] W. Al-Ahmad, K. Al-Fagih, and K. Khanfar, "A Taxonomy of an IT Project Failure: Root Causes.," *Int. Manag. ...*, vol. 5, no. 1, p. 14, 2009.

[12] R. Schmidt, K. Lyytinen, M. Keil, and P. Cule, "Identifying software project risks: An international Delphi study," *J. Manag. Inf. Syst.*, vol. 17, no. 4, pp. 5–36, Mar. 2001.

[13] F. B. Zainuddin, R. B. A. Arshah, R. B. Mohamad, R. B. Mokhtar, R. B. A. Hamid, and N. A. Bin Ahmad, "Reviewing the Challenge and Practices of Human Factor Involvement in Requirement Specification Validation," *Adv. Sci. Lett.*, vol. 24, no. 10, pp. 7322–7327, Oct. 2018.

[14] F. Binti Zainuddin, R. Bin Abdullah Arshah, and R. Binti Mohamad, "Software visual specification for requirement specification validation," in *ACM International Conference Proceeding Series*, 2018, pp. 66–71.

[15] K. Ewusi-Mensah, "Software Development Project Failures," in *Software Development Failures*, 2019.

[16] P. Brereton, B. A. Kitchenham, D. Budgen, M. Turner, and M. Khalil, "Lessons from applying the systematic literature review process within the software engineering domain," *J. Syst. Softw.*, vol. 80, no. 4, pp. 571–583, 2007.

[17] B. Kitchenham, O. Pearl Brereton, D. Budgen, M. Turner, J. Bailey, and S. Linkman, "Systematic literature reviews in software engineering - A systematic literature review," *Information and Software Technology*, vol. 51, no. 1. Elsevier B.V., pp. 7–15, 2009.

[18] D. L. Hughes, N. P. Rana, and A. C. Simintiras, "The changing landscape of IS project failure: an examination of the key factors," *J. Enterp. Inf. Manag.*, vol. 30, no. 1, pp. 142–165, 2017.

[19] T. O. A. Lehtinen, M. V. Mäntylä, J. Vanhanen, J. Itkonen, and C. Lassenius, "Perceived causes of software project failures - An analysis of their relationships," *Inf. Softw. Technol.*, vol. 56, no. 6, pp. 623–643, Jun. 2014.

[20] H. G. Chen, J. J. Jiang, G. Klein, and J. V. Chen, "Reducing software requirement perception gaps through coordination mechanisms," *J. Syst. Softw.*, vol. 82, no. 4, pp. 650–655, Apr. 2009.

[21] K. Wnuk, T. Gorschek, and S. Zahda, "Obsolete software requirements," *Inf. Softw. Technol.*, vol. 55, no. 6, pp. 921–940, Jun. 2013.

[22] M. Talhe, "Critical Requirements Engineering Errors Leads to Fails Software Project," *Educ. Rev. USA*, vol. 2, no. 2, Feb. 2018.

[23] Neetu Kumari.S, A. S. Pillai, N. Kumari.s, and A. S. Pillai, "A study on the software requirements elicitation issues - Its causes and effects," in *2013 3rd World Congress on Information and Communication Technologies, WICT 2013*, 2014, pp. 245–252.

[24] S. Besrour, L. B. A. Rahim, and P. D. D. Dominic, "A quantitative study to identify critical requirement engineering challenges in the context of small and medium software enterprise," in

2016 3rd International Conference on Computer and Information Sciences, ICCOINS 2016 - Proceedings, 2016, pp. 606–610.

[25] J. Miller, "A total benefits strategy is a valuable approach in HR outsourcing," *Employ. Relat. Today*, vol. 34, no. 4, pp. 55–61, 2008.

[26] S. Besrour, L. B. A. Rahim, and P. D. D. Dominic, "Investigating requirement engineering techniques in the context of small and medium software enterprises," *2016 3rd Int. Conf. Comput. Inf. Sci. ICCOINS 2016 - Proc.*, pp. 519–523, 2016.

[27] G. Huzooree, "A Systematic Study on Requirement Engineering Processes and Practices in Mauritius," *Int. J. Adv. Res. Comput. Sci. Softw. Eng.*, vol. 5, no. 2, pp. 40–46, 2015.

[28] Y. Lu and T. Käkölä, "A dynamic life-cycle model for the provisioning of software testing services," *Syst. Sci. Control Eng.*, vol. 2, no. 1, pp. 549–561, 2014.

[29] U. A. Raja, "Empirical studies of requirements validation techniques," *2009 2nd Int. Conf. Comput. Control Commun. IC4 2009*, 2009.

[30] M. Kamalrudin and S. Sidek, "A review on software requirements validation and consistency management," *Int. J. Softw. Eng. its Appl.*, vol. 9, no. 10, pp. 39–58, 2015.

[31] D. Torre, Y. Labiche, M. Genero, and M. Elaasar, "A systematic identification of consistency rules for UML diagrams," *J. Syst. Softw.*, vol. 144, pp. 121–142, Oct. 2018.

Acknowledgments
This work was supported in part by University Malaysia Pahang under Grant RDU RDU191802-2.

IOP Conf. Series: Materials Science and Engineering **769** (2020) 012051 doi:10.1088/1757-899X/769/1/012051

Meta-Requirement Mapping Model

Muhamad Idaham Umar Ong[1,*], Mohamad Ariff Ameeden[1]

[1]Faculty of Computing, College of Computing and Applied Sciences, University Malaysia Pahang, Gambang, 26300, Malaysia

* Correspondence: idaham@ump.edu.my

Abstract. The challenges faced by software vendor in competing for tenders in todays are massive and are desperately in need of intervention by the kind people of the research community. With very small resources available at their disposal, it is a daunting task them in the pursue to gain new software development projects each time new tenders are coming out. The objective of this publication is to propose an initial model that utilizes meta-requirement as a medium to validate the User Requirement Specification document that is included in the RFT. Looking forward, an algorithm will be developed for realizing the associations that was define in the propose model.

1. Introduction

For software vendors, software project doesn't just come falling from the sky. For high profile software project to exist, an intensive and complex task related to procurement is being executed by software purchaser. At the end of the journey, a Request for Tender (RFT) is produced. Now onwards, comes the challenging task faced by the software vendor to produce software project proposal that will be able to convince the software purchaser to choose them as the software provider. Other similar RFT terms that are being used to worldwide are Express of Interest (EOI), Request for Information (RFI), Request for Proposal RFP) or Request for Quotation (RFQ). Some of the common content in an RFT are; (1) Tender procedure, (2) Requirements and (3) Contract forms. In this publication, we will be focusing more on item no 2. Similar term that are being used to represent item no. 2 could be User Requirement Specification (URS), User Requirement, User Specification etc.

Once the RFT is being released, the next course of action will be the task of the software vendors to prepare a proposal that will then be evaluated by the software purchaser team in selecting the so-called best proposal. The main challenge of software vendor is in the proposal presentation session. They must make sure that the materials that are prepare for the session is enough to convince the panels but also must show that their solution is the best. For a software development tender, software vendors can use different approaches and methods available in the market in order to achieve their target. Based on the requirements that is included in the tender document, software vendors will be including a prototype of the requested software in order to show the evaluation panel the outcome. But at the end of the day, panels will also look for the uniqueness of the software vendor solution and the worthiness of the vendors themselves.

Meta-requirements as describe by researchers is the most basic form of requirement. Its main role in ISDT as the main artefact that describe the most basic function that a system must have, well in theory that is. The definition of ISDT is an initial draft of establish relationships between components of a system to achieve a specific result. Design theory must address the question of how to combine components and relationships to make subsystem and how to combine subsystems and relationships to make a systems [1].

The objective of this publication is to propose an initial model of formalizing USR and meta-requirement. The remaining five sections are as follows. Section 2 presents the related literature of the study. Section 3 describe briefly on a proposed model of mapping URS towards Meta-Requirement. Finally, Section 4 concludes and outlines some possible future works

2. Related Literature

This section will describe the related

2.1. Challenges Faced by Software Vendors in Request for Tender

Usually in the event of presenting software development project proposal for the evaluation panel, the use of prototype is a way to increase the panel's level of confidence towards the software vendor. Prototype are usually based on past successful project or it could an out of the wild combination of GUI and functions from a few different software projects. For a software vendor with strong background with similar past successful project, this decision of whether to take on the project or not would be a snap.

In a scenario where specifications are already included as part of the tender document that describe a detail list of thousands of specifications of the desired end-product, it is quite a daunting challenge where producing a highly convincing prototype that will meet and fulfil the specifications of the desired end-product. This is the task that will be led by Sales staff with the help and advice from a team [2] that consist but not limited to; (1) Sales staff, (2) Project Manager, (3) Technical specialists (Requirements Engineer), (4) Implementers, (5) Finance and commercial specialist, and (6) Legal experts. Some of the limitations [3]–[6] that are imposed to these team are; (1) Time. This is a case by case scenario whereby the timeframe of between the advertisement of the RFT and project proposal preparation is quite short, (2) Resources. In current situation where the is no guarantee that the software vendor will be awarded with the project, not much resources will be invested in partaking this effort and (3) Competitive. Depending on the RFT itself, the level competitiveness will vary. It is worth mentioning that even large will invest their time in small projects. Figure 1 dictates the three main target of requirements engineer.

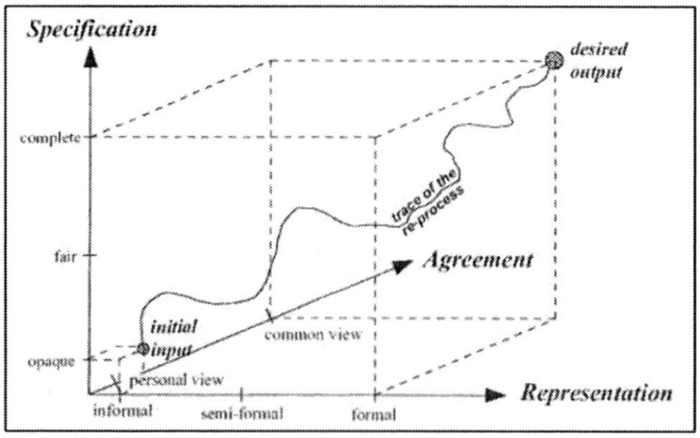

Figure 1. The Requirements Engineering (RE) Process within Three Dimension [7]

Based on the limitation shown on the previous paragraph, the task of Requirements Engineer has somewhat seemed absurd. In the scenario stated, Sales staff and Project Manager will be relying on the input from the requirement engineer to ensure that the proposal will be up-to-par with the expectation of the decision maker. Based on Fig. 1, requirements engineer main target is the achieve the desired output of a software development project by balancing the three main aspect of a software project [5]; (1) Specification, deals with the degree of requirements understanding at a given time (2) Representation, copes with the different representations (informal and formal languages, graphics, sounds etc.) used for expressing knowledge about the system. Within RE there are three categories of representations. and (3) Agreement, deals with the degree of agreement reached on a specification.

With the current attention of a Requirement Engineering is briefly illustrated in Fig. 1, it is acknowledged that with the combination of the situation of proposal presentation of a tender with the standard of work that needed to achieve by Requirement Engineers is quite challenging. Hundreds or even thousands of specifications that are included in the RFT document are also representing the almost finalize user requirement. Also, keep in mind that Requirement Engineers will need to abide with the general guideline of reviewing the quality aspect of the requirements, as shown in Table I.

Table 1. Characteristic of Good Quality Requirement [8]

No	Characteristic	Description
1	Cohesiveness	Showing the cohesiveness of a requirement and the other parts of the requirements that will be working together with that requirement
2	Completeness	The entire requirements specification should be complete and contain all relevant requirements and ancillary material depending on the standard and define template and also individual requirements should also be complete
3	Consistency	Requirements must be consistent to its source and other related requirements
4	Correctness	Individual requirements should be correct in consideration of semantics, organization business objective and requirements syntactic
5	Currency	Individual requirements must be up-to-date and not obsolete based on changes that will occur during the whole development timeline.
6	Customer/User Orientation	Individual requirements should be oriented around the needs of the customers and users if they are to be understandable and validatable
7	External Observability	Individual requirements should only specify behaviour or characteristics that are externally observable
8	Feasibility	Requirement must be able to be implemented within the scope of project budget, schedule, software, hardware and other constraints
9	Lack of Ambiguity	Requirements must be clear with no room for misinterpretation.
10	Mandatory	Questioning the nature of mandatory requirements and differentiating the requirements that is on the "wish list"
11	Metadata	Individual requirements should have that will characterizes each of them. Metadata can include but is not limited to acceptance criteria, allocation, assumptions, identification, prioritization, rationale, schedule, status, and tracing information
12	Relevance	Each requirement must be relevant and within the scope of business, application or component
13	Usability	Each requirement must be understandable and usable by stakeholders in the scope of control, evaluation, controlling cost and schedule, architectural significant, implementation and testable.
14	Validatability	Individual requirement must fulfil the need and desire of the primary stakeholders. Ensuring each requirement is accordance to the customer representatives wanted and needed.
15	Verifiability	Requirements must be verifiable against its source and associated

standards, guideline, template etc.

If the thousands of specifications that are included are being taken lightly, then software development work will projects suffer failure due to user requirements related problem such as changing requirements, unclear, ambiguous and unusable requirements and misunderstood user requirements and the failure to freeze requirements [9]–[13]. Cost and delivery overruns are resource or economic factors. Cost overruns and missed delivery can result in project termination. Publication by [14] includes increase of costs and timeline, actual project expenditures and delivery below the estimates and insufficient budget. These indirect factors may be the reason for the overrun. Failure can also be due to time and delivery below the estimates related to estimation issues in project management. Finally, the depletion of funds can result in project termination.

2.2. Meta-Requirement in Information System Design Theory (ISDT)

In looking for alternate solution of solving Requirements Engineer task of conducting earlier requirements validation of the tender specification, our view is focused on the utilization of meta-requirements. The source of meta-requirements is being heavily discuss in the area of Information System Design Theory. The breakdown of the origin of ISDT is that it originated from Design Research Science (DRS) and focused in the field of Design Theory (DT).

In 1992, a work by J. Walls, Widmeyer, & El-Sawy [1] is the first to include the phrase "Meta-requirement" in his work with relation to the field of ISDT. The future work that comes afterword in relation meta-requirement majorly in the field of ISDT, highly cited his work. Meta-requirement that was presented in his work is quite green, whereby future researchers has made improvement in the meta-requirement is being presented.

Käkölä, Koivulahti-Ojala, & Liimatainen [15]–[17] develops the product aspects of the ISDT for the class of Requirements and Release Management Systems (RRMS) with the involvement of meta-requirement. Advancement has been done thorough this work whereby the researcher introduces a different way of presenting meta-requirement. As part of the knowledge in DT, Meta-requirement could be the next big thing that is needed by Requirement Engineer as a guidance for them to do a quick check on the tender's user specification. It acts as the main theory of how the user specification should be. With such problem faced by Requirement Engineers, this would be an alternative solution and a chance for the software vendor to be unique in the sense of presenting their proposal.

3. Formalization of User Requirement Specifications

The word requirement plays a crucial role in the life of a software developer. Requirement is the heart and soul of a system. Development of a comprehensive and highly desirable software product will be achieved if the listed requirements are fulfilled. The task of studying, creating, checking and finalizing requirements are being uphold by person who called themselves Requirements Engineer. The process starts by collecting requirements, usually originated from the system owner with the intention of producing a comprehensive documentation that will contain a list of all agreed behaviours that the system will behave based on the boundaries that are defined, best known as User Requirements Specification (URS). But this is easier said than done. Requirements usually represent itself in an unstructured form and it is highly difficult for requirements engineer to settle down with their task even during the stage of delivering the finished system to the owner. Figure 2 show the list of activities that are par-take by Requirement Engineer and its close relation towards requirement validation.

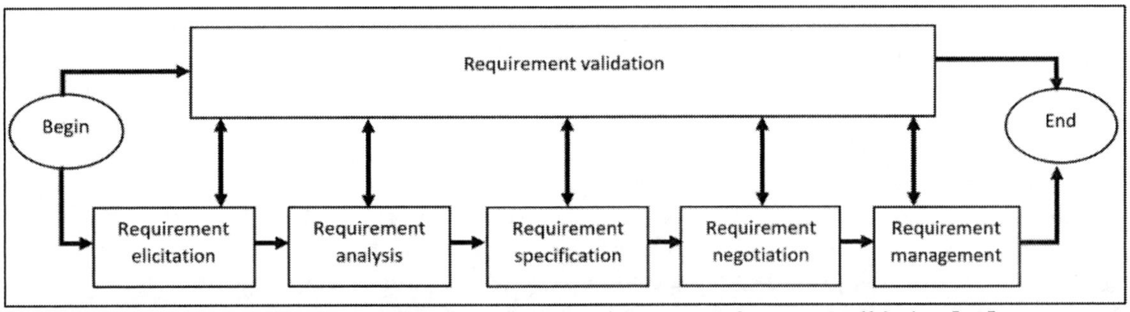

Figure 2. Requirement Engineering Relation to Requirement Validation [18]

URS document are provided earlier during the biding of tender. Different from URS that is being collect and analyse by the development team during the analysis phase, resources have already been secure through project acquisition and can be assigned during the project timeline. But for provided URS during biding of tender, normal software development organizations are not keen in investing to much resources in something that is not certain going to be theirs. The main concern of URS matter is the textual user specifications. These will be used in the meta-requirement mapping model and being considered as the crucial input of the model. Further discussion of the model will be describe in the upcoming paragraph.

3.1. Description on design theory

The main objective of introducing meta-requirements to URS is to propose an alternative solution in requirements validation. By systematically mapping the elements in both URS and meta-requirements in ISDT, the hope is to be able executing a higher abstract result of validating requirement at the earliest stages of the software project. With consideration on the consistency aspect of the URS and meta-requirements.

Based on the scope of research, it is decided that some parts of ISDT components is irrelevant. ISDT contains two parts; (1) Product hypotheses and (2) Process hypotheses. Part 2 of the ISDT component is removed from the propose meta-requirement analysis model based on the need of the model itself. Kernel Theories will also be taken out as the common ISDT will be to accommodate less than 5 kernel theories per-model. The initial model will rely mainly on the user specification product-aspect as the main input and doesn't concern on the user specification process-aspect, as shown in Figure 3.

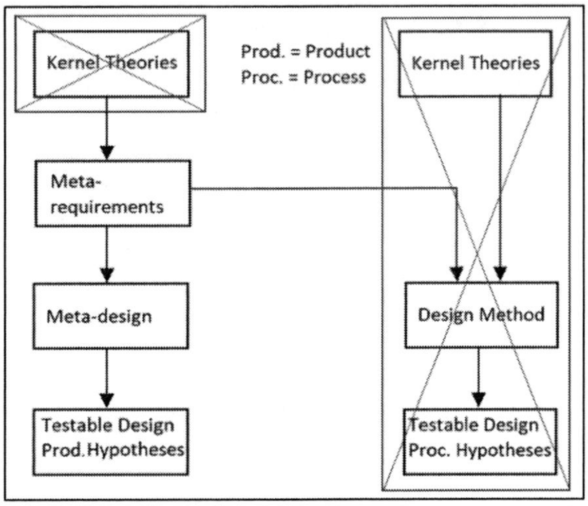

Figure 3. Excluded Components of Information System Design Theory

IOP Conf. Series: Materials Science and Engineering **769** (2020) 012051 doi:10.1088/1757-899X/769/1/012051

The remaining component of the ISDT will be carried forward in the purpose of mapping meta-requirements, meta-design and testable design product hypotheses to URS and also its requirements validation.

3.2. Mapping User specification in URS and Information System Design Theory
Based on previous paragraph, a proposed the model for the purpose of mapping meta-requirements and URS has been formulated. Figure 4**Error! Reference source not found.** illustrate the abstract level of the model. As mention in previous paragraph, the model will be focusing more on the User Specification in the USR itself. With association of "User Requirement Specification" and "Text" are generalization.

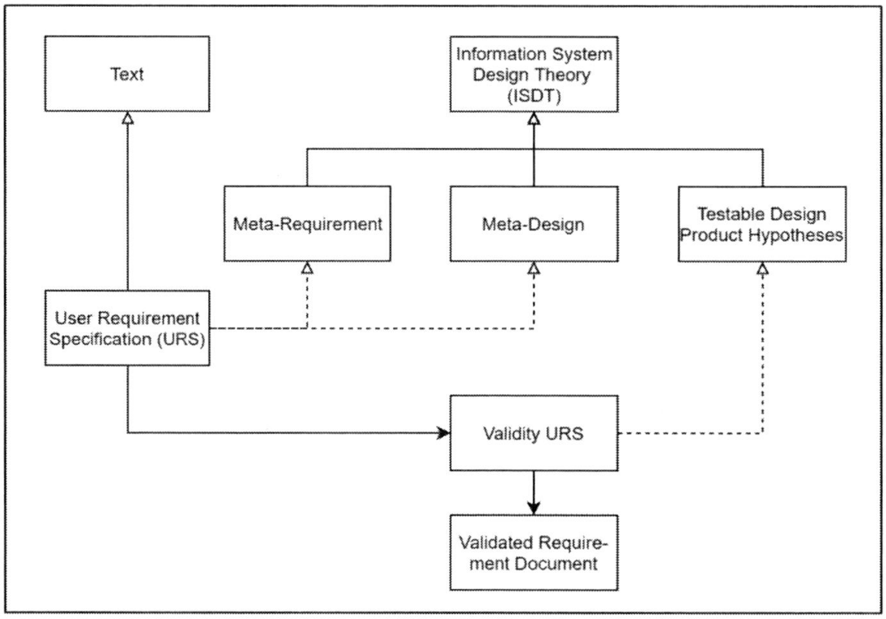

Figure 4. Meta-requirement and User Specification Mapping Model

Realization/Implementation of the "User Requirement Specification" towards the suitable "Meta-Requirement" and "Meta-Design" and "Validity User Requirement Specification" towards "Testable Design Product Hypotheses" will ensure that the model would able to achieve the objective of "Validated Requirement Document". The three main component of ISDT in Figure 4 will be interrelated, where a set of Meta-Requirement will be mapped to a set of Meta-Design where it could be uniquely mapped one-to-one or more. And finally, the meta-designs will also be mapped to the Testable Design Product Hypotheses in a similar fashion as Meta-Requirement to Meta-Design. This will enable the validation of URS in the sense of conformity and consistency of the URS vs the ISDT.

All of this will be done based on the development of an algorithm with the assistance of Object Constraint Language (OCL) that will support the needed mapping of the proposed model. Model will be focusing towards the measuring the consistency of the User Requirement Specification with the assistance of ISDT.

4. Discussion and conclusion

The proposed model of URS and Meta-requirement mapping is an effort towards producing an earlier validation of URS in the scope of consistency. With the current problems faced by software vendor in the event of an RFT, it is with high hope that the proposed model will become the pioneer of future research in utilizing meta-requirements and its contribution in the knowledge of requirement engineering.

Future research will be focusing more on the effort of realizing the correct method of creating those association fully executable. Keeping our eye on the horizon, the method of Object Constraint Language (OCL) is the method of choice and appropriately lean toward achieving the research objective. Further research will be conducted on the implementation of OCL with the purpose of realizing all the define associations in the proposed model will be next course of action.

References

[1] J. G. Walls, G. R. Widmeyer, and O. A. El Sawy, "Building an Information System Design Theory for Vigilant EIS," *Inf. Syst. Res.*, vol. 3, no. 1, pp. 36–59, 1992.
[2] D. Nickson, *Bids, Proposals and Tenders : Succeeding with effective writing.* British Informatics Society, 2012.
[3] Y. Lu and T. Käkölä, "An Information System Design Product Theory for integrated Order, Transportation and Warehouse Management Systems," *Proc. Annu. Hawaii Int. Conf. Syst. Sci.*, pp. 3717–3726, 2013.
[4] T. Käkölä, M. Koivulahti-Ojala, and J. Liimatainen, "An Information Systems Design Theory for Integrated Requirements and Release Management Systems," 2009.
[5] M. Fatehah and V. Mezhuyev, "Design and process metamodels for modelling and verification of safety-related software applications in smart building systems," in *ACM International Conference Proceeding Series*, 2018, pp. 60–64.
[6] M. I. U. Ong, M. A. Ameedeen, and I. E. Kamarudin, "Meta-requirement method towards analyzing completeness of requirements specification," in *Advances in Intelligent Systems and Computing*, 2019, vol. 881, pp. 444–454.
[7] K. Pohl, Klaus, Pohl, and Klaus, "The three dimensions of requirements engineering: A framework and its applications," *Inf. Syst.*, vol. 19, no. 3, pp. 243–258, Apr. 1994.
[8] D. Firesmith, "Specifying good requirements," *J. Object Technol.*, vol. 2, no. 4, pp. 77–87, 2003.
[9] L. Wallace, M. Keil, and A. Rai, "Understanding software project risk: a cluster analysis," *Inf. Manag.*, vol. 42, no. 1, pp. 115–125, Dec. 2004.
[10] Y. K. Dwivedi *et al.*, "IS/IT project failures: A review of the extant literature for deriving a taxonomy of failure factors," in *IFIP Advances in Information and Communication Technology*, 2013, vol. 402, pp. 73–88.
[11] A. A. Alshazly, A. M. Elfatatry, and M. S. Abougabal, "Detecting defects in software requirements specification," *Alexandria Eng. J.*, vol. 53, no. 3, pp. 513–527, 2014.
[12] W. Al-Ahmad, K. Al-Fagih, and K. Khanfar, "A Taxonomy of an IT Project Failure: Root Causes.," *Int. Manag. ...*, vol. 5, no. 1, p. 14, 2009.
[13] R. Schmidt, K. Lyytinen, M. Keil, and P. Cule, "Identifying software project risks: An international Delphi study," *J. Manag. Inf. Syst.*, vol. 17, no. 4, pp. 5–36, Mar. 2001.
[14] K. Ewusi-Mensah, "Software Development Project Failures," in *Software Development Failures*, 2019.
[15] A. Salo and T. K. Käkölä, "Groupware support for requirements management in new product development," *J. Organ. Comput. Electron. Commer.*, vol. 15, no. 4, pp. 253–284, 2005.
[16] Y. Lu and T. Käkölä, "An information system design product theory for the abstract class of integrated requirements and delivery management systems," *Proc. Annu. Hawaii Int. Conf. Syst. Sci.*, pp. 3677–3686, 2014.
[17] P. Forselius and T. Käkölä, "An information systems design product theory for software project estimation and measurement systems," *Proc. 42nd Annu. Hawaii Int. Conf. Syst. Sci. HICSS*, pp. 1–10, 2009.
[18] S. Maalem and N. Zarour, "Challenge of validation in requirements engineering," *J. Innov. Digit. Ecosyst.*, vol. 3, no. 1, pp. 15–21, Jun. 2016.

Acknowledgments

This work was supported in part by University Malaysia Pahang under Grant RDU RDU191802-2.

Factors Determination MOOCs Continuance Intention: A Proposed Conceptual Framework

N S Abdul Rahman, N S Wan Zainal Adli, A F Mat Raffei, and N S N Ismail

Faculty of Computing, Universiti Malaysia Pahang, 26300 Kuantan, Pahang, Malaysia

Email: shamsiah@ump.edu.my

Abstract. Massive Open Online Courses (MOOCs) recently start to gain its popularity at most universities and the number of MOOC has increased. Despite its popularity, previous study shown that low course completion rate and limited amount of studies in examining antecedents which influence students' continuation intention in using MOOCs. Thus, this study aims to propose a conceptual framework to figure out the factors determinant of MOOCs continuance intention among students in Malaysian higher education. By reviewing the related studies, factors namely perceived ease of use, perceived usefulness, attitude, motivation and continuance intention were identified.

1. Introduction

In recent years, MOOCs have emerged tremendously. Since MOOCs are mostly open and free to all, they have attracted millions of users' enrollment [1], which overcomes the structural limitation of traditional education system [2], improve the modern education, disseminate science around the globe and equip new informant of data [3-4]. In Malaysia, the implementation of MOOCs has been outlined in the preliminary Malaysian Education Blueprint for Higher Education under the 10th chapter called "Globalized Online Learning". As mentioned by Ab Jalil et al. [5] in Mohamad and Ab Rahim [6], Ministry of Education advocates Malaysia's intention to leverage on MOOCs as a way to improve quality and widen access to education.

In past decades, MOOCs have become one of the education revolution with a massive number of participants since its existence in 2012. This is proven as in the year of 2012, a total of 160000 enrolled in the courses. However, how far is the student's continuation intention to use MOOCs in perspective of learning has not yet been identified [7]. Even though the number of the students who enrolled in 2012 is 160000, the total of students considered completed the courses are only 20000. It shows that the success rate of the student completion is only 1/8 whereas the drop-out rate is the other 7/8. Based on the past reviews of MOOCs are focused on identifying trends and categorizing MOOCs studies rather than the purpose of using MOOCs in the midst of students [7] and it is infrequent for continuation intention in using MOOCs being the point for this research [8]. Thus, the continuance intention use of MOOCs is a huge problem as nowadays trends is to learn at a scale is more important than the education itself [9] and this could be contributed to the high number of the drop-out rate among the students who enrolled.

Although MOOCs is gaining its popularity and it is also quite popular among the Malaysian students and worldwide students, there is only a minimal number of study that investigate the antecedents which effect the adoption of MOOCs among the students [10]. As UMP students also consist of different races and religion, previous researcher also suggest to include different cultural background is interesting as it can help to generalize the findings [10-11]. Recently, the different factors related to attrition has been highlight as the focus to research on MOOCs [12] and there are fewer studies that have been done to figure out the factors [13]. This prove that there is not much research done about factors that influence students to use MOOCs for learning and this is why this topic is chosen to research on. Therefore, this research aims to propose a conceptual framework which capable to determine which antecedents will affect MOOCs continuance intention among students at universities. This paper is organized into several parts as follows: part two: describes about MOOCs, Technology Acceptance Model (TAM) and related work in prior study, section three: describes about research methodology, section four: describes about proposed conceptual framework and development of hypotheses, section five: describes about discussion, and last section number six: conclusion and recommendation.

2. Literature Review

2.1 MOOCs

In order to support the information age and revolution changes among society, several alternatives have been made in order to increase the popularity of higher education sector [14] and MOOC is one of the example for that. Since 2012, MOOCs has gained so much popularity and it is one of the best platform as it doesn't charge any fee, it can be used by all generations to study at any time, places and they learn by themselves using the modules that available [11]. While for every person that completes the modules, they can obtain a certificate with a minimal fee is charge for it [15]. It's been 11 years since MOOCs is introduced to the public and 7 years since MOOCs is popular all around the world. The growth of MOOC is increasing rapidly across the year from 2012 to 2018 for the number of subjects that is offered to the students. However, as the number of students that use MOOCs is increasing, the rate of the dropout is also increasing, thus this brings the biggest question of what is the reasons behind the increasing rate of dropout students. Two kinds of MOOCs were introduced which is known as xMOOCs and cMOOCs. xMOOCs which is an instruction based with all contents are predefined by the instructor and have a stronger tie with older established educational technology approaches [9, 11, 16]. Meanwhile, cMOOCs is a connectivity based with the student content and course materials is derived during the course [11, 16]. MOOCs Malaysia is found to be the following the x-MOOCs as most of Malaysian students use MOOCs under the supervision of their lecturer and also register under their university account.

2.2 TAM

Since this research is aiming to determine the antecedents related to students' continuance intention in using MOOCs in learning, TAM is found to be one of the simplest yet versatile model to show the relationship of every factors that is chosen as in finding the factor that influence MOOCs. TAM is the model that is chose to use among others technology acceptance theories as TAM is the frequently adopted theory and most influential theory for individual's acceptance information system [17]. Davis [18] is the first to proposed the TAM which contains the antecedents mainly for user motivation such as perceived usefulness, attitude towards technology and perceived ease of use. The example for the original model is shown in the Figure 1. In a diverse settings and samples, TAM has been found reliable to use [19-20]. However, previous researcher has recommended for new research to provide a stronger model of context and setting with extend TAM with other variables [21-22].

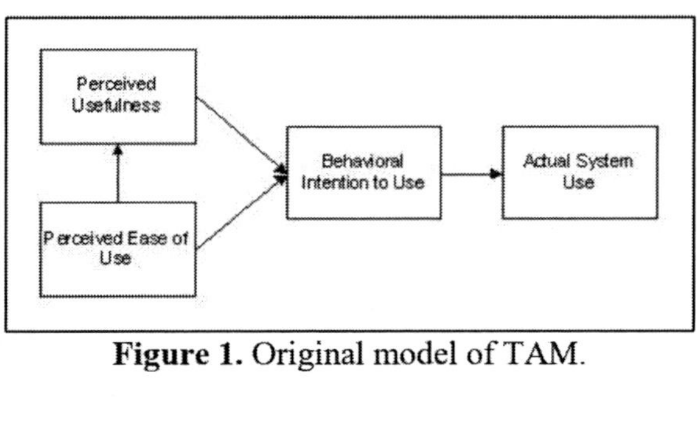

Figure 1. Original model of TAM.

2.3 Related Work in Prior Research

In order to identify the determinant antecedents related to students' continuance intention in using MOOCs, a systematic literature review was conducted where 15 publications from Science Direct, IEEE, Web of Science and Springer database were found to be related to this research. The literature founds (refer Table 1) that TAM was the most frequently and predominantly applied model in confirming the utilization of information system in various field [23].

From the prior studies, it shows that the most frequently introduced variables related to continuance intention are perceived usefulness (PU) [8, 10, 14, 24-31], perceived ease of use (PEOU) [10, 14, 24-26, 28-29, 31], attitude (AT) [10, 13, 24, 26, 31] and motivation (M) [10, 11, 13, 29]. Thus, this four variables (PU, PEOU, AT and M) will use for this research as dependent variables to identify the continuance intention for using MOOCs for learning.

Table 1. Related work by prior research.

No.	Author	Aim	Theory	Factors
1.	Scherer et al. [24]	To identify how far does the teacher accept the technologies in their teaching practice.	TAM	PU, PEOU, AT, behavioral attention.
2.	Joo et al. [25]	To examine students' motivation in using K – MOOCs.	TAM	Self – determination, PEOU, PU, satisfaction, continuance intention.
3.	Jung & Lee, [14]	To investigate learners' engagement and persistent in using MOOCs.	TAM	Academic self – efficacy, teaching presence, PU, PEOU, learning engagement, learning persistent.
4.	Khan et al. [11]	To examines the antecedents that influence students' acceptance of using MOOCs.	TTF	Social recognition, social influence, perceived autonomy, perceived competence, behavioral intention, reputation.
5.	Lemay et al. [26]	To predict behavioral intention and use of technology.	TAM	AT, behavioral intention, facilitating conditions, fidelity, PEOU, PU, self-efficacy, subjective norms.
6.	Tawafak et al. [8]	To develop a model in improving assessment techniques and student satisfaction.	TAM	Effectiveness, behavioral intention, support assessment, academic performance, student satisfaction, continuance intention, PU
7.	Mohamad et al. [27]	To propose the level of users' Internet Self-efficacy would affect the factors in identifying students continuance intention.	-	Usefulness, enjoyment, internet self-efficacy, continuance intention.
8.	Fianu et al. [32]	To examine the antecedents which influence MOOC adoption in Ghana universities.	UTAUT	Performance expectancy, effort expectancy, social influence, facilitating conditions, computer self-efficacy, system quality, instructional quality, MOOC usage, MOOC usage intention.
9.	Khan et al. [11]	To find the antecedents that influence the usage of MOOCs adoption.	TTF	Social motivation, social recognition, social influence, self-determination, behavioral intention
10.	Jun & Lee [28]	To investigate learners' engagement and persistence in MOOCs.	TAM	Learning engagement, learning persistence, self-efficacy, teaching presence, perceived usefulness, perceived ease of use.
11.	Wu & Chen [10]	To investigate the continued use of MOOCs.	TAM, TTF	PEOU, PU, AT, M, openness, reputation, continuance intention.
12.	Hakami et al. [29]	To identify students' perspectives on factors that will be affecting the continuance intention towards MOOCs.	TAM	PU, PEOU, continuance intention, perceived reputation, intrinsic motivations.
13.	Junjie [30]	To determine influential factors on continuance intention for online learning using MOOCs.	ECM	Confirmation, PU, satisfaction, continuance intention.
14.	Yang & Su [31]	To investigate the correlation between perceptions, behavioural intentions, and actual behaviour in practicing course distributed using MOOCs platform.	TAM, TPB	PEOU, PU, AT, subjective norm, perceived behaviour control, behaviour intention.
15.	Zhou [13]	To research on antecedents that affects student's decision to use MOOCs.	TPB	AT, M, behavioral control, subjective norms, controlled.

Notes:
TTF - Task-technology fit **ECM** - Expectation confirmation model **TPB** - Theory of planned behaviour **UTAUT** - Unified theory of acceptance and use of technology **PEOU** – Perceived ease of use **PU** – Perceived usefulness **AT** – Attitude **M** - Motivation

3. Methodology

As suggested by Fisher et al. [33], the framework for evaluating MOOCs continuance intention among university students is derived from a systematic literature review by following guidelines from Sekaran [34]. The guideline consists of five components which are: (i) identifying and defining variables; (ii) discussing the relationships between variables; (iii) specifying causal relationship between variables; (iv) explaining the reason of the relationship existence and (v) developing a schematic diagram.

Various studies have reported the most commonly used variables and their relation with MOOCs continuance intention. The chosen variables that are found to be applicable for this research are then defined and the relationship between the variables explained in the literatures are also discussed in order to construct the hypotheses for this study. The hypotheses guided the development of the proposed conceptual framework to outline the correlation between all the antecedents with the continued use MOOCs.

4. Conceptual Framework and Hypotheses

The conceptual framework is proposed and developed according to constructs and relevant causal links taken from previous studies. From a systematic literature review done on the factors that contribute to acceptance of MOOCs, the constructs that display most frequent significant correlation with continued use MOOCs intention are from the TAM which are PU, PEOU, followed by attitude, and motivation factor [29, 30, 35, 36]. Therefore, this research decides to develop the framework based on TAM. Figure 2 illustrates the proposed framework for evaluating MOOCs continuance intention in this research. Next, the relationship between the chosen variables are discussed.

4.1 Perceived Usefulness (PU)

Davis [18] derived the concept of perceived usefulness from TAM. Perceived usefulness is also known as the users' assumption according to new technology will aid users to know better or perform effectively [18]. By using selective system, it would increase and improve job performance and would reflect the users' subjective assessment [37]. Perceived usefulness considered as an antecedent that was proven that effecting attitude and according to several studies, it also has a significant effect on the continuance use [1, 38-39]. Thus, perceive usefulness is one of the important variable which suggest ways to help the learners recognize and enhance their learning persistence and the value of learning. Given these considerations, we intend the first hypotheses as following:

H1. Perceived usefulness appropriately influences students' continued use intention in using MOOCs.

4.2 Perceived Ease of Use (PEOU)

Perceived ease of use considered as a standard level for individual who believed that when technology is used, they are freely from doing efforts [24]. For this study context, either directly or indirectly, perceived ease of use keep affect the learners' intention in accepting MOOCs. Much research has implemented TAM in e-learning and has discovered that perceived ease of use have considerable effects individual continued use intention. According to Wu & Chen [10] in their study on MOOCs in China, perceived ease of use in using MOOCs affects the learners' learning persistence indirectly through the perceived usefulness while learning persistence consequently considered affected by perceived usefulness. Given these considerations, we intend the second hypotheses as following:

H2. Perceived ease of use appropriately influences students' continued use intention in using MOOCs.

4.3 Motivation (M)

By referring to the originality of TAM model, it is appropriate in order to analyze the motivation variable that may interfere the accepting of MOOCs among students'. The motivation variable has been addressed in the MOOCs situation and was found to be a predictor of the student acceptance and continuance of MOOCs [29]. In the rapid phenomenon of massive open online learning, motivation was considerable as great value to determine the antecedents that influence student acceptance to use MOOCs in learning. In keeping with this, we intend the third hypotheses as following:

H3. Motivation appropriately influences students' continued use intention in using MOOCs.

4.4 Attitude (AT)

Attitude refers to the degree of the individual either perceived positive or negative feeling that is related to MOOCs. User often thinks that they have control over the use of it and this leads to the intention of use is increasing due to their attitude that they perceived. Past research have proved the continuance intention to use technology is predicted by the attitude [10, 13, 31]. Thus, we intend the fourth hypotheses as following:

H4. Attitude appropriately influences students' continued use intention in using MOOCs.

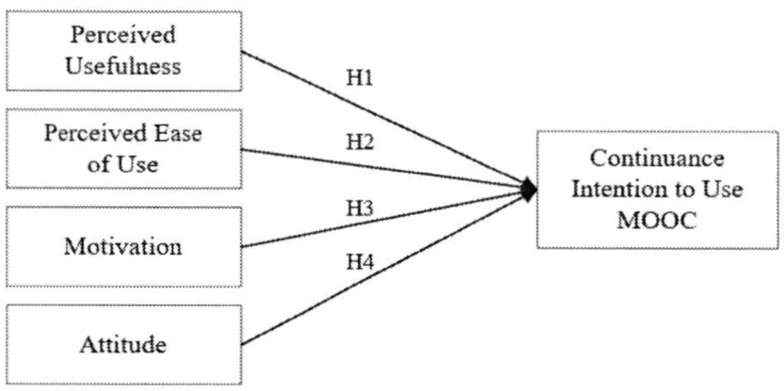

Figure 2. Proposed Conceptual Framework.

5. Discussion

Theoretically, this study contributes to the general knowledge by enhancing the understanding regarding MOOCs by proposing a conceptual framework on students' continued use intention in using MOOCs. Determinant antecedents in proposed framework covered are perceived usefulness, perceived ease of use, motivation, attitude and continued use intention. Both primary TAM constructs namely perceived usefulness and perceived ease of use are generally proven as the highest influential predictors for students' continued use intention in using MOOCs even in different context. This is not a surprise for an already well-recognized acceptance model. Our results share a number of similarities with prior studies which indicated that perceived usefulness proved to be more imperative predictor for students' continuance intention towards MOOCs [29, 30, 40]. The result also consistent with previous studies [10, 29, 31, 35] where perceived ease of use has been found points to be as a key indicator towards students' continued use intention in using MOOCs.

Our findings corroborate with previous results in the studies conducted by Azami and Ibrahim [36] found that attitude has a significant positive between students' continued use intention towards MOOCs. The same results were found in the studies conducted by Wu and Chen [10] and Yang and Su [31] whom all concluded that attitude is a significant determinant to students' continuance intention towards MOOCs. Hence, this will have supported one of the hypothesis in our proposed framework. Lastly, motivation was identified as one of the determinant antecedent for students' continued use intention in using MOOCs. This is in good agreement with previous findings [29] which marked that motivation has been conclude as one of important antecedents of students' continued use intention in using MOOCs.

6. Conclusion and Recommendation

Our intention in this paper is to propose a conceptual framework from determinant factors on MOOCs continuance intention. The factors covered are perceived usefulness, perceived ease of use, attitude and motivation which shown as the most prominent indicator from a systematic literature review done to identify the common MOOCs continuance intention factors. However, it is recommended to incorporate other factors to obtain more comprehensive understanding on students' continued intention to use MOOCs. The proposed framework hopefully will enrich the existing literatures regarding MOOCs continuance intention and provide a guideline for higher education institution and MOOCs designer in developing successful MOOCs.

Research on MOOCs continuance intention has received great attention all over the world. However, there are still a constrained number of potential exploration in the area of MOOCs that explored and examined the various antecedents which may affect continued use MOOCs intention particularly in Malaysia and to be specifically, focuses on MOOCs features. Therefore, there is a need to study in bringing the gap of the research in MOOCs continuance. As the implementation of MOOCs in Malaysia academia is still in its infancy phases, further research in this area is needed.

7. Acknowledgement

This research was supported by grant number RDU1703288, Universiti Malaysia Pahang.

References

[1] K Alraimi, H Zo, A Ciganek 2015 Understanding the MOOCs continuance: The role of openness and reputation *Computers & Education* 28-38
[2] M Dennis 2012 The impact of MOOCs on higher education *College & University* **88** 24-30
[3] S Eichhorn, G W Matkin 2016 MOOC, big data and education research *New directions for Institutional Research* **167** 27-40
[4] J Reich 2014 Rebooting MOOC research *Science* **347** 34-35
[5] H Ab Jalil, A Ismail, N Bakar, N A Azizan, K Nasir 2016 Evaluation of Malaysia Pilot MOOC (Final Report) Serdang, Malaysia: CADe UPM
[6] M Mohamad, M K I Abdul Rahim 2018 Factors Affecting MOOCs Continuance Intention in Malaysia: A Proposed Conceptual Framework *Journal of Humanities, Language, Culture and Business* **2** 61-72
[7] R Deng, P Benckendorff, D Gannaway 2019 Progress and new directions for teaching and learning in MOOCs Computers and Education 129 48–60
[8] R M Tawafak, A B Romli, A R B Arshah 2018 Continued Intention to Use UCOM: Four Factors for Integrating with a Technology Acceptance Model to Moderate the Satisfaction of Learning. *IEEE Access* **6** 66481–66498.
[9] A S Sunar, R A Abbasi, H C Davis, S White, N R Aljohani 2018 Modelling MOOC learners' social behaviours *Computers in Human Behavior* 1–12
[10] B Wu, X Chen 2017 Continuance intention to use MOOCs: Integrating the technology acceptance model (TAM) and task technology fit (TTF) model *Computers in Human Behavior* **67** 221–232
[11] I U Khan, Z Hameed, Y Yu,, T Islam, Z Sheikh, S U Khan 2018 Predicting the acceptance of MOOCs in a developing country: Application of task-technology fit model, social motivation, and self-determination theory *Telematics and Informatics* **35** 964–978
[12] W Al-Rahmi, A Aldraiweesh, N Yahaya, Y Kamin, A M Zeki 2019 Massive Open Online Courses (MOOCs): Data on higher education *Data in Brief* **22** 118–125
[13] M Zhou 2016 Chinese university students' acceptance of MOOCs: A self-determination perspective *Computers and Education* **92–93** 194–203
[14] Y Jung, J Lee 2018 Learning Engagement and Persistence in Massive Open Online Courses (MOOCS) *Computers and Education* **122** 9–22
[15] Y J Joo, H J So, N H Kim 2018 Examination of relationships among students' self-determination, technology acceptance, satisfaction, and continuance intention to use K-MOOCs *Computers and Education* **122** 260–272
[16] K F Hew, W S Cheung 2014 Students' and instructors' use of massive open online courses (MOOC): Motivations and challenges *Educational Research Overview* **12** 45-58
[17] F H Chandio, Z Irani, M S Abbasi, H A Nizamani 2013 Acceptance of online banking information systems: An empirical case in a developing economy *Behaviour & Information Technology* **32** 37-41
[18] F Davis 1989 Perceived usefulness, perceived ease of use, and user acceptance of information technology *Management Information Systems Quarterly* **13** 319-340

[19] N S A Rahman, A F M Raffei, W Al-Rahmi 2019 Understanding university students' behavioral intention to use social media for teaching and learning *International Conference on E-Learning* Universiti Malaysia Sarawak 21 – 22 August 2019 60-73
[20] W Al-Rahmi, N Alias, M S Othman, A I Alzahrani, O Alfarraj, A A Saged, N S A Rahman 2018 Use of E-Learning by University Students in Malaysian Higher Educational Institutions: A Case in Universiti Teknologi Malaysia *IEEE Access* **6** 14268-14276
[21] L Li 2010 A Critical Review of Technology Acceptance Literature Referred Research Paper PhD
[22] K Praveena, S Thomas 2014 Continuance Intention to Use Facebook: A Study of Perceived Enjoyment and TAM *Bonfring International Journal of Industrial Engineering and Management Science* **4** 25-29
[23] C-C Chang, S-W Hung, M-J Cheng, C-Y Wu 2015 Exploring the intention to continue using social networking sites: The Case of Facebook *Technological Forecasting and Social Change* **95** 48–56
[24] R Scherer, F Siddiq, J Tondeur 2019 The technology acceptance model (TAM): A meta-analytic structural equation modeling approach to explaining teachers' adoption of digital technology in education *Computers and Education* **128** 13–35
[25] Y J Joo, H J So, N H Kim 2018 Examination of relationships among students' self-determination, technology acceptance, satisfaction, and continuance intention to use K-MOOCs *Computers and Education* **122** 260–272
[26] D J Lemay, M M Morin, P Bazelais, T Doleck 2018 Modeling Students' Perceptions of Simulation-Based Learning Using the Technology Acceptance Model *Clinical Simulation in Nursing* **20** 28–37
[27] M Mohamad, M K I Abdul Rahim 2018 Factors Affecting MOOCs Continuance Intention in Malaysia: A Proposed Conceptual Framework *Journal of Humanities, Language, Culture and Business* **2** 61-72
[28] Y Jung, J Lee 2018 Learning engagement and persistence in massive open online courses (MOOCS) *Computer Education* **122** 9–22
[29] N Hakami, S White, S Chakaveh 2017 Identifying the Motivational Factors that Influence Learners' Intention to Continue to Use Arabic MOOCs *International Journal of Management and Applied Science* **3** 53-61
[30] Z Junjie 2017 Exploring the factors affecting learners' continuance intention of MOOCs for online collaborative learning: An extended ECM perspective *Australasian Journal of Educational Technology* **33** 123-135
[31] H H Yang, C H Su 2017 Learner Behaviour in a MOOC Practice-oriented Course: In Empirical Study Integrating TAM and TPB *International Review of Research in Open and Distributed Learning* **18** 35-63
[32] E Fianu, C Blewett, G O A Ampong, K S Ofori 2018 Factors Affecting MOOC Usage by Students in Selected Ghanaian Universities *Education Science* **8** 1-15
[33] C Fisher, J Buglear, D Lowry, A Mutch, C Tansley 2007 Researching and writing a dissertation; A guidebook for business students, 2nd ed. Essex, England: Prentice Hall
[34] U Sekaran 2003 Research methods for business: A skill building approach, 4th ed. Singapore: John Wiley & Sons
[35] L-Y-K Wang, S-L Lew, S-H Lau, M-C Leow 2019 Usability factors predicting continuance of intention to use cloud e-learning application *Heliyon* **5** 1-11
[36] H H R Azami, R Ibrahim 2018 A Systematic Literature Review on Acceptance of Massive Open Online Courses (MOOCs), no. May. 2018
[37] F Davis, R Bagozzi, P Warshaw 1989 User acceptance of computer technology: a comparison of two theoretical models *Management Science* **35** 982-1003
[38] A A Daneji, A F M Ayub, M N M Khambari 2019 The effects of perceived usefulness, confirmation and satisfaction on continuance intention in using massive open online course (MOOC) *Knowledge Management & E-Learning* **11** 201–214

[39] J-Y Hsu, C-C Chen, P-F Ting 2018 Understanding MOOC continuance: An empirical examination of social support theory *Interactive Learning Environments* **26** 1100-1118
[40] D A Aminu 2018 Predictive Model of Students' Continuance Intention in Massive Open Online Course among University Students. Thesis Master of Science. Universiti Putra Malaysia.

Analysis of Mamdani Method To Increase The Level Of The Advantages Of Infrastructure Development For Optimal Network

Niskarto Zendrato and M Zarlis

Faculty of Computer Science and Iinformation Technology Universitas Sumatera Utara, Medan, Indonesia

E-mail: niskarto@usu.ac.id, m.zarlis@usu.ac.id

Abstract. Network infrastructure has an important role towards national economic development. According to data from the Director General Post and Telecommunication cellular users experienced a significant increase from year to year. Efforts to improve telecommunications services have several problems because there is a tower which disturbs the comfort of the surrounding community, location who violates the rules and BTS tower that does not have permission. Location selection perspective from government and operators have differences, while the infrastructure must in a strategic location to get results optimal for improving service. Therefore, there is a need for research to formulate criteria for tower construction locations according to the level to the required optimization the method used is Mamdani to determine factor in determining the location of the tower. Then after found a factor weighting of factors was done by Mamdani's analysis. The last analysis is the formulation of criteria using Content Analysis. From the analysis 5 factors is electrical energy, the number of residents served, spatial palnning, tax cost and licensing process were generated used as a determining factor for the location of tower construction BTS (Base Transmitter Station) effectiveness.

1. Introduction

The increasingly rapid development of the era triggered the government's performance in terms of equitable development in each region in Indonesia. Infrastructure development in each region is one of the government's efforts to improve the welfare of the Indonesian people. So that to see an infrastructure development in an area that is really beneficial to improve the welfare of the people around it, it is necessary to design a computer application that can analyze the benefits of infrastructure development that has been implemented in an area that is in accordance with government objectives and shows a percentage value that represents the level of prosperity of the local people after infrastructure development was carried out.

This system is an analysis of the benefits of infrastructure development based on fuzzy logic that shows the level of feasibility of infrastructure development in an area, so that it is expected to facilitate officers working to evaluate infrastructure development in an area in determining which areas can be used as development priorities and which areas need to be evaluated re-building the project.

The last analysis is the formulation of criteria using Content Analysis. From the analysis 5 factors were generated used as a determining factor for the location of tower construction BTS (Base Transmitter Station) at. These factors are the number of people served, spatial planning, tax costs, licensing process, availability of electrical energy. From each of these factors, obtained 3 factors with the highest weight, namely availability electrical energy, in accordance with the spatial plan, and the licensing process . These three factors are the biggest influence on the selection of tower locations. Thus, the main criteria for location are obtained the construction of the transmitter tower is available electrical energy with sufficient power for BTS towers, location must be in accordance with the spatial plan and location fulfill all licensing processes. From this study obtained that adequate electrical energy is a priority for choose the location of the BTS tower

2. Methodology

Formation of Fuzzy Rules After the formation of fuzzy variables and sets, the following rules are formed (see Figure 1) in accordance with the benefit analysis data.

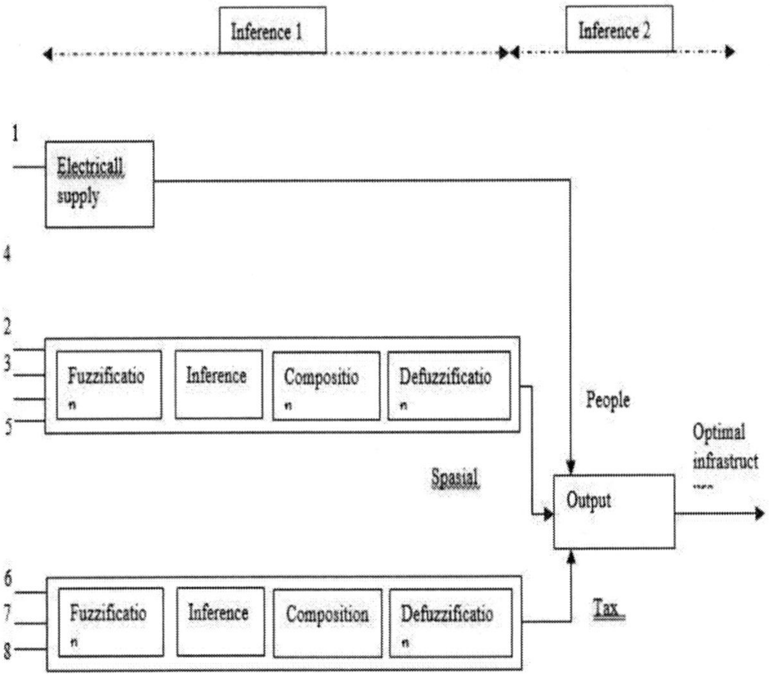

Figure 1: Formation of the fuzzy rules

Figure 2 is a picture of the feasibility classification of the analysis of the benefits of infrastructure development consisting of 2 phases of inference derived from the input set of fuzzy values. Inference 1 is the process of determining the level of effectiveness, efficiency and benefits, then the second inference determines the feasibility classification of the results of the analysis of development benefits. Savings and output data.

Simulation of the analysis of the benefits of infrastructure development using fuzzy logic with the Mamdani method (centroid) will provide convenience for officers (users) in analyzing the results of the benefits of infrastructure development in a region and classifying it into 5 conditions. Furthermore, the system will provide several outputs to the officers (users) in the form of analysis results so that this will be an evaluation and consideration material whether the development can be continued or needs to be re-evaluated.

IOP Conf. Series: Materials Science and Engineering **769** (2020) 012053 doi:10.1088/1757-899X/769/1/012053

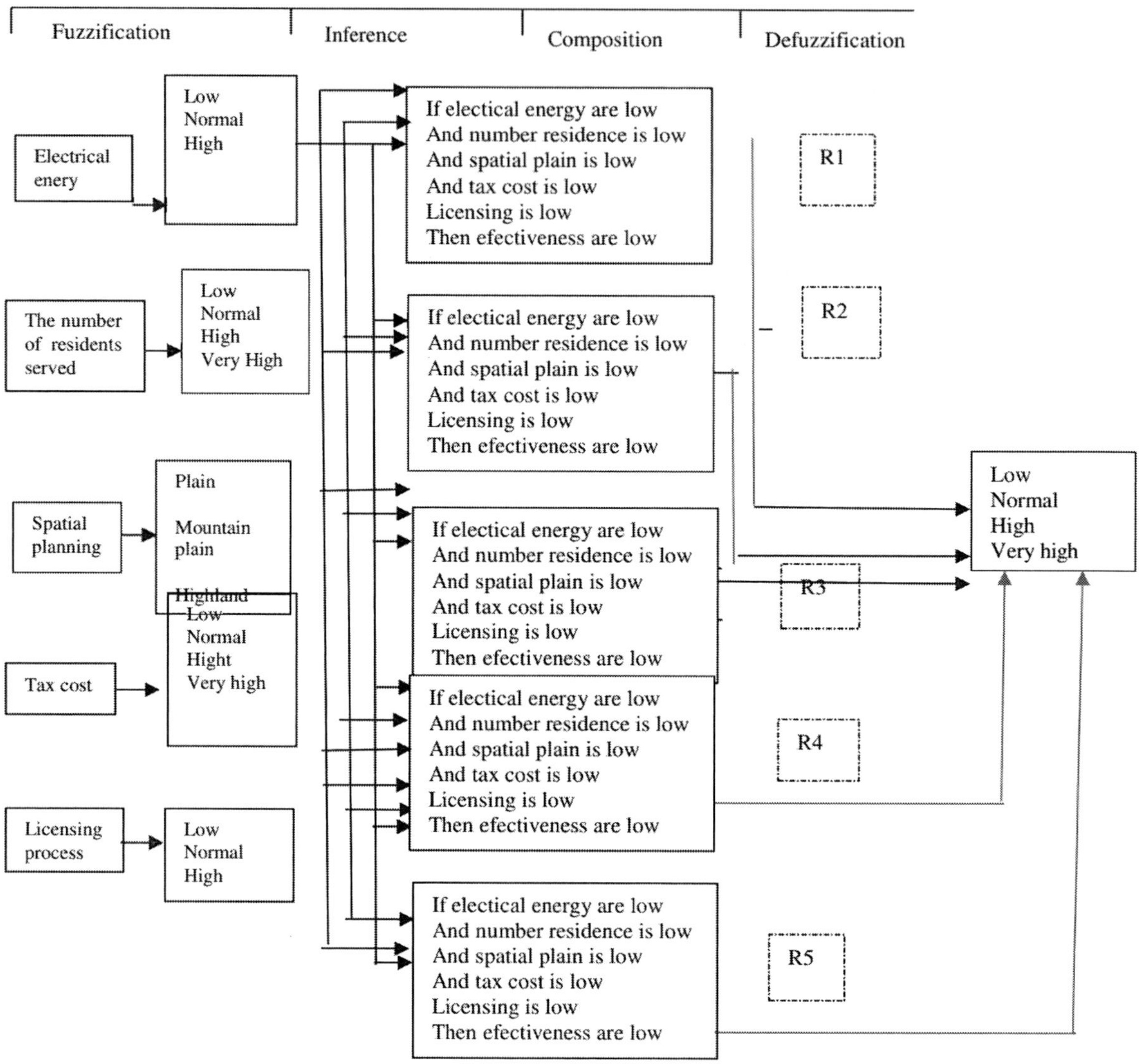

Figure 2 Inference 1 Effectiveness of infrastructure of network

2.1. *Membership function variable effectiveness*

Membership function fuzzy The membership function of the output of effectiveness can be seen in Figure 8, where the membership function consists of 4 conditions, namely low, medium, normal and high in these conditions use the trapmf type curve for low and high conditions, while for moderate and normal conditions use the trimf type curve, namely triangle. Figure 3 until figure 7 membership function fuzzy used 5 input variable with condition namely low, normal high and very high .

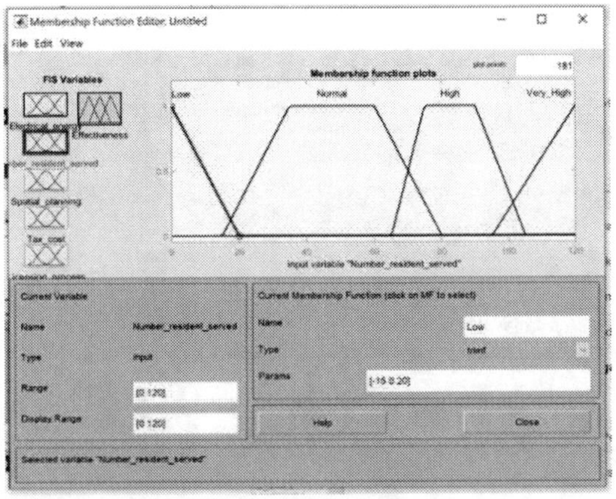

Figure 3: MF variable number of residence service

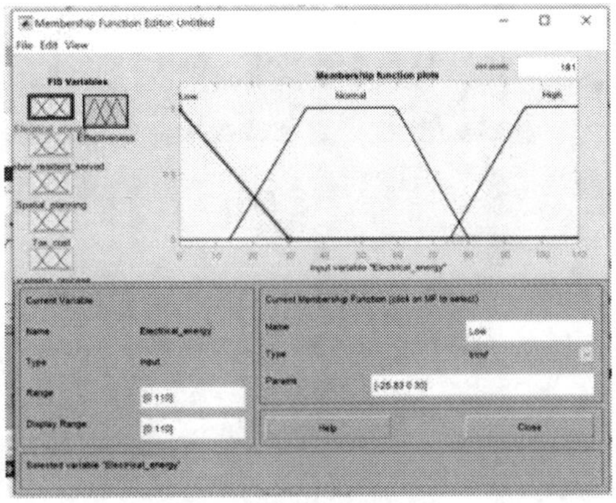

Figure 4: MF variable Electrical Energy

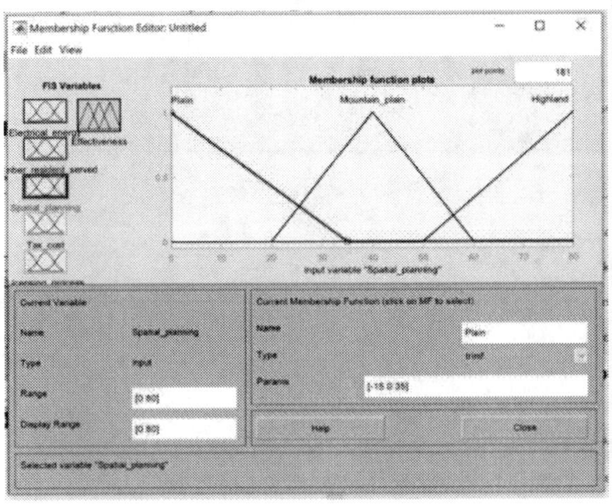

Figure 5: MF variable spatial planning

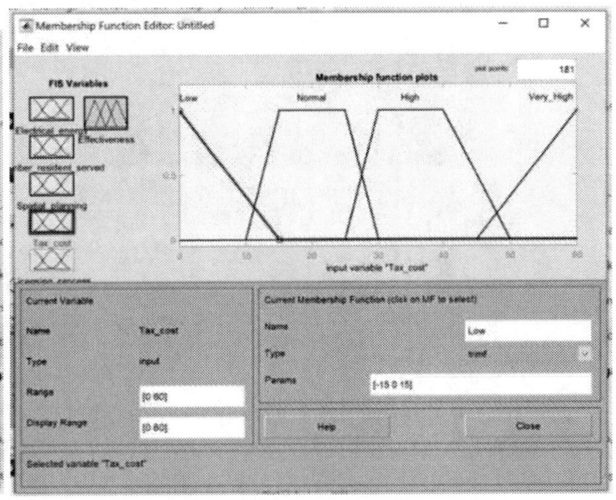

Figure 6: MF variable tax cost

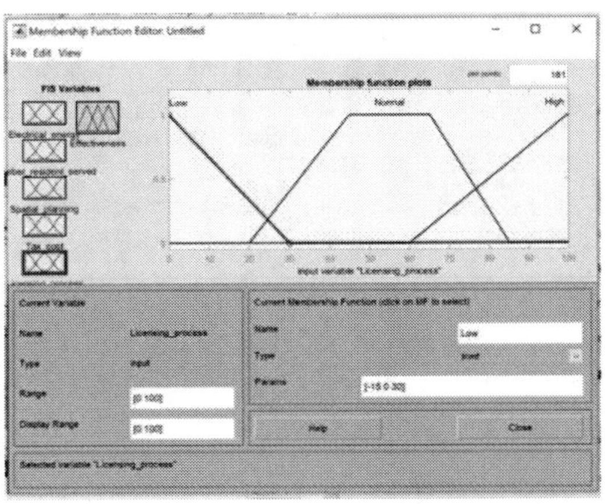

Figure 7: MF variable Licensing process

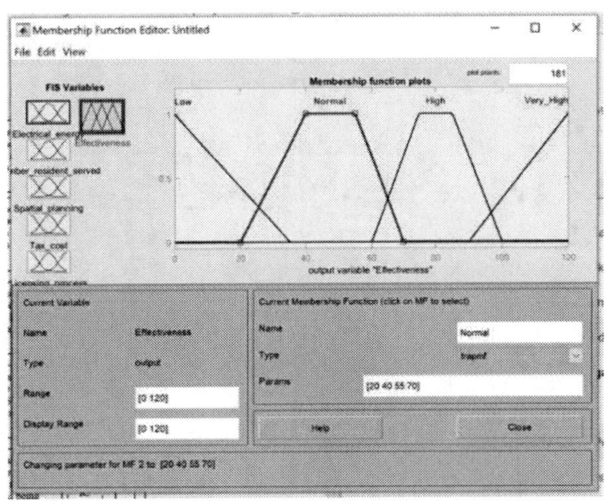

Figure 8: MF Variable output effectivenes

IOP Conf. Series: Materials Science and Engineering **769** (2020) 012053 doi:10.1088/1757-899X/769/1/012053

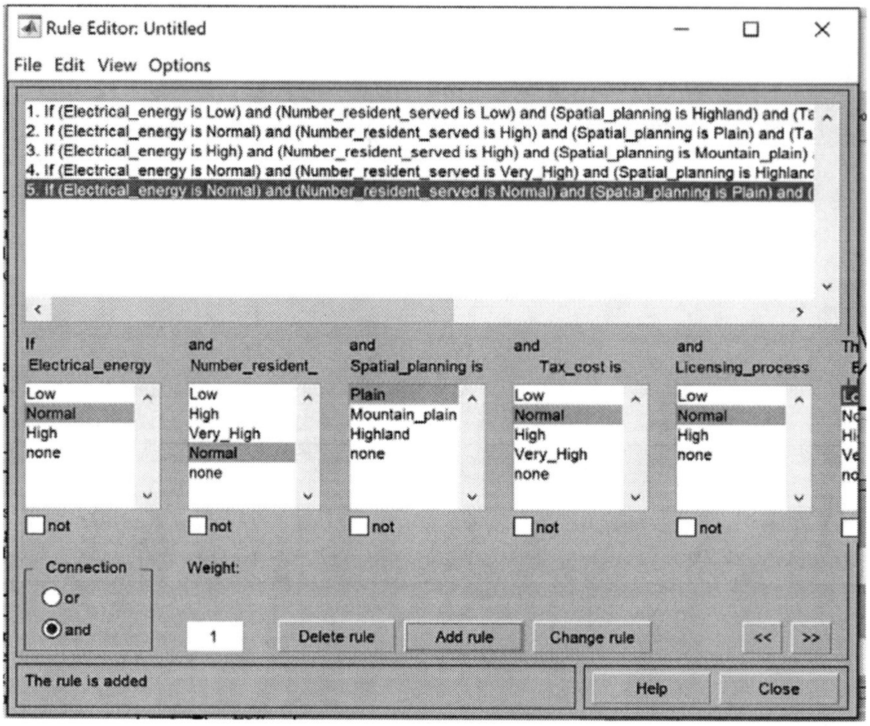

Figure 9: Rules inferences

3. Result And Discussion

The results of the simulation analysis of the benefits of infrastructure development using fuzzy logic with the mamdani method (centroid) are built using the fuzzy logic toolbox with GUI (Graphical User Interface) facilities so that it can be seen the following aspects of benefit inference Figure 3 used mamdani method.

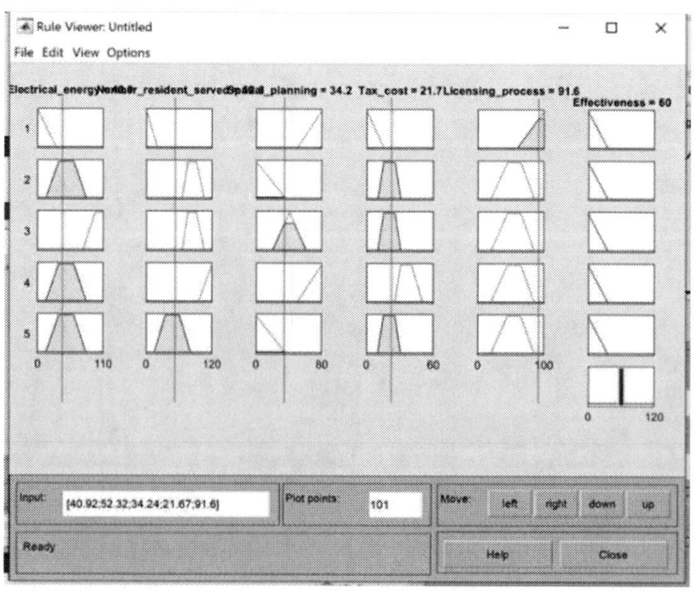

Figure 10: inference output effectiveness base on variable adjustment

data training used same data input and comparing method Mamdani and manually analysis. Based on data training data values of Mamdani

Table 1. Data training with 5 variable inputs

Variable	Training 1	Training 2	Training 3	Training 4	Training 5
Electrical energy	43,56	65,56	29,77	86,68	86,68
The number of residents served	88,1	59,52	31,09	99,52	99,52
Spatial planning	13,12	34,24	26,06	56,64	56,64
Tax cost	35,76	22,32	14,04	13,2	20,88
Licensing process	47,22	68,65	34,09	21,83	40,87
Effectiveness	60	17	15,1	16,6	15,8

4. Conclusion

The conclusions that can be drawn from this research are:

a. Variable effectiveness used to analyze SBT give advantages to build base on the 5 variable such as electrically energy, the number of residence service, tax cost and licensing process
b. Mamdani method good used to analyses the advantage of BTS than AHP
c. Base on training effectiveness stable on normal with value 15 until 17

References

[1] Sri Hartati, Imas S Sitanggang," A Fuzzy Based Decision Support System for Evaluating Land Suitability and Selecting Crops " Journal of computer science 6(4):417-424, 2010.
[2] Panduan Pelaksana PNPM PISEW TAHUN 2010
[3] " Kecerdasan buatan", http://idhaclassroom.com/2007/09/15/../kecerdasanbuatan.html. Tanggal akses 10 juli 2010.
[4] Suparman, " Komputer Masa Depan Pengenalan Artificial Intelligence", Marlan. ,Andi, Yogyakarta, 2007.
[5] Kusumadewi, S., Analisis Desain Sistem Fuzzy Menggunakan Tool Box Matlab, Penerbit Graha Ilmu, Yogyakarta, 2002.
[6] Kusumadewi, S., dan Purnomo, H., "Aplikasi Logika Fuzzy untuk Pendukung Keputusan", Penerbit Graha Ilmu, Yogyakarta, 2004
[7] IGunaidi Abdia Away, " MATLAB Programming", Informatika, Bandung, 2010
[8] Kusumadewi, S., Analisis Desain Sistem Fuzzy Menggunakan Tool Box Matlab, Penerbit Graha Ilmu, Yogyakarta, 2002.
[9] Efraim Turban, Jay E. Aronson, Ting Peng Liang," Dicision Support System and Intelligent Systems (Sistem Pendukung Keputusan dan Sistem Cerdas)", ANDI, Yogyakarta 2005
[10] Dr. Eng. Agus Naba " Belajar Cepat Fuzzy Logic menggunakan MATLAB" ANDI, Yogyakarta 2009
[11] Rofiqoh Etika Amalin dan Ketut Dewi Martha, " Kriteris Lokasi Pembangunan Tower BTS di Kota Kediri " Jurnal Teknik ITS, 2017

IOP Conf. Series: Materials Science and Engineering **769** (2020) 012054 doi:10.1088/1757-899X/769/1/012054

Investigating the Barriers of Health Information System implementation in Malaysian Public Hospitals

Norsyaliza Abd Razak[1], Norafida Ithnin[1], Mohd Zamri Osman[2]

[1]Information Assurance and Security Research Group (IASRG), School of Computing, Universiti Teknologi Malaysia, 81310, Skudai, Johor, Malaysia
[2] Soft Computing and Intelligence System (SPINT), Faculty of Computing, Universiti Malaysia Pahang, Lebuhraya Tun Razak, 26300 Gambang, Kuantan, Pahang, Malaysia

Email: zamriosman@ump.edu.my

Abstract. The advent of information and communication technology (ICT) has shaped the world today. Due to the potential increase in technology, the introduction into the healthcare industry has been progressively investigated. In 1997, Malaysia introduced a health initiative called telehealth. However, progress in IT implementation is progressively slow. Although the global healthcare sector is increasing, the implementation of the Health Information System (HIS) in Malaysia has been slow-informed due to various barriers. These barriers are due to a series of problems related to end users and challenges for IT departments and vendors. This article therefore focuses on the current situation of HIS in selected Malaysian public hospital. In addition, an interview conducted with the end users, IT department and the system providers. The data from the interview was then converted into a summary of the table. In this context, the obstacles that delay the implementation of HIS in Malaysia have been identified primarily in terms of suppliers, financial, organizational and human aspects.

1. Introduction

The advancement of ICT has reshaping the current industry especially in the healthcare industry. ICT found to be playing important role in enhancing the patient care services. Throughout the years, tremendous inventions have been presented into the healthcare sector to enhance the hospital or clinics performance and quality services. Innovations such as Telemedicine, e-Health, m-Health, TeleHealth, Health Information System (HIS), mobile technology been employed to provide better health services to society [1]. For instance, the Electronic Health Record (HER), Health Information System (HIS) and Telemedicine are such ICT innovation that proven provide better health services to the society [2]. Healthcare industries emerged as technology grow rapidly. The evolvement is necessary to increase the quality of healthcare services and the efficiency of healthcare management. In countries such as the United Kingdom (UK), they have used the comprehensive healthcare system. This aims to improve the level of health management for people in their country. [19]

This research is conducted to provide a clear understanding of the barriers in HIS implementation, especially in the public sector. Throughout the research, there are several barriers by the Malaysian Government to successfully implement the HIS in the hospitals.

IOP Conf. Series: Materials Science and Engineering **769** (2020) 012054 doi:10.1088/1757-899X/769/1/012054

This research aims to explore the implementation of HIS in the public hospital in Malaysia, the barriers and thorough analysis of the results. The end of this study, the barriers that affecting the Malaysian HIS public hospitals can be used as a guidance to provide a successful HIS.

2. Methodology

To achieve the objective of this study, there are two phases undertaken to gather information to support the literature. The first is exploring the literature on HIS implementation in Malaysian hospital covering the challenges and issues. The focus is to identify multiple factors involved in success and challenges in the implementation of HIS. Several examples of HIS implementation in Malaysian hospital was taken. The difference among HIS implementation was gathered to differentiate between each system and vendor.

The second phase is the interview with the HIS users, vendor and healthcare practitioner. An interview was conducted to support the problem exist in the Malaysian healthcare system. It is the phase which qualitative research is being conducted and preliminary results are being achieved that support the problem statement for this research. The interviews were held with the end users, system developer and system provider. For the end users, the interview is to know the current status of HIS implementation on certain hospital and clinics in Malaysia in different background point of view. Then the interview with the system developer and system provider was conducted to investigate the current status of HIS environment in Malaysian hospitals. From the interview, it shows that hospitals in Malaysia have a different system provider and every system in the hospital conduct by different provider. This came out with the issues of the implementation of HIS in each hospital.

3. HIS Scenario in Malaysia

Implementation of the healthcare system in Malaysia has long been on the run. However, overall Malaysian health organisations is not fully implemented. Several factors contribute to this implementation delays. One of the factors is the lack of acceptance of the technology for the end users in the healthcare organisation. Besides, the system environmental factors, within organisations is hampering the implementation of the healthcare system. It also included the types of systems in the organisation, the system provider for the organisation and top management authorities.

The problem with the healthcare provider has faced today are high cost and lack of human resources. Therefore, to solve these problems, an effective and economical solution must be introduced. The solution to this situation can be the integration of information and communication technologies (ICT) in the health sector. The hospital information system is the most important ICT that is integrated in the healthcare system of Malaysia. The hospital information system claims to reduce medical error, increase efficiency, cost effectiveness and increase patient involvement in healthcare decision-making [3].

A proper HIS integration subcomponent is a major challenge facing medical information. With current network technology, it is no longer acceptable to implement standalone system subcomponents which do not communicate with one another. Connectivity and integration can eliminate duplication of data entry (for example, patient demographics are required by almost all subcomponents of HIS) and allow for the automated transmission of messages between the subcomponents (for example, transmission of the details of a laboratory test order from an order entry subsystem to a laboratory information subsystem). These benefits come at the price of network hardware and software, interface software, increased complexity of the overall system, and reliance upon network availability [4].

3.1. Malaysian HIS Implementation in the Public Hospitals

In a developing country like Malaysia, hospitals are the main healthcare providers whereby aiming HIS to be improved [5]. Starting with the Telemedicine flagship introduced in 1997 is the Malaysian initiative for providing better health care services to the society. There are three healthcare sectors in Malaysia which are public, private and Non-Government organizations (NGOs) [6]. Public hospital places more complex systems due to the patients comes from various background [7]. Those patients afford higher cost, they preferred to get medication from private healthcare. Large numbers of patients especially in the public hospitals may leads to system ineffectiveness due to longer waiting time for medication. This issue has been reported between 2000 and 2008 previously in [8].

In Malaysia, we have 139 hospitals with 18 referral and tertiary hospitals. However, not all the hospitals fully implementing the HIS [9]. The implementation of HIS in Malaysia has been carried out since 1993 under the Sixth Malaysia Plan [10]. Currently, Malaysia have three type of hospitals which categorized based on the number of available beds, specialties and the budget [11][12]. Basic Hospital Information System (BHIS) is a small size hospital, while medium size Intermediate Hospital Information System (IHIS). Meanwhile, larger size hospital known as Total Hospital Information System (THIS) where placing paperless healthcare systems in all departments. Table 1 summarizes the public hospital in Malaysia implementing HIS by its classification.

Hospital Selayang and Hospital Putrajaya was the early hospital equipped with THIS since the launched of Malaysian Telehealth Flagship Application starting with Hospital Selayang in 1997 followed by Hospital Putrajaya in 2000 [14].

Table 1. Summary of classification of Malaysian public hospital based on number of beds

Type of HIS	Name of Hospitals	Component of HIS used	Number of beds
THIS (large size)	Hospital Sultan Haji Ahmad Shah, Hospital Pandan Hospital Putrajaya, Hospital Selayang, Hospital Sedang, , Hospital Sg. Buloh, Hospital Sungai Petani, Hospital Ampang, Hospital Sultanah Zahirah, Hospital Alor Setar, and Hospital Bintulu	PMS, CAIS, LIS, PIS, RIS, PACS, AIS, FIS, SIS, PIS	> 400 beds
IHIS (medium size)	Hospital Lahad Datu and Hospital Keningau	PMS, CIAIS, LIS, PIS	> 200 beds
BHIS (small size)	Hospital Tunku Ja'afar , Hospital Kuala Penyu, Hospital Kuala Batas, Hospital Pitas, Hospital Kunak, and Hospital Setiu, Hospital Port Dickson	PMS and CAIS	< 200 beds

PMS: Patient Management System, CAIS: Clinical Access Information System, LIS: Laboratory Information System, PIS: Pharmacy Information System, RIS: Radiology Information System, PACS: Picture Archiving and Communication System, AIS: Administration Information System, FIS: Financial Information System, SIS: System Inventory System, PIS: Personal Information System

3.2. HIS Revolution in Malaysian Public Hospital

Hospital Selayang was the first paperless THIS hospital in Malaysia which managed by the Cerner Malaysia Sdn. Bhd. [13]. Hospital Selayang has implemented EMR to improve the provision of health services focusing on the management and information system of patients at the organization level. THIS was also implemented at Putrajaya Hospital and Pantai Medical Center by Kompakar eHealth Tech Sdn. Bhd. Three different THIS hospital in Malaysia where Hospital Serdang, Hospital Selayang and UKM Medical Centre as shown in Table 1. Starting in 2002, Syarikat Permodalan Kebangsaan (SPK_ Bhd had initialize the development of HIS in several hospital such as Hospital Kepala Batas, Lahad Datu and Serdang. Upon the successful achievement, SPK then continue the HIS development for remaining hospitals. Table 2 shows the hospitals in Malaysia that implemented HIS developed by SPK system vendor.

Table 2. HIS Project completed by SPK system vendor

Name of hospitals	Year completed
Hospital Kepala Batas	Dec 2004
Hospital Lahad Datu	April 2005
Hospital Serdang	June 2006
Hospital Tentera Lumut	Nov 2006
Hospital Pandan	Dec 2006
Hospital Keningau	Sept 2007
Hospital Setiu	Sept 2007
Hospital Kunak	Oct 2007
Hospital Kuala Penyu	Oct 2007
Hospital Ampang	March 2008
Hospital Sg. Buloh	May 2008
Hospital Sg. Petani	May 2008
Hospital Pekan	July 2008
Hospital Alor Setar	July 2008

On the other hand, the health system in Hospital Putrajaya also operates fully using THIS. THIS was started since 1998 with the cost of RM282 million using a system called PutraCare. PutraCare system developed by the local vendor, Kompakar Sdn. Bhd [14].

A further investigation the characteristics in three public hospitals shown in Table 3. Among three hospitals compared, Hospital Serdang required the highest budget with 80 million ringgit for implementing HIS followed by Hospital Serdang with RM64 million and UKMMC with only 6.3 million ringgit. There are several criteria of the THIS in Malaysia. One of the criteria is the budget funded for implementing the HIS. HIS in Hospital Serdang and Hospital Selayang was funded fully by the Federal Government while UMKMC funded from their organisation itself. In-house development by the UKMMC only requires RM6.3 million.

Table 3. Characteristics of HIS in three Malaysian Hospitals

Variable	Hospital Serdang	Hospital Selayang	UKMMC
System Provider	IBA health	Cerner	C-Hets
Country Origin	UK	USA	Malaysia
No. of beds	630	960	1050
Staff	1900	2600	3500
Development	2005	1999	2004
Budget	Turnkey project RM80 mil	Turnkey Project RM 64 mil	In-house development RM 6.3 mil
Human resource	IT department - 1 IT Manager, - 4 IT programmer	IT department - 1 IT Manager, - 6 support staffs	IT department - Head of IT unit - 40 technical staff - 40 domain expert

According to the study conducted by [7], [12], there are only 15.2% of the Malaysian hospitals are system-based and less than 10% of 139 public hospitals are classified as HIS. This indicates the majority of the public hospitals in Malaysia remain using a paper-based method and shows slow progress on HIS reformation. Several factors have been studied the impact of implementation of the healthcare system.

4. The Challenges in Malaysian HIS

HIS implementation in a public hospital in Malaysia is still low due to various issues. Support, human, technology and infrastructure and software limitation issue are four major challenges has been identified in HIS implementation [16]. The human context in Malaysian Public Hospitals had the highest effect of HIS adoption. This could imply the importance of human skills, experience, expertise, satisfaction and quality of information for successful adoption of HIS [17]

The issue of HIS integration has been addressed by researchers and, unfortunately, is still one of the key issues for HIS's success, which remains relevant and unresolved [18]. With so many vendors developing specific applications for different medical departments and hospitals, there are challenges as hospitals try to integrate data to enable interoperability and communication between different systems [19].

The integration of the medical system between different hospital organizations (local and regional) has been claimed to be a significant step in creating a complete HIS and provides benefits to clinicians who need inter-department collaboration in solving patient's cases [20], [21]. To ensure the successful integration of the system, organizations need the necessary financial and technological resources. However, in developing countries, poor maintenance or lack of personnel in charge of these resources is exceptionally common, and that causes information systems projects to fail significantly. In support of this argument [15], it is proposed to establish a dedicated coordination unit to act as a reference for all IT or system developers (vendors) and to provide a platform for improving future HIS integration.

5. Findings and Discussion

Results from the preliminary analysis and clarification used to stimulate an overall picture of the research domain and its proposed implications for HIS integration in Malaysian hospitals.
The preliminary result is a test that conducted for pre-case study research to support the problem statement of the research. In this research, the preliminary study was done by using unstructured interview approach with the end users and system developer in different organisations and background. The result of the interview shown in Table 4.

Table 4 shows the summary of the case study healthcare system in selected hospital and clinics that is obtained from the initial interview. The interview was conducted among users in hospital and clinics of the government sector. The respondents consist of physicians, nurse, therapist and also system developer. Four different organisation from government sector have been chosen as the case study; Hospital Tunku Jaafar Seremban (HTJS), Negeri Sembilan, Hospital Sultanah Aminah (HSA) Johor, Klinik Kesihatan Gombak (KKG) Selangor and Klinik Kesihatan Masjid Tanah (KKMT) Melaka. From the table, seven criteria have been analyzed where include the knowledge about Health Information System (HIS), implementation of HIS, health system used, system developer of the health system, health data achieved, health data saved and integration of the health system (within the organization or other organisation).

From the interview, it is shown that most of the organisation using a different system and platform. Also, most of the system has no integration. In Klinik Kesihatan Taman Ehsan Gombak, they use Teleprimary Care system (TPC).

Table 4. Summary of healthcare system case study in selected public hospitals and clinics

Respondent	HIS	System Used	System Vendor	Data Achieved	Data Saved	System Integration
Physiotherapist Hospital Tuanku Jaafar Seremban (HTJS)	No	Sistem Pengurusan Pesakit (SPP)	Heitech Padu	From System and Manual	System and Manual	No
		Sistem Maklumat Pesakit	KKM			

Physician Klinik Kesihatan Taman Ehsan Gombak	Yes	TPC (Teleprimary Care)	KKM	System	System	Partial (Medication still need to do manual)
System Developer Hospital Sultanah Aminah	No	Sistem Pengurusan Pesakit Dalam (SPPD)	KKM	From System and Manual	System and Manual	No
Nurse Klinik Kesihatan Masjid Tanah	No	Rekod Pesakit	KKM	Manual from patient card /book	Manual in Patient Card	No

Based on initial findings collected from the preliminary investigation, several issues can be highlighted. These issues are:

- Different vendors. One hospital may have different vendors that developing the health system for each department. Incompatibility for other systems to communication due to a different platform, database format and programming language.

- Multiple middleware. To achieve integration of the health system, SPP was integrated with the middleware on each of the systems. The middleware is responsible for translating the data that can be read by the SPP. Since each of the system in every department was different architecture, multiple middleware required that may lead to system complexity.

- Duplication of data. Due to achieving data from a different system using a middleware, duplication of data may exist. This is due to; previous data remain in the origin database server.

6. Conclusion

As a conclusion, the healthcare system in public Malaysia is partially implemented with most of the systems are not fully integrated. Due to some reasons such as human, financial, and different vendors make the healthcare systems have isolated each other. Hence, these factors supported by the preliminary interview from selected public clinics and hospital. After a few polite preliminary studies, it can be a guideline and starting point to the ideas and intentions for effective HIS integration for future development.

Acknowledgement
We would like to show our gratitude to Universiti Malaysia Pahang (RDU vote numberRDU1803142) for supporting this research.

References

[1] D. Malvey, D. J. Slovensky, "From Telemedicine to Telehealth to eHealth : Where Does mHealth Fit ?," pp. 19–29, 2014.
[2] A. K. Jha, D. Doolan, D. Grandt, T. Scott, and D. W. Bates, "The use of health information technology in seven nations," Int. J. Med. Inform., vol. 77, no. 12, pp. 848–854, 2008.
[3] H. W. Lee, T. Ramayah, and N. Zakaria, "External factors in hospital information system (HIS) adoption model: A case on Malaysia," J. Med. Syst., vol. 36, no. 4, pp. 2129–2140, 2012.
[4] A. M. Zakaria and D. A. Giuse, "LabTalk/2: a middleware approach to HIS integration.," Proceedings of the Annual Symposium on Computer Application in Medical Care, 1995, p. 121.
[5] M. A. Malik and H. R. Khan, "Understanding the implementation of an electronic hospital information system in a developing country: a case study from Pakistan," in Proceedings of the

Third Australasian Workshop on Health Informatics and Knowledge Management-Volume 97, 2009, pp. 31–36.
[6] Kementerian Kesihatan Malaysia, MOH, "Country Health Plan: 10th Malaysia Plan," 2011.
[7] N. I. Ismail, N. H. Abdullah, A. Shamsudin, and N. A. N. Ariffin, "Implementation differences of Hospital Information System (HIS) in Malaysian public hospitals," Int. J. Soc. Sci. Humanit., vol. 3, no. 2, p. 115, 2013.
[8] Kementerian Kesihatan Malaysia, MOH, "Annual Report 2009," Putrajaya, Malaysia, 2009.
[9] Kementerian Kesihatan Malaysia, MOH, "Senarai Hospital Kerajaan," MOH, 2013. .
[10] Kementerian Kesihatan Malaysia, MOH, "Vision and Mission: Ministry of Health Malaysia," 1997. .
[11] A. Ismail, A. T. Jamil, A. F. A. Rahman, J. M. A. Bakar, N. M. Saad, and H. Saadi, "The implementation of Hospital Information System (HIS) in tertiary hospitals in malaysia: a qualitative study," Malaysian J. Public Heal. Med., vol. 10, no. 2, pp. 16–24, 2010.
[12] N. I. Ismail and N. H. Abdullah, "An Overview of Hospital Information System (HIS) Implementation in Malaysia," 3rd Int. Conf. Bus. Econ. Res. Proceeding, no. March, pp. 1176–1182, 2012.
[13] Kemententerian Kesihatan Malaysia, KKM, "Hospital Selayang," KKM, 2018. .
[14] Utusan, "Hospital Putrajaya guna e-perubatan ciptaan tempatan," Utusan, 2001.
[15] A. M. Bulgiba, "Information technology in health care-what the future holds," Asia Pacific J. Public Heal., vol. 16, no. 1, pp. 64–71, 2004.
[16] A. Baldovin, E. Mezzetti, and T. Vardanega, "Challenges in the Implementation of MrsP," no. 214, pp. 636–642, 2017.
[17] N. Izzatty, N. Hazana, and A. Shamsuddin, "Adoption of Hospital Information System (HIS) in Malaysian Public Hospitals," Procedia - Soc. Behav. Sci., vol. 172, pp. 336–343, 2015.
[18] R. Haux, "Health information systems–past, present, future," Int. J. Med. Inform., vol. 75, no. 3–4, pp. 268–281, 2006.
[19] K. A. Wager, F. W. Lee, and J. P. Glaser, Managing? Health Care Information Systems: A Practical Approach for Health Care Executives. John Wiley & Sons, 2005.
[20] G. O. Otieno, T. Hinako, A. Motohiro, K. Daisuke, and N. Keiko, "Measuring effectiveness of electronic medical records systems: Towards building a composite index for benchmarking hospitals," Int. J. Med. Inform., vol. 77, no. 10, pp. 657–669, 2008.
[21] G. F. Anderson, B. K. Frogner, R. a. Johns, and U. E. Reinhardt, "Health care spending and use of information technology in OECD countries," Health Aff., vol. 25, no. 3, pp. 819–831, 2006.

IOP Conf. Series: Materials Science and Engineering **769** (2020) 012055 doi:10.1088/1757-899X/769/1/012055

Literature review on Functional Requirements for Function Point Analysis in Mobile Game

Nur Atiqah Sia Abdullah

Centre of Computer Science Studies,

Universiti Teknologi MARA, 40450 Shah Alam, Selangor

atiqah@tmsk.uitm.edu

Nur Ida Aniza Rusli

Centre of Computer Science Studies,

Universiti Teknologi MARA, 40450 Shah Alam, Selangor

idaaniza@gmail.com

Abstract. Functional size measurement has been widely used as the effort prediction method for software project over the past decades. Despite the success in predicting the cost, time requirement, and staff resources utilizing the amount of functionality delivered by a system, the method still needs to be improved further especially in dealing with mobile application characteristics. In the domain of mobile game application, the functional features can be challenging to identify wholly due to various application platforms, diversity of game genre, and complex game scenario. This paper presents a literature review concerning the functional requirements of the mobile game. The review includes the process of defining the research question, search strategy, and selection of study. Fifty-three articles were found in the academic libraries, and only 30 relevant studies were selected for the review. This literature review aims to provide researchers and practitioners in software measurement and gaming technology with an overview of common functional characteristics in the mobile game, which can be integrated into the functional size measurement method.

1. Introduction

Software effort estimation is one of the important activities in the software project. The determination of effort estimation in the early phase of development allows both project managers and client to plan the resources such as cost, the number of staffs, and development schedule systematically.

Since the 1950s, many effort estimation models have been proposed in the literature such as Lines of Code (LOC), Software Lifecycle Management (SLIM) and Function Points Analysis (FPA). Function Point Analysis or Functional Size Measurement (FSM) methods have gained popularity in software measurement community due to the higher accuracy and the applicability to be used in the early phases of the software lifecycle. The method provides an estimation of software size by working with the amount of functionality from user perspectives.

FSM method was originally introduced by Allen Abrecht and has been updated by International Function Point User Group Function Point Analysis (IFPUG FPA) [1]. The method is then continually adapted by other proposals such as MK II FPA [2], NESMA Function Point Analysis [3], and COSMIC FFP [4]. Although FSM methods have undergone several improvements and modifications, FSM is still considered incomplete due to the difficulties to measure the latest software characteristics and unable to adopt new factors such as mobile platforms, fast pace development process, and social features. For these reasons, software measurement researchers and practitioners have been trying to develop the effort prediction model using the FSM concept for mobile application [5].

In the domain of mobile game application, the requirements and the design process are always critical. Mobile game developers need to include more artistic content, animation, gameplay models to make a good mobile game [6]. Thus, the effort of the mobile game will always be an issue since it needs to cater to those complex requirements. This paper presents a literature review concerning the standard functionality of the mobile application, which can be contributed to the functional size and included as the factors to the effort estimation. This paper is organized as follows: Section 2 illustrates the method for the review; Section 3 shows the results from the review; Section 4 describes the analysis; Section 5 emphasizes on discussion; while Section 6 presents the conclusion from the review.

2. Review Method

2.1. Research Question

The purpose of this work is to identify the common functional characteristics of the mobile game that can be useful for the establishment of software effort estimation model. This leads to the driving research question for the literature review in this study: "What functional characteristics should be considered in the mobile game effort estimation?"

2.2. Search Strategy and Selection of Article

To identify the common feature or functional requirement of the mobile game, the study began by searching the academic libraries such as ACM, SpringerLink and IEEE databases between 2000 and 2019. The academic databases were searched with the various combinations of keyword "mobile game feature" "mobile game design" and "mobile game requirements" in the mentioned libraries.

Selection of primary study was initially assessed based on the title, abstract, and keywords. Then review process was extended by including the results and conclusion if there any insufficient information in the mentioned criteria. A total of 53 articles met the necessary requirements; however, when the articles were further analyzed, 23 articles were excluded because the mobile game requirements were not discussed properly. Thus, in total, 30 articles were selected for the review, as shown in Table 1.

Table 1. List of selected articles.

Study ID	Reference	Year	Study ID	Reference	Year
S1 [7]	Alcid et al.	2017	S16 [22]	Bartel and Hagel	2014
S2 [8]	Shen and Barakova	2017	S17 [23]	Abdullah et al.	2013
S3 [9]	Cruz et al.	2017	S18 [24]	Sintoris et al.	2012
S4 [10]	Franti et al.	2017	S19 [25]	De Macedo and Rodrigues	2011
S5 [11]	Sweetser et al.	2017	S20 [26]	Peker and Can	2011
S6 [12]	Tai	2017	S21 [27]	Zhu et al.	2010
S7 [13]	Yue and Ying	2017	S22 [28]	Chee et al.	2010
S8 [14]	Shi	2016	S23 [29]	Diah et al.	2010
S9 [15]	Mildner et al.	2015	S24 [30]	Jhingut et al.	2010
S10 [16]	Kurniati et al.	2015	S25 [31]	Shiratuddin and Zaibon	2010
S11 [17]	Willis	2015	S28 [32]	Koivisto	2010
S12 [18]	Joselli et al.	2014	S26 [33]	Fotouhi-Ghazvini et al.	2009
S13 [19]	Cao et al.	2014	S27 [34]	Chehimi et al.	2008
S14 [20]	Hussain et al.	2014	S29 [35]	Ekman et al.	2005
S15 [21]	Mikkila	2014	S30 [36]	Davidsson et al.	2004

2.3. Inclusion Crieria

The review process is continued by filtering the selected articles based on the ten criteria: (1) game player, (2) user input, (3) user interface (UI), (4) gameplay character, (5) rigid modelling, (6) structural functionality, (7) media content, (8) rewarding system, (9) leaderboard, and (10) multiplatform support.

Table 2 shows ten functional requirements of the mobile game that regarded as the inclusion criteria in this review. These requirements are important to be discussed as it frequently appeared in requirement documentation, and the occurrence of issues related to mobile game appeared in academic literature.

Table 2. Functional requirements of mobile game.

Functional Requirement	Descriptions	References
Game player	The actual human that plays the game	[31] [37]
User Input	Functional mechanism to control the direction and action of a player	[31]
User Interface (UI)	Medium of interaction between player and the game. Consists of set of screens and visual elements in the game	[22][36] [37]
Gameplay Character	Set of characters that are being controlled by the player or the game itself. It consists of player character, enemy character or dynamic object modelling	[34][38]
Rigid Modelling	Set of characters that are not affected to any forces, collision or gravity during the gameplay	[34] [25]
Media Content	Medium for immersion and engagement such as music, sound effect and animation	[27][35]
Structural Functionality	Requirements that form the base of the game and coordinate the elements such as scene management, rendering, collision detection and physics systems	[38] [39]
Rewarding system	Motivational aspect includes point, badges and achievements for completing the game	[24]
Leaderboard	Lists of players according to their completion level within the game	[9][22]
Multiplatform support	Requirement to operate and performance control of the game in mobile devices	[27] [32]

2.4. Results

This section presents the results of the reviewed articles based on review method described in the previous section.

Table 3. Results

Functional Requirement	Descriptions	Total
Game player	S1, S2, S3, S4, S5, S6, S7, S9, S13, S14, S15, S16, S17, S18, S19, S20, S22, S23, S24, S25, S26, S28, S29, S30	24
User Input	S1, S2, S3, S4, S5, S8, S9, S12, S13, S14, S17, S18, S19, S20, S23, S24, S25, S26,S28, S29, S30	21
User Interface (UI)	S1, S2, S3, S4, S5, S6, S7, S8, S9, S10, S11, S12, S13, S14, S15, S16, S17, S18, S19, S20, S21, S22, S23, S24, S25, S26, S30	27
Gameplay Character	S1, S2, S3, S4, S5, S6, S7, S8, S9, S10, S11, S13, S14, S15, S17, S18, S19, S20, S21, S22, S23, S24, S25, S26, S28, S29,	27

	S30	
Rigid Modelling	S1, S2, S3, S4, S5, S6, S7, S8, S9, S10, S11, S13, S15, S16, S17, S18, S19, S20, S21, S22, S24, S25, S26, S28, S30	24
Media Content	S1, S3, S4, S5, S6, S7, S10, S11, S12, S13, S15, S16, S17, S18, S19, S20, S21, S23, S24, S25, S26, S27, S28, S30	24
Structural Functionality	S1, S3, S5, S6, S8, S9, S10, S11, S12, S13, S15, S17, S19, S20, S23, S24, S25, S27, S28	19
Rewarding system	S1, S3, S4, S5, S6, S8, S9, S11, S12, S13, S14, S15, S16, S17, S18, S19, S20, S22, S23, S25, S26, S28, S30	23
Leaderboard	S3, S16, S19, S30	4
Multiplatform support	S1, S2, S3, S4, S5, S6, S7, S8, S9, S10, S11, S12, S13, S14, S15, S16, S18, S19, S20, S21, S22, S23, S24, S25, S26, S27, S28, S29, S30	29

2.4.1. Game Player. Table 3 shows 24 articles described their target users, the number of player and player roles. These characteristics are mostly relying upon the objectives and the nature of the game itself. For the purpose of popularity and profit-maximization goals, game developers tend to offer multiplayer game mode. On the other hand, a single-player game might also render few advantages to both the developer and players since the game appeared to be less complicated compared to multiplayer games and the game does not require an Internet connection to play. Although multiplayer games are more captivating, especially if the players may interact with each other in real time, for example, the ability to talk, fight, or exchange message via the in-built chat system. However, the complexity of the game increases significantly as the game is played in multiplayer game mode.

Bartle and Hagel [22] identified game players in four types; achievers, socializer, explorers and killers. Achievers are defined as game players that aimed to achieve points or to earn status. Socializers are the players that appreciate the social interaction system in a game; while, explorers are considered as the players that enjoy in the game activities and collect elements such as badges or any component offered by the game. Finally, killers are defined as players that thrive on the competition, battle, or fight scenes.

2.4.2. User Input. Most of the articles, as mentioned above (21), are related to user input. User input facilitates the interaction between the user and the game. In the mobile game, the control of action and direction of the player character is often done by using input touch. Input touch makes the game more accessible and more appealing to a broader game player base. However, many game developers and researchers continue to find new technology to improve the input method. The trend of user input for the mobile game has now evolved by including inputs such as voice recognition, gesture recognition, acceleration, sense movement, and GPS (location detection) in the game. For example, Cruz et al. [9] implement accelerometer technology on the fishing game. The acceleration is showed when the player throws the fishing rod. The X- and Y-axis of the accelerometer indicates the right and forward direction, respectively. Meanwhile, Joselli et al. [18] proposal combined the input of touch, gesture movement and acceleration into the requirement of MindNinja game.

2.4.3. User Interface. The key to the success of the mobile game is highly dependent on the creativity of its user interface (UI) design. Throughout the comparison in Table 3, most of the articles (27) are discussed related to this requirement. UI is visual controls that allow game developers to implement functions or presents information in a game. The behavior of UI can respond to any input and integrate media elements such as animation, music, and 3D graphics. Basically, mobile games are developed by a unit of screens such as screen of setting menu, screen of leaderboard, screen of user login and screen of game levels. Common UI design also includes buttons, scroll-bars, textboxes, and image to display information such as instructions, the objectives of the game and the game score.

2.4.4. Gameplay Character. Gameplay character is one of the core requirements in the mobile game. Table 3 shows 27 articles discussed related to the requirement of gameplay characters. Mobile games are composed of a collection of gameplay characters or objects. Each character may come in different modelling according to its genres such as action, sports, adventures, or fantasy. This gameplay character requirement consists of player characters, enemy characters, or any dynamic objects that are moving during the gameplay. Each character will have a set of attributes and functionality; for example, a player character may have attributes such as level condition, health, status, or specialty. While the enemy character may have attributes such as attack, shoot or battle; and dynamic objects may have mass and forces applied to it. Each of gameplay characters must be updated according to its behaviour, the rules and the level of the game.

Tales of Mamochi mobile game [16] provides two different characters to represent the game player; the first character is a craftsman and the second character is a wizard. The game player will be given a weapon to fight the ghost monsters in the dungeon. Another character can be seen in this game is, Mamochi, which was created to help the player. The character consists of elements such as sweet, bitter, sour, and, also the level of stamina to indicate its health during the fight.

2.4.5. Rigid Modelling. The concept of the game as a rigid modelling happened when a game was in the in-game mode; the objects were affected to any collisions, forces, or gravity. This rigid modelling applies to objects such as floor, walls, windows, or any static obstacle. However, these models can be changed in accordance with the game requirements. Table 3 shows 24 articles that are related to the rigid modelling. For example, the rigid modelling in FunCopter mobile game [25] can be represented by kitchens, bedrooms, dining rooms, and toilets which allow the remote-control helicopter to fly around in an environment.

2.4.6. Structural Functionality. Structural functionality is a requirement that coordinate elements such as scene management, rendering, collision detection, and physics systems. To embed these technologies into mobile games, most developers use the game engine as a medium to integrate all structural elements and convey realistic illustrations in a game. In the game engine, scene management is responsible for managing the game objects and organizes their relationship. It deals with displaying the game map and the location of characters. Rendering module is responsible to generate the object image such as volumetric shading, light mapping, and occlusion culling, whereas, collision detection detects the physical edge of objects and examines whether object intersects. The detection of rigid and non-rigid objects results will be used in other modules such as physic engine and scene management. The physic engine is responsible for expressing the behaviour and expression of the object such as object move, object rotation, and collision.

In this functional requirement, 19 articles have discussed related to structural functionality. For example, in the requirement of Snowboard game [30], the collision detection is implemented using sprite collision method. The sprite collision is a pixel-based collision and detects any images or sprites that overlap during the gameplay. While in Tanks War mobile game [14], the game engine is responsible for rendering the player's tank and enemy's tank and detects any collision events with bullets.

2.4.7. Media Content. In the mobile game, media content is important in getting player immersion and engagement. It can convey the mood and the liveliness of the game. Game sound, audio, music, and animation are a part of this media content. The requirements can be used to deliver information about places or objects as well as movement or other actions by other characters. Each character, places, or game objects will be provided with sound, music or animation depending on the range of the player [35]. Table 3 shows that most of the articles (24) include the media contents in the requirements for the mobile game.

2.4.8. Rewarding System. From the reviewed literatures, 22 articles consider gaming rewards as part of the requirements. The rewarding system of the mobile game includes points, badges, and achievements. It provides immediate achievement according to the current level of the gameplay. Reward systems are different according to the nature of the game and are distributed in a different medium such as images, symbols or effects. Generally, the rewarding system can be designed in terms of fixed action rewards, random rewards, immediate rewards, rolling reward, social treasures, and prizing pacing.

2.4.9. Rewarding System. The purpose of the leaderboard is to make a simple ranking for comparison. It consists of a list of players according to their success level within the game. During the gameplay, a leaderboard is displayed to all game players. The leaderboard shows other players competing in the current game as well as information regarding the other players, such as player name, player avatar, player score, recent game activities, recent game awards and chat messages. Throughout the comparison, only four articles have mentioned related to this requirement.

2.4.10. Multiplatform Support. When designing a mobile game, it is important to consider technical specifications such as the type of devices and operating systems that support the mobile game. This is a crucial factor that could increase the size of target users. This requirement manages and controls the operation of hardware and software including the input and output of the game. Nowadays, cross-platform mobile game has become a trend and game developers are seeking on implementing their mobile game in multiple platforms such as iOS, Android and Windows. Most of the articles (29) considered this platform and device functionality in their mobile game.

3. Discussion

Based on the review of existing literature regarding the functional requirements of the mobile game, a few findings have been found. The characteristics related to mobile games are not clearly defined in general and to the best of researcher's knowledge, there is no literature has been reported exploring the characteristic of mobile game application specifically to mobile game functionality. Thus, the evaluation of existing articles was found to be the key process to understand functional factors better and examine the trends associated with mobile game components. The requirements of the mobile game may differ depending on the goals and objectives of the game; however, it is clear from the review that mobile game application generally shared similar functionalities. From the review, majority of articles discussed related to the inclusion criteria except to the leaderboard criteria which mentioned by four articles only.

Second, the design of the mobile game is rather a complex process from a functional point of view. The way mobile game requirement is incorporated into software effort estimation is one of aspect deserving in future research. In the context of function point, the obtained characteristics of mobile game application can be integrated as a unit of functional process in the COSMIC FFP, cooperated as a logical file in the IFPUG FPA, or derived as a unit of measurement in FISMA or NESMA FSM method. This will open opportunities to the researchers to formulate new formal rules, integrate with modern development or other instruments with the expectation to yield more accurate results.

However, it is also crucial to clarify the mobile game characteristic will vary depending on the moment in which it was released since the technology is moving faster and continually changing. With the advent of network connectivity, the mobile game is predicted to evolve rapidly in terms of user behaviour, communication technology as well as social aspects. These trends may affect the future design of the mobile game and might influence to FSM method components as well.

4. Conclusion

This paper has presented a literature review of articles from 2000 to 2019, addressing the common functionality of mobile game applications. The initial search from scientific libraries returned with 53 articles, of which only 30 articles were selected for the review based on the defined method. The details of relevant papers were extracted according to the ten criteria including game player, user input, user interface (UI), gameplay character, rigid modelling, structural functionality, media content, rewarding system, leaderboard, and multiplatform support.

Developing an FSM model for the mobile game can be difficult due to difficulty in providing a complete picture of the functional requirement of the mobile game as the mobile game design way broader due to the diversity of scenarios, features and application environments. Thus, identifying common factors of the mobile game through literature review is considered sufficient, and each of the criteria will have a significant influence on the effort estimation.

The next stage of this research is to find a solution to apply these criteria to support function point measurement and to improve the effort estimation activity. The future work will consist of proposing a new measurement design or set of rules to extract the ten criteria into the concept of FSM model.

Acknowledgment

The authors express appreciation to Faculty of Computer and Mathematical Sciences, Universiti Teknologi MARA for sponsoring this paper.

References

[1] ISO/IEC 20926: Software Engineering - IFPUG Unadjusted FSM Method - Counting Practices Manual 2003.

[2] ISO/IEC 20968: Software Engineering - MkII Function Point Analysis - Counting Practices Manual 2002.

[3] ISO/IEC 24570: Software Engineering - NESMA Functional Size Measurement Method v.2.1 - Definitions and counting guidelines for the application of Function Point Analysis 2005.

[4] ISO/IEC 19761: COSMIC Full Function Points Manual v. 2.2 2003.

[5] Rusli N I A, and Abdullah N A S UML Point for mobile game: A measurement method for sizing mobile game design *Journal of Engineering and Applied Sciences* 2017 vol 12(3) pp 481-487.

[6] Abdullah N A S and Rusli N I A Mobile game size estimation: COSMIC FSM rules, UML mapping model and Unity3D game engine *IEEE Conference on Open System (ICOS)* 2014 pp 42-47.

[7] Alcid A S Bandril L. B. P de Guzman A. E and Lopez L. J. C Analysis, design, development, implementation and evaluation of a serious game designed to inform users on environmental issues *International Journal of Computing Sciences Research* 2017 vol. 1(1) pp. 11-23.

[8] Shen X and Barakova E. I My Drama: Story-based game for understanding emotions in context *International Conference on Intelligent Technologies for Interactive Entertainment* 2017 pp. 220-230.

[9] Cruz S, Carvalho A A A and Araujo L A game for learning history on mobile devices *Education and Information Technologies*, 2017, vol. 22(2), pp. 515-531.

[10] Franti P, Mariescu-Istodor R and Sengupta L O-Mopsi: Mobile orienteering game for sightseeing, exercising, and education *ACM Transactions on Multimedia Computing, Communications, and Application (TOMM)* August 2017 vol. 13(4) 56.

[11] Sweetser P, Johnson D, Wyeth P, Anwar A, Meng Y and Ozdowska A GameFlow in different game genres and platforms *Computer in Entertainment (CIE)* February 2017 vol. 15(3) pp. 1-24.

[12] Tai Z, and Hu F Mobile games in China: Ongoing industry transformations, emerging game genres, and evolving player dynamics *Mobile Gaming in Asia* 2017 pp 173-190.

[13] Yue W S, and Ying C Y The evaluation study of gamification approach in Malaysian history learning via mobile game application *International Conference on Advanced Learning Technologies (ICALT)* 2017 pp 150-152.

[14] Shi Y P The development of Tanks War mobile game based on Android system *MATEC Web of Conferences* 2016 vol. 63.

[15] Mildner P, Stamer N and Effelsberg W From game characteristics to effective learning games *Joint International Conference on Serious Game* Springer International Publishing 2015 pp 51-62.

[16] Kurniati A, Tanzil F and Purnomo F Game development "Tales of Mamochi" with role playing game concept based on Android *International Conference Science and Computational Intelligence* Procedia Computer Science 2015 vol 59 pp 392-399.

[17] Willis G D ANTICS: A cross-platform mobile game *Electronic Theses, Projects, and Dissertations* 2015, 139.

[18] Joselli M, Binder F, Clua E and Soluri E MindNinja: Concept, development and evaluation of a mind

action game based on EEGs *Brazilian Symposium on Game and Digital Entertainment (SBGAMES)* 2014 pp 123-132.

[19] Cao L, Chang A and Mao Y Mobile game application: The MagicTale *Computer Engineering Senior Theses* 2014 27.

[20] Hussain A, Mutalib N A and Zaino A A usability testing on JFakih learning games for hearing impairment children *5th International Conference on Information and Communication Technology for The Muslim World (ICT4M)* 2014 pp 1-4.

[21] Mikkila M Creating a mobile MMO game Retrieved from World Wide Web: https://www.theseus.fi/bitstream/handle/10024/80856/Mikkila Miso.pdf?sequence=1 2014.

[22] Bartel A and Hagel G Engaging students with a mobile game-based learning system in university education *IEEE Global Engineering Education Conference (EDUCON)* 2014 pp 957-960.

[23] Abdullah N A S, Rusli N I A and Ibrahim M F A case study in COSMIC functional size measurement: Angry Bird mobile application *IEEE Conference on Open System (ICOS)* 2013 pp 139-144.

[24] Sintoris C, Stoica A, Papadimitriou I, Yiannoutsou N, Komis V and Avouris N MuseumScrabble: Design of a mobile game for children's interaction with a digitally augmented cultural space *International Journal of Mobile Human Computer Interaction (IJMJC)* 2012 vol 2(2) pp 53-71.

[25] De Macedo D V and Rodriques M A F Experiences with rapid mobile game development using unity engine *Computer in Entertainment* 2011 vol 9(3) pp 1-12.

[26] Peker A G, and Can T A design goal and design pattern based approach for development of game engines for mobile platforms *Proceedings of 16th International Conference on Computer Games: AI, Animation, Mobile, Interactive Multimedia, Educaional and Serious Games* 2011 pp 114-120.

[27] Zhu Q Y, Jin Y, Xu C and Gen R A UML model for mobile game on the Android OS *International Conference on Advances in Engineering* 2011 vol 24 pp 313-318.

[28] Chee Y S, Tan E M and Liu Q Statecraft X: Enacting citizenship education using a mobile learning game played on Apple iPhones *Proceedings of the 6th IEEE International Conference on Wireless, Mobile and Ubiquitous Technologies in Education (WMUTE)* 2010 pp 222-224.

[29] Diah N M, Ismail M, Ahmad S and Syed Abdullah S. A. SMJawi on mobile devices with Jawi wordsearch game application *IEEE International Conference on Science and Social Research (CSSR),* 2010, pp. 326-329.

[30] Jhingut M Z, Ghoorun I M, Nagowah S D, Moloo R and Nagowah L Design and development of 3D mobile games *Proceedings of the 3rd International Conferences on Advances in Computer-Human Interactions (ACHI)* 2010 pp 119-124.

[31] Shiratuddin N and Zaibon S B Mobile game-based learning with local content and appealing characters *International Journal of Mobile Learning and Organisation* 2010 vol 4(1) pp 55-82.

[32] Koivisto E M I Mobile games 2010 *CyberGames '06: Proceedings of the 2006 International Conference on Game Research and Development* 2010 vol 223 pp 1-2.

[33] Fotouhi-Ghazvini F, Earnshaw R, Robison D and Excell P The MOBO City: A mobile game package for technical language learning *International Journal of Interactive Mobile Technologies (IJIM)* 2009 vol 3(2) pp 19-24.

[34] Chehimi F, Coulton P and Edwards R Evolution of 3D mobile games development *Personal and Ubiquitous Computing* 2008 vol 12 pp 19-25.

[35] Ekman I, Ermi L, Lahti J, Nummela J, Lankoski P, and Mayra F Designing sound for a pervasive mobile *game Proceedings of the 2005 ACM SIGCHI International Conference on Advances in Computer Entertainment Technology (ACE'05)* 2005 pp. 110-116.

[36] Davidsson O, Peitz J and Bjork S Game design patterns for mobile games *Project Report to Nokia Research Center* Finland 2004.

[37] Alfadhli S and Alsumit A Game-based learning guidelines: Designing for learning and fun *Computational Science and Computational Intelligence (CSCI)* 2015 pp 595-600.

[38] Kurkovsky S Mobile game development: Improving student engagement and motivation in introductory computing courses *Computer Science Education* 2013 vol 23(2) pp 138-157.

[39] Kleinschmidt C and Haag M Evaluation of game egine for cross-platform development of mobile serious games of health *Stud Health Technol Inform* 2016 vol 223 pp 207-214.

IOP Conf. Series: Materials Science and Engineering **769** (2020) 012056 doi:10.1088/1757-899X/769/1/012056

Mobile Learning Application Impact Towards Student Performance in Programming Subject

Nurul Farahah Abdul Halim, Danakorn Nincarean Eh Phon
Faculty of Computing, Universiti Malaysia Pahang, Pahang, Malaysia

danakorn@ump.edu.my

Abstract. Writing a program is significant basic proficiency for computer science students. However, majority of the students found it is difficult to learn and score a high mark. In an era of rapidly changing technologies, mobile learning has received increasing emphasis and widely used among educational practitioners and researchers. Thus, the objective of this study is to investigate the factors that influence students learning performance and evaluate the effectiveness of mobile learning using Learn C application in Programming subject. Prior to the factors that influence students learning outcome, a survey was conducted with 88 first year undergraduate students from Faculty of Computer Systems & Software Engineering. Next, a pre-experimental one group pretest-posttest research design was employed with 17 students as respondents to examine the impact of mobile learning application. The collected data was then analysed using thematic analysis, descriptive and inferential statistics. The results showed that there are several factors that affect students learning performance in programming that is, misunderstanding, lack of practices, poor logical thinking and problem solving, no prior knowledge, psychological disorder, less interest, ineffective teaching method and bad attitude. The paired sample t-test of student's pre-test and post-test scores demonstrate that there were no significant differences before and after the treatment through mobile learning application given. The study has gone some way in enhancing our understanding of factors that affect student in programming learning which could be used to facilitate the development of the students programming abilities.

1. Introduction

The education sector nowadays has undergone a process of change in teaching and learning styles [1]. This change can be seen from using whiteboard and chalk up to the use of mobile technology. Recently, the adoption of mobile learning has received increasing emphasis and widely used among higher education institutions. The acceptance of mobile learning among students can increase their understanding and access knowledge in freeing from the constraints of time and place [2].

Mobile learning has turn out to be a learning instrument that come with great benefit in both classroom and outdoor learning [3]. As the growth of mobile learning, there have been several investigations into the use mobile technologies in science learning to assist educators to expand their abilities [4][5]. Therefore, there are a high demand for programming subject due to a high dropout rate in higher education among Information and Communication Technology (ICT) students especially registered in computer science courses [6][7].

Writing a program is a significant fundamental skill for computing students. However, majority of the students face difficulty in learning and scoring a good mark [8]. One of the main reason of this difficulty is due to this subject is new to most of them [9]. Students interest in the subject is also important [10]. Moreover, how the teachers conduct the teaching and learning in a class also can be a influence the student to be less interested in the subject.

For many years, researchers have been looking for another possibility in teaching programming subjects. Lacking the comprehension in programming concept has reduced undergraduate's interests to pursue further investigation and self-experimentation. Computing field is facing a crisis in the current enrolment due to decreasing demand of computing as a career of choice [11] [12] and indirectly it effects on the reputation of the institute [13]. There are several factors that drag to this problems including students are loss of interest in programming course due to subject difficulties and it pushes them to the high failure and dropout rates [14]. This impact has been alarming many lectures especially from universities of developing countries [15]. However, computing lecturers have tried many efforts to overcome this problem especially approach a variety of instructional methods to assist the novice students [16].

Recent studies have discussed on the educational benefits that mobile learning application have garnered for programming learners. However, the evidences are mixed in terms of the usefulness of mobile learning application. Positive outcome with mobile based-learning have been reported [17] [18], however several studies found no difference in the outcome of utilization of mobile learning application. Therefore, in this study, the researcher seeks to determine the factors that influence students learning performance and investigate how far the effectiveness of integration of mobile learning application into programming course and find the inferences for teaching and learning practices.

2. Literature Review

Programming language is one of the main knowledge areas in the Computer Science curriculum. It covers programming paradigms, concepts, technologies, patterns and algorithm. Writing a program is considered as a vital skill that must be mastered by any individual interested in continuing a career in computer science field.

2.1. Factor that influence students learning performance in programming course

Several factors that may influence students learning outcome in programming course have been identified as shown in Table 1.

Table 1. Factor that affect students learning performance in programming

Authors	Factors that affect students learning performance
[7]	Lack of problem solving skill
[6] [19]	No prior knowledge
[6] [15]	Low motivation
[20]	Poor lecturer skill
[7]	Poor mathematical knowledge
[15]	Peer influences
[12]	Lack of future expectation

2.2. Mobile Learning

Mobile learning can be referring to the usage of mobile devices to access learning resources, to learn, collaborate and share ideas among each other at anytime and anywhere with the support of mobile devices and networks. Therefore, it considered as a fastest growing learning platform [21] across a number of dimension such as mobility of technology, learners, educator and learning [11].

In addition, academic achievement of the students is considered enhanced by mobile learning since it offers self-regulating learning environment which can nurture student engagement and provide chances to make learning process essential to their daily life activities [1]. However, the achievement must be supported by institution commitment to mobile learning and institution learning practices to the success of mobile learning [21]. There are several findings from previous studies claims that the mobile Apps has positive effect on the knowledge comprehension of the students [22] as well as support learning for the delivery of instruction [23] and majority of the students revealed favorable attitudes towards the practice and use of mobile learning apps.

2.2.1. Learn C Apps. Learn C Application is a programming tool for learning C language. The app can be access through the browser of Android device and also can be installed through Google Play. The apps consist of the following features: various notes covering overall chapter in Programming Technique subject, code example supports with pictures, various of C programs, downloadable notes for quick revision, MCQ type tests, helpful interview questions, useful links to refer to resources and quizzes. Table 2 presents a comparison characteristic between C programming language apps. Within seven apps, Learn C apps is chosen to be used as a learning tool in with the integration of mobile learning.

Table 2. Comparison characteristics between C programming language apps

Criteria	Learn C	Learn C#	C Prowess	101 C Programmin g Problems	C Programmin g	All C Program s	C Programmin g - 200+ Offline Tutorial and Examples
Navigation	Easy	Hard	Easy	Easy	Easy	Moderate	Easy
Interface	User-friendly	User-friendly	User-friendly	User-friendly	User-friendly	User-friendly	User-friendly
Content of learning	Covered	Covered	No Flowchar t and algorithm topic	None	No Flowchart and algorithm topic	None	No Flowchart and algorithm topic, control statement 1,File Processing
Connectivity method	Offline	Offline	Offline	Offline	Offline	Offline	Offline
Compiler	No	No	No	No	No	No	No
Storage size	11 Mb	5.3 Mb	3.3Mb	5Mb	2.5Mb	6.9Mb	3.6Mb
Language	English	English	English	English	English	English	English
Exercises/Quizze s	Both	Both	Both	No	No	No	No
Example of programs	Yes	No	No Flowchar t and algorithm	Yes	Yes	Yes	Yes
Developer	ReezX	SoloLear n	Faiz & Daneyal	AppsFacto	SNJ	Mahidhar Guggilam	Ravrani Software Developers
Purchase	Offer in-app	Free	Free	Free	Free	Free	Free

3. Methodology

3.1. Research Design
This study used a research design of a quantitative approach with pre-experimental design of one group pre-test post-test design. The researcher tended to use a "single group study", due to the researchers concern on threats of internal validity [24]. Any direct or indirect interaction between the respondents in an experimental or control group can invalidate the comparisons between these two groups. Furthermore, quantitative data analysis method is suitable for this study because it offers a quantitative explanation of student's feedback in programming concept [25].

3.2. Participant
The samples involved in this research were categorized into two sets. Set I included students who were involved in answering the first objective. The sample consisted of 88 first year undergraduate students from Faculty of Computer Systems & Software Engineering. These participants were carefully chosen as they were within the targeted population and just begun to learn programming subject. In set II, 17 students from set I students were involved to answer the second objective that were included in the present study.

3.3. Research Instruments
There are two instruments were employed in this study. They consisted of online questionnaire and student's performance tests. A questionnaire is adopted and adapted from the instrument developed by [27] to assess the factors, perception of the difficult programming topics and revision methods. This questionnaire aims to discover the factors that influence the learning performance of the students based on their learning experience. Prior to the effectiveness of mobile learning application, a pre-test and post-test assessment was used in this study. The test comprised of structured questions which were constructed based on the previous final exam questions. The test has been validated and verified by two subject content experts.

3.4. Procedure
Figure 1 displays the stage of the experimental design. At the beginning of the semester, the online survey is conducted among 88 participants who take the Programming Technique subject in their second semester and the Data Structure and Algorithms at the present. Approximately 5 minutes are taken by the participants to complete the survey. It seeks to examine the factors that influence student learning and analyze the difficult topics of programming concept in terms of their learning experience.

The pre-test is administered in the second week of the semester. The participants were provided one hour to complete the test. The aim of the pre-test is to assess the prior knowledge in programming concepts. From the (n=88) students, 17 students are employed for the experimental group.

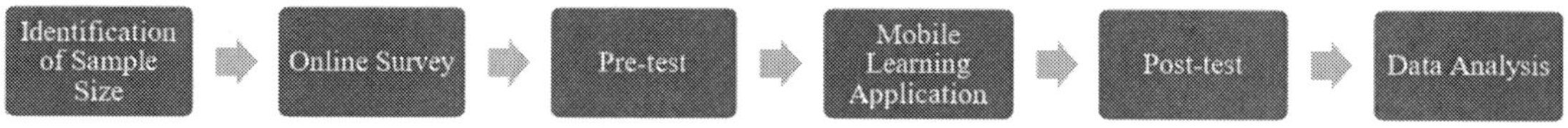

Figure 1. Experimental Design

The mobile learning application is accessible to the experimental group in the third week of the semester along with the conventional. The conventional learning is defined as when the lecturer interacts face-to-face with students while delivering instructional material. Along the treatment period, students were using mobile learning in the class as self-learning tool. They explore the mobile apps on their own while the lecturer act as facilitator. In the seventh week of the semester, the post-test is administered by experimental group to measure and correlate their learning performance. The collected data then analyzed using thematic analysis, descriptive and inferential statistics.

4. Result and Discussion

4.1. Factors that influence students learning performance in subject of Programing Technique

The first objective attempts to recognize the factors that influence students learning performance in subject of Programing Technique. Based on Figure 2, there are 65.1 % students are facing the problem of Pointer Operations and Expression topics followed by Pointer to Pointer (62.8 %), Multi-dimensional array (53.5%), Pointer variables (47.7%), Pass by value and pass by reference (46.5%), End of file and file verification (43%), Writing and reading file (37.2%), Array of character and string (37.2%), Basic operation and declare file (31.4%), One-dimensional array (31.4%), Open and close file (27.9%), Calling and return function (24.4%), User-defined function and prototype function (19.8%), Pre-processor directives and header file, Identifier Variables and Constant, Programming Keyword and Reserve Word (16.3%), Loop-while, Do-while and For (11.6%), If-else, Switch-case, continue and break (10.5%),Pseudocode and flowchart (8.1%), and Input and Output operation (4.7%). Therefore, the researcher can identify that Pointer to pointer is the most difficult topic for students to understand in programming concept.

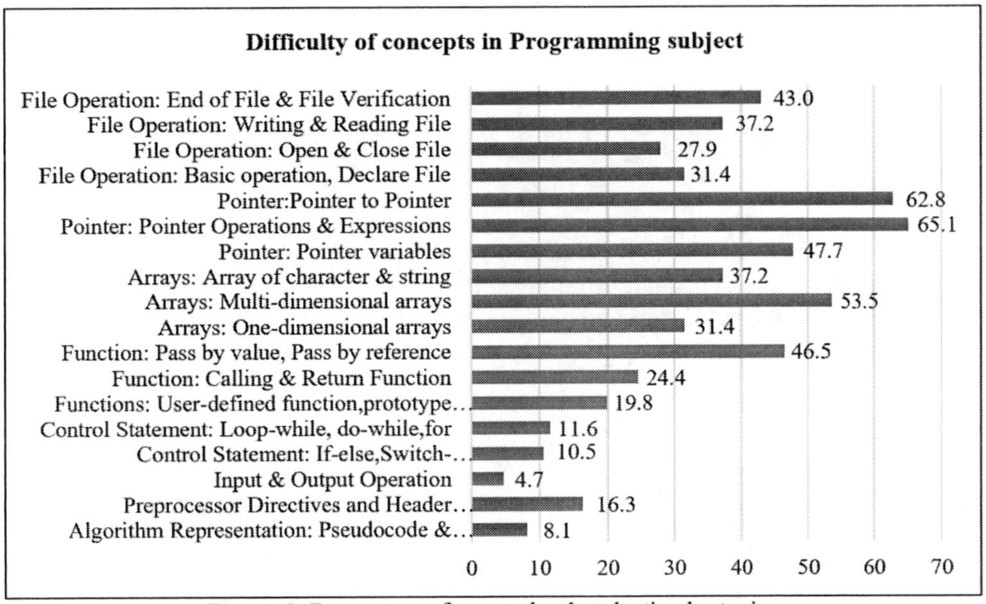

Figure 2. Percentage of respondent's selection by topics

Figure 3 shows the difficult topics with number of students. Based on the results, the number of students is higher for four critical topics in subject of Programming Technique such as Function, Array, Pointer and File Processing compared with other topics. However, only a few students think that topic of Operator and Expression is difficult. Figure 4 shows the domain factor that affect students learning performance. Compared to other domain factors, 33 % of students are misunderstanding of programming concepts, following by 22 % of students are poor in logical thinking skills and problem solving, 10% of students have no prior knowledge, 6% of students are psychological disorder, 5% of students are blaming to the ineffective of lecturers teaching method, 4% of students are less interest in programming subject, and 1% are students bad attitude such as laziness to study instead of do practices.

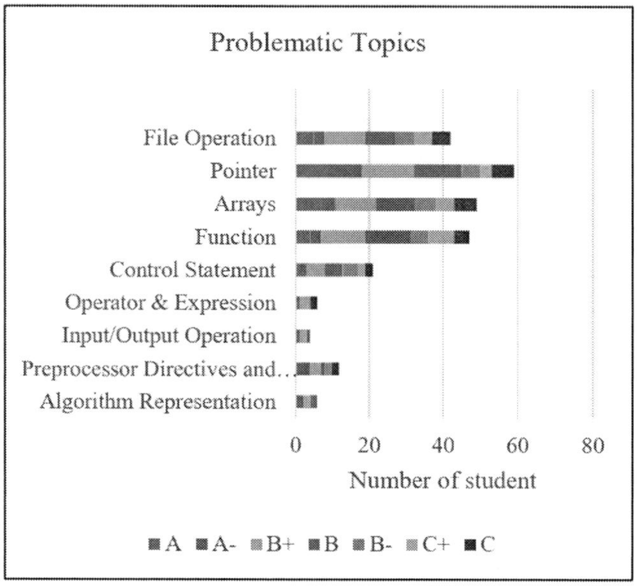

Figure 3. Topic difficulty by grade

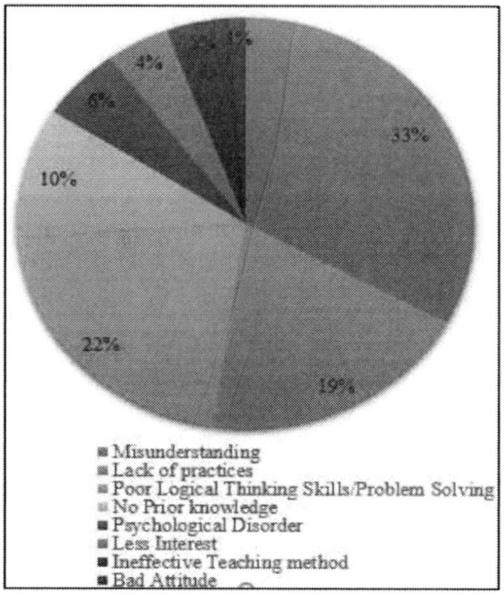

Figure 4. Factors that affect student learning performance

4.2. Effectiveness of mobile learning application in Programming Technique

Figure 5 illustrates the differences of the learning outcome before and after using mobile learning in this study. It revealed that there is a small difference between a scores in pre-test and post-test assessment. To add, the result showed that the participants understanding of the subject is increasing for 8 students while 7 students is decreasing and 2 students have a same score and no differences after using mobile learning application.

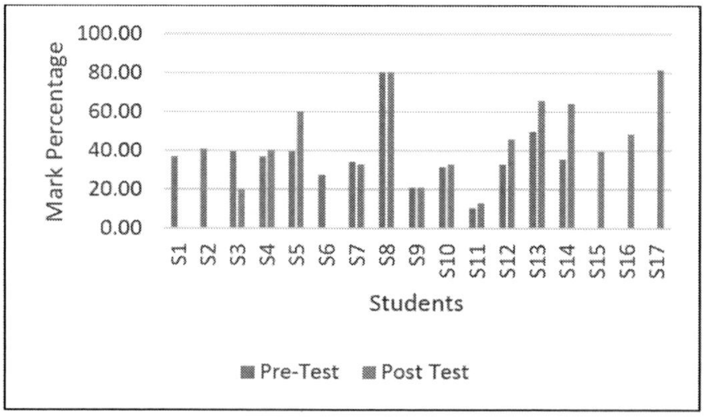

Figure 5. Comparison of pre-test score and post-test score

Prior to determine the significant difference between the mean of pre-test and post-test assessment scores, the researcher had undergone a normality test. Shapiro-Wilk used to determine whether the distribution of the data was normally distributed. The significant value of Shapiro – Wilk normality test for pre-test data is .065 and post-test data is .012. Since the value is greater than the chosen alpha level, which is concluded that the data is normally distributed.

With the proven that the data was normally distributed, the researcher continued in conducting paired-sample t-test analysis. Table 3 presents the descriptive analysis for pre-test and post-test assessment. For N=17, mean for pre-test is 30.4, standard deviation is 20.2 and standard error mean is 4.9 while mean for post-test is 38.7, standard deviation is 22.5 and standard error mean is 5.5.

Table 3. Descriptive analysis

Descriptive analysis				
	Mean	N	Std. Deviation	Std. Error Mean
Pre-Test	30.4171	17	20.21203	4.90214
Post-Test	38.6976	17	22.50843	5.45910

Table 4 showed the result for paired sample t-test for this study. The result indicated that t (16) = -1.233, p > 0.005. Because of the means of the two test scores and the direction of the p-value, the researcher can conclude that there was no statistically significant enhancement of understanding towards the programming concepts in Programming Technique for experimental group which essentially states that the intervention mobile learning application is not give a huge impact.

Table 4. Paired Sample T-Test

Paired Samples Test								
	Paired Differences							
	Mean	Std. Deviation	Std. Error Mean	95% Confidence Interval of the Difference Lower	Upper	t	df	Sig. (2-tailed)
PreTest - PostTest	-8.28059	27.69644	6.71737	-22.52078	5.95961	-1.233	16	.235

Based on the analysis performed on investigation the factor that affect students learning performance, the study found that students are posing four critical topics in programming Techniques such as Function, Array, Pointer and File Processing. Moreover, misunderstanding in learning the programming concepts bring them to the major factor to the dropout rate and then fail. The study also found that eight participants understanding of the subject in increasing while seven students is decreasing and two students have a same score.

5. Conclusion

In this investigation, the aim was to classify the factors that influence students learning in programming subject and the effectiveness of mobile learning application impact towards students' achievement in programming subject. This study has found that several factors that affect students learning performance in programming. Based on these factors, the finding can give some initial understanding and guidelines for the educators to improve the pedagogical strategy in teaching programming subject. The integration of mobile learning application in this study also revealed that one of these differences were statistically significant.in the students' performance. Learn C apps is chosen to be utilized as it covers all chapter studied in Programming Technique, an easy navigation, user-friendly in use, stand-alone apps after installation, consists both assessment including quizzes and exercises and several of example of programs. Taken together, these findings can contribute to the features that might have to be use in mobile application that can give more effect toward programming understanding. This research extends our knowledge on the factors and effect of mobile learning application in programming concept. The generalizability of these results is subject to certain limitations. For instance, the experimental group used in this study were relatively limited. Therefore, further research is needed to account for a larger sample.

Acknowledgement
The authors would like to thank the Universiti Malaysia Pahang (UMP) for the financial support through RDU Grant Vot No. RDU180361 initiated by UMP.

References

[1] Baharudin, A. F., Sahabudin, N. A., & Kamaludin, A. 2017. Behavioral tracking in E-learning by using Learning styles approach. *Indonesian Journal of Electrical Engineering and Computer Science*, **8(1),** 17-26.

[2] Burston, J. 2016 Realizing the potential of mobile phone technology for language learning, *IALLT Journal of Language Learning Technologies*, **41(2).**

[3] Sung, Y. T., Chang, K. E. and Liu, T. C. 2016 The effects of integrating mobile devices with teaching and learning on students learning performance: A meta-analysis and research synthesis, *Computers and Education*. Elsevier Ltd, **94**, pp. 252–275

[4] Crompton, H. et al. 2016 The Use of Mobile Learning in Science: A Systematic Review, *Journal of Science Education and Technology*, **25(2),** pp. 149–160

[5] Oyelere, S. S. et al. (2018) Design, development, and evaluation of a mobile learning application for computing education, *Education and Information Technologies*, **23(1)**, pp. 467–495

[6] Kori, K. et al. 2016 The Role of Programming Experience in ICT Students Learning Motivation and Academic Achievement, *International Journal of Information and Education Technology*, **6(5)**, pp. 331–337

[7] Dolgopolovas, V., Jevsikova, T. and Dagiene, V. 2018 From Android games to coding in C—An approach to motivate novice engineering students to learn programming: A case study, *Computer Applications in Engineering Education*, **26(1)**, pp. 75–90

[8] Lishinski, A., Yadav, A., & Enbody, R. 2017 Students emotional reactions to programming projects in introduction to programming: Measurement approach and influence on learning outcomes. *In Proceedings of the 2017 ACM Conference on International Computing Education Research* (pp. 30-38). ACM.

[9] Anand, V. K. *et al.* 2018 Recursive clustering technique for students performance evaluation in programming courses, in *Majan International Conference (MIC), 2018*. IEEE, pp. 1–5.

[10] Jin, L. and Su, Y. 2018 The Teaching Reform of Integration of Theory and Practice with Interest Leading and Ability Training, in *Proceedings of the 2018 International Conference on Big Data and Education*. ACM, pp. 116–120.

[11] Anohah, E., Oyelere, S. S. and Suhonen, J. 2017 Trends of mobile learning in Computing Education from 2006 to 2014: A systematic review of research publications, *International Journal of Mobile and Blended Learning (IJMBL)*. IGI Global, **9(1)**, pp. 16–33.

[12] Kinnunen, P. et al. 2018 Understanding initial undergraduate expectations and identity in computing studies, European Journal of Engineering Education. *Taylor & Francis*, **43(2)**, pp. 201–218.

[13] Papadakis, S. and Kalogiannakis, M. 2018 Using Gamification for Supporting an Introductory Programming Course. The Case of ClassCraft in a Secondary Education Classroom, in Brooks, A. L., Brooks, E., and Vidakis, N. (eds) *Interactivity, Game Creation, Design, Learning, and Innovation.* Cham: Springer International Publishing, pp. 366–375.

[14] Álvarez, A. and Larrañaga, M. 2016 Experiences Incorporating Lego Mindstorms Robots in the Basic Programming Syllabus: Lessons Learned, *Journal of Intelligent & Robotic Systems*, **81(1)**, pp. 117–129.

[15] Dasuki, S. and Quaye, A. 2016 Undergraduate Students Failure in Programming Courses in Institutions of Higher Education in Developing Countries: A Nigerian Perspective, *The Electronic Journal of Information Systems in Developing Countries*. Wiley Online Library, **76(1)**, pp. 1–18.

[16] Kunkle, W. M. and Allen, R. B. 2016 The Impact of Different Teaching Approaches and

Languages on Student Learning of Introductory Programming Concepts, *Trans. Comput. Educ.* New York, NY, USA: ACM, **16(1)**, p. 3:1-3:26

[17] Zhao, J., Wei, Y., Maideen, I., Moe, Z. K., & Nasirudeen, A. M. A. 2018 The Relationship between Smartphone use and Academic Performance in a Sample of Tertiary Students in Singapore: A Cross-Sectional Study. *i-Managers Journal of Educational Technology*, **14(4)**, 28.

[18] Demir, K. and Akpinar, E. 2018 The Effect of Mobile Learning Applications on Students Academic Achievement and Attitudes toward Mobile Learning., *Malaysian Online Journal of Educational Technology*. ERIC, **6(2)**, pp. 48–59

[19] Cazzola, W. and Olivares, D. M. 2016 Gradually Learning Programming Supported by a Growable Programming Language, *IEEE Transactions on Emerging Topics in Computing*, **4(3)**, pp. 404–415.

[20] Hijón-Neira, R. *et al.* 2017 An analysis of the current situation of teaching programming in Primary Education, in *Computers in Education (SIIE), 2017 International Symposium on*. IEEE, pp. 1–6.

[21] Alrasheedi, M., Capretz, L. F. and Raza, A. 2016 Managements Perspective on Critical Success Factors Affecting Mobile Learning in Higher Education Institutions—An Empirical Study, *Journal of Educational Computing Research*, **54(2)**, pp. 253–274

[22] Arain, A. A., Hussain, Z., Rizvi, W. H., & Vighio, M. S. 2018. An analysis of the influence of a mobile learning application on the learning outcomes of higher education students. *Universal Access in the Information Society*, **17(2)**, 325-334.

[23] Fraga, L. M. and Flores, B. B. 2018 Mobile learning in higher education, in Handbook of Research on Mobile Technology, Constructivism, and Meaningful Learning. *IGI Global*, pp. 123–146

[24] Phon, D.N.E, Rahman, M. H. A., Utama, N. I., Ali, M. B., Halim, N. D. A., & Kasim, S. 2019. The Effect of Augmented Reality on Spatial Visualization Ability of Elementary School Student. *International Journal on Advanced Science, Engineering and Information Technology*, **9(2),** 624-629.

[25] Ghazal, S. et al. 2018 Acceptance and Satisfaction of Learning Management System Enabled Blended Learning Based on a Modified DeLone-McLean Information System Success Model, *International Journal of Information Technology Project Management (IJITPM)*. IGI Global, **9(3)**, pp. 52–71

[26] Rahmat, M. *et al.* 2012 Major problems in basic programming that influence student performance, *Procedia-Social and Behavioral Sciences*. Elsevier, **59**, pp. 287–296.

A Code Profiling Model for StART

Roslina Mohd Sidek

Faculty of Computing, College of Computing and Applied Sciences,
Universiti Malaysia Pahang, Lebuhraya Tun Razak,
26300 Kuantan Pahang, Malaysia

E-mail: roslinams@ump.edu.my

Abstract: Test case selection in software testing is one of the process that support to quality in finding bug. Decreasing number of bug may increase quality of software. Many strategies have been proposed by researcher for software testing. Specially to avoid exhausted testing which all the testers take into account which may increase time and performance. One of the strategy is StART. The code profiling is a part of StART, which is a selection test cases strategy using code profiling. The strategy is based on the adaptive random testing and support with code profiling in the strategy. The aim of this paper is to show the code profiling model. The selection of domain area to test is based on the highest probability in the code profiling result. The probability of in the code profiling is calculated and the highest value to be chosen as the first area to test. The value to be chosen as a domain area for the first to be tested in SUT. This model used AspectJ as case study to show the model can be implemented for a new programming paradigm. Tetris selected as case study to show the model flow. From the case study of Tetris, the model shows the area to be tested first is in the package Gui, for aspect menu and advice after the probability value shows the highest result. For future study, this model need to test their efficiency of their strategy.

Key words: code profiling, aspect, StART, model.

1. Introduction

There are many strategies in selection test case in testing phase such as random and systematic approach. The test cases are the essence of software testing which need clarification of information should be tested. Random approach [1], [2] selects the test cases without consider any information behind the software under test (SUT). Random testing is very simple and easy strategy to apply due to unbiased [3] in a way of selection normally happened from human mistake. This reason random testing is very popular testing strategy to be used in selection the test cases when to test software. Meanwhile, systematic approach [4], [5] will use certain strategy to get the best test cases to find the error or bug [6]. The information behind the software in systematic approach is considered to select which part that need to test to get potential error. Nevertheless, both approaches have similar objectives which are to avoid more effort to test all test cases known are known as exhausted testing.

Test cases is a set of conditions or a specification of the input, condition or variables under which a tester will determine whether a system under test satisfies requirements or works correctly. Test cases underlie testing that is methodical rather than haphazard. A battery of test cases can be built to produce the desired coverage of the software being tested. Formally defined test cases allow the same tests to be run repeatedly against successive versions of the software, allowing for effective and consistent. In testing, to test all test cases is impossible. The problem is to test all combination of input is time consuming [7]. Such as, if test Internet Explore shows in [7] for Advance Option which have 53 condition produce 108,086,391,056,891,904 possible combinations of conditions.

Selection test cases using model [8] can be used random or systematic approach and even combination of both random and systematic. Actually, no standard for test case selection. Thus, selection is depending on strategy that tester try to apply during testing. Offering many strategies may support the testing process more option to be chosen and give more testing to find bug.

In this study, code profiling had been chosen as strategy that tester need to be considered. This proposed model uses AspectJ as a code to be tested. The code involved consists of core concern and crosscutting concern. The crosscutting concern is the area of study. Basically, source code can be many package, module and line of code such as [9] measuring the time spent on each line of code, code coverage or memory usage during its execution, [10] using probability calculation and focus on line of code, and package used. [11] gathered profiling information is structured in line with the application structure in terms of packages, classes, and methods. In this paper, the elementary of code is the advice. The advice consists of after, before and around.

Detail explanation for the model in code profile with the probability calculation organized according to the sections. Section 2 discuss about related work, section 3 research background, section 4 proposed model and section 5 is about discussion and section 6 conclusion and future work.

2. Related Work

In this study, code profile used to help in selection test case in software testing especially to avoid biasness [12],[10], [11]. In [12] , the author used the profiling in forecasting model selection and the accuracy. Performance has been assessed in terms of a model's ability to generate both unbiased and accurate forecasts, and accuracy has been examined using both error magnitude and directional change error criteria. In [10], the author generate code profile using statistical testing to increase the reliability of test case selection. [11] code profile generated to easily build profilers and visualize profiling information. It is also can be used to help in many areas such as to show usage model [13]. It provides the most cost- effective use of the testing budget and the only direct statistical method for making inferences concerning reliability of software in operational use.

3. Research Background

Test case selection in aspect-oriented domain still need more strategy especially in random approach. [14] researcher said the exclusive random testing may to cater this testing strategy. As a software tester to test all (known as exhausted testing) is impossible due to time limitation and expensive. That is the reason test cases to be selected wisely to be test to minimize the cost.

3.1 StART Strategy

StART is a strategy to select the test cases [10]. The strategy is implemented in aspect-oriented program domain area. So, the process model based on AspectJ code. The process code profiling involved with AspectJ code and focus on advice: after; before and around [10]. In this paper, model proposed only for code profiling in the strategy. StART also consider about the distance between test cases [15]. The whole strategy discusses in different research paper.

3.1.1 Code Profiling. This code profiling is referring to the consideration of element used in code. It has been discussed in [10]. If the code is object-oriented the classes and the method will take as code profile with using operational profile. Operational profile is to determine the execution frequencies of various thread and to use this information to select thread for system testing. One way to determine the operational profile of a system is to use decision tree. The decision tree shows the transition probabilities value. The

IOP Conf. Series: Materials Science and Engineering **769** (2020) 012057 doi:10.1088/1757-899X/769/1/012057

value for any state need is to be found or estimated of each going transition. The sum of these must be 1. The example of operational profile shown in [16].

3.1.2 Probability. There two type of probability theory in the study of software testing: the probability that a particular path of statements executes; and the other generalized to a popular industrial concept called operational profile [17]. It is a tool for the design and analysis of probabilistic and randomized algorithms.

The probability concept explain detail in [6]. It is defined as a sample space S, which is a set of elements called elementary events. The elementary event can be viewed as a possible outcome this study. The element in this study are packages, aspects and advices.

4. Proposed Models

According to the subject program as data set used throughout this paper, the main requirements are code profiling at the first step of the proposed method. This operation already discussed in [10].

Definition in this step are provided as follows: A: the number of aspect; a: the elements of A; F(a): the set of aspect in A. Given a program, let $a \epsilon A$ be an aspect. Denote that A(a) the set of aspect in A and $F(A_1) = \cup_{a \epsilon a_1} F(a), F(a_i)$ contain aspects which are in a_i. Suppose that P is a probability of aspect from 0 to 1; equivalent there is a distribution of aspect. Then the aspect size of a_1 is defined as

$$size\ (A_l) = \frac{\sum_{a \epsilon F(A_l)} P(a)}{\sum_{a \epsilon A} P(a)}$$

The highest of value will select as the first domain to be tested. Figure 1 shows the Code Profiling process model. Firstly, read the input source code. In this process, aspect is counted. Certain program has package. The package also need to be list down and counted. If the aspect found, then check whether there is advice or not. Thus, advice can be after or before or around. If more than one of each advice such as after. The name for after is *after 1, after 2* until *after n.* Same case to other advice. The process continues until finish the code. Then, probability calculation needed. In this calculation, line of code need to consider. In this study, tool used to ensure the number is correct. From this calculation the value for each of advices are determined.

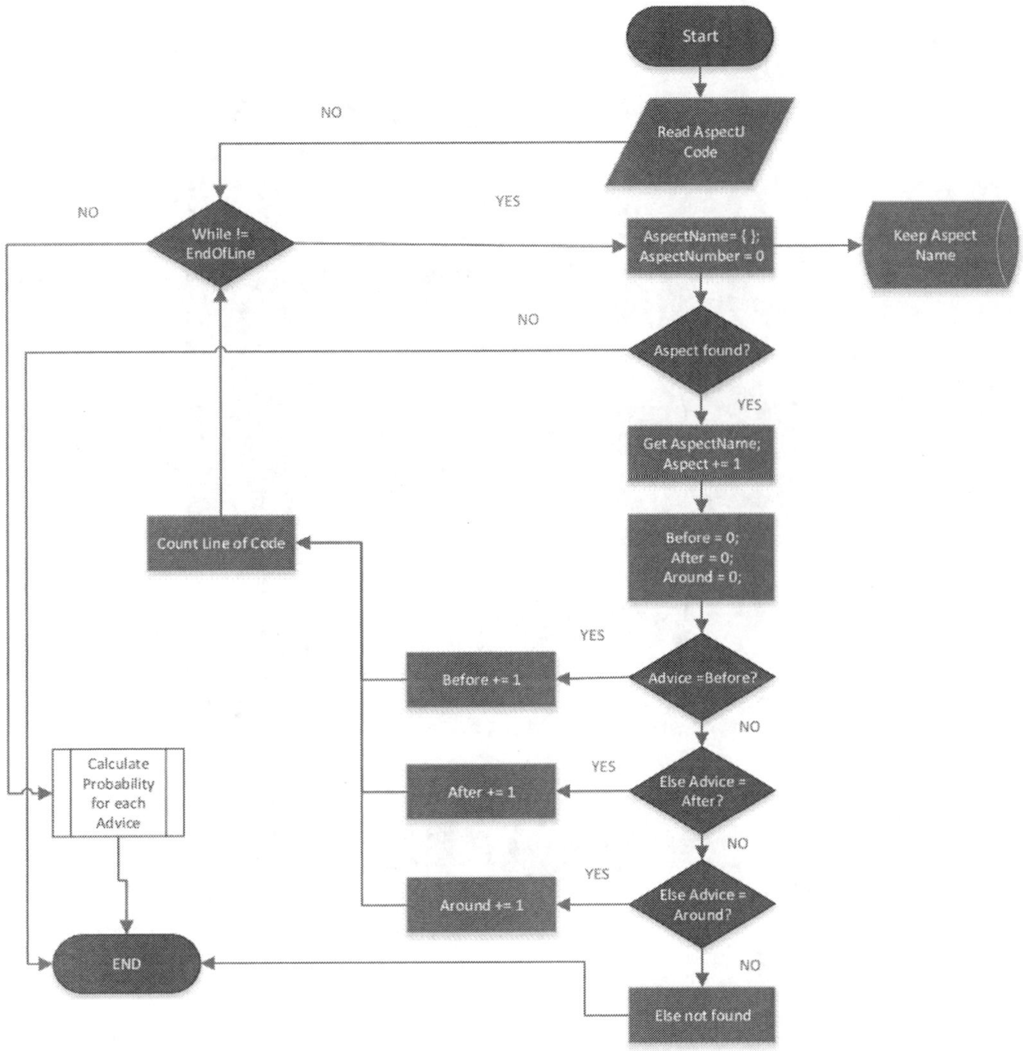

Fig. 1. Code Profiling Process Model.

A. Code profiling

According to the subject program as data set used throughout this paper, the code profiling is the first step of the proposed methods. Tetris code used as a sample in this research as case study. The Tetris code is a benchmark code from eclipse [18]. After analyzing the code using tool, there are three packages named: AspectLogic; Gui; and HighScore. For AspectLogic, two aspect involved: nextBlock with two advices; and newBlock with five advices. Package Gui, consists of two aspects: Levels with four advices; and Counters with 6 advices. Meanwhile, package HighScore consists of two aspects: Menu with two advices; and TestAspect with single advice. In Figure 2. shows the tree of Tetris code. This tree to show the profile of the code graphically. This method easy to calculate the probability for each element inside the code. The elementary of code can see clearly with package, aspect and advices.

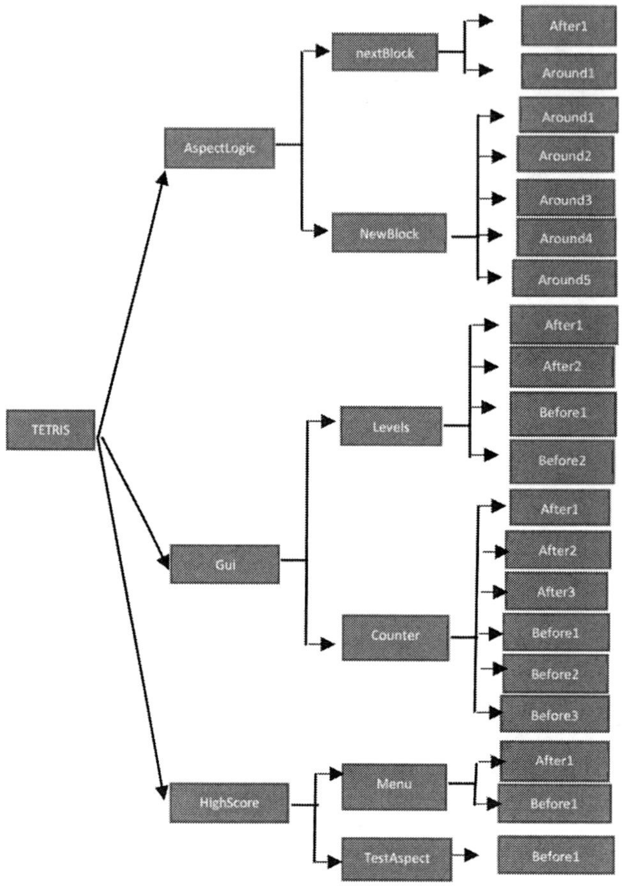

Fig 2. Tetris Tree.

B. Calculate the Probability

Then, probability calculation based on the profile analysis for the code. The calculation needed to get the highest value. The value will be used as the first domain to be tested in the software under test. The probability value shown in Table 1.

Table 1: Probability Value

CODE	PACKAGE	prob	ASPECT	prob	ADVICE	LOC	Prob(LOC)	Prob(a)	MAX
			nextBlock	1/2	after	5	0.55556	0.09	
					around	4	0.44444	0.07	
					around1	1	0.02222	0.00	
	aspectLogic	1/3			around2	23	0.51111	0.09	
			newBlock	1/2	around3	7	0.15556	0.03	
					around4	7	0.15556	0.03	
					around5	7	0.15556	0.03	
					after	3	0.2	0.03	
			levels	1/2	after	7	0.46667	0.08	
Tetris					before	2	0.13333	0.02	0.141026
					before	3	0.2	0.03	
					after	3	0.25	0.04	
	highScore	1/3			before	1	0.08333	0.01	
			counter	1/2	after	3	0.25	0.04	
					before	1	0.08333	0.01	
					before	3	0.25	0.04	
					after	1	0.08333	0.01	
			menu	1/2	after	11	0.84615	0.14	
	Gui	1/3			before	1	0.07692	0.01	
			testAspect	1/2	before	1	0.07692	0.01	

5. Discussion

One of the issues in code profiling is the elementary of profile. This work focus on advice part in AspectJ code. Advice is the part that execute the and can change the operation of code. The advice consists of after, before and around. In this strategy, object also is one of the factor will affected to the value of profile. Thus, calculation from the whole code that refer to the aspect should take into consideration for next study. The result in Figure below shows the highest is gui.menu.after with 17% from the aspect code. This code will be selected as the first area to be tested. If core concern or object involved in this profiling, the value may affect to the test cases selection.

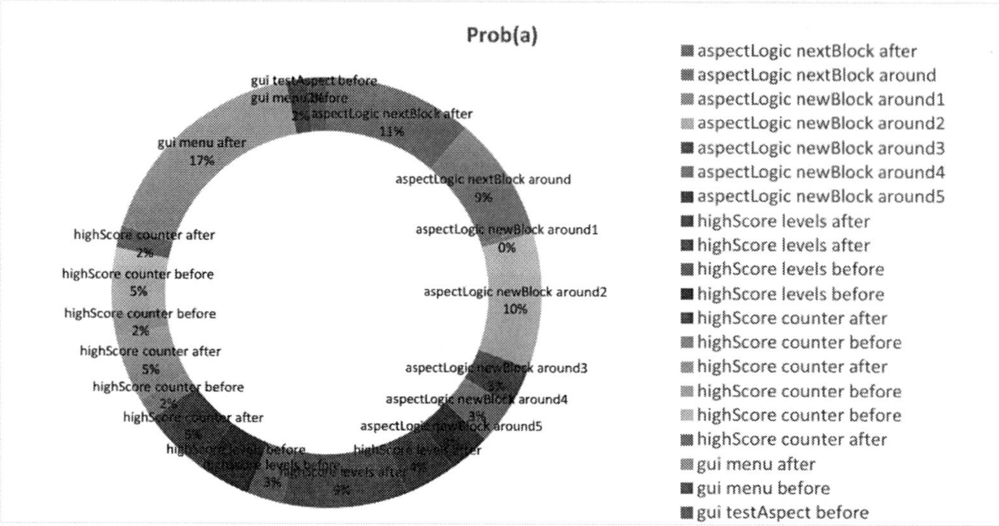

Fig 3. Percentage for aspect code in Tetris

6. Conclusion and Future Work

StART is a testing strategy for test case selection. In this strategy, information from source code is taken into account called code profiling. In this paper, shows the code profiling model support the tester to test which part the need to be test first. The model applies probability calculation to calculate elementary in AspectJ code profile.

Future work, the model can be applied to next phase of StART strategy and can be implemented to any real case study.

Acknowledgment

Appreciation conveyed to Universiti Malaysia Pahang Grant RDU170303 for financing this research.

References

[1] A. Arcuri, M. Z. Iqbal, and L. Briand, "Random Testing: Theoretical Results and Practical Implications," *IEEE Trans. Softw. Eng.*, vol. 38, no. 2, pp. 258–277, Mar. 2012.

[2] T. Y. Chen and D. H. Huang, "Adaptive Random Testing by Localization," in *11th Asia-Pacific Software Engineering Conference*, 2004, pp. 292–298.

[3] J. Mayer and C. Schneckenburger, "An empirical analysis and comparison of random testing techniques," in *Proceedings of the 2006 ACM/IEEE international symposium on International symposium on empirical software engineering - ISESE '06*, 2006, no. Empirical Software Engineering and Measurement, pp. 105–114.

[4] H. D. Mills, M. Dyer, and R. C. Linger, "Cleanroom Software Engineering Cleanroom Software Engineering," *IEEE Softw.*, 1987.

[5] S. Poulding and J. A. Clark, "Efficient Software Verification : Statistical Testing Using Automated Search," *IEEE Trans. Softw. Eng.*, vol. 36, no. 6, pp. 763–777, 2010.

[6] P. C. Jorgensen, *Software Testing: A Craftman's Approach*. 2013.

[7] "Exhaustive Testing: Definition, Principles, and Examples," *Resources Test Automation*, 2018. [Online]. Available: https://testautomationresources.com/software-testing-basics/exhaustive-testing-fundamentals/.

[8] S. Fujiwara, G. v. Bochmann, F. Khendek, M. Amalou, and A. Ghedamsi, "Test selection based on finite state models," *IEEE Trans. Softw. Eng.*, 1991.

[9] A. Rubio and F. Villar, "Code Profiling in R: A Review of Existing Methods and an Introduction to Package GUIProfiler," *R J.*, 2015.

[10] Roslina *et al.*, "A Code Profiling Using Statistical Testing in StART," *Adv. Sci. Lett.*, vol. 24, no. 10, pp. 7295–7299, 2018.

[11] A. Bergel, F. Ba??ados, R. Robbes, and D. R??thlisberger, "Spy: A flexible code profiling framework," *Comput. Lang. Syst. Struct.*, vol. 38, no. 1, pp. 16–28, 2012.

[12] S. F. Witt, H. Song, and P. Louvieris, "Statistical Testing in Forecasting Model Selection," *J. Travel Res.*, vol. 42, no. 2, pp. 151–158, Nov. 2003.

[13] gwendolyn h. Walton, J. h. Poore, and carmen j. Trammell, "Statistical Testing of Software Based on a Usage Model," *Softw. – Pract. Exp.*, vol. 25, no. January 1994, pp. 97–108, 1995.

[14] R. M. Parizi, A. Azim, and A. Ghani, "On the Preliminary Adaptive Random Testing of Aspect-Oriented Programs," in *ICSEA 2011: The Sixth Interntaional Conference on Software Engineering Advances*, 2011, no. c, pp. 49–57.

[15] M. S. Roslina, A. Azim, Abdul Ghani, S. Baharom, and H. Zulzalil, "A Preliminary Study of Adaptive Random Testing Techniques," *Int. J. Inf. Technol. Comput. Sci. (IJITCS)*, vol. 19, no. 1, pp. 116–127, 2015.

[16] J. D.Musa, "Operational Profiles in Software Reliabilitv Engineering," *IEEE Comput. Soc.*, vol. 10, no. 2, pp. 14–32, 1993.

[17] Paul C. Jorgensen, *Software Testing A Craftsman's Approach*. 2014.

[18] S. Apel, "How aspectj is used: An analysis of eleven AspectJ programs," *J. Object Technol.*, 2010.

IOP Conf. Series: Materials Science and Engineering **769** (2020) 012058 doi:10.1088/1757-899X/769/1/012058

Preventive Self-Care Serious Games for Diabetes: A Game Design

S N Ihsan, T A A Kadir, R Benedict and A R Abdullah

Faculty of Computing, College of Computing and Applied Sciences, Universiti Malaysia Pahang, Kuantan, Pahang, Malaysia
tuty@ump.edu.my, rafiqabdullah263@gmail.com and benmouse7@gmail.com.

Abstract. Diabetes is a significant factor that contributes to obesity. In recent years, there has been an increasing number of patients, including adults and children diagnosis of diabetes. Extensive awareness has been carried out for educating people to prevent from diagnosed with diabetes such as sharing the information through electronic media, social media, forum by medical experts, and others. However, less implementation of awareness is given through entertainment computing, such as serious games to attract people to learn in a fun way. In order to overcome the problem, a serious game to prevent diabetes through awareness was developed, namely as *Grab2BeHealthy* to educate people in choosing healthy food. The primary purpose of this study is to design the game architecture for this serious game using SGDA framework. There are a few literature studies addressed on game design framework and the implementation of a serious game in the healthcare industry, especially for awareness and diseases prevention. The game design architecture was developed using game design framework as a guideline to meet the objective of prevention. The prototype development has been tested by 50 players to proof the game design architecture.

1. Introduction

Extreme intake of calories will eventually lead to a build-up of fat in the human body, and that can lead to diseases such as diabetes. A statistics report from the *World Health Organization (WHO)* by 2016 shows that 422 million adults have diabetes. 3.7 million people dying due to the combination of diabetes and high blood glucose and 1.5 million deaths caused by diabetes [1]. Practical approaches are available to prevent type 2 diabetes and the complications of premature death from all types of diabetes. These can be including policies and practices across whole populations within specific settings (school, home, workplace) to ensure everyone become acquainted exercising regularly, eating healthily, avoiding smoking, and controlling blood pressure and lipids [2]. According to reports by the *International Diabetes Federation (IDF)* in 2017, there were over 3,492,600 cases of diabetes in Malaysia [3]. Primary prevalence of diabetes and related risk factor can classify into four categories, which are diabetes, overweight, obesity, and physical inactivity. However, we can manage to cure diabetes through monitoring (blood glucose level), proper diet (by following dietitian advice) and patients/guardians' motivation (to motivate diabetic to have the urge of managing their disease)[4]. One of the methods in preventing the diseases is giving an awareness to increase exposure or knowledge on the factor of being suffered by diabetes and how to avoid it. According to the *American Diabetes Association (ADA)*, the Standards of Medical Care in Diabetes provides the clinical practice recommendations with the components of diabetes care, treatment goals and guidelines, and tools to evaluate the quality of therapy. They have provided the standards of medical care in diabetes, including nutrition therapy, physical inactivity, smoking cessation and psychological issues [5]. Once the patient diagnoses as a diabetic's patients, the most challenging part of the treatment plan is controlling the type of food to take and follow an advice meal plan [5]. Hence, knowledge of identifying healthy food is very important to prevent diabetes. Self-management education in controlling the people with diabetes and sharing knowledge programs is an example of appropriate awareness for people with prediabetes to receive education and support to maintain the behaviours that can prevent or delay the development of type 2 diabetes [6].

Brian Oldenburg et al. summarise the technology used recently in improving and managing diseases can consider the potential and evidence base for using contemporary technology applications and platforms within the new communications landscape to improve the prevention and management of lifestyle-related chronic diseases in the future [7]. Nowadays, serious games are widely used in healthcare industries. The implementation of serious games in increasing the awareness is proof by several examples to educate people to know the diseases such as preventing of neglected diseases [8] and empowering people with down's syndrome [9]. Development teams, including the game researcher and practitioner, play an essential role to improve technology applications to overcome public health challenges. The development of serious games can help youth learn about healthy eating and physical activity to support efforts to prevent obesity [10]. Managing diabetes prevention using serious games gives an exciting way to learn knowledge [11]. It is a tough challenge to develop the concept of self-learning applications to educate the public on preventing diseases. However, the implementation of concept can be applying through serious games due to the elements in the serious games are focusing on the learning principle embedded with entertainment elements.

This study briefly describes an overview and literature study related to the serious games design assessment framework in Section 2. In Section 3, we proposed our game design architecture for preventive self-care serious games mapping with elements in a serious game design assessment framework. The prototype development and games flow are explained in Section 4. In Section 5, we analyse the player's feedback and presented the User Acceptance Test (UAT) result and finally, in Section 6, final discussion and conclusion being address.

2. Literature Review

There are a few industries currently created a serious game for the professional purposes more than entertainment such as defence, education, scientific exploration, healthcare, management, politics and many more for learning purposes [12]. Their intended aim is focusing on understanding the knowledge of related field either as a piece of information or upskilling gap. Learning activity or pedagogy using gamification is a challenge that can improve player's understanding or practical skills while they experience it by themselves [13] and the learning approach by gamification technology adapt knowledge with fun elements to helps the learner understand the process with skills. In the twenty-first century, healthcare industries such as medical training, information sharing, and rehabilitations are applying serious games in various activity for pre-treatment, treatment, and post-treatment [14]. As indicated by Drummond, Hadchouel, and Tesniere, serious games technology in the healthcare industry is a useful tool with two main objectives, which are can motivate and educate the player [15].

There are a few serious games developed for healthcare in preventing and educating the patient. Nutrition Rush [16] is a serious game that aims to teach healthy food with nutrition. It is a preventing and awareness games created using Jump and Runs mechanics to increase player interest to play the games interactively which embedded the elements of task-based learning in the games. Another example of games application in healthcare is iLift [17] developed for nursing personnel to train them for lifting and transfer techniques to prevent lower-back injuries in healthcare. This game will not only cause the player to use correct techniques for static lifting and transferring but also revealed the necessity for future social system development, especially related to intervention acceptance using fantasy game genre and apply practice and feedback as a learning technique. A serious game for self-management healthcare designed to older adults offers the home system that implies the serious games for older adults using machine learning models for exercises recognition and remote activity supervision [18]. The goals of learning are to minimise the physical activity and offers patients to use and simulates kinesiology exercise, answer the quizzes to improve their understanding and give feedback. This game implements intelligent tutor principles for giving feedback to the player. Its aims for preventing and self-manage the activity using exergames as a game's rules. Teaching children on eating healthy habits is a hard task for the parents. Hence, the serious games, namely as Yummy Tricks [19], is developed to educate them on this. This game genre apply to the serious game is a puzzle and apply to learn from mistakes as a principle of learning. The application tested to the three (3) children in a different stage of the game. The result shows that the proposed application was successfully engaged children on playing while learning good healthy habits. Findings of listed serious games proof that this technology can beneficially help healthy lifestyle for peoples. It can improve learner understanding and replace the conventional method of learning for prevention and awareness of diseases.

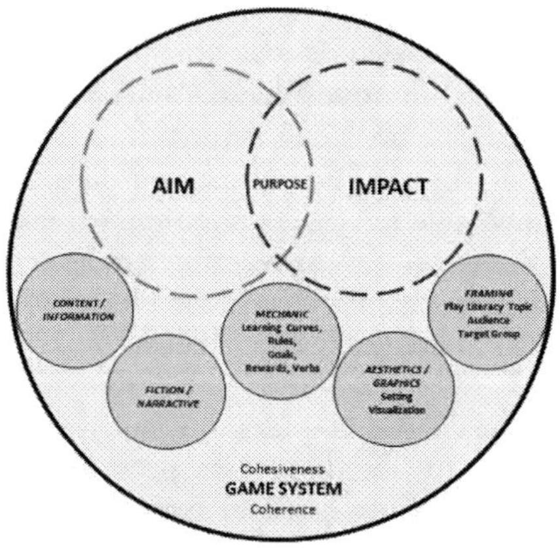

Figure 1. Serious Game Design Assessment (SGDA) Framework [22]

Generally, there are a few game design frameworks for the development serious games such as Mechanics, Dynamics & Aesthetics (MDA) framework [20], scenario-based game design framework [21], the six "I's" of SEG design framework [22], Serious Games Design Assessment (SGDA) Framework [23] and many more.

MDA formal approach to understanding games (bridge gap between game design, game criticism, and technical game research. It consists of three (3) components which are Mechanics for game rules, Dynamics for game systems and Aesthetic for fun. The fundamental of this framework is games are like artefacts than media. Content of the games is the behaviour and streams out towards the player through emotional responses evoked by the player when they interact with the game's systems. Unfortunately, the implementation of this framework is weak when applying to the education line since there are no intelligence or coherence interaction with AI logic with the gameplay. The scenario-based game design framework is used to develop a system architecture and the associated tools for efficient scenario-based game development. It consists of three (3) parts which are conceptual level, technical level and practical level. The conceptual level in this framework explained the static game configuration and game environment. It can be comprising into four (4) subsystems: the gameplay, the learner world, the teacher world and the game management world. The technical level is a tooling system implemented in game development. The architecture of this level comprises four (4) types of tools which are location builder, object builder, role builder and scenario builder. Finally, on top of the conceptual level and technical level to develop serious games, the practical level should incorporate to support the development of complex serious games. The levels cover the topics of game structure, feedback and game representation. It is used to reduce the complexity of feedback and representation of game design. However, the framework will be more established with the evaluation of learning outcome for certain scenario-types — the six "I's" of SEG design framework derived from developing and testing educational games. The component consists of Identity, Informed Teaching, Increase Complexity,

Interactivity, Immersion and Identity. It is based on combination from research and theory in education and psychology with instructional technology and learning science.

Finally, the fourth game design framework, Serious Games Design Assessment (SGDA) framework are designed for constructive structure to develop purposed-based games. It consists of seven (7) components which are purpose, content or information, fiction or narrative, mechanics, aesthetic or graphics, framing and game systems. Figure 1 shows the serious games assessment design framework. According to this framework, game design starts with the game purpose with the combination of aim and impact of playing the games to player. The aim should reflect directly to the purpose of playing and its topic. Next component in the framework is all the information visible to the player, facts and data offered to display or show to the player while playing the games. The third components are game mechanics where the developer must consider the learning curve of playing the games. The games method are invoked by the games agent for interacting the game world. It also uses for the establishment of the rules of operation in the games including rewarding systems, obstacle or challenge and the winning condition. The game mechanics also translate the verbs or action in-game world into the rules and algorithm. Fourth, the fiction and narrative flow are used to create a situational space and relate to the game purpose. The context of fictional includes the characters in-game space, settings, narrative story, backstories, scenario and problems. The narrative story could be a linear story, simulate, represent issues or created by the player. Fifth, the aesthetics and graphics is a basic element to introduce the game purpose and impact on the player. Usually, it is representing the audio-visual language which is the aesthetic characteristics, imagery, style preference, artistic media and computer games technique. It also conceptualized and visualize to display important elements to the player. Example of the important elements is the content through descriptive information, fiction as a game character, framing to focus the target group, game settings and game mechanics as instruction and rewarding systems. Besides the five (5) key components: purpose, content & information, mechanics, fictions & narrative and aesthetics & graphics, the framing are the important components to specify the target group of players. It is used to control the play literacy of the game purpose and the broader topic of content or information. The framing can essentially influence the gameplay experience. Finally, the coherence and cohesiveness of the game system used to show how it holistically relates to each other between the game purpose and all the components. It also worked as an integral entity to shape the gameplay experience. The massiveness of game system leads to the conflict in a gameplay.

Game design framework is essential to implement in the development of the serious games to ensure the achievable of learning objective with experiencing the game world in a fun way. Unfortunately, some of the game design frames is not suitable to implement with specific issues.

3. Game Design

The serious game architecture for prevention and self-care awareness of diabetes or known as *Grab2BeHealthy* was designed using Serious Games Design Assessment (SGDA) framework to guide on the prototype development. The development process embedded the adoption concept of software development life cycle in order to help in improving the comprehensiveness of the requirements and satisfaction of player [24]. Figure 2 shows the game design architecture for *Grab2BeHealthy*. The game design architecture of *Grab2BeHealthy* refers to the seven (7) elements in design framework, which are purpose, content & information, game mechanics, fiction & narrative, aesthetics & graphics, framing and game systems.

Figure 2 shows the game design architecture of Grab2Behealthy serious games with guidance from the SGDA framework. The primary purpose of the development is to give awareness on balancing diet to the player by choosing the suitable food for healthy life. The main aim or mission of this game is differentiating healthy and unhealthy food. After they complete the mission the games should help the player to choose and categorize healthy food in their daily life. The heart of the serious games is the content and information. This game offer tips on differentiating the food, healthy food, unhealthy food, timer, score and the total score to the player as a fulfilment of content and information. Next element is a game mechanics which are roles as a mechanical part in a game. This element creates rules using algorithm to control the action in a Grab2BeHealthy such as how to avoid from unhealthy food, catching or grabbing the healthy food using basket, collecting the suitable food and moving the basket using mouse controller. It is including the operation of rewarding system while collecting the correct food, obstacle and challenge, the difficulty of each level by speed and how to calculate the score to win the games.

In order to create a narrative story and the fiction of serious games, the arrangement of character, games setting, game stories, backstories and main problem have to clearly identify. The main problem created in Grab2BeHealthy is people usually take more unhealthy food in daily life which cause obesity and diseases. Based on the problem, scenario has been identified and created, where people quickly get unhealthy food. It is to create stories where people usually experiencing the situation. Next is the aesthetic and graphics which are fundamental to introduce the game purpose and give experience to the player through visual graphics and head-up display (HUD) in games interfaces. Grab2BeHealthy starts with the image of the impact if people not choosing healthy food in life which are overweight and obesity. It also shares the tips on choosing the food before starts playing the games. Fact and information related to the healthy food also display in between each level to remind and keep the player on track of the purpose of playing the games. Music and graphics are embedding and design in a 2D to keep player exciting with the mood of challenging.

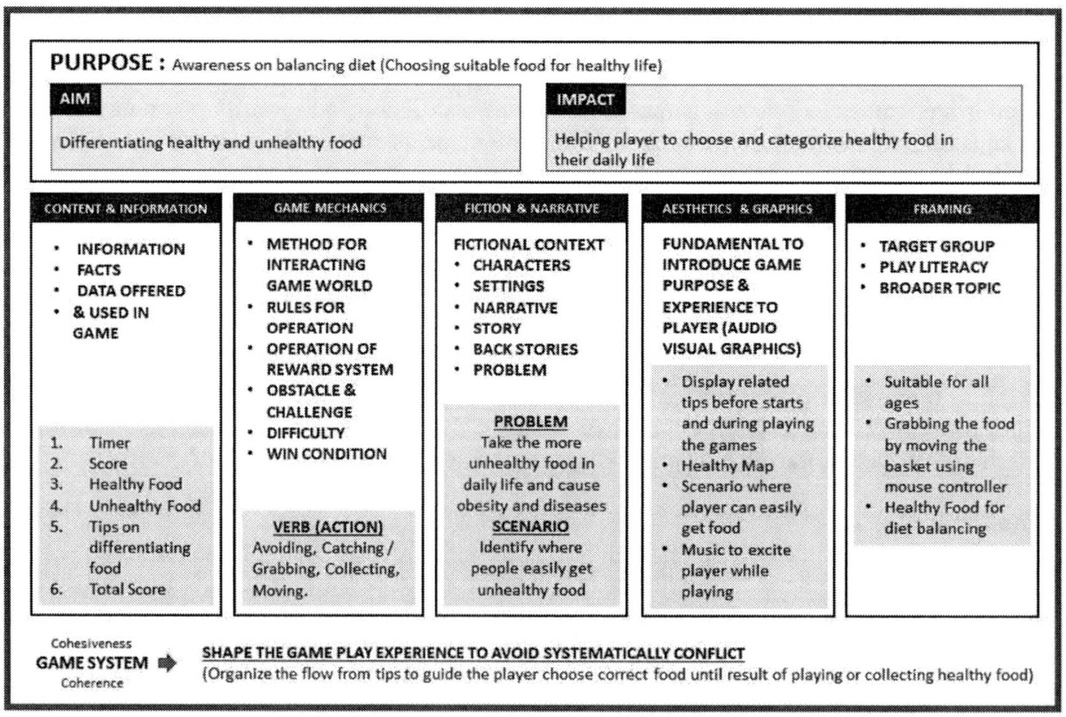

Figure 2. Grab2BeHealthy Games Design Architecture

Framing of the development is important to create the scoping area of game purpose. Grab2BeHealthy is targeted to all suitable ages. The main play literacy is grabbing or collecting the healthy food by moving the basket using mouse controller to get score within time given and the difficulty of each level. The broader topic of this games is healthy food for diet balancing. Overall cohesiveness and coherence of the game system are used to shape the gameplay experience in order to avoid conflict systematically. The cohesiveness of this games is clearly identified to organize the flow from tips to guide the player until the result of playing or collecting healthy food.

4. Prototype Development

The prototype development is based on the effect cause unbalance diet or taking more unhealthy food than healthy food. To start the games, the player must read the guideline or list of food that can choose to collect rewards score. Given guidelines are works as awareness to the player to differentiate healthy and unhealthy food. Two (2) types of information as in Figure 3 shows the guidelines and knowledge for diabetes prevention.

In the beginning, the player must choose the category then the player will see 2 pages of instruction food that they should grab and shouldn't grab. The games continue with level 1, and there is a lot of food combined with healthy and unhealthy food drop in a slow mode. The player must recall the information given on the previous page and grab healthy food. It is a compulsory requirement where the player must expose with either direct or indirect knowledge as an awareness. The mission started when the player was aiming to grab food, as shown in the instruction earlier. 5 points of marks will be adding to the score result as a reward in level 1 and minus 3 points from score result if the player grabs the unhealthy food. Games will continue to level 2, where the challenge will increase with the increase of speed in a food drop. Each collected unhealthy food will minus 4 points, and 5 points of mark will be adding as a reward if the player collects the healthy food successfully. The player will continue to level 3 with a challenging obstacle where the food drop in a fast mode.

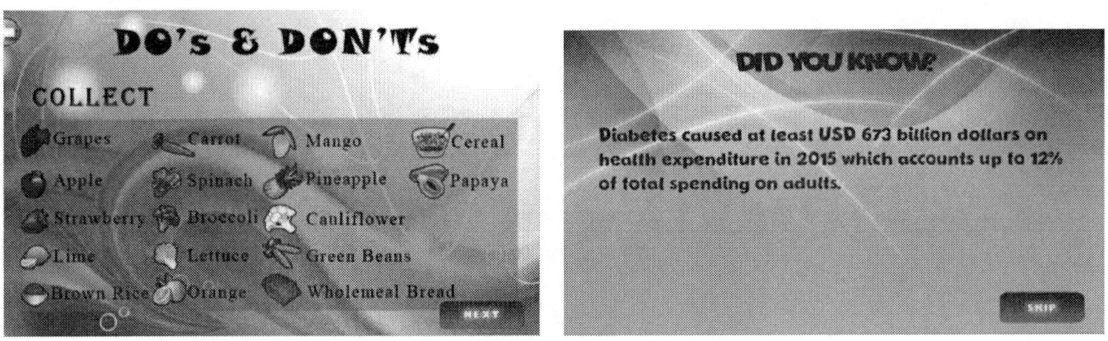

Figure 3. Guidelines and knowledge for diabetes prevention.

Figure 4. Example of the game environment (scenario) in *Grab2BeHealthy*

The penalty of 5 points will be deducted from the current score if the player grabs the healthy food and they will get another 5 points of marks if they catch good food for health. The process of counting the marks is a mechanical process in the game engine to increase the challenge of the games. In between each level, the player must read tips as a piece of additional knowledge to them to prevent and awareness regarding diabetes. 60 seconds are given to the player to grab the food for each level. The games share feedback on reward's result and achievement after completing the games. Figure 4 shows examples of the scenario in the game environment.

5. Result

The games were tested and accepted by 50 participants with basic computer knowledge (22 Male and 28 Female) and various age range from 18 to 50 years. 68% from them are exposed to diabetes knowledge either from media, books and family history. The results show that 52% of the tester agree with this application suitable to use for awareness purposes while the other two motivation, which is 28% suitable for prevention and consultation purposes. 20% from the tester agree that this application can be used for teaching purposes.

User acceptance test result was collected and categorized into components in game design architecture. From the pie chart in Figure 5, it is apparent that 60% of player agrees with the gameplay mission while they played the games. However, another 40% disagree with the aims. Interestingly, most of the player agrees with the arrangement and design for the game engine, which contains game mechanics and challenge. Of the overall response to the questionnaire, 96% was reported to agree with rewards and penalty given to the player, and another disagrees with their own reason. The most surprising feedback is 98% of the player agree with the challenge in the games. The feedback on game engine shows in Figure 6.

Figure 5. User Acceptance for gameplay in Mission

On average, gameplay for feedback in Fig.7 shows that only 46% of player agrees with the overall prototyped development game. However, the concept of games using game design architecture mostly satisfied based on overall feedback as they get benefit after playing the games. Table 1 shows the feedback on overall prototype development using game design architecture and agrees with the alternative platform using serious games to educate the consumer on a healthy diet for diabetics' patient.

Table 1. *Grab2BeHealthy* Games Characteristics

Feedback	Agree	Disagree
Players get the benefit after playing the games.	50	0
Alternative awareness platform to educate the consumer on a healthy diet for diabetics' patient.	50	0

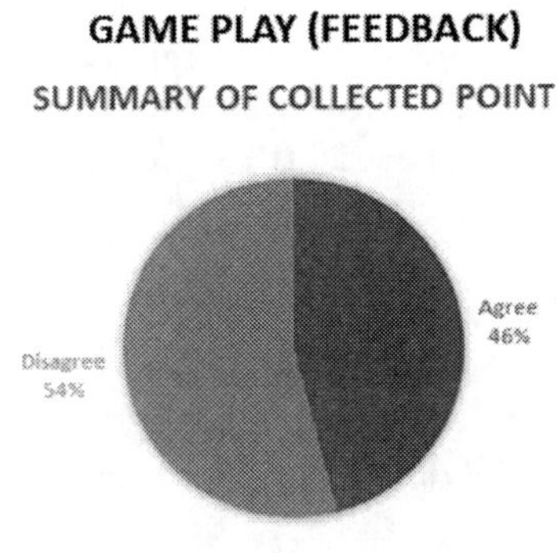

Figure 6. User Acceptance for gameplay in feedback

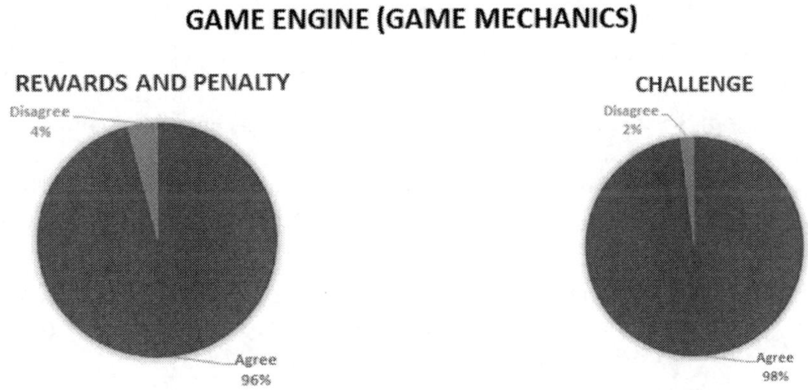

Figure 7. User Acceptance for the game engine in game mechanics

6. Conclusion

In this study, game design architecture for preventive self-care serious game for diabetes by differentiating healthy and unhealthy food was developed to utilizes the compulsory elements in the framework, such as the purpose, content & information, game mechanics, fiction and narrative, aesthetics & graphics, framing and the game system. The implementation of the framework into the design helps player to increase the understanding and attract various age range to understand the facts of healthy food in a fun way. This application used to educate and give awareness to people related to diet and differentiate healthy and unhealthy food using technology tools.

The application suitable for various range of age and implement puzzle games as a game's genre that suitable to the learning technique such as memorization, feedback, practice, and others. The repetition of activity in this game can encourage the player to memories the information compared to reading. Based on the survey result, all tester agree that they get to benefit from the application after playing *Grab2BeHealthy*. There are a few reasons how people have diabetes such as lack of physical activity, smoking cessation and depression. Self-care preventive management helps to minimize the risk of getting diabetes. The application is not limited to the information learning but increasing the human activity that leads to the reducing risk of diabetes. The current implementation of awareness, such as article, video sharing, and graphics illustration is one-way information. The interactive application gives them more enjoyable and improves memorization, especially on the do and don'ts. We aim for the data monitoring enhancement of the games by embedding the physical activities to increase the understand and awareness of diabetes to the public.

Acknowledgements
We would like to thanks to the development team of this research for their effort in designing and developing serious games. This research was awarded Silver Medal during Creation Innovation Technology and Research Exposition (CITREx 2018) at University Malaysia Pahang and funded by the Universiti Malaysia Pahang (UMP) under UMP Postgraduate Research Grant Scheme (PGRS190387).

Reference

[1] World Health Organization 2016 Global Report on Diabetes *WHO Press. World Heal. Organ.* vol. 978 pp. 4–4

[2] World Health Organization 2016 Global Report on Diabetes Executive Summary *WHO Press. World Heal. Organ.* pp. 4–4

[3] I. D. F. (IDF) Accessed: 14-Jan-2019 IDF Western Pacific members *International Diabetes Federation (IDF) 2017.* [Online]. Available: https://www.idf.org/our-network/regions-members/western-pacific/members/108-malaysia.html

[4] Lokman A S and Zain J M 2009 Designing a Chatbot for Diabetic Patients *International Conference on Software Engineering & Computer Systems (ICSECS'09)* pp. 19–21

[5] Care D and Suppl S S 2018 Lifestyle management: Standards of medical care *Diabetes 2018 Diabetes Care* vol. 41 pp. S38–S50

[6] Care D and Suppl S S 2018 Prevention or Delay of Type 2 Diabetes: Standards of Medical Care in *Diabetes Diabetes Care vol. 41* pp. S51–S54

[7] Oldenburg B, Taylor C B, O'Neil A, Cocker F, and Cameron L D 2015 Using New Technologies to Improve the Prevention and Management of Chronic Conditions in Populations *Annu. Rev. Public Health*, vol. 36 no. 1 pp. 483–505

[8] Masoodian M, Luz S, Cesario M, Cesario R R, Rogers B, and Borges D A 2015 A serious game for improving community-based prevention of neglected diseases *Proc. - IEEE Symp. Comput. Med. Syst.* vol. 2015–July pp. 286–291

[9] Bourazeri A, Bellamy-Wood T, and Arnab S 2017 EnCity: A serious game for empowering young people with Down's syndrome, *2017 IEEE 5th Int. Conf. Serious Games Appl. Heal. SeGAH*

[10] Ruggiero L 2015 Diabetes Prevention and Management: What Does a Serious Game Have to Do with It? *Games Health*, vol. 4, no. 5, pp. 333–334

[11] Kahol K 2011 Integrative gaming: A framework for sustainable game-based diabetes management *J. Diabetes Sci. Technol.*, vol. 5, no. 2, pp. 293–300

[12] Wattanasoontorn V, Boada I, García R, and Sbert M 2013 Serious games for health, *Entertain. Comput.*, vol. 4, no. 4, pp. 231–247

[13] De Freitas S and Liarokapis F 2011 Chapter 2 Serious Games: A New Paradigm for Education? No. *Serious Games and Edutainment Applications. Springer-Verlag London Limited*

[14] Korhonen, Tanja, Kemppainen J, Kajaani, Koskela, and Kyösti 2017 A Multidisciplinary Approach To Serious Game Development *Health Sector 11th Mediterr. Conf. Inf. Syst. (MCIS), Genoa, Italy*

[15] Drummond D, Hadchouel A, and Tesnière A 2017 Serious games for health: three steps forward *Adv. Simul.*, vol. 2, no. 1, p. 3

[16] Baranyi R, Steyrer B, Lechner L, Agbektas G N, Lederer N, and Grechenig T 2017 NutritionRush-A serious game to support people with the awareness of their nutrition intake *IEEE 5th Int. Conf. Serious Games Appl. Heal. SeGAH*

[17] Kuipers D A *et al 2016* iLift : A health behaviour change support system for lifting and transfer techniques to prevent lower-back injuries in healthcare *Int. J. Med. Inform.*, vol. 96, pp. 11–23

[18] Codreanu I A and Florea A M 2015 A Proposed Serious Game Architecture to Self-Management HealthCare for Older Adults *Symb. Numer. Algorithms Sci. Comput. (SYNASC), 2015 17th*

Int. Symp. on. IEEE

[19] Inglés-Camats G, Presno-Rivas M M, Antonijoan M, Garcia-Panella O, and Forrest T 2012 Yummy tricks: A serious game for learning healthy eating habits *Stud. Health Technol. Inform.*, vol. 172, pp. 185–190

[20] Hunicke R, Leblanc M, and Zubek R 2004 MDA: A formal approach to game design and game research *Proceedings of the AAAI Workshop on Challenges in Game AI*, p. Vol 4 No 1

[21] Westera W, Nadolski R J, Hummel H G K, and Wopereis I. G. J. H. 2008 Serious games for higher education: A framework for reducing design complexity *J. Comput. Assist. Learn.* vol. 24, no. 5, pp. 420–432

[22] Annetta L A 2010 The 'I's' Have It: A Framework for Serious Educational Game Design *Rev. Gen. Psychol.* vol. 14, no. 2, pp. 105–112

[23] Mitgutsch K and Narda A 2012 Purposeful by Design ? A Serious Game Design Assessment Framework *Proceedings of the International Conference on the foundations of digital games* pp. 121–128.

[24] Ihsan S N and Kadir T A A 2018 Adoption of Software Development Life Cycle (SDLC) Model in Games Development Framework for Serious Games Applications *Adv. Sci. Lett.*, vol. 24, no. 10, pp. 7300–7304

A Model of the Best Private University Selection in Aceh

Taufiq[1], Muhammad Zarlis[2], Herman Mawengkang[3], Saib Suwilo[3]

[1]Doctoral Student in Computer Science, Universitas Sumatera Utara, Medan, Indonesia
[2]Department of Computer Science, Universitas Sumatera Utara, Medan, Indonesia
[3,4]Department of Mathematics, Universitas Sumatera Utara, Medan, Indonesia

E-mail: taufiqstmt@yahoo.com, m.zarlis@usu.ac.id

Abstract. The Integrated Population Decision Support Model in the village needs to be developed according to the rules of Law number 6 of 2014 with a decision information system model that can assist one in making accurate and targeted decisions in choosing a university. Many problems can be solved using a model, one of which is the Decision Making System, for the Determination of Human Resources in an area. There are several models that have been built including Analytical Hierarchy Process (AHP). which is the most widely used research method in solving multi-criteria problems, such as in the Determination of Human Resource Selection. This research model uses Analytical Hierarchy Process (AHP) in determining the University for a superior place of learning for human resources. In determining the University, there are several basic decision-making, including the place of the educational model, the model of expertise / skills taught, reputation, quality, teaching and learning process. The final results in this study are the priority results of university criteria, which are sorted from highest to lowest, so that prospective student users who wish to continue their higher education in choosing a regional university (rural) can easily make decisions by determining the university in the village which, and finally this model is almost 80% accurate.

Keywords: Model, Analytical Hierarchy Process, University

1. Introduction

The countryside has always been a dream of every human being for a place to live because it is cool and can breathe free and fresh air, and free from noise, therefore every individual in the village needs a reliable / good place to study. Our country has many private developing universities. Unlike state universities, private universities are more established, which makes us confused to determine which is the best private university in rural areas in an area, so the object of research is in one of the provinces in Aceh, because it has a specialty in managing government administration. Therefore this research is thinking carefully in order to be the best decision and it is not wrong in determining the choice to continue to a higher education level, namely the University in the village.

The problem that many of us find now in a university is that it is not accredited, costs soar, lecturers are not reliable so that later they will graduate graduates who are not qualified. The consequence that will arise is the name of the university that is not good because it is not able to produce qualified skilled graduates. In determining the best university, information about the university's reputation, university accreditation, departmental, financial and output accreditation is needed in a private university.

In the mandate of law No. 6 of 2014 concerning Villages, must strive to develop independence and prosperity of the community by increasing knowledge, attitudes, skills, behavior, abilities, awareness, and utilizing resources through the determination of policies, programs, activities and assistance in accordance with the essence of the problems and priority needs of rural communities, then the village community should take an active stance to prepare everything related to the implementation of the Act - the village, one of which is about information systems choosing private universities in rural areas that aim to open information on the implementation of good and reliable governance to the public to make a choice

IOP Conf. Series: Materials Science and Engineering **769** (2020) 012059 doi:10.1088/1757-899X/769/1/012059

2. Research Methodology

Research methodology used in solving problems:

1. Library Studies
   At this stage the literature study is carried out, namely collecting reference materials from books, articles, journals, papers, and on the internet in making applications in the field.
2. Data Collecti on
   At this stage the data were collected in the form of data about the data of Private Universities in Aceh.
3. Problem Analysis
   At this stage an analysis will be made of the application of the selection of the best private tertiary institution in the Aceh countryside.
4. The system is tested
   At this stage the results of the analysis of the solution to the solution in the application of the selection of the best private universities and the design of the application resulting in an easy-to-use interface for determining the best university.

3. Results And Discussion

3.1 Pairwise comparison weight.

Pairwise comparisons are carried out based on the rules for assessing the weight of criteria in the following table 1:

Table 1: Scale of Pairwise Comparison Assessment

Level of importance	Definition
1	Both elements are very important
3	One element is a little more important than the other elements
5	One element of the essence of the tool is very important compared to the other elements
7	One element is really more important than the other
9	The absolute element is more important than the other elements
2,4,6,8	Middle values between two consecutive assessments

3.2 Pairwise Comparison Matrix

Pairwise comparison matrix is conducted for comparison assessment between one criterion with other criteria, namely Reputation criteria with accreditation criteria, reputation criteria with lecturer quality criteria, reputation criteria with learning process criteria, reputation criteria with facilities, quality accreditation lecturers, accreditation criteria with learning process, accreditation criteria with facilities, lecturer quality criteria with learning process, criteria for learning process with facilities. The results of the assessment are in table 2.

Table 2: Making Criteria Comparison Matrix

Goal	Reputation	Accreditation	Lecturer Quality	Learning Process	Amenities
Reputation	1	3	5	7	9
Accreditation	0,333333	1	3	5	7
Lecturer Quality	0,2	0,333333	1	3	5
Learning Process	0,142857	0,2	0,333333333	1	3
Amenities	0,111111	0,142857	0,2	0,333333333	1
Total	1,787302	4,67619	9,533333333	16,33333333	25

After entering the data in table 2 above, the division value of the number of columns is divided by the number of each column and sums the element values of each column. From the values of the criteria matrix elements above, the number of elements in each column are:
Number of Column 1: 1 + 0.333333 + 0.2 +0.142857 + 0.111111 = 1.787302
The number of Column 2: 3 + 1 + 0.333333 + 0.2 + 0.142857 = 4.67619
Number of Column 3: 5 + 3 + 1 + 0.333333333 + 0.2 = 9.533333333
Amount of Column 4: 7 + 5 + 3 + 1 + 0.333333333 = 16,33333333
Number of columns 5: 9 + 7 + 5 + 3 + 1 = 25

3.3 Criteria Value Matrix

Consideration of consideration for pairwise comparisons is synthesized to obtain overall priorities. The things done in this step are:

1. Add the values of each matrix column
2. Divide each value from the column by the corresponding column to obtain the normalization of the matrix.
3. Add the values of each row and divide them by the number of elements to get the average value.

Table 3: Criteria value matrix

Reputation	Accreditation	Lecturer Quality	Learning Process	Amenities	Sum	Priority vector
0.559503	0.641548	0.524475524	0.428571429	0.36	2.514097	0.502819496
0.186501	0.213849	0.314685315	0.306122449	0.28	1.301158	0.260231588
0.111901	0.071283	0.104895105	0.183673469	0.2	0.671752	0.134350441
0.079929	0.04277	0.034965035	0.06122449	0.12	0.338888	0.067777667
0.062167	0.03055	0.020979021	0.020408163	0.04	0.174104	0.034820809

In this matrix Reputation column and reputation row 0.559503 are obtained from the reputation row reputation column value divided by the number row value and reputation column in table 2, this process is done until the quality row quality column. While the amount is the sum of 0.559503 + 0.641548 + 0.524475524 + 0.428571429 + 0.36 = 2.5149097. For the values in the priority column obtained from the values in the column number divided by the number of criteria, in this study there are 5 criteria.

3.4 The Sum of Each Line Matrix

The summing matrix of each row is obtained from the multiplication of priority values in table 3 with a paired comparison matrix in table 2.

Table 4: Criteria Consistency Matrix

Reputation	Accreditation	Lecturer Quality	Learning Process	Amenities	Sum	Bobot
0,502819496	0,780694763	0,671752203	0,474443668	0,313387281	2,743097	5,455431689
0,167606499	0,260231588	0,403051322	0,338888334	0,243745663	1,413523	5,431790266
0,100563899	0,086743863	0,134350441	0,203333001	0,174104045	0,699095	5,203520323
0,071831357	0,052046318	0,04478348	0,067777667	0,104462427	0,340901	5,029698779
0,055868833	0,037175941	0,026870088	0,022592556	0,034820809	0,177328	5,09259353

Value of 0.502819 on line Reputation of reputation column is obtained from the highest priority value available in table 3. multiplied by reputation row reputation value in table 2. The value of 0.167606 in the reputation column accreditation line is obtained from the highest priority value in table 3multiplied by the value of the accreditation column of the reputation column in table 2. and so on until the facility line and facility column. While the number column in table 4 is obtained by summing the values on each row in the table. Like 2.743097 in the column number is the sum of 0.502819 + 0.780694763+ 0.671752203+ 0.474443668+ 0.313387281. n (number of criteria) = 5. Random Index (IR) = 1.12. Calculate max λ (number / n) = 26,21303459 / 5= 5,242606918. CI ((λ max-n) / n) = 5.242606918 - 5 / 5-1 = 0.0606517295. CR (CI / IR) = 0.0606517295 / 1.12 = 0.0541533299. From the above calculation, produce a CR value of <0.1, so that the calculation of the consistency ratio can be accepted. For the calculation of the sub-criteria of each criterion carried out in the same way.

3.5 Alternative Comparative Matrix based on University Reputation criteria

Pairwise comparison matrices are conducted for the comparison of one alternative with another based on criteria. For alternative comparisons based on reputation criteria can be seen in table 5.

Table 5: Alternative pairing matrix based on University Reputation criteria

		1	2	2	3	3	4	5	5	6	7	7
		Almuslim	Jabal Ghafur	Iskandar Muda	Abulyatama	Muhammadiyah Aceh	Serambi Mekkah	Teuku Umar Meulaboh	Gajah Putih	Gunung Leuser	Ubudiyah Indonesia	Sains Cut Nyak Dhien
1	Almuslim	1	2	2	3	3	4	5	5	6	7	7
2	Jabal Ghafur	0,5	1	1	1,5	1,5	2	2,5	2,5	3	3,5	3,5
2	Iskandar Muda	0,5	1	1	1,5	1,5	2	2,5	2,5	3	3,5	3,5
3	Abulyatama	0,33333	0,666667	0,666667	1	1	1,3333333	1,66666667	1,66667	2	2,33333	2,333333333
3	Muhammadiyah Aceh	0,33333	0,666667	0,666667	1	1	1,3333333	1,66666667	1,66667	2	2,33333	2,333333333
4	Serambi Mekkah	0,25	0,5	0,5	0,75	0,75	1	1,25	1,25	1,5	1,75	1,75
5	Teuku Umar Meulaboh	0,2	0,4	0,4	0,6	0,6	0,8	1	1	1,2	1,4	1,4
5	Gajah Putih	0,2	0,4	0,4	0,6	0,6	0,8	1	1	1,2	1,4	1,4
6	Gunung Leuser	0,16667	0,333333	0,333333	0,5	0,5	0,6666667	0,83333333	0,83333	1	1,16667	1,166666667
7	Ubudiyah Indonesia	0,14286	0,285714	0,285714	0,42857	0,4286	0,5714286	0,71428571	0,71429	0,8571429	1	1
7	Sains Cut Nyak Dhien	0,14286	0,285714	0,285714	0,42857	0,4286	0,5714286	0,71428571	0,71429	0,8571429	1	1
	Total	3,76905	7,538095	7,538095	11,3071	11,307	15,07619	18,8452381	18,8452	22,614286	26,3833	26,38333333

After determining the alternative comparison matrix value based on reputation criteria, the next is to calculate the priority vector value in the same way as calculating the priority vector criteria in table . Priority vectors for alternative matrices based on reputation criteria can be seen in table.

IOP Conf. Series: Materials Science and Engineering **769** (2020) 012059 doi:10.1088/1757-899X/769/1/012059

Table 6: Priority vector alternative comparison matrix based on Reputation criteria

0,26532	0,265319	0,265319	0,26532	0,2653	0,265319	0,26531901	0,26532	0,265319	0,26532	0,265319015	0,265319015
0,13266	0,13266	0,13266	0,13266	0,1327	0,1326595	0,13265951	0,13266	0,1326595	0,13266	0,132659507	0,132659507
0,13266	0,13266	0,13266	0,13266	0,1327	0,1326595	0,13265951	0,13266	0,1326595	0,13266	0,132659507	0,132659507
0,08844	0,08844	0,08844	0,08844	0,0884	0,0884397	0,08843967	0,08844	0,0884397	0,08844	0,088439672	0,088439672
0,08844	0,08844	0,08844	0,08844	0,0884	0,0884397	0,08843967	0,08844	0,0884397	0,08844	0,088439672	0,088439672
0,06633	0,06633	0,06633	0,06633	0,0663	0,0663298	0,06632975	0,06633	0,0663298	0,06633	0,066329754	0,066329754
0,05306	0,053064	0,053064	0,05306	0,0531	0,0530638	0,0530638	0,05306	0,0530638	0,05306	0,053063803	0,053063803
0,05306	0,053064	0,053064	0,05306	0,0531	0,0530638	0,0530638	0,05306	0,0530638	0,05306	0,053063803	0,053063803
0,04422	0,04422	0,04422	0,04422	0,0442	0,0442198	0,04421984	0,04422	0,0442198	0,04422	0,044219836	0,044219836
0,0379	0,037903	0,037903	0,0379	0,0379	0,0379027	0,03790272	0,0379	0,0379027	0,0379	0,037902716	0,037902716
0,0379	0,037903	0,037903	0,0379	0,0379	0,0379027	0,03790272	0,0379	0,0379027	0,0379	0,037902716	0,037902716

3.6 Vector Priority Ranking

The calculation steps for the Overall Composite Weight are as follows:

1. The weight column is taken from the Priority Vector column and the criteria matrix.
2. The other eleven columns (alternative campus names) are taken from the Priority Vector column, the five criteria matrix are Reputation, Accreditation, Quality of Lecturers, Learning Process and Facilities.
3. The Composite Weight line is obtained from the number of cell multiplication results above with Weight

Table 7: Priority Vector Ranking Results

	Reputation	Accreditation	Learning Process	Lecturer Quality	Amenities	Results
Almuslim	0,26532	0,028571	0,033333	0,02564	0,0222	0,14783
Jabal Ghafur	0,13266	0,085714	0,066667	0,05128	0,0444	0,10299
Iskandar Muda	0,13266	0,085714	0,066667	0,05128	0,0444	0,10299
Abulyatama	0,08844	0,057143	0,1	0,07692	0,0667	0,08031
Muhammadiyah Aceh	0,08844	0,057143	0,1	0,07692	0,0667	0,08031
Serambi Mekkah	0,06633	0,114286	0,066667	0,05128	0,0889	0,07862
Teuku Umar Meulaboh	0,05306	0,142857	0,066667	0,10256	0,1111	0,08363
Gajah Putih	0,05306	0,057143	0,1	0,10256	0,1111	0,06581
Gunung Leuser	0,04422	0,085714	0,133333	0,12821	0,1333	0,07579
Ubudiyah Indonesia	0,0379	0,142857	0,166667	0,15385	0,1556	0,09447
Sains Cut Nyak Dhien	0,0379	0,142857	0,1	0,17949	0,1556	0,08725
Weight	0,5028195	0,2602316	0,1343504	0,0677777	0,0348208	

Based on the table above, a decision can be made that those who have the highest score are Almuslim, which is 0.14783. finally it can be concluded that the Best Private University in Aceh is the University of Almuslim.

3.7 Display Admin Login

This page is used for data security. Admin who will use the system must login first by filling in the username and password according to the conditions of the system. Inputting data from the login form must match. If it does not match, the system will not respond and cannot proceed to the following menu. The admin login format can be seen in the following picture:

3.8 Page Analysis with AHP Method

This page is a page where the user determines the best university by choosing a comparison of values between criteria1 and criteria 2 so as to provide reports in the form of ranking or ranking of the best universities. The appearance can be seen in the following picture:

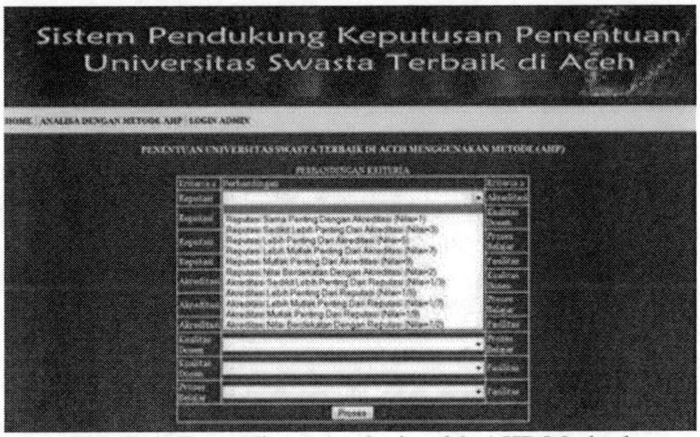

Figure 1: Page Views Analysis with AHP Method

3.9 Admin page view

This page is the scope of all forms that are in the Admin login system, including alternative menu, criteria menu, alternative-criteria menu, change password menu and exit menu. This page can also be used to add, edit, and delete all data in the system. The appearance can be seen in the picture below.

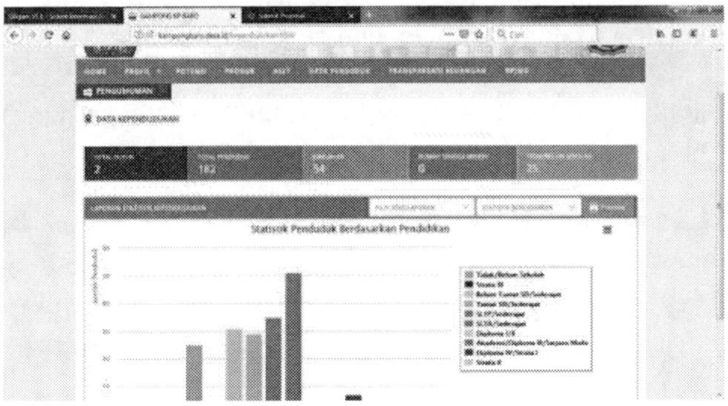

Figure 2: Admin Page Display

This page can also be used to add, edit, and delete item data. The appearance can be seen in the picture below.

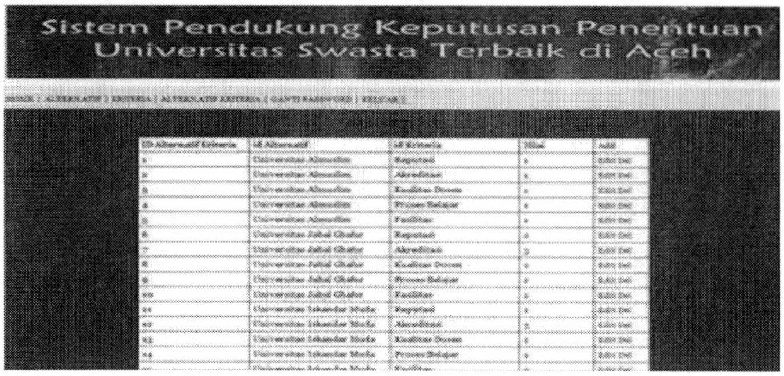

Figure 3: Alternative Page Views-Criteria

4. Conclusion

The use of the AHP method in this application can solve problems from 11 University objects and 5 criteria, resulting in input into the decision making process of determining the best private university in Aceh. This decision support application provides user convenience in getting advice on the best private universities in Aceh. So that prospective students can determine the best university.

References

[1] Kusrini, 2007. *Konsep dan Aplikasi Sistem Pendukung Keputusan*, Andi, Yokyakarta.
[2] Sante´, I., N. Pacurucu, M. Boullo´n, A. M. Garcı´a, and D. Miranda. 2016. "An Open Source GIS-based Planning Support System: Application to the Land Use Plan of La Troncal, Ecuador." Transactions in GIS 20:976–90.
[3] Sarkar, S., J. S. Dyer, C. Margules, M.Ciarleglio,N. Kemp, G. Wong, D. Juhn, and J.Supriatna.
[4] Soemarno. 1991. Studi Perencanaan Pengelolaan Lahan di Sub DAS Konto, Malang. Disertasi, Jurusan Pengelolaan Sumberdaya Alam dan Lingkungan (PSL), Institut Pertanian Bogor.
[5] Soemarno. 1991a. Implementasi Model Tujuan Ganda dalam Pengelolaan Daerah Aliran Sungai. Studi Kasus di Sub DAS Pinjal, Kabupaten Malang. Jurnal Universitas Brawijaya, Vol. 3 No.2, Hal. 41-60.
[6] Saaty, T.L., 1980, *The Analytic Hierarchy Process*, McGraw-Hill, New York.
[7] Saaty, T. L. 1996. Decision Making withDependence and Feedback The Analytic NetworkProcess. RWS Publications: Pittsburgh.
[8] Wahyuni, 2015. Sistem Pendukung Keputusan Penentuan Universitas Swasta Terbaik Di Aceh Menggunakan Metode *Analytical Hierachy Process*, Skripsi, Progam Studi Teknik Informatika. Bireuen.
[9] Taufiq, Maryana, M.Daud Nurdin. Model Informasi Desa Terintegrasi Berbasis Android. Unimal Pres 2017 Lhokseumawe.
[10] Taufiq Modul Aplikatif Sistem Informasi Adminitasi Gampong (SiAgam V.1) Penerbit Umuslim 2017 Bireuen
[11] Wang, Z., Glynn, P. W., Ye, Y. 2014. Likelihood robust optimization for data-driven problems. arXiv: 1307.6279v3 [math.OC]. Robyn M. Dawes. 1988. Rational Choice in an Uncertain World. Hartcourt Brace Jovanovich, Publishers.
[12] Xue Ling, Zhu Yeping, Xue Yan RAEDSS: An integrated decision support system for regional agricultural economy in China School of Government, Peking University, Beijing, 100871, Beijing, ChinaInstitute of Agricultural Information, Chinese Academy of Agricultural Sciences 100081, Beijing, China
[13] Yao, J., Zhang, X., Murray, A. T. 2017. Spatial optimization for land-use allocation: Accounting for sustainability concerens. International Regional Science Review, 1-22. 2017."DevelopinganObjectivesHierarchyforMulticriteriaDecisionsonLandUseOptions, with a

IOP Conf. Series: Materials Science and Engineering **769** (2020) 012059 doi:10.1088/1757-899X/769/1/012059

Case Study of Biodiversity Conservation and Forestry Production from Papua, Indonesia." Environment and Planning B: Urban Analytics and City Science 44:464–85.

[14] Machine Learning for Vehicle Plate Number Recognition DIP DIP DIP Researcher: D.Sihombing, H.A. Nugroho, S. Wibirama (2015) A. Prasetyo, S. Wibirama, N.A. Setiawan (2016) Machine learning Machine learning for traffic monitoring Researcher: D.A. Kurniawan, S. Wibirama, N.A. Setiawan (2016)

[15] Pros and cons Source: Introducing Deep Learning with Matlab (Mathworks, 2018) Summary • Artificial intelligence has been used in so many applications. Practically, we are surrounded by AI technologies, in so many forms. • Machine learning is a sub-field of AI that focuses on giving ability to a computer to learn from data without explicitly being programmed by a human. • Deep-learning is a new form of neural network research. It is now the most popular algorithm of machine learning technology used in various companies.

IOP Conf. Series: Materials Science and Engineering **769** (2020) 012060 doi:10.1088/1757-899X/769/1/012060

Requirement Prioritization Based on Non-Functional Requirement Classification Using Hierarchy AHP

Thant Z. Win, Rozlina Mohamed and Jamaludin Sallim
Faculty of Computing, College of Computing and Applied Sciences, Universiti Malaysia Pahang, 26300 Kuantan Pahang Malaysia

E-mail: rozlina@ump.edu.my

Abstract. Requirement prioritization is a process in requirement engineering, which is a part of software development life cycle (SDLC). Requirement is prioritized due to constraints such as budget, time and resource allocation. Requirements of software is often classified as functional requirements (FR), and non-functional requirement (NFR). In order to produce a high-quality software, both requirement must be considered during requirement prioritization process. Various prioritization techniques have been invented, and Analytical Hierarchical Prioritization (AHP) is the most popular technique that has been cited. However, AHP does not support the NFR and unscalable. Meanwhile, Hierarchy-AHP has been introduced unto increase the scalability of AHP by using hierarchical requirements as input. Nevertheless, hierarchy-AHP does not meant for NFR and experimental result for increasing the scalability is not received significant attention. Thus, we intend to use NFR with large dataset on hierarchy-AHP. Aim of this paper is an exploration of hierarchy-AHP experimenting on RALIC dataset. Our major findings are: (i) NFR can be used hierarchy-AHP with minor process amendment, and (ii) hierarchy-AHP able to reduce pairwise comparison which is up to 97.33% for 403 number of requirements, compared to original AHP.

Keyword. Requirements prioritization, Non-functional requirement, Analytic Hierarchy Process (AHP), Hierarchy AHP.

1. Introduction

Requirement engineering is one of the most important and fundamental stage in SDLC in any software process model. Requirement engineering is the process to collect the software requirement from user, analyses and document are known as requirement engineering. Requirements prioritization is an activity for identifying the most important requirements for a system [1]. Requirement prioritization is in either software validation or software analysis process which is part of software requirement engineering [2]. Determining the accurate requirement which is selected from massive amount requirements is a crucial step. Requirement is prioritized based on many aspects such as significance, cost, time, risk and volatility. Various scale can be used in prioritization activity such as a ratio scale or ordinal scale [3].

2. Related work

The related works in this research is divided into 4 sub-sections.

2.1. Analytic Hierarchy Process

Analytic Hierarchy Process (AHP) is initially developed by Prof. Thomas L. Saaty. AHP is widely known in software prioritization as a structured technique for organizing and analyzing requirements to be prioritized. Prioritization decision is made based on mathematics and psychology [5]. In this technique, all requirements will be compared in pairwise. Each comparison is compare based on a ratio scale [6]. Then, the ratio scale of the pairwise comparison is put on a matrix. The matrix is used to compute the relation significance of each requirement.

Thus, total number of pairwise comparison in AHP is up to $n \times (n-1)/2$, where n is number of requirements. Therefore, the entire pairwise comparison will dramatically increase proportion to the increment number of requirements [7]. As a result, AHP is *unscalable* due to inappropriate for a large number of requirements. This is because, number of requirements stipulate the number of pairwise comparisons that need to be made [8].

2.2. Hierarchy AHP

Normally, requirements for a large scale software project are structured. Thus, structured requirements can be broken-down into hierarchy of interrelated requirements. The hierarchically structured requirements are arranged where the most generalized requirement is placed on top of the hierarchy, while the more specific requirements are place on the lower level of the hierarchy. This hierarchically structured requirements are becoming the input unto AHP in stages, based on the hierarchy level. In contrast with input unto original AHP, requirements are in flat structure. By using hierarchy AHP, not all requirements are compared in pairwise. The pairwise comparison is only occurred between requirements in the same hierarchy level.

Using a hierarchical AHP reduces the required number of pairwise comparison. As a result, hierarchy AHP is more scalable than original AHP.

2.3. Hybrid Assessment Method

Hybrid Assessment Method (HAM) is used to support a prioritization ranking for NFR. It is used to aid software engineers to decide which qualities are more important to be satisfied and, in case of conflict, which should be prioritized over others and HAM recommend the best architecture by ranking architectural alternatives considering the stakeholder's views [11]. HAM is an efficient method combines one single pairwise comparison matrix for a decision making to determine the weights of criteria with one typical weighted decision matrix to prioritize the alternatives [11]. The consistency problems associated to scale and the prioritization method, geometric scale is used in HAM. The geometric scale is used for evaluating the criteria and the geometric mean for defining the alternative rankings.

Advantages of HAM are: (1) consistent, (2) strong and (3) flexible. HAM allows trade-offs between criteria importance, for example; non-functional requirement resulting in the automatic determination of their relative importance (weights). Disadvantages of HAM are: (1) does not support group decision making, and (2) requires technical knowledge about the context domain for providing judgments for relative importance of criteria, in addition to provide appraisals for the alternatives [11].

2.4. Non-Functional Requirements

Non-functional requirements are the system properties and limitations for example reliability, performance usability, efficiency and space requirements. Non-functional requirement is expressed to increase the accuracy of FR. Three parts of NFR are: product requirements, organization requirements and external requirements [12]. NFR may affects the overall architecture of a system rather than the individual components. NFR is typically identify or limit characteristics of the system as an entire such as performance, security, or usability. Type of NFR are: Efficiency Requirement, Security Requirement, Dependability Requirements, Performance Requirements, Space Requirements, Portability Requirements, Modifiability Requirements, Reliability Requirements, and Usability Requirements [12]. Table 1 represented comparative for some requirement prioritization.

Table 1. Comparative of requirement prioritization techniques

Requirement Prioritization Techniques	Advantages	Disadvantages	References
AHP	It is use for complex decision making and usage for pairwise comparison to compute the relation importance of each requirement. By decreasing complex decisions pairwise comparison AHP supports decision makers to get best decision.	Time consuming for pairwise comparison when use the large amount of requirement. Thus this technique is unscalable .	[5, 7, 13, 14]
Hierarchy AHP	Reduce number of pairwise comparison	Requires initial works to disseminate requirements into hierarchical structured No experimental results reported.	[20, 21]
HAM	Combining one single pairwise comparison decision matrix. Thus, it is simple, efficient, consistent, strong and flexible method.	Does not support group decision making. Requires technical knowledge about the context domain to provide the importance judgments for criteria compared.	[11]

3. Methodology

Methodology in this research is consists of four main process, study the existing requirement prioritization techniques, pairwise comparison of non-functional requirement for hierarchy (a), and pairwise comparison at low level of requirement hierarchy (a), prioritized non-functional requirement for each functional requirement.

3.1. RALIC dataset

Dataset is a collection of data. Most commonly a data set corresponds to the contents of a single database table, or a single statistical data matrix. The RALIC project was a software development originated to substitute the existing access control systems at UCL and consolidate the new system with library access and borrowing. RALIC stands for Replacement Access, Library and ID Card. The RALIC project datasets were collected by Soo Ling Lim at the University College London (UCL) [4, 19].

3.2. Pairwise comparison

Analytic Hierarchy Process technique uses a pair-wise comparison matrix to compute the relative significance of each requirement. The total number of comparisons to make with AHP are n × (n-1)/2 (where n is the number of requirements) at each hierarchy level. After making pairwise comparison for hierarchy (a) lowest to highest (a.1 to a.3), calculate the average the crossing for every row of requirement and got the weights result for each requirement row prioritized list which was highest weight was first prioritized. Figure 1 represented the research methodology flow for this paper.

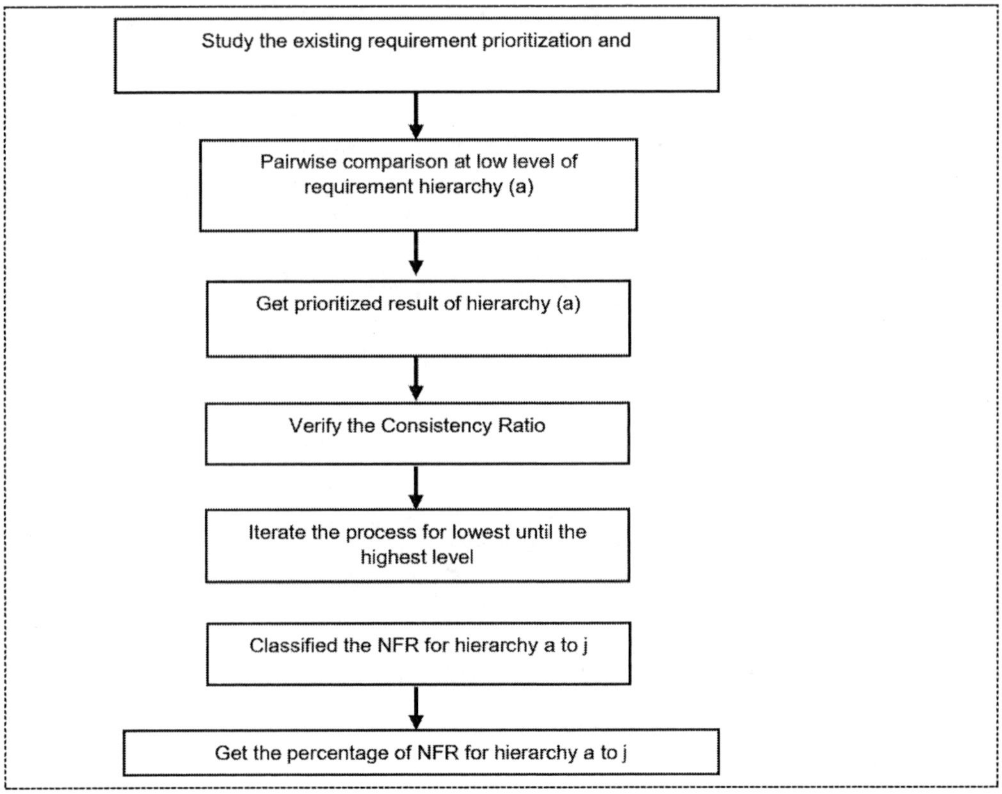

Figure 1. Research methodology

3.3. Consistency Ratio Check

The consistency of comparisons is assessed by using the Eigenvalue (Max) (λ) after all pairwise comparisons have been finished. A consistency index (CI) is then calculated with the eigenvalue, via the following equation 1: After getting pairwise comparison weights value for requirement, verify the consistency ratio check accuracy for pairwise comparison using Equation 1 of consistency index check for consistency ratio (CR). Where (λ) is Eigenvalue (Max) and n is number of requirements.

IOP Conf. Series: Materials Science and Engineering **769** (2020) 012060 doi:10.1088/1757-899X/769/1/012060

$$CI = (\lambda - n) / (n - 1) \qquad (1)$$

where CI is a Consistency index

λ is an Eigenvalue (Max)

n is a Number of requirements

n represents number of requirements consistency and can be checked by considering the consistency check, of *CI* with the appropriate value of Random Index (*RI*), Equation (2) of consistency ratio check, *CR* for Consistency Ratio, *CI* for Consistency Index and *RI* for Random Index. If consistency ratio value is (0.1) below, pairwise comparison is accurate and consistency ratio value is more than (0.1) need to make pairwise comparison again. According to the Table 2 shows the random index value for consistency check.

$$CR = (CI) / (RI) \qquad (2)$$

where CR is a Consistency ratio

CI is a Consistency index

RI is a Random index

Table 2. Random index value.

n	*1*	*2*	*3*	*4*	*5*	*6*	*7*	*8*	*9*	*10*
RI	0	0	0. 58	0.9	1. 12	1. 24	1. 32	1. 41	1. 46	1. 49

4. Implementation

The implementation includes two parts, which are: comparison matrix for using RALIC dataset hierarchy (a) non-functional requirement, and part B is determine the non-functional requirement for hierarchy (a) for this study.

4.1. Pairwise Comparison Using Ralic Dataset Hierarchy (a) Non-functional Requirement

AHP pairwise comparison using RALIC dataset requirement hierarchy (a). Table 3 describes the ratio scale for AHP technique.

Step 1, compare the hierarchy (a) requirement number 45 requirements need to compare for pairwise comparison. Table 6 illustrations the comparison of RALIC dataset hierarchy (a) for non-functional requirement.

Step 2, when assign the actual value into upper triangular matrix, reciprocal value for lower triangular matrix automatically appeared in pairwise comparison table. Pairwise comparison for RALIC dataset hierarchy (a) illustrated in Table 6.

Step 3 comparison hierarchy (a), divide each element of hierarchy (a) the matrix with the sum of its column and total sum of each column is 1 and the normalized the row by row of the matrix by averaging across the rows, got the weights result.

Step 4 calculate the consistency measure for consistency index, and calculate the consistency ratio (CR).

Table 3. Ratio scale of analytic hierarchy process.

Scale	*Definition*
1	Equal
3	Moderate
5	Strong
7	Very strong
9	Extreme
2, 4, 6, 8	Intermediate values

Table 4. Pairwise comparison for sub-hierarchy (a.1).

a.1	*a.1.1*	*a.1.2*	*a.1.3*	*a.1.4*
a.1.1	1	3	5	3
a.1.2	0.33	1	4	7
a.1.3	0.20	0.25	1	9
a.1.4	0.33	0.14	0.11	1
Sum	**1.87**	**4.39**	**10.11**	**20**

a.1	*a.1.1*	*a.1.2*	*a.1.3*	*a.1.4*	Weights
a.1.1	0.54	0.68	0.49	0.15	0.47
a.1.2	0.18	0.23	0.40	0.35	0.29
a.1.3	0.11	0.06	0.10	0.45	0.18
a.1.4	0.18	0.03	0.01	0.05	0.07
Sum	**1**	**1**	**1**	**1**	1

Table 5. Pairwise comparison for sub-hierarchy (a.3).

a.3	*a.3.1*	*a.3.2*	*a.3.3*	*a.3.4*	*a.3.5*	*a.3.6*
a.3.1	1	6	5	7	3	9
a.3.2	0.167	1	3	4	3	6
a.3.3	0.20	0.33	1	3	3	2
a.3.4	0.14	0.25	0.33	1	5	7
a.3.5	0.33	0.33	0.33	0.2	1	9
a.3.6	0.11	0.17	0.50	0.14	0.11	1
Sum	**1.95**	**8.08**	**10.17**	**15.34**	**15.11**	**34**

a.3	*a.3.1*	*a.3.2*	*a.3.3*	*a.3.4*	*a.3.5*	*a.3.6*	Weights
a.3.1	0.51	0.74	0.49	0.46	0.20	0.26	0.44
a.3.2	0.09	0.12	0.30	0.26	0.20	0.18	0.19
a.3.3	0.10	0.04	0.10	0.20	0.20	0.06	0.12
a.3.4	0.07	0.03	0.03	0.07	0.33	0.21	0.12
a.3.5	0.17	0.04	0.03	0.01	0.07	0.26	0.10
a.3.6	0.06	0.02	0.05	0.01	0.01	0.03	0.03
Sum	**1**	**1**	**1**	**1**	**1**	**1**	1

Table 6. Pairwise comparison for hierarchy (a).

a	*a.1*	*a.2*	*a.3*
a.1	1	5	3
a.2	0.2	1	7
a.3	0.33	0.14	1
Sum	**1.53**	**6.14**	**11**

a	*a.1*	*a.2*	*a.3*	Weights
a.1	0.65	0.81	0.27	0.58
a.2	0.13	0.16	0.64	0.31
a.3	0.22	0.02	0.09	0.11
Sum	**1**	**1**	**1**	1

IOP Conf. Series: Materials Science and Engineering **769** (2020) 012060 doi:10.1088/1757-899X/769/1/012060

4.2. Determined the Hierarchy (a) For Non-Functional Requirement

In this part determined the hierarchy (a) for non-functional requirement for RALIC dataset. According to the Table 4, pairwise comparison for sub-hierarchy (a.1) and Table 5 shows, pairwise comparison for sub-hierarchy (a.3). Table 6, described the pairwise comparison for hierarchy (a).

Table 7. Determined the NFR for hierarchy (a).

Hierarchy (a)	*Weights*	*Type of NFR*
a.1	0.58	Efficiency
a.2	0.31	Security
a.3	0.11	Usability

5. Result

The results of requirement prioritization based on non-functional requirement classification using hierarchy analytic hierarchy process (Hierarchy AHP). The result of hierarchy (a) for non-functional requirements is efficiency (60%) which is highest prioritized in hierarchy (a). The outcome weights result and percentage of pairwise comparison using RALIC dataset hierarchy (a) for non-functional requirement list described in Table 7.

Figure 2 described the percentage of NFR for hierarchy (a) dataset and Figure 3 indicated the percentage of hierarchy (a to j) for non-functional requirements. For hierarchy (a), efficiency requirement is the highest percentage of non-functional requirement. And then, for the all hierarchy (a to j) of the highest percentage of non-functional requirement was Portability requirement which reached (92%) in hierarchy (e) RALIC dataset.

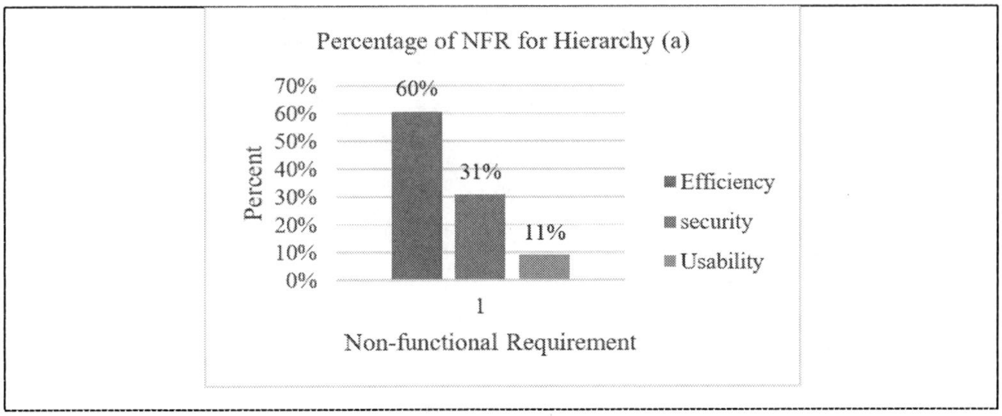

Figure 2. Percentage of NFR for hierarchy (a).

6. Conclusions

A Prioritization process can be completed based on several features such as significance, budget, time, risk and instability. Requirement prioritization techniques are important and help to right decisions making in order to develop software products. The study focus on hierarchy analytic hierarchy process pairwise comparison matrix calculation using existing RALIC dataset requirement and classified the hierarchy (a) for non-functional requirement. The outcome of hierarchy (a) consistency ratio accepted for accuracy. As well as, the percentage of efficiency requirement is the highest percentage for non-functional requirement in hierarchy (a). When making the pairwise comparison for hierarchy (a to j) requirements input is *(15, 051)*, in contrast, hierarchically pairwise comparison for (a to j) requirements input is *(403)*. Therefore, hierarchically comparison can reduce the number of requirements *(97.33 %)* by using hierarchy AHP. Secondly, AHP technique had weakness when validate consistency ratio check for the accuracy calculation accepted for 10 number of requirements only because Random Index *(RI)* has value for 10 number of requirements only. Consistency Ratio = Consistency Index / Random Index *(CR = CI / RI)* in that formula when calculate the accuracy formula able to accept for 10 requirements and then not allow to more than 10 number of requirements. Therefore, this paper segregated and prioritized the non-functional requirements using RALIC dataset hierarchy (a) requirements.

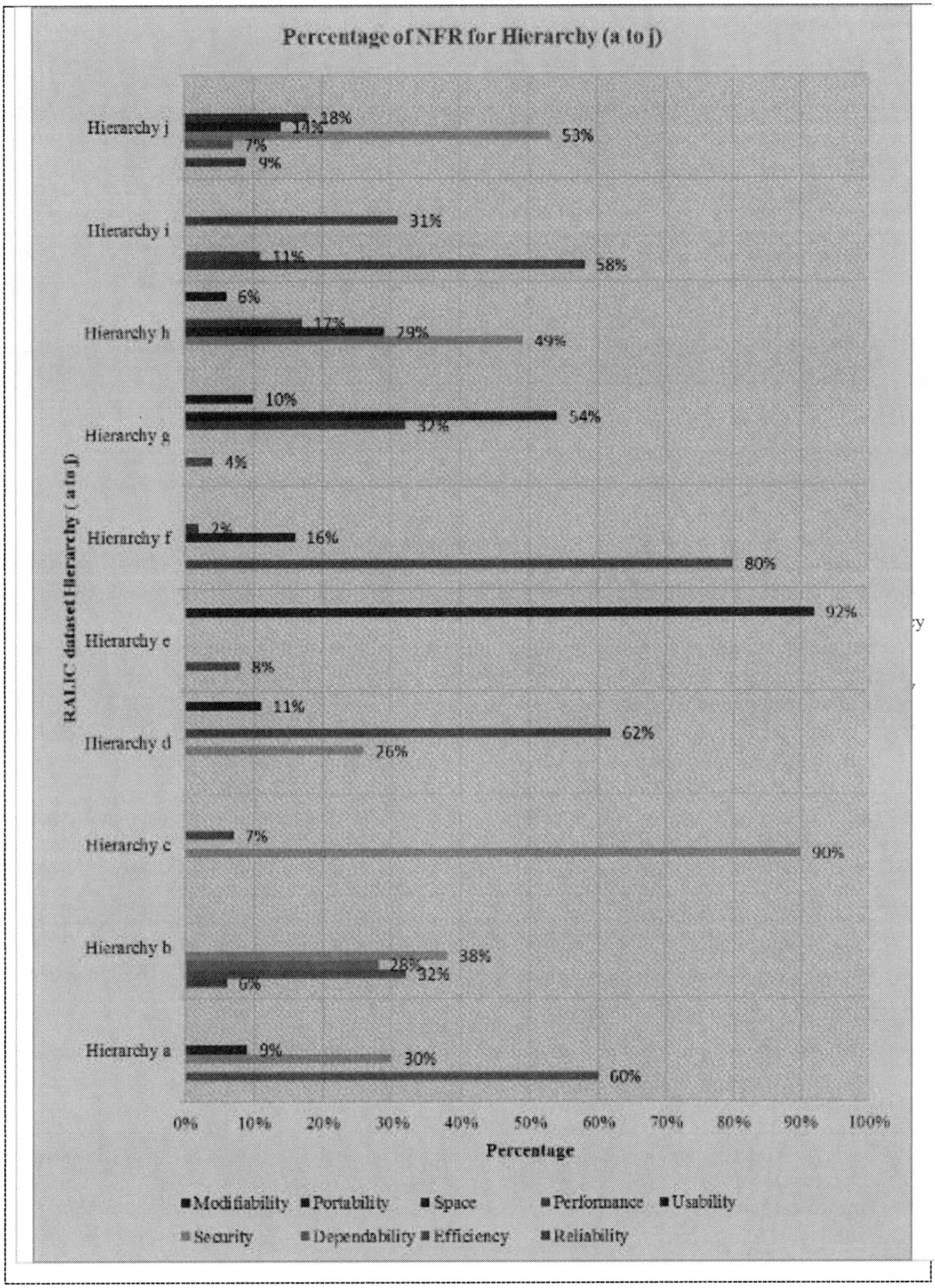

Figure 3. Percentage of NFR for hierarchy (a to j).

Acknowledgment

The work presented in this paper is supported by Universiti Malaysia Pahang Grant: RDU182207-2.

References

[1] Sommerville, I., Introduction to software Engineering. Addison-Wesley, 2007.
[2] Vestola, M., A Comparison of Nine Basic Techniques for Requirements Prioritization. Helsinki University of Technology, 2010.
[3] Lehtola, L., Providing value by prioritizing requirements throughout software product development. Licentiate thesis, Helsinki University of Technology, 2006.
[4] Lim, S.L. and A. Finkelstein, StakeRare: using social networks and collaborative filtering for large-scale requirements elicitation. IEEE Transactions on Software Engineering, 2012. 38(3): p. 707-735.
[5] Saaty, T., The Analytic Hierarchy Process, McGraw-Hill, New York, 1980. There is no corresponding record for this reference.
[6] Karlsson, J. and K. Ryan, A cost-value approach for prioritizing requirements. IEEE software, 1997. 14(5): p. 67-74.
[7] Berander, P. and A. Andrews, Requirements prioritization, in Engineering and managing software requirements. 2005, Springer. p. 69-94.
[8] StephFam. A List of Requirements Prioritization Techniques You Should Know About. 2016 August 18, 2016 [cited 2017 2/1/2017].
[9] Berander, P. and P. Jönsson, Hierarchical cumulative voting (hcv)—prioritization of requirements in hierarchies. International Journal of Software Engineering and Knowledge Engineering, 2006. 16(06): p. 819-849.
[10] Voola, P. and A.V. Babu, Comparison of requirements prioritization techniques employing different scales of measurement. ACM SIGSOFT Software Engineering Notes, 2013. 38(4): p. 1-10.
[11] Ribeiro, R.A., et al., Hybrid assessment method for software engineering decisions. Decision Support Systems, 2011. 51(1): p. 208-219.
[12] Sommerville, I., Software engineering. 2010: Pearson.
[13] Karlsson, J., C. Wohlin, and B. Regnell, An evaluation of methods for prioritizing software requirements. Information and Software Technology, 1998. 39(14): p. 939-947.
[14] Khan, K.A., A systematic review of software requirements prioritization. 2006, Blekinge Institute of Technology.
[15] Berander, P., Evolving prioritization for software product management. 2007.
[16] Leffeingwell, D. and D. Widrig, Managing software requirements. A use case aproach. 2003, Addison-Wesley.
[17] Brackett, J.W., Software requirements. 1990, DTIC Document.
[18] Mustafa, B.A. and A. Zainuddin. An experimental design to compare software requirements prioritization techniques. in Computational Science and Technology (ICCST), 2014 International Conference on. 2014. IEEE.
[19] Lim, S.L., *Social Networks and Collaborative Filtering in Large-Scale Requirements Elicitation.* 2010
[20] Khan, J.A, et al., An Evaluation of Requirement Prioritization Techniques with ANP, International Journal of Advanced Computer Science and Applications, Vol. 7, No. 7, pp. 419-429. 2016
[21] J. Karlsson, C. Wohlin and B. Regnell, "An Evaluation of Methods for Prioritizing Software Requirements", Information and Software Technology, Vol. 39, No. 14-15, pp. 939-947, 1997-98

IOP Conf. Series: Materials Science and Engineering **769** (2020) 012061 doi:10.1088/1757-899X/769/1/012061

A Review of User Authentication Model for Online Banking System based on Mobile IMEI Number

Waleed A. Hammood[1], Ruzaini Abdullah[1], Omar A. Hammood[1], Salwana Mohamad @ asmara[1], Mohammed A. Al-Sharafi[1], Ali Muttaleb Hasan[1]

[1]Faculty of Computing (FKOM)
College of Computing and Applied Sciences
Universiti Malaysia Pahang 26300
Kuantan, Pahang, Malaysia

engwaleed54@yahoo.com, ruzaini@ump.edu.my, omer_almajeed@yahoo.com, salwanamohamad@ump.edu.my, ma_shrafi@yahoo.com, alimatlab65@yahoo.com

corresponding author: engwaleed54@yahoo.com

Abstract. The traditional banking system, such as transactions over the counter using bank book is enhanced by adding the elements of electronic banking, where nowadays all transaction can be done over the network, because of globalization and the advancement of information and communication technology. However, significant threats to this technology also come in parallel, such as from banking frauds. This is why building trust among banking users by providing security mechanisms is very important. In the electronic banking, security mechanisms mainly focus on proving a secure environment for the online transaction, especially user authentication. Many researchers have been proposed various models for the online user access authentication for the banking industry. Most of the researches are based on traditional the form of username and password, but with the varying mechanism of password forms. For example, password in arbitrary value, password in arbitrary value through secured channel, as well as biometric like fingerprint recognition, voice recognition, or retina recognition. In this paper we reviewed the current existing user access for online banking based on mobile phone, the functions of the SIM card, the characteristics mobile of the SIM card are described. Cloning, a threat to the SIM card is also described. Findings show that all the security model for online user access contains password in the form of value, biometric, or PIN. Thus, none of the existing user access proposed the idea of the user access based on International Mobile Equipment Identity (IMEI) number to strength the security of the user access. The IMEI is a 15- or 17-digit code that uniquely identifies mobile phone sets.

1. Introduction

Online security is more complicated than computer security, the hackers always try to hack social network account, personal bank account, and e-mail account [1]. According to Nattakant [2] has discussed the phishing that includes sending of e-mails pretending in the form of legal financial institutions to the receiver and request private information such as username and password. it may potentially to collect cash from e-account by hacking. These hackers till now chopped many bank's web site and collect hugs amount of money through their technological efficiency, the e-transaction is fully dependent by online banking. So, the mainly depends on e-banking security is the protection of software, called software security. In this modern world, the traditional banking transaction is changing from using bank book over the counter to the transaction over the internet [3], as a result of technological advancement. So, clients' of the banks can perform transaction online by using the computer as well as mobile phone, which eases the transaction because clients' can make their banking regardless of time and place. That is why providing security to online banking transaction is the main concern in the modern banking system [4]. There are many security incidents reported in the news. In August of 1995, it was reported that there was a security incident, which involved attackers breaking into their system [5, 6] which estimate a $10 million lost. It was the primary successful penetration by a hacker into the arrangement of the $10 million dollars illicitly moved, $400,000 was as yet not found. In August 2000, British police have captured three men regarding an endeavour of deceitful in Egg banking system [5]. As per the report, the bank was the target of an effort to get cash through deceitful records however no cash was stolen, then the Egg authority strongly said that none of their computer systems had been breached.

In April 2010, a fire alarm company in Arkansas lost more than $110,000 because of a hacking incident [7]. Hackers hack into the system of the company and stole its online banking credentials and drained its payroll account It was seen that through the span of the past couple of days, somebody had affirmed two bunches of finance instalments one for $45,000 and another for $67,000. A couple of days after the fact, Melanie Eakel, CEO of JE Frameworks Inc., was educated by the bank about the Web Address that was utilized to process the payments, the username and password of the online banking [8]. Different security policies or frameworks have been deployed to banking systems to secure financial transaction for their online users. For example, there are banks that provide their own application to their clients to make the transaction. There are also banks that provide two-way verification method by sending a code to the clients' mobile phone, called SMS-based authentication [9] while the transaction is made.

The banking frauds reported, it is clear that we need to strengthen the online banking system. Experts and researchers have published many types of research on banking security, especially in a user authentication scheme. Most of the papers describe new security policy or framework for the banking transaction. For example, digital watermarking scheme Alam et al [10], digital identity Chowdhury & Noll [11], and PKI-SIM Card scheme Xue et al [12]. In this paper, we studied authentication methods used in various banking operations of user authentication for online banking system.

2. Literature review

As a result of technological advancement, the current traditional banking system and financial services are being transformed to the electronic banking system [13, 14], which is call online banking system. Using this system, users can now perform transactions, pay the bill, and shop online without any hassle, which makes lives easier. While online banking system provides hassle-free banking to the users, security to banking transaction for users is the main concern. Many technologists are trying to enhance the security of the bank[15]. The concern is to make user authentication system more secure [16]. Authentication is the process by which the end user application is identified and authorized so that, the user can have access to the system or application [17, 18]. The authentication aim is to verify that the certain information provided is a request from a given individual to be authentic. [19]. Generally, some of the information security management (ISM) systems follow the guidelines of the International Electro-Technical Commission (IEC) 27001 standard and International Organization for Standardization (ISO). [20]. At the particular level of Internet banking security, there seems to be no common standard that can be followed by all Web banking services, but there are certain set of guidelines for specific contexts. Two of these are the Data Security Standard (DSS) for Payment Card Industry (PCI), and the VISA requirements for payment security [20]. According to Poetta et al [21], the existing user authentication can be summarized as follows:

- Password, which can be any arbitrary secret value.
- Password through secured channel.
- SMS Code – one-time password sent by SMS.
- PIN – limited confidential value.
- Grid Card – the client inserts these codes sequentially into the process with a paper card with payment.

- PKI – Public key cryptography-based encryption scheme.
- Token - Trustworthy tool for authentication.

From the above existing user authentication mechanisms, they have been classed in three groups of categories, based on the factors that are [22]:

2.1. Single Factor Authentication Mechanisms

This is the traditional security method that requires user ID and password [22] to logging into the web, mobile phone application, or system. It is also called Single Factor Authentication (SFA) [22]. The security in SFA is totally depended on end-user's creativeness of providing the password, because the password should be trickier making it harder to guess. We have classified SFA in two categories. They are:

- Password – arbitrary secret value

This is the traditional security method that requires user ID and password [22] for logging into the web, mobile phone application, or any system. Hence, the users should take more precautions [22, 23] to choose the password. They should create a strong password, which contains alphabet, numerical value, and special characters.

- Password through secured channel (for example SSL)

Like any other web application, Web banking systems use The Hypertext Transfer Protocol (HTTP) for data transfer [20]. Regarding data transfer through HTTP, all communications must be secured. That is why the Secured Socket Layer (SSL) or Transport Layer Security (TLS) protocol (i.e. HTTP Secure over SSL or HTTP over TLS (HTTPS) must be used with POST method for sending data as POST method is more secured than GET method of HTTPS.

Clients can access to a website in a both secured or insecure manner. When clients have the opportunity to access to the website in a secure manner, the clients at that point be confirmed in certain designs. At the point when the client goes from an public page through a secure part, the web server begins SSL and secures this sort of communication. The difference between SSL and Non-SSL connection between user and bank website is shown in Fig 1.

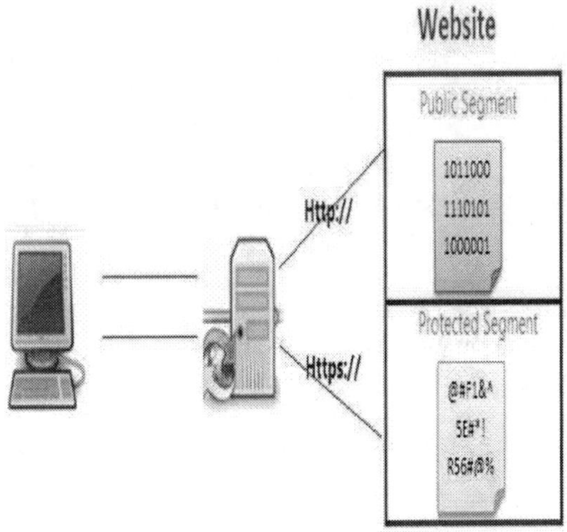

Fig. 1. Difference between SSL and Non-SSL connection adapted from [24].

In figure 1, it is shown that user is visiting a website. In Http://, the connection between user and server is a public segment, which means data is transmitted through the channel only on binary format. On the other hand, in Https://, the connection between the server and the user is protected segment or secured channel. Data is transmitted in a encrypted format through this secured channel.

The following Fig 2. is showing the overall process of mutual authentication for both client and server to each other in an SSL.

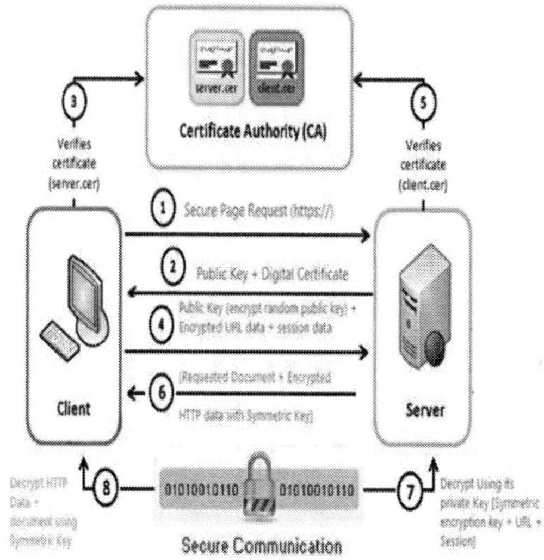

Fig. 2. Steps of SSL connection between user and bank server.

Asymmetric encryption is SSL's basic building block where the public key is freely shared to encrypt the message that the private key can only decrypt. SSL hires a third-party agency, a Certificate Authority (CA) to classify one or both ends of the transactions. The previous diagram shows how it works..

- A browser must send a request to interface with the server and requests a protected page (more often than not a report).
- The server of the web sends the public key with the user's labeling certificate.
- The browser tests if the certificate issued by the CA it trusts. The customer contrasts the data in the certificate and the data got from the site and checks every one of the subtleties. Assuming this is the case, the browser demonstrates the virtue of the server certificate by demonstrating a green lock and the customer continues.
- The browser creates a symmetric random encryption key and then encodes it to the server's open key. At last, it sends it to the server alongside the encoded URL and other encoded HTTP information.
- The web server decodes the approaching packet utilizing its private key and uses the symmetric key to decode the HTTP and URL information that was produced randomly at the user side.
- The requested file from the server is then sent back to the user together with other info rmation encrypted with the symmetric key.
- Eventually, using the symmetric key, the browser decrypts the packet and safe handshaking is created.

Figure 3 shows a typical SFA log in procedure that normally has user name and password fields for users to enter their confidential information.

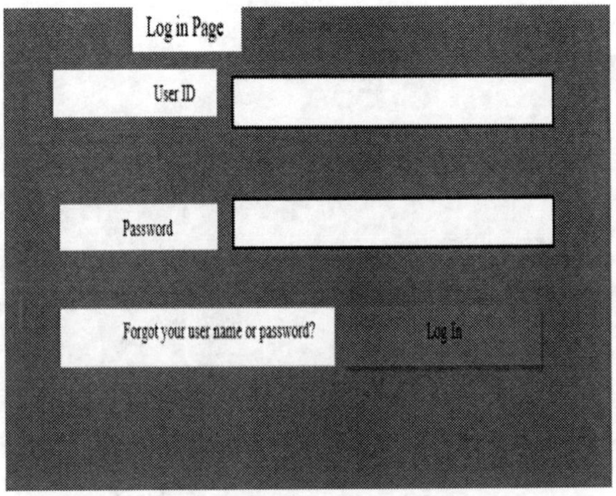

Fig. 3. A typical SFA login procedure.

2.2. Two-Factor Authentication Mechanisms (TFA)

Two-Factor Authentication is an end-user protection system having to go through following two factors. In this mechanism, user supplies or provides two types of identification, the one of which is normally a physical token, like a unique key generator or a smartphone, and the other of which is regularly something memorized by the user, for example, a security code or PIN or a token or a password. In this unique circumstance, the two elements included that you have and something you know. TFA methods are described below:

- SMS Code - One Time Password (OTP) Send by SMS

To fall in the TFA class, the user should have a devicelike Grid Card or Smart Card or "what you have" anduser should knowsomething (PIN or password) or "what you know". In this method, userhas something (e.g. Mobile phone) torecieve a PIN or a Code and User knows something (PIN or Code) after receiving them in their mobile phone. For this, user need toregister his or her phone number on their accounts [25, 26].

OTP is a disposable password, which is impossible to find out because of its dynamic feature. That is why this password can not be used again onceit is used next time [27]. Thus, it is valid only short time, which lasts for onesession or transaction [22]. This method is a combination of time-synchronous method and the event-synchronous method [27]. This method is widely used nowadays. A new code is generated in every specific time interval and it changes the code by increasing the countervalue for reattempt in each time interval.

An OTP log in process is shown in Fig 4. In this method, the user will visit the bank's log-in page and will be asked to enter phone number to receive OTP after entering his or her user ID. The number will be verified in the server. Then, the user will receice a short code in their given mobile phone number. Thus, user will enter this given code in the password box in the user log-in page.

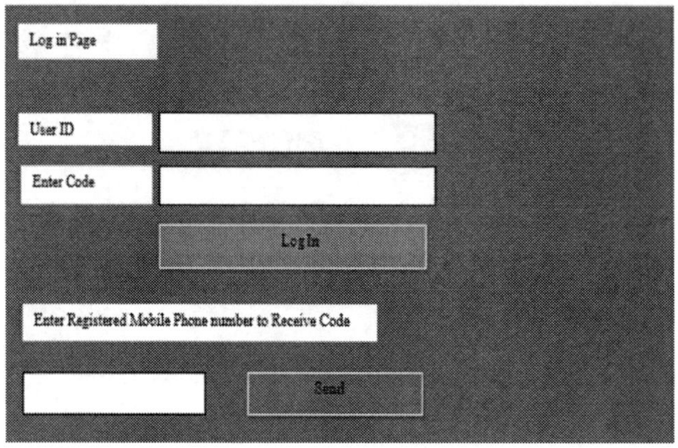

Fig. 4. OTP Log in process.

- Grid Card – paper card with transaction codes, user enters these codes in sequence into the system

This is a combination of paper card or device and something, such as Personal Identification Number (PIN) number or password or secret phase that the user knows [22, 28]. The use of Grid Cardin our daily life is withdrawing money from Automate Teller Machine (ATM) system. The user need to insert his or her debit card or credit into the ATM machine and then type their PIN. This process can be classified in two steps Grid Card - user has something and PIN - user knows something. This authentication method can also be brought to the online or Internet world, especially for online banking transaction. In modern Grid Card, there is a smart chip, which contain user information stored in the bank server, has been pasted. The Flowchart of Grid Card using PIN method is shown in Fig 5. below. In This Flowchart, we have been assumed that all banks share each other's users' information if itis their debit card.

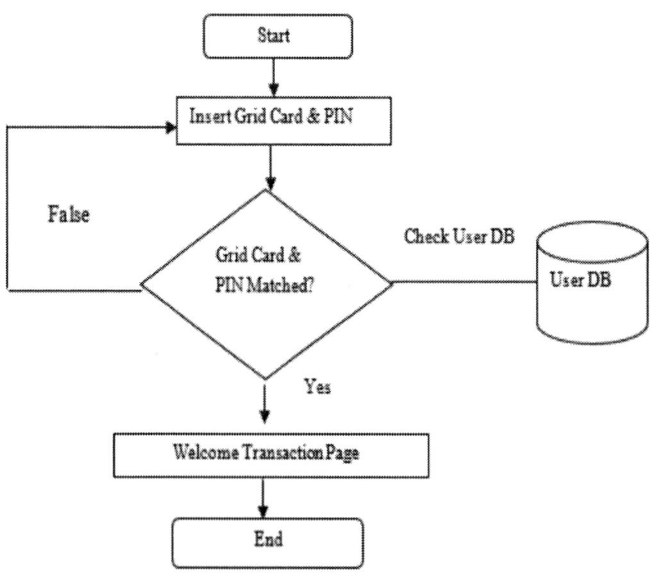

Fig. 5. Flowchart of Grid Card & PIN method.

- Token - trustworthy authentication device

Token is a small pocket-size device for authentication, which display passcode on Liquid Cristal Display (LCD) [22, 29]. It generates password based on an algorithm [25] and is increasingly used in banking industry for user authentication. If users try to log into the bank's internet banking system, a username and password created by the token will be requested from the protected web page. The token also has a PIN which must be inserted in token whenever a code is needed.

It can also provide an electronic signature code based on the key issued by the internet banking application's protected website. Consequently, it has two functions: Provide a login password (option 1 from the main menu) and a signature token if requested as part of a web page transaction (option 2 from the token main menu).

The token will have a registration number and is connected to the account of the user. For other accounts, it cannot be used to create passwords. Last, authentication system based the Public Key Cryptography (PKI) password through secured channel, in this mechanism, password is sent through secured channel using Public Key Cryptography.

2.3. *Multifactor Authentication*

Multi-Factor Authentication (MFA) or strong authentication mechanism [22, 30] authenticates users for transaction such as make a payment, fund transfer, or log into the system by providing multi-level security mechanism [31]. In addition to two factor authentication mechanism, where user provides two level of identification; for example, physical token like smart card or debit card and a security code or PIN or password, in MFA, an additional authentication method is used to verify users. The additional authentication method can be biometric verification process such as iris recognition, finger print scanning, facial recognition, or voice ID. MFA method is shown in Figure 6 below:

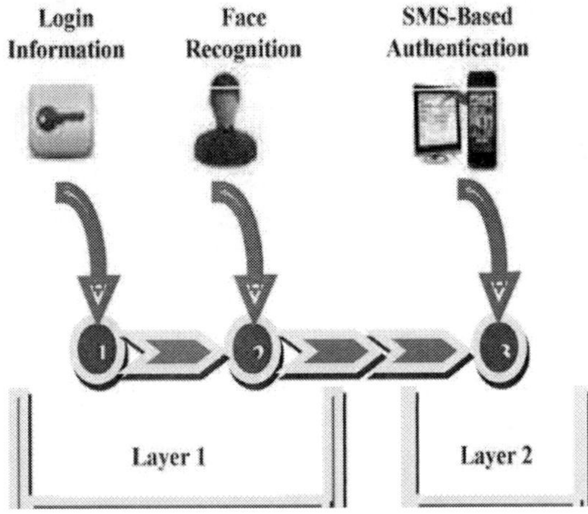

Fig. 6. MFA using face recognition process adapted from [32].

In the figure above, there are two layers for MFA. In step one of first layer , users will enter their typical user name and password. After they enter those credentials and values, the system will recognize users' face in step two. And lastly, in step 3 in layer two, users will enter the PIN number that is received in their mobile phone after completing Layer 1.

3. Security in Online Banking System

This section reviews the existing work in online banking, particularly the security procedures on user authentication, using mobile phones. Alam et al and McCarthy et al [10, 24] described mobile banking process - opening bank account using mobile phone, checking account balance, and making transaction through mobile phone. And later, they proposed a two-level security scheme for transaction by using: biometric verification, and digital watermarking.

In their proposed scheme, they divided the transaction method in two steps. The first step covers the security between Client and Mobile Server by sending SMS from their registered mobile number saved in the bank and then user will be prompted to record their own voice, which will be forwarded to and stored in the banking server. So, when the user wants to make any transactions later on, their given voice will be verified with the stored voice in the server, and if the voice is matched, they will be able to make transaction. In the second step, they cover the security between Mobile Station and Bank Server by securing the stored data in the server using digital water making technique. To do so, they used two data hiding techniques namely the standard Least Significance Bit (LSB) method and Bit Least Significance Bit (BLSB) method. In this method all processes of user verification or authentication is performed using the registered mobile number of clients. However, this method is susceptible to mobile Subscriber Identity Module (SIM) card cloning. Another work proposed by Xue et al [12] is an authentication scheme called Public Key Infrastructure - Subscriber Identity Module (PKI - SIM), which generates authentication key independently both in mobile phone and in the authentication server. This is a good authentication scheme for online transaction, but still has some weaknesses, especially SIM card cloning. The summary of literature review is shown in the table 1:

Table 1. The existing security scheme and framework for the clients' authentication for online transaction.

Authors	Description
[10]	Propose two way security scheme for user verification and authentication using mobile SIM card.
[12]	Proposed PKI-SIM card based authentication method.
[9]	Online banking services: user authentication and data origin authentication. SMS based authentication by sending an OTP.
[11]	Described a model "My Digital Identity" using mobile SIM card.

4. Conclusion and Future Work

The traditional banking transaction is changing from using bank book over the counter to the transaction over the internet, as a result of technological advancement. This paper reviewed the existing framework for online user access for the bank. It was also shown the classification of the existing framework based on factors. As mobile phone is the main character of this research. It also gives primary idea about cloning SIM card, analysed the previous research works related to online user access authentication system for bank, building a trust among banking users by providing security mechanisms is very important. The previous work found that all the security model for online user access contains password in the form of value, biometric, or PIN. But it is time to get out of this current trend. Experts and researchers have published many researches on banking security, especially in user authentication scheme. Most of the papers describe new security policy or framework for the banking transaction. For example, digital watermarking scheme [10, 33], digital identity [34] , and PKI-SIM Card scheme [35]. However, to best of our knowledge, none of the literature reviews have been performed that focuses on Subscriber Identification Module (SIM) card cloning attack for the online clients' authentication for the banking industry. So, future research should be focused on user authentication system or model without any password based on mobile IMEI number.

References

[1] Sadekin, M.S. and A. Shaikh, *Security of E-Banking in Bangladesh.* Journal of Finance and Accounting, 2016.

[2] Utakrit, N., *Security awareness by online banking users in Western Australian of phishing attacks.* 2012.

[3] Adrian, T. and A.B. Ashcraft, *Shadow banking: a review of the literature*, in *Banking Crises*. 2016, Springer. p. 282-315.

[4] Al-Otaibi, S., et al., *The satisfaction of Saudi customers toward mobile banking in Saudi Arabia and the United Kingdom.* Journal of Global Information Management (JGIM), 2018. 26(1): p. 85-103.

[5] Lee, A.J.Y., et al., *Customers' Satisfaction Towards Online Banking In Malaysia: A Primary Data Analysis*. 2017, UTAR.

[6] Yang, Y.-J. *The security of electronic banking.* in *Proc. Nat. I International Systems Security Conference*. 1997.

[7] Gunson, N., et al., *User perceptions of security and usability of single-factor and two-factor authentication in automated telephone banking.* Computers & Security, 2011. 30(4): p. 208-220.

[8] Hiltgen, A., T. Kramp, and T. Weigold, *Secure internet banking authentication.* IEEE Security & Privacy, 2006. 4(2): p. 21-29.

[9] Panja, B., et al. *Cybersecurity in banking and financial sector: Security analysis of a mobile banking application.* in *2013 International Conference on Collaboration Technologies and Systems (CTS)*. 2013. IEEE.

[10] Alam, S.B., et al. *A secured electronic transaction scheme for mobile banking in Bangladesh incorporating digital watermarking.* in *2010 IEEE International Conference on Information Theory and Information Security*. 2010. IEEE.

[11] Chowdhury, M.M. and J. Noll. *Distributed identity for secure service interaction.* in *2007 Third International Conference on Wireless and Mobile Communications (ICWMC'07)*. 2007. IEEE.

[12] Yin, X., et al. *An improved dynamic identity authentication scheme based on PKI-SIM card.* in *2009 5th International Conference on Wireless Communications, Networking and Mobile Computing*. 2009. IEEE.

[13] Priya, R., V. Tamilselvi, and G. Rameshkumar. *A Novel algorithm for Secure Internet Banking with finger print recognition.* in *2014 International Conference on Embedded Systems (ICES)*. 2014. IEEE.

[14] Al-khatib, A.A. and W.A. Hammood, *Mobile malware and defending systems: Comparison study.* International Journal of Electronics and Information Engineering, 2017. 6(2): p. 116-123.

[15] de Oliveira Malaquias, F.F. and Y. Hwang, *An empirical investigation on disclosure about mobile banking on bank websites.* Online Information Review, 2018. 42(5): p. 615-629.

[16] Rosener, D.K., W.O. Brown, and E.L. Reuss, *User authentication system and method.* 2016, Google Patents.

[17] Basavala, S.R., N. Kumar, and A. Agarrwal. *Authentication: An overview, its types and integration with web and mobile applications.* in *2012 2nd IEEE International Conference on Parallel, Distributed and Grid Computing*. 2012. IEEE.

[18] Omar A. Hammood, M.N.M.K., Muamer N. Mohammed and Waleed A. Hammood *Issues and Challenges of Video Dissemination in VANET and Routing Protocol: Review.* Journal of Engineering and Applied Sciences, 2017. 12(11): p. 9266-9277.

[19] Conklin, A., G. Dietrich, and D. Walz. *Password-based authentication: a system perspective.* in *37th Annual Hawaii International Conference on System Sciences, 2004. Proceedings of the*. 2004. IEEE.

[20] Martino, A.S. and X. Perramon. *A model for securing E-banking authentication process: antiphishing approach.* in *Services-Part I, 2008. IEEE Congress on*. 2008. IEEE.

[21] Peotta, L., et al., *A formal classification of internet banking attacks and vulnerabilities.* International Journal of Computer Science & Information Technology, 2011. 3(1): p. 186-197.

[22] Basavala, S.R., N. Kumar, and A. Agarrwal. *Authentication: An overview, its types and integration with web and mobile applications.* in *Parallel Distributed and Grid Computing (PDGC), 2012 2nd IEEE International Conference on*. 2012. IEEE.

[23] Alghathbar, K. and H. Mahmoud. *Noisy password security technique.* in *Internet Technology and Secured Transactions, 2009. ICITST 2009. International Conference for*. 2009. IEEE.

[24] McCarthy, C. and A.N. Zincir-Heywood. *An investigation on identifying SSL traffic.* in *2011 IEEE Symposium on Computational Intelligence for Security and Defense Applications (CISDA)*. 2011. IEEE.

[25] Lupu, C., V.-G. Gaitan, and V. Lupu. *Security enhancement of internet banking applications by using multimodal biometrics.* in *Applied Machine Intelligence and Informatics (SAMI), 2015 IEEE 13th International Symposium on*. 2015. IEEE.

[26] Arshah, R.A., W.A. Hammood, and A. Kamaludin, *An Integrated Flood Warning and Response Model for Effective Flood Disaster Mitigation Management.* Advanced Science Letters, 2018. 24(10): p. 7819-7823.

[27] You, H.-N., et al., *A Study on the Two-channel Authentication Method which Provides Two-way Authentication using Mobile Certificate in the Internet Banking Environment.* The Journal of Korean Institute of Communications and Information Sciences, 2011. 36(8B): p. 939-946.

[28] Hammood, O.A., et al., *RESP: Relay Suitability-based Routing Protocol for Video Streaming in Vehicular Ad Hoc Networks.* International Journal of Computers, Communications & Control, 2019. 14(1).

[29] Hammood, O.A., et al., *The VANET-Solution Approach for Data Packet Forwarding Improvement.* Advanced Science Letters, 2018. 24(10): p. 7423-7427.

[30] Hammood, O.A., M.N.M. Kahar, and M.N. Mohammed. *Enhancement the video quality forwarding Using Receiver-Based Approach (URBA) in Vehicular Ad-Hoc Network.* in *2017 International Conference on Radar, Antenna, Microwave, Electronics, and Telecommunications (ICRAMET)*. 2017. IEEE.

[31] Hashim, E.W.A., M.O.A. Hammood, and M.T.I. Al-azraqe, *A Cloud Computing System based Laborites' Learning universities: Case study of Bayan university's Laborites-Erbil.* BOOK OF PROCEEDING, 2016: p. 538.

[32] Mohammed, M.M. and M. Elsadig. *A multi-layer of multi factors authentication model for online banking services.* in *Computing, Electrical and Electronics Engineering (ICCEEE), 2013 International Conference on*. 2013. IEEE.

[33] Alam, S.B., et al. *A secured electronic transaction scheme for mobile banking in Bangladesh incorporating digital watermarking.* in *Information Theory and Information Security (ICITIS), 2010 IEEE International Conference on*. 2010. IEEE.

[34] Chowdhury, M.M. and J. Noll. *Distributed identity for secure service interaction.* in *Wireless and Mobile Communications, 2007. ICWMC'07. Third International Conference on*. 2007. IEEE.

[35] Xue, Y., et al. *An improved dynamic identity authentication scheme based on PKI-SIM card.* in *Wireless Communications, Networking and Mobile Computing, 2009. WiCom'09. 5th International Conference on*. 2009. IEEE.

Acknowledgment

UMP Postgraduate Research Grants Scheme, PGRS190386, funds this research.

Analysis of energy consumption in a wireless sensor network using fuzzy memberships function

Wan Isni Sofiah Wan Din[1], Mritha Ramalingam[2] and Nuramera Abdullah[3]

Faculty of Computing, Universiti Malaysia Pahang, 26300, Gambang, Pahang, Malaysia[1, 2, 3]

sofiah@ump.edu.my[1]

Abstract. The use of sensors in every application as well as in today's life has become a huge demand. Since that, the public sphere is starting to move to mobile applications that are easy to access. For example, child monitoring at school, health monitoring, tracking system, fire detection and so on. Thus, the Wireless Sensor Network (WSN) is the preferred choice to meet these needs. However, despite the enthusiasm of building various applications using sensors, the WSN itself is still hampered by limited battery usage. Due to some applications that require long battery life, various types of research have been conducted to address this problem. One of these is the involvement of artificial intelligent (AI) in extending the life of a battery on a sensor. Fuzzy Logic (FL) is one of the preferred AI that have been chosen by researchers to be implemented with WSN especially to protract the lifetime of WSN. In Fuzzy Logic, there are three membership functions that need to investigate the capability of it towards the WSN applications. The proposed approach is by combining different types of Fuzzy Logic membership functions which are Triangular with Gaussian, Gaussian with Trapezoidal and Trapezoidal with Triangular to get the best results for analysing the use of sensor batteries. The parameters involved for the cluster head selection are communication cost, centrality and residual energy as Fuzzy inputs. This approach will use an existing algorithm which is Multi-Tier Algorithm (MAP) and this is a part of the MAP enhancement towards the WSN lifetime. The results will compare, discuss and analyst the number of dead nodes and energy usage of the sensor node during data transmission. In conclusion, through this approach, it able to prolong the lifetime for the sensor network since the proposed technique can reduce the energy usage of the sensor nodes.

1. Introduction

In the past few years, the use of sensors has become wider due to the advances in technologies such as in vehicles, manufacturing, wireless communication, agriculture, healthcare and many more. A WSN is known as tiny devices that are used for monitoring and reporting the environmental conditions like pressure, sound, humidity, image and temperature. Typically, a WSN comprises of hundreds to thousands of sensor nodes and there must allocate at least one or more base stations in sensing area for user to access the data. The sensor nodes will collect all the data or information such as videos, audios, humidity, temperature, smoke and any others relevant data. After that, all the data will be placed at base station and user can retrieved the data from that. The base station will act as interference or in other names, an intermediary between users and network where it will observe and analyze the data after gathering them from the sensor nodes. The application of WSN that has been used nowadays in our daily life like the water quality detection, intrusion detection, tracking for military application, forest fire detection and recent development is in air pollution monitoring application [1] [2].

Unfortunately, there is a major constraint of WSNs where it has limited energy resource inside the batteries and it will be impossible in replacing the batteries when WSN is implemented far from the power source. Thus, it gave a bad impact on the lifetime of the network since it cannot complete the monitoring task. Due to this issue, the solution must be addressed and one of the effective methods that can be used to optimize the usage of energy and prolong the lifetime of the network is called clustering [3]. Clustering is one of the method to organizes all the distribute sensor nodes in numerous clusters and for every cluster that exist in the network, it must have at least a head of cluster. The cluster head may be assigned by the members of the cluster and the cluster head will become the intermediary between the base station and the other nodes. The cluster head will gather all the information that the other nodes have sent to them and forward the data to the base station [4] [5].

In [6] the researcher proposed a new clustering algorithm that forms a multitier network or often called as "n-tiers" topology to lengthen the lifetime of WSN. The two-tier technique has been designed. The author used Fuzzy Logic technique to choose the cluster head [6]. For the input parameters, they used remaining energy power, the central of the sensor node and also the cost of the communication. From aforementioned parameters, each of that will be combined to satisfy the characteristic of cluster head such as combination of energy with centrality, or energy with cost or cost with energy and centrality. In each tiers, it must locate cluster head as it responsible to collect the data and transmit to the base station. For the data transmission, it involves another two representatives in the network. The sensor node will send the data to their respected cluster head. Cluster head will pass the data to its primary node and then primary node will find the nearest secondary cluster head before all the data safely arrived at base station. Hence, this paper proposed three combinations of fuzzy logic membership function such as Gaussian with Triangular, Triangular with Trapezoidal and Trapezoidal with Gaussian. All these three combination will be run into multi-tier network topology to identify which combinations can enhanced the lifetime of sensor network.

2. Literature review

There are a lot of studies done specifically for energy consumption and optimization issues in Wireless Sensor Network. Many studies had proposed using Artificial Intelligent (AI) techniques in order to solve this main issue that always occurs in WSN. For example, the author in [7] has proposed the Genetic Algorithm (GA) to minimize the communication distance between the sensor nodes and cluster heads at the same time thus enhanced the WSN lifetime. In their work, a cluster-based method is used where there are two phases proposed known as set-up phase and steady-state phase. In the set-up phase, the clusters are created and not changed in the whole network. The selection of the new cluster head is done by selecting a node that has the highest residual energy if the current cluster head has residual energy below average after being checked at the end of each round. This cluster-based and GA method was proven helps to expand the lifetime of wireless sensor network after the simulation is done.

According to Gupta et al. [8], a fuzzy logic approach is proposed to depend on three parameters - energy, centrality and the concentration to choose a suitable node as the cluster head in every cluster. On that paper, fuzzy logic will help in minimizing the energy used in every node at once increase the lifetime of the network in WSN by using three parameters - energy level in each node, the nodes centrality and the nodes concentration. In order to get a better result, fuzzy logic can blend the different parameters that are proposed together to produce the best result. There is a research where the author proposed an enhancement for clustering in wireless sensor network which also implemented Fuzzy Logic [9]. They used three parameters which are energy of the sensor, the range to the base station and also the concentration of the sensor nodes.

Meanwhile, according to a study by Rostami and Marzieh et al [10], a new approach has been introduced for clustering WSN based on particle swarm optimization (PSO) that aim to extend the network lifetime by using fitness function. This study proposed a new cost function for the selection of

IOP Conf. Series: Materials Science and Engineering **769** (2020) 012062 doi:10.1088/1757-899X/769/1/012062

cluster heads that combined different criteria that affects the energy efficiency in cluster heads. There are five parameters that were used in this study and they are residual energy, the distance of cluster head from each other, distance from the base station, density and lastly, the intra-cluster distance. In this study, all the sensor nodes used power control to adjust the power transmission that is dependent on the distance to the receiver. For clustering, the distribution of particles is done randomly and nodes with the best point are appointed as the cluster heads and the other member nodes will form clusters. All of the decisions such as the cluster head selection and clustering were performed in the base station as the proposed algorithm is a centralized algorithm. After the simulation is done, the result showed that the proposed method is way effective in terms of energy consumption and network lifetime compared to other methods that have been used in the previous studies such as LEACH, CHEF and PSO-MV [11].

The Multi-tier Protocol (MAP) is another approach that has been introduced for selecting the optimal cluster head in WSN. It used Fuzzy Logic approach with three suitable parameters to identify the cluster head. From three input with Fuzzy Logic, it will produced one output and will be ranked to identify the potential nodes to become a cluster head [6].The transmission between the sensor node and cluster head using multi-hop clustering technique. In MAP, each of the network tiers will have a cluster head, for example, the two-tier network clustering, the cluster head in second-tier will responsible to collect and aggregate the packet from other sensor nodes in the same tier, then the information will be forward to the nearest cluster head in a first tier. The process of cluster head in first-tier will be same, the packet will collect from the sensor node in first-tier members and the cluster head in the second tier, and then it will directly send to the base station. It will keep continuing the process until the cluster head loss of energy, then the new cluster head will be select based on K-Optimal formula [12].

3. Methodology

The proposed algorithm is stimulated by using MATLAB tools. Table 1 shows the preliminary design setup for the whole network and values of metrics for the simulation process. Two-tier network design was used in order to compare the result with the previous research [13].

Table 1. Preliminary Design Setup

Parameter	Value
Number of Tiers Used	2
Network Diameter	100M x 100M
Number of Nodes Used	100
Base Station Coordination	(50,50)
Energy Used (Joule)	(1)
Data (Bits)	2000, 4000, 6000,8000

Figure 1 shows the proposed two-tier network design. The location of the base station had been set with coordinate (50, 50). The sensor area measured for tier 1 radius from the base station is 25 meters meanwhile for tier 2 is 50 meters.

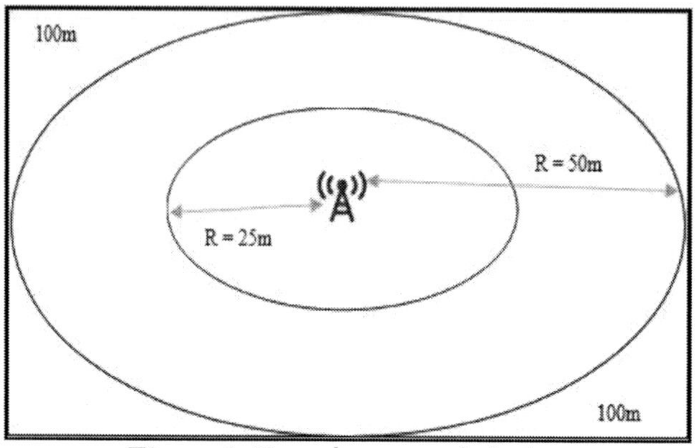

Figure 1. Two-tier network design

For the proposed approach, three parameters have been identified which are energy, centrality and communication cost. All these three parameters will used by 100 sensor nodes and three type of Fuzzy Logic membership function will involve to select the cluster head. The Fuzzy Logic membership function were Triangular, Gaussian and Trapezoidal. From the three input, Fuzzy Logic will only produce one output. The purpose of blending the parameters with different combinations of the membership function is to identify the best combination that can produce the output hence improving the energy efficiency for cluster head selection and prolonging the lifetime of the wireless network.

Figure 2 shows the Fuzzy inference system for three parameters proposed. As mentioned previously, all three parameters will be blended into Fuzzy Logic to produce an output. For each parameter, it will have three fuzzy sets. Centrality was presented by Far (F), Near (N) and Satisfactory (S). Meanwhile, for residual energy and communication cost, Low (L), Medium (M) and High (H) were used.

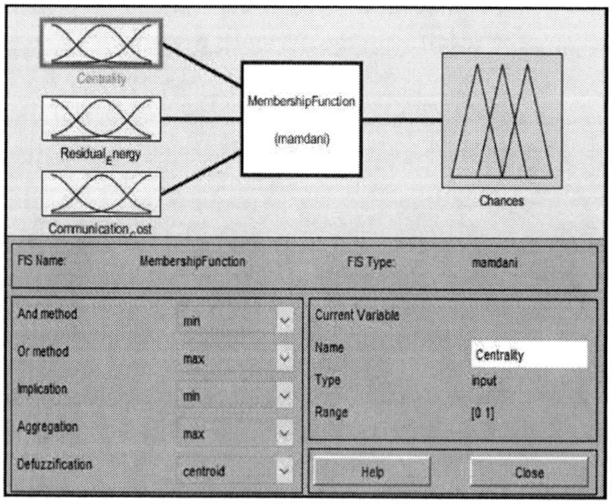

Figure 2. Fuzzy Inference System (FIS) for three parameters

Figure 3 shows the fuzzy sets that represent the parameter of centrality which is Far (F), Near (N) and Satisfactory (S). For Far, Near and Satisfactory, the value is (0-0.5, 0-1 and 0.5-1) respectively. For example to the respective sensor, if the centrality of the sensor node is near to the base station, the residual energy is high and the communication cost low, the chances of the sensor to become a cluster head is high compare to the other sensor nodes. Therefore, for all the sensor nodes, the output chances will be ranked and the sensor with the higher chances will become a cluster head.

Since there are three parameters that are involved in this study, so each input variable will have three linguistic states. Therefore, it specifies the total number of fuzzy inference rules which are a total of 27 rules (3x3x3 =27). Table 2 shows the total of 27 rules for the three parameters – residual energy, communication cost and centrality. The output of the variables indicates the chances of each sensor nodes to be the cluster head. The higher the output, the higher the chances and possibility of the sensor nodes to be the cluster head.

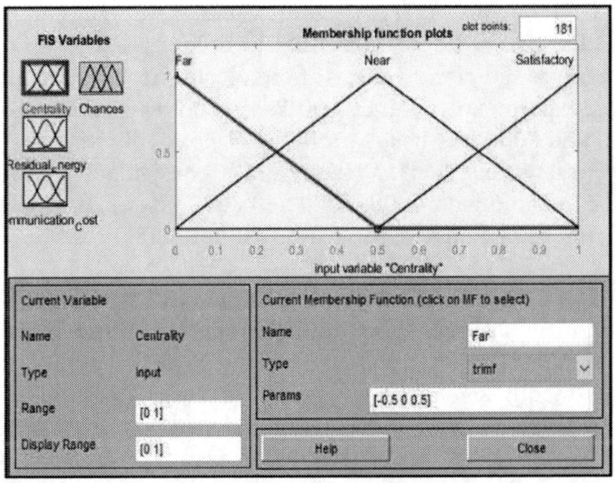

Figure 3. Membership function for parameter centrality

Table 2. Fuzzy rules [6]

Rule	Centrality	Residual Energy [6]	Communication Cost [6]	Output Chances [6]
1	Far	High	High	Normal
2	Near	High	High	Normal
3	Satisfactory	High	High	Strong
4	Far	High	Medium	Normal
5	Near	High	Medium	Strong
6	Satisfactory	High	Medium	Strong
7	Far	High	Low	Strong
8	Near	High	Low	Strong
9	Satisfactory	High	Low	Strong
10	Far	Medium	High	Weak
11	Near	Medium	High	Normal
12	Satisfactory	Medium	High	Normal
13	Far	Medium	Medium	Normal
14	Near	Medium	Medium	Normal
15	Satisfactory	Medium	Medium	Strong
16	Far	Medium	Low	Normal
17	Near	Medium	Low	Strong
18	Satisfactory	Medium	Low	Strong
19	Far	Low	High	Weak
20	Near	Low	High	Weak
21	Satisfactory	Low	High	Normal
22	Far	Low	Medium	Weak
23	Near	Low	Medium	Normal
24	Satisfactory	Low	Medium	Normal
25	Far	Low	Low	Normal

26	Near	Low	Low	Normal
27	Satisfactory	Low	Low	Strong

4. Result and discussion

4.1. Analysis of dead node for different data bits

This experiment was run by using MATLAB tool where the combination of the fuzzy membership function is implemented in Fuzzy Logic Controller in order to select the cluster head in the two-tier network. In order to measure the effectiveness of the cluster head, the analysis will be done based on data transmission. This experiment involves 100 sensor nodes and the data used in this experiment were 2000, 4000, 6000 and 8000 bits based on LEACH protocol [14] [15]. The maximum iteration is at maximum 20000 iterations. In order to maintain the consistency of the result for the lifetime of sensor nodes, we have compared with 4000, 6000 and 8000 bits of data. Table 3 shows the iteration of the dead node on different bits of data starting from 2000 to 8000 for the combination of MF that has been proposed in this study.

Based on Table 3, it was observed that the combination of Trapezoidal and Gaussian MF survived longer for every testing that has been done on different data bits compared to the combination of Triangular and Trapezoidal MF. Trapezoidal and Gaussian MF has better performance because the Gaussian MF provided smooth transitions and provided a way to stimulate the maximum number of rules in the rule base thus an accurate representation of input and output relationship is achieved [16]. In addition, with Trapezoidal MF where its width can be adjusted and at once produced a desirable result. When the width of Trapezoidal MF is adjusted to a higher number, the greater the result would be. It was concluded that, the bigger the data bits, the faster the nodes to die which caused the network lifetime to be shorter. It can be seen at 2000 data bits the range of dead node iteration was at 9000th to 10000th and above while, for 8000 data bits, the range of dead node iteration was literally at 3000th and below which was about 25% to 27 % less than 2000 data bits. The combination of Trapezoidal and Gaussian is proven to be efficient and helps to prolong the lifetime of the wireless sensor network.

Table 3. Dead node for each combination using different data bits

MF Combination	2000		4000		6000		8000	
	First	Last	First	Last	First	Last	First	Last
Triangular + Trapezoidal	7448th	10121st	3724th	5061st	2482nd	3374th	1862nd	2530th
Triangular + Gaussian	7448th	9844th	3724th	4932nd	2482nd	3288th	1862nd	2466th
Trapezoidal + Gaussian	7620th	10507th	3810th	5254th	2540th	3502nd	1905th	2627th

4.2. Analysis of energy used in node 50th using different data bits

Figure 4 shows the graph of energy possessed in 50th node in Joule versus the number of node in 2000 bits and 8000 bits of data for every combination of membership function proposed which were Triangular and Trapezoidal, Triangular and Gaussian and Trapezoidal with Gaussian. There are 3 areas that have been marked in these graphs that represent the highest energy possessed based on the combination of MF proposed.

From Figure 4, it was observed that the energy possessed for 2000 and 8000 bits' data are almost in the same pattern. For the first 2000 bits' data, the total energy possessed recorded to be higher compared to 8000 bits. The total energy possessed for 8000 bits is recorded as the lowest because, at the beginning of the iteration, the nodes used a lot of energy to send bigger bits of data. This can be seen in Table 4 where the energy was recorded for 3 areas that indicated the combination of MF proposed. This explained why the energy in 50th node is lower because the nodes before have consumed a lot of energy for data transmission in the network.

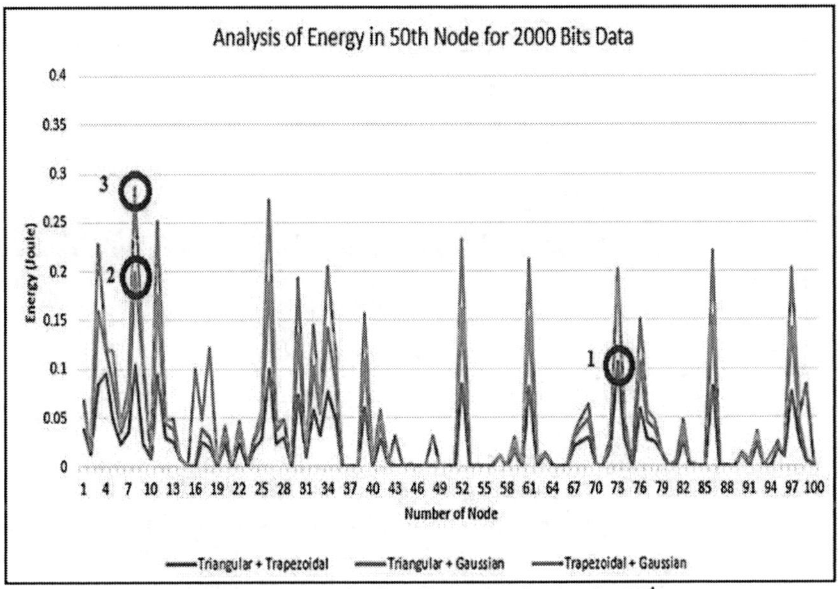

Figure 4. Analysis of energy for node 50th

Table 4. Energy Consumption for Different Area

Area	The energy in 2000 bits	The energy in 8000 bits
1	0.0953706599998195	0.0952539200000434
2	0.093994300000205	0.093859200000019
3	0.08860572000202	0.08863592000002

4.3. Analysis Of Data Sent By 100 Nodes During Data Transmission

During the data transmission phase, every sensor node will send the different size of data to its appointed cluster head. The cluster head then will gather all the data before sending it directly to the base station. In this study the data from sensor node in tier one is sent to its own cluster head and then, the cluster head will compress, aggregate and analyse before sending it to the PCH appointed in its own cluster. The PCH will send the data that is received from a cluster head in tier 1 directly to the base station. Meanwhile, in tier 2 all node will send their data to the SCH in their own cluster and SCH will compress and aggregate all the data that have been received and transmit them to the nearest PCH in tier one. After PCH received all the data from SCH, it will instantly send the data to the base station. Figure 5 showed the graph of the number of data versus the number of nodes where all data sent for 100 sensor nodes for 4000 bits is recorded and plotted. As we can see, there are several nodes recorded to have data sent above 350000000. The highest data sent for the combination of Triangular Trapezoidal and Triangular Gaussian are at the same which were 3500000000 bits at node 28th. Meanwhile, the highest data sent recorded to be at 48th node which for the combination of Trapezoidal and Gaussian was 361096000 bits.

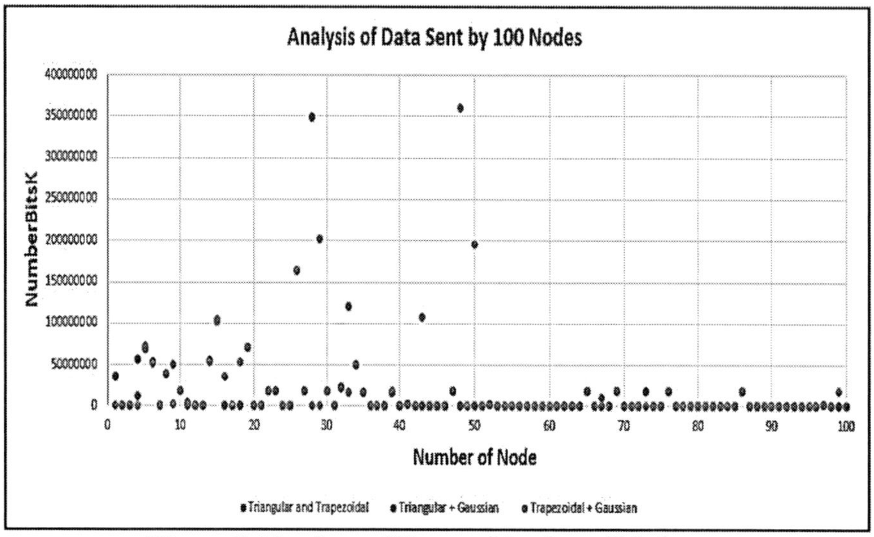

Figure 5. Numbers of Data vs Number of Nodes

Figure 6 shows the total of data sent by all 100 sensor nodes to the selected cluster head in the clusters. For the combination of Triangular Trapezoidal, it showed that they had the lowest total number of data sent which was only 1633692000 bits followed by 1635408000 bits for the combination of Triangular Gaussian function. While, the combination of Trapezoidal and Gaussian function recorded the highest total number of data sent, 1651532000 which was 0.54% higher than the total data of the combination of Triangular and Trapezoidal function. The total number of data sent in this combination recorded the highest because the iteration can sustain up to 5000 iterations. This means the longer the iteration, the more data that can be sent by every sensor node in the network.

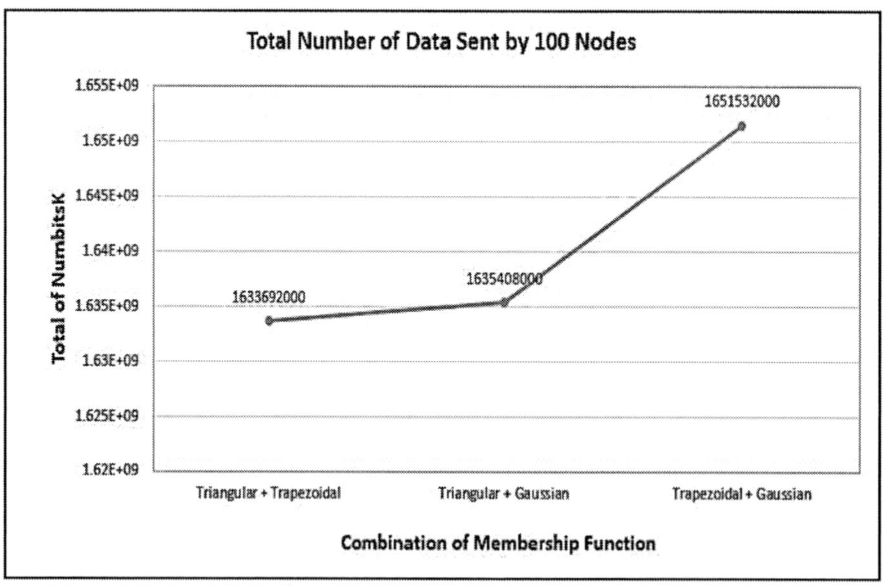

Figure 6. Total Data for each Combination of Membership Function

5. Conclusion

The cluster head selection successfully implemented using Fuzzy Logic approach. Three parameters have been used which are energy, centrality and communication cost. From these three parameters, with the Fuzzy Logic approach it can produce one output and the highest output will be selected as a highest chances for the sensor nodes to become a cluster head. There are also three Fuzzy Logic membership function that involved in this proposed method which are Triangular, Trapezoidal and Gaussian. Based on that, we combined the membership function as; Triangular and Trapezoidal, Triangular and Gaussian and lastly Trapezoidal with Gaussian. The effectiveness of the cluster head gave a big impact on network lifetime [17]. Hence, it is important to make sure which combination of fuzzy membership function helps to prolong the network lifetime [18] [19]. In this study, the effectiveness of cluster head selected is measured based on data transmission between the nodes in the cluster. It was observed that, the smaller the data, the longer the network would survive. By using data up to 6000 bits it is proven to shorten the network lifetime as the energy used by the nodes higher. The nodes can sustain up to 10000 iterations when using 2000 bits' data which is 20 % longer than using 8000 bits' data. In this study, it was observed that the combination of Trapezoidal and Gaussian function is the best to use as it can support up to 1000 of iterations. Besides, this combination also has the highest total number of data sent by all 100 nodes which were 0.54 % higher compared to the combination of Triangular Trapezoidal and Triangular Gaussian. Thus, it concluded that the implementation of Fuzzy Logic with the help of the combination of Trapezoidal and Gaussian membership function has allowed the nodes use their energy effectively and helps in prolonging the network lifetime.

Acknowledgement

The author would like to thank UMP RDU Grants (RDU-180364) for providing financial support and facilities in this study. Also gratefully, we would like to acknowledge and thank the Faculty of Computing and Systems Network & Security (Sysnet) research group UMP that support this research.

References

[1] J. A. Nayak, "Improving the Network Life Time of Wireless Sensor Network using EEEMR Protocol with Clustering Algorithm," Int. J. Electron. Electr. Comput. Syst. IJEECS, vol. 6, no. 6, pp. 342–346, 2017.

[2] K. Okokpujie, M. Odusami, E. Noma-Osaghae, O. Modupe, S. John, and O. Oluwatosin, "A Smart Air Pollution Monitoring System," Int. J. Civ. Eng. Technol. (IJCIET, vol. 9, no. 9, pp. 799–809, 2018.

[3] A. A. Qasem, A. E. Fawzy, M. Shokair, W. Saad, S. El-Halafawy, and A. Elkorany, "Energy Efficient Intra Cluster Transmission in Grid Clustering Protocol for Wireless Sensor Networks," Wirel. Pers. Commun., vol. 97, no. 1, pp. 915–932, 2017.

[4] J. Anzola, J. Pascual, G. Tarazona, and R. González, "A clustering WSN routing protocol based on k-d tree algorithm," Sensors (Switzerland), vol. 18, no. 9, pp. 1–26, 2018.

[5] D. R. Das Adhikary and D. K. Mallick, "An Energy-Aware Unequal Clustering Algorithm using Fuzzy Logic for Wireless Sensor Networks," J. ICT Res. Appl., vol. 11, no. 1, pp. 56–77, 2017.

[6] W. I. S. W. Din, S. Yahya, R. Jailani, M. N. Taib, A. I. M. Yassin, and R. Razali, "Fuzzy logic for cluster head selection in wireless sensor network," in AIP Conference Proceedings, 2016, vol. 1774, p. 050006.

[7] M. Iqbal, M. Naeem, A. Anpalagan, A. Ahmed, and M. Azam, "Wireless Sensor Network Optimization: Multi-Objective Paradigm," Sensors, vol. 15, no. 7, pp. 17572–17620, 2015.

[8] C. P. Gupta and A. Kumar, "Optimal Number of Clusters in Wireless Sensor Networks with Mobile Sink," Int. J. Sci. Eng. Res., vol. 4, no. 8, pp. 1706–1710, 2013.

[9] T.-T. Nguyen, J.-S. Pan, S.-C. Chu, T.-K. Dao, and V.-C. Do, "Improved Performance of Wireless Sensor Network Based on Fuzzy Logic for Clustering Scheme," Adv. Smart Veh.

Technol. Transp. Commun. Appl. VTCA 2018. Smart Innov. Syst. Technol. Springer, vol. 128, pp. 104–113, 2019.

[10] A. S. Rostami, M. Badkoobe, F. Mohanna, H. Keshavarz, A. A. R. Hosseinabadi, and A. K. Sangaiah, "Survey on clustering in heterogeneous and homogeneous wireless sensor networks," J. Supercomput., vol. 74, no. 1, pp. 277–323, 2018.

[11] A. A. R and O. H. I, "Performance Evaluation of Selected Distance-Based and Distribution-Based Clustering Algorithms," Int. J. Softw. Eng. Comput. Syst., vol. 4, no. 2, pp. 38–48, 2018.

[12] A. A. Rahman, M. N. M. Kahar, and W. I. S. W. Din, "Distance-based thresholds for 2-tier relay nodes selection in WSN," Comput. Stand. Interfaces, vol. 66, no. April 2019.

[13] N. K. Jassi and S. S. Wraich, "An enhanced K-Means clustering technique with Hopfield Artificial Neural Network based on reactive clustering protocol," in 2014 5th International Conference - Confluence The Next Generation Information Technology Summit (Confluence), 2014, pp. 821–825.

[14] S. Vasuhi and V. Vaidehi, "Target tracking using Interactive Multiple Model for Wireless Sensor Network," Inf. Fusion, vol. 27, no. Supplement C, pp. 41–53, 2016.

[15] M. Mirzaie and S. M. Mazinani, "Adaptive MCFL: An adaptive multi-clustering algorithm using fuzzy logic in wireless sensor network," Comput. Commun., vol. 111, no. Supplement C, pp. 56–67, 2017.

[16] S. Ramakrishnan and S. P. Shyry, "Fuzzy-based cluster head election mechanisms for wireless sensor networks: A review," in 2017 International Conference on Information Communication and Embedded Systems (ICICES), 2017, pp. 1–6.

[17] A. A. Allahham and M. A. Rahman, "A Smart Monitoring System for Campus Using Zigbee Wireless Sensor Networks," Int. J. Softw. Eng. Comput. Syst., vol. 4, no. 1, pp. 1–14, 2018.

[18] C. Wu, T. Yoshinaga, X. Chen, L. Zhang, and Y. Ji, "Cluster-Based Content Distribution Integrating LTE and IEEE 802.11p with Fuzzy Logic and Q-Learning, "IEEE Comput. Intell. Mag., vol.13, no.1, pp 41-50, 2018.

[19] A. Saxena et al., "A review of clustering techniques and developments," Neurocomputing, vol. 267, pp. 664–681, 2017.

IOP Conf. Series: Materials Science and Engineering **769** (2020) 012063 doi:10.1088/1757-899X/769/1/012063

Applying Heuristics Evaluation to Improve the Usability of Malaysia Accounting Training Management System

Zuraidah binti Mat Harun[1], Azma binti Abdullah[2], Kiriyadhatshini a/p Gunaratnam[2]

[1] MAMPU Cyberjaya, Sepang Selangor, Malaysia
[2] Faculty of Computing (FKom), Universiti Malaysia Pahang, Pahang, Malaysia

mszuraidah@mampu.gov.my, azma@ump.edu.my, kiriyadatshini@gmail.com

Abstract. The website is a connection of communication between the organization and their customers. By using web applications customers can obtain information and deal with the organization. The user interface and the interface design of web-based application affect the interest and attention of the customers. The consideration of web navigation, web usability and web accessibility are compulsory to make a website was designed in a user-friendly format. This study aimed to evaluate system based on issues of interface design with focusing on human-computer interaction and user interface design which are error human handling and accessibility. The questionnaire based on questions from the User Interface Usability Evaluation (User Interface Usability Evaluation) was conducted to obtain feedback on the system developed. The findings and results of the questionnaire was used to assess whether the training management system design interface both in terms of handling and ease of access error information (error information handling and accessibility).

1. Introduction

Website become a common place that any information about the business of the organization was shared and displayed. The website become as a connection of communication between the organization and their customers. The user interface and the interface design of web based application affect the interest and attention of the customers [1][2]. Interface can be describing as a layout of the website. User interface can be deliberated as all aspects of information system which related to the users. Interface design for website must be made properly to get positive perception of the user on the contents and services provided in the website. It also must be prepared to facilitate the user's needs and make sure that the interface has elements of an easy to access, easy to understand, and easy to use [3][4].

Wells et al. [5] mention that website design with attractive visual make positive perception to the customer. The consistency, aesthetic, and the attractiveness of the website's appearance such as images, colors, fonts, shapes, animations, and layout reflect the visual design of a website [6]. Selection of the interface elements must be consistent and predictable because it will help in the task completion, efficiency and satisfaction [7].

In this study, Training Management System(TMS) for National Accounting Institute was evaluate using heuristic evaluation criteria. The module evaluates include the functioning for training information details, apply for training offered, training status and reports modules. The design issues that evaluated were error information handling and accessibility.

The rest of this paper is organized as follows. Section 2 presents background and related work, while section 3 gives the explanation research methodology, Section 4 presents experiment and Section 5 discuss about results of the study.

2. Background and Related Work

2.1. Definition

In order to establish the scope and get an overall picture of the adopted terms used in interaction design and human computer interaction in the web-based system or application and to ensure uniformity throughout this paper, we present the following definitions, organized in alphabetical order:

- **Human computer interaction (HCI)**: Comprises the study of human-computer interaction with the overall objective of informing the design of more humanly acceptable technology. HCI focuses on empirical methods to identify the most credible and valid interface design rules. From the HCI view, several main problems arise in the design of web-based training, the instrument, the customers, the activities and the environment [8].
- **Interaction design**: Connecting the digital environment with the human being. This link enables users to use digital products and devices, from desktop computers, phones and watches to websites, applications and games [9].
- **Interface design**: Refers to User Interface Design, which focuses on predicting what customers might need to do and ensuring that the interface has components that are simple to access, comprehend and use. Selection of the interface elements must be consistent and predictable because it will help in the task completion, efficiency and satisfaction [7]. The interface elements are including input controls, navigational components, informational components and containers.
- **Information architecture (IA)**: Focuses on organizing, structuring, and labelling content in an effective and sustainable way. The goal is to help users find information and complete tasks.
- **Visual design**: Focuses on the aesthetics of a site and its related materials by strategically implementing images, colors, fonts, and other elements.

2.2. Related Work

Existing Training Management System that have been taken into account and analyzed for their available features in comparison with the development of new TMS development system is IPN (Institut Perakaunan Negara). IPN was established to provide accounting and financial training to civil servants in Malaysia. Based on the revision of this system, there are some drawbacks found as listed in Table 1 which encloses the whole system.

Table 1. The list of issues in the existing TMS.

Issues	Explanation	Suggestions
Long main page display	Need to scroll down to get all information in the main page	The main page should display in exact page without any scroll
Unorganised Menu	In main page, menu stated by displaying as image but in other pages, its display in texts.	Menu must state in one place either on top or other places to make it easy to access
Redundant Information	Menu and other information were	Menu and other information must be

	displayed in same area and make its replace each other	displayed in organized place such as menu on top of page and other information in the middle of page
Blank Space	There are some blank spaces	The blank space must be filled with some images, texts or any information to make it utilized
Maximised Window Screen	The window screen is not maximized	The size of windows screen must be design as 100% on their width to make it display maximized with any resolution of the device setup

A study of the existing system is also done based on the principles of usability heuristics based on stated principals [10]. The main aspect that flawed throughout the system is the design. The organization and consistency of design has failed to follow the usability heuristics for user interface design such as consistency and standards, as well as aesthetic and minimalist design.

The Table 2 shows the revision of the existing system based on the principles of usability heuristics.

Table 2. The existing TMS based on the principles of usability heuristics.

Principles	**Details**	**Availability in Existing TMS (Yes / No)**
Long main page display	Need to scroll down to get all information in the main page	No
Match between system and the real world	Use the users' language, with words, phrases and concepts familiar to the user	No
User control and freedom	Support undo and redo	No
Consistency and standards	Users should not have to wonder whether different words, situations, or actions mean the same thing	No
Error prevention	A careful design is better than good error messages to prevent a problem from occurring in the first place	No
Recognition rather than recall	Minimize the user's memory load by making	No

	objects, actions, and options visible	
Flexibility and efficiency of use	Allow users to tailor frequent actions	No
Aesthetic and minimalist design	Dialogues should not contain information which is irrelevant or rarely needed	No
Help users recognize, diagnose, and recover from errors	Error messages should be expressed in plain language, precisely indicate the problem, and constructively suggest a solution	No
Help and documentation	Provide help and documentation which such information should be easy to search, focused on the user's task, list concrete steps to be carried out, and not be too large	No

These issues were considered as the initial requirements of the new TMS development. The results from the analysis were used for enhancement and upgrading the quality of the systems based on human computer interaction (HCI).

3. Research Methodology

In this process, a research methodology has been introduced to improve Training Management System in IPN. This research methodology divided into three phases which are analysis, design & development and evaluation as shown in Figure 1.

Figure 1 demonstrates the phases of Research Methodology for this research. It starts with phase 1: analysis on the existing system, design issues, Human Computer Interaction and web application. literature review was conducted based on these topics. Followed by phase 2: designing and developing new TMS based on design issues and implement usability heuristic to produce new TMS embedded with error information handling and accessibility. Research ends with phase 3: evaluation based on questionnaire given to expert users.

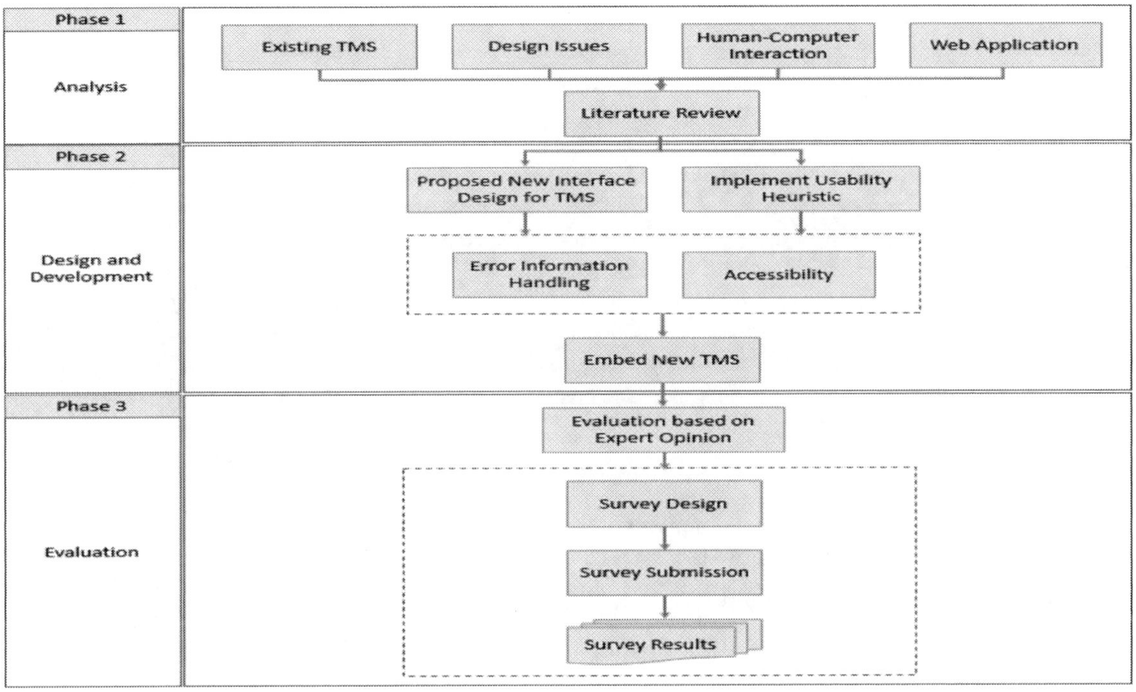

Figure 1. Research Methodology Phases

4. Experiment

In TMS, emphasis was made to ensure that the Error Information Handling taken into account in the implementation of this system. Certain features were implemented to minimize user errors when entering information and allows users to reduce errors include incorrect information. It is done by providing a helpful message to users and restrict their choice. For example, the user can simply choose the date of appointment prior to today's date as the applicant can only apply after he was appointed. Another, limited choice of the courses that are enrolled in the system which courses can only be selected via the dropdown list.

The implementation of the errors information handling in the web based system development involved the design of the web pages. In this system, design is based on the elements shown in Table 3.

Table 3. The error information handling elements.

Element	Function	Implementation on the system
Avoid the error occurring in the first place	Avoid a potential error	Date-picker for Training Date: the user should not be able to select a date prior to today
Show the user the field where the error has occurred	The error message shown below the relevant element	Highlight the message text with red colour
Use inline validation	Correct the error immediately	Inline validation enables the customer

		decrease the waiting time before the mistake occurs and decreases the opportunity of data loss in the form
Be polite – don't blame the user	The message use word "please" and point out how to fix the error	The message should never place blame on the user
Be precise – clearly describe the issue, and how to resolve it	The message is specific	The message specifies the issue to be solved by the user
Reduce the work required to fix the problem	Keep the data that the user has entered	Display the key-in information to see if it appears incorrect and attempt to solve it

The error information handling elements which are related to error prevention heuristic principle are being highlighted in detailed on how the system is preventing the users from being distracted by the errors occurring while usage. Apart from that, this table is highlighting the failure of the developer to implement principle of recognition rather than recall by not displaying the entered data by user. Meanwhile, the elements of the error information handling applied in the TMS shown in Table 4.

Table 4. The elements errors information handling applied in TMS

Element	**Implementation on the system**
Avoid the error occurring in the first place	Date-picker for Tarikh Lantikan: the user should not be able to select a date today or after to avoid a potential error because the date must come first before they apply for the courses
Show the user the field where the error has occurred	The error message shown below the relevant field which highlight the message text with red colour
Be polite – don't blame the user	The message for the error is by using the word "Sila…"

Accessibility is a significant component of the site design and should be regarded throughout the phase of growth. There is a need to guarantee that all prospective consumers, including disabled individuals, can access the data readily. Accessibility focuses primarily on how a disabled individual can access or benefit from a site, system or application. Accessibility is an significant component of the site design and should be taken into account throughout the design phase. By making the website accessible, it is necessary to guarantee that all prospective customers, including persons with disabilities, have a good user experience and can access the data readily [9]. By implementing web accessibility best practices, it also can improve the usability of the site for all users [10]. The best practice that is applied in this system are as shown in Table 5.

Table 5. The accessibility elements

Content	**Best Practice**
Consistent Navigation	Clear and logical designs will benefit the user
Keyboard Navigation	The availability of keyboard navigation and hotkey shortcuts for standards keyboard control
Prioritize Text Clarity	Increase clarity of letters and clarity of text blocks
Colour Code	A colour code can make faster and more efficient communication
Explanatory Link Text	Include link descriptions that can be understood independently

There are three elements of accessibility that has been implemented in TMS are shown in Table 6 which are consistent navigation, prioritize text clarity and colour code.

Table 6. The elements accessibility applied in TMS

Element	**Implementation**
Consistent Navigation	Fix menus on the pages
Prioritize Text Clarity	Use suitable letters and text blocks for clarity
Colour Code	Choose colour code which can make efficient communication with the user

The questions about User Interface Usability Evaluation were picked from the website of Gary Perlman at http://garyperlman.com/quest/. Only 22 related questions were selected, for the purpose of this study on error information handling and accessibility. These questions were group in four (4) which are System Usability, User Interface Satisfaction, Accessibility and User Guidance. The questions in the questionnaire are shown in Table 7.

Table 7. User Interface Usability Evaluation

Questions		Scale of Selection							
System Usability		**1**	**2**	**3**	**4**	**5**	**6**	**7**	
This system is easy to use	strongly disagree								strongly agree
It was simple to use this system	strongly disagree								strongly agree
The system gives error messages that clearly tell me the problems	strongly disagree								strongly agree
Whenever I make a mistake using the system, I recover easily and quickly	strongly disagree								strongly agree
It is easy to find the Information I needed	strongly disagree								strongly agree
The information provided for the system is easy to understand	strongly disagree								strongly agree
The organization of information on the system screens is clear	strongly disagree								strongly agree
The interface of this system is pleasant	strongly disagree								strongly agree
User Interface Satisfaction		**1**	**2**	**3**	**4**	**5**	**6**	**7**	
Reading characters on the screen	Hard								Easy
Sequence of screens	Confusing								Very clear
Position of messages on screen	Inconsistent								Consistent
Prompts for input	Confusing								Clear
Error messages	Unhelpful								Helpful
Accessibility		**1**	**2**	**3**	**4**	**5**	**6**	**7**	
Does it provide default values in certain field?	strongly disagree								strongly agree
Is the shifting among windows easy?	strongly disagree								strongly agree
Is the menu selection by pointing?	Bad								Good
Is the menu selection by keyed entry (key-in)?	Bad								Good

Does it require minimal cursor positioning?	Bad								Good
Does it require minimal steps in sequential menu selection?	Bad								Good
User Guidance		**1**	**2**	**3**	**4**	**5**	**6**	**7**	
How helpful is the error message?	Bad								Good
Does it provide CANCEL option?	Bad								Good
Are erroneous entries displayed?	Bad								Good

5. Results

The respondents for the survey were selected from the existing users of the TMS whose are IPN's Staffs. They are 10 expert users whose position are Accountants (2 users), Accountant's Assistants (4 users) and office clerks (4 users). The findings of result were divided into four (4) group of feedback, which are feedback on System Usability, User Interface Satisfaction, Accessibility and User Guidance.

Feedback on System Usability are based on eight (8) questions. All these questions must be answered by the scale of 1 (strongly disagree) to 7 (strongly agree). The findings show that most users strongly agree with all the question. For the question of "The organization of information on the system screens is clear", 100% of the respondents strongly agree that the new TMS has it. There are five (5) questions to get feedback on User Interface Satisfaction. These questions must be answered by the scale of 1 to 7. The scale of selection is different for each question. For Question "Reading characters on the screen" the scale from 1 (hard) to 7 (easy). For Question "Sequence of screens" the scale from 1 (confusing) to 7 (very clear). For Question "Position of messages on screen" the scale from 1 (inconsistent) to 7 (consistent). For Question "Prompts for input" the scale from 1 (confusing) to 7 (clear). For Question "Error messages" the scale from 1 (unhelpful) to 7 (helpful). The finding shows that more than 50% of the users give positive feedback. For the question "Reading characters on the screen", 80% respondents choose it easy.

Feedback on Accessibility are based on six (6) questions. These questions must be answered by the scale of 1 to 7 but the scales are whether 'strongly disagree' to 'strongly agree' or 'bad' to 'good'. The finding shows that more than 80% respondents agree and replied good for all questions except on the question of "Does it provide default values in certain field? ". Feedback on User Guidance are based on three (3) questions. All these questions must be answered by the scale of 1 (bad) to 7 (good). For all questions, 70% and above respondents replied "good" for all questions. There is one respondent give nearly negative feedback on all three questions.

The results of the questionnaire to the expert users were point out that the characteristics or elements of error handling information and accessibility that have been implemented in the new system has achieved the expected goals to improve the system. Although not one hundred percent agree with such features but it gives positive feedback where more than 50% of respondents admit it. The summary of the results was show in Table 8.

Table 8. The Respondents' Feedback

Question	**Respondents' Feedback**
System Usability	
This system is easy to use	80% strongly agree
It was simple to use this system	60% strongly agree
The system gives error messages that clearly tell me the problems	50% strongly agree
Whenever I make a mistake using the system, I recover easily and quickly	40% strongly agree
It is easy to find the information I needed	70% strongly agree
The information provided for the system is easy to understand	90% strongly agree
The organization of information on the system screens is clear	100% strongly agree
The interface of this system is pleasant	50% strongly agree
User Interface Satisfaction	
Reading characters on the screen	80% easy
Sequence of screens	80% very clear
Position of messages on screen	70% consistent
Prompts for input	50% clear
Error messages	70% helpful
Accessibility	
Does it provide default values in certain field?	70% strongly agree
Is the shifting among windows easy?	70% strongly agree
Is the menu selection by pointing?	60% good
Is the menu selection by keyed entry (key-in)?	60% good
Does it require minimal cursor positioning?	50% good
Does it require minimal steps in sequential menu selection?	40% good
User Guidance	
How helpful is the error message?	40% good
Does it provide CANCEL option?	70% good
Are erroneous entries displayed?	60% good

Based on feedback from respondents, more than 50% of them gave positive feedback on the TMS. This shows that the system has been developed to have the characteristics of good interface design. There are a small number of respondents giving negative feedback.

6. Conclusion

In this paper, we have presented the details Usability Heuristics Evaluation for National Accounting Training Management System followed by the discussion of adopted terms used in interaction design and human computer interaction in the web-based system or application as well as the existing TMS system that was analysed and stated the improvement to be done in the proposed development. The research methodology, results and surveys have been shared in order to provide a clear version of what is the root of this process and the expected outcome as well.

The development of the new TMS that takes into the elements of error information handling and accessibility have been implemented as described. Evaluation through questionnaires to actual users of the system showed that most of our users agree that this system has been improved in terms of interface design.

A further research must be done on the system because the system is not covered all the function in TMS. More studies need to be done on other considerations design issues such as internationalization. This system can be improved by using a bilingual and can be accessed through the mobile application.

7. References

[1] Ng, W.S., Enhancing the Quality of Educational Website Design through Assessment for Learning Strategies, in Design Solutions for Improving Website Quality and Effectiveness. 2016, IGI Global. p. 24-51.

[2] Hujainah, F., Bakar, A., Binti, R., & Abdulgabber, M. A. (2019). Investigation of Requirements Interdependencies in Existing Techniques of Requirements Prioritization. *Tehnički vjesnik*, *26*(4), 1186-1190.

[3] Pham, N. and Y. Zhao, The Role of User Interface Design in a Digital Document Reader. 2018.

[4] Abdullah, A., Khatan, H., Abdulleteef, M., Bakar, R. A., & Hamid, R. A. (2018). A Tool for Capturing Safety Properties from Critical System Specifications Written in SOFL. *Advanced Science Letters*, *24*(10), 7311-7316.

[5] Wells, J.D., J.S. Valacich, and T.J. Hess, What signal are you sending? How website quality influences perceptions of product quality and purchase intentions. MIS quarterly, 2011: p. 373-396.

[6] Stone, R.W., L.J. Baker-Eveleth, and D.M. Eveleth, The Role of Aesthetics Influencing User Satisfaction with Recruitment Websites. International Journal of Humanities and Social Sciences, 2018. 10(1): p. 76-92.

[7] Garrett, J.J., Elements of user experience, the: user-centered design for the web and beyond. 2010: Pearson Education.

[8] Errity,A.,Human-computer interaction. An Introduction to Cyberpsychology, (2016): p.241.

[9] CarrieC., https://designshack.net/articles/graphics/interaction-design-what-is-it-and- how-can-you-use-it/ viewed 28-04-2019, (2015).

[10] Nielsen,J,http://www.designprinciplesftw.com/collections/10-usability-heuristics-for- user-interface-design viewed 28-04-2019,(2013).

[11] Ellcessor, E., < ALT="Textbooks">: Web accessibility myths as negotiated industrial lore. Critical Studies in Media Communication, 2014. 31(5): p. 448-463.

[12] Norman, K.L., Cyberpsychology: An introduction to human-computer interaction. 2017: Cambridge university press.

Acknowledgments
This work was supported in part by UMP under grant number RDU190189.

IOP Conf. Series: Materials Science and Engineering **769** (2020) 012064 doi:10.1088/1757-899X/769/1/012064

A Review of Cloud Computing Implementation in ASEAN Countries

Asyran Zarizi Bin Abdullah[1], Wan Isni Sofiah Wan Din[2], Zalili Binti Musa[3], Razulaimi Bin Razali[4]

Faculty of Computing, Universiti Malaysia Pahang, 26300, Gambang, Pahang, Malaysia[1, 2, 3]
Faculty of Computer Science & Mathematics, UiTM Cawangan Pahang Kampus Raub Raub, Pahang, Malaysia[4]

sofiah@ump.edu.my

Abstract. Cloud Computing is a technology that allows services of the communication, file transfer, infrastructure and platform usage to the user through the internet. In the era of globalisation, each of the countries raced to growth cloud technology as the main part of their industrial development. Similarly, with ASEAN countries, they start using cloud technology as a core part of the industrial revolution 4.0 (IR 4.0) to increase their economy and raise productivity. ASEAN already started using cloud computing, but some country has slow progress due to several challenges such as the broadband quality, lack of cloud knowledge and security issues. They require some strategies to address this problem in order to prove the ASEAN can be competitive in the use of technology as it can open up better industrial opportunities. This article reviews the published literature on the implementation of cloud computing in ASEAN. The author systematically analysed a total of 37 selected articles. There were 22 of the articles selected through restrictive search criteria while the other 15 were selected based on unrestrictive search criteria. The review shows the cloud implementation, expose the benefit and challenge to grow a cloud computing in general, and focusing the challenge of ASEAN in cloud computing development. This paper concludes with some recommended future works to construct the theory and perform empirical research.

1. Introduction

Cloud Computing is computing based on the Internet [1]. Past previous years, people need to purchase physical software and install in the computer for them to use it. Nowadays, with cloud computing its allow people to use the same software via the internet. While user uploads their picture on Facebook, checking account balance on a smartphone, send an email to another and watching favourite live sport streaming on the laptop, cloud computing allows unrestricted technology that simplifies user tasks in everyday life. There is a Cloud Computing definition mention by others shown in Table 1.

Table 1: Cloud Computing definition

Reference	Definition
[2]	Cloud Computing is the computing technologies that provide services such as infrastructure, platforms, and applications through the internet. It is parallel and distributed computing system.
[3]	Cloud computing defined as the delivery of IT infrastructure and applications as a service on-demand to individuals and organizations via the internet.
[4]	Cloud computing is a mechanism that allows an organisation or industry to focus on its core business or the activity by outsourcing its resources in IT.

Cloud Computing has issued as powerful platforms for companies or individual for hosting and delivering services over the Internet [5]. The term "cloud" describes a large scale of objects, and for this platform, it explains about publicly stored user data at an unknown location and can be accessed at any time and anywhere [6]. One of the benefits of cloud computing is to allow user to use applications online without need for software installation, maintenance and physical data storage [7]. The reason for the user requires the cloud is because all the data can be processed virtually through the Internet, instead of using physical hardware and software. Due to many benefits, Cloud Computing was adopted by industry such as Microsoft (Azure), Google (App Engine), and Amazon (AWS) as the main platform for cloud business. Nowadays cloud computing is growing rapidly, and it leads to several issues about freedom and creativity. Cloud user do not have freedom of their data since the user does not know and cannot access physical storage, and user needs approval from the administrator to install new applications and achieve specific tasks. Another issue is about security, which exposes the user data to attackers using a lot of attack techniques [8]. Otherwise, efficient energy usage in a natural technology environment, which will save the emission of carbon and reduce some electricity consumption is one of the cloud benefits [9].

This paper focuses on cloud computing, current implementation in ASEAN, the benefits, risk and challenge. The selected countries are Singapore, Malaysia, Philippines, Thailand, Indonesia and Vietnam. It organised as follows, the history of cloud will discuss in part two. Part three discusses the migration of cloud computing and how to implement cloud computing referred to a model of cloud. The material and method explained in part four. Then, discuss the current adopted in ASEAN and the challenge faced by ASEAN in part five. Lastly is a conclusion of the paper and a future discussion of the cloud.

2. Review of Cloud Computing

In networking, typically, the cloud symbols utilised to represent a wide network connection with computing tools. As John McCarthy said in 1961, "Computation may someday be organised as a public utility", and now a public utility that mentions by him already exists, and it is called the Internet [10]. Cloud computing in the modern perception appeared around 1996 with the earliest acknowledged by Compaq [11]. The particular popularisation with the term could be traced to 2006 when Amazon released its cloud product which is Amazon's Elastic Compute Cloud (EC2) as a commercial web service that allowed small companies and individuals to rent computers on which to run their computer applications [12].

The history of the computer business, as shown in Figure 1, there is some useful software that has been created. Starting in the 60s, the call central systems is first systematic as electronic data processing. Mainframe computers are first used by businesses for computation, like census data processing, statistics and financial transaction processing. Then in 80s client-server technology was used to make all data processed will store by centralised on a server, but still in standalone mode.

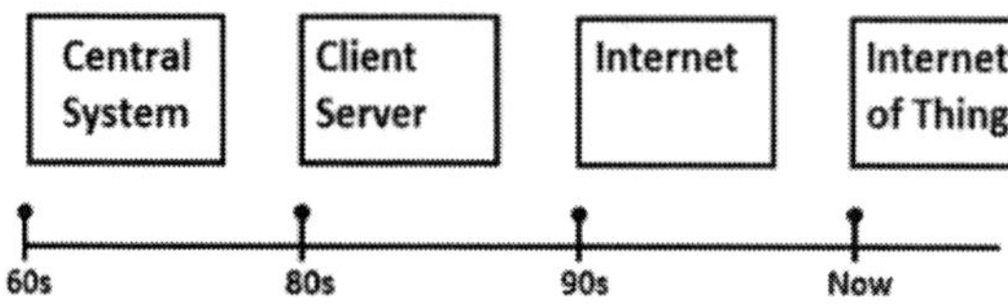

Figure 1: History of Cloud Computing

The 90s is an era of the Internet which all the data will store in server storage that can be accessed via the Internet. The term "cloud computing" was first used by Information System Professor, Ramnath Chellappa [13] on their talk in 1997. Now is the Internet of Thing (IoT) revolution, which is not only computer used to process and send a data, but there are more devices such as smartphone, smart car, smart home, and smart factory that can remotely control their device using the Internet at any time and anywhere. Unfortunately, this technology needs a big space of storage to store data, and better solution is the Cloud Computing [14].

Today, cloud computing is seen as the main technology to driving the success against Industry Revolution 4.0. Similarly, with ASEAN, that has a plan to deploy a cloud computing in order to assist the region in moving forward within a short period of time. Based on ASEAN ICT Master Plan 2020 [15] they aim to transform the ASEAN economy to the ASEAN digital economy that is secure, sustainable and transformative in order to compete with the world economy [16]. The three big giant of cloud provider (Amazon AWS, Google App Engine and Microsoft Azure) also opens their branch in Singapore [17] to strengthen the ASEAN market and provide the best services in the region, this explains the good growth of cloud computing in the ASEAN global environment.

3. Material and Method

The published literature reviewed in this paper were collected through two different ways, namely restrictive and unrestrictive search criteria. Articles were collected from three sources, namely ISI Web of Science, SCOPUS and Google Scholar. The keywords “cloud computing AND issues” and “cloud computing AND review” were used when the search was conducted using the ISI Web of Science and SCOPUS. On the other hand, the search using Google Scholar used the keyword “cloud computing”. Both SCOPUS and ISI Web of Science searches were restricted for resources published from 2000 to 2018, while Google Scholar was for all years. The articles yielded from the searches were then refined to ensure that only the related ones were selected for review. In general, the search yielded articles which may be divided into three broad categories, namely cloud implementation, benefit and risk. As a result, 22 articles were selected from restrictive search criteria, and 15 more articles were selected from unrestrictive search criteria.

Cloud computing has several characteristics which provide low-cost service, high reliability, sustainability, multi-tenancy, independence of device and location, and data security, which make cloud is the best choice for a budget company to start a business with minimum cost [18]. In 2019, RightScale was surveyed over 786 IT professional about acceptance and implementation of cloud computing on their organisation in the year of 2019 [19]. Figure 2 shows the migration of cloud usage from public and private cloud to hybrid cloud.

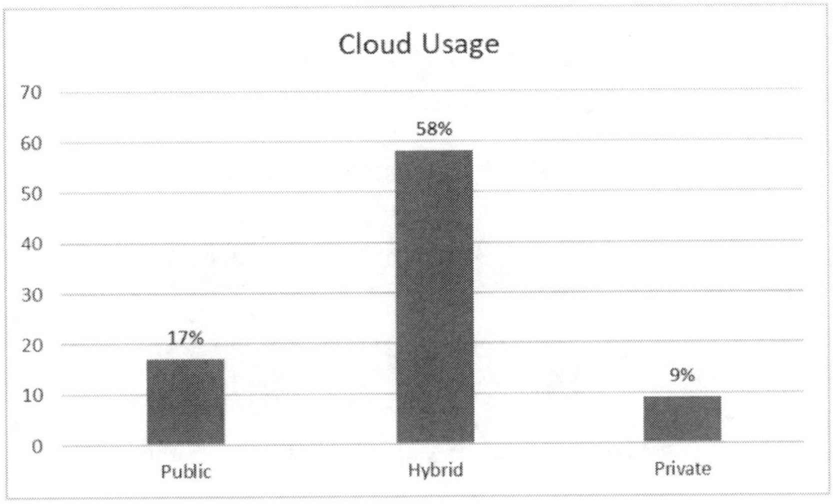

Figure 2: Cloud Usage for 2019

Mostly, hybrid cloud adopting by an organisation or a company, due to a lower-cost cloud implementation compare to the private cloud, and it provides better security compared to the public cloud [20], [21]. The reason for this migration because the user needs both which is security as well as cost-effectiveness and the hybrid cloud is the answer [22]. In 2015, there is 81% cloud user using a public and private cloud and the drastic changing on 2016 show that 57% of cloud user (additional of 38%) in 2015 started using the hybrid cloud, it is because of slow performance in the public cloud and difficult to access a data due to a high-security implementation in a private cloud. Otherwise, the hybrid is greater flexible and most effective model compares to other clouds [21].

In ASEAN, around 360,288,700 million internet user and estimated around 40% of them use cloud computing, but to become a digital world economy, ASEAN facing some challenges. There will be a discussion about cloud computing, focusing on implementation and challenges in ASEAN countries, then some suggestion to overcome the issues.

4. Related Works

There are some studies about cloud implementation. Based on NIST, cloud computing has a three different stage, which contains five essential characteristics, three delivery models and three deployment models, as shown in Figure 3 [21]. Cloud Deployment Model is a service provided by the cloud, and it grouped into three different types, each type will address a different set of business requirement. Cloud Delivery Model is a reference model on which of the Cloud Computing based, also about user selection that described based on the appropriate service, and its dependence on a factor such as management of distributed data, availability of the suitable program, or developing an application.

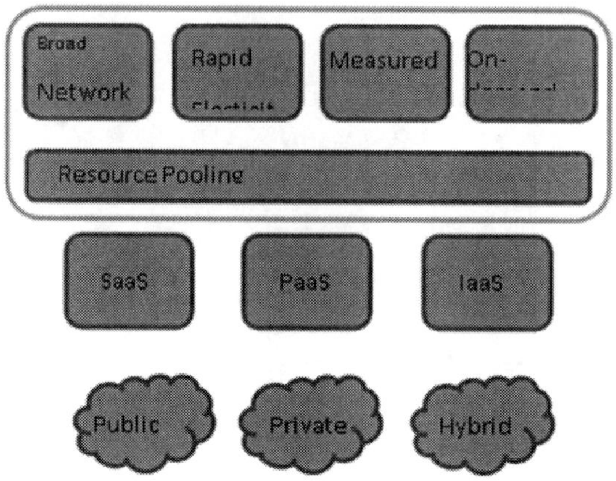

Figure 3: The NIST cloud computing definitions

4.1. Cloud Deployment Model
Cloud Computing Deployment Model as Figure 4 determines a way of the user to access to the cloud, and it will depend on business and the type of services that user need, it might be sorted into three types based on the deployment:

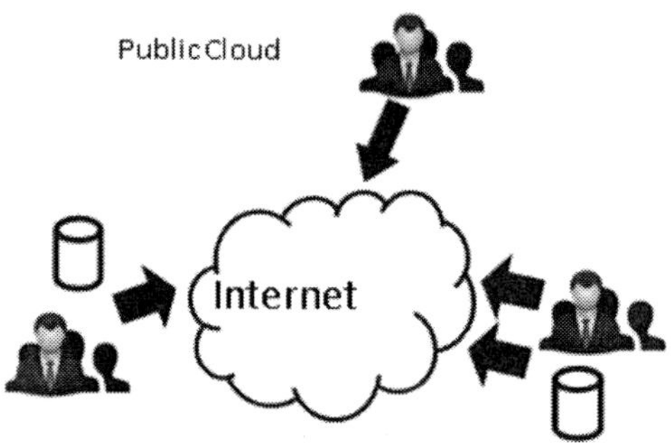

Figure 4: Cloud Deployment Model

Public Cloud: The Cloud provider will manage and control a resource located in the provider facilities, and it is available to the general public via the Internet. It could be no cost and offered on a pay-per-usage product [21]. Usually, the public cloud user has limited control to the application, storage and security implementation. If the user needs more access, storage and better security, the user needs to pay or subscribe some amount from a cloud provider.

Private Cloud: Generally, an organisation use the private cloud to have its own data centre (DC), and only registered members can access. The user has permission to customise the cloud because the organisation has full access to manage and control its own cloud [21]. Private cloud is suitable for a high-security implementation since it has a restricted number of users behind a firewall [23].

Hybrid Cloud: This flexible cloud is a combination between Public and Private cloud provides more deployment option for better performance [24], [25]. It will use a function of a private cloud when to deploy sensitive data, and it will use a public cloud when to deploy a non-sensitive data. For example, an organisation will use public cloud applications such as Google Email (Gmail) for backup data but continues to maintain a private cloud when performing sensitive user data.

4.2. Cloud Delivery Model
Cloud computing describes the applications, hardware and system software provided as services via the internet [26]. It uses to improve business flexibility and provide a lower cost of engineering

infrastructure resources with a pay-per-use. The term used to describe a service is "as-a-service" or "aaS", and XaaS is a term used to refer to any cloud-based services. "X" means anything services provide through the cloud [27]. The most common services that provide by cloud computing known as Software as a Service (SaaS) offer a cloud application to the user, then Infrastructure as Service (IaaS) offers infrastructure and hardware on the web, and Platform as a Service (PaaS) offers a development of application via the web [26]. This delivery model provides a different level of users that can integrate directly to a service. There are three main services provided by cloud delivery model:

Infrastructure as a Service (IaaS): The cloud services provide physical infrastructures such as a server, storage, and other networking devices to cloud user [28]. This service will directly perform by network architecture or network engineer that responsible for configuring and setting up a physical data centre for use by cloud user. The organisation will take a service based on the need, and user also can subscribe if the current services not enough for them. Amazon S3 that provides big data storage for an organisation is an example of IaaS.

Platform as a Service (PaaS): This service will provide a facility for application developer or programmer to develop an application through the web. Usually, PaaS is comprised an Operating System, Programming Language, Database and web server [29]. Then the developer can execute their program with no cost and easy to develop because no need to think either the software compatible or not with a hardware, for example, Amazon Aurora that provide MySQL and PostgreSQL as a database platform and Google that provides cloud operating system services namely Chromium OS.

Software as a Service (SaaS): SaaS is a software distribution model that provides access to a user through web-based services [26]. This platform allows end-user to use an application without the need to buy a license or pay for additional hardware. Suitable for a small organisation that runs a low-cost business, because this service will wipe out all possibilities for organisations to manage the installation, software updating, license setup and maintenance. For example, Google provides several SaaS services such as a Google Doc, Spreadsheets, Slide and Form that can be accessed and shared with another, anywhere and anytime through the internet.

5. Discussion

There have several issues about cloud computing, which is a benefit and a risk-based on a survey conducted by RightScale [30] in January 2017, a total of 1002 respondents from technical professional background already answered, they discover several benefits and the challenge of using Cloud Computing.

5.1. Benefit of Cloud Computing

Cloud computing is a technology to deploy and delivery services; many of benefit related to the cloud will be determined as a new. According to Figure 5, shows there is a top benefit of cloud implementation in 2017 [30] which is, faster access to the infrastructure, greater scalability, higher availability, faster time to market and business continuity.

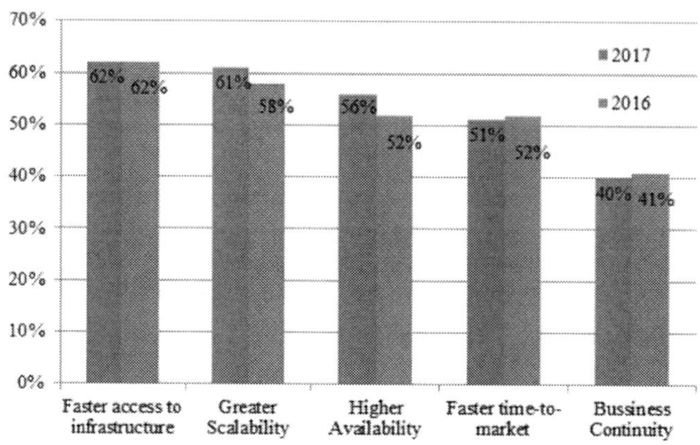

Figure 5: Shown cloud top benefit of Cloud Computing in 2017

Faster access to Infrastructure: Nowadays, a user needs faster access to infrastructure resources across multiple devices, network and storages. When the needs grow, and some resource is added to the environment, the packet will take a long time for delivery, and it makes slower business growth. The higher percentages of the survey show the importance of faster access to infrastructure to ensure the reliable delivery of cloud computing services is manageable, scalable and easy for a user to develop and market their application.

Greater Scalability: Scalability is an important factor in a technology environment, it shows an increase of 3% in 2017 from 58% in the previous year, this explains the ability to increase the scale without implying increased costs [31], and also decrease time to market, which is the meantime, use to develop or deliver a cloud service. Now, the internet becomes popular, this increase due to the additional cloud user where, the request from the user will make the server receive more traffic and if there is no solution to handle, the site will not available (breakdown) to access by the user.

Higher Availability: According to [32] measurement of higher availability in the cloud services is based on accessible 99.999% of the time, its mean 5.26 minutes of downtime per year [33], based on the formula shown in Figure 6. The application, tools or data provided by cloud services can be accessed anytime and anywhere. It is also mentions about reliability, where the cloud provider will always provide technical support 24/7 at 365 days without fail. An increase from 2016 (52%) to 2017 (56%) show the importance of high availability in a cloud provider, because nowadays in the paperless workplace, most of the data will be processed by computer, it needs downtime protection in order to make user data can be accessible as possible.

Availability = MTBF /(MTBF + MTTR)
*MTBF (Mean Time between Failure)
*MTTR (Mean Time to Repair)

Figure 6: Availability formula

Faster time-to-market: The average length of time has been taken in the product development process from product idea to the complete finished product, and it is also a critical element of time-based competition. The organisation can speed up and get the job done by the service or product to market more rapidly. Based on Figure 5, it decreases 1%, due to a restrictive control process that applies a certain level of security in the cloud, and it will make the high latency in data processing and affect the time taken to market.

Business Continuity: Cloud computing is network dependent, it needs an internet connection to operate, and without the internet, the cloud cannot deliver a service to users [34]. The most important part of making the business grow is, when the occurrence of the disaster, user still can access the data on a cloud. The cloud still can operate a service, but a reduction of 1% shown that the trusted is the main issue, where the file management of the organization need to manage by cloud providers, where there is not the owner, and it maybe can be accessed by unauthorized way. Some company not ready because of the vulnerabilities of interception during transmission of data.

5.2. Risk of Cloud Computing
The current acceptance of cloud computing is associated with many challenges. Users are still in doubt about the use of cloud computing, especially in some aspects that recognised by cloud maturity which declare as a major challenge [30], such as lack of resources, security, managing cloud spend, compliance and governance/control shown in Figure 7.

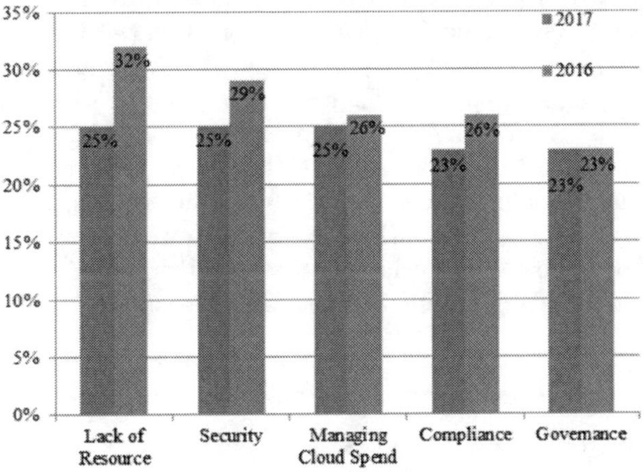

Figure 7: Shown cloud top challenge of Cloud Computing in 2017

Lack of Resource: The lack of resource here is about knowledge and expertise, according to a McAfee on September 2016 [19] and supported by Forbes on 2017, the lack of security skills causing almost half of IT organisation and engineering firms slowdown of their cloud adoption plan. Cloud starting the huge adoption around 2016 and it brings more workloads to the cloud, at that time, not many IT organisations are capable of solving cloud issues, and it needs to refer to cloud expert. Some provider focusing on internal training and skill improvement to overcome this issue and the result in 2017 has shown the significant decrease in that challenge.

Security: The security is the big challenge in Cloud Computing. IT organisation will increase 46% budget in 2015 to spend on security implementation. A report from RightScale [30] said the security issues increase by 4% from 25% in 2016. Due to the cloud still new, there are vulnerability and issues to address [35]. The cloud provider needs to have the ability to control and overcome a security problem to prevent the cloud from beginning attacked. The addition of a high level of security implementation and strategy on previous years has made a changing on cloud security, and it starts to be trusted by users and the strategy cause the decrease of security issues in 2017.

Managing Cloud Spend: Although cloud computing implementation is more cost-efficient compared to the traditional way, it also contains a wasted spend on cloud cost. The cloud user is wasting 30-45% of their spend in cloud computing [30]. An organisation that uses a hybrid or private cloud spends money to set up a data center that focuses on server, infrastructure, power requirements and networking [36]. Optimisation a cost of cloud was applied in 2017 to reduce wasted spending in cloud implementation.

Compliance: Cloud compliance becomes an issue for cloud users; it's also a reason for several companies still using traditional server-based rather than using cloud computing. The user still argues about security, data storage location, the integrity of data, availability of access and so on. The Service Level Agreement (SLA) which is a contract on the particular services between the cloud provider and cloud user [37], used to fix the compliance argument and make cloud start to gain user trust.

Governance: Cloud computing provides governance services for applying policy and rules to control and manage a system [38]. The main function is to secure a service, assets, remote facility and data that provide by cloud. Cloud Governance has some good factor where the optimisation can leverage to cloud to prevent cloud wasted, and the profitability showed the good governance applies in cloud computing will make the highest profit compare to the poor governance.

5.3. ASEAN Challenge

Compare to other countries, ASEAN started quite late for adopter cloud computing development because of several challenges such as cost, social-economy, network connectivity, and broadband quality. Now the awareness of adoption of cloud computing in ASEAN shows a good improvement. Based on a report from the Asia Cloud Computing Association (ACCA) about Cloud Readiness Index on 2018 presents that Asia is leading the world in Cloud Computing readiness and ASEAN countries such as Singapore, Malaysia, Philippines, Thailand, Indonesia and Vietnam overcome the challenge as a catalyst to their industrial revolution 4.0 development. Figure 8 shows the cloud computer readiness index of ASEAN country in 2018 [13]. In ASEAN, Singapore leading with a 76.6% score follows by Malaysia with 61%, and the Philippines with 53.6%, then Thailand 50.6%, Indonesia 49.4% and Vietnam 41%. The result in Figure 8 influenced by various factors and one of it is an ability of network connectivity to support the development of cloud computing. The good internet speed will make the development of cloud computing become faster.

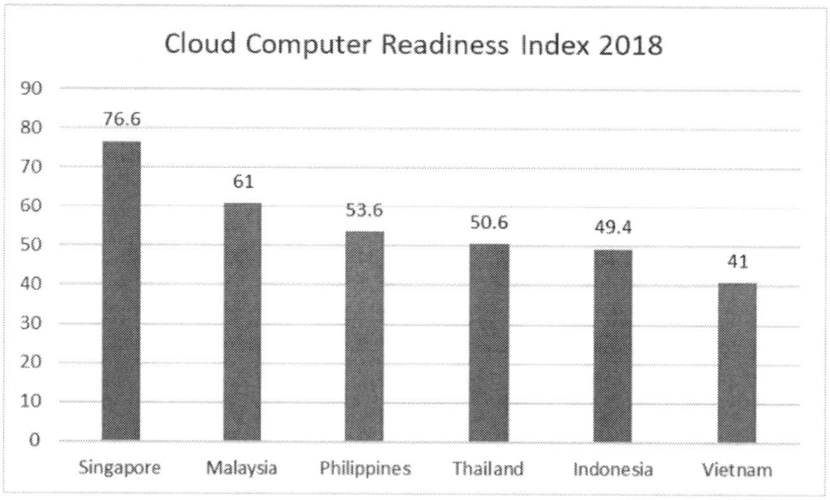

Figure 8: ASEAN Cloud Computing Readiness Index

Table 2 shows the average connection speed of the internet in ASEAN. Refer to Table 2, Singapore has a 181.47Mbps in 2018 and 199.62Mbps in 2019, which increase 18.15Mbps in a year followed by Thailand with 76.62Mbps in 2019, the lowest in Indonesia with 16.65 Mbps in 2019. Malaysia shows the highest increase in internet connection with a different of 39.72Mbps from previous average speed.

Table 2: Average connection speed of internet in ASEAN 2018 and 2019 [39-40]

Country	Global Rank		ASEAN Rank		Ave. Speed (Mbps)	
	2018	2019	2018	2019	2018	2019
Singapore	1	1	1	1	181.47	199.62
Thailand	34	27	2	2	47.35	76.62
Malaysia	58	36	3	3	27.43	67.15
Vietnam	60	72	4	4	25.28	27.60
Philippines	88	101	5	5	17.32	19.51
Indonesia	94	111	6	6	15.18	16.65

Figure 9 shows the affordability of cloud user to subscribe to Internet service on fixed broadband. Based on Figure 8, 9 and Table 2, show that the highest internet speed connection will make the country smoothly develop and deploy a cloud computing, but the lower connection speed with the high cost will make vice versa. ASEAN announces in AIM2020 (2016 – 2020), planning to develop the cloud computing environment in the private and public sectors [15]. Some of ASEAN country has slow development progress compare to others, because they have a different problem and challenge.

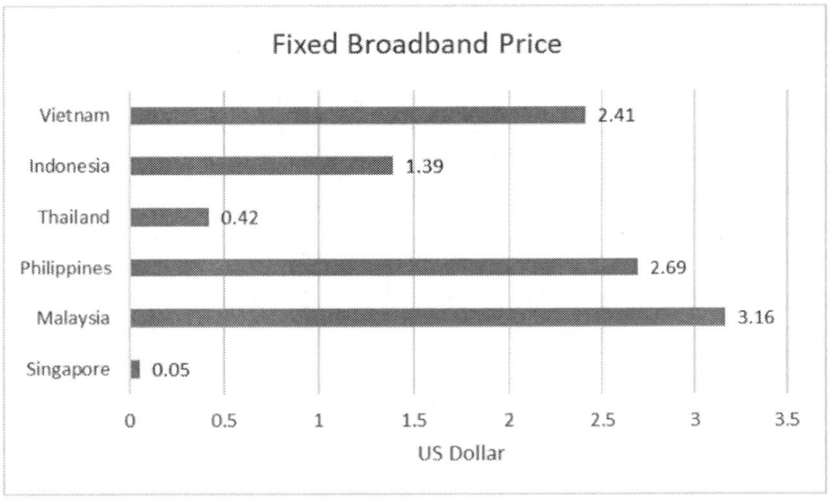

Figure 9: Fixed broadband Internet price in 2018 [39]

Table 3 shows the challenge of each country which are Malaysia, Indonesia, Singapore, Philippines, Thailand and Vietnam.

Table 3: Challenge of ASEAN country to develop a cloud computing

Country	Challenge
Malaysia [13], [41]	• Lack of physical infrastructure. • Lack of competent cloud engineers, knowledgeable cloud lawmakers, and service consumers. • Security and privacy issues.

Indonesia [13], [42]	• Low quality of internet. • Security issues. • Low-speed internet connection in rural areas. • Slow user adoption. • Users anxiety to use cloud computing. • Lack of knowledge of the cloud business model. • Low partnership opportunity.
Singapore [13]	• Lack of international connectivity. • Lack of freedom of information.
Philippines [13], [43], [44]	• Low quality of broadband and mobile access to the internet. • Lacks a standard for ICT handling and exchange. • Most expensive broadband connectivity.
Thailand [13], [45]	• Lack of international connectivity. • Lack of comprehensive software for cloud development. • Only a few domestic cloud providers.
Vietnam [13], [46]	• Lack of freedom of information. • High risk of Data Center. • Lack of explicit a cloud strategy.

6. Conclusion

Nowadays, cloud technology was used as the main medium to store a data as individual (small storage) or organisation (large scale of data), a cloud used as an online application that allows a user to do their job anywhere and anytime. Cloud computing is about a time, cost, and performance for applying effective technology. It is a good comparison between traditional server-based and cloud computing in term of effectiveness. Cloud computing saves time to set up a data centre, cost of power and electricity and save 50% emission of carbon on earth. Every day the usage of cloud increased, and it will continue to increase in the next few years, but, recently security, reliability and scalability have arisen as issues, and it causes some organisation still not using a cloud service. It is a difficult part to manage because of virtualisation and the resources it is. ASEAN countries also have a challenge to ensuring that cloud computing will be operating well in their country. They need a strategy to fix the issues, to fulfil the requirement of Industries Revolution 4.0. The suggestion is to empower the ICT infrastructure, including a broadband quality, provide enhancement of security and assurance to the user, and transform the ASEAN economy to a digital economy in order to overcome the highlight challenges.

7. Acknowledgement

The author would like to thank UMP RDU Grants (RDU-180364) for providing financial support and facilities in this study. Also gratefully, we would like to acknowledge and thank the Faculty of Computing and Systems Network & Security (Sysnet) research group UMP that support this research.

References

[1] H. A. S. Ahmed, M. H. Ali, L. M. Kadhum, M. F. Bin Zolkipli, and Y. A. Alsariera, "A review of challenges and security risks of cloud computing," J. Telecommun. Electron. Comput. Eng., vol. 9, no. 1–2, pp. 87–91, 2017.

[2] A. Shihab Ahmed and M. F. Zolkipli, "Data Security Issues in Cloud Computing: Review," Int. J. Softw. Eng. Comput. Syst., vol. 2, no. February, pp. 58–65, 2016.

[3] P. K. Senyo, E. Addae, and R. Boateng, "Cloud computing research: A review of research themes, frameworks, methods and future research directions," Int. J. Inf. Manage., vol. 38, no. 1, pp. 128–139, 2018.

[4] C. A. B. de Carvalho, R. M. de C. Andrade, M. F. de Castro, E. F. Coutinho, and N. Agoulmine, "State of the art and challenges of security SLA for cloud computing," Comput. Electr. Eng., vol. 59, pp. 141–152, 2017.

[5] A. Shawish and M. Salama, "Inter-cooperative Collective Intelligence: Techniques and Applications," vol. 495, pp. 39–68, 2014.

[6] N. Chidambaram, P. Raj, K. Thenmozhi, and R. Amirtharajan, "Enhancing the Security of Customer Data in Cloud Environments Using a Novel Digital Fingerprinting Technique," Int. J. Digit. Multimed. Broadcast., vol. 2016, 2016.

[7] S. Ouf and M. Nasr, "The Cloud Computing: The Future of BI in the Cloud," Int. J. Comput. Theory Eng., vol. 3, no. 6, pp. 750–754, 2011.

[8] S. Aldossary and W. Allen, "Data Security, Privacy, Availability and Integrity in Cloud Computing: Issues and Current Solutions," Int. J. Adv. Comput. Sci. Appl., vol. 7, no. 4, pp. 485–498, 2016.

[9] S. K. Garg, C. S. Yeo, and R. Buyya, "Green cloud framework for improving carbon efficiency of clouds," Lect. Notes Comput. Sci. (including Subser. Lect. Notes Artif. Intell. Lect. Notes Bioinformatics), vol. 6852 LNCS, no. PART 1, pp. 491–502, 2011.

[10] K. Alisetty, A. Professor, and K. E. Balachandrudu, "Cloud Computing & Grid Computing Environments," Int. J. Innov. Eng. Technol., vol. 3, no. 2, pp. 214–218, 2013.

[11] C. Science, Y. S. P. Mahavidyalaya, and P. Medinipur, "Cloud Computing in brief," IOSR J. Comput. Eng., vol. 18, no. 6, pp. 101–103, 2016.

[12] V. Kishore, G. G. G. Bhavani, and M. Ramakrishna, "Simplifying Complex Tasks : Cloud Computing," Int. J. Comput. Sci. Inf. Technol., vol. 3, no. 5, pp. 4988–4996, 2012.

[13] ACCA, "Cloud Readiness Index 2018," Asia Cloud Comput. Assoc., pp. 1–46, 2018.

[14] B. James, The enterprise cloud, Brian Ande. Unitedd States of Amarice: O`Reilly, 2015.

[15] S. Jakarta, "The Asean ICT Masterplan 2020," Jakarta, Indonesia, 2015.

[16] P. M. Tehrani, "Feasibility Of Sustainable Development," vol. 13, no. 4, pp. 559–576, 2016.

[17] D. Sikeridis, I. Papapanagiotou, B. P. Rimal, and M. Devetsikiotis, "A Comparative Taxonomy and Survey of Public Cloud Infrastructure Vendors," pp. 1–20, 2017.

[18] C. Gong, J. Liu, Q. Zhang, H. Chen, and Z. Gong, "The characteristics of cloud computing," Proc. Int. Conf. Parallel Process. Work., pp. 275–279, 2010.

[19] Flexera, "Cloud Computing Trends: 2019 State of the Cloud Survey," 2019.

[20] G. Sivi and T. Narayanan, "A Review on Matching Public , Private , and Hybrid Cloud Computing Options," Int. J. Comput. Sci. Inf. Technol. Res. ISSN, vol. 2, no. 2, pp. 213–216, 2014.

[21] S. Goyal, "Public vs Private vs Hybrid vs Community - Cloud Computing: A Critical Review," Int. J. Comput. Netw. Inf. Secur., vol. 6, no. 3, pp. 20–29, 2014.

[22] K. Kaur, "A Review of Cloud Computing Service Models," Int. J. Comput. Appl., vol. 140, no. 7, pp. 15–18, 2016.
[23] D. R. K. R. . Hemamalini. B. H.1, Dr. Suresh L.2 and 1., "A Survey on Cloud Computing," Ijmcr, vol. 1, no. 10, pp. 270–272, 2013.
[24] R. Balasubramanian, "Security Issues : Public vs Private vs Hybrid Cloud Computing," Int. J. Comput. Appl., vol. 55, no. 13, pp. 35–41, 2012.
[25] D. R. Malhotra and P. Jain, "How to Choose an Economic Cloud Deployment Model," Int. J. Comput. Trends Technol., vol. 4, no. 8, pp. 2607–2614, 2013.
[26] C. C. Rao, M. Leelarani, and Y. R. Kumar, "Cloud: Computing Services And Deployment Models," Int. J. Eng. Comput. Sci., vol. 2, no. 12, pp. 3389–3392, 2013.
[27] M. S. Narasimha and V. S. Murthy, "A Study on Cloud Computing Services," ICT Crit. Infrastruct. Proc. 48th Annu. Conv. Comput. Soc. India, no. November, 2016.
[28] R. Sharma and R. K. Trivedi, "Literature review: Cloud Computing –Security Issues, Solution and Technologies," Int. J. Eng. Res., vol. ISSN, no. 34, pp. 221–225, 2014.
[29] J. F. Zhao and J. T. Zhou, "Strategies and methods for cloud migration," Int. J. Autom. Comput., vol. 11, no. 2, pp. 143–152, 2014.
[30] B. Clement, "State Of The Cloud Report," United States of Amarice, 2017.
[31] R. Dabra, T. Kumar, and P. Dhiman, "Performance Analysis of Scale-out Cloud Architecture In Multiple Geographical Regions Using Cloud Sim 2 . Why Worry About Cloud," vol. 2, no. 4, pp. 33–37, 2013.
[32] A. Kanso and Y. Lemieux, "Achieving High Availability at the Application Level in the Cloud," 2013 IEEE Sixth Int. Conf. Cloud Comput., pp. 778–785, 2013.
[33] M. Toeroe, N. Pawar, and F. Khendek, "Managing application level elasticity and availability," Proc. 10th Int. Conf. Netw. Serv. Manag. CNSM 2014, pp. 348–351, 2015.
[34] Australian Government, "Cloud Computing Strategic Direction," Inf. Secur., pp. 1–43, 2011.
[35] Columbus, L. "State of Cloud Adoption and Security," Forbes, 2017.
[36] L. Columbus, "Computerworld's 2015 Forecast Predicts Security, Cloud Computing And Analytics Will Lead IT Spending," 2014.
[37] G. Ramachandra, M. Iftikhar, and F. A. Khan, "A Comprehensive Survey on Security in Cloud Computing," Procedia Comput. Sci., vol. 110, no. 2012, pp. 465–472, 2017.
[38] A. Greenberg, J. Hamilton, D. A. Maltz, and P. Patel, "The cost of a cloud: research problems in data center networks," ACM SIGCOMM Comput. Commun. Rev., vol. 39, no. 1, pp. 68–73, 2008.
[39] Stephen Chin, "Singapore Wins Internet Speed Race," The Asean Post, 2018.
[40] Jason Thomas, "Are ASEAN's Internet Speeds World Class?," The Asean Post, 2019.
[41] L. D. Introna and D. Wood, "Picturing algorithmic surveillance: The politics of facial recognition systems," Surveill. Soc., vol. 2, no. 2–3, pp. 177–198, 2004.
[42] S. Baller, S. Dutta, and B. Lanvin, "The Global Information Technology Report 2016 Innovating in the Digital Economy," Geneva, Switzerland, 2016.
[43] S. Abolfazli, Z. Sanaei, A. Tabassi, S. Rosen, A. Gani, and S. U. Khan, "Cloud Adoption in Malaysia: Trends, Opportunities, and Challenges," IEEE Cloud Comput., vol. 2, no. 1, pp. 60–68, 2015.
[44] I. S. I. S. S. Mangula, I. Van De Weerd, and S. Brinkkemper, "Adoption of the Cloud Business Model in Indonesia: Triggers, Benefits, and Challenges," Int. Conf. Inf. Integr. Web-based Appl. Serv., pp. 54–63, 2012.
[45] Electronic Government Agency, "Study on Cloud Computing on ASEAN for ICT Framework and Policy Recommendations Initiative," 2015.
[46] D. K. M. V. Dr. Eduardo Araral; Dr. Kris Hartley, Dr. Reuben Ng, "A Cloud for Doing Good : A Technology Revolution for All in ASEAN," Singapore, 2017.

Coinhive's Monero Drive-by Crypto-jacking

Azizah Binti Abdul Aziz[1], Syahrulanuar Bin Ngah[2], Yau Ti Dun[3], Tan Fui Bee[4]

[1,2]Systems Network & Security (SYSnet), Faculty of Computing, Universiti Malaysia of Pahang, 26300, Gambang, Kuantan Pahang, Malaysia

[3,4]SyaArmy R&D Team, SysArmy Sdn Bhd, Wisma Zelan, No., 12,, 1, Jalan Tasik Permaisuri 2, Bandar Tun Razak, 56000 Kuala Lumpur, Malaysia

E-mail: azizah.official@gmail.com, syahrulanuar@ump.edu.my

Abstract. This paper provides study of behaviour on a drive-by crypto-jacking, specifically Coinhive's Monero. Crypto-jacking is a form of cyber threat where a host machine's processing power hijacked thru infected website to solve cryptographic puzzles as an unwitting participant. The sample study share host machine and network behaviours when host visited websites that have been embedded with Coinhive's Monero related script. These data collection of behaviour can be utilize to identify and detect Coinhive's Monero mining activities in networks.

Keywords: Coinhive's Monero, cryto-jacking, drive-by

1. Introduction

Cryptocurrency is digital currency that uses cryptography for security purpose, hence the term 'Crypto' in cryptocurrency. The first idea of cryptocurrency was circulated in the internet from a write up by Satoshi Nakamoto, an anonymous author of paper titled "Bitcoin: A Peer-to-Peer Electronic Cash System" that proposed a system for electronic transactions that would use individual host machine's resources to perform work that creates a "coin" and verifies the transfer of ownership of these virtual coins without being controlled by regulatory [1] One of common currency used are Monero (XMR) and Bitcoin (BTC).

Crypto mining in other hand refer to act of utilizing host machine's processing power to solve cryptographic puzzle, validate transactions and generate profit. As the value of cryptocurrency increases exponentially, the need of mining them also increases. Figure 1 shows the increases value and volume of cryptocurrency over the years [2]. Thus, cybercriminals resorted to illegal act of crypto-jacking to acquire more profit from crypto mining. Crypto-jacking is define as form of cyber threat where a host machine's processing power is hijacked to mine cryptocurrency as an unwitting participant [3].

These illegal activities may cause harm to organization's network as mining activities always resulted in workstation become slow or freeze that will hinder normal business activities, consuming power that will resulted in money loss and can also be used as "door" for something more malicious that may lead to data loss and cyber-attack. Online services have always been a target of internet attackers with new sophisticated tactics [4]

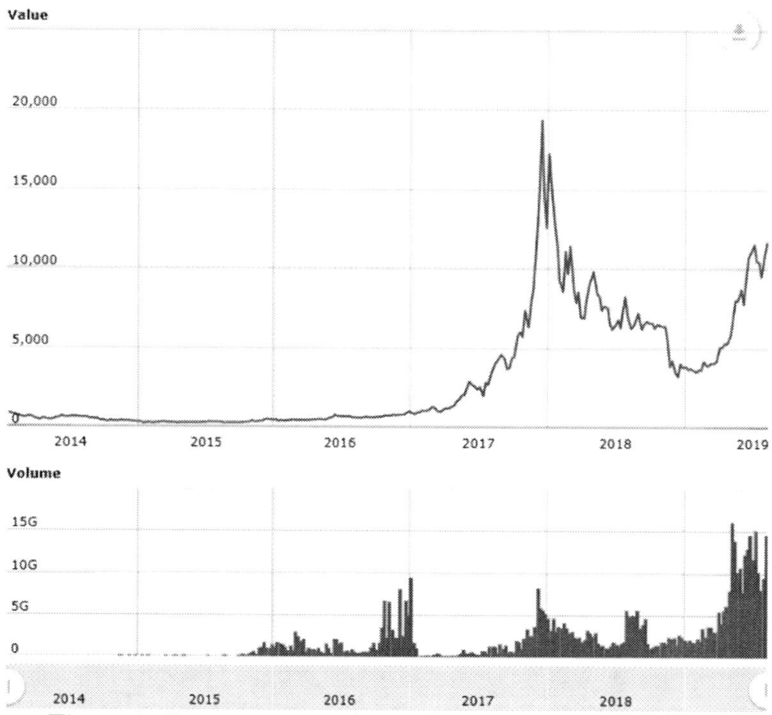

Figure 1. Crytocurrency value and volume over the years

2. Coinhive's Monero

Monero is one of the cryptocurrencies used worldwide besides Bitcoin. Due to its high level of privacy where its transactions are virtually untraceable by outsider, Monero have been one of choices for cybercriminal to do transaction.

One of the cryptocurrency mining services that mines for Monero is Coinhive. Coinhive provide its users with JavaScript that can be embedded in webs sites. The purposes of this JavaScript are to do mining by using websites' visitors' resources. This will eventually lock up visitors' browser and drains the device's battery.

Based on report produced by security firm, Malwarebytes [3], their telemetry identified usage of silent Coinhive JavaScript of 3M/day during the period of January 10 to February 6, 2018. Silent version of Coinhive refer to JavaScript version that enable script owner to utilise visitor host machine resources to do mining without their consent.

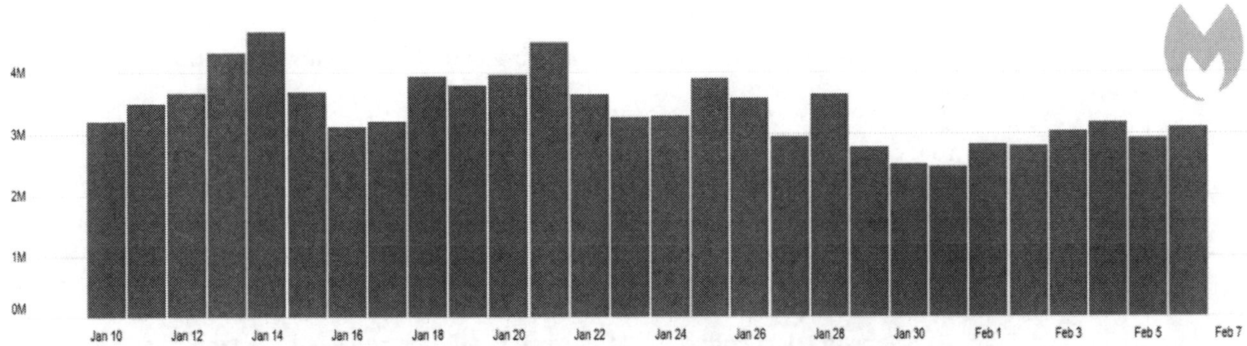

Figure 2. Usage statistics for the silent version of Coinhive

Researcher at Krebs On Security considered Coinhive as the top malicious threat to web users as they found nearly 32,000 websites running Coinhive's JavaScript miner code as for March 2018 [5].

3. Crypto-jacking thru drive-by compromise

'Drive-by compromise' refer to exploit technique to gain access of a system and its resources when user visited compromised websites while browsing. Cybercriminals embedded malicious code modules in websites and it will be deployed in user's browser when they visit the compromised website. This method often leaves no traces of malicious files (file-less), thus making it difficult to be detected by end point protection (EPP) such as antivirus [6]. This method often observed been leveraged by cybercriminals to launch crypto mining modules on users' machines to hijacked its resources. Figure 3 shows typical anatomy of drive-by compromise.

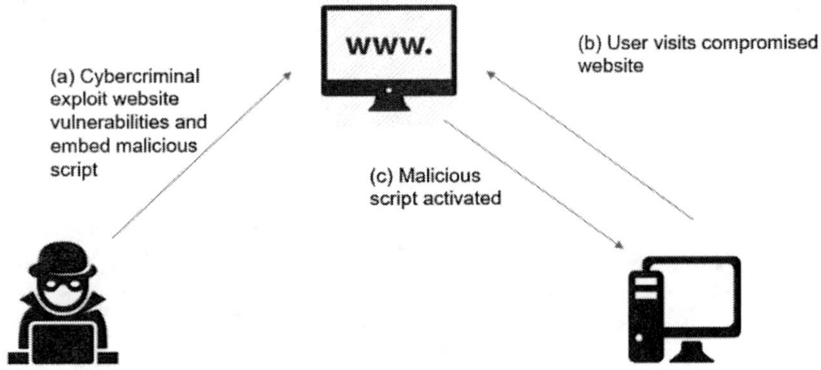

Figure 3. Typical anatomy of drive-by compromise

Even though file-based detection is near impossible for drive-by crypto-jacking, there are alternative detection method that are more feasible to identify and detect drive-by crypto-jacking which is behavior-based detection.

4. Research Methodology

Detecting mining activities similar to this is becoming a challenging task due to increasing connectivity of systems that gives greater access to the internet and makes it easier for cyber criminals to do illegal activity that gives benefit to them.

Thus, to overcome this problem, analysing mining behaviour will help to established the pattern of behaviour where it then can be utilized to detect mining activities in a network.

In this study, several work stages have been identified for the methodology for the research. Initially, an investigation on the cryptomining activities and how it contributes to cyber risk are done.

a) Stage 1: Preliminary study. This initial study will start by reviewing on how Coinhive's Monero works and collecting sample of infected website.
b) Stage 2: Design controlled environment. In this stage, designs of controlled environment architecture and applications are done that are appropriate for mining behaviour analysis environment.
c) Stage 3: Testing the samples. This stage involves with visiting the suspected websites and collects relevant data.
d) Stage 4: Analysing data collected. At this stage, all data collected is analysed and grouped into appropriate group based on the findings.

5. Sample case Coinhive's Monero mining behaviour

In order to identify behavior pattern of crypto-jacking via Coinhive, we leverage 53 public websites that have been reported as source of Coinhive' Monero distribution for our sample case. These websites all have "coinhive" string in their source page. Traffic and host machine behavior are observed starting from the before, during and after user visited the website.

In this sample case, for host machine, Windows 10 with Chrome Ver. 68.0.3440 are used as browser to surf the websites and Intel ® Core ™ i7-5500u CPU, 2.40GHz is used as CPU. Different browser application and CPU may have different result. Whereas, Juniper Firewall SRX and Cisco ASA firewall are used to monitor the traffics.

5.1. Host level

Immediately when host visit the infected websites, we observed sudden increment of CPU and memory usage for browser application. With average increment of CPU up to 99.4% and memory utilization of 532.7 MB. Figure 4 shows differences of CPU utilization before and during host visited websites with Coinhive's Monero, whereas Figure 5 shows differences of memory before and during host visited the websites.

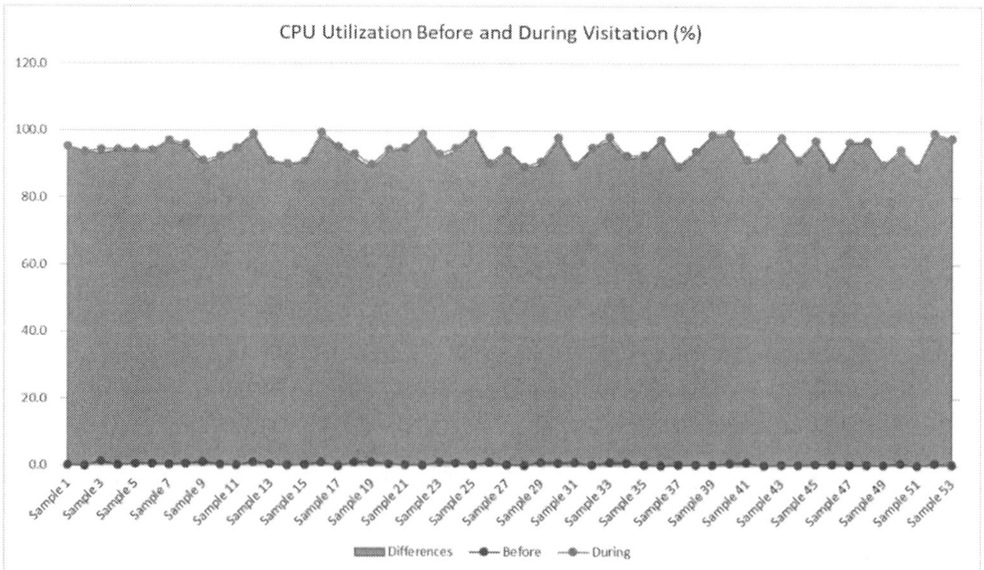

Figure 4. CPU utilization before and during host visiting the websites.

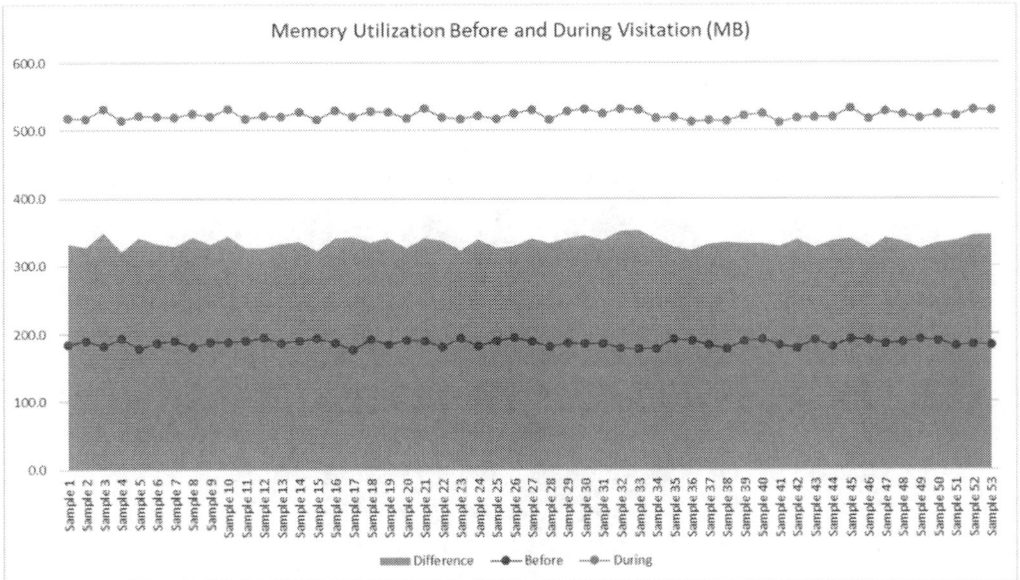

Figure 5. Memory utilization before and during host visiting the websites

75.47% of the sample case shows high utilization of CPU and memory continues even after the browser windows that used to visit the websites closed. We found that this is due to hidden browser windows that have been initiated during the infection phase under the taskbar. The hidden window's coordinates vary based on each user's screen resolution. Figure 6 shows sample of hidden browser as observed.

Figure 6. Hidden browser as launched during mining activities.

Figure 7 shows percentage breakdown where mining activities continues even after the main browser closed.

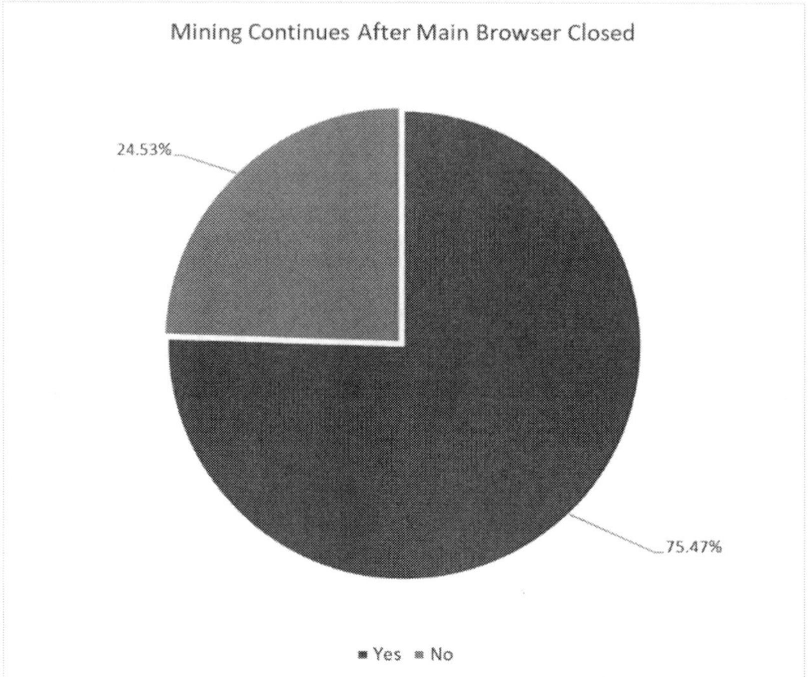

Figure 7. Percentage breakdown for host that continue mining activities even after the main browser have been closed.

5.2. Network level

In controlled environment, as host visited the infected websites, we are able to observe two (2) group of behaviour, where we labelled as Group A and Group B. For Group A, we observe multiple persistent outbound traffic toward the infected websites. This match with typical cryptocurrency mining activities. Host visited the websites, JavaScript on the websites triggered and mining started. Refer Figure 8.

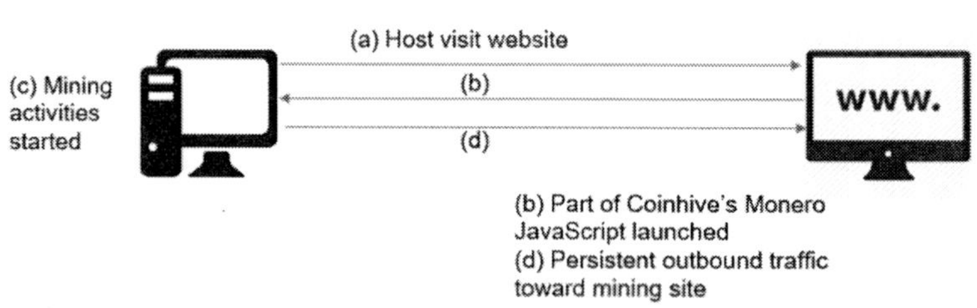

Figure 8. Anatomy of Group A Coinhive's Monero mining behaviour

Whereas for Group 2, instead of persistent outbound traffic to the visited website, we observe outbound traffic from host toward advertisement sites via content delivery network (CDN). Host visited websites, JavaScript triggered and redirect host to another sites. These redirected sites hosted another JavaScript that started the mining. Process from direction to host started mining create another hidden browser. At the time of research, we observed multiple persistent outbound traffic toward CDN, where host as hosts were mining at the mining sites. Refer Figure 9 for the anatomy

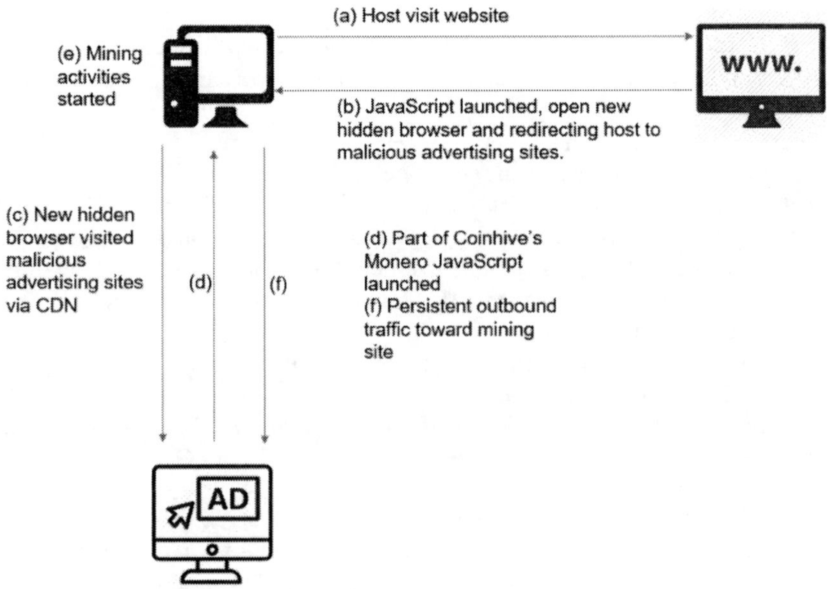

Figure 9. Anatomy of Coinhive Monero mining via advertistment sites

Table 1 shows summarize characteristic for group A and B

Table 1. Summarize characteristic for both group

Group A	Group B
• JavaScript that triggered mining resides on the visited websites • Persistent outbound traffic toward the infected websites from host machine.	• JavaScript that triggered mining does not resides on the visited websites • Persistent outbound traffic toward the redirected sites.

Based on the sample tested, instead of embedding JavaSript in the website like traditional drive-in exploit, cybercriminal leverage the technology of CDN and hosted the mining JavaScript in other sites, mainly advertisement sites for mining purposes. Hosts were redirected to these mining sites when visiting the infected websites. Thru this way, cybercriminal can maintain their infrastructure longer and able to infect large scale of host with minimum changes to crypto-jacking infrastructure.

6. Conclusion

Detecting mining activities is not just tedious but sometimes impossible and costly. Identifying these behaviors may assist to increase protection against organization. Based on the behavior identified, set of rules can be implement at network and host level to either preventing or detecting these activities.

Acknowledgments
The work is funded by Universiti Malaysia Pahang under Grant ID RDU1703294. The authors would like to thank all reviewers for their constructive comments, and Universiti Malaysia Pahang for the support.

References

[1] Berentsen, A. (2019). Aleksander Berentsen Recommends "Bitcoin: A Peer-to-Peer Electronic Cash System" by Satoshi Nakamoto. 21st Century Economics, 7–8. doi:10.1007/978-3-030-17740-9_3

[2] Bitcoin price | index, chart and news | WorldCoinIndex. (2019). Retrieved from https://www.worldcoinindex.com/coin/bitcoin

[3] The state of malicious cryptomining. (2018, September 04). Retrieved from https://blog.malwarebytes.com/cybercrime/2018/02/state-malicious-cryptomining/

[4] Ahmed, A. A. (2015). Investigation Model For Ddos Attack Detection In Real-Time. International Journal of Computer Systems & Software Engineering, 1(1), 93–105. doi: 10.15282/ijsecs.1.2015.8.0008

[5] Krebs, B. (2018). Who and What Is CoinHive. Website https://krebsonsecurity.com/2018/03/who-and-what-is-coinhive.

[6] Hong, G., Duan, H., Yang, Z., Yang, S., Zhang, L., Nan, Y., … Qian, Z. (2018). How You Get Shot in the Back. Proceedings of the 2018 ACM SIGSAC Conference on Computer and Communications Security - CCS '18. doi:10.1145/3243734.3243840

[7] Musale, M. (n.d.). Hunting For Metamorphic JavaScript Malware. doi:10.31979/etd.ky84-3b5q

[8] Cimpanu, C. (2018). Over 1.65 Million Computers Infected With Cryptocurrency Miners in 2017 So Far. Retrieved from https://www.bleepingcomputer.com/news/security/over-1-65-million-computers-infected-with-cryptocurrency-miners-in-2017-so-far/

[9] Eskandari, S., Leoutsarakos, A., Mursch, T., & Clark, J. (2018). A First Look at Browser-Based Cryptojacking. 2018 IEEE European Symposium on Security and Privacy Workshops (EuroS&PW). doi:10.1109/eurospw.2018.00014

IOP Conf. Series: Materials Science and Engineering **769** (2020) 012066 doi:10.1088/1757-899X/769/1/012066

Cybersecurity Risk Assessment of the University of Northern Philippines using PRISM Approach

Bryan Irvin Lamarca

Management Information Systems, University of Northern Philippines, Vigan City, Philippines

bijlamarca@unp.edu.ph

Abstract. The University of Northern Philippines (UNP) rely on critical infrastructure systems to promote innovation and efficiency in fulfilling its core mandate - deliver quality education. Like any environment, UNP's cyber environment is vulnerable to security risks which dampen the privacy and safety of stakeholders, the security of assets, and the confidentiality of business proceedings. To proactively address these and other potential risks, this study reviewed existing risk management frameworks used across governments and selected one to be utilized for improving organizational cyber policies and risk mitigation procedures and practices – PRISM, a model to identify and implement cybersecurity risk management tailored towards the problems and needs of the university. Results showed that at least half of the risk areas have poor preparedness level, stemming from the lack of institutionalization of knowledge and solutions, assessing unusual behavior, and proactive and enterprise risk management. Most risk areas seemed to be prioritized, had allocated resources, and has reactive risk management in place. GAP Analysis was also conducted in conjunction with the results of PRISM assessment to better steer the university in the right direction. The study concluded that PRISM is a tool that aided the university in laying down a formalized groundwork for cybersecurity.

1. Introduction

The surge of sophisticated attacks on Higher Education Institutions' (HEIs) cyber environment poses threat, risks, and challenges, thus, prompting awareness of cybersecurity implications on a national and international scale. Organizations, specifically, State Universities and Colleges (SUCs) must thwart these attacks, improve their defenses, and put into place risk mitigation measures.

Cybersecurity's definition has many iterations; however, it mainly revolves around: defensive methods used to detect and thwart would-be intruders[1], safeguarding of computer networks and the information from penetration and malicious damage of disruption[2], and reducing the risk of a malicious attack to software, computers, and networks[3]. Also, it is a collection of tools, policies, security concepts, security safeguards, guidelines, risk management approaches, actions, training, best practices, assurance and technologies that can be used to protect the cyber environment and organization and user's asset[4]. To name a few, data breach, insecure application user interface (API), cloud abuse, malware attack, loss of data, hacking, single-factor passwords, zero-day vulnerabilities, and insider threat are some of the most common cyber-attacks that organizations have experienced on their operations.

In the Philippines, the 2004 National Cyber Security Plan addresses the cyber aspect of critical infrastructure protection focusing on Information and Communication Technology (ICT). The plan focuses on better measures to respond to cyber threats. The National Cybersecurity Plan 2022 targets the institutionalization, adaption, and implementation of globally recognized Security Governance and Risk Management standards by the public and private sector. Also, the plan mandates all government and public systems undergo the National Security System assessment which covers Philippine networks connected through government networks, domestic, and international5. Currently, both the Department of Information and Communication Technology (DICT) and National Privacy Commission (NPC) are the governing branches responsible for the enactment of cybersecurity and data privacy initiatives respectively. With UNP being funded by the government, it must abide by the following cyber/information security laws and standards: National CyberSecurity Plan 2022 (NCSP 2022)[5,] 2011-2016 National Security Policy[6], R.A. 8792 (E-Commerce Act)[7], R.A. 9775 (Anti-Child Pornography Act of 2009)[8], R.A. 9995 (Anti-Photo and Video Voyeurism Act of 2009)[9], R.A. 10173 (Data Privacy Act of 2012)[10], R.A. 10175 (Cybercrime Prevention Act of 2012)[12], M.O. 37, s2001[13], E.O. 810, s2009[14], A.O. 39, s2013[15], and Draft A.O. on Social Media Use[16].

Like any other organization, UNP has experienced more than one form of a cybersecurity threat or incident. A crucial organizational component to respond to cybersecurity threats is Cybersecurity Risk Management (CSRM). Risk management is concerned primarily with hardening a system component to avoid its failure; resilience management is concerned with maintaining the system's critical functionality by preparing for adverse events, absorbing stress, recovering the critical functionality, and adapting to future threats[17]. The organization to mitigate cyber-related risks in the university is still at its infancy. With the persistence of the university to comply with the rules and policies set forth by R.A. 10173 and R.A. 10175, securing data and managing cyber risk must now be an integral element and be prioritized within SUCs. Other factors that impede compliance including but not limited to lack or limited funding for the allocation of resources, prioritization of initiatives, saturated technology standards to follow and implement, and lack of institutionalized knowledge and solutions. The conduct or cybersecurity risk assessment is the first step to incorporate cybersecurity preparedness and responsiveness into the university's standard operating procedures. Further, this study potentially contributes to the literature of cybersecurity in the country and applications of the methodologies in institutions similar in nature.

2. Research Methodology

The research utilized the Input-Process-Output methodology, focusing on selecting an appropriate risk assessment framework and a baseline set of controls, apply an overlay based on a targeted assessment of threats unique to the university and conduct the risk assessment. ICT domain experts with the use of existing artifacts such as strategies, policies, best practices, inventory of assets, and other various sources will address key risk assessments using the selected framework. Further, GAP Analysis will be performed with the output of the assessments in each area to accumulate deeper insight.

3. Related Literature

Frameworks exist for decision-makers as guidance documents detailing risk assessment processes and transforming them into actionable mitigation strategies. To manage institutional risks, the applicability of primary risk-based standards, models, and frameworks are further discussed below.

Table 1. Cybersecurity Frameworks

Framework	Details
Framework for Improving Critical Infrastructure Cybersecurity (Cybersecurity Framework led by NIST)	Focusing on using business drivers to guide cybersecurity activities and considering cybersecurity risks as part of the organization's risk management processes. NIST consists of three parts: a) the Framework Core is a set of cybersecurity activities, outcomes, and informative references that are common across sectors and critical infrastructure; b) elements of the Core provide detailed guidance for developing individual organizational Profiles; and c) through the use of Profiles, the Framework will help an organization to align and prioritize its cybersecurity activities with its business/mission requirements, risk tolerances, and resources. The Tiers provide a mechanism for organizations to view and understand the characteristics of their approach to managing cybersecurity risk, which will help in prioritizing and

	achieving cybersecurity objectives[18].
Control Objectives for Information and Related Technologies (COBIT)	Created by Information Systems Audit and Control Association (ISACA), COBIT 5 includes an add-on related to information security and assurance and is based on the following five key principles for governance and management of enterprise IT: a) meeting stakeholder needs to align business and IT goals; b) end-to-end coverage of the enterprise that covers all functions and processes, both internal and external; c) a single, integrated framework that integrates various other frameworks and can serve as an overarching framework; d) a holistic approach that accounts for organizational processes, culture, structures, etc.; and e) Separate governance and management to de-conflict roles and responsibilities.[19]
ISO 27000	The ISO/IEC 27000 family of standards will help organizations manage the security of assets such as financial information, intellectual property, employee details or information entrusted to an organization by the third parties[20]. ISO/IEC 27001 is the best-known for providing requirements for an information security management system (ISMS).
Decision Framework for Cyber Security Risk Assessment PRISM Approach	PRISM, which stands for Prioritize, Resource, Implement, Standardize and Monitor, is a cybersecurity evaluation model will help cyber decision-makers and agencies create tailored approaches to risk management and better communicate the impact of investments in security resources on reducing targeted cyber risks. This model is also used to explore gaps in current processes to steer an organization in the right direction to resolve risk management and cyber risks specific to an organizational strategy and its functions.[21]
CORAS (Construct a platform for Risk Analysis of Security Critical Systems) methodology	CORAS, a model-based method for security risk analysis that provides a customized graphical language for threat and risk modelling has been developed through both empirical investigations and a series of industrial field studies (projects financed by the Norwegian Research Council and the EU) is based on international standards for risk management[22]
OCTAVE (Operationally Critical Threat, Asset and Vulnerability Evaluation)	OCTAVE is a security framework for determining risk level and planning defenses against cyber assaults, primary focusing on: a) helping organizations minimize exposure to likely threats, determine the likely consequences of an attack and deal with attacks that succeed; and b) leveraging the experience and expertise of people within the organization[23]. The first step is to construct profiles of threats based on the relative risk that they pose. The process goes on to conduct a vulnerability assessment specific to the organization.
DREAD MS (Damage Potential, Reproducibility, Exploitability, Affected Users, Discoverability)	Being part of a system for risk-assessing computer security threats previously used at Microsoft, DREAD provides a mnemonic for risk rating security threats using five categories used as initials[24]. Initially, DREAD was proposed for threat modeling, but it was discovered that the ratings are not very consistent and are subject to debate. Although currently used by OpenStack and other corporations, DREAD was abandoned by its creators.
CRAMM (CCTA Risk Analysis and Management Method)	The CRAMM is a risk analysis method developed by the British government organization CCTA (Central Communication and Telecommunication Agency), now renamed the Office of Government Commerce (OGC). A tool having the same name supports the method: CRAMM. The CRAMM method is rather difficult to use without the CRAMM tool. The first releases of CRAMM (method and tool) were based on best practices of British government organizations. At present, CRAMM is the UK government's preferred risk analysis method, but CRAMM is also used in many countries outside the UK. CRAMM is especially appropriate for large organizations, like government bodies and industry[25].

NCSP 2022 compels government agencies to manage their risks by conducting three stages of risks assessment: identification, analysis, and evaluation. Guidelines for implementing risk management is to be formulated by the government through DICT. Also, the plan advises that security system must be compliant with the requirements relevant to IS0 27000. Further, NPC recommends the use of ISO/IEC 27002 control set as a minimum standard to assess any gap in an agency's control framework. As to the time of this writing, there is no reference to the existing literature as on what standard cybersecurity assessment framework to be utilized by SUCs in the country. For the preceding reasons, this study leveraged on PRISM. Also, sophisticated use, implementation, and applicability issues of the other models and required extensive knowledge by domain experts are needed to implement the other frameworks. Utilization of some of the frameworks are for compliance and certification, not to mention the other frameworks are already discontinued and out of use. Further, PRISM caters to complement the profile of government agencies' shortcomings, heavily focusing on prioritization and strategic decisions, reactive and ad hoc approaches, the inadequacy of resources, lack of standard practice, and monitoring capabilities. Through the PRISM decision model, an agency leader continuously Prioritizes, evaluates and allocates adequate Resources, Implements, Standardizes, and Monitors an agencies' cybersecurity posture, preparedness, and responsiveness. The model aligns with the NIST Cybersecurity Framework[26].

4. PRISM

When implementing PRISM, domain experts of the university will follow the following multi-step process: 1) Assess the likelihood of an attack and the impact of damage across the enterprise and prioritize major risk areas and attack vectors; 2) Annually review individual cyber factors to determine the agency's current stage of preparedness and responsiveness to risk; 3) Examine and rate the resource allocations to reducing cybersecurity risk; 4) For each risk area, assign a risk score and a preparedness level; and 5) Recommend improvements for each component of the PRISM model, standardize solutions, and monitor for compliance.[26] PRISM also includes scorecards to aid agencies to prioritize and identify the precise risk vectors, leverage existing standards to resource and implement a risk-mitigation strategy[26]. The ICT domain experts composed by the web administrator, network administrator, data protection officer, compliance officers for privacy, cybersecurity personnel, head of the Management Information Systems, and other key personnel addressed the risk assessments along: Area 1 & 2 (Components and Governance), Area 3 & 4 (Data and Information), Area 5 (Applications and Tools), Area 6 (Networks), Area 7 (Storage Devices) and Area 8 & 9 (Operational Products and Services).

5. Results and Discussion

The risk assessment reveals the results of each area in Table III. Specific domains in each area are prioritized and allotted resources due to the University's 6th Governance Initiative Towards Quality and Excellence: Modernization of facilities with an emphasis on ICT. Currently, main efforts are being redirected to the procurement of an enterprise enrolment system. All areas have "Reactive" Risk Management activities or protocols in place such as the basic implementation of physical and electronic security controls to mitigate risks, standard open-source/freeware firewall, anti-virus, and other security software is installed on all computing devices, and securing the network and Wi-FI. Also, appropriate access controls are set for employees. For Incident Response and Reporting, cybersecurity issues are handled by DICT CERT and data privacy issues are forwarded to NPC.

Table 2. Risk drivers analysis

			P	R	I			S	M		
Area	Risk Factors	Risk Factor relevant to Cybersecurity	Prioritization	Resources	Stage 1 "Reactive" Risk Management	Stage 2 "Proactive" Risk Management	Stage 3 "Enterprise" Risk Management	Knowledge & Solutions institutionalized	Constantly Assess Unusual Behavior	Risk Rating Level	Preparedness Level
Summary Risk Driver Analysis:											
1	Internal/External	Adequacy of Actions/Activities	7/0	4/3	4/3	4/3	0/7	3/4	3/4	3	F
2			11/2	9/4	13/0	2/11	0/13	6/7	5/8	3	F
3			4/2	5/1	6/0	1/5	0/6	4/2	1/5	2	P
4			4/2	5/1	6/0	1/5	0/6	4/2	2/4	2	P
5			5/4	6/3	6/3	1/8	0/9	0/9	0/9	2	P
6			4/1	3/2	5/0	0/5	1/4	1/4	4/1	3	P
7			4/4	2/6	1/7	1/7	0/8	0/8	4/4	1	P
8			2/5	0/7	7/0	0/7	0/7	0/7	0/7	1	P
9			8/0	6/2	8/0	5/3	0/8	4/4	8/0	3	F
Yes Subtotal:			49	40	56	15	1	22	27		
No Subtotal:			20	29	13	54	68	47	42		

LEGEND: risk rating level and preparedness level. 1=Poor, 2=Poor, 3=Fair, 4=Fair, 5= Good, 6=Very good, 7=Excellent

IOP Conf. Series: Materials Science and Engineering **769** (2020) 012066 doi:10.1088/1757-899X/769/1/012066

Most areas do not have knowledge and solutions institutionalized. Existing ICT organizational structures in the university is focused on maintaining information systems, network, and tech support. The Data Privacy Act of 2012 led to the Board Resolution 165 s2018– the approval of the Privacy Manual and the creation of Data Security and Technology Standards Division last December 19, 2018 with provisions of filling up the positions of Data Protection Officer, Compliance Officers for administrative and academic privacy, Legal Adviser, and Cybersecurity Threats and Risk Officers. The inclusion of cybersecurity personnel was influenced by Section IV A of M.O. 6, s2017 by the DICT-establishment of Government Computer Emergency Response Team (GCERT) in government agencies. However, the three cybersecurity personnel positions are not filled up in the current time being as there were restrictions in hiring. The lack of ICT plantilla positions not only contribute to the high turnover rate of ICT personnel but also impedes on the recruit and retaining of certified IT, IT risk, and information security talent. Regular assessment and review of policies, internal audit, certifications were not yet conducted because the oversight organization is still new. Also, the Data Breach Response Team is yet to be established.

Stage 2 "Pro-active" Risk management has 15 "Yes" across the areas. Employee access to data and information and authority to install software are limited. Regular update of all applications is not conducted by all employees. Data privacy and/or cybersecurity awareness campaign, seminars, training and workshops were provided for employees focusing on identify potential reputation risks, recognizing social engineering, protect against online fraud, protecting against malware, safe web browsing, strong password policies, scams, and fraud, verifying the identify of telephone information seekers, and minimize and safeguard printed materials with sensitive information. A layered approach to guard against malicious software is not practiced. To pass a policy, said policy must be presented to the Administrative Council and approved by the Board of Regents. Policies and procedures for every stage of the data life cycle (use, records retention policy, disposal) is already in effect while policy development and management is yet to focus on security roles and responsibilities, network security, website security, email, mobile devices, facility security, payment cards, email usage policy, social media policy, employee Internet usage policy, and design cybersecurity policies, training, and testing. A collective view of the Project Procurement Management Plan (PPMP) of offices reveals that procurement is focused on office supplies, software, network infrastructure, hardware, and peripherals. There was no indication of procurement for security essentials. Prioritization of investments and contrasting organizational functions impede on the protection of information, computers, and networks from cyber-attacks. The university's procurement procedures are governed by the Republic Act No. 9184, otherwise known as the Government Procurement Reform Act. Competitive bidding could impact the quality of procured goods, infrastructure projects, and consulting services concerning cybersecurity investments.

Most vectors do not assess unusual behavior. Industry practices such as the following are not implemented: monitoring security breaches, security features of software and applications used, regularly testing and evaluation of the effectiveness of security measures, and encryption, authentication, monitor the network and review logs, implementing access controls that only permit required administrative or management services (SNMP, Network Time Protocol, Secure Shell, FTP, Trivial File Transfer Protocol, RDP, SMB), the use of Enterprise Firewall, VPN, Service Auditing, File Auditing, and Intrusion Detection Systems, VLAN, and utilize Microsoft Baseline Security Analyzer (MBSA). Detecting, authenticating, and inventory of unknown devices is not practiced on an institutional level and data protection and loss prevention programs with related monitoring are not fully implemented.

At least half of the areas (Data, Information, Application, Tools, Networks, and Operational Products) have "Poor" preparedness level, the rest have "Fair" preparedness level. The result is due to the lack of segregate standard network traffic from management traffic, Endpoint Security, security engineering, vulnerability assessment, network intrusion detection system (NIDS), Password "salt and peppering", authorization, authentication, and two-factor authentication, Virtual Private Networks, Application Whistling, securing internal network and cloud services, Enterprise Firewalls, Anti-virus

IOP Conf. Series: Materials Science and Engineering **769** (2020) 012066 doi:10.1088/1757-899X/769/1/012066

software, and encryption. Also, hardening network management devices by testing patches, turning off unnecessary services on routers and switches, and enforcing strong password policies is not practiced. Best practices are followed by ICT personnel. However, these are not published for institutional consumption. The absence of Operations Security (OPSEC) led to several incidents of infiltration and infection of WannaCry Ransomware on office computers and web defacement. In a commercial context, OPSEC is the process of denying hackers access to any information about the capabilities or intentions of a business by identifying, controlling and protecting evidence of the planning and execution of activities that are essential for the success of operations. This aids in minimizing the effects of a cybersecurity incident while quickly restoring operations and security. Physical security measures such as access procedures of agency personnel and visitors, design of office space/workstation, guarding and storage of ICT equipment, securing the area of work, and protecting output are hardly observed.

Areas 7 (Storage Devices) and 8 (Operational Products) garnered the lowest risk rating level of 1. Making backup copies of essential business data and information is a constant practice. However, there are no policies relating to storage devices and products that are approved and published other than the Privacy Manual. The Bring Your Own Device (BYOD) policy is being drafted as of the current time frame. Also, protocol and product standards for cybersecurity technology and operational guidelines for the selection, implementation, and management of cybersecurity technologies was inexistent. Further, reporting procedures for lost or stolen equipment and ensuring all devices are wiped clean before disposal is also non-existent. However, suppliers enforce integrity checks of the supply chain to validate hardware and software authenticity.

99% of the areas do not have Stage 3 "Enterprise" Risk Management. The assessment revealed the lack of presence of use enterprise solutions such as authentication and cryptographic technologies as appropriate to protect certain types of sensitive data where feasible, vulnerability scanning, security scanning, penetration testing, risk assessment, security auditing, posture auditing, SSH, Enterprise Firewall, VPN, Public Key Infrastructure, SSl/TLS Encryption, Service Auditing, Intrusion Detection Systems, Enterprise Vulnerability assessment and ethical hacking. Also, little to partial implementation of R.A. 8792, R.A. 9995, R.A. 10173, R.A. 10175, M.O. 37 s2001, E.O. 810 s2009, A.O. 39 s2013, PNS ISO/IEC 270001:2005, and PNS ISO/IEC 27002:2005, FOI Act, Executive Order No. 608, Series of 2007, and NAP Guidelines General Records Disposition Schedule Common to All Government Agencies Series 2009 impacts enterprise risk management. Expenses for the compliance or implementation of some of the laws are required to be included in the annual budget of the university.

To have a partial atomistic view of the deficiencies and what else has to be accomplished, the treatment of GAP Analysis on the different risk areas further aids in ensuring the desired direction of the University in terms of cybersecurity, as observed in Table 3 for Networks. Area 6 yielded the score of 2 as the lowest risk rating level and 4 as the highest. Resource allocation for networks is evident as the deployment of campus network infrastructure nears halfway to completion. Also, a decent internet connection is provided to the different units. However, the campus backbone was prioritized, therefore, network security is limited to open source applications. Monitoring of unusual behavior and lack of standardization of knowledge and solutions seems to manifest on all risk areas. All risk areas possess the same deficiencies. Codependent components such as enterprise security solutions, hardware, and software are proposed to resolve problems in all risk areas. Further, hardening the security should involve audits and assessments, implementing policies, and observing the best networking practices.

Table 3. Area 6: Networks

Risk Factors	Current Status	Deficiency	Action Plan
Internal Structure	Basic physical security measures (e.g. proper network cabling) to protect internal infrastructure; Known topologies are maximized; Risk Rating Level of 2, which translates to Poor preparedness level; Risk area is not a priority; Unusual behavior is not constantly assessed; Knowledge and solutions are not institutionalized.	Observing network security practices are not evident such as: monitoring physical and electronic inventory of network and other devices, disabling unencrypted remote admin protocols used to manage network infrastructure (e.g., Telnet, File Transfer Protocol [FTP]), disabling unnecessary services (e.g., discovery protocols, source routing, Hypertext Transfer Protocol, Simple Network Management Protocol [SNMP], Bootstrap Protocol), using SNMPv3 (or subsequent version), Secure access to the console, auxiliary, and virtual terminal lines, implementing robust password policies, using the strongest password encryption available, protecting routers and switches by controlling access lists for remote administration, restricting physical access to routers and switches, periodically testing security configurations against security requirements, and protecting configuration files with encryption or access controls when sending, storing, and backing up files.	To achieve preparedness level of "Very Good", risk areas must be installed or preconfigured with moderate-to-advanced security tools to combat viruses/malware, purchase licensed Enterprise solutions to harden network security against viruses/malware such as endpoint security, security engineering, network intrusion detection system (NIDS), network intrusion protection system (NIPS), application whistling, firewalls, and anti-virus software. Also, administrators must Segment and segregate networks and functions, monitor, detect, prevent, and correct incoming/outgoing data flow and bandwidth speed, track data usage, segregate standard network traffic from management traffic, implement acceptable use policy for internet, perform vulnerability assessment and penetration testing, conduct regular security audits and assessments, utilize virtual private networks (VPNs) to securely extend a hosts or the network, apply encryption to all management channels, encrypt all remote access to infrastructure devices, manage all administrative functions from a dedicated, fully patched host over a secure channel and implement patch and update management policy. Harden network management devices by testing patches, turning off unnecessary services on routers and switches, enforce strong password policies, monitor the network and review logs. Implement access controls that only permit required administrative or management services. Also, the procurement committee must maintain strict control of the supply chain, require resellers to enforce integrity checks to validate hardware and software authenticity.
External Infrastructure	Basic security measures are in place; Principles of least privilege and a need-to-know basis for network details implemented; Risk Rating Level of 3, which translates to Fair preparedness level; Knowledge and solutions are not institutionalized; Resources are not allocated for the risk area.		
Connectivity	Utilization of DHCP connectivity; Network configured to adequately manage and protect network traffic flow; Risk Rating Level of 4, which translates to Fair preparedness level; Knowledge and solutions are not institutionalized.	No conduct of network vulnerability testing of basic network attacks such as DoS, packet sniffing, phishing, and DNS spoofing; Bandwidth is monitored, but not controlled.	
Communications	Network communication is adequately managed; Risk Rating Level of 3, which translates to Fair preparedness level; Knowledge and solutions are not institutionalized; Resources are not allocated for the risk area.	Limitation and control of network ports, protocols, and services; Encryption to all management channels is not utilized	
Architecture	Resource allocated for the backbone of campus network; Almost half of the campus is connected locally; Risk Rating Level of 4, which translates to Fair preparedness level;	Private virtual LANs is not utilized/practiced to isolate a user from the rest of the broadcast domains; Virtual routing and forwarding (VRF) technology to segment network traffic is not utilized.	

6. Conclusion and Future Work

To overcome existing and potential cybersecurity incidents, the University must conduct a cybersecurity risk assessment. As the results show, PRISM is a tool that coincides and highlights the university's weaknesses and shortcomings, revealing the areas to be top priorities for risk mitigation. The assessment also revealed the university's strengths – most areas seemed to be prioritized, most areas have resources allocated, and most areas do have existing reactive risk management. It is crucial to implement appropriate security management practices and controls on all the areas assessed. Also, the university must ensure appropriate steps to improve on the pro-active risk management, enterprise risk management, the institutionalization of knowledge and solutions, and the constant assessment for unusual behavior on all risk areas. This paper concludes that risk assessment is just the first step in the path of being a cyber resilience institution, and it is a necessity. GAP Analysis alongside PRISM can be used to ensure the desired performance and the direction of the University. The results of PRISM assessment and GAP Analysis can be presented during the strategic planning of the University annually to enforce solutions to address the deficiencies identified. Both results can impact the procurement procedure by identification and prioritization of crucial assets. Further, the insights could influence policymaking along ICT, data privacy, and cybersecurity. For now, the groundwork has already been laid. Forward-looking, the results will serve as inputs for a succeeding larger study, focusing on creating a cybersecurity framework for SUCs - a tailor-made framework with proactive strategies, policies, and procedures guided by national cybersecurity strategy - a constituted effort to traverse the ever-evolving landscape of the cybersecurity.

References

[1] R. A. Kemmerer, "Cybersecurity," Proceedings of the 25th international conference on Software engineering. IEEE Computer Society, 2003.
[2] J. A. Lewis, "Cybersecurity and critical infrastructure protection," Center for Strategic and International Studies, 2006.
[3] E. Amoroso, "Cyber Security," New Jersey: Silicon Press, 2006.
[4] ITU, "Overview of Cybersecurity," Recommendation ITU-T X.1205. Geneva: International Telecommunication Union (ITU), 2009.
[5] DICT, NATIONAL Cybesecurity Plan 2022, 2019.
[6] NSC, National Security Policy
[7] LawPhil, Republic Act No. 98792
[8] LawPhil, Anti-Child Pornography Act of 2009 (R.A. 9775)
[9] DICT, Anti-Photo and Video Voyeurism Act
[10] NPC, Data Privacy Act of 2012
[11] LawPhil, Cybercrime Prevention Act of 2012
[12] DICT, Inventory of Published Policies
[13] DICT, Cybersecurity, 2019.
[14] Official Gazette of the Republic of the Philippines, Executive Order No. 810, s. 2009
[15] iGovPhil Program, Government Web Hosting Service (GWHS)
[16] DICT, Draft AO on Social Media Use
[17] Z. A. Collier, D. DiMase, S. Walters, M. M. Tehranipoor, J. H. Lambert, and I. Linkov, "Cybersecurity standards: Managing risk and creating resilience," Computer, 47(9), 70-76, 2014.
[18] Cybersecurity, Critical Infrastructure. "Framework for Improving Critical Infrastructure Cybersecurity." Framework 1: 11, 2014.
[19] ISACA, COBIT 5: A Business Framework for the Governance and Management of Enterprise IT, 2012.
[20] ITGOVERNANCE, The ISO 27000 family of information security standards, 2017.
[21] S. Friedman, "A new model for cyber risk management", 2018.
[22] SourceForge, The CORAS Tool - The CORAS Method, 2012.
[23] C. J. Alberts, A. J. Dorofee, J. F. Stevens, and C. Woody, "Introduction to the OCTAVE Approach," 2003.
[24] OPENSTACK, OSSA
[25] Z. Yazar, "A Qualitative Risk Analysis and Management Tool – CRAMM," 2002.
[26] R. Goel, J. Haddow, and A. Kumar, "Managing Cybersecurity Risk in Government: An Implementation Model," 2018.

An open cloud-based platform for the creation and delivery of smart applications and services

Heba M. Fadhil[1], Mohamed I. Younis[2*], AbdulRahman A. Alsewari[3], Mohammed Abd-Alrazzaq[1], Ameer Toran Mahdi[1], Alaa A. Alomoush[2]

[1]Information and Communication Department, University of Baghdad, Baghdad, Iraq

[2]Computer Engineering Department, University of Baghdad, Baghdad, Iraq

[3]Faculty of Computing, Universiti Malaysia Pahang,Pahang, Malaysia

*younismi@coeng.uobaghdad.edu.iq

Abstract. With the increasing integration of computers and smartphones into our daily lives, in addition to the numerous benefits it offers over traditional paper-based methods of conducting affairs, it has become necessary to incorporate one of the most essential facilities into this integration; namely: colleges. The traditional approach for conducting affairs in colleges is mostly paper-based, which only increases time and workload and is relatively decentralized. This project provides educational and management services for the university environment, targeting the staff, the student body, and the lecturers, on two of the most used platforms: smartphones and reliable web applications by cloud hosting. The services include project management, attendance marking, various notifications and alerts, files and resources, a grading system, and an assignment management system. This paper also aims to ultimately digitalize most of the information inside the college which is a great challenge. The proposed system defines five actors, namely: an admin; who is the person with the highest authority within the system, the head of the department, the department coordinator, the project coordinator, the lecturers, and the students. Each of those actors has its privileges and the interaction among participants via the cloud. The mobile platform targets the students exclusively and interacts with the web platform. Finally, this paper stresses the benefits of the proposed system and gives the avenues for future development.

Keywords. Smart Campus, Web Portal, Mobile Application, Cloud, IoT.

1. Introduction

Ever since computers and the World Wide Web have been introduced into organizations, they have offered exponential growth and realized possibilities which were thought to be extremely difficult before. Management systems have especially benefited from this. The traditional paper-based methods consume too much time; reports need to be drafted and delivered by hand, meetings need everyone's physical presence, information and news need to be distributed manually, and simple operations on data need much time to go through all of that paperwork [1].

Online management systems provide an environment in which all of the above-mentioned work is reduced to simple operations, and make some of the operations completely automated, with the bonus of that they can be done anywhere as long as the person is connected to the network [2].

Colleges are organizations which can especially benefit from these advantages, especially in Iraq. The amount of paperwork and waiting involved in doing even the simplest of tasks is immense. Moreover, there is relatively poor communication between the faculty and the students which suffer greatly from human error. Furthermore, the existing members are reluctant to adapt to more technologically advanced methods, and still prefer the traditional one, partly because most existing systems are never implemented completely and fail to address the needs of the faculties and students directly, or are too complicated for them to use [3].

Various College Management Systems have existed since the '90s; each providing a customized solution for the college it is used in. In existing systems, the majority of the college, maintaining the records and other information is the annual process. Taking an existing system keen on deliberation, concise on the necessities they anticipate and so on. All of these require more work and more time [4]. The project focuses mainly on colleges and the synchronization of all the sparse and diverse information on the regular college schedule. Usually, students face problems getting the correct notifications at the right time, sometimes important notifications such as campus interviews, training and placement events, holidays and special announcements. Smart Campus attempts to mind the gap between students, teachers, and college administrators. Therefore, in the environment, like a college campus, the information (e.g., oral communication, notices, files) can be reached for communication and made available for the participants through the android devices. The use of architectural Model-View-Controller (MVC) makes both the development maintenance and corrective maintenance easier because of the separation of concerns in the development of an application (e.g., data management, GUI) from and web service which requires fast and efficient maintenance of the application.

Implementing a cloud-based smart campus system, which is tailored for the needs of the currently existing faculty and students, is a major step towards solving the above-mentioned issues and guaranteeing a future that is committed to technology and its benefits.

The objective of this paper is to provide a central database for all of the information within the campus, and provide a fast and reliable communication method between the student body and the faculty using smartphones, both Android and IOS, and the web portal, with the automation of some of the tasks within the campus.

2. Literature Survey

The smart campus is an emerging field of study which attracts researcher. Many related works try to design and implement smart campus system from different points of view such as infrastructure, services, and usability.

Moreno et al [5] provided a series of services implemented on intelligent campus for energy efficiency. Services implemented include internal settlement assessment, energy consumption forecasting, convenience, and energy conservation. Arshad et al [6] designed a mixed label system to label the contents and devices of the campus environment to become smart internet-based objects. In a plan developed, features contain scalability and security for campus contents and Internet objects. Xiang et al [7] proposed a campus information dissemination target to university. A model of intelligent campus been proposed by Atif and Mathew [8], which integrate learning resources in the real world within a social network on the campus level. Liu et al [9] introduced a smart digital campus combining both internet of things and cloud computing technologies and illustrated the complications that can be detected after the creation of intelligence application platforms. Younis et al [10] proposed a secure graduation certificates issuing system (GCIS). They attempt to provide the graduation certificates in a secure server based on digital signature. However, the proposed system is used inside the examination committee of a department and do not scale with college and university. They stress the need of reliable web portal through a cloud hosting.

3. College Software

The college software is a cloud-based and an enterprise resource planning (ERP) software which provides an online solution for computerization of administrative activities of a college campus. This software handles academic and non-academic activities. The education landscape is facing continuously changes for academic and non-academic sectors as a result of digitalization era. To facilitate this, institutions must shift the gear toward the state-of-the-art innovations and technologies. The software enables colleges to modernize the education process easier and provide a unified framework where learning and teaching can thrive.

Cloud-based college ERP provides a total online solution for computerization of all administrative activities, which includes admission, student administration, attendance, grading, etc. This software has many modules to execute multiple operations on the college campus. Each module can be tailored according to the requirements of the college.
Following are the benefits of college management system software:

- reduces manual workload
- increased productivity
- effective communication between teachers & students
- reliable backup & recovery system
- minimal data redundancy
- cost-effective

The college management software is a vital requirement to help participants to reach effective management for complex operations. While implementing an ERP it's very important that it should fulfill institute requirement and generate expected results. Implement a well suitable, flexible and innovative education ERP. The system proposed in this project explores the following tasks:

1. Project management: This includes the process of assigning teams to the projects and managing the project tasks until completion.
2. Attendance marking: This includes marking the student's attendance performance and sending alerts when required.
3. News and notifications: This includes the college newsletter, mentioning important events within the campus and sending notifications when needed.
4. Files and resources: This includes the various textbooks and study materials that the students need.
5. Grading system: This includes grading the students until the end of the course and delivering their grades on time.
6. Assignment management system: This includes placing assignments for the students and getting their answers for grading.

3.1. MVC

The MVC is an architectural pattern that provides the separation of concerns principals in developing a GUI application. In MVC design an application consists of three components: the model, the view, and the controller as seen in figure.1 MVC is an industrial-standard framework that enables the developers to provide extensible and expandable products [11].

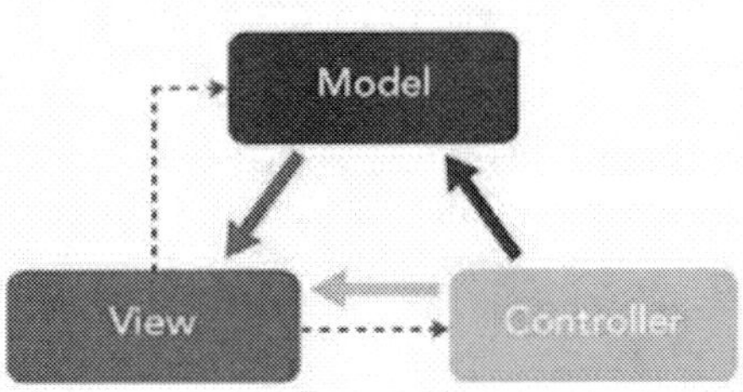

Figure 1. MVC components interaction.

3.1.1. Model
The Model represents the logical way that the user interacts with the data. The Model is responsible for the data movement between the View and Controller components. For instance, a Customer object retrieves information from a table in a database, manipulates or renders it for displaying or updates the records.

3.1.2. Controller
Controllers process the interaction logic between the Model and the View components. For instance, a Customer controller takes the inputs from the Customer View and updates the database through the Customer model. Alternatively, the controller is re-used to view the Customer data. As such, the controller provides two-way interfaces.

3.1.3. View
The View component includes all UI components that are physically the end-user interacts with them. Examples include radio-buttons, drop-down menus, text areas, etc.

3.2. Technologies and Frameworks

The project's web portal portion was developed using Laravel for its backend and Vuejs for its frontend. The mobile application portion was developed using React Native. Some features required a Nodejs server.

3.2.1. React Native
React Native is an open-source tool released by Facebook community in 2015 [12]. The React Native is widely used by the developers for developing mobile applications target to iOS and Android. It uses the same declarative approach to constructing user interfaces as React for the web. React Native also aims to reduce many native development inefficiencies. Rather than deal with the slow process of write, compile, deploy and debug, which can cripple development on larger native apps, React Native allows you to simply refresh the app after making a change without the slow compile and deploy steps, just like on the Web! This makes for a much-improved developer experience and, unlike normal native development, React Native allows you to share far more code across platforms. However, Facebook points out that React Native is not intended to be a write once, run anywhere solution. They acknowledge that each platform has its own look, feel, and capabilities. Instead, React Native allows you to leverage common technologies to build across multiple platforms. They call this learn once, write anywhere [12].

A React Native application runs the low-level mobile platform code through a JavaScript "bridge" layer to interact with the phone for accessing all its functionality and presenting a UI. There is a JavaScript engine interface that exists on both iOS and Android. This is a low-level layer that then interacts with the OS for the phone to do the same things that other platform languages on the phones do.

3.2.2. Vuejs
Vue is a popular library with rich features that has easy adoption with official extension libraries. The core library of Vue (e.g., React) manipulates the view layer. On the other hand, the extension libraries like Vuex and Router, responsible for data management and routing respectively [13] [14] [15]. The key design principles of Vue are as follows:

The key design principles of Vue are as follows:

- Focus: Vue has opted for a small, focused API, and its sole purpose is the creation of UIs
- Simplicity: Vue's syntax is terse and easy to follow

IOP Conf. Series: Materials Science and Engineering **769** (2020) 012067 doi:10.1088/1757-899X/769/1/012067

- Compactness: The core library script is ~25 KB minified, making it smaller than React and even jQuery
- Speed: Rendering benchmarks beat many of the main frameworks, including React
- Versatility: Vue works well for small jobs where you might normally use jQuery, but can scale up as a legitimate SPA solution.

3.2.3. Node.js

Node is a run-time environment for JavaScript with platform-independent and open-source functionalities. Node.js is widely used in programming command-line tools outside a browser, unifying both the client-side as well as server-side into the industrial standard JavaScipt programming. Besides, it is used for deploying networked services. JavaScipt is dominant the web programming, when combined with the Node's simplicity, facilitate the integration of the on-the-shelf third-party modules [14].

3.2.4. Laravel

Laravel is a PHP open-source framework based on MVC. It is used for building robust web applications with elegant syntax and rich features. The current version of Laravel is 5.5. Examples of famous web applications and services written by Laravel involve, but not limited, customer portal, image cropper, social network, and RESTful APIs [13] [16].

3.2.5. Socket.io

Socket.IO is a JavaScript package for real-time enables the developers to write bi-directional streams to handle the communication between client-sides and server-sides [17].

3.2.6. SQL

A database is a set of structural tables that can be accessed or updated without knowing the details of the tables, by using industrial standard application programming interface (API). Structured Query Language (SQL) is dominant query language in a relational database development [18]. SQL statements enable the developers to write quires for gathering data and reports generation.

3.3. Design Methodology

Throwaway Prototyping Model is a powerful framework to identify the requirements for a project when the needs are poor and ambiguous. Prototyping is considered as proof by doing. The end-user inters a loop of satisfaction to complete the project (i.e., Close-Ended Prototyping as shown in figure 2). Thus, Throwaway Prototyping Model methodology is adopted for the following reasons:

1. Users participate in the development cycle.
2. The users get experience about the system under the development.
3. Bugs can be detected at the early stages.
4. Faster feedback from users to assist in better developments.
5. Missing features will be identified smoothly.
6. Confusing can be resolved
7. Difficult functions can be refined through the step-wise-refinement process.
8. Provides fast implementation for an incomplete working application.
9. Simplify Verification and Validation (V&V) for software testing.

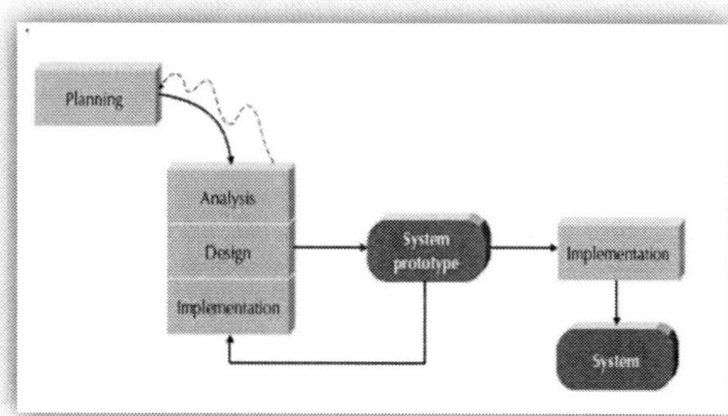

Figure 2. Throwaway prototyping methodology.

4. Implementation and Results

The following subsection discuss the system design, platforms, services and describe the mobile applications.

4.1 System Design

In the proposed system, the students can access both the web portal and the mobile application. The Admin, as well as other faculties, can access the web portal only. The participants need to log in to use the applications. The applications have privileged to access the database. The database is implemented in the cloud and shared through the internet as shown in figure.3 Various user privileges' types are enlightened as seen in figure 4.

1. **Student:** After the students' login, they can access various services and applications like: voting, read the notice, read the blogs published by the staff, read the manuals, etc.
2. **Lecturer**: After the lecturers' login, they can write the post as well as notes on the blog, view blogs, making an election for voting, upload grades, and absents of students, resources, and other services.
3. **Department head:** the department head manages all department options; he can add courses, subjects and assign them to lecturers, add students and manage their accounts.
4. **Admin:** The admin can manage all departments of the college in every detail. In addition, he/she has permission to add the department head and all departments' accounts and control their management.
5. **Projects committee:** it a group of lecturers has permissions to assign projects to students after they get votes of the project from students. In addition, the department head can create these groups.

4.2 System Platforms

- **Mobile Application**: Only students use mobile applications. There are many applications for student involves projects tasks, assignments, resources, attendance, and grades.
- **Web Portal**: All actors can access to the web portal. The services inside the web portal are depending on the privilege of the actors as shown in figure 5.

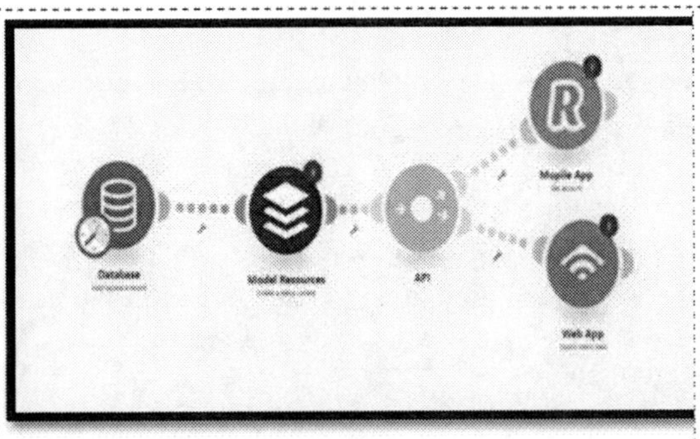

Figure 3. System architecture with modal resources of mobile and web application.

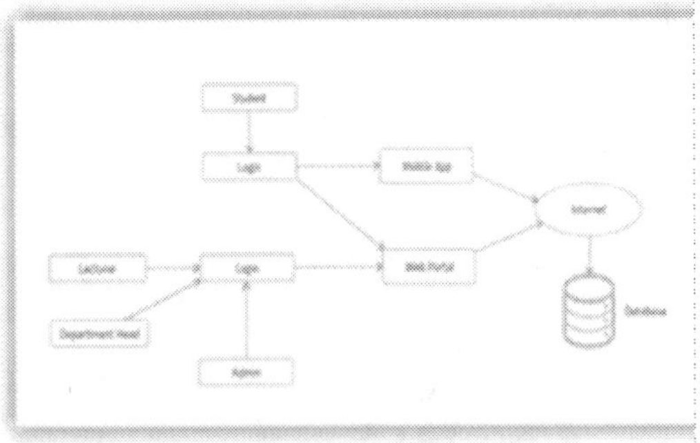

Figure 4. System access architecture.

4.3 System Services

4.3.1 Projects Management

This module manages projects of all stages and departments of the college, so department head can upload the projects and assign lecturers for them firstly, then students can vote on projects with a partner for each one. After that, the projects committee can give students their projects based on their votes as shown in figure 6. So each project has its supervisors and student, supervisors can upload tasks of projects to the student, when students complete their tasks return feedback to the supervisor, see figure 7.

4.3.2 Grades

Lecturer can upload marks of exams and student get their marks directly, when all exams have completed, lecturer can calculate the midterm mark automatically by clicking on **"publish midterms"** button, an algorithm will run to collect all exams and calculate then midterm grade, after that the **Department head** get a notification to approve the midterms of the stage students as shown in figures 8 and 9.

4.3.3 Attendance

Lecturer uploads the hours of absents of his subject for each student, and then alerts will be calculated automatically depending on hours of the subject see figures 10 and 11.

4.3.4 Files and Resources

Lecturers and staff upload files on cloud and student can download them. The files include the lectures, sheets, and books of subjects as shown in figure 12.

4.3.5 Assignments

Students upload homework's and assignments as pdf or images for lecturers, each lecturer can see only his assignments as shown in figure 1).

4.3.6 Blogs and news:

The staff from lecturers and department heads even admins can post news or events. Admin can post for one department or all and one stage or all stages because he has full permissions for post, department head and lecturer can post for only his department to one stage or all stages, Students just see the posts for their departments, so they cannot see the posts for other departments as shown in figure 14.

IOP Conf. Series: Materials Science and Engineering **769** (2020) 012067 doi:10.1088/1757-899X/769/1/012067

4.4 Mobile Application

The cross-platform mobile application was developed using React Native and relies on Expo Notifications for notification services. It consists of a login page and 5 swappable tabs, as shown below in figure 15.

4.4.1 Login

The student provides his email and password to log in. There is also a link to the web portal where the user can register a new account. The users can choose whether if they want to stay logged-in the next time they use the app. In case the users forget their respective passwords, they are redirected to the web portal to acquire a new password via email see figure 16.

4.4.2 Login Main page

The main page is shown in figure 17 consists of a search function and a button to the notifications list in the header, and the five mentioned tabs.

4.4.3 Study

The study contains four basic services:

1. Grades: The users will receive their grades of a subject once they are uploaded to the web portal by the lecturer, along with a notification.
2. Assignments: Once an assignment is assigned to the students via the web portal, they will receive a notification. The application will show the description of the assignment along with the deadline. The students can choose to submit their answer from the application, which is only possible before the assigned deadline as shown in figure 18.
3. Project Tasks: The students will receive the tasks assigned by the project supervisor with its description, and have the option to submit a prompt when the task is completed as shown in figure 19.
4. Resources: The resources show the files (Textbooks, PDFs, Images) of the subjects of the course the student is currently enrolled in, which can be downloaded.

4.4.4 Profile

The profile tab shows the profile of the student including his information as shown in figure 20.

4.4.5 Bus Order

The bus order system connects to Socket.io running on a Node.js server. Each student can submit a single order when he or she wants and the order increments the total number of orders as shown in figure 21. The transportation department then can inform the students that a bus is on the way, resetting the number of orders and still keeping the students from ordering. After the bus makes a lap around the campus, the transportation department cancels, letting the students order again as shown in figure 22.

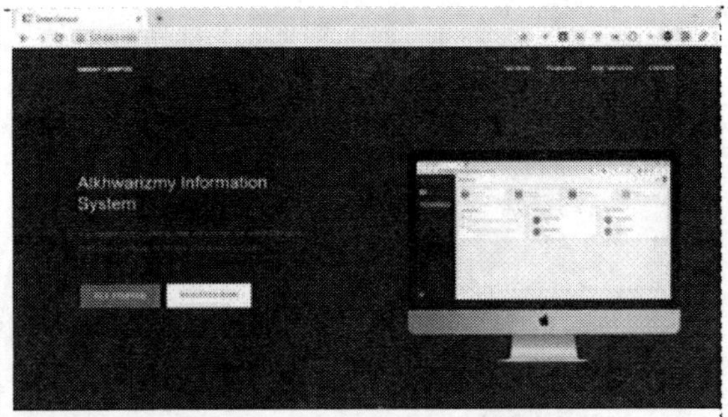

Figure 5. Web portal homepage.

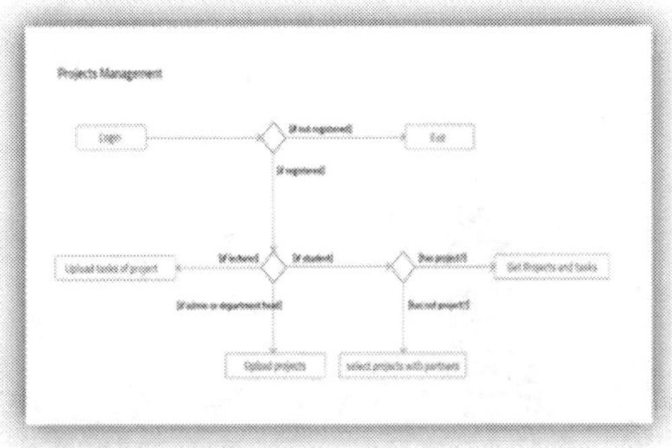

Figure 6. Projects management.

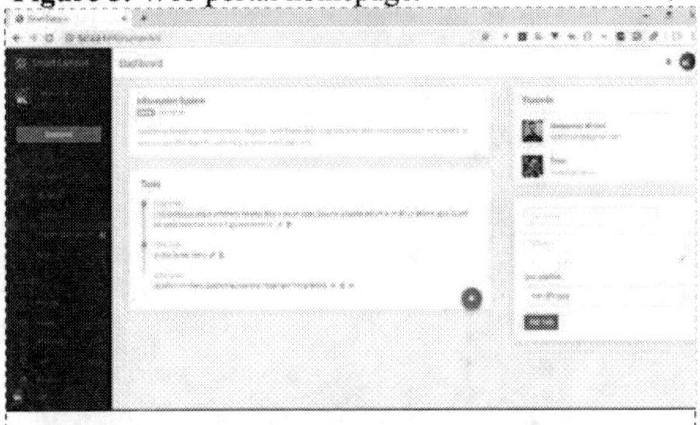

Figure 7. A project with tasks, each task has deadline.

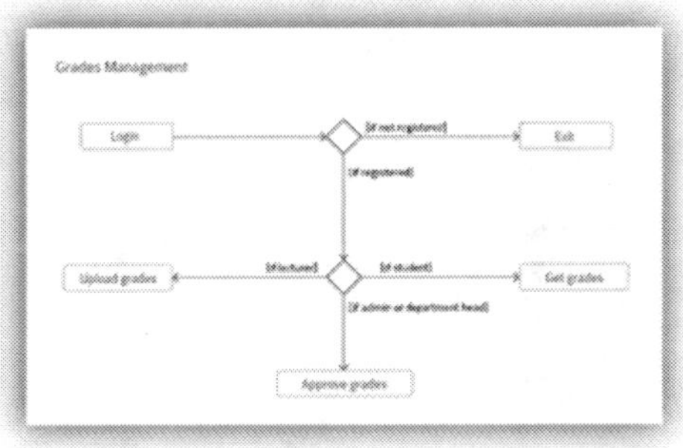

Figure 8. Grades management.

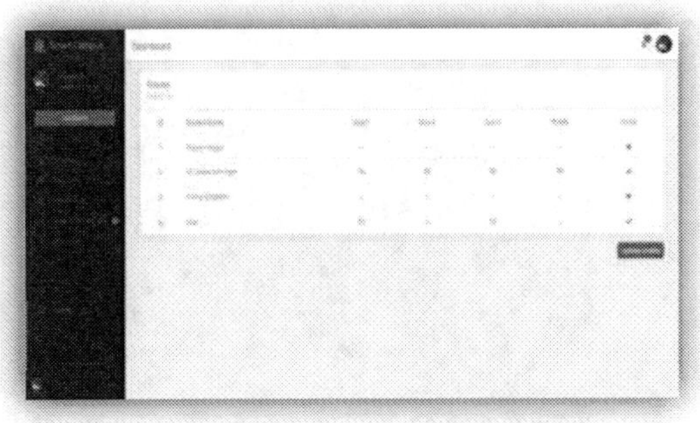

Figure 9. Grades of students with the approval of admin on it.

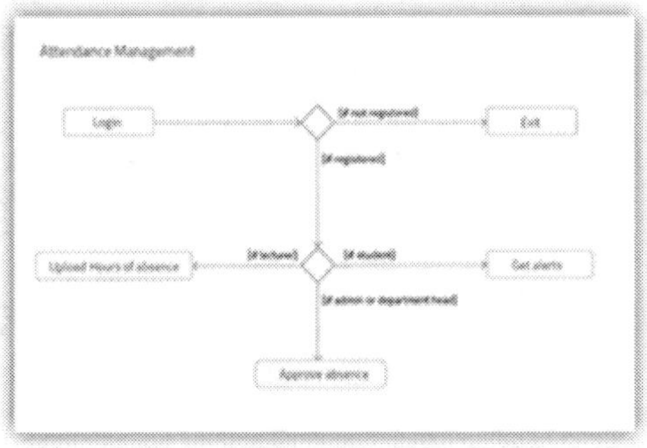

Figure 10. Attendance management.

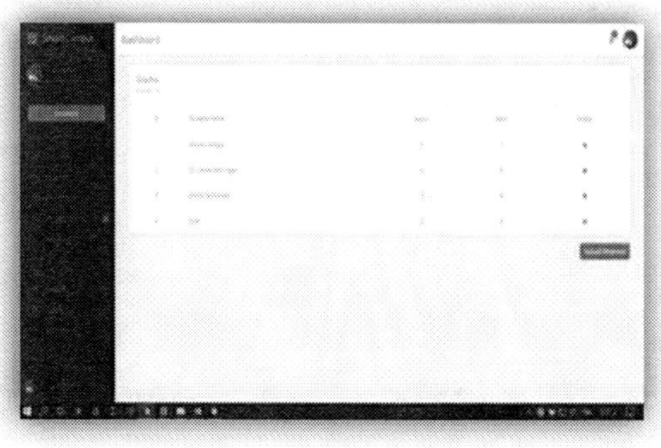

Figure 11. Attendance of students with the approval of admin on it.

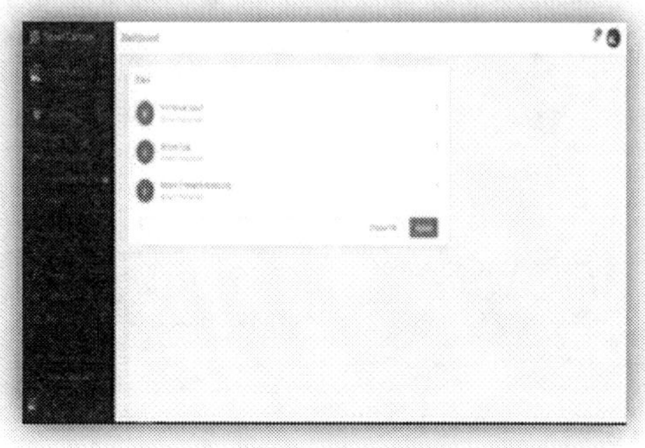

Figure 12. Files upload from lecturer to students.

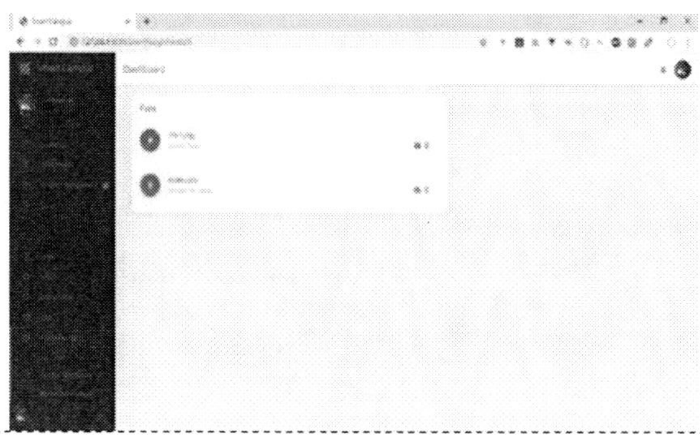

Figure 13. Assignments module.

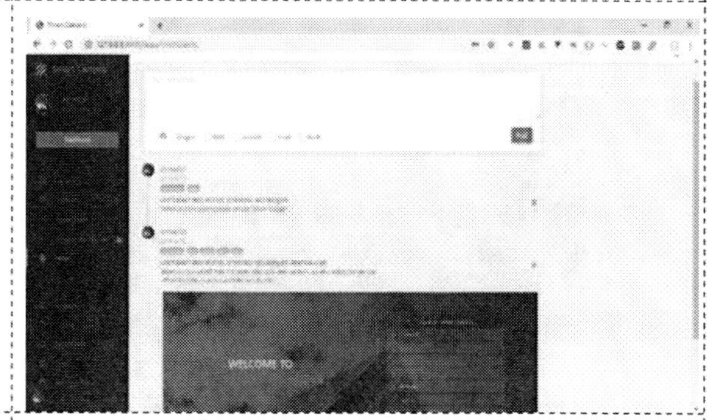

Figure 14. Blog and news module.

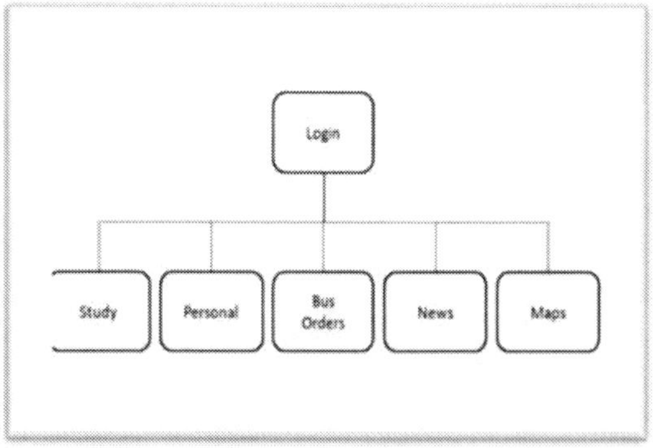

Figure 15. Mobile application pages.

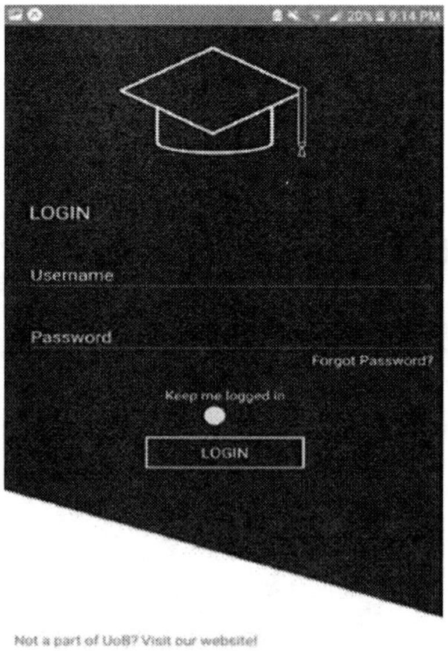

Figure 16. Login page.

Figure 17. The main page.

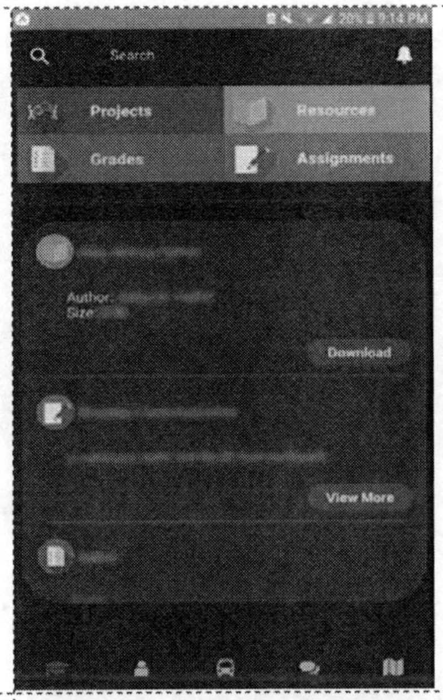

Figure 18. Uploading an assignment.

Figure 19. Project tasks.

Figure 20. Profile.

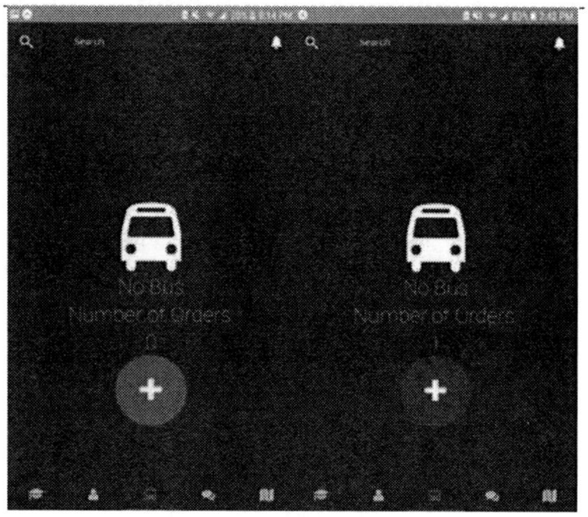

Figure 21. Before and after a student places a bus order.

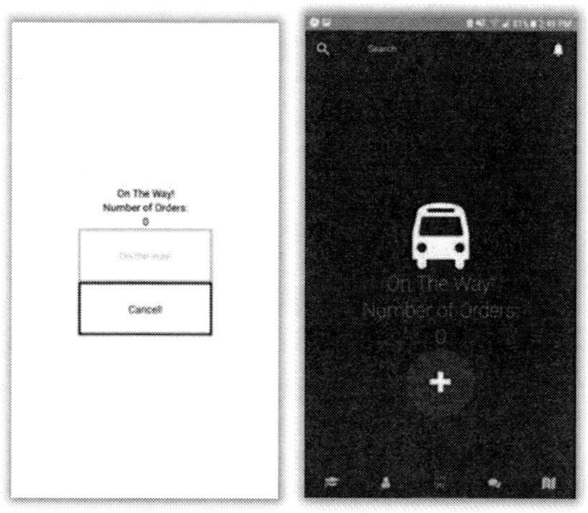

Figure 22. The right picture represents the application running in the transportation department and the left picture shows that the bus is on the way.

5. Conclusions

This paper has been discussed the development of an open, cross, cloud-based platform for the creation and delivery of smart applications and services. The proposed system has the following advantages when compared to the literature review. First, it provides cross-platform functionality and follows the industrial design strictly. Second, the proposed reduces the time and physical resources required to reach information and thus, facilitates the working beyond both place and time barriers. Third, the proposed system provides an easy way to maintain the database records based on the reliability of the cloud, as a result, the system is considered a reliable system. Fourth, provides an easy interaction among the participants. Besides, the system is open, that can be extended horizontally by providing other services and applications and can be expanded vertically by adding hierarchies hosting. Finally, and more to the points, all participants improve their familiarities with the institution roles significantly and the quality of the institution will be improved as a result. From this perspective, there are multiple avenues for the future-works. One avenue is to add healthcare services and applications. Another avenue is to add payment and security-aware applications. Finally, the database could be available for machine learning researchers to perform statistics and machine learning methodologies to improve the skills and the rank of the University.

Acknowledgments

This research is funded by, UMP (RDU190334): A Novel Hybrid Harmony Search Algorithm with Nomadic People Optimizer Algorithm for Global Optimization and Feature Selection, and (FRGS/1/2018/ICT05/UMP/02/1) (RDU190102): A Novel Hybrid Kidney-Inspired Algorithm for Global Optimization Enhance Kidney Algorithm for IoT Combinatorial Testing Problem.

References

[1] Magno-Tan, M.J., Crisostomo, A.V., Villaflor, B. and Faller, J.C. "Cloud-Based College Management Information System for Universities". *International Journal of Information and Education Technology*, *4*(6), p.508, 2014.

[2] Bilawane, Shilpa, Jambhulkar, Pranali and Naidu, Kokila "Information System Based On College Campus". *International Journal of Engineering and Computer Science*, 4, pp. 2319-7242, 2015.
[3] Rajesh Shah, Makhan Kumbhkar, "Cloud-Based College Management Information System for Autonomous Institute" *International Journal of Application or Innovation in Engineering & Management (IJAIEM))*, 5 (5), May 2015.
[4] Sagar Rajebhosale, Shashank Choudhari, Sachin Patil, Akshay Vyavahare, Sanket Khabiya, "SMARTCAMPUS – An Academic Web Portal with Android Application", International Research Journal of Engineering and Technology, 3 (4), 2016.
[5] M. V. Moreno et al." Applicability of Big Data Techniques to Smart Cities Deployments", IEEE Transactions on Industrial Informatics, 13(2):800809, Apr. 2017.
[6] S. Arshad, M. A. Azam, S. H. Ahmed, and J. Loo," Towards Information Centric Networking (ICN) Naming for Internet of Things (IoT): The Case of Smart Campus". International Conference on Future Networks and Distributed Systems (ICFNDS), July 2017.
[7] Y. Xiang , D. Chang ,B. Chen ," A smart university campus information dissemination framework based on wechat platform". *In: LISS*, vol 2013, pp 927–932,2015.
[8] Y. Atif, S. Mathew and A. Lakas, "Building a smart campus to support ubiquitous learning", *Journal of Ambient Intelligence and Humanized Computing*, 6 (2), pp. 223-238, 2014.
[9] Y. Liu, W. Zhang and P. Dong, "Research on the Construction of Smart Campus Based on the Internet of Things and Cloud Computing", *Applied Mechanics and Materials*, pp. 3213-3217, 2014.
[10] Mohammed Issam Younis, Hayder Faez Abdulkareem, Hamid Mohammed Ali, "Construction of Graduation Certificate Issuing System Based on Digital Signature Technique". *Journal of Engineering*, 21(6), pp. 15-36, 2015.
[11] MVC hierarchy, https://www.tutorialspoint.com/mvc_framework.
[12] Eric Masiello, Jacob Friecmann, Mastering React Native, Packt, 2017.
[13] Anthony Gore, Full-Stack Vue.js 2 and Laravel 5, Packt Publishing, 2017.
[14] Ethan Brown, Web Development with Node and Express, Orielly Publishing, 2014.
[15] VueJS Framework Documentation https://vuejs.org/v2/guide/.
[16] Laravel Framework Documentation https://laravel.com /docs/5.7.
[17] Mohammed Issam Younis, Taif Fadhil Hussein, "Design and Implementation of a Contactless Smart House Network System". *International Journal of Electrical and Computer Engineering* 8(6), pp. 4663-4672, 2018.
[18] Aniqa Rehman, Saba Latif, Nazir Ahmad Zafar, "Non-Deterministic Formal Modeling of Registration System towards Smart Campus", 12th International Conference on Mathematics, Actuarial Science, Computer Science and Statistics (MACS), pp. 1-6, 2018.

IOP Conf. Series: Materials Science and Engineering **769** (2020) 012068 doi:10.1088/1757-899X/769/1/012068

Region of Interest-based Tamper Detection and Lossless Recovery Watermarking Scheme on MRI and X-ray Medical Images

Jessie Ooi[1], Hui-Liang Khor[2] and Siau-Chuin Liew[1]
[1] Faculty of Computing, University Malaysia Pahang, Pahang, Malaysia
[2] Department of Computing And Information Technology, Tunku Abdul Rahman University Collage,Pahang, Malaysia

Abstract. The advancement of the technology has introduced many applications that will ease human daily process and improve the way people connect to each other. One of the applications that has been introduced to the user and vastly improve our life is the healthcare system. Health care system allows the medical practitioner to aware of the patient's previous health condition and documents their current. It also allows the medical practitioner to communicate with each other or get some advice from their peers and colleague even if they are far from each other physically. To ensure the authenticity of an image when it is transfer through the health care system, watermarking has been implemented. Through the process of watermarking the images, the authenticity of the images can be ensured. In this study, a ROI-based tamper detection and recovery watermarking scheme has been introduced. The proposed scheme has been tested out on the MRI and XA images. The PSNR value for MRI is 49.09 for the sequential mode and 49.07 in parallel mode. On the other hand, PSNR for XA images is 52.68 in the sequential mode and 52.68 in the parallel mode.

1. Introduction

With the increasingly common in application of healthcare system in a medical facility, communications between multiple medical partitional from different locations has become possible. However, this has brought the issue of medical images authentication to light. Especially when these images were shared between different parties. In certain fields such as military and healthcare, authenticity of the documents received, or transfer is always the one of the most important factors due to their sensitive nature. For example, military often involved with intelligence gathering while the documents transfer via the healthcare system are used often used for the patient's diagnostic.

To enhance and verified the authenticity of the transferred data, digital watermarking has been employed. Digital watermarking is a type of data hiding technology [1]. It allowed the original data to be protected while provide an extra layer of security to the data. Watermarking is often carried out in spatial domain. The intensity of the randomly selected pixel will be modified during the watermarking process[2]. Aforementioned, exchanging medical images or documents between different parties/ hospitals has become more and more common due to the advancement of the technology. Therefore, it has increased the risk of medical images/ documents attacked or modified by third parties. Any tampering that has been done on the medical documents may lead to the false diagnostic of the patients. Medical images such as ultrasound, X-ray, MRI can be easily tampered or modified by using basic image processing tools. Therefore, it is very important to focus on the authenticity of the documents and also increase the difficulties for the documents to be modified. Another important aspect for the watermarking process is its reversibility. The sensitive nature of medical images required the watermark to prove their authenticity. However, they also need to have the ability to recover themselves from the watermarked images as they are required to be view in their original state. Hence, the reversibility of the watermark is highly regarded.

Medical images often been divided into 2 different regions, ROI (region of interest) and RONI (region of non-interest). ROI is the area of images where important data or the diagnosis information is located. On the other hand, RONI often hold less or no significant information. In our previous work, a ROI-based tamper detection and recovery watermarking scheme (ROI-DR) has been proposed [3]. The proposed approach has been tested on the ultrasound medical images. The result proved that the proposed scheme is able to achieve a good result in imperceptibility with peak signal-to-noise ratio (PSNR) values approximately 48dB. The proposed scheme has also show that it is robust enough to fight against various type of tampering. Most importantly, the tampered ROI can be recovered to its original form. The proposed scheme was tested on a ultrasound medical images, while it is able to achieve a good result, it has lead us to one question where if the proposed scheme is able to be applied in other type of medical images such as MRI and X-Ray (XA). In this paper, we will discuss about the application and the result of the proposed scheme when it is applied on different type of medical images. In this paper, past study will be discussed in the section 2, and the methodology used will be discussed in section 3. Section 4 will discuss about the result and the last section will be the summary of this study.

2. Literature review

To address the authentication issue of the medical images, watermarking has been implemented. Through the implementation of watermarking in the medical images, unauthorized modification can be prevented. During the watermarking process, tampered localization was able to detect and locate modification of pixel values of the image. It is also able to deduct the motivation of a tampering process and if the modification is legitimate. Liew [4] proposed a tamper localization and recovery scheme by dividing the image into ROI and RONI. ROI is the significant part of medical images that used by doctors to diagnose the patient, and RONI is the area outside the ROI. The original Least Significant Digits (LSBs) that are removed in watermark embedding process were stored in RONI after compression. Gran [5] proposed a watermarking scheme that uses lossless compression which allows larger ROI to be used. These researches that has been mentioned were conducted on a single frame ultrasound image which is impractical in the real world. Therefore, digital watermarking on multi-frames medical images is proposed. However, several researches have proven that a multi-frames watermarking required a longer processing time. To speedup multi-frames watermarking processing time, a parallel computation in multicores watermarking processing on medical images is introduced.

Recent research about parallel processing has demonstrated its potential in improving medical image processing performance [6]. Parallel processing or computing is a type of computation approach that allows multiple processes or calculations to be carried out simultaneously. This approach is based on the concept of separating a complicated process into a few smaller processes. These smaller processes will then be solved concurrently. Parallelism has been employed for many years, mainly in high-performance computing (i.e., supercomputers such as Cray, and PC- based cluster and grid computing), but the more recent rise of multicore processor technology such as Intel Quad-Core technology has allow server or similar system to take advantages of its processing power [7]. The advantages of the parallel computing have attracted the researcher to implement parallel processing into the watermarking process. Hosny, Darwish, Li and Salah [8] proposed an approach of that make use of the parallel multi-core CPU and GPU for a fast and robust medical image watermarking. The result has shown that the application of parallel multi-core CPU and GPU has resulted in a huge amount of time reduction in the overall watermarking process. Glowacz and Pietroń [9] has proposed a watermarking algorithm by exploring the mixed CPU and GPU processing approach. It has been discovered that parallelization played an important part in the reducing the computation time by making an optimal use of the resources.

While the previous methods address the issue of speed in a watermarking process, there is other issue that has not been addressed. Medical images often are very sensitive in nature. Hence, the originality of the images needed to be preserved. Therefore, reversibility has become one of the important factors in the watermarking process. In certain area, modification in the images are not allowed regardless of the size and type of modification. Any form of modification will affect the authenticity of the images hence render the images to be unusable [10]. The main objective of reversible watermarking is to embed the information without any kind of distortion. Catalin and Coltuc[11] proposed a local prediction based reversible watermarking. This reversible watermarking provides a good result. However, it requires extra computing cost for each pixel a least square predictor in a square block center on the pixel. Hence, a longer processing time is required. Later, Catalin and Coltuc [12] revisit a previous reversible watermarking scheme proposed earlier. The adaptive pixel paring consider only pixels with similar predictions errors is introduced. By increasing the number of pixel pairs where both the pixels are embedded and decrease in the number of shifted pixels, the proposed watermarking scheme were able to outperforms the state-of-art low embedding bit rate schemes. Pakdanan and Saryazdi [13] has proposed a reversible watermarking based on error prediction in Hadamard domain. The proposed method divides the original image into block and transform them to Hadamard domain. Value of error between the original and the predicted coefficient is computed. Proposed method shows that it provides a higher capacity and quality as compared to other well-known methods.

3. Methodology

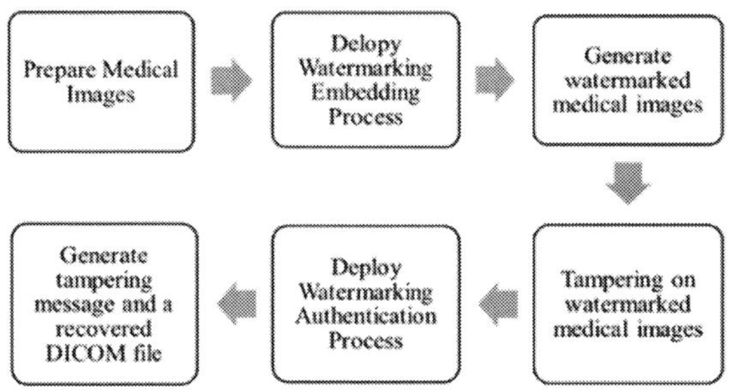

Figure 1. Watermarking process flow

Figure 1 shows the process flow of the proposed watermarking scheme. The proposed watermarking scheme will be implemented in both sequential and also parallelly in order to speed up the watermarking process. When the process is carried out parallelly, the watermarking scheme will subdivide the medical images based on the number of core available and hence perform the all the watermarking process in parallel. This process is shown in Figure 2.

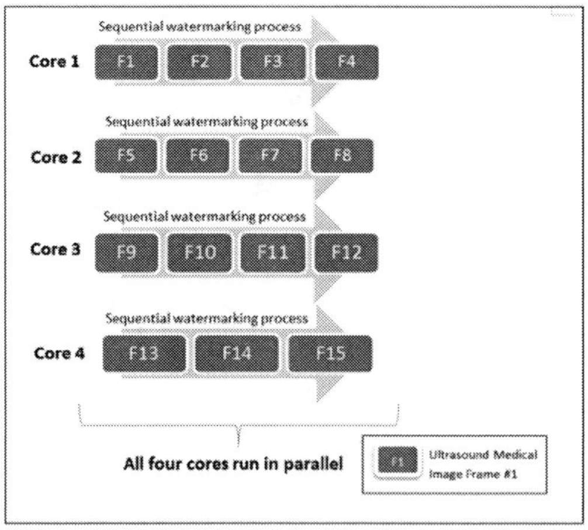

Figure 2. Multi-frame watermarking scheme

Three different type of medical images are selected in this study, MRI, XA and CT scan. While these medical images are different in nature, they are all 8 bits per pixel. The properties of these images is shown in Table 1.

Table 1. Medical Images Properties

Medical Image Sample	Image dimension in pixels	Bits per pixel	Number of frames	ROI size definition [x, y, width, height]
MRI.dcm	640 x 476	8	16	[132, 135, 284, 149]
XA.dcm	512 x 512	8	12	[140, 127, 226, 149]
CT.dcm	512 x 512	8	1	[135, 124, 206, 168]

The entire watermarking and authentication process started with the image preparation process. Each medical image is divided into one main section of ROI and 5 different sections for RONI. ROI and RONI will be predefined before the start of watermarking embedding process. ROI is the section where it holds the important information about the image where RONI is basically contained less to none information.

The main process of the watermarking embedding is adding ROI bits into RONI area to produce a watermarked medical image. These added bits will later be used in the watermarking authentication process.

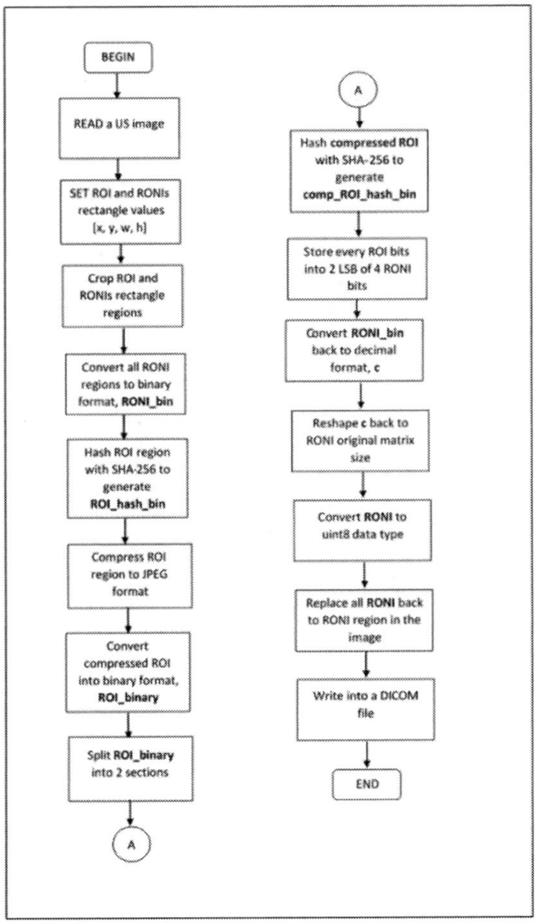

Figure 3. Proposed watermarking embedding process

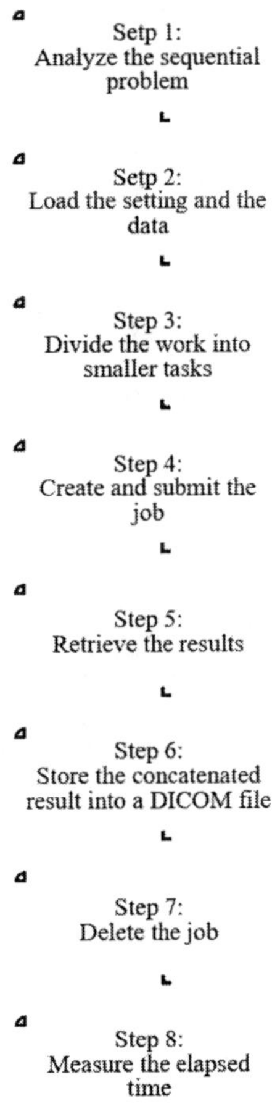

Figure 4. Main algorithm flow of parallel watermarking process

Figure 3 shows the proposed watermarking embedding process. RIO region was hashed with SHA-256 and stored as ROI_hash in ROI 3. This process is carried out in both parallel and sequential modes. The multi-frame ultrasound medical images were loaded into a quad-core processor. While these images were loaded into different core, the sequential watermarking process will be carried out in each core parallelly. The main algorithm for the parallel watermarking is shown in Figure 4. The methodology of this study was described in details in our previous work [3] .

4. Result

In our last study, the proposed scheme has been deployed onto the ultrasound medical images. To test our proposed approached with different type of images, the scheme is carried out on the MRI and XA images. The PSNR and MSE values after the watermarking process for the MRI and XA are shown in Table 2 and Table 3.

Table 2. PSNR and MSE Value after ROI-DR multi-frames watermarking embedding process on MRI image

	ROI-DR Multi-Frames Watermarking Embedding Process			
MRI.dcm	**Sequential mode**		**Parallel Mode**	
Frame_No	**PSNR**	**MSE**	**PSNR**	**MSE**
1	49.34	35.64	49.00	37.45
2	49.00	37.45	49.00	37.45
3	49.11	35.11	49.11	35.11
4	49.12	35.51	49.12	35.51
5	49.06	34.51	49.06	34.51
6	49.02	34.79	49.02	34.79
7	49.01	35.85	49.01	35.85
8	49.07	36.39	49.07	36.39
9	49.02	35.86	49.02	35.86
10	48.96	35.65	48.99	35.69
11	49.10	36.09	49.10	36.09
12	49.15	36.21	49.15	36.21
13	49.15	36.31	49.16	36.33
14	49.10	36.88	49.10	36.88
15	49.10	37.50	49.10	37.50
16	49.11	37.42	49.11	37.42
Average	**49.07**	**36.07**	**49.07**	**36.19**

Table 3. PSNR and MSE Value after ROI-DR multi-frames watermarking embedding process on XA image

	ROI-DR Multi-Frames Watermarking Embedding Process			
XA.dcm	**Sequential mode**		**Parallel Mode**	
Frame_No	**PSNR**	**MSE**	**PSNR**	**MSE**
1	52.70	38.96	52.70	38.96
2	52.69	39.54	52.69	39.54
3	52.65	39.69	52.65	39.69
4	52.65	39.80	52.65	39.80
5	52.65	39.74	52.65	39.74
6	52.72	39.93	52.72	39.93
7	52.68	39.90	52.68	39.90
8	52.69	39.87	52.69	39.87
9	52.69	39.87	52.69	39.87
10	52.71	239.90	52.71	39.90
11	52.68	39.89	52.68	39.89
12	52.67	39.86	52.67	39.86
Average	**52.68**	**39.75**	**52.68**	**39.75**

Based on the result it can be concluded that the PSNR values for MRI and XA are both well above the minimum acceptable PSNR value which is 30dB. When the two group of images were compared, it can be notice that PSNR value for the XA is better than the PSNR value for MRI. One of the reasons for this is the ROI size for XA is smaller than the ROI size of the MRI. Hence, there are fewer embedding bits in the RONI.

Table 4 Speedup Factor for ROI-DR multi-frames watermarking embedding process on different medical images modalities

Medical Image Modalities	Number of frames	Elapsed time (seconds) Sequential	Parallel	Speedup factor= Sequential / Parallel
MRI	16	16.57	18.50	0.90
XA	12	10.94	16.10	0.68
US	15	14.30	12.60	1.14

Table 4 shows the speedup factor for DOI-DR multi-frames watermarking embedding process on different medial images modalities. The speedup factor is calculated by dividing the elapsed time of the process in the sequential mode with the parallel mode. The result shows the speedup factor for MRI and XA is smaller than 1. This result might due to the frame size of the MRI and XA as it is smaller than an ultrasound image. Therefore, they are unable to take advantages the proposed multi-frames watermarking scheme.

In order to test the robustness of the medical images towards any tampering, the MRI, XA and CT has been tampered. This process was completed by adding "Paper and Salt" noise to the ROI. A portion of the images has been cloned and flipped vertically into the ROI.

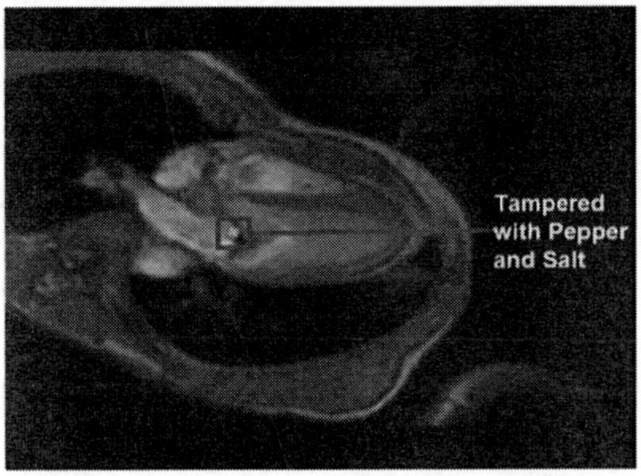

Figure 5 Watermarked MRI tampered by "Pepper and Salt" noise

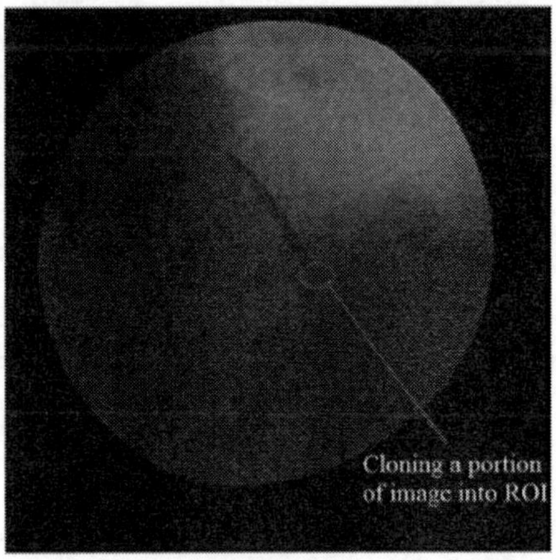

Figure 6 Watermarked XA tampered by "Pepper and Salt" noise

IOP Conf. Series: Materials Science and Engineering **769** (2020) 012068 doi:10.1088/1757-899X/769/1/012068

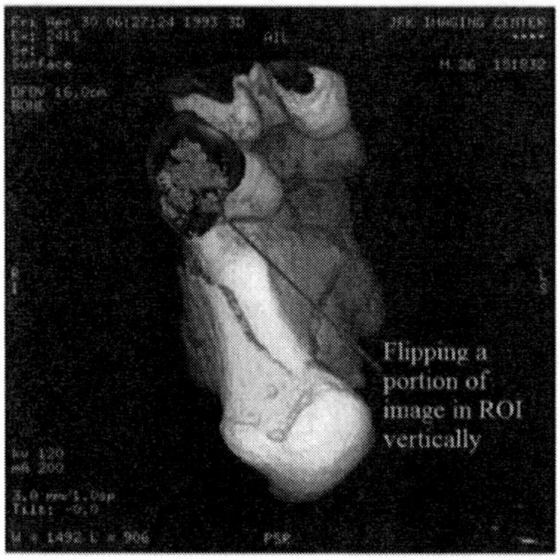

Figure 7 Watermarked CT tampered by "Pepper and Salt" noise

The time needed for the authentication process has been recorded. In Table 5, it can be noticed that the elapsed time in the watermarking authentication process for different medical images is very efficient which only took less than 0.5 seconds. At the same time, it is also able to detect any form of tampering and recovery regardless of the image type. However, it has been discovered that the elapsed time for the parallel watermarking authentication process is quite high which render it not suitable to be adapt. This might largely due to the frame size of the medical image that are too small to take advantages of the parallel watermarking process.

Table 5. Speedup factor for ROI-DR multi-frames watermarking authentication process on different medical images

Measurement	Result			
	MRI	**XA**	**CT**	**US**
Elapsed time in watermarking authentication process (seconds)	0.46	0.29	0.43	0.34
Able to detect tampering	Yes	Yes	Yes	Yes
Able to produce exact ROI recover	Yes	Yes	Yes	Yes
Able to generate a recovered DICOM file	Yes	Yes	Yes	Yes

Table 6. Speedup factor for ROI-DR multi-frames watermarking authentication process on different medical images

Medical Image Modalities	Number of frames	Elapsed time (seconds)		Speedup factor= Sequential / Parallel
		Sequential	**Parallel**	
MRI	16	3.28	16.90	0.19
XA	12	1.69	12.70	0.13
US	15	5.70	9.30	0.61

5. Conclusion

The proposed scheme similarly is able to receive a fast authenticating result when it is applied to different medical images. The robustness of the proposed scheme was also been tested and has been proven to be able to detect any tampering and has the ability to recover the ROI. However, due to the small size of the MRI and XA, the elapsed time for the parallel watermarking scheme are quite high.

IOP Conf. Series: Materials Science and Engineering **769** (2020) 012068 doi:10.1088/1757-899X/769/1/012068

The small size of these images does not allow them to take advantages of the parallel processing. Future study needs to be carried out in order to investigate this issue.

Acknowledgments
This research is financially supported by Universiti Malaysia Pahang Research Grant RDU1703280.

References

[1] K. Parasuraman and G. Deeparani, "Reversible image watermarking using interpolation technique," 2014 Int. Conf. Electron. Commun. Comput. Eng. ICECCE 2014, vol. 5, pp. 200–205, 2014.
[2] N. Nikolaidis and I. Pitas, "Robust image watermarking in the spatial domain," Signal Processing, vol. 66, no. 3. pp. 385–403, 1998.
[3] H. L. Khor, S. C. Liew, and J. M. Zain, "Region of Interest-Based Tamper Detection and Lossless Recovery Watermarking Scheme (ROI-DR) on Ultrasound Medical Images," J. Digit. Imaging, vol. 30, no. 3, pp. 328–349, 2017.
[4] S.-C. Liew, S.-W. Liew, and J. Mohamad Zain, Tamper Localization and Lossless Recovery Watermarking Scheme with ROI Segmentation and Multilevel Authentication, vol. 26. 2012.
[5] G. Badshah, S.-C. Liew, J. Mohamad Zain, and M. Ali, "Watermark Compression in Medical Image Watermarking Using Lempel-Ziv-Welch (LZW) Lossless Compression Technique," J. Digit. Imaging, vol. 29, 2015.
[6] G. Allan and G. S. Almasi, Highly parallel computing. 1988.
[7] H. L. Khor, S. C. Liew, and J. M. Zain, "A review on parallel medical image processing on GPU," 2015 4th Int. Conf. Softw. Eng. Comput. Syst. ICSECS 2015 Virtuous Softw. Solut. Big Data, no. January 2018, pp. 45–48, 2015.
[8] K. M. Hosny, M. M. Darwish, K. Li, and A. Salah, "Parallel multi-core CPU and GPU for fast and robust medical image watermarking," IEEE Access, vol. 6, pp. 77212–77225, 2018.
[9] A. Głowacz and M. Pietroń, "Implementation of Digital Watermarking Algorithms in Parallel Hardware Accelerators," Int. J. Parallel Program., vol. 45, no. 5, pp. 1108–1127, 2017.
[10] J. J. Fernandez, "Reversible Image Watermarking : A Review," vol. 4, no. 4, pp. 738–746, 2018.
[11] I. C. Dragoi and D. Coltuc, "On local prediction based reversible watermarking," IEEE Trans. Image Process., vol. 24, no. 4, pp. 1244–1246, 2015.
[12] I.-C. Dragoi and D. Coltuc, Adaptive Pairing Reversible Watermarking, vol. 25. 2016.
[13] Z. Pakdaman, S. Saryazdi, and H. Nezamabadi-pour, "A prediction based reversible image watermarking in Hadamard domain," Multimed. Tools Appl., vol. 76, no. 6, pp. 8517–8545, 2017.

IOP Conf. Series: Materials Science and Engineering **769** (2020) 012069 doi:10.1088/1757-899X/769/1/012069

An Improved Image Steganography Scheme Based on RDWT and QR Decomposition

Ke-Huey Ng[1], **Siau-Chuin Liew**[2] **and Ferda Ernawan**[3]

[1,2,3]Faculty of Computing, Universiti Malaysia Pahang, Gambang, Kuantan, Malaysia.

kehuey2809@gmail.com[1], liewsc@ump.edu.my[2], ferda@ump.edu.my[3]

Abstract. This paper demonstrates an improved image steganography that hides grayscale secret image into grayscale cover image using RDWT and QR decomposition. The proposed scheme made use of the human visual system in the information hiding process. Both cover and secret image are being divided into non-overlapping image blocks with identical block size. Entropy values generated from every image block will be sorted from the lowest value to the highest value. The embedding process starts by embedding the secret image block with lowest entropy value into the cover image block with lowest entropy value. The process goes on until all image blocks have been embedded. Embedding secret image into cover image according to the entropy values causes differences that HVS can less likely to detect because of the small changes on image texture. The proposed scheme gives higher PSNR values and better image quality.

1. Introduction

Image steganography is the science of information hiding. The purpose of steganography is to prevent outsiders to discover the existence of hidden data so it can hardly be detected and further manipulated. Its goal is to protect the secret data. It can be in any forms, such as audio, video, text or image files. In this paper, it focuses on hiding image into another image. In general, an image steganography scheme can be represented as below [1].

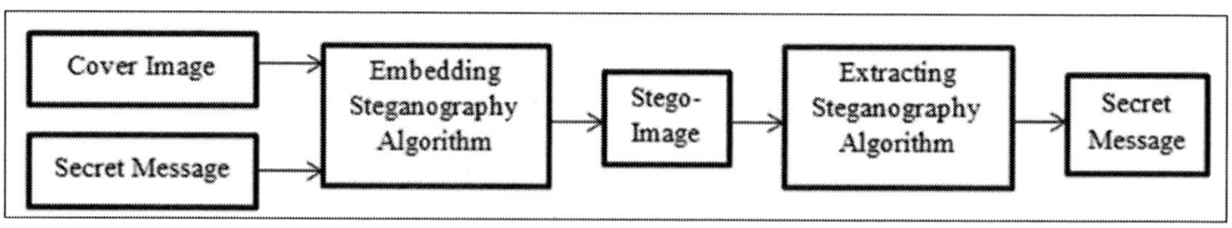

Figure 1. Basic Steganography Process

In the context of image steganography, there are two (2) domains which are spatial and transform domain. Spatial domain involves the direct bitwise manipulation [2] whereas transform domain focuses on the transformed image manipulation, which means the original cover image will be changed or transformed first before embedding secret message. Spatial domain is easier to be developed as compared to transform domain. It requires shorter computational time. However, it is more vulnerable to attacks. The reason is that it embeds secret information into cover image directly without transforming the cover image itself, this can cause the secret information to be destroyed easily if the stego image has been attacked.

Thus, transform domain is more preferable because it ensures a certain level of robustness as it withstands against attacks such as geometric attacks and compression. The secret information should still be present and can be detected regardless of the attacks done to stego image.

2. Literature Review

2.1. Frequency Domain Transform Techniques

To be able to achieve good quality and robustness to attacks, embedding in transform domain is much more efficient than embedding in spatial domain [2].The most commonly used frequency-domain transform methods include the Discrete Fourier Transform (DFT), Discrete Cosine Transform (DCT), Discrete Wavelet Transform (DWT) and Redundant Discrete Wavelet Transform (RDWT). Frequency-domain methods are more widely applied as compared to spatial-domain method.

In information hiding schemes, DFT coefficients gives only modest results and is fragile to attacks, especially sensitive to JPEG and MPEG attacks. Capacity and lack of HVS models are also the drawback of DFT [3] whereas DCT based techniques are robust against simple image processing operations but they are hard to implement, takes more computational time and cost and weak against geometric attacks.

Discrete Wavelet Transform (DWT) is a modern technique popularly utilized in digital image processing. The transforms are based on wavelet of limited duration and different frequency. The wavelet transform decomposes the image into three directions. Most of the image energy concentrates at LL band. Hence, embedding in other sub-bands would lower the quality of image. Hiding information in the transform domain is generally more robust and less perceptible [4,5].

As compared to DCT as adopted by Lai [6], DWT has more advantages. As it does not suffer from blocking artefacts, it takes less computational time. Therefore, information hiding techniques based on wavelets are more robust against attacks than those based on DCT [7].

Comparing DWT and DCT, DWT has better energy compaction and presents a sparse time-frequency but there is a major disadvantage of DWT which is the poor directional selectivity and lack of shift variance.

Redundant Discrete Wavelet Transform (RDWT) is a shift invariance property. Let $f(n)$ be the input signal and $f'(n)$ be its reconstructed version. $h[-k]$ and $g[-k]$ are low pass and high pass analysis filters while $h[k]$ and $g[k]$ are corresponding low pass and high pass synthesis filters. c_j and d_j are output coefficients at level j . RDWT avoids down and up sampling of coefficients. During image extraction process, DWT produce inaccuracy because of its shift variances property [8]. Many information hiding schemes apply RDWT to overcome the shirt variance problem of DWT [9]. It removes the down-sampling operation from DWT to produce an over-complete representation of the frequency coefficients [10]. Also, as compared to DWT, RDWT is more robust [11]. Besides that, RDWT helps to enhance embedding capacity because its sub-bands have the same size of the original image [12].

2.2. Matrix Factorization Techniques

There are several matrix factorization methods such as singular value decomposition (SVD), Schur decomposition, QR decomposition, LU decomposition etc. resulting from solutions of linear equation [13].

SVD of an matrix A with dimensions m x m is given by

$$A = USV^T \tag{1}$$

With SVD's good stability, applying it to an image does not cause noticeable change on the appearance [14, 15].

Schur decomposition of a real matrix A results in two matrices U and D such that

$$A = U \times D \times U' \tag{2}$$

It is suggested to be used in digital image processing as it requires less computational cost than SVD [16].

QR decomposition or QR factorization [17] is decomposition of matrix into an orthogonal matrix and triangular matrix. Any real matrix A can be expressed as

$$[Q, R] = qr(A) \tag{3}$$

LU factorization is almost similar to QR decomposition. However, QR decomposition has been proven to be more precise for least square problems. LU factorization can only be applied to square matrices whereas QR factorization can be applied to both square and rectangular matrices [12].

When comparing SVD with QR decomposition, the latter requires less computational complexity. Another feature of QR decomposition is the resistance to some signal processing operations, such as filtering, lossy compression and noise addition. QR decomposition could also solve the major issue of SVD which is its false positive problem [4].

2.3. *Arnold Transform*

Arnold Transform is widely used in image permutation. It is also called the Cat Face transfer, and it is given by

$$\begin{pmatrix} x' \\ y' \end{pmatrix} = \begin{pmatrix} a & b \\ c & d \end{pmatrix} \begin{pmatrix} x \\ y \end{pmatrix} (mod\ N) \tag{4}$$

Arnold transform is often being applied on involved images to improve the security of information hiding schemes.

2.4. *Entropy*

There are several ways to measure the texture of a digital image. Entropy is the most suitable way to measure the image's texture content. Texture provides measure of properties of an image such as regularity, coarseness and smoothness. An image that is perfectly flat will have an entropy value of zero. Entropy can be defined as the statistical measure of randomness [18].

According to Shannon's definition, the entropy of a grayscale image is given by the following mathematical relation:

$$E_1 = -\sum_{i=0}^{L-1} p_i \log p_i \tag{5}$$

If the entropy value is high then it is considered to have more details [19].

2.5. *Performance Evaluation*

To measure the quality of a digital image, PSNR is one of the popular metrics to use. It is done by analyzing the mean squared error value between the Cover and the stego image. The higher the PSNR value, the smaller the possibility of visual attack by human eyes [20]. PSNR is being presented as [21]

$$PSNR\ (in\ dB) = 10log_{10}\left(\frac{255^2}{MSE}\right) \tag{6}$$

and

$$MSE = \frac{\sum_{i=1}^{N}(C_i - C'_i)^2}{N} \tag{7}$$

On the other hand, SSIM is also used as a metric to measure the similarity between two images. It is being done by

$$\mathrm{SSIM}\ (x, y) = \frac{(2\mu_x\mu_y + c_1)(2\sigma_{xy} + c_2)}{(\mu_x{}^2 + \mu_y{}^2 + c_1)(\sigma_x{}^2 + \sigma_y{}^2 + c_2)} \tag{8}$$

$$c_1 = (k_1 L)^2 \tag{9}$$

$$c_2 = (k_2 L)^2 \tag{10}$$

The higher the SSIM index is, the better the visual quality of the stego image with respect to the cover image [21].

2.6. *Related Work*

The spatial domain methods are less robust and use simple embedding process. When the issues of imperceptibility and robustness are concerned, the transform domain techniques are a better option [2], especially when they are being tested against geometric attacks.

Recently, many hybrid transform domain methods are being developed to improve the robustness. While individual transform domain techniques such as DFT, DCT and DWT are good enough to improve the properties as mentioned in 2.6, combining them with Singular Value Decomposition (SVD) will further improve the image quality of the scheme. One of the features of SVD is that it helps to achieve good transparency and robustness.

In 2013, Anumol and Anusudha [22] presented a robust scheme based on discrete wavelet transform and singular value decomposition. The paper chose to use DWT over DFT and DCT as the DFT and DCT are full frame transforms and hence any changes in the transform coefficient affects the entire image. In the embedding process, HL and LH sub-bands are used instead of LL and HH band to improve the robustness and imperceptibility. The proposed method is robust against various types of attacks.

In 2016, Mansi and Vijay [12] proposed an image steganography scheme using QR decomposition and RDWT. Artefacts due to variation in energy distribution caused by shifts in input signal have been solved by RDWT. QR decomposition requires less computational complexity and eliminates false positive issue. The proposed work has shown improvement in terms of robustness, capacity and imperceptibility.

In 2016, Taha et al. [15] evaluated the performance of both RDWT-SVD and DWT-SVD schemes. RDWT is a shift invariance property. Many schemes based on RDWT have been proposed to overcome the shift variant problem of DWT. It removes the down-sampling operation from DWT to produce an over-complete representation of the frequency coefficients. Performing SVD on images is computationally costly. By combining SVD with DWT or RDWT, less computational effort is required to produce better performance. Both schemes showed robust against all attacks, RDWT-SVD is better than DWT-SVD, especially for geometrical attacks.

In 2017, Shaoli Jia et al. [4] introduced a scheme based on DWT and QR Decomposition for colour images. The main advantage of using DWT is that it better takes into account the local image characteristics at different resolution levels which can significantly improve the robustness of hidden message. Arnold transform is applied on the message to ensure security. The work used QR decomposition instead because SVD requires greater computational complexity.

In 2018, Poonam et al. [23] presented a scheme for grayscale images based on DWT-SVD. DWT provided better robustness and visible transparency as compared to DCT and DFT. This paper illustrated an improvement in imperceptibility as well as robustness against attacks.

In 2018, Ferda and Muhammad [24] proposed a block-based RDWT-SVD method using human visual system (HVS) characteristics. This scheme presents an embedding method by examining the coefficients in the first column of U vectors. The proposed scheme can avoid the false-positive problem faced by many other RDWT-SVD schemes during extraction process. The hidden image is scrambled by Arnold transform for security purpose. Compared to existing methods, the method achieved better robustness and imperceptibility under various attacks.

In 2018, Divya and [25] developed a false-positive-free scheme based on shuffled SVD (SSVD) and RDWT. The scheme embeds hidden message on the cover image using redundant discrete wavelet transform (RDWT) and chaotic mapping. Chaotic mapping is achieved by Shuffled singular value decomposition (SSVD). SSVD enhanced the quality of reconstructed image by breaking an image into a set of ensemble images. The proposed work eliminates the false-positive problem that usually found in other RDWT-SVD methods.

Inspired by above-mentioned previous works, an improved image steganography scheme based on RDWT and QR decomposition is proposed to embed grayscale image into grayscale cover image by analysing the image texture using entropy.

Table 1. Comparison between different image steganography methods

Year	Ref.	Features	Advantages
2013	[22]	-DWT -SVD -Embed in HL and LH	-Improved robustness -Improved imperceptibility
2016	[12]	-RDWT -QR factorization -Arnold transform	-Less computational complexity -Avoids false positive issue
2016	[15]	-RDWT -SVD	- Removes the down-sampling operation from DWT -Less computational effort -Robust against geometrical attacks
2017	[4]	-DWT -QR Decomposition -Arnold transform	-Less computational complexity -Improved robustness of hidden message
2018	[23]	-DWT -SVD	-Better robustness than DCT and DFT -Improved imperceptibility
2018	[24]	-RDWT -SVD by examining orthogonal matrix $U_{2,1}$ and $U_{3,1}$. -Arnold transform	- Remove up and down-sampling of coefficients. -Avoid false-positive problem -Higher imperceptibility -Higher robustness
2018	[25]	-RDWT -SSVD -Chaotic mapping	-Avoid false-positive problem -Robust against geometrical attacks

3. Methodologies

This section explains the detailed embedding and extraction process of the proposed work. The proposed scheme made use of the image texture by calculating entropy values for every image block for both cover and secret images before embedding process (refer to Figure 2) takes place. Figure 3 shows the extraction process of proposed method.

The main contribution of the proposed work is the use of entropy in the embedding process. Before embedding takes place, both cover image and secret image will be segmented into non-overlapping blocks with the same block size. Every image block will be computed to produce an entropy value. Then, all values will be sorted in ascending order, from the lowest entropy value to the highest entropy value. This applies to both cover and secret image.

Starting from the block with the lowest entropy value, RDWT will be applied on the blocks. The LL sub-band of cover image block will then be decomposed by QR decomposition. The secret image information is embedded by modifying the R value of the LL sub-band of cover image. After that, inverse QR decomposition is performed to get the modified LL sub-band. Them, inverse RDWT will be applied to get the modified image block. This process will continue until all secret image information is being embedded. Finally, a stego image will be formed by combining all modified image blocks.

The reason to use RDWT instead of other transform techniques is that, RDWT removes the coefficients' up-sampling and down-sampling that exists in DWT. Also, RDWT can increase robustness and provides more embedding capacity. On the other hand, QR decomposition provides better imperceptibility and avoids false positive issue, as compared to SVD.

Entropy identifies the texture of a digital image. The higher the entropy value of an image block, the more details it contains. By utilizing the use of entropy during embedding process, the cover image block with the lowest entropy value will be embedded with the secret image block with the lowest entropy value. The minor changes in the particular image block will not cause significant effect on the entropy values. Now, imagine the secret image block with the highest entropy value, which contains more details, is embedded into the cover image block with the lowest entropy. This process will cause more changes in the image texture as compared to the previous case, which will indirectly affect the imperceptibility of the scheme.

The proposed work embeds secret image into cover image by following the steps shown in Figure 2. By considering the order of entropy values of all image blocks, higher imperceptibility is achieved. As the image block with lowest entropy value will remain having the lowest entropy value as it is embedded with another image block, which, in this case, the secret image block with the lowest entropy value. This will not be accomplished if the process replaces any of the image block with entropy value of different order instead of the same order because the image texture will be changed in a different manner that causes more differences.

Figure 3 demonstrates the extraction process which extracts secret image information from the image location that was set during the embedding process. There is no need for re-calculating of entropy values for every image blocks.

3.1. Embedding Process

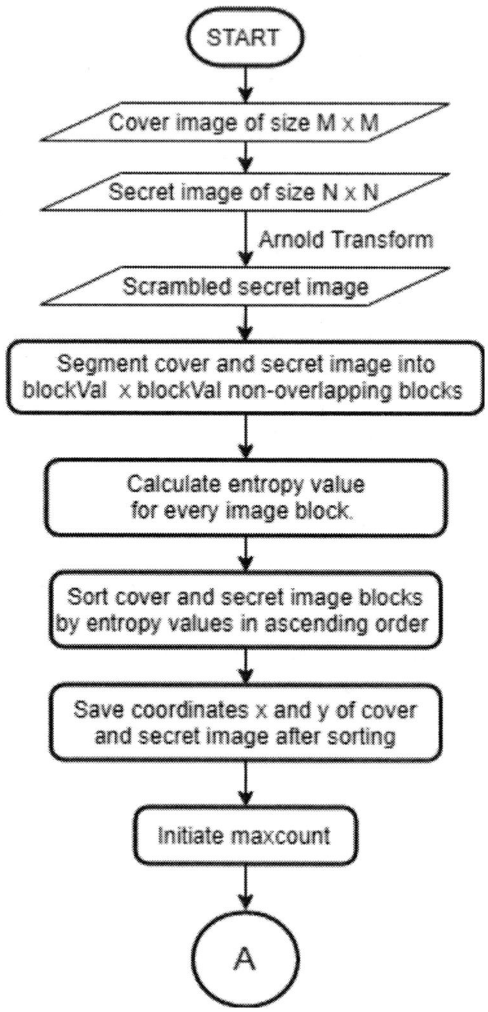

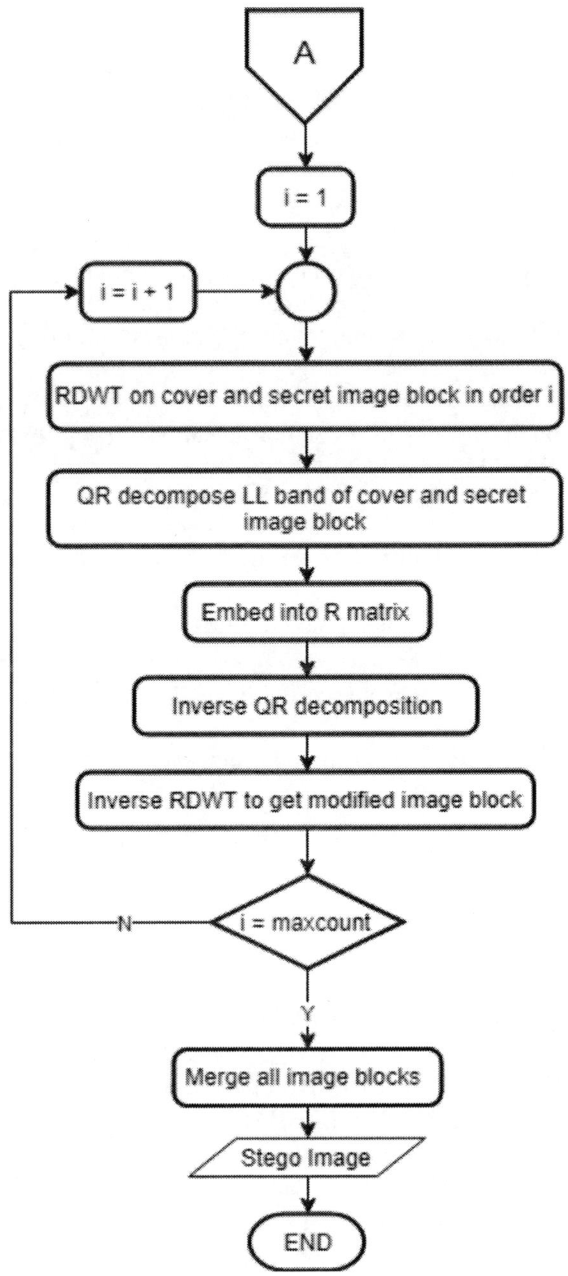

Figure 2. Embedding Process of Proposed Method

The embedding algorithm:

Input: Cover Image; Secret Image

Pre-processing:

Step 1: Segment cover image, c of size M x M into blocksize x blocksize non-overlapping blocks.

Step 2: Apply Arnold transform on secret image.

$$S' = \mathrm{arnold}(S, \mathrm{key}) \tag{11}$$

Step 3: Segment secret image, s of size N x N into 4 x 4 non-overlapping blocks.

Step 4: Calculate entropy for each cover image blocks.

Step 5: Sort the cover image blocks according to entropy values from lowest to highest.

Step 6: Save coordinates of x_c and y_c after sorting.

Step 7: Calculate entropy for each secret image blocks.

Step 8: Sort the secret image blocks according to entropy values from lowest to highest.

Step 9: Save coordinates of x_s and y_s after sorting.

Secret Image Embedding:

Step 10: Select cover image blocks from the lowest entropy value to the highest.

Step 11: Each selected blocksize x blocksize cover image block is transformed by 1-level RDWT.

$$[\mathrm{LL}_c, \mathrm{LH}_c, \mathrm{HL}_{c,} \mathrm{HH}_c] = \mathrm{rdwt}(\text{cover image block}) \tag{12}$$

Step 12: Select secret image blocks from the lowest entropy value to the highest.

Step 13: Each selected blocksize x blocksize secret image block is transformed by 1-level RDWT.

$$[\mathrm{LL}_s, \mathrm{LH}_s, \mathrm{HL}_{s,} \mathrm{HH}_s] = \mathrm{rdwt}(\text{secret image block}) \tag{13}$$

Step 14: The LL subband of cover image block is decomposed by QR decomposition.

$$[Q_{c,}, R_c] = \mathrm{qr}(\mathrm{LL}_c) \tag{14}$$

Step 15: The LL subband of secret image block is decomposed by QR decomposition.

$$[Q_{s,}, R_s] = \mathrm{qr}(\mathrm{LL}_s) \tag{15}$$

Step 16: The secret image is embedded by modifying the R value of the LL_c.

$$R_{st} = R_c + \propto R_s \tag{16}$$

Step 17: Perform inverse QR decomposition to get the modified LL_c to form LL_{st}.

$$\mathrm{LL}_{st} = Q_c \times R_{st} \tag{17}$$

Step 18: Apply inverse RDWT to get the modified image block

$$\text{Modified image block} = RDWT^{-1}(\mathrm{LL}_{st}, \mathrm{LH}_c, \mathrm{HL}_{c,} \mathrm{HH}_c) \tag{18}$$

Step 19: Repeat step 10-18 until all secret image blocks have been embedded.

Post-processing:

Step 20: Merge all modified image blocks to form Stego Image.

Output: Stego Image

3.2. Extraction Process

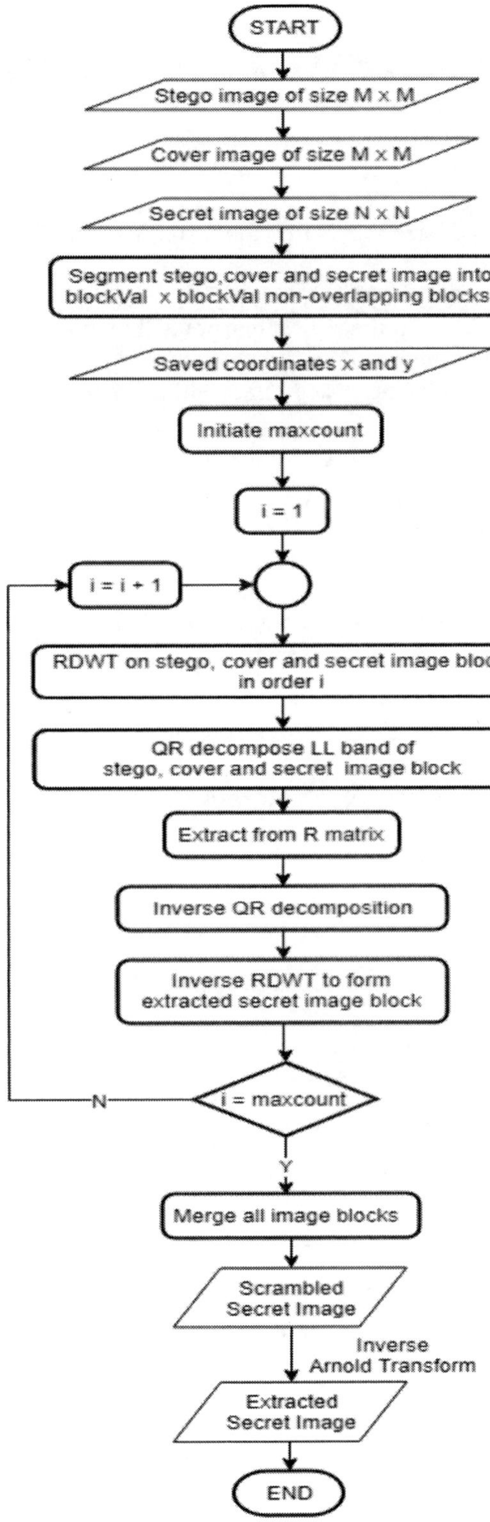

Figure 3. Extraction Process of Proposed Method

IOP Conf. Series: Materials Science and Engineering **769** (2020) 012069 doi:10.1088/1757-899X/769/1/012069

The extraction algorithm:

Input: Stego Image; Cover Image; Secret Image

Pre-processing:

Step 1: Segment stego image , st of size M x M into blocksize x blocksize non-overlapping blocks.

Step 2: Segment cover image, c of size M x M into blocksize x blocksize non-overlapping blocks.

Step 3: Segment secret image, s of size N x N into blocksize x blocksize non-overlapping blocks.

Secret Image Extraction:

Step 4: Select stego image blocks from the lowest entropy value to the highest according to the x_c and y_c coordinates obtained.

Step 5: Each selected blocksize x blocksize stego image block is transformed by 1-level RDWT.

$$[\mathrm{LL}_{st}, \mathrm{LH}_{st}, \mathrm{HL}_{st,}\mathrm{HH}_{st}] = \mathrm{rdwt}(\text{stego image block}) \tag{19}$$

Step 6: Select cover image blocks from the lowest entropy value to the highest according to the x_c and y_c coordinates obtained.

Step 7: Each selected blocksize x blocksize cover image block is transformed by 1-level RDWT.

$$[\mathrm{LL}_{c}, \mathrm{LH}_{c}, \mathrm{HL}_{c,}\mathrm{HH}_{c}] = \mathrm{rdwt}(\text{cover image block}) \tag{20}$$

Step 8: Select secret image blocks from the lowest entropy value to the highest according to the x_s and y_s coordinates obtained.

Step 9: Each selected blocksize x blocksize secret image block is transformed by 1-level RDWT.

$$[\mathrm{LL}_{s}, \mathrm{LH}_{s}, \mathrm{HL}_{s,}\mathrm{HH}_{s}] = \mathrm{rdwt}(\text{secret image block}) \tag{21}$$

Step 10: The LL subband of stego image block is decomposed by QR decomposition.

$$[Q_{st,}, R_{st}] = \mathrm{qr}(\mathrm{LL}_{st}) \tag{22}$$

Step 11: The LL subband of cover image block is decomposed by QR decomposition.

$$[Q_{c,}, R_{c}] = \mathrm{qr}(\mathrm{LL}_{c}) \tag{23}$$

Step 12: The LL subband of secret image block is decomposed by QR decomposition.

$$[Q_{s,}, R_{s}] = \mathrm{qr}(\mathrm{LL}_{s}) \tag{24}$$

Step 13: The secret image is extracted by

$$R_{s2} = (R_{st} - R_c) / \propto \tag{25}$$

Step 14: Perform inverse QR decomposition to form LL_{s2}.

$$\mathrm{LL}_{s2} = Q_s \times R_{s2} \tag{26}$$

Step 15: Apply inverse RDWT to get the extracted secret image block.

$$\text{Extracted secret image block} = RDWT^{-1}(\mathrm{LL}_{s2}, \mathrm{LH}_{s}, \mathrm{HL}_{s,}\mathrm{HH}_{s}) \tag{27}$$

Step 16: Repeat step 4-15 until all secret image blocks have been extracted.

Post-processing:

Step 17: Merge all extracted image blocks to form image S2.

Step 18: Apply inverse Arnold transform on image S2 to get secret image S_{ext}.

$$S_{ext} = \mathrm{inarnold}(\mathrm{S2}, \mathrm{key}) \tag{28}$$

Output: Extracted Secret Image

IOP Conf. Series: Materials Science and Engineering **769** (2020) 012069 doi:10.1088/1757-899X/769/1/012069

4. Results

To test the effect of utilizing entropy values in the embedding process, M.S. Subhedar's method [12] is used to compare with proposed scheme. Both methods used RDWT and QR decomposition, the difference is that proposed scheme embeds the secret image blocks into cover image blocks according to blocks' entropy values. The secret image block with the lowest entropy value will be embedded into the cover image block with the lowest entropy value. The process continues until all secret image blocks have been embedded from lowest to highest entropy values.

Table 2. Comparison of PSNR and SSIM values between proposed scheme and M.S. Subhedar's scheme

Secret Image	M.S. Subhedar		Proposed Scheme	
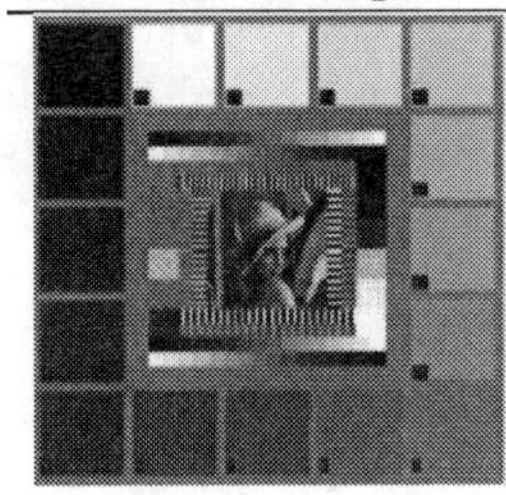			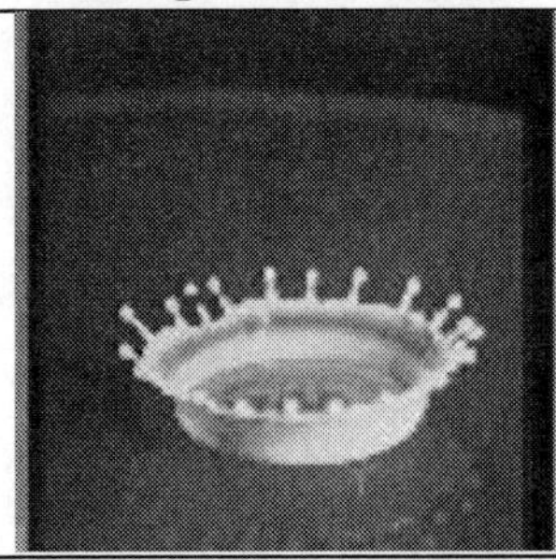	
	PSNR	54.8019	PSNR	57.3025
	SSIM	0.9975	SSIM	0.9995

	PSNR	49.2608	PSNR	57.3105
	SSIM	0.9971	SSIM	0.9996

	PSNR	49.0369	PSNR	57.3093
	SSIM	0.9963	SSIM	0.9997

PSNR	50.1480	PSNR	57.3333
SSIM	0.9966	SSIM	0.9996

The comparison results have shown significant improvement in terms of imperceptibility by comparing PSNR and SSIM values. By using entropy values to determine the embedded position of secret image information, the quality of image will be increased.

5. Conclusion

In transform domain, DWT is a popular method in information hiding techniques but it has one major flaw, which is the shift variant characteristic which arises from using the down-sampling operation. To overcome the shift problem, RDWT has been introduced. Moreover, RDWT can increase robustness. Also, RDWT provides more embedding capacity than DWT.

There are many matrix decomposition methods that can be combined to develop a better steganography schemes. Among all, QR decomposition and SVD is the most popular one as they help achieving better imperceptibility. By addressing the false-positive errors, modifying the elements in the orthogonal U matrix instead of direct embedding into the singular value S could solve the problem. When comparing the usual SVD which embeds in singular values with QR decomposition, the latter provides better imperceptibility and avoids false positive issue.

Therefore, an improved image steganography scheme based on RDWT and QR decomposition is proposed to embed grayscale image into grayscale cover image by analyzing the image texture using entropy.

By analyzing the image texture before embedding, secret image will be embedded according to the entropy values. Both cover and secret image are being segmented into non-overlapping blocks with identical block size. Then, entropy values generated from every image block will be sorted from the lowest value to the highest value. The embedding process starts by embedding the secret image block with lowest entropy value into the cover image block with lowest entropy value. The process goes on until all image blocks have been embedded. Embedding secret image into cover image according to the entropy values causes differences that HVS can less likely to detect because of the small changes. The proposed scheme gives higher PSNR values and better image quality.

Many different experiments will be carried out in the future in order to provide better analysis. Future work concerns deeper analysis of how the proposed scheme may affect imperceptibility, security and robustness by testing it using loads of different cover and secret images. Robustness test will also be carried out to see how proposed scheme reacts under different attacks.

References

[1] Thiyagarajan, P.; Aghila, G.; Prasanna, Venkatesan V., "Stego-Image Generator (SIG) – Building Steganography Image Database," CDBR-SSE Lab Department of Computer Science, Pondicherry University, Puducherry 605014, 2012.

[2] Prerna Gupta, Girish Parmar, "Image Watermarking using IWT-SVD and its Comparative Analysis with DWT-SVD," in International Conference on Computer, Communications and Electronics (Comptelix), 2017.

[3] Serdean C.V., Tomlinson M., Wade G.J., Ambroze A.M., "Protecting intellectual rights: Digital

watermaking in the wavelet domain," in Trends and Recent Achievements in Information Technology, 2002. K. Elissa, "Title of paper if known," unpublished.

[4] Shaoli Jia, Qingpo Zhou and Hong Zhou, "A Novel Color Image Watermarking Scheme Based on DWT and QR Decomposition," Journal of Applied Science and Engineering, pp. 193-200, 2017.

[5] L.-Y. Hsu, H.-T. Hu, "Robust blind image watermarking using crisscross inter-block prediction in the DCT domain," J. Vis. Commun. Image R., pp. 33-47, 2017.

[6] Lai, C.C, "An improved SVD-based watermarking scheme using human visual," *Optical Commununication,* pp. 938-944, 2011.

[7] Serdean C.V., Tomlinson M., Wade G.J., Ambroze A.M., "Protecting intellectual rights: Digital watermaking in the wavelet domain," in *Trends and Recent Achievements in Information Technology*, 2002.

[8] Bradley, A.P., "Shift-invariance in the discrete wavelet transform," in *Proc. VIIth Digital Image Computing: Techniques and Applications*, 2003.

[9] Hien T.D., Nakao Z., Chen Y.W., "RDWT domain watermarking based on independent component analysis extraction," *Advance Software Computing,* pp. 401-414, 2006.

[10] Nasrin M. Makbol, Bee Ee Khoo, Taha H. Rassem, "Block-based discrete wavelet transform singular value decomposition image watermarking scheme using human visual system characteristics," *IET Image Process,* pp. 34-52, 2016.

[11] F. JE., "The redundant discrete wavelet transform and additive noise," *Signal Processing Letters,* pp. 629-632, 2005.

[12] M.S. Subhedar, V.H. Mankar, "Image steganography using redundant discrete wavelet transform and QR factorization," *Computers and Electrical Engineering,* pp. 406-422, 2016.

[13] Q. Su et al., "Embedding color watermarks in color images based on Schur decomposition," *Optics Communications 285,* pp. 1792-1802, 2012.

[14] Sunil et al., "An Improved Image Steganography based on 2-DWT-FFT-SVD on YCBCR Color Space," in *International Conference on Trends in Electronics and Informatics*, 2017.

[15] Taha H. Rassem, Nasrin M. Makbol, and Bee Ee Khoo, "Performance evaluation of RDWT-SVD and DWT-SVD watermarking schemes," in *International Conference on Advanced Science, Engineering and Technology (ICASET)*, 2016.

[16] G.H. Golub, C.F. Van Loan, "Matrix Computations," Johns Hopkins University Press, Baltimore, 1989.

[17] C.-J. Ahn, "Ahn CJ . Parallel detection algorithm using multiple QR decompositions with permuted channel matrix for SDM/OFDM.," *IEEE Transactions on Vehicular Technology,* pp. 2578-2582, 2008.

[18] Manoj Kumar, Gursewak Singh, "Block based Image Steganography using Entropy with LSB and 2-bit Identical Approach," *International Journal of Computer Applications (0975 – 8887),* 2017.

[19] Swati Bhargava and Manish Mukhija, "Hide Image and Text Using LSB, DWT and RSA Based on Image Steganography," *ICTACT JOURNAL ON IMAGE AND VIDEO PROCESSING,* pp. 1940-1946, 2019.

[20] Hossain et al., "Variable Rate Steganography in Gray Scale Digital Images Using Neighborhood Pixel Iriformation," in *Proceedings of 2009 12th International Conference on Computer and Information Technology (ICCIT 2009)*, Dhaka, Bangladesh, 2009.

[21] I.J. Kadhim, P. Premaratne and P.J. Vial et al., "Comprehensive survey of image steganography: Techniques, Evaluations, and trends in future research," *Neurocomputing,* pp. 299-326, 2019.

[22] Anumol Joseph, K. Anusudha, "Robust watermarking based on DWT SVD," *International Journal of Signal & Image Processing,* 2013.

[23] Poonam, Shaifali M.Arora, "A DWT-SVD based Robust Digital Watermarking for Digital Images," in *International Conference on Computational Intelligence and Data Science*

(ICCIDS), 2018.

[24] F. Ernawan, M. N. Kabir, "A block-based RDWT-SVD image watermarkingmethod using human visual system characteristics," *The Visual Computer,* 2018.

[25] J L Divya Shivani, Ranjan K. Senapati, "False-positive-free, Robust and Blind Watermarking Scheme based on Shuffled SVD and RDWT," *Journal of Advanced Research in Dynamical and Control Systems,* 2018.

Acknowledgments
The authors sincerely thank Universiti Malaysia Pahang, Malaysia, for providing financial support for this work through UMP Postgraduate Research Grants Scheme (PGRS170372).

Reversible Watermarking Technique for Fingerprint authentication based on DCT

Mohamed Lebcir[1], Suryanti Awang[1], and Ali Benziane[2]

[1]Soft Computing & Intelligent Systems Research Group (SPINT), Faculty of Computing, University Malaysia Pahang, UMP, Lebuhraya Tun Razak 26300, Kuantan, Pahang, Malaysia.

[2]Faculty of science and Technology, University of Djelfa, Algeria.

mlebcir78@gmail.com

Abstract. In this paper, a new reversible and blind fingerprint image watermarking technique based on the differential method and discrete cosine transform (DCT) domains is presented. The focus is to increase the security of the fingerprint image in authentication systems. Two DCT-transformed sub-vectors are employed to embed the bits of the watermark sequence in a differential scheme. The original sub-vectors are acquired by the DCT transform on the host fingerprint image. In the extraction process, a minor variance between the sub-vectors that correspond to the watermarked fingerprint image directly allows access to the embedded watermark sequence; therefore, the extraction process doesn't require an original fingerprint. The original fingerprint image is then recovered from the watermarked fingerprint image based on the reversible watermarking technique. The similarity between the reversible fingerprint image and the original is considered, and we could extract minutiae points from it without a problem. The proposed technique is evaluated using 80 fingerprint images from 10 persons for each from FVC2002 fingerprint database. Eight fingerprint images have been taken from each person to be used as the template then the process was followed by embedding the watermark into each fingerprint image. The experimental results validate the proposed method able to give promising results in preserving the fingerprint image security.

Keywords: *Fingerprint, Fingerprint image watermarking; Frequency Domain; DCT; Reversibility*

1. Introduction

Automated systems with the implementation of Artificial Intelligence techniques widely used in many areas to enhance the quality of the system [1,2,3]. It's becomes more popular from late 1990 until now. An authentication system based on fingerprint image also does not left behind in this trend. A fingerprint image indicates the process of impressing the ridges and valleys pattern on the fingertip surface; the combination of their minutiae points then determines the uniqueness of a fingerprint [4,5] which permits the use of the fingerprint images in the user authentication process due to their uniqueness [6]. Thus, it is essential to protect the fingerprint's authenticity. In order to avoid issues of compromised fingerprint recognition system, three different research domains, namely fingerprint biometric for authentication, cryptosystem for confidential data and reversible watermarking for the integrity.

Digital watermarking could be used to check a fingerprint sample authenticity [6]. This approach is used to embed the watermark data into the original fingerprint image in the purpose of maintaining fingerprint ownership security. As an additional protection layer, we may encrypt the watermark data before the watermark is embedded [7]. Thus, the operator's key role is preventing attackers to gain access to the watermark. The reversible watermarking technique extracts minutiae points from fingerprint image watermarked without problem.

We may classify the watermarking techniques into three embedding domains after watermark pattern embedding location of the domain [7]. The domains are spatial domain, frequency domain and the multiresolution domain. Spatial domain of embedding fingerprint watermark directly into the pixels of the host image. A common technique used for watermarking in the spatial domain involves the substitution of the least significant bits (LSBs) with more significant bits in the image pixel [8]. The technique has several advantages such as less complication coupled with high payload. However, they are not sufficiently robust against the common forms of image processing attacks such as compression. In addition, this approach is highly vulnerable to piracy attacks as the watermark can be easily modified during the process of watermarked image transmission between a sender and a receiver.

Whereas, frequency domain denotes the rate of change in the values of pixel at the spatial domain. In the frequency domain, insertion of watermarks is made into the image transform coefficients. An inverse transformation is used to rebuild the watermarked image [9]. The commonly used watermarking transformation techniques in frequency domain are the Fast Fourier Transform (FFT) technique, and the Discrete Cosine Transform (DCT) [10]. The use of DCT algorithms for embedding watermark into fingerprint image is sufficiently for compression attacks. Thus, in this paper, DCT is implemented that based on reversible watermarking technique to enhance the security of the fingerprint recognition system for authentication purposes.

The next section of this paper will discuss about related work that combined watermarking and fingerprint biometrics for enhancing the security on fingerprint recognition system, followed by our watermarking technique methodology and ends with experimental results and discussion.

2. Related work

There are some notable researches in a combination of watermarking and fingerprint biometrics for enhancing the security purpose in fingerprint recognition system. A fragile authentication technique for watermarking fingerprint images was proposed by Pankanti and Yeung [11]. In their research, an authentication key was used to embed a watermark into a fingerprint image in the spatial domain. The method proposed has the ability pinpoint any part of the fingerprint image which has been altered. Their watermarking approach does not present observable loss to the process of the fingerprint authentication. However, the technique is not suitable for protected fingerprint sample transmission over network which is not secure. This is due to the high vulnerability of embedded watermarks towards attackers.

In another research, a method was presented by Gunsel et al. [12] and Uludag et al. [13]. This method safeguards the orientation of the quantized gradient in the domain of the embedded watermark. Two techniques for fingerprint image watermarking in the spatial domain were described by Gunsel et al. The first approach makes use of gradient orientation analysis during the process of embedding the watermark. Hence, none of the features extracted by making use of gradient information is altered. Likewise, the second technique conserves the singular points within the fingerprint image. Therefore, there is no disruption to the classifications of the fingerprint image that is watermarked using this approach. In similar manner, Uludag et al. put forward a spatial method for embedding watermarking fingerprint images without ruining the features of the fingerprint. With this approach, the characteristic features of the fingerprint are first detected. The watermark is then embedded into the fingerprint image in such a way that the fingerprint features are not distorted by the watermark. It is questionable if this method could trigger potential changes in the features of the fingerprint after the incorporation of the watermark data, or whether the features shall be retained. However, the entire single points as well as features of the fingerprint which were extracted using gradient information are secured within the fingerprint image.

In a different research, incorporation of a facial image was proposed to be hidden in the fingerprint image by A. K. Jain et al. [14]. This method involves inserting a user face's eigenface coefficients in the fingerprint image. Hence, during the process of decoding, the recovered facial image may be used to ascertain the originality of the image. In their research, the information from the user face was primarily used to verify the fingerprint image. A secret key is randomly generated and one-bit stream of the coefficients of the eigen face is then inserted into the pixels of selected fingerprint image. The process of embedding the face is in the spatial domain. Thus, it does not depend on the original image in order to extract the watermark (blind). The main disadvantage of this approach is it mainly works in the spatial domain, whereby, the embedded watermark is highly vulnerable to attack. This is because spatial domain makes use of direct values of image pixels.

In a similar approach, a scheme of watermarked spatial fingerprint which utilizes eigenfaces was proposed [15]. The watermark was embedded into the fingerprint through steganography in the 'sailboat' image. The solution was aimed at inscribing black squares around the minutiae in order to denote the ones that need to be dodged during the watermarking phase. After evaluation, this technique was proven to reveal good performance especially towards authentication and protection of fingerprint minutiae following the watermarking process. However, there is the challenge of computational complexity. This is mainly due to the long duration of time would be required to process a combined watermarking and steganography procedure. Other than that, this approach is the spatial embedding domain that was not sufficiently resistant to factors such as compression, noise as well as filtering attacks. This is associated with the pixels were changed thereby sacrificing the image quality.

A new watermark embedding method has been presented by Bousnina et al.; it consists of generating a secret key that can be used to designate the pixels to be watermarked. The fingerprint minutia points, thus, would be protected and security would be enhanced along the extraction stage of the watermark. The following step consists of inserting the face features (watermark) into the fingerprint image (cover); the fingerprint minutia points, thus, would be protected via Orthogonal Locality Preserving Projections (OLPP) method [16]. Although this approach resistance efficiency was not checked to confront other potential attacks such as geometric attacks, it was able to prevent the intentional attacks where the attacks would acquire the secret key in watermark extracting.

Ghany et al.'s based on wavelet approach, on the other hand, consists of embedding DNA data based multi-bit watermark into fingerprint images using DWT [17]. This method proved its robustness and straightforwardness, yet the original fingerprint template is required at the stage of extraction (non-blind). This method proved to be robust with high imperceptibility; however, the need for the original fingerprint when extracting the watermark decreased its security.

Another method was proposed by Abraham et al.; the technique is simple and blind, and it uses DWT in watermarking and in detecting digital image [18]. This method is characterised as blind which implies the unnecessity of the original fingerprint template along the extraction stage (blind). The embedded information robustness of in this approach proved to be acceptable.

Alkhathami et al's approach proposed a watermarking algorithm which can be embedded with the unique minutiae into fingerprint images so as to protect it from possible effects of the embedded watermark [19]. A new Dual-tree Complex Wavelet Transform (DTCWT) is used in this technique which had the advantages of preserving the fingerprint minutiae from potential embedded watermark effects. This technique, however, is non-blind; the watermark extraction process requires the the original image.

Alkhathami et al., adding to the last algorithm, introduced two watermarking techniques [20,21] in which both DWT and DCT techniques features are combined. These two techniques consist of combining the fingerprint of owner's grayscale facial image and other identification details in the model of binary text image into the fingerprint image [20]. While, in [21], Discrete Cosine Transform (DCT) algorithm is used in which two watermarks are combined for embedding into fingerprint images. These techniques achieved successful payload capacity without corrupting the minutiae points. Through this technique, more authentication factors will be added based on the watermark messages. However, this technique is characterised as non-blind due to the necessity for the original fingerprint image during the watermark data extraction stage.

Therefore, based on the related works, we can see that most of the researchers focus on non-blind watermarking. However, the problem with non-blind is that it prone to lack of security to the fingerprint image because original fingerprint image is needed in extracting the watermark data. Thus, we come out with our proposed solution in this paper that will utilize blind watermarking in frequency domain with reversible watermarking technique.

3. Methodology

The proposed algorithm is extended from the idea of Benoraira et al [22] that used the hybrid of DWT and DCT algorithms in line with the differential embedding technique. The technique of differential embedding depends on two DCT transformed sub-vectors to ensure the blind extraction of the watermark. Furthermore, the proposed algorithm added the Reversible Watermarking Technique for Fingerprint authentication based only on DCT blind that Benoraira et al. not used it in their method.

To embed watermark into fingerprint image, the reversible watermarking algorithm that based on Discrete Cosine Transform (DCT) is used in this proposed method. The transform DCT permits to divide the image into different blocks and easily embedding the watermark into selected blocks. In addition, this algorithm based on differential embedding technique that depends in two DCT transformed sub-vectors to ensure the blind extraction of watermark. Further, new idea has been added in this algorithm, could recovering the original fingerprint image from the watermarked fingerprint image based on reversible watermarking technique. Thus, could extracting minutiae points easily from reversible fingerprint image that seem as copy than the original.

The most common DCT definition of a 1-Dimension sequence x of length N is:

$$c(k)=\alpha(k)\sum_{i=0}^{N-1} x(i)\cos\left[\frac{(2i+1)}{2N}\right] \tag{1}$$

where

$$\alpha(k)=\begin{cases}\sqrt{2/N} & for \quad k=0 \\ \sqrt{2/N} & for \quad k=1,2..N-1\end{cases}$$

The corresponding inverse transformation for 1- Dimension sequence x of length N is:

$$x(i)=\sum_{k=0}^{N-1} \alpha(k)c(k)\cos\left[\frac{(2i+1)k\pi}{2N}\right] \tag{2}$$

DCT based watermarking is split into three levels: low, middle and high. The lower frequencies sub band includes the image most important visual parts where the majority of the signal energy lies. Compression and noise attacks delete the image high frequency component. The watermark is embedded through changing the middle frequency sub-band coefficients. This, therefore, will not affect the image visibility and the compression will not remove the watermark [23]. DCT is applied on the fingerprint image to obtain two sub-vectors for purpose watermarking and reversibility process based on embedding technique. The extraction is therefore blind due to differential embedding of the watermark in the resulting two DCT transformed sub-vectors.

The flowchart that shown in in figure 1 represents fingerprint authentication system using reversible watermarking technique. There are 3 major processes; watermarking embedding process, watermarking extracting process and reversibility watermarking process in the respective systems. The watermarking embedding system begins with performing a fingerprint scan using a fingerprint scanner to acquire a fingerprint image and the extracted features are stored in the database to be used during the feature matching. The embedding process is implementing on the fingerprint image that contains valid features as a host image and the bipolar {-1,1} binary sequence W of size L as a watermark. Using DCT and the differential embedding method which will be explained in detail later. Here, the watermark is embedded into fingerprint image for further use in the fingerprint-based authentication system that the possibly attacked. The next process, the extracted watermark is compared with the original watermark. If the extracted watermark is not genuine, the verification process is rejected, and the process is ended. Otherwise, the reversible watermarking system to extract new features from reversible watermarked fingerprint image, the new extracted features from reversible watermarking system are sent with the genuine features in the database to achieve matching together and the process is ended with the approved verification.

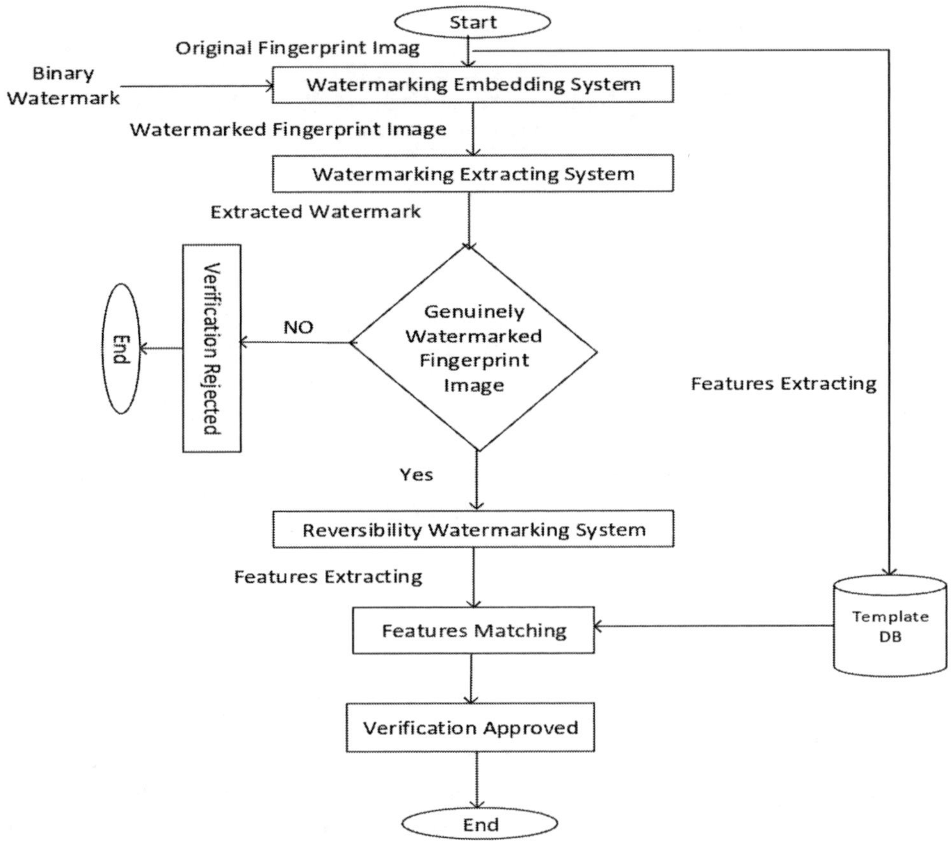

Figure 1. Fingerprint authentication system using reversible watermarking technique

The main steps of the proposed watermark embedding procedure can be described as follows:

3.1. Watermark Embedding Process:

This is the first process in proposed method. The proposed watermark embedding process is shown in figure 2 and the embedding process can be described as follows:

• **Step 1**: perform a zigzag scanning on coefficients matrix which obtained from receiving original fingerprint image F_o . The aim is to convert it into a vector of complex coefficients $Z(n)$ to easily process of watermarking. Where $n=1...N$, and N size of coefficients matrix.
• **Step 2:** the vector of coefficients z is decomposed into two (correlated) sub-vectors z_1 and z_2 to allow applying differential embedding technique (will explain it on details in next steps) as follow:

$$z_1(k) = Z(2k) \quad (3)$$

$$z_2(k) = Z(2k-1) \quad (4)$$

where $k = 1 \ldots N/2$.

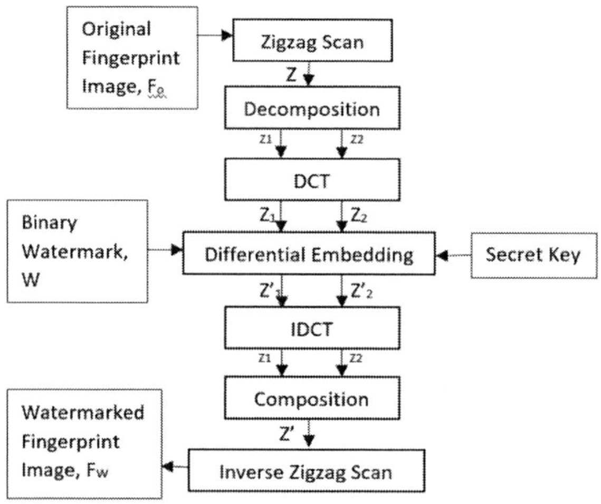

Figure 2. The embedding procedure of proposed reversible watermarking method

• **Step 3:** DCT is performed on z_1 and z_2 to produce their DCT transformed forms Z_1 and Z_2 which permit an image to be decomposed into various blocks in such way that making it much easier to embed the watermark into selected blocks.

$$Z_1 = DCT(z_1) \quad (5)$$

$$Z_2 = DCT(z_2) \quad (6)$$

• **Step 4**: The watermark (binary sequence W of size L) sequence bits $W(i)$ for $i = 0,2..L-1$, is inserted into the transformed vectors Z_1 and Z_2 using a differential embedding method that is a mathematical equation will describe in equations 7 and 8. It allows to include the watermark randomly into the transformed vectors Z_1 and Z_2 which obtained from DCT transform to use in extraction stage without need the original data (blind). This will output two changed (by watermarking) sub-vectors Z'_1 and Z'_2 as follows:

$$Z'_1(i') = Z_1(i') + \alpha W(i) \quad (7)$$

$$Z'_2(i') = Z_2(i') - \alpha W(i) \quad (8)$$

where α is the gain factor and i' are the random places into the highly strength band of Z_1 and Z_2 are the embedded watermark bits. These places are the elements of a vector r which can be generated the usage of a random permutation operation:

$$i' = r(i) \quad (9)$$

$$r = RandPerm(S, a, b) \quad (10)$$

where S is the seed of the associated Pseudo Random Number Generator (PNRG), a and b are the beginning and the ending places of the highly strength band applied to put in the watermark. Therefore, the user's secret key is $K=(S,a,b)$, which prevents the watermark from tempering or unauthorized access by attackers

• **Step 5:** The inverse DCT $ICDT$ is performed on Z'_1 and Z'_2 using the following equations 11 and 12. The purpose of performing the $ICDT$ is to reconstruct vector of complex coefficients Z embedded by watermark.

$$z'_1 = IDCT(Z'_1) \tag{11}$$

$$z'_2 = IDCT(Z'_2) \tag{12}$$

• **Step 6:** the two changed sub-vectors z'_1 and z'_2 is combined in the composing step using the opposite operation in equations 3 and 4 in order to produce the modified vector Z' to reconstruct complex coefficients vector embedded by watermark.

• **Step 7**: The modified vector is converted into the matrix employing the inverse of the zigzag scanning operation to reconstruct coefficients matrix embedded by watermark, to attain the watermarked fingerprint image F_W.

There are small modifications in original coefficients matrix. During the embedding process, the coefficients of the fingerprint image are modified but still can obtained almost the original fingerprint image without affected their features that had been due to the embedded watermark.

3.2. Watermark Extraction Process:

The watermark extraction process follows the same steps as the embedding process until Step 3 where the extraction is taking place as shown in figure 3.

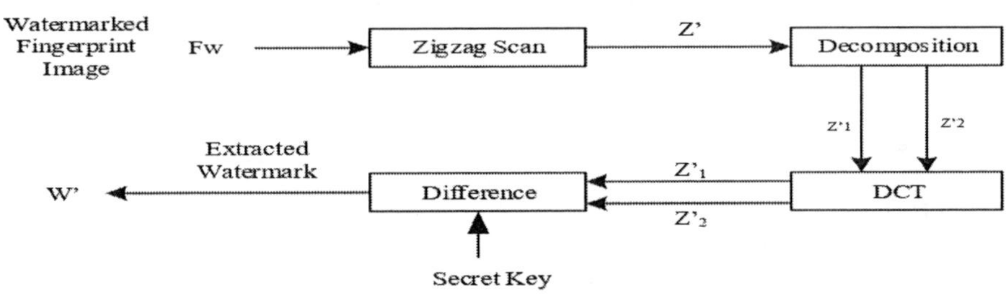

Figure 3. The extraction procedure of proposed reversible watermarking method

• **Step 1:** perform a zigzag scanning on coefficients matrix which obtained from watermarked fingerprint image F_W. The aim is to convert it into a vector of complex coefficients $Z'(n)$ to easily process of watermarking. Where $n=1..N$, and N size of complex coefficients matrix.

• **Step 2:** Decompose the vector of coefficients Z' to two related sub-vectors z'_1 and z'_2 is performed next.

• **Step 3:** perform DCT on z'_1 and z'_2 to produce their DCT-transformed forms Z'_1 and Z'_2.

If the enter fingerprint image is the watermarked one, and by using analogy with the embedding operations, the extraction operations will deliver two sub-vectors Z'_1 and Z'_2. Consequently, the difference between them has a proportional dating with the watermark sequence W:

$$\Delta Z(i) = Z'_1(i') - Z'_2(i') = 2\alpha W(i) \tag{13}$$

with $i = 1..L$; and i' are the random places in which the watermark bits are embedded. These places are determined clearly by using recreating the vector r , the usage of the user-selected secret key and the random permutation as shown in equations 7 and 8.

Finally, because the difference $\Delta(i)$ may vary from $+1/-1$ values, therefore, a difficult challenge characteristic is implemented on it to be able to recover the source bits of the watermark:

$$W'(i) = \begin{cases} +1 & if \quad \Delta Z(i) \geq 0 \\ -1 & otherwise \end{cases} \tag{14}$$

After that, a comparison between tow watermarks W and W' is performed. If there is a similarity between it, then begins work of extracting features from watermarked fingerprint image through reversible watermarking process for using in fingerprint-based authentication system.

3.3. Reversible Watermarking Process:

The reversible watermarking process begin as the same step 4 in the embedding process, where we replace the original watermark with the extracted watermark obtained from watermark extraction process as shown in figure 4.

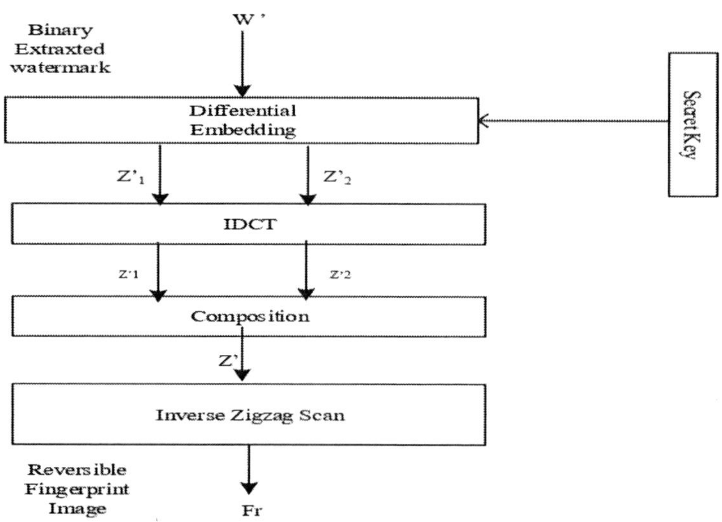

Figure 4. The reversibility procedure of the proposed reversible watermarking method

• **Step 1**: The extracted watermark (binary sequence W' of size L) sequence bits $W'(i)$ for $i = 0,2..L-1$, is inserted into the transformed vectors Z_1 and Z_2 using a differential embedding method that is a mathematical equation will describe in equations 21 and 22. It allows to include the watermark randomly into the transformed vectors Z_1 and Z_2 which obtained from DCT transform to use in extraction stage without need the original data (blind). This will output two changed (by watermarking) sub-vectors Z'_1 and Z'_2

Step 2: The inverse DCT IDCT is performed on Z'_1 and Z'_2 using the following equations 9 and 10. The purpose of performing the IDCT is to reconstruct vector of coefficients Z' embedded by extracted watermark.

IOP Conf. Series: Materials Science and Engineering **769** (2020) 012070 doi:10.1088/1757-899X/769/1/012070

Step 3: The two changed sub-vectors z'_1 and z'_2 is combined in the composing step using the opposite operation in equations 27 and 28 in order to produce the modified vector z' to reconstruct coefficients vector embedded by extracted watermark.

Step 4: The modified vector is converted into the matrix employing the inverse of the zigzag scanning operation to reconstruct coefficients matrix embedded by extracted watermark, to attain the reversible fingerprint image F_r.

There are small modifications compared of original coefficients matrix. Thus, during the second embedding process by extracted watermark, the coefficients of the reversible fingerprint image obtained are modified but it is very close to the original, we could extract minutia without problem.

4. Results and discussion

Several experiments in this section have been conducted aimed at evaluating the performance of the adopted watermarking technique. The performance of the reversible fingerprint image-watermarking algorithm is evaluated by employing a benchmark fingerprint image database FCV2002-DB2 and compare it with previous works that had used the same database. To conduct this experiment, 80 fingerprint images have been taken from 10 persons from which the template is set using eight fingerprint images of each person; after that, the watermark is embedded into each of the taken fingerprint images.

Firstly, we compare the input original fingerprint images in databases against its corresponding watermarked fingerprint images to find positive matching results between them for minutia points and watermark data. Secondly, we embed the extracted watermarked into watermarked fingerprint image based on embedding equation that shown in equations 21 and 22 to obtain the reversible fingerprint image. Finally, we compare the reversible fingerprint image against its corresponding original fingerprint image to extract minutia easily without problem. It is because the reversible fingerprint image is very close to the original, as proved from the experiments results. In order to evaluate the obtained reversible and watermarked of fingerprint image quality, we utilised the peak signal to noise ratio (*PSNR*) [16].

Fingerprint recognition tools is used to measure the rate of similarities between fingerprints features. The purpose of the similarity measurement is to find a positive score of minutia points matching between original fingerprint images in databases and its corresponding which the possibly attacked and distorted. *PSNR* represents the ratio noise between the fingerprint image before and after operations of watermarking and its reversibility. This ratio can be used to determine the percentage of deformation on the original image caused by the embedding of the watermark. The high ratio of *PNSR* means the watermarked and reversible of fingerprint image quality is high, thus, the watermarked and reversible image is similar to the original image after operations of watermarking and its reversibility means good imperceptibility. figure 5 shows PSNR values obtained from our experiments versus the gain factor α to purpose embedding coefficients.

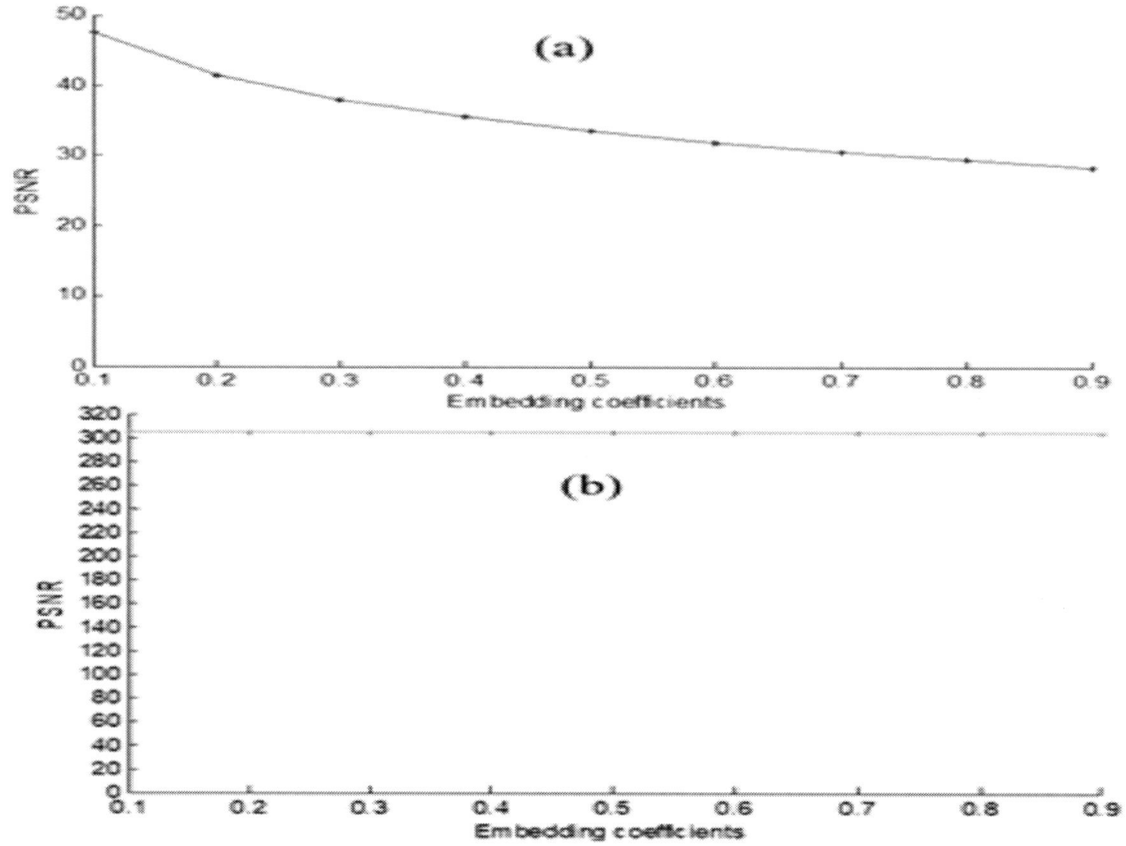

Figure 5. PSNR fingerprint watermarked image (a), PSNR reversible fingerprint image (b) for the proposed reversible watermarking method

Looking at figure 5(a), the higher α values make lower PSNR of the watermarked fingerprint images. In contrast, for reversible fingerprint image, the PSNR value increases with α values until it stabilizes at value (α=0.2) as show in figure 5(b).

The DCT with the differential extracting method is used to extracting watermark embedded into fingerprint image watermarked which may be affected due to watermarking process, the extracted watermark is compared with its corresponding original watermark which not need the original fingerprint image. We used the bit-correct ratio (BCR) in the process of evaluating the extracted watermark quality by measuring the rate of the similarity between the original and the extracted watermarks w and w' respectively. The bit-correct ratio (BCR) is the ratio of correct extracted bits to the total embedded bits number. It is worth mentioning that since BCR generates further values detailed scale, it is getting widely spread and used. [17]. Thus, it was chosen as to measure the similarity of watermark data that have used as sequence binary in the proposed method, if the rate of BCR is equal to 100%, it's means that the extracted watermark is equal to the original. That means, the technique is achieving stronger robustness against watermarking process. figure 6 shows BCR rates of the extracted watermark obtained from our experiments versus the gain factor α. Based on that figure, the BCR is getting better at α value of 0.2 and higher.

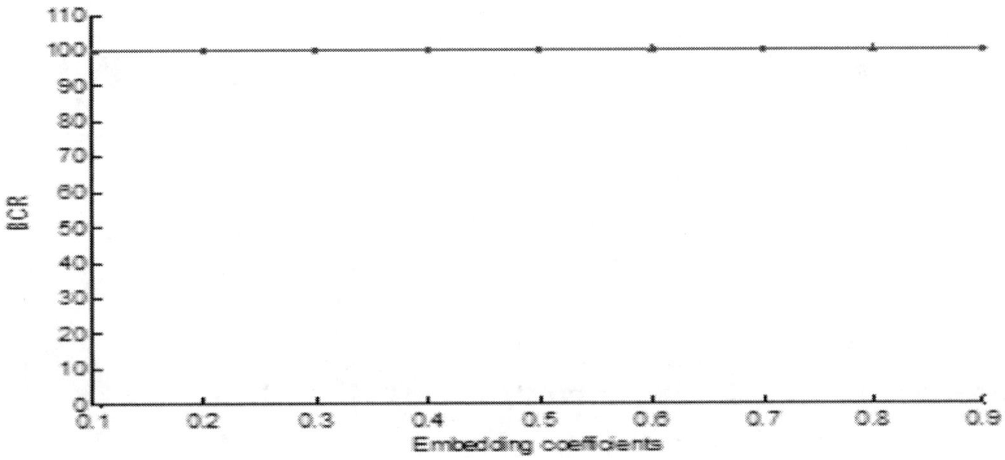

Figure 6. BCR extracted watermarked of the proposed reversible watermarking method

4.1. Gain Factor Selection:

Other than above experiment, several experiments are conducted to observe the invisibility, robustness and reversibility. These experiments are under the gain factor selection experiment. The aim of this experiment is to observe if the proposed technique able to increase the security of the fingerprint recognition system. In order to select the suitable values of the gain factor in equations 5, 6, 20 and 21 that fulfil both the invisibility, the robustness and reversibility requirements of the watermarking. In addition, due to the random permutation value in equations 8 and 23, the experiment is deployed 20 times with what is agreed with our random algorithm for embedding the watermark data in order to obtain good stable results. Further, embedding the watermark data in various location to increase the security against intentional attacks. In the proposed algorithm based on databases, FCV2002-DB2 which were tested on 80 fingerprint images from 10 persons and there are 8 fingerprint images for each person. As shown in figure 5 and 6, the best trade-off between visual quality and watermark robustness is achieved with the value of $\alpha = 0.2$. Therefore, value of $\alpha = 0.2$ is used as the default value of the gain factor. figure 7 shows the original, the watermarked image and reversible image along with their corresponding histograms in databases FCV2002-DB2. It is clearly noticeable that watermarking process reversibility has not affected of the fingerprint image visual quality as shown in each if the histogram graphs. Furthermore, values of *PNSR* and Matching score that obtained in experiments as shows in Table 1, prove that the fingerprint image has not been affected by reversibility of watermarking process.

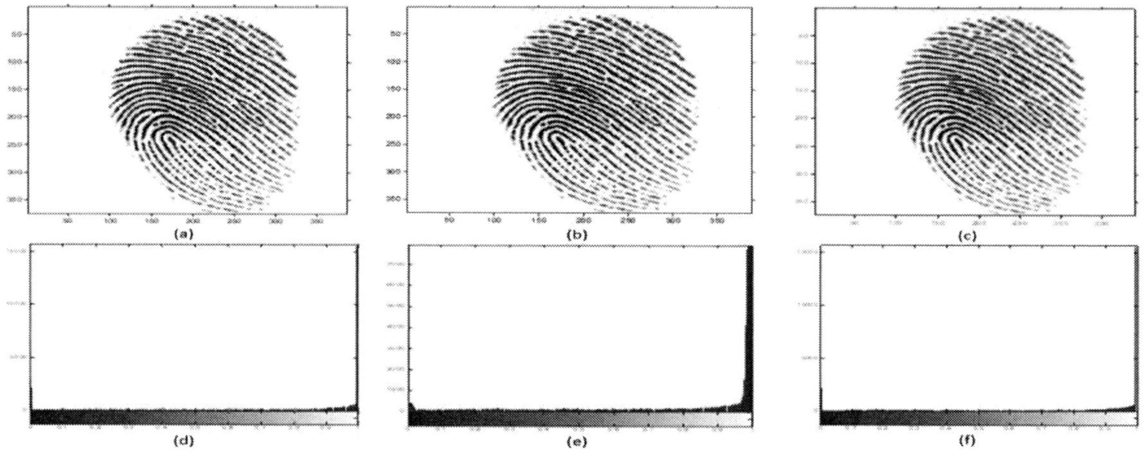

Figure 7. The original cover image (a), watermarked image (b), The reversible watermarked image (c), histogram of the original cover (d), histogram of the watermarked image (e) and histogram of the reversible watermarked image (f).

The average of *PSNR* values, *BCR* rates and the matching score values between the original fingerprint images and its corresponding in database is shown in Table 1. Looking at that table, *PSNR* results obtained for reversible and watermarked fingerprint image from our experiments are acceptable, where, proposed method achieved over 47 dB for *PSNR* value of the watermarked image and 301dB for reversible image in database. Whereas, the *BCR* rates of the extracted watermark and the minutia score matching of watermarked fingerprint images without attacks are almost achieved 100% for all experiments.

Table 1. PSNR values, BCR rate and the Matching score value of the proposed reversible watermarking technique

DB	PSNR watermarked image	PSNR reversible image	BCR	Matching score
FVC2002-DB2	*47,53*	*301,67*	*99,61*	*1*

5. Conclusions

Briefly, this paper introduces a blind, robust but simple technique of reversible watermarking based on DCT domains. A differential technique is performed on both transformed sub-vectors during the embedding process stage in order to fulfil the extraction of the watermark by using the difference of the corresponding watermarked sub-vectors. Then, the extracted watermarked in another embedding process based on differential technique is used to obtain a reversible fingerprint image. The reversible fingerprint image is very close to the original, as proved from the experiments results. All in all, the results obtained from the experiments confirms that the adopted technique achieves a high visual quality of the watermarked and reversible fingerprint image marking over 47 dB for PSNR per watermarked image, and over 301dB for reversible fingerprint image. Thus, extracting minutiae will be easy without problem. Besides, the achieving high strong of security in terms of correlation of extracted watermark data as the accepted results obtained of *BCR* values of extracted watermark and score matching of minutiae points for fingerprint image (cover image). The proposed technique will add a factor of authentication to the authorization process by combining both identity of the user and biometric features with a high security when access to the fingerprint recognition system.

Acknowledgment

We would like to show our gratitude to Universiti Malaysia Pahang (RDU vote number RDU180380) for supporting this study.

References

[1] Awang, S., Azmi, N. M. A. N. (2018). Vehicle Counting System Based on Vehicle Type Classification Using Deep Learning Method. In IT Convergence and Security 2017 (pp. 52-59). Springer, Singapore

[2] Al-Saffar, A., Awang, S., Tao, H., Omar, N., Al-Saiagh, W., Al-bared, M. (2018). Malay sentiment analysis based on combined classification approaches and Senti-lexicon algorithm. PloS one, 13(4), e0194852.

[3] W. M. Syahrir, A. Suryanti, N. N. Ain, N. N. (2007, September). Egg_s Bloodspot Detection Using Image Processing. In Second International Conference on Innovative Computing, Informatio and Control (ICICIC 2007) (pp. 424-424). IEEE.

[4] A.K. Jain, S. Prabhakar, L. Hong, S. Pankanti, Filterbank-based fingerprint matching, IEEE Trans. Image Process. 9 (2000) 846–859.

[5] S. Awang, J. Sulaiman, J., N. Noor, K. Mohd, L. Bayuaji (2017). Comparison of accuracy performance based on normalization techniques for the features fusion of face and online signature. Advanced Science Letters, 23(11), 11233-11236.

[6] G. Balamurugan, A Fingerprint Based Reversible Watermarking System For The Security Of Medical Information, 2016 World Conf. Futur. Trends Res. Innov. Soc. Welf. (Startup Conclave). (2016) 1–6. doi:10.1109/STARTUP.2016.7583901.

[7] A. Noore, R. Singh, M. Vatsa, M.M. Houck, Enhancing security of fingerprints through contextual biometric watermarking, Forensic Sci. Int. 169 (2007) 188–194.

[8] S.M. Mousavi, A. Naghsh, S.A.R. Abu-Bakar, Watermarking techniques used in medical images: a survey, J. Digit. Imaging. 27 (2014) 714–729.

[9] M. Lebcir, S. Awang, Fingerprint Image Watermarking in Spatial , Frequency and Multiresolution domain : Techniques and challenges, 2018. http://umpir.ump.edu.my/id/eprint/23029/7/Fingerprint Image Watermarking in Spatial1.pdf.

[10] F.Y. Shih, S.Y.T. Wu, Combinational image watermarking in the spatial and frequency domains, Pattern Recognit. 36 (2003) 969–975.

[11] M.M. Yeung, S. Pankanti, Verification watermarks on fingerprint recognition and retrieval, J. Electron. Imaging. 9 (2000). doi:10.1117/12.344704.

[12] B. Gunsel, U. Uludag, A. Murat Tekalp, Robust watermarking of fingerprint images, Pattern Recognit. 35 (2002) 2739–2747. doi:10.1016/S0031-3203(01)00250-3.

[13] U. Uludag, B. Günsel, M. Ballan, A Spatial Method for Watermarking of Fingerprint Images., in: PRIS, 2001: pp. 26–33.

[14] L.R. Haddada, I.H. Trimech, N.E. Ben Amara, A biometric watermarking approach of fingerprint images by DLDA Gabor face features without altering minutiae, IPAS 2016 - 2nd Int. Image Process. Appl. Syst. Conf. (2017) 1–6. doi:10.1109/IPAS.2016.7880135.

[15] A. K. Jain, U. Uludag, R.-L.H.R.-L. Hsu, Hiding a face in a fingerprint image, Object Recognit. Support. by User Interact. Serv. Robot. 3 (2002) 756–759. doi:10.1109/ICPR.2002.1048100.

[16] N. Bousnina, S. Ghouzali, M. Lafkih, O. Nafea, M. Mikram, W. Abdul, D. Aboutajdine, Watermarking for protected fingerprint authentication, in: 2016: pp. 127–131.

[17] K.K.A. Ghany, G. Hassan, A.E. Hassanien, H.A. Hefny, G. Schaefer, A.R. Ahad, A Hybrid Biometric Approach Embedding DNA Data in Fingerprint Images, (2014) 1–5.

[18] J. Abraham, V. Paul, A blind watermarking method for fingerprinting digital images, in: 2016 Int. Conf. Data Min. Adv. Comput., IEEE, 2016: pp. 145–149. doi:10.1109/SAPIENCE.2016.7684124.

[19] M. Alkhathami. (2015). Watermarking techniques for genuine fingerprint authentication.

[20] M. Alkhathami, F. Han, R. Van Schyndel, Fingerprint authentication enhancement using combination of DWT and DCT, Proc. 2015 10th IEEE Conf. Ind. Electron. Appl. ICIEA 2015. (2015) 387–392. doi:10.1109/ICIEA.2015.7334144.
[21] M. Alkhathami, F. Han, R. Van Schyndel, Fingerprint image protection using two watermarks without corrupting minutiae, Proc. 2013 IEEE 8th Conf. Ind. Electron. Appl. ICIEA 2013. (2013) 1151–1155. doi:10.1109/ICIEA.2013.6566540.
[22] Benoraira, A., Benmahammed, K., & Boucenna, N. (2015). Blind image watermarking technique based on differential embedding in DWT and DCT domains. Eurasip Journal on Advances in Signal Processing, 2015(1). https://doi.org/10.1186/s13634-015-0239-5.
[23] K. Ramani, E. V Prasad, S. Varadarajan, DWT Based Watermarking For Biometric Data, Indian J. Inf. Sci. Technol. 3 (2007) 35–41.

Transforming Malware Behavioural Dataset for Deep Denoising Autoencoders

Mohd Razif Bin Shamsuddin, Fakariah Hani Hj Mohd Ali, Mohd Shahril Bin Zainol Abidin
Fakulti Sains Komputer & Matematik, Universiti Teknologi MARA, Malaysia

Corresponding author's e-mail address: razif@fskm.uitm.edu.my

Abstract. This research is a part of a major research on automation of malware identification using Deep Denoising Autoencoders. Malicious software, or in short called malware refers to any software designed to cause damage to a single computer, server, or computer network. This malware term includes all kind of malicious software such as computer virus and spyware. All these malicious malware behaviour is monitored, logged and recorded using a cuckoo sandbox with the help of an x86 hosted supervisor software. The intent of recording the malware behaviour is to understand the pattern of behaviour of each known malware family. This collected data will be further trained to a Deep Denoising Autoencoders to automate the identification process of new malware within the identified malware families. However, the raw behaviour data is not suitable for an optimum training process. This paper will discuss the process of transforming the text based behavioural dataset to a more suitable dataset for deep learning purposes. At the end of the research a cleaned bit string format that should represent a unique malware behaviour will be produced.

1. Introduction

This paper will explain the process of transforming malware executable dataset into bit-string format that is suitable for Machine Learning input. Explanation and appropriate figure are shown to give clear look on each process that will be implemented throughout this research paper.

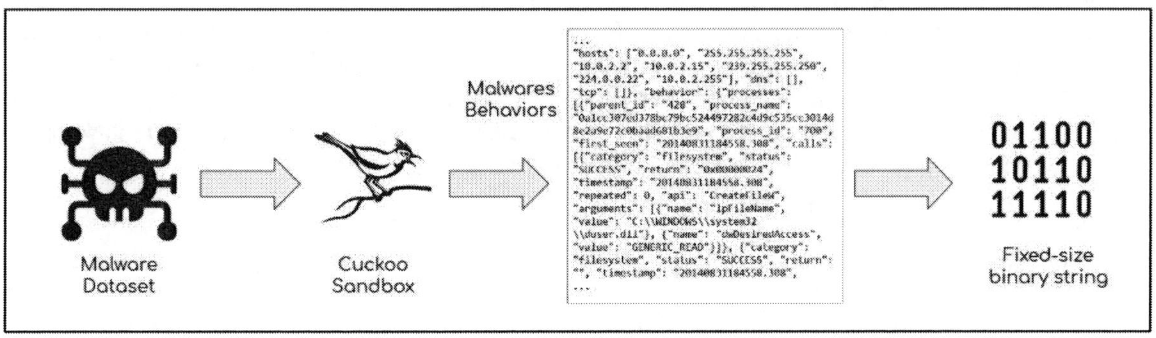

Figure 1. Transformation of malware dataset into bit-string format data

As the malware dataset consists only binary executables (raw format), several processing steps need to be done to gain useful data representation from every each of the malware samples. Behaviors extraction technique was chosen as this shown to be very reliable to generalize malware and their harmful actions. Moreover, generalization of malware through its behaviors proved to be a popular method among literature, and it allows Machine Learning to predict new unseen malware even if the malware has been modified to evade traditional detection system [1][7].

2. Malware Family Scopes

The scopes of this study comprised the following families of malware:

2.1. Cerber (Ransomware)

Cerber ransomware was first seen in the mid of 2015. Cerber is categorized as a Ransomware-as-aService (RaaS), where the developer will rent the ransomware functionality with split profit between the buyers and the sellers. Among many others, it is a type of ransomware that displays ransom message after it has infected the user's computer. It encrypts files and adds an extension ".cerber" which means that particular files have been encrypted. Furthermore, like any other high-profile malware, it has a "malware factory" form; which means it uses code obfuscation technique that can automatically generate large volumes of unique-hash malware variants from the original malware code [2][4].

2.2. Cryptowall (ransomware)

CryptoWall is a ransomware family that is designed to use sophisticated encryption (RSA-2048) algorithm to make files inaccessible on the targeted computers. Malware researchers spotted the first version of ransomware in 2013. Since then, crypto-virus was updated several times, with additional functionalities (such as stealth mode and deletion of volume shadow copies) have been added in order to avoid ongoing antivirus detections [2][4].

2.3. GandCarb (ransomware)

GandCrab ransomware was discovered as Ransomware-as-a-Service (RaaS) near the end of January 2018 and soon became the most popular and widely used ransomware of the year. GandCrab was potentially born from the need to monetize encrypted data from organizations further by customizing the ransom notes based on the victim's profile and the type of encrypted data. As a result, demand for GandCrab ransom could be between $600 and $700,000 per victim. This behavioral change has probably led to a significant increase in revenue for cybercriminals, especially since they started to deliver it as-a-service [2][4].

2.4. Petya (ransomware)

Petya is a ransomware encryption family that discovered for the first time in March 2016. The malware targets Microsoft Windows systems which it infects the master boot record to execute a payload that encrypts the file system table on a hard drive and prevents Windows from booting. Petya has evolved in several ways. One such example can be seen in June 2017, where a new variant of Petya (dubbed as NotPetya by researchers) was used at a global scale cyberattack and could propagates itself via computer worm technique [2][4].

2.5. Sality (computer botnet)

As the only non-ransomware type malware inside this project scope, Sality surely has its own reputation. Sality is a family of malware which infects file on Microsoft Windows systems and made its first debut in 2003. It has gone through several evolutions and advancement over the years to become a dynamic, enduring and full-featured form of malicious code. Because of its continued development and capabilities, Sality is one of the most complex and formidable forms of malware to date [4][5].

2.6. Wannacrypt (ransomware)

WannaCrypt is a ransomware computer-worm that targets the family of Microsoft Windows operating systems. It was first discovered on May 2017, targeting Microsoft Windows system and has affected more than 230,000 computers in 48 hours to over 150 countries. The ransomware demanded payment to unlock the infected system. WannaCrypt caused ambulances to be diverted, shutting down nonemergency services, nabbed machines at Telefonica in Spain, and affected airlines and ministry [2].

3. Cuckoo Sandbox to Record Malware Behavioural data

To obtain the malware behaviours in the form of text data, computer sandbox has been chosen to articulate this task. A sandbox is a computer controlled-environment in which it can monitor program behaviours through several logging mechanisms such as API tracing, system-call logger and man-inthe-middle network traffic recorder. Among every sandbox that has been released into the internet, Cuckoo Sandbox is the one that shown to be the most potential, as it is constantly get updated with new features over time. Furthermore, Cuckoo Sandbox could monitor executable behaviours until the kernel level, where it is the lowest point of every operating system. Using this feature, it helps computer sandbox to monitor any executable behaviours even if the malware tries to hide by injecting into any kernel modules [3].

To achieve these feet, Cuckoo Sandbox has been set up together with the help of an x86 hosted supervisor software. Oracle VirtualBox has been chosen as it is open source and has great support from the developers of Cuckoo Sandbox themselves. To fully utilize host resources for more efficient and less consuming analysis, a cluster of three sandboxes have been set up. These three sandboxes have the hardware specifications which are Windows 7 x64 bit and 2GB of memory access. An operating system in which Windows 7 has been chosen as this OS is popular among malware analysts and can run most of modern malware without much hassle as current Windows version put many protections in order to thwart malware infections.

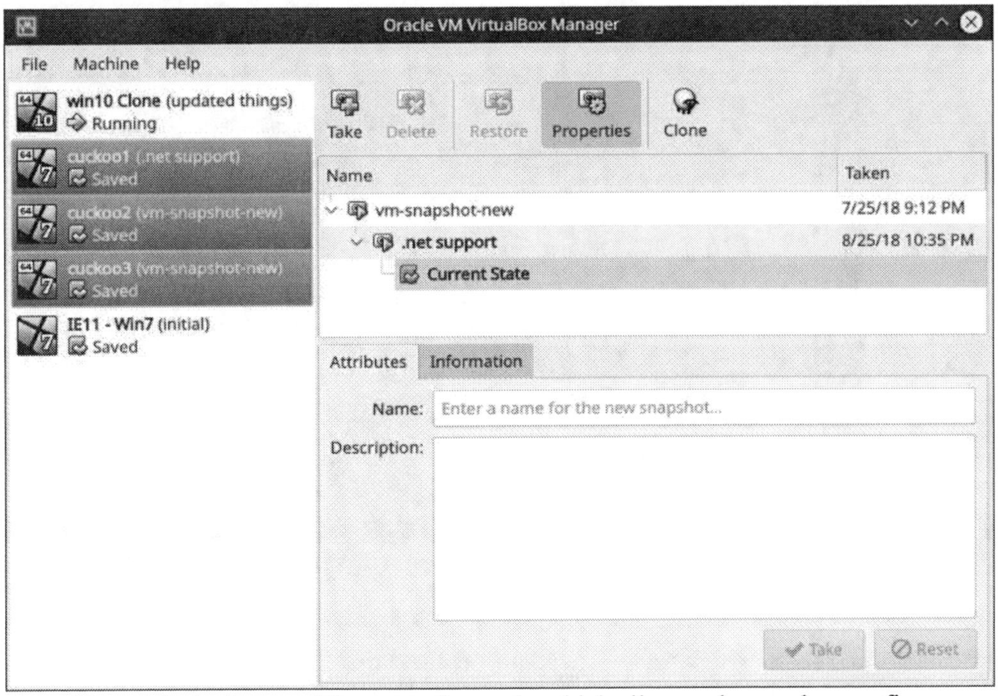

Figure 2. A cluster of three sandboxes in which all named as cuckoo prefix

In order to let Cuckoo Sandbox uses these three sandboxes, further configuration files have been written to point out which virtual machines that it should use when processing and analyzing malware. Using `cuckoo` command-line interface program, all malware has been submitted into Cuckoo Sandbox's system, in which it uses queue technique to process malware which has been submitted first into the system.

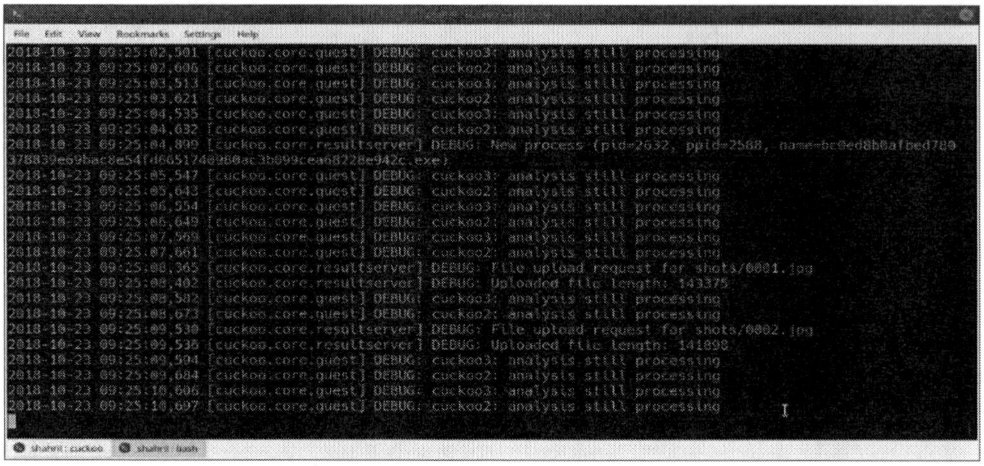

Figure 3. Cuckoo Sandbox interface while analyzing executable.

Figure 4. Executable runtime's behaviors produced by Cuckoo Sandbox.

By using these clustering and sandboxing techniques, Cuckoo Sandbox was able to analyze 70 executables in one-hour timeframe, speeding up the time required to analyze the malware and its behaviors. For every executable that has been analyzed by Cuckoo Sandbox, it will produce a report file containing the executable behaviors while it is running inside the computer sandbox. The report produced has the format of JavaScript Object Notation (JSON), in which this format can be easily parsed by common JSON parser.

However, as the recorded behaviors report file is in the JSON format, there is still another preprocessing step that needs to be done. The problem with this report file is that for every malware recorded behaviors, the output of the report file is going to be varied, depending on how much behaviors the malware spits out. This data representation cannot be fed into Machine Learning algorithms as most of them expected the size of the input to be fixed. One such example is neural network; it requires the input to be a vector of fixed-size or otherwise cannot proceed because the different number of neurons in the input layer does not match with input vector size.

4. Using n-Gram Extraction Method to Produce Fixed Sized Binary Files for Machine Learning

As explained in the previous section, raw malware data representation cannot be fed into Machine Learning algorithms as it is. A raw malware data behaviour contains many text based information such that is not suitable for Machine Learning training process. Thus, a method to transforms varying sized behavioural report file into sets of fixed-size binary string is needed. In order to process the datasets, a method from Natural Language Processing (NLP) is used to complete this task. One such method is by using n-gram extraction. The simplest form of n-gram extraction is *1*-gram extraction (or called as unigram). 1-gram is an algorithm which finds the most frequent words in text sample [4][5][6].Below are the steps taken for the unigram's data transformation process.

4.1. Step 1, For all report files, count all occurrences of unigram words.

Treat all reported JSON file as normal text file. Split the text into unigrams by doing text tokenization such as splitting by spaces and removing word which size is less than or equals to three. Build a dictionary which for every unigram, count its occurrences.

4.2. Step 2, Sort all words by their frequency.

Sort the dictionary with respect to their occurrences, using Python's sort function in which it uses Tim sort algorithm to speed up the sorting operation.

4.3. Step 3, Choose top 10,000 unigram words.

Split the dictionary and choose top 10,000 most frequent unigrams. As the dictionary has first been sorted out, this operation can be done quickly.

4.4. Step 4, Map each executable's unigrams with top 10,000 unigrams.

By having top 10,000 unigrams, bit-string vector can be created by mapping every unigram inside each executable and mark it 1 if the unigram exists or 0 if otherwise.

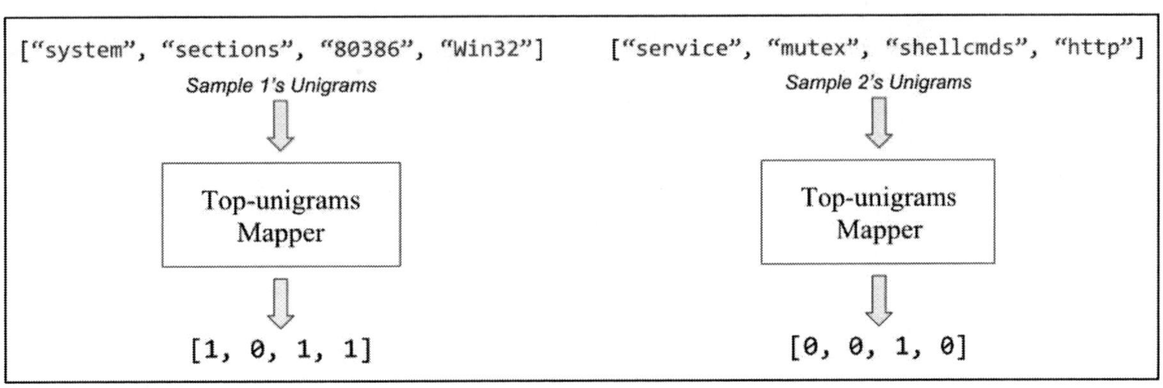

Figure 5. Example of bit-string conversion based on samples' unigrams.

All the processes listed above have been done with the help of Python programming languages. The Python script will parse all reports, tokenize every text, and find top 10,000 most occurrences unigram. Developing the script is technically challenging as with every report averagely sized more than 50 megabytes, it is time-consuming to process all 6,000 executable samples. Many optimizations have been

IOP Conf. Series: Materials Science and Engineering **769** (2020) 012071 doi:10.1088/1757-899X/769/1/012071

applied such as using Python's dictionary data structure which has a great algorithmic complexity to speed up the occurrences counting process.

Once the script has done processing, for every executable sample it will generate a bit string representation of ones and zeros as its data. These ones and zeros are treated as the executable's features and will have fixed-size of 10,000 in length. By having this representation of input, machine learning algorithms such as neural network will not have a problem dealing with it for learning purposes.

5. Data Pre-Processing Result

As explained earlier, several methods have been applied to the malware samples to convert the file executable representation into features that can be understood by machine learning algorithm.

The first process manage to capture all behavioural data from the malware executable samples. Using 6,000 executable samples, all of them have been submitted into Cuckoo Sandbox for dynamic analysis. It took roughly one minute for each executable sample, thus approximately 100 hours of processing time to successfully examine the malwares and capturing their behaviours.

However, as mentioned previously in Section 3 where each JSON report file going to be varied depending on the behaviors spitted out by the malware thus incompatible with machine learning that expecting input data to be all the same. To solve this problem, pre-processing steps need to be applied for all these JSON behaviors file. Each figure should have a brief caption describing it and, if necessary, a key to interpret the various lines and symbols on the figure.

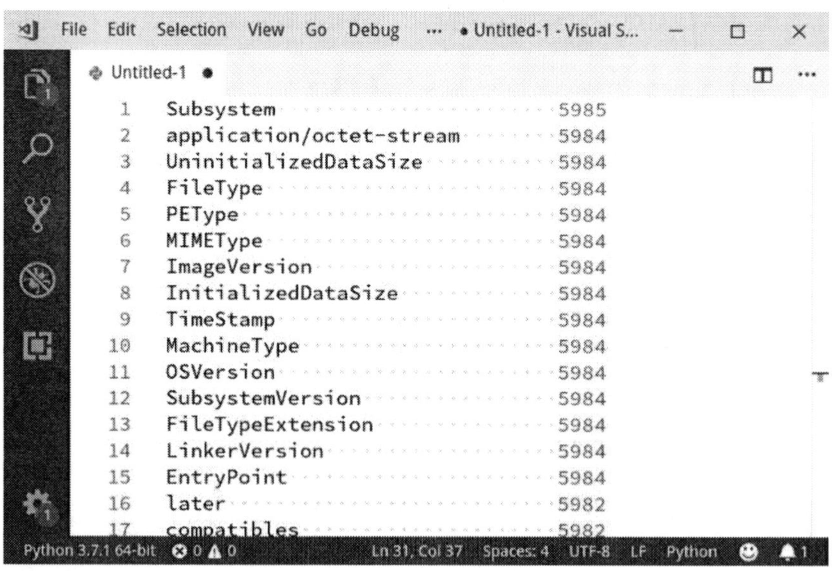

Figure 6. List of behaviours file produced by Cuckoo Sandbox

By applying Python script coded with techniques mentioned in Section 3.6, every malware's behaviors have been sampled, unigram extracted, and top 10,000 unigrams chosen based on their repetition frequency. These top 10,000 unigrams are then marked as a feature-map, where each of malware behaviors' unigrams have been mapped to either one or zero based on their availability in the top frequent unigrams. With having these transformed files, initial visual differences can be done to have an initial look on how our data has been transformed. Below is an example of two completely different malwares and their families on the bit-level representation. Figure 7 shows data visualization that give an idea about their differences and how machine learning algorithms such as deep Denoising Autoencoders can exploit this non-similarity to create generalized form of malware signatures which can represent these bit-string behaviors on much lower dimensions.

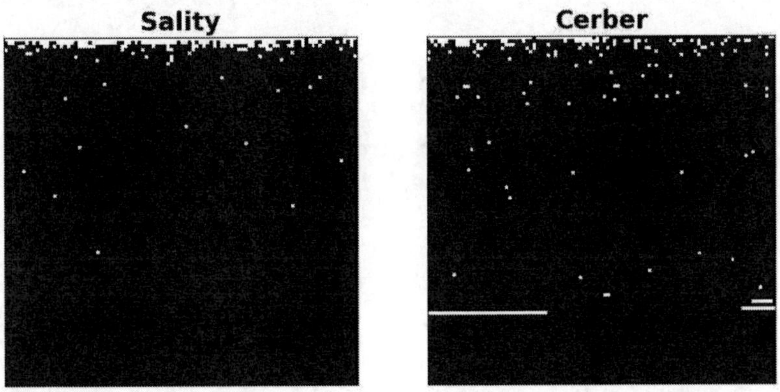

Figure 7. Bit-string visualization of two malware samples on bit-level

Acknowledgments
The authors are grateful to the sponsor of the research Ministry of Education, Universiti Teknologi MARA and the Research Management Centre (RMC) UiTM Shah Alam for the support under the national Fundamental Research Grant Scheme FRGS/1/2019/ICT02/UITM/02/9.

References

[1] Rieck K, Trinius P, Willems C, & Holz_aff2n3 T. (2011). Automatic Analysis of Malware Behavior Using Machine Learning. *Journal of Computer Security*, 19(4), pp 639–668.

[2] Nieuwenhuizen, D. (2017). *A behavioural-based approach to ransomware detection.* https://labs.mwrinfosecurity.com/assets/resourceFiles/mwri-behavioural-ransomware detection-2017-04-5.pdf

[3] Claudio Guarnieri. (n.d.). Cuckoo Sandbox - Automated Malware Analysis. Retrieved June 22, 2018, from https://cuckoosandbox.org/blog/206-interim-release

[4] Daniele U,Leonardo A, Roberto B. (2019) Survey Of Machine Learning Techniques for Malware Analysis. *Computers & Security* . Volume 81, pp 123-147.

[5] Şerif B Mehmet B Y, Can Y A. (2019). A multi-dimensional machine learning approach to predict advanced malware. *Computer Networks*. Volume 160, pp 118-129.

[6] Sen C, Minhui X, Lingling F, Shuang H, Lihua X, Haojin Z, Bo L. (2018). Automated poisoning attacks and defenses in malware detection systems: An adversarial machine learning approach. *Computers & Security*. Volume 73, pp 326-344.

[7] Wei Z, Feng G. (2019) multi-level deep learning system for malware detection. *Expert Systems with Applications*. Volume 133, pp 151-162.

IOP Conf. Series: Materials Science and Engineering **769** (2020) 012072 doi:10.1088/1757-899X/769/1/012072

Types of anti-phishing solutions for phishing attack

Siti Hawa Apandi, Jamaludin Sallim and Roslina Mohd Sidek

Faculty of Computing, College of Computing and Applied Sciences, Universiti Malaysia Pahang, Gambang, Kuantan, Pahang, Malaysia

E-mail: sitihawa.apandi@gmail.com

Abstract. Nowadays, many people use Internet to do online activity. This scenario exposed them to danger in Internet which is phishing attack. In order to solve phishing attack, the anti-phishing solutions are needed. Based on our review, there are still lacks of articles that review on the types of anti-phishing solutions in detail. In this paper, a general idea of phishing attack and anti-phishing solutions is presented. The phishing attack can be classified into two categories which are social engineering and malware-based phishing attack. The anti-phishing solutions can be differentiating into two types which are phishing prevention and phishing detection. Compared to phishing prevention, the phishing detection is more important to solve the phishing attack. In phishing detection, there are two categories which are user awareness and software detection. The software detection has two methods which are traditional and automatic. There are two types for automatic method of software detection which are public phishing detection toolbars and academic phishing detection / classification schemes. Based on the comparison of all types of phishing detection, the academic phishing detection / classification schemes are more useful for phishing detection. For future work, we can do more research on the academic phishing detection / classification schemes that utilize deep learning to see its potential of accuracy to detect phishing websites.

1. Introduction

The people lifestyle has changed with the arrival of the Internet. Nowadays, the people spent a lot of time on the Internet. The people surf Internet by using smartphone, laptop or computer. The common things that people do with the Internet are online shopping, emails, surfing social media, chatting and many more. The people also use Internet to access information in the World Wide Web, or simply called as Web.

Internet users are exposed to cybercrime. The cybercrime is also referred as computer crime. It is a crime that involves computer and network. One of the cybercrime is phishing. The phishing is an act to trick Internet users to visit the phishing website for the purpose to obtain the personal information of Internet users such as username and passwords to login into the website. The attacker of phishing is known as phisher. The phisher exploit the retrieved information of the Internet users for their own gain such as financial gain, sell the stolen identities of the Internet users, fame and notoriety [1].

The most targeted for the phishing website are financial and online payment companies. The reason for why the Internet user fall the phishing attack is that the phishing website look similar with the legitimate website [2]. There is increasing number of phishing attack even though there are many research have been done to solve the phishing attack [3]. The anti-phishing solutions for the phishing attack can be divided into two types which are phishing detection and prevention.

The remainder of this paper is organized as follows. In Section 2, the phishing attack is explained. This is followed by the discussion of anti-phishing solutions in Section 3. Lastly is the conclusion in Section 4.

2. Phishing attack

One of the popular cybercrime is phishing attack. It was first discovered in 1996 [3]. The term phishing sound similar with fishing. This is because the idea of phishing attack is same as the idea of fishing where someone will throw out a bait for the user to catch the bait [2, 4, 5]. The phisher lures the Internet user to visit the phishing website in order to steal the personal information of the user. Figure 1 shows the life cycle of phishing attack.

There are three components in the phishing attack which are medium, vector and technical approaches. For the medium of phishing, there are three medium which are internet, voice and short messaging service (SMS). The internet is commonly used medium for phishing because it has opened a huge chance for the phisher to deploy the phishing attack. The vector means a place to launch the phishing attack. The vector for internet are email, website and social networks [4]. The technical approaches for the phishing attack can be differentiate into two category which are social engineering and malware-based phishing attack [6]. The social engineering exploit the user emotion of fear to lose something valuable, thus the user reveal the personal information to the phisher [4]. For the malware-based phishing attack, it secretly install malicious programs in order to give the phisher access to the user computer [6].

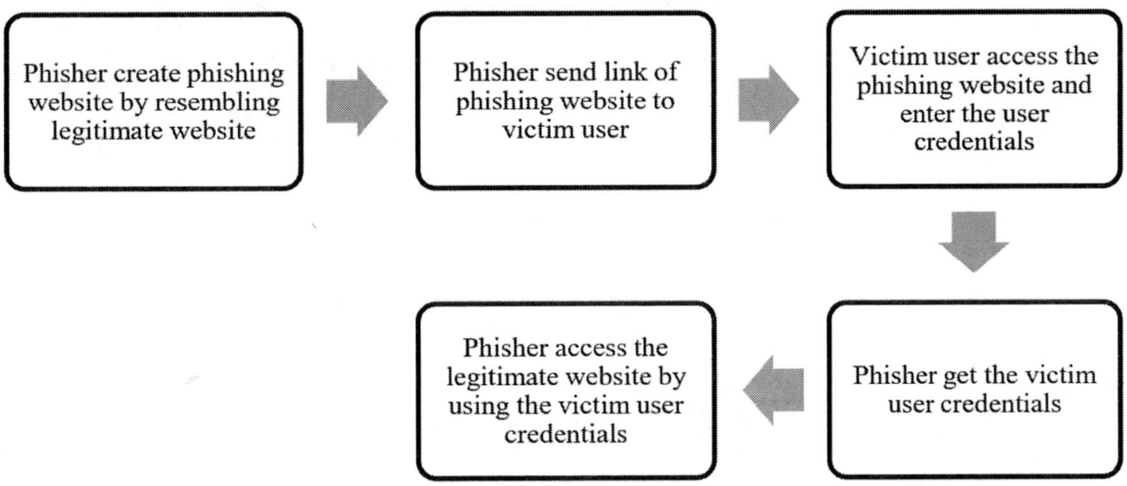

Figure 1. Life cycle of phishing attack [2].

3. Anti-phishing solutions

There are many tools exists to solve phishing attack. The phishing attack solutions also known as anti-phishing solutions. Figure 2 shows the types of anti-phishing solutions. There are two types for anti-phishing solutions which are phishing prevention and phishing detection. For phishing detection, it can be differentiate into two categories which are user awareness and software detection. There is traditional and automatic method of the software detection. For automatic of the software detection, there are two categories which are public phishing detection toolbars and academic phishing detection / classification schemes. Each of the anti-phishing solutions is briefly described.

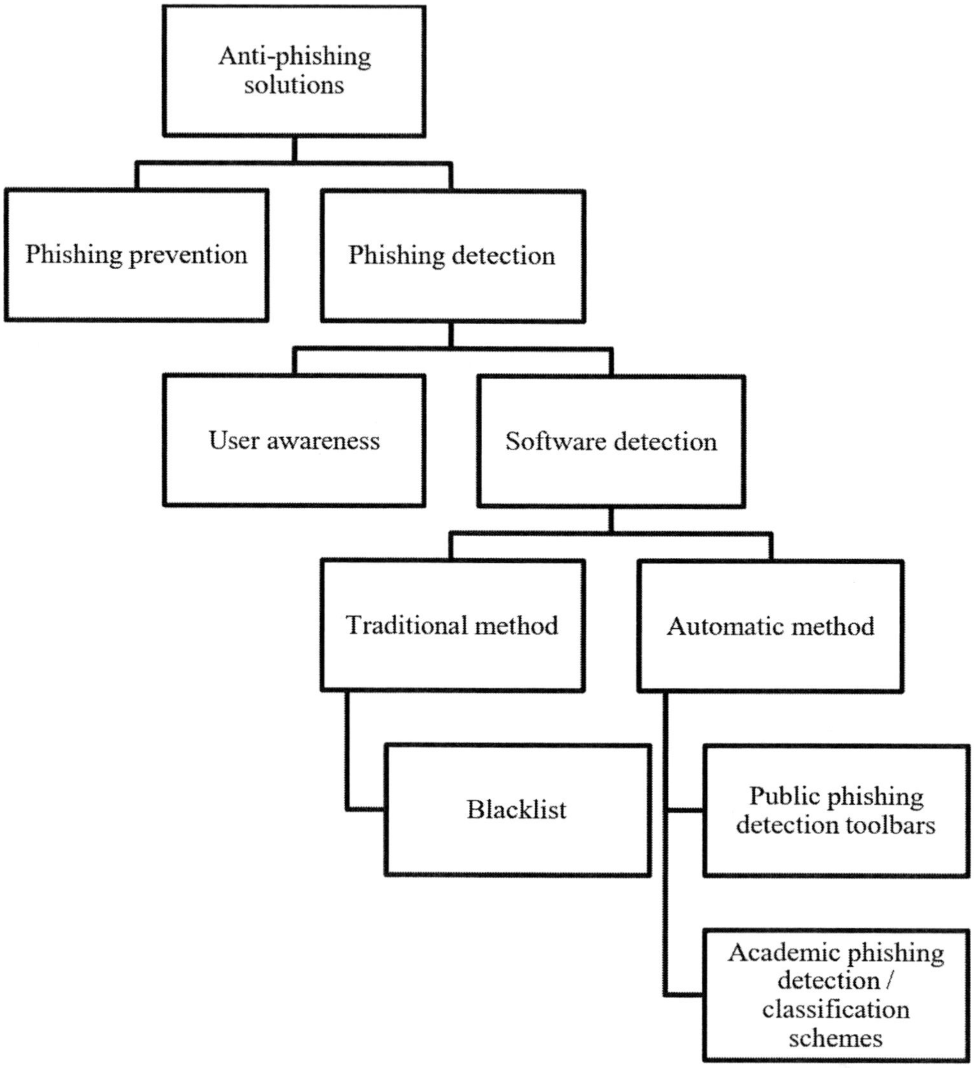

Figure 2. Types of anti-phishing solutions.

3.1. Phishing prevention

In order to prevent phishing attack, the phishing prevention is introduced by providing an extra layer of security when the user login into the website. The extra layer of security is via two-factor authentication, which is a process to confirm the user identity before the user is granted to access its login account in the website.

Figure 3 shows an example of two-factor authentication via SMS. When the user has entered username and password to login into the website, a verification code will be sent to the user's registered mobile phone number via SMS. Then, the user need to enter the verification code before the user can login into the website. The verification code only can be used in a short time before it is expired.

Even though the phishing prevention can provide extra security to prevent the phishing attack, it may hard to be implementing. An extra device is required for the user to get the verification to access into the website. This can cause extra cost to implement the extra security for the phishing attack.

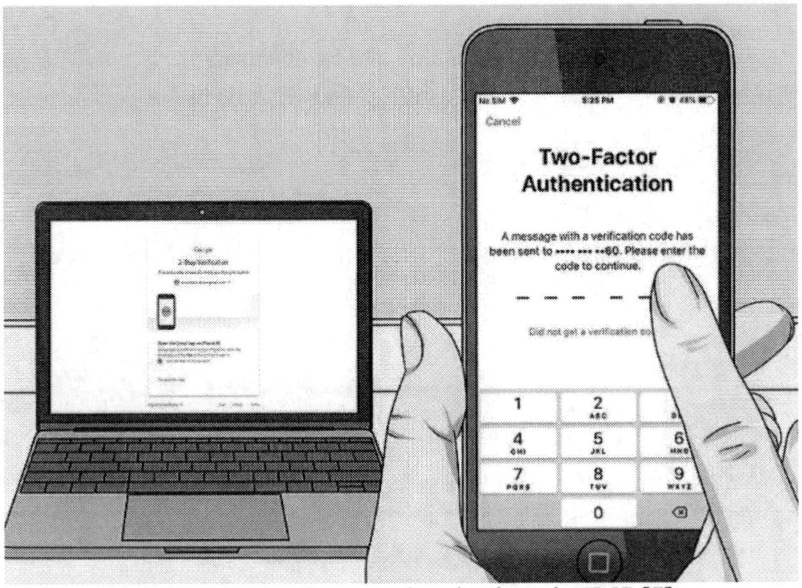

Figure 3. Two-factor authentication via SMS [7].

3.2. Phishing detection

There are two categories for phishing detection which are user awareness and software detection as shown in Figure 2. The user awareness is for to educate users so that they are able to identify phishing attempts targeted at them. The users need to be careful when visiting the web page, for example by checking at the web page URL first. Even though the user has being careful, there is a chance that the user can be deceived by the phisher to visit the phishing web page [3, 8].

Therefore, software detection is introduced to use for distinguish whether the website is legitimate or phishing. The software detection can be differentiating into two methods which are traditional and automatic method as shown in Figure 2 [2].

For traditional method of software detection, the blacklist is used to manage the list of phishing websites, which are manually entered and updated in the system. The advantage of blacklist is it has high accuracy. The drawback of blacklist is it lacking to identify the phishing website that has short lifetime [1, 2]. Furthermore, if no one report about the phishing website, then the blacklist cannot detect the phishing website [9].

For automatic method of software detection, it can be classified into two categories which are public phishing detection toolbars and academic phishing detection / classification schemes as shown in Figure 2. The automatic method of software detection use combinations of heuristic and blacklist based approach. The heuristic based approach examines contents of the website. There are three types of heuristic based approach which are surface level content, textual content and visual content. The heuristic of surface level content means by examine at the URL of website. The heuristic of textual content means by examine the terms or words in the website. Lastly, the heuristic of visual content means by examine the layout of website [1]. Next, the public phishing detection toolbars and academic phishing detection / classification schemes is briefly described.

The purpose of the public phishing detection toolbars is to detect and blocking the phishing website. The user can see these toolbars as a web browser extension. A security warning is displayed to alert the Internet user when the user visits the phishing website. There are two types of security warning which are passive warning and active warning. For passive warning, it does not block the content of the website and just show the warning to notify the user about the phishing attack as shown in Figure 4. While for active warning, it block the content of the website, thus make the user unable to view the website as shown in Figure 5. There are many public phishing detection toolbars that are freely available. Some of them are Google Safe Browsering, McAfee SiteAdvisor, Netcraft Anti-

Phishing Toolbar and many more [1, 10]. This is a useful tool as a protection for the Internet users who has lack knowledge to identify the phishing attack themselves. For public phishing detection toolbars, it is hard to know on the algorithm used for phishing detection.

Figure 4. Passive warning in Internet Explorer (IE) [10].

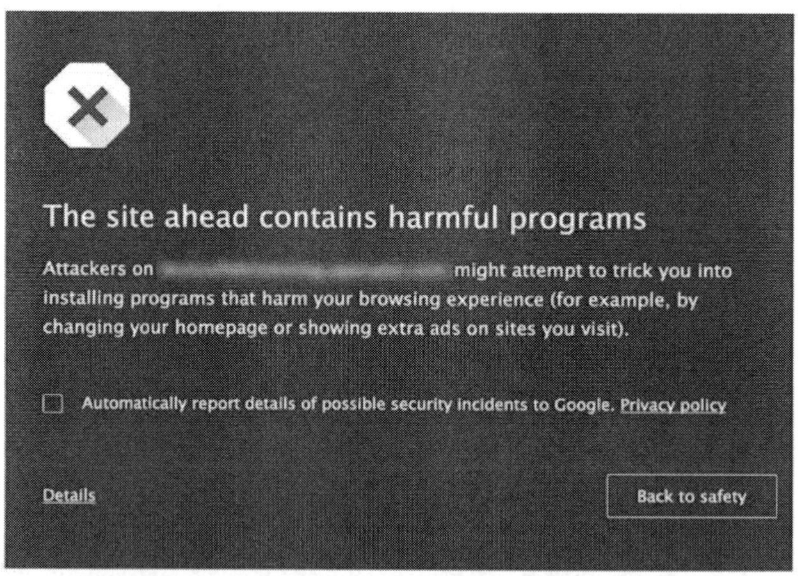

Figure 5. Active warning in Google Safe Browsering [1].

The purpose of the academic phishing detection / classification schemes is to identify and classify whether the website is legitimate or phishing. It utilizes Artificial Intelligence (AI) method which uses supervised learning classification algorithms to do binary classification of website whether it is legitimate or phishing [11]. The algorithm used includes machine learning and deep learning. There are many types of machine learning and deep learning algorithms. The common machine learning algorithms used for the phishing detection techniques are Support Vector Machine, Logistic

IOP Conf. Series: Materials Science and Engineering **769** (2020) 012072 doi:10.1088/1757-899X/769/1/012072

Regression and Bayesian-based classifiers [1]. The machine learning algorithm has a limitation which is it become ineffective when dealing with a large scale datasets [12]. In order to improve the performance of the academic phishing detection / classification schemes, the deep learning algorithm is introduced. Compared to machine learning algorithm, the deep learning algorithm is able to handle large scale datasets [1, 8]. However, the deep learning algorithm need more time for training [8]. The comparison performance of phishing website detection by using machine learning and deep learning algorithm can be seen in [13], where it shows deep learning algorithm namely deep belief network achieve high accuracy which is 0.94 compared to the accuracy of the other types of machine learning algorithm which are Random forest achieve 0.92 and J48 achieve 0.89. The challenging task to build algorithm for the academic phishing detection / classification schemes is feature extraction or also known as feature engineering. The website features that are useful for the phishing detection need to be selected in order to improve the algorithm accuracy. The evaluation metrics used to measure the performance of the academic phishing detection / classification schemes are accuracy, precision, recall and F1-score.

Based on the discussion of the phishing detection, we have summarized some important point regarding the category of phishing detection as shown in Table 1.

Table 1. Comparison on category of phishing detection.

Category of phishing detection	User awareness	Software detection		
		Traditional method	Automatic method	
		Blacklist	Public phishing detection toolbars	Academic phishing detection / classification schemes
Aim	Detect phishing attacks based on the user knowledge	Manage list of phishing website	Detect and block phishing website	Identify and classify whether the website is legitimate or phishing
Characteristics	User need to carefully check the URL website before visit the website	Blacklist based approach	Combinations of heuristic and blacklist based approach	
			Little information about algorithm used to detect phishing attack	Algorithm used to detect phishing attack by using machine learning or deep learning
			Freely available to use	Not available for public use
Advantage		High accuracy	Real-time system that determine whether the website is legitimate or phishing	
				Reduce the number of false positive

				rates
Disadvantage	Need to educate the user about phishing attack	Unable to detect zero-hour phishing attacks which mean attacks that were not seen previously	High false positive rates	
	Human error or ignorance to detect phishing attacks	Easy for the phisher to escape from phishing detection	User ignore the security warning displayed	
		Require frequent update on the dataset		
		Too slow to handle large scale datasets		

4. Conclusion

This paper has discussed about phishing attack and anti-phishing solutions. This paper can be used as a guideline for the researchers who are interested to know about the anti-phishing solutions exists in order to solve the phishing attack. There are two types of anti-phishing solutions which are phishing prevention and phishing detection. We need to do phishing detection first before we can do phishing prevention. If the phishing attack is not detected first, then it is no use to do the phishing prevention. After the phishing attack has been detected, only then the phishing prevention can be applied. This shows the importance of the phishing detection compared to the phishing prevention. There are many categories of the phishing detection which are user awareness, traditional method of software detection, public phishing detection toolbars and academic phishing detection / classification schemes. Based on the comparison of the categories of the phishing detection, the academic phishing detection / classification schemes are more useful in order to detect the phishing attack. For the future work, we would like to explore more on the academic phishing detection / classification schemes that utilizes on deep learning. We choose to focus on deep learning as the phishing detection technique because it is the recent technique being used and it has shown the improved performance in terms of accuracy of phishing detection compared to the machine learning.

5. Acknowledgment

The work presented in this paper is supported by Universiti Malaysia Pahang Grant: RDU182207-2. This work also partially supported by Adnuri SMA Research Center (M) Sdn. Bhd.

References

[1] Z. Dou, I. Khalil, A. Khreishah, A. Al-Fuqaha, and M. Guizani, "Systematization of Knowledge (SoK): A systematic review of software-based web phishing detection," *IEEE Communications Surveys & Tutorials,* vol. 19, pp. 2797-2819, 2017.

[2] R. Das, M. Hossain, S. Islam, and A. Siddiki, "Learning a deep neural network for predicting phishing website," Degree of B.Sc. in Computer Science, Brac University, 2019.

[3] G. Varshney, M. Misra, and P. K. Atrey, "A survey and classification of web phishing detection schemes," *Security and Communication Networks,* vol. 9, pp. 6266-6284, 2016.

[4] K. L. Chiew, K. S. C. Yong, and C. L. Tan, "A survey of phishing attacks: their types, vectors and technical approaches," *Expert Systems with Applications,* vol. 106, pp. 1-20, 2018.

[5] I. Qabajeh, F. Thabtah, and F. Chiclana, "A recent review of conventional vs. automated cybersecurity anti-phishing techniques," *Computer Science Review,* vol. 29, pp. 44-55, 2018.

[6] H. Thakur and S. Kaur, "A survey paper on phishing detection," *International Journal of Advanced Research in Computer Science,* vol. 7, 2016.

[7] w. Staff. (2019). *How to Prevent Phishing*. Available: https://www.wikihow.com/Prevent-Phishing

[8] O. K. Sahingoz, E. Buber, O. Demir, and B. Diri, "Machine learning based phishing detection from URLs," *Expert Systems with Applications,* vol. 117, pp. 345-357, 2019.

[9] A. I. Abunadi, "Anti-phishing Model for Phishing Websites Detection: Using Pruning Decision Tree," Universiti Teknologi Malaysia, 2013.

[10] M. Khonji, Y. Iraqi, and A. Jones, "Phishing detection: a literature survey," *IEEE Communications Surveys & Tutorials,* vol. 15, pp. 2091-2121, 2013.

[11] V. Ra, B. G. HBa, A. K. Ma, S. KPa, P. Poornachandran, and A. Verma, "DeepAnti-PhishNet: Applying deep neural networks for phishing email detection," in *Proc. 1st AntiPhishing Shared Pilot 4th ACM Int. Workshop Secur. Privacy Anal.(IWSPA)*, 2018, pp. 1-11.

[12] L. Safae, B. El Habib, and T. Abderrahim, "A Review of Machine Learning Algorithms for Web Page Classification," in *2018 IEEE 5th International Congress on Information Science and Technology (CiSt)*, 2018, pp. 220-226.

[13] M. K. Verma, S. Yadav, B. K. Goyal, B. R. Prasad, and S. Agarawal, "Phishing Website Detection Using Neural Network and Deep Belief Network," in *Recent Findings in Intelligent Computing Techniques*, ed: Springer, 2019, pp. 293-300.

IOP Conf. Series: Materials Science and Engineering **769** (2020) 012073 doi:10.1088/1757-899X/769/1/012073

A Comparative Analysis on Artificial Intelligence Techniques for Web Phishing Classification

Tengku Balqis binti Tengku Abd Rashid, Jamaludin bin Sallim and Yusnita binti Muhamad Noor

Faculty of Computing, College of Computing and Applied Sciences, Universiti Malaysia Pahang, 26300 Kuantan Pahang Malaysia

E-mail: jamal@ump.edu.my

Abstract. Over the last few years, the web has been expanded to serve millions of users for various purposes all over the world. The web content filtering is essential to filter offensive, unwanted web content from web pages, reduced inappropriate content to prevent access to content which could compromise the network and spread malware. It also to tightened network security where web content filtering adds a much-needed layer of security to the network by blocking access to sites that raise an alarm. However, there are lack of comparison between classification techniques in previous studies in order to find the best classifier for the web page classification and the analysis related to it. Thus, the purpose of this study was to apply web page classification techniques and their performances is compared as it is the initial step in data mining before going to web filtering. In this project, three classifiers called Artificial Neural Network, J48 Decision Tree and Support Vector Machine were used to web phishing dataset in order to find the best possible classifier with small computational efforts that will give the best result in classifying the web page.

Keywords. Web page classification, Artificial Neural Network, SVM (Support Vector Machine), J48 Decision Tree.

1. Introduction

In a matter of time, the web has changed the way people do business and communicate by becoming a very powerful platform that influence everyday life. According to [1], growth of internet user from 1995 until today increase by 55% and for internet host, it growth more than almost double the amount a decade prior where the statistic shows around 1.03 billion internet hosts were available on the DNS by January 2018 [2].

From the statistic we can see the rapid growth of the Internet usage as source of information which make the classification of web pages becoming a significant process and challenging task to do as the number of web pages also increasing [3]. Web page classification is important especially in education institute for web content filtering to control the access to unwanted data, inappropriate web pages and to manage the traffic [4]. Classification is the most crucial stage in web filtering as it is the initial stage of the process.

According to [5], classification can be explained as organized learning problem where the labelled dataset is used to train classifier to before applying it to new dataset for categories determination. Web page classification based on [6] are tasks to determine either a certain web page is in a category or categories which also known as web page categorization. It is in the area of machine learning that can provide ways to classify web pages to certain categories. There are many ways and approaches to classify web pages and this project will be focusing on classification algorithms and techniques used for classifying web pages.

Classification related problems arise as the rapid growth of Internet host and web pages happen. First problem discovered is institutions are lacking in terms of web page classification and identification without the content of web page. Most of institution still can't classify and identify the web page can be access accurately as the classifier used is selected randomly. This happen because lack of comparison between classification techniques used in order to find the best classifier for the web page classification and the analysis related to it [7].

Besides that, another problem arises is about manual classification of web pages where the classification process of web page is done manually. According to [8], manual classification is costly and need a very intensive labor. Other than it is cost more than automated process [9], have a heavy labor-intensive process, manual classification is an extremely time-consuming process that does not guarantee with the latest results [10]. Manual classification process also has limitation where it is not always reliable and adequate as it need for automated keywords prediction and tags recommendation through machine learning [11].

Lack of reliability in existing studies of web page classification also one of the problems discovered as lot of research discussing more about way to improve the existing classification algorithms and lot of the research is focusing on web page content rather than discovering new features for classifying web pages without the content of web pages [12].

Based on the problem as explained above, in the next section, which is section 2 will summarize the related technique used in previous studies. Section 3 will explain about selected techniques for this study. Section 4 is about the methodology, Section 5 will discuss about the result and will summarize the conclusion in Section 6.

2. Related works

[13] using URL based approach for web page classification where they don't need to download the whole content of the web page to do the classification. They proposed an automated way of learning universal dictionary as through this way the challenge to manage a big scale data can be avoided with the training set is made independent. They also applying this technique on another dataset gain search results gained from Google and the final result of the research by implementing the universal dictionary rather than to use dataset-specific term dictionary show that the difference is not statistically significant, but the classification process is faster, and the bandwidth can be saved as they use URL features instead of taking the web page content.

[14] stated that automatic web classifier is required urgently in this fast growth of information in world wide web as manual process is time-consuming. In this research they proposed a way to reduced dimensionality of thousands of inputs by introducing two novel feature selection approaches for web page classification. As classifier need to manage massive number of web pages with thousands features and categories, it faces many problems in the implementation. So, it is important to reduce the dimensionality of the data for classification process work smoothly without facing problems. To decrease the number of input dimensionality they proposed a fuzzy ranking analysis paradigm together with a novel relevance measure and discriminating power measure (DPM). For the result of this research, analysis on fuzzy ranking show that this technique has successfully validate uncertain behavior of each relevance measure and DPM method capable in setting up a better classifier as it can decrease amount of redundancy and noise features.

Next, [15] recommended automatic recognition method using classification rule by combining content, structure and uniform resource locator (URL) attributes for news web page classification. They are focusing on classifying new web page as it is different from other web pages to retrieve only useful news web pages. In this research they used Naïve Bayes algorithm to classify news articles from other non-news articles. Based on their research show that Naïve Bayes algorithm give competent accuracy of classification with different dataset.

Also, [16] proposed that to overcome web page classification problem it is important to use the most efficient way in order to select best feature and reducing the feature space. In this research, for web page classification best features selection they used firefly algorithm (FA) for features subset selection and J48 classifier for selected features fitness evaluation. The dataset is accurately classified in a short time by applying the FA algorithm. The result in this research show that the process of web page classification is done faster with high quality classification after reducing the features by removing unnecessary features.

In their study, [17] explained about phishing website classification based on URL and their feature extraction process. First, they extract the features of website URL then analyze the techniques used for feature selection and classification to detect the phishing website. As phishing technology has becoming more complicated and advanced, the anti-phishing program can be pass easily by them. So, in this research they proposed to increase the number of URL features extracted and to analyze the techniques for feature selection that have not been used yet in phishing websites detection based on URL. They choose Naïve Bayes and SMO algorithms for the experiment and the result show that SMO algorithms have the best accuracy in feature selection and classification ofphishing website.

Next [18] pointed out about the need of web page classification to get efficient indexing, search and retrieval as web information repository has increase drastically. In this research they use SVM classifier for classification to build web classifier with highest accuracy using structure information including META tags, title, descriptions of links and alternative texts of images. Based on their research show that there is improvement in term of accuracy by combining the full text with structure information compared to traditional methods and the compatibility of SVM classifier with web classification is very high.

[19] in their research propose a novel approach for web page classification that uses the HTML information present in a web page for its classification. There are many ways of achieving classification of web pages into various domains. This paper proposes an entirely new dimension towards web page classification using Artificial Neural Networks (ANN). The presence of additional information provided by the HTML tags and the hyperlinks gives the researchers idea of exploring new techniques for representing Web sites for automatic classification. As conclusion they proposed a solution for web page classification using HTML elements of a web page. The proposed model will provide the necessary web page classification technique for fast and efficient working of the search engines. Further, it is also expected to obtain results with high classification accuracy.

Next [20] have proposed a technique for web page categorization using artificial neural network (ANN) through automatic feature extraction is proposed. The main objective behind this task is to provide an efficient way for classification of web pages. This will facilitate the different search engines to classify the web pages more efficiently and also to provide a rich web directory. The end user will also be facilitated to find the web page of their desired classes. From [21], they firstly research the SVM classifiers for Web page classification with different features based on same data set. On the Web page presentation, the HTML structure is considered and separately tested on different combination. Then they study the classification performance of SVM classifier based on the polynomial kernel function and the radius basis function (RBF) kernel function. Then, NB classifier is used to investigate the different performance of the SVM classifier and NB classifier changing with the number of the feature dimension. Finally, the results are proved that the SVM classifier has better performances.

Based on [22], three different data mining algorithms have been discussed for the analysis of anti-phishing website data sets. Theses algorithms are Random Forest (RF), Nearest Neighbor Classification (NNC), Bayesian Classifier (BC). The Random Forest shows around 68 percentage of successful result when the training data is split to 75 percentages. The Nearest Neighbor Classification technique gives better and accurate result when the checking conditions are less. The result of Bayesian Classification shows the accuracy rate is around 88 percentages for finding the phishing websites. With the comparison of all these algorithms, the Bayesian classification is more accurate and shows fast response to the system.

IOP Conf. Series: Materials Science and Engineering **769** (2020) 012073 doi:10.1088/1757-899X/769/1/012073

[23] stated that medical professionals need a reliable prediction methodology to diagnose hematological data comments. There are large quantities of information about patients and their medical conditions. In this paper they are studying the various classification algorithms. Their main aims are to show the comparison of different classification algorithms using Waikato Environment for Knowledge Analysis or in short, WEKA and find out which algorithm is most suitable for user working on hematological data. The best algorithm based on the hematological data is J48 classifier with an accuracy of 97.16% and the total time taken to build the model is at 0.03 seconds. Naïve Bayes classifier has the lowest average error at 29.71% compared to others.

[24] investigated the features selection aiming to determine the effective set of features in terms of classification performance. They compare features selection and classification methods in order to determine the least set of features of phishing detection using data mining. Experimental tests on large number of features data set have been done using Information Gain and Correlation Features set methods. Further, five data mining algorithms Naïve Bayes, KNN, Random Forest, SVM and j48 have been used to classify the web phishing data set, analyze the results and identify the efficient technique to classify the web page phishing data set. In this research work, web page's phishing data sets are used to analyze the various classification techniques and find out the efficient classifier. The research show that Random forest model shows better performance than KNN, SVM, J48 and Naïve Bayes classification models.

3. Selected techniques

Based on the related work as explained and analyzed in section 2, there are three (3) techniques were selected based on their advantages which are Artificial Neural Network, J48 Decision Tree and Support Vector Machine. Each of this technique is explained in the next section.

3.1. Artificial Neural Network

Neural networks try to replicate biological systems which is human brain where in the human brain there are synapses; connected neurons through several points. In biological systems, the changes of synaptic connections strength in response to impulse is where the learning process happened. ANN has adopted this biological analogy. Neuron and unit are the basic computation unit in an ANN that is connected by arranging them in different kinds of architectures and perceptron is the most basic architecture of the neural network that have a set of input nodes and an output node where the input units pass a set of inputs to output unit [25].

ANN is collection of highly interconnected processing elements or neurons that working together to solve specific problems such as related to pattern recognition or data classification through their ability to derive the meaning behind complicated or imprecise data. This is especially for extracting and detecting complex patterns and trends that are above the capability of humans or other computer techniques [26] and the result of using ANN in various field have showed many positive result with high probability to save time resources between input predictor and known output responses. They can track changes happen on signals over time with ability to continuously adapting to new data. Their ability to adapt continuously on new data allows them to track changes in a signal over time and can handle appropriately some problem that can't be solve by conventional statistical technique like adaptability to learn from arbitrary. Also ANN able to classify multi source data as they are non-parametric classifiers [27].

3.2. J48 Decision Tree

J48 algorithm is the improved version of the C4.5 or optimized implementation of the C4.5. The Decision tree will be the output of J48. A Decision tree is similar to tree structure where it has different nodes, such as root node, intermediate nodes and leaf node. Every node of the tree holds a decision and the result is led by the decision as name is decision tree. Decision tree partitioned the data set input space in mutually exclusive areas with every area describing or elaborating its data points by having a label, a value or an action [23]. According to [28], decision tree mechanism is transparent and easy to follow a tree structure for the decision making. A decision tree is a tree structure consisting of internal and external nodes connected by branches. An internal node is a decision-making unit that evaluates a decision function to determine which child node to visit next. The external node, on the other hand, has no child nodes and is associated with a label or value that characterizes the given data that leads to its being visited.

3.3. Support Vector Machine

Support Vector Machine is a machine learning technique based on Statistical Learning theory. SVM has been proved to be very effective in dealing with high-dimensional feature spaces, the most challenging problem of other machine learning techniques due to the so-called curse of dimensionality [29]. Support vector machine are basically binary classification algorithms. The basic support vector machine takes a set of input data and predicts, for each given input, which of two possible classes forms the output, making it a non-probabilistic binary linear classifier [30]. SVM has been constantly evolving to apply its excellent learning ability to many types Classification, information processing and so on [31].

SVM is one of machine learning methods, which based on statistical theory will automatically find methods utilizing hyperlink to improve precision, methods based on ontology and subject oriented methods with high dividing capacity. Classifier produced by these support vectors can maximize distances between categories. As a result, SVM gets high precision for classification. But if the training set is large, time consumedby SVM is too long [32]. When using SVM, first transforming data into high dimensional space may convert complex classification problem into simpler problem that can use linear discriminant function. Secondly, SVM provides the most useful information for classification [27].

4. Methodology

4.1. Research framework

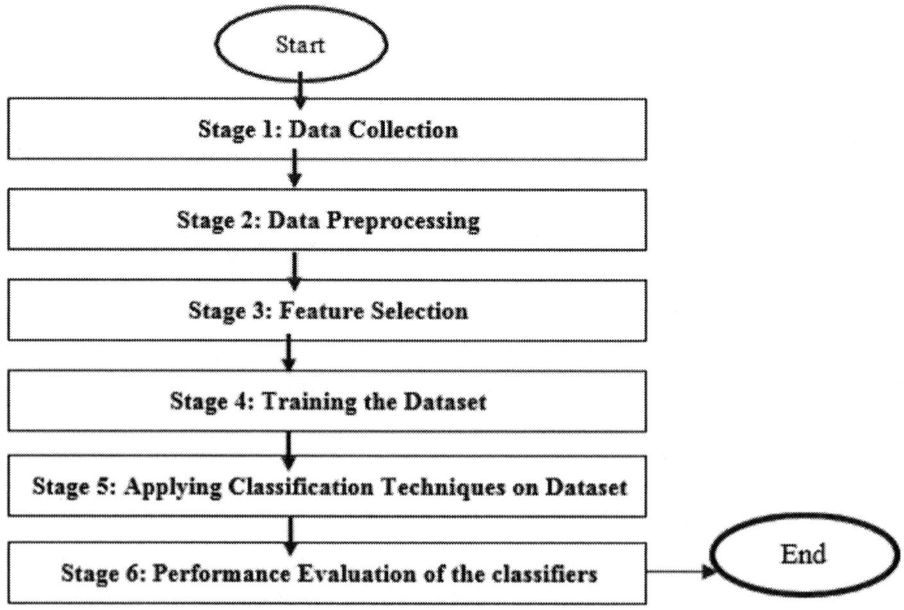

Figure 1. Research Framework

IOP Conf. Series: Materials Science and Engineering **769** (2020) 012073 doi:10.1088/1757-899X/769/1/012073

The process to evaluate the performance of classifier is composed of six steps: data collection, data pre-processing, feature selection, data training and applying the classifier techniques on the dataset which will be described in the following sections. **Figure 1** illustrates the research framework that consists of five main steps. The initial stage for evaluating classifier is to collect the data. The second step is data pre-processing which includes. The third, fourth and fifth step are the feature selection process followed with training the dataset and applying classification techniques on the dataset. The details of these stages will be elaborated in the following sections.

4.2. Data collection

The dataset used as shown in **Table 1** is from UCI repository. The researcher faced challenges about the unavailability of reliable training datasets. In fact, this challenge faces any researcher in the field. However, although plenty of articles about predicting phishing websites have been disseminated these days, no reliable training dataset has been published publicly, may be because there is no agreement in literature on the definitive features that characterize phishing webpages, hence it is difficult to shape a dataset that covers all possible features. In this dataset, we shed light on the important features that have proved to be sound and effective in predicting phishing websites.

Table 1. Dataset Information

No	Name of Features	Rules
1.	Having_ip_Address	If the Domain Part has an IP Address → Phishing Otherwise→ Legitimate
2.	URL_Length	If URL length<54 → Legitimate else if URL length ≥54 and ≤75 → Suspicious otherwise→ Phishing
3.	Shortinig_Service	If TinyURL → Phishing Otherwise→ Legitimate
4.	Having_At_Symbol	If URL Having @ Symbol→ Phishing Otherwise→ Legitimate
5.	Double_slash_redirecting	IF The Position of the Last Occurrence of "//" in the URL > 7→ Phishing Otherwise→ Legitimate
6.	Prefix_suffix	If Domain Name Part Includes (-) Symbol → Phishing Otherwise → Legitimate
7.	Having_sub_Domain	If Dots in Domain Part=1 → Legitimate Else if Dots in Domain Part=2 → Suspicious Otherwise→ Phishing
8.	SSLfinal_State	If Use https and Issuer is Trusted &and Age of Certificate≥ 1 Years → Legitimate Else if Using https and Issuer Is Not Trusted → Suspicious Otherwise→ Phishing
9.	Domain_registeration _length	If Domains Expires on ≤ 1 years → Phishing Otherwise→ Legitimate
10.	Favicon	If Favicon Loaded from External Domain→ Phishing Otherwise→ Legitimate
11.	Port	If Port # is of the Preferred Status→ Phishing Otherwise→ Legitimate
12.	HTTPS_token	If Using HTTP Token in Domain Part of the URL→ Phishing Otherwise→ Legitimate

IOP Conf. Series: Materials Science and Engineering **769** (2020) 012073 doi:10.1088/1757-899X/769/1/012073

13.	Request_URL	If % of Request URL <22% → Legitimate Else if %of Request URL≥22% and 61%→ Suspicious Otherwise→ Phishing
14.	URL_of_Anchor	If % of URL Of Anchor < 31% → Legitimate Else if % of URL Of Anchor ≥ 31% And ≤ 67% → Suspicious Otherwise→ Phishing
15.	Links_in_tags	If % of Links in "<Meta>","<Script>" and "<Link>" <17% → Legitimate Else if % of Links in "<Meta>", "<Script>" and "<Link>" ≥ 17% And ≤ 81% → Suspicious Otherwise→ Phishing
16.	Server from Handler (SFH)	If SFH is "about: blank" Or Is Empty → Phishing Else if SFH Refers to A Different Domain→ Suspicious Otherwise → Legitimate
17.	Submitting_to_email	If Using "mail()" or "mailto:" Function to Submit User Information → Phishing Otherwise → Legitimate
18.	Abnormal_URL	If the Host Name Is Not Included in URL → Phishing Otherwise→ Legitimate
19.	Redirect	If number of Redirect Page ≤ 1 → Legitimate Else if number of Redirect Page ≥ 2 &And <4 → Suspicious Otherwise → Phishing
20.	On_Mouseover	If onMouseOver Changes Status Bar → Phishing It Does't Change Status Bar → Legitimate
21.	Right Click	If Right Click Disabled → Phishing Otherwise → Legitimate
22.	popUpWidnow	If Pop-up Window Contains Text Fields → Phishing Otherwise → Legitimate
23.	Iframe	If Using iframe→ Phishing Otherwise → Legitimate
24.	Age_of_domain	If Age Of Domain ≥ 6 months → Legitimate Otherwise → Phishing
25.	DNSRecord	If no DNS Record for The Domain → Phishing Otherwise → Legitimate
26.	Web_traffic	If Website Rank < 100,000 → Legitimate Website Rank > 100,000 →Suspicious Otherwise → Phishing
27.	Page_Rank	If PageRank < 0.2 → Phishing Otherwise → Legitimate
28.	Google_Index	If Webpage Indexed by Google → Legitimate Otherwise → Phishing

IOP Conf. Series: Materials Science and Engineering **769** (2020) 012073 doi:10.1088/1757-899X/769/1/012073

29.	Links_poiniting_to_page	If number Of Link Pointing to The Webpage = 0 → Phishing Else if Of Link Pointing to The Webpage > 0 and ≤ 2 → Suspicious Otherwise → Legitimate
30.	Statistical_report	If Host Belongs to Top Phishing IPs or Top Phishing Domains → Phishing Otherwise → Legitimate

4.3. Data pre-processing

In the third phase, data need to be prepared for modeling tools. Therefore, features should be selected and extracted. For the purpose of classification, documents are represented by their features. A feature is simply a decimal value, a measure of a given aspect of a document. For instance, the frequency of a word in the document could be a feature. Note that features don't necessarily have a decimal part. They could be Boolean, integer, or some sort of label. But all of these can be represented as a decimal number. When considered as a whole, these features form a vector of decimal numbers. The length of the vector is equal to the number of features chosen to model the documents.

4.4. Feature selection

Feature selection is pretty important as it can reduce both the data and the computational complexity. It can also get more efficient and find out the useful feature subsets. The raw data collected is usually large, so it is desired to select a subset of data by creating feature vectors that feature subset selection is the process of identifying and removing much of theredundant and irrelevant information possible. This results in the reduction of dimensionality of the data and thereby makes the learning algorithms run in a faster and more efficient manner [33].

High dimensional data consists of features that can be irrelevant, misleading, or redundant which increase search space size resulting in difficulty to process data further thus not contributing to the learning process. Feature selection is the process of selecting best features among all the features that are useful to discriminate classes. Feature selection algorithm (FSA) is a computational model that is provoked by a certain definition of relevance [34].

For this research project, a single-attribute evaluator that doesn't evaluate a subset is used. It evaluates each attribute individually. This can help to eliminate irrelevant attributes, but it can't remove redundant attributes because it's only looking at individual attributes, one at a time. We applied AttributeSelectedClassifier using Ranker search method to get fair evaluation. The ranking search method doesn't really search; it just sorts them into rank order of the evaluation. It sorts attributes according to their evaluation and can specify the number of attributes to retain. The default is to retain them all, or to discard attributes whose evaluation falls below a certain threshold or can specify a certain set of attributes that need to ignore.

4.5. Training the dataset

Training Dataset is a dataset to be used in machine learning algorithm to train the model. Creating a train and test split of dataset is one method to quickly evaluate the performance of an algorithm on related problem. The training dataset is used to prepare a model, to train it. The chosen classification techniques which is ANN, J48 DT and SVM will be trained using the training data. In this research 75% of data will be used for training and the rest will be used for testing and evaluating the model as both processes can't be done with the same data to get the best reliable estimates of the models' performance. Comparing test vs. training performance help to avoid overfitting. If the model performs very well on the training data but poorly on the test data, then it's over fit.

4.6. Applying classification techniques
Applying classification techniques on dataset is where the process implementation of techniques is done on the selected dataset. The techniques will be applied on the dataset using WEKA. WEKA tool has widely used from Machine Learning researchers and proving its strength in classification as it provides many algorithms for these tasks that can be used on our datasets. Two cases will be tested to evaluate the classification techniques performance which is with and without feature selection. Three different classification techniques will be testes where the techniques were chosen based on their different nature of work in order to compare their efficiency in terms of accuracy in both cases with and without feature selection.

4.7. Performance evaluation
This section is how the results obtained by applying the classifier on dataset is compared and analyzed in order to evaluate their performance which classifier have the highest accuracy for web page classification. The experiments in this research are evaluated using the standard metrics of accuracy, precision, recall and f- measure for Web Classification. These were calculated using the predictive classification as in **Table 2**, known as Confusion Matrix.

Table 2. Predictive Classification

		PREDICTED	
		IRRELEVANT	RELEVANT
ACTUAL	IRRELEVANT	**TN**	**FP**
	RELEVANT	**FN**	**TP**

TN (True Negative) = Number of correct predictions that an instance is irrelevant
FP (False Positive) = Number of incorrect predictions that an instance is relevant
FN (False Negative) = Number of incorrect predictions that an instance is irrelevant
TP (True Positive) = Number of correct predictions that an instance is relevant

Precision – The proportion of the predicted relevant pages that were correct:

$$\text{Precision} = TP / (FP + TP) \quad (1)$$

Recall – The proportion of the relevant pages that were correctly identified

$$\text{Recall} = TP / (FN + TP) \quad (2)$$

F-Measure – Derives from precision and recall values:

$$\text{F-Measure} = (2 \times \text{Recall} \times \text{Precision}) / (\text{Recall} + \text{Precision}) \quad (3)$$

The F-Measure was used, because despite Precision and Recall being valid metrics in their own right, one can be optimized at the expense of the other. The F- Measure only produces a high result when Precision and Recall are both balanced, thus this is very significant.

5. Result and discussion
In this experiment, the classification task will be examined in order to classify instances and predict their category in web phishing problem; instances will be classified to legitimate or phishing website. Two cases have been tested to evaluate the classification performance; with and without feature selection. Different three types of classifiers have been tested; classifiers were chosen due to their different nature of work, to compare their efficiency in terms of accuracy in Phishing Website Prediction problem, in both cases with and without feature selection. Experiments were done using test set as testing mode.

5.1. Experiment with Full Features

Full features included all instances and category in web phishing were evaluated to get their percentage of accuracy. **Table 3** shows result of three types of classifiers from this full features experiment and **Figure 2** shows the same result in figure to make it more clear and easy to understand.

Table 3. Accuracy result for experiment with full features

Classifiers	Precision	Recall	F-Measure
Artificial Neural Network	0.963	0.963	0.963
J48 Decision Tree	0.953	0.953	0.953
SVM	0.937	0.937	0.937

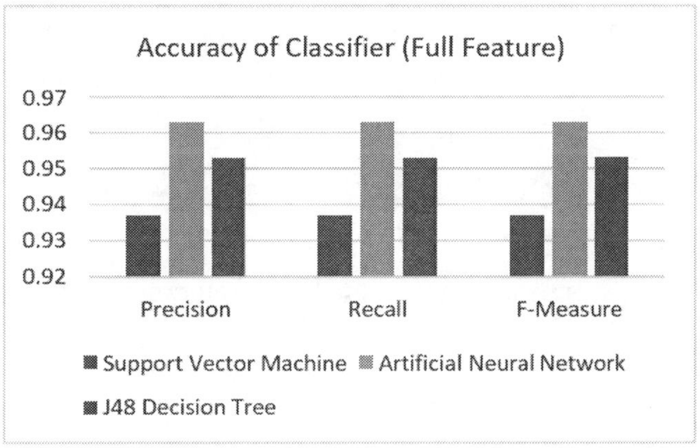

Figure 2. Graph of full features accuracy result

The above **Table 3** and **Figure 2** reveal that the classification process in the web page phishing data set, Artificial Neural Network classification holds highest F-Measure value (0.962), also followed by J48 classifier's F-Measure value which is 0.956 and Support Vector Machine is 0.934 with the lowest value.

As we observe, all classifiers accuracy without feature selection had better results than using feature selection as a preprocess step before classifying instances. The performance of the Artificial Neural Network classifier and J48 algorithms were almost similar on the data set, whereas the SVM classifier is slightly lower in accuracy.

5.2. Experiment with reduced features

Reduced features included some instances and some category in web phishing were evaluated to get their percentage of accuracy. **Table 4** shows result of three types of classifiers from this full features experiment and Figure 3 shows the same result in figure to make it more clear and easy to understand.

Table 4. Accuracy result for experiment with reduced features

Classifiers	Precision	Recall	F-Measure
Artificial Neural Network	0.962	0.962	0.962
J48 Decision Tree	0.956	0.956	0.956
SVM	0.934	0.934	0.934

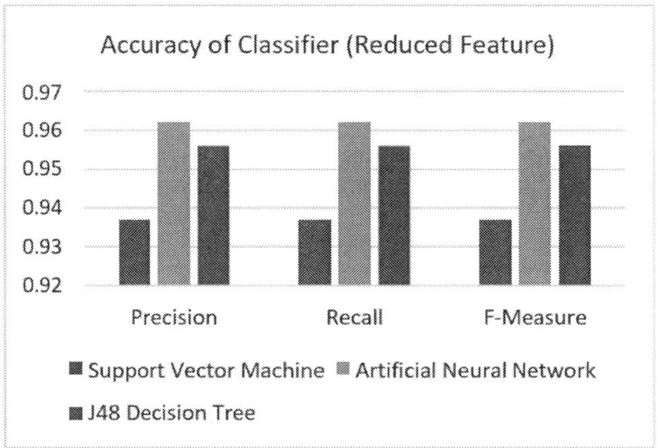

Figure 3. Graph of reduced features accuracy result

6. Conclusion

As a conclusion, we have met our objective which is to evaluate and investigate three selected classification algorithms based on Weka. Classification technique showed significant performance in Phishing Website Prediction. Three algorithms have been used and compared in terms of accuracy. The feature selection pre-processing was considered to observe the performance of classifiers with a minimal number of features. The obtained results showed that the classifiers are doing good without eliminating features in the tested dataset. But there is a tradeoff between the accuracy and the consumed time in the prediction process. Where if more accuracy was required, it's better to use the classification techniques without feature selection. And if the best performance was the target it's better to use the feature selection process in this dataset. From the experiment, it was found that the Artificial Neural Network Classifier had the highest accuracy and the Support Vector Machine had the least one. In the future, more classification techniques can be compared, with different measures, and more datasets can be used, in addition to the feature extraction from a number of phishing websites then many classification techniques can be applied for the prediction process.

Acknowledgement

A big appreciation to Universiti Malaysia Pahang for financial support under research grant RDU190310 to publish this research paper and a big thanks to all research team.

References

[1] "Internet Growth Statistics 1995 to 2019 - the Global Village Online," 2019. [Online]. Available: https://www.internetworldstats.com/emarketing.htm. [Accessed: 20- Mar-2019].
[2] "Number of internet users worldwide 2005-2018," 2019. [Online]. Available:

https://www.statista.com/statistics/273018/number-of- internet-users-worldwide/. [Accessed: 20-Mar-2019].

[3] P. Vinod and P. Prajapati, "Comparative Study of Web Page Classification Approaches," *Int. J. Comput. Appl.*, vol. 179, no. 45, pp. 6–9, 2018.

[4] S. S. Modi and S. B. Jagtap, "Web Content Filtration Using Different Web Mining Techniques in Educational System: An Overview," *Int. J. Adv. Res. Comput. Sci. Softw. Eng.*, vol. 7, no. 3, pp. 135–139, 2017.

[5] T. Mahmoud, T. Abd-El-Hafeez, and D. El-Deen, "A Design of an Automatic Web Page Classification System," *Br. J. Appl. Sci. Technol.*, vol. 18, no. 6, pp. 1–14, 2017.

[6] B. Choi and Z. Yao, Web Page Classification. In: Chu W., Young Lin T. (eds) *Foundations and Advances in Data Mining*. Studies in Fuzziness and Soft Computing, vol 180. Springer, Berlin, Heidelberg.

[7] A. Siddiqui, M. Adnan, R. A. Siddiqui, and T. Mubeen, "A comparative study of web pages classification methods applied to health consumer web pages," *2015 2nd Int. Conf. Comput. Technol. Inf. Manag. ICCTIM 2015*, pp. 43–48, 2015.

[8] A. S. Patil and B. V. Pawar, "Automated Classification of Web Sites using Naive Bayesian Algorithm," *Proc. Int. MultiConferenceEng Comput. Sci.*, vol. I, 2012.

[9] R. K. Roul, "An Effective Approach for Web Document Classification using the Concept of Association Analysis of Data Mining," *Citeseer*, vol. 3, no. 10, pp. 483–491, 2012.

[10] D. Xhemali, C. J. Hinde, and R. G. Stone, "Naïve Bayes vs. Decision Trees vs. Neural Networks in the Classification of Training Web Pages," *IJCSI Int. J. Comput. Sci. Issues*, vol. 4, no. 1, pp. 16–23, 2009.

[11] Y. Yang, P. Liu, L. Ding, B. Shen, and W. Wang, "ServeNet: A Deep Neural Network for Web Service Classification," pp. 1–17, Jun. 2018.

[12] Z. Lu, "Web Page Classification Using Features from Titles and Snippets," no. May, 2015.

[13] R. R. and C. Aravindan, "An Effective and Discriminative Feature Learning for URL Based Web Page Classification," pp. 1374–1379, 2019.

[14] C. M. Chen, H. M. Lee, and Y. J. Chang, "Two novel feature selection approaches for web page classification," *Expert Syst. Appl.*, vol. 36, no. 1, pp. 260–272, 2009.

[15] S. K. Dwivedi and C. Arya, "News web page classification using url content and structure attributes," *Proc. 2016 2nd Int. Conf. Next Gener. Comput. Technol. NGCT 2016*, no. October, pp. 317–322, 2017.

[16] E. Sarac and S. A. Ozel, "Web page classification using firefly optimization," in *2013 IEEE INISTA*, 2013, no. September, pp. 1–5.

[17] S. Moein, "Feature Extraction and Classification," *Med. Diagnosis Using Artif. Neural Networks*, pp. 159–169, 2014.

[18] K. He and C. Li, "Structure-Based Classification of Web Documents Using Support Vector Machine," 2016.

[19] P. Manchanda, S. Gupta, and K. K. Bhatia, "On The Automated Classification of Web Pages Using Artificial Neural Network," *IOSR J. Comput. Eng.*, vol. 4, no. 1, pp. 20–25, 2012.

[20] S. M. Kamruzzaman, "Web Page Categorization Using Artificial Neural Networks," *Training*, p. 4, 2010.

[21] X. Weimin, B. Hong, H. Weitong, and L. Yuchang, "Web page classification based on SVM," *Proc. World Congr. Intell. Control Autom.*, vol. 2, pp. 6111–6114, 2006.

[22] D. R. Gupta, "Comparison of classification algorithms to detect phishing web pages using feature selection and extraction," *Int. J. Res.*, vol. 4, no. 8, pp. 118–135, 2016.

[23] M. N. Amin and M. A. Habib, "Comparison of Different Classification Techniques Using WEKA for Hematological Data," no. 3, 2015, pp. 55–61.

[24] S. Nandhini and V. Vasanthi, "Extraction of Features and Classification on Phishing Websites using Web Mining Techniques," vol. 5, no. 4, pp. 1215–1225, 2017.

[25] C. C.Aggarwal, Data Classification: Algorithms and Applications. CRC Press, 2015.

[26] V. A. Zilpe and M. Atique, "Web Usage Mining Using Neural Network Approach: A Critical Review," *Int. J. Comput. Sci. Inf. Technol.*, vol. 3, no. 1, pp. 3073–3077, 2012.
[27] H. Bouali and J. Akaichi, "Comparative study of different classification techniques: Heart disease use case," *Proc. - 2014 13th Int. Conf. Mach. Learn. Appl. ICMLA 2014*, pp. 482–486, 2014.
[28] C. Nasa and S. Suman, "Evaluation of Different Classification Techniques for WEB Data," *Int. J. Comput. Appl.*, vol. 52, no. 9, pp. 34–40, 2012.
[29] W. . Awad, "Machine Learning Algorithms in Web Page Classification," *Int. J. Comput. Sci. Inf. Technol.*, vol. 4, no. 5, pp. 93–101, Oct. 2012.
[30] K. Wisaeng, "A Comparison of Different Classification Techniques for Bank Direct Marketing," *Int. J. Soft Comput. Eng.*, vol. 3, no. 4, pp. 116–119, 2013.
[31] W. Huang and H. You, "Web Page Classification Algorithm Based on Semi-Supervised Support Vector Machine," *Proc. 2018 2nd IEEE Adv. Inf. Manag. Commun. Electron. Autom. Control Conf. IMCEC 2018*, no. Imcec, pp. 2144–2148, 2018.
[32] J. Wang, H. Cai, B. Xu, and L. Jiang, "CUCS: A web page classification algorithm for large training set," *Proc. - 2008 IFIP Int. Conf. Netw. Parallel Comput. NPC 2008*, pp. 440–445, 2008.
[33] T. Z. Phyu and N. N. Oo, "Performance Comparison of Feature Selection Methods," *MATEC Web Conf.*, vol. 42, p. 06002, Feb. 2016.
[34] S. Khalid, T. Khalil, and S. Nasreen, "A Survey Of Feature Selection And Feature Extraction Techniques in Machine Learning," pp. 372–378, 201.

IOP Conf. Series: Materials Science and Engineering **769** (2020) 012074 doi:10.1088/1757-899X/769/1/012074

A Survey on Current Malicious JavaScript Behavior of infected Web Content in Detection of Malicious Web pages

Wan Nurulsafawati Wan Manan, Mohd Nizam Mohmad Kahar and Noorlin Mohd Ali

Faculty of Computing, College of Computing and Applied Sciences, Universiti Malaysia Pahang, 26300 Kuantan Pahang Malaysia

E-mail: safawati@ump.edu.my

Abstract. In recent years, the advance growth of cybercrime has become an urgent issue to the security authorities. With the improvement of web technologies enable attackers to launch the web-based attacks and other malicious code easily without having prior expert knowledge. Recently, JavaScript has become the most common attack construction language as it is the primary browser scripting language which allow developer to develop sophisticated client-side interfaces for web application. This lead to the growth of malicious websites and as main platform for distributing malware or malicious script to the user's computer when the user access to these webpages. Initial act and detection on such threats early in a timely manner is vital in order to reduce the damages which have caused billions of dollars lost every year. A number of approaches have been proposed to detect malicious web pages. However, the efficient detection of malicious web pages previously has generated many false alarm by the use of sophisticated obfuscation techniques in benign JavaScript code in web pages. Therefore, in this paper, a thoroughly survey and detailed understanding of malicious JavaScript code features will be provided, which have been collected from the web content. We conduct a thorough analysis and studies on the usage of different JavaScript features and JavaScript detection technique systematically and present the most important features of malicious threats in web pages. Then the analysis will be presented along with different dimensions (features representation, detection techniques analysis, and sample of malicious script).

1. Introduction

The dramatic rise of the Internet technology has given various impact to our daily lives. This technology become the most popular medium of communication and information collection. Web based attack is happening daily where, JavaScript has become the most common language used by attackers. From the G Data Security report, most of the common malware threats are launch by the attacker coming from web sites and not only from the executable file [1].

As reported from the previous years, the most remarkable security breach is, Ransomware attack, which caused billions of dollar lost and identity breach or stealing user's money. Attackers have used many techniques to launch malicious attack through websites. Figure 1 below illustrates some examples of attack classification.

Figure 1: Malicious web attacks classification.

Malicious web crime can be categorized into several types of attacks and drive by downloads are become the most common surreptitious attack on the Internet [19]. While, cross sites scripting (XSS), code injection, URL redirection, malvertising, and many more become the most predominant attack and threats on web pages nowadays.

JavaScript is the most common programming language used in web development. This scripting language is a lightweight and object-based client-side scripting language which add many features and effect to the end user experience. This common web language was used by the web developer to develop interactive content for a web pages [1] and making the web user experience the dynamic and better control of web interaction. Besides, HTML and CSS are the common construction language which used in the web browser in developing the interactivity in web pages.

Detecting and filtering web pages URLs based on a blacklist is a simple and powerful technique. However it is difficult to notice malicious threats on websites and unknown malicious URLs on the web. Besides, it is difficult to keep up to date with newly created malicious URLs for blacklisting [2]. This is because a new URL can be generated easily and a massive amount of web content is generated every day by both legitimate and malicious users.

In this research, we have categorized and suggested a set of JavaScript features from the thoroughly literature studies of other researcher. Features which have been considered in this paper aim at describing a classes of web pages or web content components that are frequently used attackers for conducting attack in website. Besides, these proposed features was collected based on various comparisons from different researchers' point of view thus making the features expected to correctly classify the malicious websites with a high level of precision rate.

This paper is organized as follows. We presented the related works on previous research in detecting malicious websites in section two. In the section 3, we have discussed the characteristic and features representation in identifying malicious website. Last section conclude the proposed idea and future works.

2. Related works

This section surveys related approaches malicious JavaScript detection and classification, related applications of statistical method, machine learning and other techniques.

IOP Conf. Series: Materials Science and Engineering **769** (2020) 012074 doi:10.1088/1757-899X/769/1/012074

2.1. Malicious web pages detection

Variety of approaches have been proposed by previous researchers in detecting the malicious JavaScript in web pages. In authors [4] a novel approach was proposed to detect the malicious JavaScript Code. They have developed a system which used a number of JavaScript features and combining the anomaly detection with emulation. This machine learning classifier was used to classify whether the web pages is benign or malicious.

R.Wang et al [5], suggested a hybrid analysis technique to detect malicious web pages. According to the researcher detecting the malicious activity in web pages using static analysis only was become difficult. Therefore, researcher has combined the dynamic analysis in order to achieve high performance result. In their studies, static analysis utilized the classification algorithm in machine learning while dynamic analysis method checked the web pages that are directly from the web browser to determine whether they contain malicious shell codes during execution.

Another work by Canali et al [6] has designed a fast and reliable filter, which also use static web page analysis based on machine learning for detecting malicious web pages. The features being used in the analysis were derived from the HTML, JavaScript code, and corresponding URL. The author also claim that the system was faster than the where processing time about 0.27 s/page

In [7] authors have proposed five features that capture different characteristics of script for detecting malicious JavaScript. They are, execution time, external referenced domains and calls to JavaScript functions. From the experiment results, it depicted that a combination of selected features is able to detect malicious JavaScript code successfully with a precision of 0.979 and a recall of 0.978

Another well-known method which used by [8] in detecting the malicious web is through the blacklist-based method. Blacklisting is a technique where it was compiled through crowdsourcing technique, and comes from various sources of repository or database. A blacklist is a repository which contains a various list of known malicious URLs. This method particularly used by filtering systems against web-based malware infection and also widely used in search engines and browsers toolbars. However there were there are several limitation, the URL list need to be updated regularly as the web environment is changing at rapidly and unable to detect unknown or zero-day delay malicious web URLs.

2.2. Critical Analysis of Malicious Features Selection

Malicious features is one of the biggest threats in web crime activities. In this paper, different features of malicious has been studied and proposed to help researcher to decide between malicious and benign web pages. In particular, the selection of features is important to prove better performance of the machine learning classifier as well as the accuracy of the result. We focus on content based features which are obtained upon downloading or opening the website.

HTML and JavaScript features are based detail content about the whole web pages. The most common characteristic of malicious JavaScript features are not human-readable, and the character frequency of JavaScript codes being used in the content [9], [10]. While, [11] have used 77 features, among which 45 are new features in their dataset preparation.

This paper [12] implement the vector space representation in extracting semantic features which normally use during classification and clustering. Benign webpages were collected from Alexa (2013), while malicious web pages collected mostly from Phishtank (2013). Furthermore, researcher also used lexical features of URLs, visual features which are based on images and web pages that contain many links to collect information.

We have conducted a thoroughly studies on the other researchers work in order to propose and come out with a set of malicious JavaScript features that are more representative and structured which can be refer by researcher and security practice in future. Besides, this proposed features also could help in detecting malicious web pages as well as any web threats especially zero-day attacks.

With the dynamic behavior in nature of JavaScript, it executes certain programming behaviors at runtime, using different technique provided by both interpreter and browser. Figure 2 below present an example JavaScript code which attempts to execute the malicious pages where, script elements was stored and pointed to external document at runtime [13].

```
<script type="javascript">
document.write("<iframe width='0' height='0'
src='www.xxx.com '> </iframe>");
</script>
```

Figure 2: Example of malicious JS code (1)

While figure 3 show the example built in JavaScript function of eval( ) and unescape( ) where hacker uses these two function to decode and encode string and write the output on the webpage [14]. This JavaScript function is the most used by attackers in inserting malicious code and activity in the web pages.

```
< script > document.write(unescape("%3C%73%63%72%69%70%
    74%20%6C%61%6E%67%75%61%67%65%
    3D%76%62%73%63%72%69%70%74%3E")) < /script >
...
eval("arrNum = 0 : ReDimTempStr(0) : strLength
    = Len(NbjHrXYekZCCM ...")
```

Figure 3: Example of malicious JavaScript located in functions: eval() and unescape()

Another research done by [15], developed a technique based on JavaScript timing attacks for stealing information from the victim machine and from the sites the victim visits during the attack. While [7] proposed a method for detecting malicious JavaScript code based on five features which are: execution time, external referenced domains and calls to JavaScript functions (avgExecTime, maxExecTime, funcCalls, totalUrl, extUrl).

A malicious URL is more likely to be developed with a malicious URL [5]. As discussed in [16], the malware behavior is more resource-consuming than a trusted application. So we have emerge with an analysis within web page and web content only in detecting malicious web pages. This will help researcher in shorten the analysis time and process in detecting the malicious or benign web pages.

Table 1 below summarize the proposed features which have been done and implement by various researcher in detecting malicious web page content. We collected several features which commonly been used by most of the researcher and also few features only can be implemented also in defining the web page whether benign or malicious. Most of the proposed features are from content based features which are primarily from the HTML content and JavaScript method and function used in the web pages.

Table 1: JavaScript string methods type features (1)

JavaScript Method Types	Examples of Features	[17]	[11]	[5]	[18]	[10]	[7]	[6]
JavaScript String methods	search()		/			/		
	charAt()		/					
	charCodeAt()		/			/		
	Concat()		/					
	indexOf()		/					
	substring()		/	/		/		
	replace()		/			/		
	Split()		/	/		/		
	toString()		/					
	document.write()	/	/					
	Window.location()		/					

Based on Table 1, we categorized several features based on JavaScript strings and method. These feature was classified based on several factor and consideration. The most common consideration where the features representation was denote by most of the attackers in their malicious web pages found in dataset.

Most of this strings and method category are quite vulnerable and frequently used by an attackers to spread the malicious script. Based on thoroughly analysis from other researchers [11, 10] suggest that this features should be considered when detecting a malicious web pages.

Table 2: JavaScript string methods type features (2)

JavaScript Method Types	Examples of Features	[17]	[11]	[5]	[18]	[10]	[7]	[6]
JavaScript Global Function	encodeURL()				/			
	decodeURL				/			
	parseInt()	/	/		/	/		
	parseFloat()				/			
	encodeURIComponent()				/			
	encodeURIComponent()				/			
	escape()	/	/	/	/	/		/
	eval()	/	/	/	/	/		/
	unescape()	/	/	/	/	/		/
	Document.cookie				/			
	fromCharCode()	/	/		/	/		/
	replace()		/	/				
	exec()			/		/		

From the table 2 above, W3Schools have categorized another most popular JavaScript features are from Global Function category. This features are the most common list being used, which are vulnerable and exploited by attackers for injecting this functions with malicious script.

Table 3: JavaScript string methods type features (3)

JavaScript Method Types	Examples of Features	[17]	[11]	[5]	[18]	[10]	[7]	[6]
JS DOM HTML	setInterval()				/	/		/
	setTimeout()	/	/		/	/		/
HTML tags	<iframe>	/	/	/				/
	Number of hidden elements			/				/
	Unequal HTML tags			/				/
HTML Global Event	onload		/			/		/
	onerror		/			/		/
	onunload		/			/		/
	onmouseover		/					/
	onbeforeunload		/			/		/
	addEventListener(), attachEvent(), dispatchEvent()							/

While from the table 3 above we have summarize based on content based HTML types of features. This features was achieved upon downloading the web pages. Most of the researcher [6, 10, 11] proved that this features are important to be considered in detection the malicious web pages.

As mention by Canali et al, [3] classify that this features was very helpful in detection of an early threats of malicious script in web pages. Most of the content based features are derived from HTML and the JavaScript.

Another features have been categorized was other features which are selected by researcher based on pattern or based on their own heuristics. Table 4 below depicted and example of features used.

Table 4: Other features (4)

JavaScript Method Types	Examples of Features	[17]	[11]	[5]	[18]	[10]	[7]	[6]
Others Features	No of Characters in JavaScript		/	/				/
	The ratio between keywords and words	/	/	/				/
	No of blank spaces	/	/	/				/
	Average length of Words	/	/	/				/
	No Hex Values	/	/					/
	String sequence in document	/	/					/
	Length the script string	/	/					/
	No of plugins and ActiveX controls	/	/	/				/
	No of classid	/	/					
	No of suspicious tag string	/		/				/
	avgExecTime,						/	
	maxExecTime						/	
	funcCalls						/	
	totalUrl,						/	
	extUrl						/	

The selection of these features in consideration with content-based feature, which are related to JavaScript and HTML elements. This features primarily contribute to how the patterns of the malicious script that inserted by the attackers during the malicious execution in the web pages.

3. Conclusion

Malicious website has evolved nowadays with various and sophisticated techniques by attackers. Various techniques with difference malicious JavaScript characteristic features was proposed by other researcher in detecting malicious website.

However, some of the features proposed only focus on certain detection threats, such as, malicious URL, obfuscation issues and malicious JavaScript in web pages only. Besides, the total number of the chosen features in each researcher also differ based on the how the researcher declare the features itself. For instance, the tagging and naming of the features sometime repeated or convey the same meaning with the already declare features. The biggest issues in defining the features is detecting the obfuscated malicious JavaScript as attacker used various obfuscation techniques.

Therefore, we conducted a thorough study on the malicious code (JavaScript, HTML & other related code) on web pages and propose the most preferred and more representative which being used by others researchers. By using various features hopefully able to detect the malicious activity or threat in website accurately and precisely with minimum duration of detection.

4. Acknowledgement

The authors would like to thank the supervisor and co-supervisor for their helpful discussions and suggestions. This work was supported by Postgraduate Research Grant Scheme (PGRS), University Malaysia Pahang,PGRS 190388.

References

[1] R. B. Hauke Gierow, "Malware in 2018: The danger is on the web | G DATA Blog," 2018. [Online]. Available: https://www.gdatasoftware.com/blog/2018/09/31037-malware-figures-first-half-2018-danger-web. [Accessed: 08-Nov-2018].

[2] N. Bielova, "Survey on JavaScript security policies and their enforcement mechanisms in a web browser," J. Log. Algebr. Program., vol. 82, no. 8, pp. 243–262, 2013.

[3] M. Akiyama, T. Yagi, and M. Itoh, "Searching structural neighborhood of malicious URLs to improve blacklisting," Proc. - 11th IEEE/IPSJ Int. Symp. Appl. Internet, SAINT 2011, pp. 1–10, 2011.

[4] M. Cova, C. Kruegel, and G. Vigna, "Detection and analysis of drive-by-download attacks and malicious JavaScript code," Proc. 19th Int. Conf. World wide web - WWW '10, p. 281, 2010.

[5] R. Wang, Y. Zhu, J. Tan, and B. Zhou, "Detection of malicious web pages based on hybrid analysis," J. Inf. Secur. Appl., vol. 35, pp. 68–74, 2017.

[6] D. Canali, M. Cova, G. Vigna, and C. Kruegel, "Prophiler : A Fast Filter for the Large-Scale Detection of Malicious Web Pages Categories and Subject Descriptors," Proc. Int. World Wide Web Conf., pp. 197–206, 2011.

[7] G. Canfora, F. Mercaldo, and C. A. Visaggio, "Malicious JavaScript Detection by Features Extraction," e-Informatica Softw. Eng. J., vol. Vol. 8, no. nr 1, pp. 65–78, 2014.

[8] A. Y. Daeef, R. B. Ahmad, Y. Yacob, and N. Y. Phing, "Wide scope and fast websites phishing detection using URLs lexical features," 2016 3rd Int. Conf. Electron. Des. ICED 2016, pp. 410–415, 2017.

[9] P. Likarish, E. Jung, and I. Jo, "Obfuscated malicious javascript detection using classification techniques," 2009 4th Int. Conf. Malicious Unwanted Software, MALWARE 2009, pp. 47–54, 2009.

[10] G. Canfora and C. A. Visaggio, "A set of features to detect web security threats," J. Comput. Virol. Hacking Tech., vol. 12, no. 4, pp. 243–261, 2016.

[11] D. R. Patil and J. B. Patil, "Detection of Malicious JavaScript Code in Web Pages," Indian J. Sci. Technol., vol. 10, no. 19, pp. 1–12, 2017.
[12] H. B. `Kazemian and S. Ahmed, "Comparisons of machine learning techniques for detecting malicious webpages," Expert Syst. Appl., vol. 42, no. 3, pp. 1166–1177, 2015.
[13] G. Lu and S. Debray, "Automatic simplification of obfuscated JavaScript code: A semantics-based approach," in Proceedings of the 2012 IEEE 6th International Conference on Software Security and Reliability, SERE 2012, 2012, pp. 31–40.
[14] Y. T. Hou, Y. Chang, T. Chen, C. S. Laih, and C. M. Chen, "Malicious web content detection by machine learning," Expert Syst. Appl., vol. 37, no. 1, pp. 55–60, 2010.
[15] Dennis Fisher, "JavaScript and Timing Attacks Used to Steal Browser Data | Threatpost | The first stop for security news," 2013. [Online]. Available: https://threatpost.com/javascript-and-timing-attacks-used-to-steal-browser-data/101559/. [Accessed: 10-Dec-2018].[Online]
[16] M. F. Zolkipli, "An Approach for Malware Behavior Identification and Classification," 2011 3rd Int. Conf. Comput. Res. Dev., vol. 1, pp. 191–194, 2010.
[17] W.-H. Wang, Y.-J. Lv, H.-B. Chen, and Z.-L. Fang, "A Static Malicious Javascript Detection Using SVM," Proc. 2nd Int. Conf. Comput. Sci. Electron. Eng. (ICCSEE 2013), no. Iccsee, pp. 214–217, 2013.
[18] Z. S. 3 Javad Hajian Nezhad 1, Majid Vafaei Jahan 3,∗, Mohammad-H. Tayarani-N 2, "Analyzing New Features of Infected Web Content in Detection of Malicious Web Pages," ISC Int'l J. Inf. Secur., vol. 9, no. 2, pp. 63–83, 2017.
[19] N. Provos, D. Mcnamee, P. Mavrommatis, K. Wang, and N. Modadugu, "The Ghost In The Browser Analysis of Web-based Malware," *Proc. first Conf. First Work. Hot Top. Underst. Botnets*, vol. 462, p. 4, 2007.

Review of Hybrid Analysis Technique for Malware Detection

Yus Kamalrul Bin Mohamed Yunus[1], Syahrulanuar Bin Ngah[2]

[1,2]Faculty of Computing, Universiti Malaysia Pahang, 26300 Gambang, Kuantan, Pahang, Malaysia

E-mail: yus.kamalrul96@gmail.com, syahrulanuar@ump.edu.my.

Abstract. Malware is a problem spread out worldwide. Current techniques to analyze these malware are static analysis technique and dynamic analysis technique. Later, the two analysis technique is combined into a technique called hybrid analysis technique. This paper discusses on the current analysis technique and introduces a new approach towards the hybrid analysis technique by introducing memory analysis technique into it. The expected outcome of producing memory analysis technique in hybrid analysis technique will be discussed in the latter part of this paper.

1. Introduction

Malware is the term used by security professionals and is commonly used in the Information Technology (IT) field in regards to any malicious software that enters a system without the authorization of the computer administrator and/or user of the said system. This term emerges after combining the word "Malicious" and "Software" thus creating the word malware. This type of attack is growing in volume and evolving in complexity. With the current explosion of Internet of Things and Internet of Everything, threats from malware is increasing and require efficient monitoring and detection of the affected device. Effective analysis method is needed to match with these data-intensive systems and its scale. Currently, there are many types of malware detector had been created by security companies and researchers; these detectors heavily depend on the techniques being implemented in them.

Several types of threads that is referred as malware is trojan, ransomware and botnet [1]. Trojan is a type of malware that mislead user of its true intent and is often disguised as a legitimate computer program. The term is derived from the famous story of a wooden horse used by ancient Greek to penetrate the city of Troy. Once activated, the Trojan malware can delete, block, modify and copy data, also it can disrupt computer and network performance. To date, Trojans can perform many actions, therefore, it can be further categorized as backdoor Trojan, exploit Trojan, rootkit Trojan, Distributed Denial of Service (DDOS) Trojan, and but not limited to proxy Trojan. Ransomware on the other hand is a type of malware that threatens and at the same time block users from accessing their data on their system by encrypting it [2]. Most of the time, the attacker will demand ransom in cryptocurrency in order to release the files back to the victim. It is divided into several types such as scareware ransomware, locker ransomware and crypto ransomware [2]. In recent years, the use of ransomware as an attack vector has grown internationally. Attacker generates a key pair and place the public key in the ransomware. The ransomware is then released to the victim. To carry out the attack, the malware generates a random asymmetric key while it proceeds to encrypt victim's data and in certain cases render the system unusable. It puts up a message to victim that includes the cyphertext and steps on how to pay the ransom. Victim will then send the cyphertext back to the attacker with the amount of cryptocurrency the attacker demand. The attacker then deciphers the cyphertext with a private key, and sends the asymmetric key to the victim. The data is decrypted using the key provided. At times, upon receiving their ransom, the attacker did not send any key to victim resulting victim with monetary loss alongside data loss. The last example is botnet, botnet is a collection of internet connected device such as PC, IOT device or mobile phones where its security has been compromised and its control is handled by malicious third parties. The usual communication channel used by botnet is either Internet Relay Chat (IRC) or Hyper Text Transfer Protocol (HTTP). Once infected, victims machine will be used to carry out massive DDOS attacks, email spam to millions of users and generate fake internet traffic. Botnet is a commodity for cyber criminal because it is rented out for a variety of purposes.

Malware analysis technique came in three classifications, static analysis technique, dynamic analysis technique and hybrid analysis technique. These techniques have been around for many years and are being used by security analyst and professional to detect malware in a device. This three analysis technique will be discussed in more depth in the next sections. Figure 1.1 shows types of malware analysis technique and the methods it uses.

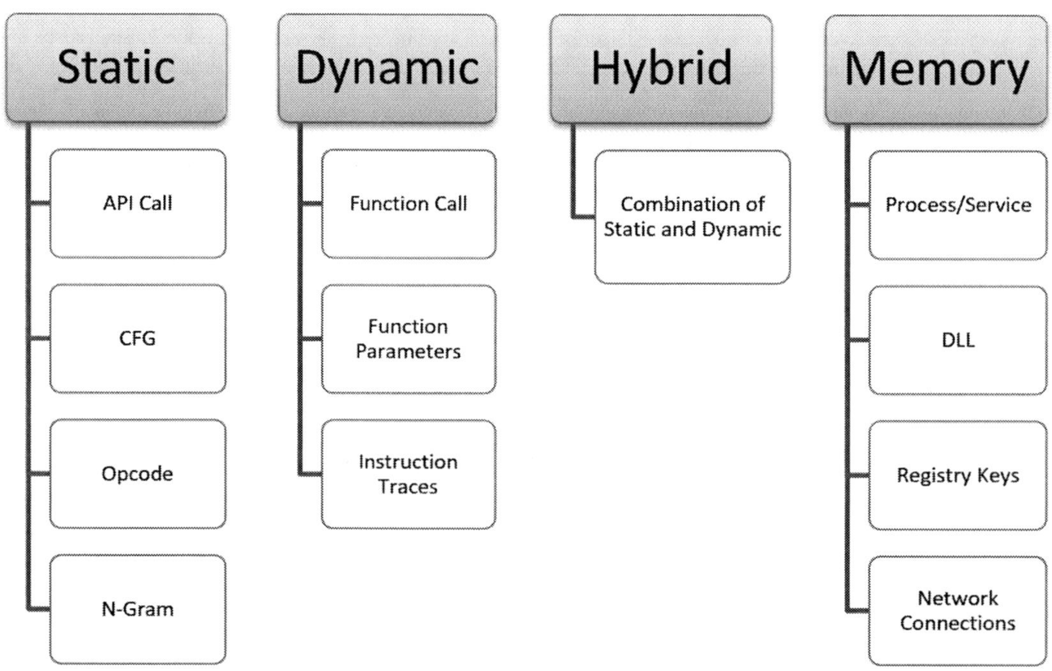

Figure 1.1 Malware Analysis Techniques

The aim of this paper is to discuss the current malware analysis technique also the approach of combining hybrid analysis technique with memory analysis technique. The proposed method in this paper will be adopted as the future work by the authors.

2. Static Analysis Technique

This analysis technique works by analyzing the code sections of the malware in order to gain knowledge on how the malware is operating [3]. This analysis procedure is done without executing the malware. Using reverse engineering tools, the malware is dissected in order to analyze its code and later, it can be

IOP Conf. Series: Materials Science and Engineering **769** (2020) 012075 doi:10.1088/1757-899X/769/1/012075

re-build. Due to not executing the malware, static analysis technique consumes fewer resource and time [4]. The different tools used to perform this analysis are debugger, dissembler, decompiler and source code analyzer. The purpose of analyzing the code is to identify the inner working of the malware; the unique algorithm in the malware will help identify the malware and can be analyzed further to eliminate the malware.

This analysis technique can be further broken down into three approaches, signature based approach, permission based approach and heuristic approach. Signature based approach are used mostly on commercial anti-malware product or on-the shelf product. It extracts the malware pattern and creates a unique signature that represents the extracted malware. A code, executable or program is flagged as a malware if the signature matches with the existing malware signature. The major drawback of this technique is that it only detects existing or known malware type, if a new malware is introduced in the system where the signature is not yet available, this approach will not detect the malware. Also due to limited library of signature, most of malware is undetected.

The second approach is permission based approach. This approach is used to control by approving or denying access right of any application. When a program is installed, as default settings, it will not have any access towards data stored in the device and does not have any impact towards system security.

The last approach searches for the instruction those are not present in the infected program or code rather than searching for a certain signature. This makes finding new variant of malware easier. Some of the technique in heuristic approach is file based heuristic, rule based heuristic and generic signature analysis.

The drawback of using this analysis technique is code obfuscation. Malware that uses code obfuscation will not be detected by this technique as it can't find the malware signature and instructions. Aside from that, malware that uses anti-debug and anti-disassembly will not be detected using this method.

3. Dynamic Analysis Technique

The analysis process is done during the execution of the malware code or the infected file. The malware code or infected file is analyzed in a sandbox environment or virtualized environment such as VMware player, Oracle Virtual Machine (VM) VirtualBox or Hyper-V environment. The environment can be a simulator, emulator, sandbox and more. The analysis can be done by monitoring certain components which is function calls, information flow, function parameter and tracing instructions. Malware or infected file need to be analyzed in an invisible environment due to some malware came with anti-virtual machine and anti-emulator technique [5]. When the malware code detects the environments, it will behave normally and will not execute any malicious activity.

A little different from static analysis technique, this analysis technique can be broken down into two approaches, anomaly-based approach and taint analysis. In order to detect malicious behavior of applications, this approach uses machine learning. Existing malware features are used to train a model for unknown malware. The tools provide deep analysis, as a result, it require more resource to operate. The drawback is application must be installed in a system in order to detect malicious software. Also, if a legitimate application invokes more system call than the usual call set in the model, the application can be falsely detected as malware.

User input can be very dangerous if it is not properly checked. Taint approach checks for variable that user input can modify. A scientific technique called "dynamic taint analysis" used by one of the application that uses this approach [4]. Information of interest will be marked as "taint". It stays with information when used. The tracking system then tracks the movement of that taint.

The drawback of using this method is it consumes resources while being time consuming. Other than that, malware that incorporated anti-virtual machine and anti-emulator will slip pass through this method.

4. Hybrid Analysis Technique

This analysis technique is introduced to overcome the limitation available in both static and dynamic analysis technique. It starts by analyzing the signature of any malware code and continue by combining it with other behavioral pattern parameters to enhance malware analysis [3]. Due to this reason, it overcomes both the shortcoming of static and dynamic analysis technique. This increases the ability in detecting malicious software correctly. In the same time, this analysis technique has almost all of the strength of static and hybrid technique.

In detecting malware in android application, mobile sandbox is an example of hybrid analysis technique. Static analysis will analyze the APK file, user permission and identifying suspicious code. While the dynamic analysis using emulator will be used to run the suspicious APK file to check the application behavior. Figure 4.1 shows the mobile sandbox.

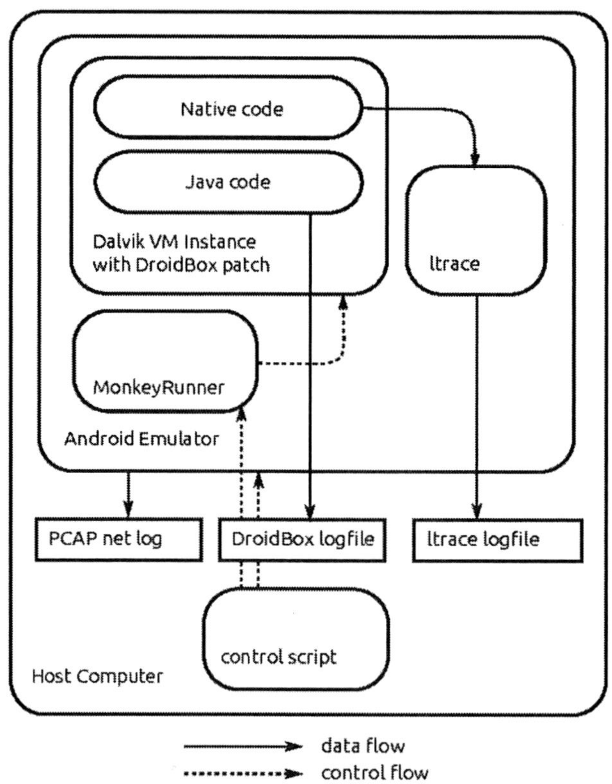

Figure 4.1 Mobile Sandbox

The limitation of using this method is this technique is resource hungry. This method is time consuming as it is running both static and dynamic technique as hybrid technique.

5. Proposed Method

Proposed method is combination of hybrid analysis technique with memory analysis technique. Although hybrid analysis technique analyze the signature of malware and then combine it with behavioral pattern parameter but it only analyzes the malware that is stored on the disk and the same malware runs in the memory. Malware that sits and run in the memory is either not checked or high probability that their signature and behavioral pattern is not the same as the classic malware that sits in the hard disk and run in the memory. Previously, computer hijacker will send malware and open a backdoor into unsuspecting user computer and start mining bitcoin, a type of cryptocurrency, without user knowledge.

Earlier 2019, ESET security team had found out that attacker injected malicious JavaScript code to website ads. Now, the website contains malicious ads. User who visited the website will have their system used for Monero or Dash cryptocurrency mining as bitcoin mining will alert user because it will skyrocket processor usage. This is all happening in the memory as the browser is running in the memory and all tabs and page is in the memory.

Memory analysis technique is able to examine malware code outside of function normal scope [3]. It uses the image of memory to analyze operating system, running programs and the general state the computer. In digital forensics on regards of memory, there are two steps which is memory acquisition and memory analysis. In acquisition, memory image is obtained by acquisition of volatile memory to non-volatile memory [8]. In analysis, the memory image is analyzed thoroughly while looking for malicious activity in the image. Memory analysis especially offline memory analysis approach is a process that can be repeated and can be followed to acquire precise digital evidence for research and court use [9]. Researchers in regards of memory forensics had proposed several approaches such as trigger-based memory analysis approach and Application Programming Interface (API) trigger-based memory analysis approach. Memory forensics technique able to monitor behavior of Malware in regards to API hooking, Dynamic-link Library (DLL) injection and hidden processes.

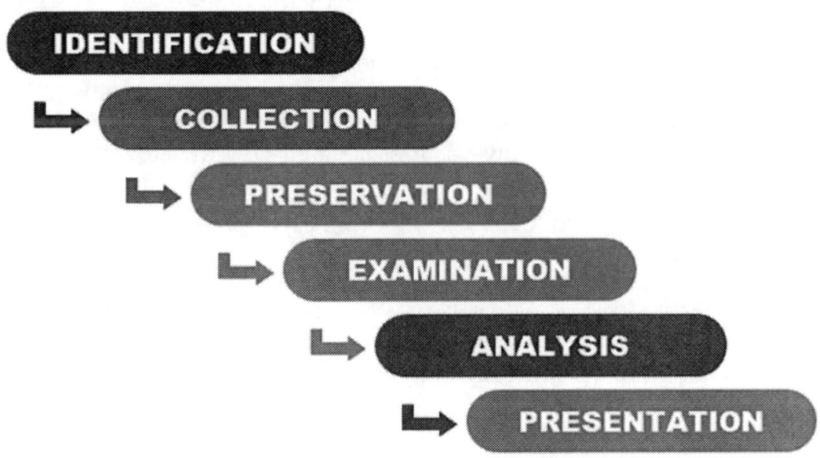

Figure 5.1 Memory Acquisitions for Digital Forensics

Malware uses API hooking to interrupt function calls. Most common type of API hooking are Import Address Table (IAT) hooks, inline API hooks, Interrupt Descriptor Table (IDT) hook, System Service Dispatch Table (SSDT) hook and Input/Output Request Packet (IRP) hook. DLL injection inserts malicious code into a legitimate process. DLL injection categories are classic DLL, registry modification injection, and Windows hooking function injection.

The process of the proposed technique starts by capturing the contents of memory by using forensics tool. After the capturing process, the memory contents will be further analyzed by using hybrid analysis technique. The process if memory capture is done manually as the automation of this process is still being explored.

Malware programs nowadays frequently contain checking criteria. This will determine the existence of certain directories or files on a machine. When the checking criteria are met, the malware runs part of its code matching with the criteria. Other type of malware may need internet connection or that a specific mutex object does not exist in order for it to execute its code [6].

There are several problems that need to be addressed during the research process. One of the problems is steps to capture the memory contents before some of the malware starts to wipe its tracks and disappear from the system. The steps to address the problem are still being explored. Other than that, anti-forensics tools are being executed within the malware to prevent memory acquisition and analysis, some went into extra lengths by planting fake artifacts in-memory [7].

The steps mentioned earlier were just a discussion on how the process will be done theoretically as the research on implementing the proposed technique will be done in future work. Therefore, during the execution phase of the future work, majority of the steps discussed will be revised to improve the execution and implementation of the proposed technique while at the same time conducting analysis on the effectiveness and performance of the proposed technique.

Expected outcome from this proposed technique is a more robust malware analysis method. The combination of hybrid analysis technique and memory analysis technique will certainly increase resource usage but it will detect more malware than only using hybrid analysis technique. This new method will be expected to detect more malware than hybrid analysis technique alone.

6. Conclusion

Malware is critical threat to a system either it is a computer mobile phone or IOT device. This paper introduces briefly on three types of malware, Trojan, ransomware and botnet. Other than that, this paper also briefly discuss on three analysis technique available for malware analysis which is static analysis technique, dynamic analysis technique and hybrid analysis technique. The drawback of each analysis technique is also provided briefly.

New proposed method for malware analysis technique is introduced. The method proposed to work by combining hybrid analysis technique and memory analysis technique. While proposing the new method, a brief introduction to memory analysis technique is also provided. Malware change their signature frequently has present many open research on how to detect them and what will be the best way to detect them. These researchers help in providing information to security professionals and researchers to tweak current analysis technique and helps in a new analysis technique creation process.

Acknowledgments

The work is funded by Universiti Malaysia Pahang under Grant ID RDU1703294. The authors would like to thank all reviewers for their constructive reviews, and Universiti Malaysia Pahang for the support.

References

[1] Jyoti Landage, Prof. M. P. Wankhade,Malware and Malware Detection Techniques : A Survey, International Journal Of Engineering Research & Technology (IJERT) Volume 02, Issue 12 (2013)
[2] Sultan, Hirra. A Survey On Ransomware: Evolution, Growth, And Impact. International Journal of Advanced Research in Computer Science vol 9, No. 2 (2018).
[3] Rao, V.R., & Hande, K. A comparative study of static , dynamic and hybrid analysis techniques for android malware detection (2017).
[4] Rami Sihwail,Khairuddin Omar and Khairul Akram Zainol Ariffin,"A Survey on Malware Analysis Techniques: Static, Dynamic, Hybrid and Memory Analysis," International Journal on Advanced Science, Engineering and Information Technology, vol. 8, no. 4-2, pp. 1662-1671, (2018).
[5] Tahir, R. A study on malware and malware detection techniques. International Journal Education and Management Engineering (IJEME) 8(2), 20–30 (2018)
[6] A. Moser, C. Kruegel and E. Kirda. Exploring Multiple Execution Paths for Malware Analysis. 2007 IEEE Symposium on Security and Privacy *(SP '07)*, Berkeley, CA, 2007, pp. 231-245 (2007).
[7] Teller, Tomer and Adi Hayon. Enhancing Automated Malware Analysis Machines with Memory Analysis. (2014).
[8] Arends, Jan, and Ing Kerstin. "Malware Analysis." (2018).

[9] Sali VR, Khanuja HK. RAM Forensics: The Analysis and Extraction of Malicious Processes from Memory Image Using GUI Based Memory Forensic Toolkit. In2018 Fourth International Conference on Computing Communication Control and Automation (ICCUBEA) 2018 Aug 16 (pp. 1-6). IEEE.

AUTHOR INDEX

Institute of Physics
Dirac House, Temple Back
Bristol BS1 6BE UK

ISSN: 1757-8981
ISBN 978-1-7138-4707-6

CURRAN ASSOCIATES INC.
proceedings.com